CHEMICAL PETROLOGY

Robert F. Mueller
Surendra K. Saxena

CHEMICAL PETROLOGY

with applications to The Terrestrial Planets and Meteorites

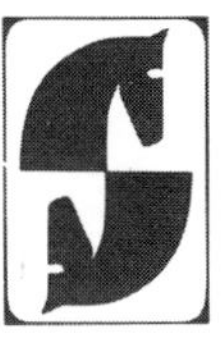

Springer-Verlag
New York Heidelberg Berlin

Robert F. Mueller
Goddard Space Flight Center
National Aeronautics and
Space Administration
Greenbelt, Md. 20771

Surendra K. Saxena
Department of Geology
Brooklyn College
City University of New York
Brooklyn, N.Y. 11210

Library of Congress Cataloging in Publication Data

Mueller, Robert Francis, 1923–
Chemical petrology.

Bibliography: p.
Includes index.
1. Rocks, Igneous. 2. Rocks, Metamorphic. 3. Geochemistry. I. Saxena, Surendra Kumar, 1936– joint author. II. Title.
QE461.M75 552′.06 77-26049

Printed in the United States of America.

ISBN 0-387-90196-5 New York
ISBN 3-540-90196-5 Berlin Heidelberg

Cover photo courtesy of NASA

Designer: Edmée Froment

To Hans Ramberg, who showed us how to do it

Preface

Chemical petrology is essentially the physical chemistry of rocks and associated fluids, although it also borrows heavily from such other sciences as mineralogy. In terms of fundamentals it is firmly grounded in chemical thermodynamics and kinetics. In its treatment of terrestrial environments it grades imperceptably into sedimentology, geochemistry, and geophysics and in extraterrestrial environments into cosmochemistry. It is one of the most important branches of planetology and meteoritics.

The unity of approach of thermodynamics and kinetics to processes in these diverse environments is stressed in this book by numerous examples which have been chosen to illuminate different aspects of the subject. Thus we have discussed in some depth such problems as the genesis of layered basic complexes, calc-alkaline batholiths, chondritic meteorites, and the surface–atmosphere interaction of the planet Venus because these are important and because they are particularly good illustrations of the chemical petrology approach. Considerable attention also has been devoted to volcanic processes. In our treatment of metamorphism in particular, an attempt has been made to correlate and integrate the vast number of recent experimental, theoretical, and field studies. However, we have not attempted a comprehensive survey of all known rock types or occurrences, nor did we review all the diverse opinions and conclusions on the origins of controversial rocks. Instead we have chosen to stress interpretations we regard as following most directly from the evidence. For readers who may not agree with all our interpretations we hope that there remains enough of the unbiased fundamental information and theory to aid them in reaching their own conclusions.

Because of our belief in their importance we have heavily stressed the thermodynamic properties of solid solutions and have attempted to illustrate how these are related to such factors as temperature, pressure, gravitational potential, composition of associated fluids, and rates of reaction and diffusion. We have also placed considerable emphasis on the process of oxidation-reduction since interest in this subject has developed so greatly recently. A particularly intriguing subject is the interaction between dissolved oxygen, hydrogen, water, and the silicate components of rock melts, and the bearing these interactions have on the origin of such rocks as the calc-alkali suite.

It is apparent that chemical petrology provides many tools for the evaluation of new concepts and ideas such as the recently proposed models for global tectonics as well as comparative studies of the planets. Ideas on the global distribution and movement of rocks and magmas need to be tested for their thermodynamic and kinetic compatibility. Similar arguments apply to the interpretation and comparison of the vastly different surface features and atmospheres of the planets Mercury, Venus, Earth, and Mars. Indeed we present evidence that these diverse features of the terrestrial planets follow fairly directly from the operation, in varying degrees, of kinetic constraints on the tendency toward thermodynamic equilibrium.

Progress in chemical petrology has been enormous in recent years, especially in the experimental investigation of oxidation and hydration effects in systems of complexity approaching natural rocks and magmas. Much progress has also been made in the experimental and theoretical study of the properties of solid and liquid solutions. We hope that this book reflects these advances. Although there have also been some important studies in the application of chemical kinetics to petrology, experiments along this line have been few. Consequently, in order to give the subject its due emphasis, we have been required to employ more theoretical interpolation than in the case of equilibrium studies.

This book was made possible by the work of a large group of highly talented investigators who carried out the field, laboratory, and theoretical investigations to which we refer. We hope that we have represented their results and ideas accurately and fairly here. Much of the background work for this book was done while one of us (R.F.M.) was at the University of California, San Diego and the University of Chicago and while the other (S.K.S.) was at the University of Uppsala and the Goddard Space Flight Center. We wish to acknowledge the support of these institutions. Much of the material has in one way or another been incorporated into classes in petrology taught by us. We are particularly indebted to Professor Ralph Kretz for a critical review of the manuscript and for many helpful suggestions. We are also indebted to Doctor E. J. Olsen and Professor Somdev Bhattacharji for review of parts of the manuscript and to Professor P. J. Wyllie for help in the publication of this book. Mona Saxena's able technical assistance in preparing the manuscript was invaluable.

January, 1977

Robert F. Mueller
Surendra K. Saxena

Contents

1 The energetic basis

The foundation of the energetic approach to mineral systems is provided by the laws of thermodynamics. These laws are broad generalizations that apply to all forms of matter, and in principle no system is too complex to yield to their application if the required data are available. However, in many problems of mineral chemistry in which complex phases are involved, the equations of thermodynamics cannot be directly applied to obtain quantitative results. The reason for this is that they give us relations only in their implicit forms and what we require are the equations of state of the phases that lead to explicit functions linking the intensive variables such as temperature, pressure, and composition. These equations of state may be based on empirical data or upon theoretical models, which usually have their foundation in statistical mechanics. Thus we shall be concerned with deducing activity coefficients that give us a measure of the deviation of a phase from the simplest model—that of an ideal solution.

This two-fold aspect of energetics is also displayed in the equations of modern kinetic theory. Here the thermodynamic properties of the activated state play an important role; however, the quantum mechanical basis is in this case more evident in the model. Unfortunately, the theory of rate processes has as yet only limited quantitative application in mineral systems.

In the following exposition it is assumed that the reader is familiar with the fundamental definitions and derivations of the thermodynamic potentials and has some acquaintance with the theory of rate processes. We therefore proceed with a discussion of the most useful relations.

General relations

The following general relations and definitions are given here for reference:

$$\left(\frac{\partial G}{\partial T}\right)_{P,n} = -S \tag{1.1}$$

$$\left(\frac{\partial G}{\partial P}\right)_{T,n} = V \tag{1.2}$$

$$\left[\frac{\partial(G/T)}{\partial T}\right]_{P,n} = -\frac{H}{T^2} \tag{1.3}$$

$$\left(\frac{\partial H}{\partial T}\right)_{P,n} = C_P \tag{1.4}$$

$$\frac{1}{V}\left(\frac{\partial V}{\partial T}\right)_{P,n} = \alpha \tag{1.5}$$

$$-\frac{1}{V}\left(\frac{\partial V}{\partial P}\right)_{T,n} = \kappa \tag{1.6}$$

In the above relations the following terminology applies: G is the Gibbs free energy; H is the enthalpy; S is the entropy; T is the absolute temperature; P is the pressure; V is the volume; C_P is the heat capacity at constant pressure; α is the coefficient of thermal expansion; and κ is the coefficient of compressibility. The subscript n indicates that all composition variables are to be held constant.

For one mole of any phase of variable composition we have the Gibbs–Duhem relation:

$$\bar{S}dT - \bar{V}dP + \sum_i X_i d\mu_i = 0 \tag{1.7}$$

in which $\bar{S}$ and $\bar{V}$ are the molar entropy and volume, respectively, X_i is the mole fraction of the component i, and μ_i is its chemical potential. At constant pressure and temperature we have

$$\sum_i X_i d\mu_i = 0 \tag{1.8}$$

Equations of state and chemical potential of pure gases, liquids, and solids[1]

General relations

The chemical potential μ of any pure phase may be expressed in terms of the molar enthalpy $\bar{H}(T, P)$ and the entropy $\bar{S}(T, P)$:

$$\mu(T, P) = \bar{H}(T, P) - T\bar{S}(T, P) \tag{1.9}$$

In this expression the letters with a bar are used to indicate *one mole of a pure phase*, and (T, P) indicates that the quantities are functions of the temperature and pressure. In other sections the bar on the symbols for the molar property are not used.

From the general relation

$$d\bar{H} = \left(\frac{\partial \bar{H}}{\partial T}\right)_P dT + \left(\frac{\partial \bar{H}}{\partial P}\right)_T dP \tag{1.10}$$

we obtain by integration

$$\bar{H}(T, P) = \bar{H}(T^0, P^0) + \int_{T^0}^{T} C_p(T, P^0)dT + \int_{P^0}^{P}\left[\bar{V} - T\left(\frac{\partial \bar{V}}{\partial T}\right)_P\right]dP \tag{1.11}$$

where the superscript ^{0} indicates standard P and T. Since

$$\left(\frac{\partial \bar{H}}{\partial T}\right)_P = C_p$$

we obtain

$$\left(\frac{\partial \bar{H}}{\partial P}\right)_T = \bar{V} - T\left(\frac{\partial \bar{V}}{\partial T}\right)_P$$

In these expressions the molar volume $\bar{V}$ is a function of the temperature.[2]

The molar entropy $\bar{S}$ as a function of T, and P may also be evaluated in a similar way using

$$d\bar{S} = \left(\frac{\partial \bar{S}}{\partial T}\right)_P dT + \left(\frac{d\bar{S}}{\partial P}\right)_T dP \tag{1.12}$$

to give

$$\bar{S}(T, P) = \bar{S}(T^0, P^0) + \int_{T^0}^{T} \frac{C_P(T, P^0)}{T} dT - \int_{P^0}^{P}\left(\frac{\partial \bar{V}}{\partial T}\right)_P dP \tag{1.13}$$

Gases

We shall first apply these expressions to the perfect gas, which has the equation of state

$$PV = RT \tag{1.14}$$

If we differentiate Equation (1.14) and substitute the result into Equation (1.11), we see that the last term of

[1] The development in this section is to a large extent based on Prigogine and Defay (1954).

[2] For a derivation of this expression the reader is referred to Prigogine and Defay (1954, pp. 24 and 161).

this expression drops out. This is in conformity with the definition of a perfect gas as one in which the internal energy and enthalpy depend only on the temperature. If we now substitute the same Equation (1.14) into Equation (1.13), we obtain

$$\bar{S}(T, P) = \bar{S}(T^0, P^0) + \int_{T^0}^{T} \frac{C_P(T)}{T} dT + R \ln P^0 - R \ln P \quad (1.15)$$

If P^0, the standard state of the gas, is taken as 1 atm, the term $R \ln P^0$ becomes zero. Substitution of Equations (1.15) and (1.11) into Equation (1.9) then gives

$$\mu(T, P) = \mu(T) + RT \ln P \quad (1.16)$$

which gives the chemical potential of a perfect gas in terms of the chemical potential $\mu(T)$ in its standard state at 1 atm.

Expression (1.16) is not generally applicable; however, if P is regarded as a *fugacity*, the relation is correct for nonideal gases as well. The pressure of a nonideal gas is related to the fugacity f by a *fugacity coefficient* $\gamma(T, P)$ as follows:

$$f = P\gamma \quad (1.17)$$

Sometimes nonideal gases also are discussed in terms of the *compressibility* factor $z = PV/RT$. To apply this we use Equation (1.2) as follows:

$$\left(\frac{\partial \mu}{\partial P}\right)_T = V$$

Integration then gives

$$\mu(T, P) = \mu(T) + \int_1^P V dP = \mu(T) + RT \int_1^P \left(\frac{z}{P}\right) dP$$

Consequently we obtain the equivalent of Equation (1.16) for a nonperfect gas:

$$\mu(T, P) = \mu(T) + RT \ln P + RT \int_1^P \left(\frac{z-1}{P}\right)_T dP \quad (1.18)$$

Comparison with Equation (1.17) shows that the last term in Equation (1.18) is equivalent to $RT \ln \gamma$.

Condensed phases

Previously we defined the coefficients of thermal expansion and compressibility. If these quantities are known as a function of T and P for any substance, the equation of state of that substance is determined. In the case of a perfect gas, α and κ have very simple forms, but for solids and liquids these functions are usually too complex to determine very exactly. However, for certain ranges of pressure and temperature and certain substances, α and κ may be regarded as independent of T and P. If κ is assumed to be independent of T, Equation (1.6) integrates to

$$V(T, P) = V(T, P^0) \exp[-\kappa(P - P^0)] \simeq V(T, P^0)[1 - \kappa(P - P^0)] \quad (1.19)$$

In the range in which κ becomes important $P - P^0$ is usually very large so that we may set $P^0 = 0$. If it is assumed that α is independent of temperature and pressure, and if Equations (1.5) and (1.19) are substituted into Equations (1.11) and (1.13), we obtain the expressions for the molar enthalpy and entropy for a condensed phase

$$\bar{H}(T, P) = \bar{H}(T^0, 0) + \int_{T^0}^{T} C_P(T, 0) dT + P\bar{V}(T, 0)(1 - \alpha T)(1 - \tfrac{1}{2}\kappa P) \quad (1.20)$$

and

$$\bar{S}(T, P) = \bar{S}(T^0, 0) + \int_{T^0}^{T} \frac{C_P(T, 0)}{T} dT + \alpha P\bar{V}(T, 0)(1 - \tfrac{1}{2}\kappa P) \quad (1.21)$$

The corresponding expression for the chemical potential of a pure condensed phase therefore becomes

$$\mu(T, P) = \mu(T) + P\bar{V}(T, 0)(1 - \tfrac{1}{2}\kappa P) \quad (1.22)$$

For most petrologic calculations the effects of the compressibility and the thermal expansion may be ignored relative to the first-order effects of temperature, pressure, and composition. Therefore Equation (1.22) assumes the simpler form

$$\mu(T, P) = \mu(T) + P\bar{V}(T, 0) \quad (1.23)$$

Phases of variable composition

Activities and activity coefficients

The general expression for the chemical potential of a component i in any multicomponental phase is

$$\mu_i = \mu_i^0 + RT \ln a_i \quad (1.24)$$

In this expression μ_i^0 is the same quantity obtained in Equations (1.22) or (1.23)[3] and a_i is the relative activity. A major concern throughout much of this book will be

[3] The superscript ^{0} is used here to indicate a pure substance.

the relation of a_i to the temperature, pressure, and the total composition of a given phase. The implicit form of this relation may be expressed as

$$a_i = f(T, P, c_j \cdots) \tag{1.25}$$

in which c_j refers to the concentration of the various components. For the case in which the standard state is pure i, Equation (1.26) takes the form

$$a_i = X_i \gamma_i \tag{1.26}$$

where X_i is the mole fraction of i in the solution. The quantity γ_i is the activity coefficient of the component i and in general is also a function of T, P, and the total composition of the phase.

In the analysis of mineral assemblages the standard state frequently chosen is the pure end member of a given solid solution series. Thus pure forsterite (Mg_2SiO_4) is the standard state for the magnesian component of olivine. The problem here, as for all solutions, is the deduction of explicit expressions for the activity coefficients. The forms that these expressions take are governed by our information of the statistical mechanics of the solutions. Fundamental to this problem is the identification of the statistical unit. In the case of simple atomic or molecular solutions in which the interaction energies are not very dependent on composition, the problem is relatively simple. The same may be said for many of the silicate solid solutions in which we can readily determine that mixing of similar ions occurs. Thus we are assured by x-ray evidence that Mg^{2+} and Fe^{2+} ions mix to form the intermediate members of the olivine series. When this mixing occurs perfectly randomly, and when the unlike particles interact with each other to the same extent as the like particles, Equation (1.26) assumes the form

$$a_i = X_i \tag{1.27}$$

which is a definition of the ideal or perfect solution.

In petrologic problems we frequently work with complex silicate melts and require a knowledge of the activities of their components. The problem then is of considerably greater complexity because the statistical mechanics of such systems are very poorly known. Usually we cannot readily identify the statistical units, although it has been shown that such melts consist, in part, of charged ions or complexes (Tomlinson, 1953). Our only recourse then is to make use of macroscopic variables such as the total concentration of a given independently variable component of the melt. This approach enables us to deduce the corresponding macroscopic activity coefficients although we do not know the microscopic characters such as the numbers of each given complex, etc. We shall return to this interesting subject later when we discuss silicate melts in detail.

The simplest explicit expression for γ_i of Equation (1.26) is provided by the regular solution model, which was introduced by Hildebrand (1929).[4] This model provides a particularly simple expression for the case of the binary solutions. Thus if 1 and 2 refer to the components, we have

$$a_1 = X_1 \exp\left(\frac{W}{RT} X_2^2\right) \tag{1.28}$$

$$a_2 = X_2 \exp\left(\frac{W}{RT} X_1^2\right) \tag{1.29}$$

For the regular solution, as strictly defined, W is a constant independent of T and P. Then the quantities WX_1^2 and WX_2^2 represent the partial molar heats of mixing. However, the status of W is sometimes not known, a situation that arises in the study of naturally occurring mineral assemblages. If such systems formed under essentially isothermal and isobaric conditions over a substantial composition range, we may write the forms

$$a_1 = X_1 \exp(\alpha X_2^2) \tag{1.30}$$

$$a_2 = X_2 \exp(\alpha X_1^2) \tag{1.31}$$

in which α is an unknown function of T and P but is independent of the compositional changes.

Regular solution models for ternary solutions also exist (Prigogine and Defay, 1954), and in some cases they have proved useful in the interpretation of mineral solid solutions (Mueller, 1961b; Saxena, 1973, p. 24).

The regular solution is the simplest model that exhibits several of the interesting features of natural systems. One of the most important of these is the phenomenon of unmixing or exsolution. Unmixing occurs when a cooling solution becomes unstable and breaks up into two solutions of differing composition. This occurs when the activity coefficients greatly exceed unity. For the regular solution the critical point is given by the condition $(W/RT) = 2$. It is informative to consider the mean molar Gibbs free energy of a regular solution which may be derived from Equations (1.24), (1.28), and (1.29). Thus we obtain

$$\bar{G} = X_1\mu_1^0 + (1 - X_1)\mu_2^0 + [X_1 RT \ln X_1 + (1 - X_1)RT \ln(1 - X_1)] + WX_1(1 - X_1) \tag{1.32}$$

[4] For a lucid explanation of the statistical mechanical basis for this model, the reader is referred to Gurney (1949).

The term enclosed by square brackets is the negative of the free energy of mixing and is always negative. The last term may be either positive or negative depending on the sign of W. This term is also referred to as the excess free energy. A plot of $\bar{G}$ against X_1, for various values of T, is useful in illustrating various points concerning the stabilities of the solution. For a detailed discussion of this aspect,' the reader is referred to Saxena (1973, Chapter 3).

The constant W is sometimes referred to as the "interchange energy." A regular solution with one energy constant W results in a "symmetrical nonideal solution." For such binary solutions the two limbs of a solvus are symmetric in relation to the intermediate composition. Many crystalline solutions, however, are not symmetric, and their excess free energy requires an expression with two constants as described by Thompson (1967):

$$G^E = [W_{G1}(1 - X_1) + W_{G2}X_1]X_1(1 - X_1) \tag{1.33}$$

Where W_{G1} and W_{G2} are the energy parameters commonly referred to as Margule's parameters. Both the parameters W_{G1} and W_{G2} are independent of the compositional changes but may vary as a function of P and T.

Another type of solution of great importance for certain petrogenic environments is the aqueous electrolyte. The conventions of standard states and solution models for these systems differ considerably from those discussed above. Because a discussion of these conventions and models would require too much space, the reader is referred to the appropriate chemical literature.[5] However, the conventions and the corresponding notation as presented below will be referred to when required.

The concept of the electrolytic solution assumes particular importance in the investigation of mineral systems in the low-temperature aqueous environment and has great bearing on the problems of chemical weathering, chemical sedimentation, ion-exchange adjustments of clay minerals, and diagenesis in sediments.

The activity of a dissolved electrolyte is defined as equal to the product of the activities of its ions at all concentrations. Thus for NaCl we obtain

$$a^{s}_{\mathrm{NaCl}} = a_{\mathrm{Na}}a_{\mathrm{Cl}} \tag{1.34}$$

Similarly for $BaCl_2$ we have

$$a^{s}_{\mathrm{BaCl_2}} = a_{\mathrm{Ba}}a^2_{\mathrm{Cl}} \tag{1.35}$$

In these expressions the superscripts refer to the solution.

Another quantity in use for electrolyte solutions is the mean activity, which takes the following form for NaCl and $BaCl_2$

$$a_{\pm} = (a_{\mathrm{Na}}a_{\mathrm{Cl}})^{1/2} \qquad a_{\pm} = (a_{\mathrm{Ba}}a^2_{\mathrm{Cl}})^{1/3} \tag{1.36}$$

In order to relate these activities to the concentrations (here expressed as *molal* quantities) we also make use of ionic activity coefficients. Thus the complete expression for the activity of NaCl becomes

$$a^{s}_{\mathrm{NaCl}} = c_{\mathrm{Na}}c_{\mathrm{Cl}}\gamma_{\mathrm{Na}}\gamma_{\mathrm{Cl}} \tag{1.37}$$

in which we have set

$$a_{\mathrm{Na}} = c_{\mathrm{Na}}\gamma_{\mathrm{Na}} \qquad a_{\mathrm{Cl}} = c_{\mathrm{Cl}}\gamma_{\mathrm{Cl}} \tag{1.38}$$

However since the ionic activity coefficients γ_{Na} and γ_{Cl} are not measurable quantities, we also define a mean activity coefficient. For NaCl this is

$$\gamma_{\pm} = (\gamma_{\mathrm{Na}}\gamma_{\mathrm{Cl}})^{1/2} \tag{1.39}$$

and for $BaCl_2$

$$\gamma_{\pm} = (\gamma_{\mathrm{Ba}}\gamma^2_{\mathrm{Cl}})^{1/3} \tag{1.40}$$

In general, if an electrolyte with the formula $A_{\nu_+}B_{\nu_-}$ dissociates into ν_+ cations and ν_- anions we have the following expressions

$$\begin{aligned} a_{\pm} &= [(a_{\mathrm{A}})^{\nu_+}(a_{\mathrm{B}})^{\nu_-}]^{1/\nu_+ + \nu_-} \\ &= [(c_{\mathrm{A}}\gamma_{\mathrm{A}})^{\nu_+}(c_{\mathrm{B}}\gamma_{\mathrm{B}})^{\nu_-}]^{1/\nu_+ + \nu_-} \\ &= [c_{\mathrm{A}}^{\nu_+}c_{\mathrm{B}}^{\nu_-}]^{1/\nu_+ + \nu_-}\gamma_{\pm} \end{aligned} \tag{1.41}$$

An interesting part of the theory of electrolytes is concerned with the Donnan membrane and osmotic equilibria. These equilibria are of considerable importance in the study of the reactions of clay minerals with aqueous solutions in geochemical environments of weathering, sedimentation, and diagenesis or early metamorphism. The general theory was first proposed by Donnan and Guggenheim (1932), although much previous work by Donnan and others had preceded this paper. The theory of Donnan and Guggenheim treats the compressibilities as constants, however. The equilibria of both ions and solvent must be considered. For the neutral solvent, which we index as u, we equate the chemical potentials as the two different sides of a membrane.

$$\mu'_u(T, P, c) = \mu''_u(T, P, c) \tag{1.42}$$

Here the single and double primes indicate the two different sides of the membrane. We shall treat only the isothermal case. Since each chemical potential is of the form of Equation (1.22), we obtain for the osmotic pressure

$$P' - P'' = \frac{RT}{[V_u]} \ln \frac{c''_u\gamma''_u}{c'_u\gamma'_u} \tag{1.43}$$

[5] A comprehensive and lucid exposition of standard states is to be found in Klotz (1950).

in which $V_u = V_u^*[1 + \frac{1}{2}\kappa_u(P' + P'')]$, V_u^* being the molar volume under standard conditions.

An analogous derivation for the electrolyte leads to the following expression for the osmotic pressure

$$P' - P'' = \frac{RT}{[V_s]} \ln \frac{(c''_+)^{\nu_+}(c''_-)^{\nu_-}(\gamma''_\pm)^{\nu_+ + \nu_-}}{(c'_+)^{\nu_+}(c'_-)^{\nu_-}(\gamma'_\pm)^{\nu_+ + \nu_-}} \tag{1.44}$$

in which $V_s = (\nu_+ V_+) + (\nu_- V_-)$. We shall return to this interesting subject later.

Electrode potentials[6]

It is frequently convenient to relate ionic concentration to easily measurable electrode potentials such as the oxidation potential and the pH. The oxidation potentials refer to half-cell reactions, which are compared with the standard half-cell reaction

$$H^+ + e \longrightarrow \tfrac{1}{2}H_2$$

The potential of this reaction has been arbitrarily set equal to zero. The standard free-energy change for any half-cell reaction is related to the standard potential E^0 by the relation

$$\Delta G^0 = \nu_e E^0 F \tag{1.45}$$

in which ν_e is the number of electrons involved in the reaction and F is the faraday constant equal to 23,060 cal/V-gm equivalent.

Reactions and equations of equilibrium

One of the most useful concepts in the treatment of the complex equilibria of chemical petrology is that of the reaction. From each reaction there follows an equation of equilibrium that places a restriction on the variables of the system. Obviously the greater the number of reactions that can be found, the more restricted the system is. Thus the simultaneous solution of all the equations of equilibrium leads to the elimination of variables and the expression of certain variables in terms of others. In the rare case of invariance as many equations as variables exist, and the system is completely determined.

The general form of the reaction may be written as

$$\sum \nu_R R \rightleftharpoons \sum \nu_P P \tag{1.a}$$

[6] For a detailed discussion of this subject, the reader is referred to standard texts such as Klotz (1950).

In this expression R and P refer to the reactant and product components, and ν_R and ν_P are the corresponding stoichiometric coefficients. The reactant and product components may all be confined to one phase; in this case the reaction is homogeneous. But in general some of the components will be dissolved in solids or fluids, depending upon the coexisting phases.

It is sometimes convenient to write the equation of equilibrium for a given reaction in its exponential form, that is, in terms of an equilibrium constant that is a function of the temperature and pressure only. It is also convenient to affix the index of the reaction as a subscript to the equilibrium constant so that a given constant may always be assigned to its reaction and be identifiable no matter how it occurs. This practice will generally be followed. Thus we write for the equilibrium constant corresponding to Reaction (1.a)

$$K_{(a)} = \frac{\Pi X_P^{\nu_P} \gamma_P^{\nu_P}}{\Pi X_R^{\nu_R} \gamma_R^{\nu_R}} \tag{1.46}$$

in which $K_{(a)} = \exp(-\Delta G_a^0/RT)$.[7] Here the Greek letter Π stands for the product of all the mole fractions and the corresponding activity coefficients raised to powers of their stoichiometric coefficients. This form of the equation of equilibrium is convenient when dealing with isothermal–isobaric equilibria.

If the change in the equilibrium constant as a function of temperature or pressure is sought, we may derive the following general relations from Equations (1.2) and (1.3):

$$\left(\frac{\partial \ln K}{\partial T}\right)_P = \frac{\Delta H^0}{RT^2} \tag{1.47}$$

and

$$\left(\frac{\partial \ln K}{\partial P}\right)_T = -\frac{\Delta V^0}{RT} \tag{1.48}$$

It is interesting at this point to examine a specific case. For this purpose it will serve to consider the equilibrium between Ca-pyroxene, hematite, calcite, quartz, and a fluid phase. This assemblage of minerals is of common occurrence in many contact metamorphic rocks. Let us consider the reaction involving hedenbergite, the iron component of Ca-pyroxene. Normally the amount of the component in Ca-pyroxenes in association with hematite

[7] When referring to the equilibrium constant of a reaction in another chapter, the chapter number is added to the subscript, i.e. $K_{(3.a)}$.

is small, and we may determine why this is the case. The reaction is

$$\underset{\textit{Ca-pyroxene}}{CaFeSi_2O_6} + \underset{\textit{gas}}{CO_2} + \underset{\textit{gas}}{\tfrac{1}{2}O_2} \rightleftharpoons \underset{\textit{hematite}}{\tfrac{1}{2}Fe_2O_3} + \underset{\textit{calcite}}{CaCO_3} + \underset{\textit{quartz}}{2SiO_2} \quad (1.b)$$

Here the phase in which the component occurs has been written under the component. In the analysis of such reactions it is necessary to identify the thermodynamically significant components, that is, the species, whether they be chemical elements or compounds that are independently variable. If only Reaction (1.b) is considered, the components of the system are seen to be five in number. However, if we consider the case in which $CaFeSi_2O_6$ is mutually dissolved with $CaMgSi_2O_6$, we must add the latter as a component. The actual choice of components is arbitrary since only their number counts. We may, for example, choose the elements, Ca, Mg, Fe, Si, O, and C, or the oxides CaO, MgO, FeO, Fe_2O_3, SiO_2, and CO_2. We shall anticipate a later derivation of the phase rule by writing

$$v = (n + 2) - \phi \quad (1.49)$$

Here v is the variance or number of degrees of freedom of the system, n is the number of components, and ϕ is the number of phases. The number 2 represents the pressure and temperature variables. The number of phases involved in Reaction (1.b) is obviously five since one would certainly expect CO_2 and O_2 to mix in all proportions under geologic conditions. The variance is therefore seen to be three. Thus at constant temperature and total pressure any variation in the proportions of the gaseous species should be reflected in the composition of the Ca-pyroxene. To illustrate in detail we write for the condition of equilibrium that the total change in free energy for the reaction must be zero:

$$\Delta G_{(b)} = 0 = 2\mu_Q + \mu_{Cal} + \tfrac{1}{2}\mu_{Hem} - \mu_{Hed} - \mu_{CO_2} - \tfrac{1}{2}\mu_{O_2} \quad (1.50)$$

where Q is quartz, Cal is calcite, Hem is hematite, and Hed is hedenbergite.

If now equations of type (1.16) for the gaseous components and of type (1.23), (1.24), and (1.26) for the solid components are substituted into (1.50), we arrive at the useful relation

$$\Delta G_{(b)} = 0 = \Delta G^0_{(b)} + P(2V_Q + V_{Cal} + \tfrac{1}{2}V_{Hem} - V_{Hed}) - RT(\ln P_{CO_2} - \tfrac{1}{2}\ln P_{O_2}) - RT\ln X^{Ca\text{-}Px}_{Hed}\gamma^{Ca\text{-}Px}_{Hed} \quad (1.51)$$

where

$$\Delta G^0_{(b)} = 2\mu^0_Q + \mu^0_{Cal} + \tfrac{1}{2}\mu^0_{Hem} - \mu^0_{Hed} - \mu^0_{CO_2} - \tfrac{1}{2}\mu^0_{O_2}$$

The quantities P_{CO_2} and P_{O_2} may be regarded as either pressures or fugacities if we remember that the former applies only when the species behave as perfect gases. At this stage, however, we disregard this complication.

We note that the second term on the right of Equation (1.51) includes the molar volumes of the solid phases, the third term expresses the effect of the gaseous components as partial pressure, and the last term takes into account the solid solubility of hedenbergite in the pyroxene solution.

Alternatively we may express the equation of equilibrium in terms of the equilibrium constant:

$$K_{(b)}(T, P) = \frac{1}{X^{Ca\text{-}Px}_{Hed}P_{CO_2}P^{1/2}_{O_2}\gamma^{Ca\text{-}Px}_{Hed}} \quad (1.52)$$

This form best exhibits the functional relations of the compositional variables at constant temperature and total pressure. As will be shown later, the activity coefficient $\gamma^{Ca\text{-}Px}_{Hed}$ is equal to unity in this case since Ca-pyroxene forms virtually ideal solutions of its magnesium and iron end members. It is plain, therefore, that increasing P_{O_2} or P_{CO_2} calls for a decrease in the iron content of the pyroxene under these conditions, in agreement with observation.

It is instructive to return now to the general form of the equation of equilibrium corresponding to Reaction (1.a). For the logarithmic form we obtain

$$\Delta G_{(a)} = 0 = \Delta G^0_{(a)} + P[\sum \nu^c_P V^c_P - \sum \nu^c_R V^c_R] + RT[\sum \nu^g_P \ln P^g_P - \sum \nu^g_R \ln P^g_R] + RT[\sum \nu^c_P \ln X^c_P\gamma^c_P - \sum \nu^c_R \ln X^c_R\gamma^c_R] \quad (1.53)$$

In this equation the superscripts c and g refer to the condensed and gaseous components, respectively. The terms are in the same order as in Equation (1.51).

Consideration will show that certain difficulties are encountered when an attempt is made to apply Equation (1.53) to rock systems. Immediate difficulties arise, for example, when the P^g_P of the third term are desired as functions of the partial pressures. In systems as complex as rocks, we cannot in general know the full extent of the types of species that were present during crystallization. Also, usually insufficient chemical data exist on the interactions between these species. For example, we do not even know the deviations from ideality of mixtures of H_2O and CO_2 for petrogenic ranges of pressure and temperature. However, even if we assume that all the gases behave ideally, some difficulties still remain. Thus, for example, P in the second term is frequently taken as equal to the total gas pressure, which is in turn assumed to equal the pressure of the rock overburden. Thus we have

$$P = \sum (P^g_{\mathrm{R}} + P^g_{\mathrm{P}}) \tag{1.54}$$

This means that each set of equations of type (1.53) involves the simultaneous solution of transcendental equations whose solutions are only implicit. Such equations may, of course, be solved by various methods of approximation.

An even greater difficulty arises when P cannot be taken as equal to the sum of the partial pressures, as, for example, when the pressure of the overburden is in excess to the gas pressure. It may even happen that the pressure has no simple relation to the overburden, as when large horizontal stresses exist. In any case, we may then have

$$\sum (P^g_{\mathrm{R}} + P^g_{\mathrm{P}}) < P \tag{1.55}$$

In this situation P appears as an extra independent variable as the phase rule (see next section) takes the form

$$v = (n + 3) - \phi \tag{1.56}$$

The Gibbs phase rule

When all the phases of a system are in equilibrium, the chemical potential of each component must be the same in each phase. Thus in the system represented by Reaction (1.b) we have

$$\mu_{\mathrm{CO_2}}\,(\text{gas}) = \mu_{\mathrm{CO_2}}\,(\text{calcite})$$
$$\mu_{\mathrm{SiO_2}}\,(\text{Ca-pyroxene}) = \mu_{\mathrm{SiO_2}}\,(\text{quartz}) \qquad \text{etc.} \tag{1.57}$$

When the conditions of type (1.57) do not exist, the components are transferred from phases in which they have a high chemical potential to those in which this quantity is low until equality is attained. Relations such as Equation (1.57) are therefore the foundation of the Gibbs phase rule as given in Equations (1.49) and (1.56). There are $\phi - 1$ separate relations such as Equation (1.57) for each component or $n(\phi - 1)$ total relations of this type. There is also for each phase a relation of type $\sum X_i = 1$ relating the mole fractions of the components i. Thus there are in all $n(\phi - 1) + \phi$ equations of condition. In addition to the n composition variables there are such variables as the temperature (one variable for the total system) and the various pressures that bear on the various phases. As we have seen, these latter can be different for different phases. Additional degrees of freedom are introduced by the influences of external fields, but these complications will not be considered now. If only one pressure variable is considered, we may obtain Equation (1.49) by subtracting the total number of equations from the total number of variables:

$$v = n\phi + 2 - [n(\phi - 1) + \phi] = (n + 2) - \phi \tag{1.49}$$

An extension to arrive at Equation (1.56) is obvious.

Effect of the gravitational field

General

Of all the possible influences on mineral equilibria the gravitational field assumes one of the most ambiguous roles. This ambiguity results largely from the fact that differentiation can only have occurred on a global scale in the solid material of the earth since it is obvious that most crustal materials are not in chemical or mechanical equilibrium with respect to the earth's field.

Current views on the gross structure of the earth lead us to suspect that some differentiation has in the past produced shells of varying density and chemical content with respect to the center of the earth. It seems reasonable to suppose that some of this differentiation was chemical in character, that is, it involved the independent movement of particles by means of diffusion.

It is our purpose to show here that the existence of a gravitational field implies the existence of chemical potential gradients of a type that gives rise to continuous variations in the composition of phases equilibrated to this field. This problem in relation to the earth was discussed by Ramberg (1947, 1948b) and by Brewer (1951).

The general equation for the change of the chemical potential of a component i in a phase of variable composition in the gravitational field is

$$d\mu_i = \left(\frac{\partial \mu_i}{\partial T}\right)_{P,\phi,X} dT + \left(\frac{\partial \mu_i}{\partial P}\right)_{T,\phi,X} dP + \left(\frac{\partial \mu_i}{\partial \phi}\right)_{T,P,X} d\phi + \sum_{j}^{j-1} \left(\frac{\partial \mu_i}{\partial X_i}\right)_{T,P,\phi,X_{j-1}} dX_i \tag{1.58}$$

The mole fraction of i is related to the mole fractions of the $j - 1$ other components as follows:

$$X_i = 1 - \sum_{i=1}^{j-1} X_j \tag{1.59}$$

The quantity ϕ in Equation (1.58) is the gravitational potential and is equal to the integral of the gravitational acceleration through the height h. In a relatively constant field such as obtains in the earth's upper regions we have

$$d\phi = g dh \tag{1.60}$$

in which g is the constant of acceleration.

Binary solutions in the gravitational field

It is instructive to consider the behavior of a binary solution composed of the species A and B. Equation (1.58) then gives for the chemical potential of A

$$d\mu_A = \left(\frac{\partial \mu_A}{\partial T}\right)_{P,h,X_A} dT + \left(\frac{\partial \mu_A}{\partial P}\right)_{T,h,X_A} dP + \left(\frac{\partial \mu_A}{\partial h}\right)_{T,P,X_A} dh + \left(\frac{\partial \mu_A}{\partial X_A}\right)_{T,P,h} dX_A \tag{1.61}$$

The chemical potential is defined as

$$\mu_A = \mu_A^0 + RT \ln X_A \gamma_A + M_A g h \tag{1.62}$$

where M_A is the molar weight of A. If δ_{AB} is the density of the solution, we also have

$$\delta_{AB} = \frac{M_{AB}}{V_{AB}} = \frac{M_A X_A + M_B(1 - X_A)}{V_A X_A + V_B(1 - X_A)} \tag{1.63}$$

in which M and V refer to the individual and mean molar weights and volumes. The pressure is related to the height as

$$dP = -\delta_{AB} g dh \tag{1.64}$$

From Equations (1.1), (1.2), and (1.62) we obtain

$$\left(\frac{\partial \mu_A}{\partial T}\right)_{P,h,X_A} = -s_A \qquad \left(\frac{\partial \mu_A}{\partial P}\right)_{T,h,X_A} = V_A$$

$$\left(\frac{\partial \mu_A}{\partial h}\right)_{T,P,X_A} = M_A g$$

and

$$\left(\frac{\partial \mu_A}{\partial X_A}\right)_{T,P,h} = \frac{RT}{X_A} + RT \frac{\partial \ln \gamma_A}{\partial X_A}$$

In order for equilibrium to obtain we must have a uniform temperature throughout the body; therefore $dT = 0$ and Equation (1.61) becomes

$$d\mu_A = 0 = \left(M_A g - V_A \frac{M_{AB}}{V_{AB}} g\right) dh + \frac{RT}{X_A} dX_A + RT \frac{\partial \ln \gamma_A}{\partial X_A} dX_A \tag{1.65}$$

An interesting case to which we may apply Equation (1.65) is the regular solution which is defined by the relations (1.28) and (1.29). Thus we obtain

$$\frac{\partial \ln \gamma_A}{\partial X_A} = -\frac{2W}{RT}(1 - X_A) \tag{1.66}$$

which we substitute in Equation (1.65) and get

$$\frac{dX_A}{dh} = \frac{[V_A(M_{AB}/V_{AB}) - M_A]g}{(RT/X_A) - 2W(1 - X_A)} \tag{1.67}$$

It is helpful at this point to introduce a slight simplification. We therefore let $V_A = V_B = V_{AB}$. This simplification is justified in many petrologic problems in which the members of certain ion pairs have similar radii though they may differ considerably in atomic weight. Equation (1.67) then becomes

$$\frac{dX_A}{dh} = \frac{(M_B - M_A)(1 - X_A)g}{(RT/X_A) - 2W(1 - X_A)} \tag{1.68}$$

We now consider a column of the regular solution in a gravitational field. In such a column the ratio A/B will vary uniformly from top to bottom if equilibrium is obtained. To show this we subdivide the column into multilayers[8] so thin that any one has an essentially constant composition. Any two of the multilayers may be compared. We designate the lowest of these α and the highest as $\beta (h\beta - h\alpha \equiv \Delta h)$, and integrate Equation (1.68) between these limits to obtain

$$\frac{X_A^\beta}{1 - X_A^\beta} = \frac{X_A^\alpha}{1 - X_A^\alpha} \exp\left[\frac{(M_B - M_A)g\Delta h}{RT}\right] \times \exp\left[\frac{2W}{RT}(X_A^\beta - X_A^\alpha)\right] \tag{1.69}$$

If $W = 0$ the second exponential factor becomes 1 and we obtain the ideal-solution case. In this event we see that if $M_B > M_A$ and $\Delta h > 0$, the species A is concentrated in the upper multilayer β. However, if $W < 0$, corresponding to attraction between A and B which exceeds the attraction between the like species, we see that X_A^β might approach X_A^α in spite of the large difference in mass. We shall return to this subject later.

Kinetics

The preservation of rocks formed under environments remote from the earth's surface is almost entirely due to the failure of these rocks to react chemically with sufficient

[8] It is obvious that the multilayers α and β cannot be considered phases in the ordinary sense since a column of solution is really a continuum of varying composition.

rapidity to adjust to their new environment, since virtually no plutonic mineral assemblages are stable at the surface. Thus we owe the very existence of chemical petrology to kinetic factors.

As applied to the general case of heterogeneous systems, kinetics has a three-fold aspect involving the fundamental processes of diffusion, reaction, and nucleation. These processes involve the movement of individual particles: atoms, molecules, ions, or complexes. However, kinetics is also involved in the interpretation of mass movements such as the hydrodynamic flow of gases and liquids and the plastic flow of solids, since the latter movements are the result of the directed flow of individual particles. Diffusion is also closely related to the flow of heat and electricity, and the soret effect is probably of considerable petrologic importance.

Theory of absolute reaction rates

For the fundamentals of the theory the reader is referred to the major work of Glasstone, *et al.* (1941). A very lucid short discussion may also be found in Darken and Gurry (1953). Here we merely present the outlines of the theory and give the fundamental equations for reference.

All forms of atomistic or chemical processes are assumed to involve an intermediate or *activated* state in which the reactants assume a form which is a precurser of the products. Thus we obtain a more general form of Reaction (1.a):

$$\sum \nu_R \mathrm{R} \rightleftharpoons A \longrightarrow \sum \nu_P \mathrm{P} \qquad (1.c)$$

in which A represents the activated complex. It should be noted that in the first step the reactants are regarded as being in equilibrium with the activated complex. Therefore, following the usage of the thermodynamics of equilibrium processes, an equilibrium constant may be written in terms of the activities of the reactants and activated complex as follows:

$$K^\dagger = \frac{a^\dagger}{\Pi a_R^{\nu_R}} = \left(\frac{c^\dagger}{\Pi c_R^{\nu_R}}\right)\left(\frac{\gamma^\dagger}{\Pi \gamma_R^{\nu_R}}\right) \qquad (1.70)$$

Here the dagger indicates the activated complex. The concentrations c are usually expressed in molecules or moles per unit volume, and the other symbols are as previously defined.

The second step of Reaction (1.c) involves the observed rate of the reaction which is directly proportional to the concentration of the activated complex. From the wave mechanical theory this rate is found to be

$$\text{Rate of reaction} = r = c^\dagger\left(\frac{RT}{Nh}\right) \qquad (1.71)$$

In this equation R and T are as usual the gas constant and absolute temperature, respectively, N is the Avogadro number, and h is Planck's constant. It should be observed that the factor in parentheses is independent of the nature of the system and contains T as the only variable. Relation (1.71) also shows that the rate of reaction at any given temperature depends only on the concentration of the activated complex. For ordinary temperatures the term RT/Nh is approximately 10^{13} or 10^{14} sec^{-1}. The reciprocal of this term, Nh/RT, is then the average lifetime of the activated complex. If $c^\dagger$ is substituted from Equation (1.70) into Equation (1.71) we obtain the relation

$$r = \left(\frac{RT}{Nh} K^\dagger \frac{\Pi \gamma_R^{\nu_R}}{\gamma^\dagger}\right) \Pi c_R^{\nu_R} \qquad (1.72)$$

The term in the parentheses of Equation (1.72) is referred to as the specific rate constant[9] K.

The equilibrium constant $K^\dagger$ may also be expressed in terms of the standard Gibbs free energy of formation of the activated complex $\Delta G^\dagger$ from the reactants in their standard states. $\Delta G^\dagger$ may also be expressed as a function of the entropy and enthalpy. Thus we have

$$\begin{aligned} K^\dagger &= \exp(-\Delta G^\dagger/RT) \\ &= \exp(-\Delta H^\dagger/RT)\exp(\Delta S^\dagger/R) \end{aligned} \qquad (1.73)$$

The second factor in Equation (1.72), the product of the concentrations, determines the order of the reaction. Thus the bimolecular reaction $\mathrm{A} + \mathrm{B} \rightarrow \mathrm{C}$ *may* be representable by $r = Kc_A c_B$, in which the time rate of formation of the species C is proportional to the concentrations of A and B. However, the order observed need not coincide with the molecularity of the reaction as written. The order of a given reaction can only be established experimentally.

Diffusion

Classical diffusion theory is based on Fick's laws, which date to 1855. The mathematical expression of Fick's first law is

$$J = -D\frac{\partial c}{\partial s} \qquad (1.74)$$

Here J is the flux of particles which cross 1 cm^2/sec of a surface oriented perpendicular to the diffusion direction s. D is known as the coefficient of diffusion and dc/ds is the concentration gradient. The minus sign

[9] This development omits the transmission coefficient, which is equivalent to a correction factor.

occurs to make J positive, since c decreases with increasing s in the direction of diffusion. It is obvious that Equation (1.74) can apply only to the special case in which the concentration gradient is only a function of the direction s. In the general case in which gradients exist in x, y, and z directions of coordinates, Fick's law assumes the form

$$\mathbf{J} = -D\nabla c \tag{1.75}$$

Fick's second law may be derived from Equation (1.74) by a consideration of the variation of $\mathbf{J}$ with s.[10] This leads to the expression

$$\frac{\partial c}{\partial t} = D\frac{\partial^2 c}{\partial s^2} \tag{1.76}$$

For the case of the steady state we have $dc/dt = 0$; thus dc/ds = constant. Numerous solutions to these equations for nonsteady state conditions and various geometries are given in the literature. Most of these solutions assume that D is independent of the composition. However, this is generally not the case, and in most petrologically significant materials the deviation may be expected to be large. This is especially well brought out in the case of solutions that show miscibility gaps, as we shall presently illustrate.

We may also derive Equation (1.74) from the theory of reaction rates. Thus if λ_s is the distance between successive equilibrium positions of the diffusing particle, we get for the flux in the forward direction

$$J_{\mathrm{f}} = c\lambda_s K \tag{1.77}$$

Here c is as usual the concentration and K is the specific rate constant. Now the concentration in the adjacent equilibrium position is $c + \lambda_s(\partial c/\partial s)$ to a first approximation. Consequently the flux in the backward direction is

$$J_{\mathrm{b}} = \left(c + \lambda_s\frac{\partial c}{\partial s}\right)\lambda_s K \tag{1.78}$$

The difference between Equations (1.77) and (1.78) is the resultant flux in the forward direction:

$$J = -\lambda^2 K\frac{\partial c}{\partial s} \tag{1.79}$$

Thus by comparison with Equation (1.74) we obtain

$$D_1^0 = \lambda^2 K \qquad \text{(for ideal solutions)} \tag{1.80}$$

It may be noted that Equations (1.77) and (1.78) have a form similar to Equation (1.72) for the case of the first-order reaction; the rate of the diffusion "reaction" is proportional to the concentration.

The interesting case of diffusion in nonideal systems may also be treated according to the theory of absolute reaction rates if certain simplifying assumptions are made. We note that the specific rate constant K is the same for both directions of flux in the ideal systems. This is not true for nonideal systems. For the latter the expression corresponding to Equation (1.79) becomes

$$J = c\lambda_s K_{\mathrm{f}} - \left(c + \lambda\frac{\partial c}{\partial s}\right)\lambda_s K_{\mathrm{b}} \tag{1.81}$$

In this expression K_{f} and K_{b} are the specific rate constants for the flux in the forward and backward directions, respectively.

The difference between ideal and nonideal systems may be illustrated by a plot of the Gibbs free energy against the reaction (diffusion) coordinate. Figure 1.1 is a plot of the Gibbs free energy difference between an initial equilibrium position A and a final equilibrium position B separated by a distance λ_s. The full curve gives the free energy variation in the ideal system, for which the equilibrium positions are at the same level and the free energy of activation is $\Delta G_0^{\ddagger}$. Thus diffusion is controlled solely by the concentration gradient. The dashed curve shows the relations for a nonideal solution in which the free energy of activation for diffusion to the right is $\Delta G_0^{\ddagger} + \frac{1}{2}\Delta G$ and to the left $\Delta G_0^{\ddagger} - \frac{1}{2}\Delta G$. The

1.1 Gibbs free energy relations for ideal and nonideal systems. $\Delta G_0^{\ddagger}$ is the standard free energy of activation for diffusion. (After Glasstone *et al.*, 1941).

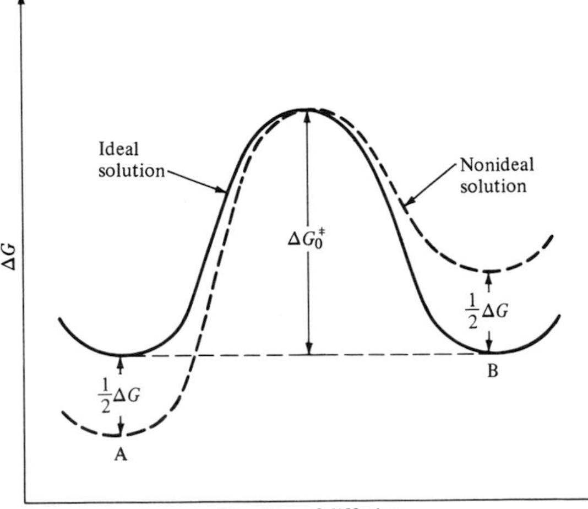

[10] See, for example, the derivation given in Darken and Gurry (1953).

term ΔG represents the difference in excess free energy between the two equilibrium positions and is therefore related to the change in the activity coefficient as follows:

$$\Delta G = \lambda_s RT \frac{\partial \ln \gamma_i}{\partial s} \tag{1.82}$$

The change in the excess free energy with the distance s is of course brought about by the statistical difference in concentration between the two positions. If the activity coefficient in any part of the system exceeds the critical value, the solution becomes metastable or unstable and unmixing will result. Equation (1.81) may be written[11] in terms of the activity coefficient and mole fraction as follows:

$$J_i = -D_i^0\left(1 + X_i \frac{\partial \ln \gamma_i}{\partial X_i}\right)\frac{\partial c}{\partial s} \tag{1.83}$$

in which D_i^0 is the coefficient of diffusion for the ideal solution.

It is interesting to examine Equation (1.83) for the case of diffusional instability. The stability condition for a binary solution consisting of components 1 and 2 are

$$\frac{\partial \mu_1}{\partial X_2} < 0, \quad \frac{\partial \mu_2}{\partial X_1} < 0$$
$$\frac{\partial \mu_1}{\partial X_1} > 0, \quad \frac{\partial \mu_2}{\partial X_2} > 0 \tag{1.84}$$

For component 1 we have

$$\frac{\partial \mu_1}{\partial X_1} = \frac{RT}{X_1} + RT \frac{\partial \ln \gamma_1}{\partial X_1} \tag{1.85}$$

We see that if the last term assumes a large enough negative value, $(\partial \mu_1/\partial X_1) < 0$, and the solution becomes unstable. Under these conditions a negative diffusion occurs. That is, particles move against the concentration gradient to form nuclei of the exsolved phase. This may be illustrated for the case of the regular solution by making use of the activity coefficients in Equations (1.30) and (1.31):

$$\frac{\partial \ln \gamma_1}{\partial X_1} = -2\alpha(1 - X_1) \tag{1.86}$$

We see that for $\alpha > 0$ this expression is always negative, and that for $\alpha = 2$ we have

$$1 + X_1 \frac{\partial \ln \gamma_1}{\partial X_1} = 0$$

and unmixing is imminent. These relations are illustrated in Figure 1.2, a plot of the activity in a solution in which

[11] The derivation is simple although somewhat long. It is given in the book by Glasstone *et al.* (1941).

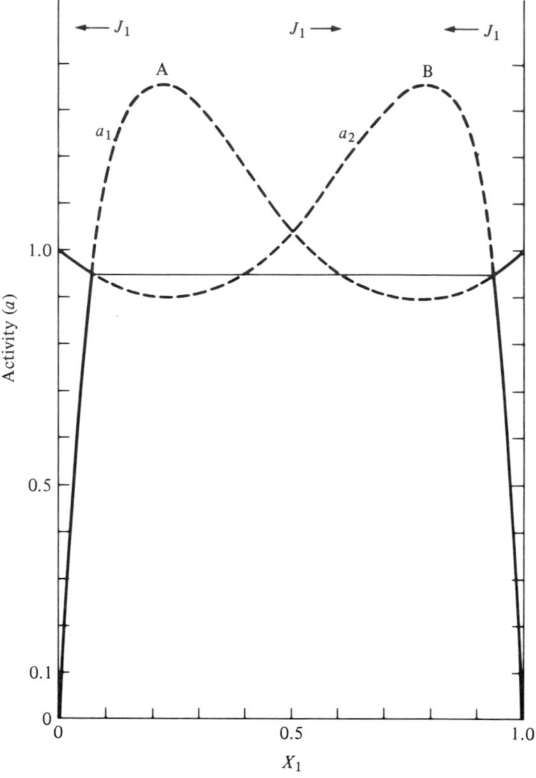

1.2 Miscibility gap for a binary solution with $\alpha (= W/RT)$ equal to 3.0. The metastable and unstable parts of the activity curves are shown as broken lines. a_1 and a_2 are activities. J_1 shows direction of diffusion.

$\alpha = 3$. In this figure the metastable and unstable parts of the activity curves are shown as dashed lines and the miscibility gap by a solid horizontal line. The regions to the left of A and to the right of B are regions of stability and metastability. In the unstable region between A and B diffusion against the concentration gradient initiates unmixing.

Nucleation

The consideration of diffusional instability leads naturally to the problems of nucleation and the growth of nuclei. The subject is conveniently approached through the study of successively more complicated cases. We may, for example, choose the following sequence:

1. Nucleation of liquid from a vapor in a one component system
2. Nucleation of crystals from a vapor in a one component system

3. Nucleation of one liquid in another liquid in a binary system exhibiting a miscibility gap
4. Nucleation in a binary solid solution with a miscibility gap
5. Nucleation in multicomponental heterogeneous systems

Item 5 is the case most often encountered in petrologic systems.

In the formation of liquid drop embryos from a one-component vapor phase the condition of supersaturation must be fulfilled because the first small embryos to form have a high surface energy which makes them unstable even under a small degree of supersaturation. The magnitude of the necessary supersaturation pressure P_i required to stabilize an embryo of radius r_i, surface energy σ, molar volume V_E, and saturation pressure P_∞ is given by the following expression:

$$RT \ln \frac{P_i}{P_\infty} = \frac{2\sigma V_E}{r_i} \tag{1.87}$$

which is discussed by Dunning (1955).

If we consider the equilibrium between a species A and the nucleus composed of i particles of A we may write the reaction for the embryo formation as

$$i\mathrm{A} \rightleftharpoons \mathrm{A}_i \tag{1.d}$$

It is evident that Reaction (1.d) requires a fluctuation or negative diffusion to bring the i molecules together. The equilibrium constant for Reaction (1.d) may be written as

$$K_{(\mathrm{d})} = \frac{c_{\mathrm{A}_i}}{c_{\mathrm{A}}^i} = \exp\left(-\frac{\Delta G_i^0}{RT}\right) \tag{1.88}$$

in which c_{A_i} and c_{A} are the concentrations of the embryos and the A particles, respectively, and ΔG_i^0 is the standard free energy of formation of an embryo. For a special droplet at rest it may be shown that

$$\Delta G_i^0 = i(\mu_\infty^L - \mu_i) + 4\pi\sigma r_i^2 \tag{1.89}$$

in which μ_∞^L is the chemical potential of A in the liquid drop of infinite size (macroscopic drop) and μ_i is the chemical potential of a molecule in the supersaturated vapor.

The critical nucleus for the growth of a droplet is the embryo for which ΔG_i^0 attains a maximum. To show this we substitute $i = 4\pi r_i^3/3V_E$ into Equation (1.89) and differentiate with respect to r_i. At the maximum $r_i = r_m$, and we have

$$r_m = \frac{2\sigma V_E}{\mu_i - \mu_\infty^L} = \frac{2\sigma V_E}{RT \ln (P_i/P_\infty)} \tag{1.90}$$

Comparison of Equation (1.90) with (1.87) shows that there is a critical size for each degree of supersaturation. The expression for ΔG_i^0 in terms of r_m then is

$$\Delta G_i^0 = 4\pi\sigma\left(r_i^2 - \frac{2}{3}\frac{r_i^3}{r_m}\right) \tag{1.91}$$

Figure 1.3 shows a plot of ΔG_i^0 against r_i. It is apparent that at first with increasing r_i of the embryo the r_i^2 term dominates because of the surface energy, but that at $r_i = r_m$ this effect is overcome by the increased size of the embryo. As a consequence any embryo which has a radius exceeding r_m can grow because ΔG_i^0 is decreasing.

The continued growth of an embryo may be expressed by reactions of the following type:

$$\begin{aligned} \mathrm{A}_i + \mathrm{A} &\longrightarrow \mathrm{A}_{i+1} \\ \mathrm{A}_{i+1} + \mathrm{A} &\longrightarrow \mathrm{A}_{i+2} \quad \text{etc.} \end{aligned} \tag{1.e}$$

The kinetics of nucleation in condensed systems has been treated by Turnbull and Fisher (1949). They conclude that if c_{A_i} represents the steady state concentration of nuclei, then the "forward" rate r_f, or the rate for reactions of the type (1.e) may be written as

$$r_\mathrm{f} = c_{\mathrm{A}_i}(a_1 i^{2/3})\left(\frac{RT}{Nh}\right)\exp\left(-\frac{\Delta G_1^\ddagger}{RT}\right) \tag{1.92}$$

Here $(a_1 i^{2/3})$ is the number of "A" molecules in contact with the nucleus and $\Delta G_1^\ddagger$ is the free energy of activation. The rate of the reverse reaction $\mathrm{A}_{i+1} \rightarrow \mathrm{A}_i + \mathrm{A}$ may be written as

$$r_\mathrm{b} = c_{\mathrm{A}_{i+1}}(a_2 i^{2/3})\left(\frac{RT}{Nh}\right)\exp\left(-\frac{\Delta G_2^\ddagger}{RT}\right) \tag{1.93}$$

1.3 Plot of the standard free energy (ΔG_i^0) against radius of embryo r_i. r_m corresponds to maximum ΔG_i^0.

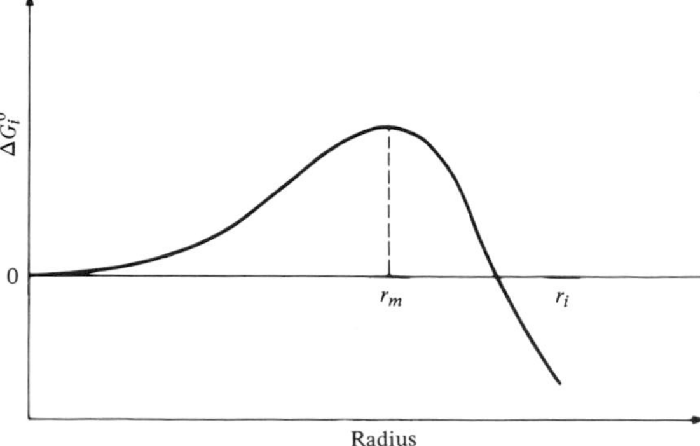

The net rate of formation of the A_{i+1} nuclei is therefore

$$r = r_f - r_b \tag{1.94}$$

An expression analogous to Equation (1.89) may also be written for the free energy of formation of a crystalline embryo:

$$\Delta G_i^0 = i(\mu_\infty^0 - \mu_i) + \sum_n \sigma_n O_n \tag{1.95}$$

Here μ_∞^0 is the chemical potential of A in an infinitely large crystal and μ_i is the chemical potential in the vapor. O_n is in this case a structural factor for each of the n faces of the crystal embryo and is proportional to the area of the face.

2 Equilibrium of mineral systems

One of the major objectives of chemical petrology is knowledge of the heterogeneous reactions in the chemical systems of nature. Considerable progress in this direction has been made in the application of methods of experimental phase equilibria and in the theoretical interpretation of observational and experimental data. In this chapter we shall be particularly concerned with applications of theory to some well known systems and its extension to certain less known but important natural systems. In the analysis of the more complex systems a particular advantage of the theoretical-analytical approach lies in the manner in which the pertinent variables are displayed in the derivations and the unambiguous way in which limits, boundary conditions, and assumptions may be kept track of. Another advantage of the analytical approach is that many diagrams may be dispensed with, and those that are used represent simple surfaces or volumes in multi-componental space.

The writers believe that the following sequence of mineral systems is a fair representation of those likely to be encountered in nature over the broad range of conditions from weathering to magmatism. The examples have been chosen because of their importance or because they illustrate some particular type of variable interaction. In the presentation of these diverse systems an attempt is made to exhibit the unifying theme of the equations of equilibrium which are always of the same form although the phases involved may be quite different in their physical characteristics.

One-component transitions

This type of reaction is of special significance because it is univariant in pressure and temperature. That is, the coexistence of two phases fixes one of these variables uniquely in terms of the other.

The best example of this type of system is that in which the component is a pure element, as in the transition

$$\textit{graphite} \rightleftharpoons \textit{diamond} \tag{2.a}$$

which was discussed by Berman and Simon (1955). An almost equally good example is the system SiO_2 which was extensively studied by Mosesman and Pitzer (1941). The latter system is effectively one componental because of the strong binding between silicon and oxygen.

Particular interest attaches to such simple univariant transitions which remain "decoupled" from the more complex assemblages in which they occur. Coupling occurs when a participant in a simple transition interacts in some way with other components. Such is probably the case with certain of the high temperature forms of SiO_2. Among the more complex silicates, coupling is the rule rather than the exception, as represented by such transitions as

$$\textit{orthorhombic pyroxene} \rightleftharpoons \textit{monoclinic pyroxene}$$

in which solid solution of Mg and Fe^{2+} plays an important role. It is obvious that coupling may destroy the value of a transition as a geothermometer or a geobarometer.

If we use Reaction (2.a) as an illustration, a convenient general form for the equilibrium equation is

$$RT \ln K_{(a)}(T, P) = \mu_C^{0G} - \mu_C^{0D} - \int_0^P \Delta V_{(a)} dP = 0 \tag{2.1}$$

in which μ_C^{0D} and μ_C^{0G} are the standard chemical potentials of carbon diamond and graphite, respectively, and $\Delta V_{(a)}$ is the volume change for Reaction (2.a). In this case it is necessary to take into account the effect of the compressibilities because of the high pressures involved. According to Berman and Simon (1955) the compressibility of diamond may be considered a constant, and the analogue of Equation (1.22) for its chemical potential may be used. However, the variation of the compressibility of graphite is considerable; a graphical method was employed to evaluate the contribution of graphite to the last term in Equation (2.1). The term $\mu_C^{0G} - \mu_C^{0D}$ may, of course, be evaluated from standard thermochemical data. The results of the calculation are

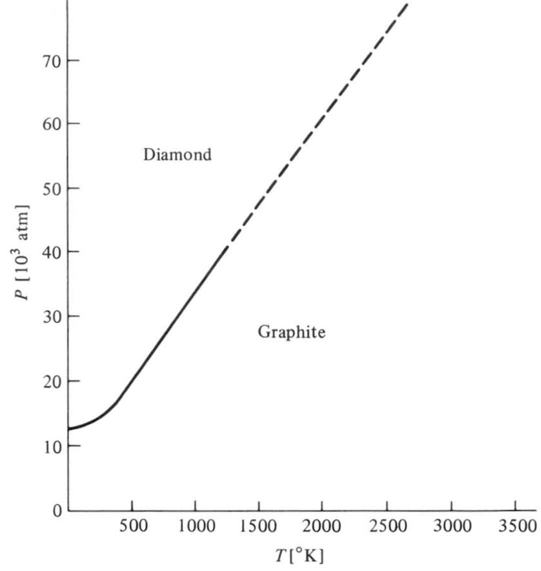

2.1 Univariant graphite-diamond equilibrium curve calculated by Berman and Simon (1955; see also Bundy, 1968).

shown in Figure 2.1. For a recent experimental and theoretical treatment of graphite-diamond equilibrium, see Bundy (1968).

Univariant transformations of solids

We now consider a type of reaction which is a degree more complex than the simple transition. A well known example is

$$\underset{\textit{albite}}{\tfrac{1}{2}NaAlSi_3O_8} + \underset{\textit{nepheline}}{\tfrac{1}{2}NaAlSiO_4} \rightleftharpoons \underset{\textit{jadeite}}{NaAlSi_2O_6} \tag{2.b}$$

A theoretical treatment of this reaction was given by Adams (1953). The volume change amounts to -34.9 cm^3, so pressure will greatly affect the equilibrium. However, the previous work of Yoder and Weir (1951) showed that in this case the combined effect of the compressibilities and thermal expansion had a *relatively* small effect and could safely be disregarded. The equation of equilibrium thus assumes the form

$$\ln K_{(b)}(T, P) = -\Delta G^0_{(b)} - P\Delta V_{(b)} = 0 \tag{2.2}$$

for the case where albite, nepheline, and jadeite are all stoichiometric compounds. According to Adams the thermochemical data yield the following expression for the standard free energy change

$$\Delta G^0_{(b)} = -6120 + 14.7T \qquad (2.3)$$

Thus Equation (2.2) becomes

$$-6120 + 14.7T = -P\Delta V_{(b)} \qquad (2.4)$$

where $\Delta V_{(b)}$ is regarded as a constant.

It should be kept in mind that the direct applicability of this equilibrium to rocks is severely limited. The presence of calcium or potassium would require the consideration of the energetics of the distribution of these elements among the three phases involved, and this problem is a difficult one because of the nonideal characters of the solid solutions. Compared to Equation (2.4), the experimental determination of the equilibrium curve by Newton and Kennedy (1968) yielded the expression P[bar] = $25T$[°C] − 4000, which represents a considerable shift toward higher pressure.

Univariant reactions involving gases

We turn now to a higher stage of complexity, that of the heterogeneous reactions involving a single gas phase. As an illustration we take the well known wollastonite reaction:

$$\underset{calcite}{CalO_3} + \underset{quartz}{SiO_2} \rightleftharpoons \underset{wollastonite}{CaSiO_3} + \underset{gas}{CO_2} \qquad (2.c)$$

The thermodynamics of this reaction have been considered in detail by Danielson (1950) for the case of stoichiometric crystalline phases, to which we confine our discussion here. Following our development in Equations (1.17), (1.46), and (1.52), we may write the equilibrium equation as follows:

$$\ln K_{(c)} = \ln f_{CO_2} \qquad (2.5)$$

And using Equations (1.17) and (1.18) we obtain

$$-\Delta G^0_{(c)} - P\Delta V_{(c)} = RT \ln P_{CO_2} + \int_1^P \left(\frac{z-1}{P}\right) dP \qquad (2.6)$$

where $-\Delta G^0_{(c)} - P\Delta V_{(c)} = RT \ln K_{(c)}$ and $\Delta V_{(c)}$ is as usual the difference in volume between the solid reactants and products of Reaction (2.c). Since $\Delta G^0_{(c)}$ is known from thermochemical data and the integral term may be evaluated from the universal properties of gases, it is possible to solve Equation (2.6) by successive approximations.

Reaction (2.c) was used by Ramberg (1946) and Danielson (1950) to illustrate the effect of rock pressure.

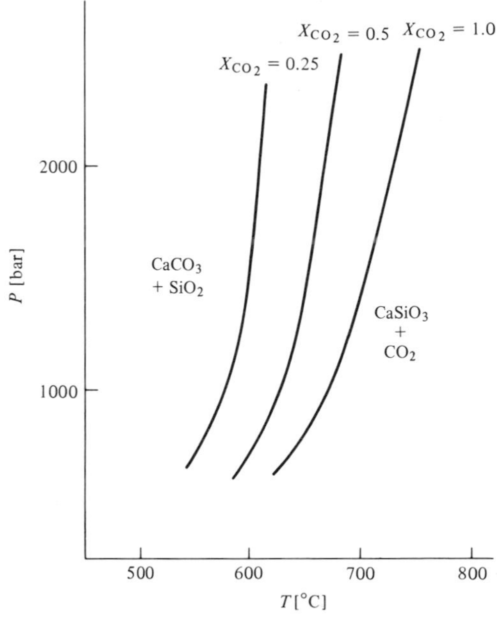

2.2 Calcite-wollastonite equilibrium as a function of P, T, and X_{CO_2}. (Data from Greenwood, 1967).

Figure 2.2 shows the wollastonite-calcite equilibrium curves. The curves shift as a function of the mole fraction of CO_2 in the fluid phase. It is obvious that the presence of another gas increases the total pressure on the solid phases while reducing the partial pressure of CO_2. Thus the reaction proceeds at a lower temperature than when the other gas is not present, an effect similar to that of rock pressure.

The system CaO-MgO-CO_2

This system is of great interest from a theoretical standpoint and, in addition, provides us with a very useful geologic thermometer, since the system CaO-MgO-CO_2 is of reasonably wide occurrence as such (in limestones) or is effectively decoupled when it occurs as a subsystem in more complex rocks. The essentials of the system were worked out by Graf and Goldsmith (1955) with contributions by Harker and Tuttle (1955).

Figure 2.3 shows the dolomite decomposition curve as represented by Reaction (2.f). This curve is of the same character as those of Figure 2.2 except that it has a nonlinear scale for plotting convenience. Originally all the minerals which enter into the decomposition reaction

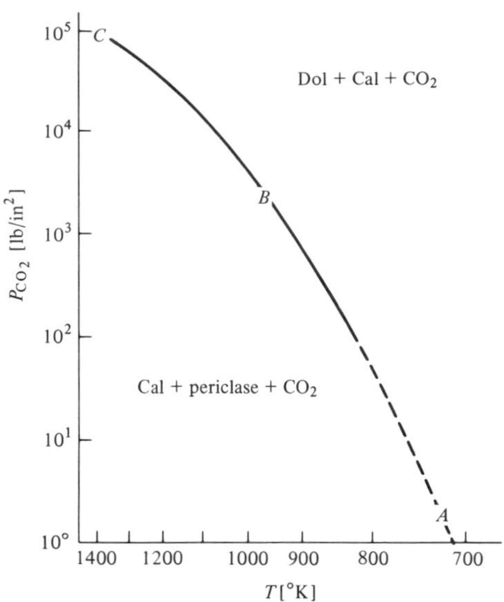

2.3 System CaO-MgO-CO_2. The curve *A-B-C* is a projection of the space curve *A-B-C* of Figure 2.5. (Data after Graf and Goldsmith, 1965).

2.4 System $CaCO_3$-$MgCO_3$. The curve is a projection along the P_{CO_2} axis as shown in Figure 2.5. (Data after Graf and Goldsmith, 1958, and Goldsmith and Heard, 1960).

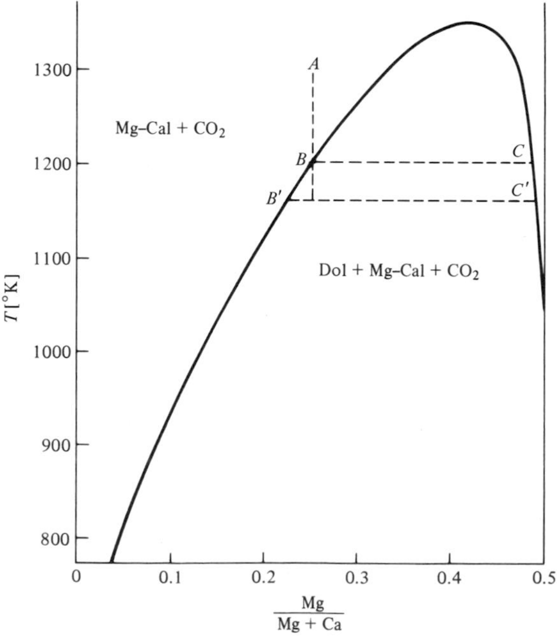

were regarded as stoichiometric, and a qualitatively similar curve was computed from thermochemical data by Hall and Markus (1952). However, it was subsequently found that the composition of calcite in equilibrium with dolomite is a function of the temperature. The empirical relationship is shown in Figure 2.4. It may be seen from this figure that a Mg-calcite of the composition A when cooled to B on the curve reacts to form a dolomite of a composition represented by C [Reaction (2.e)]. Further cooling would give rise to a more Ca-rich calcite (B') and a coexisting dolomite (C').

Figure 2.5 shows that the curves of Figures 2.3 and 2.4 are merely projections of three-dimensional relations in the ternary system. The curve of Figure 2.3 is here represented by the space curve *ABRCSD*, which defines a trough formed by the intersecting divariant surfaces. Figure 2.4 is of course a cross section of the steepest of these surfaces and appears on the upper surface of the block diagram. The steepness of this surface is a consequence of the small volume change involved in Reaction (2.e) (see below), so that pressure has little influence on the equilibrium according to Relation (1.2). This characteristic enhances the system's usefulness as a geothermometer.

The other divariant surface represents the equilibrium between solid solutions of Mg-calcite, periclase, and CO_2 gas [Reaction (2.d)], and is more complex. Figure 2.5 shows isocomposition lines of Mg-calcite on this surface.

2.5 Block diagram of the system CaO-MgO-CO_2. The construction is for the data of Graf and Goldsmith (1955, 1958) and Goldsmith and Heard (1960). See text for further explanation.

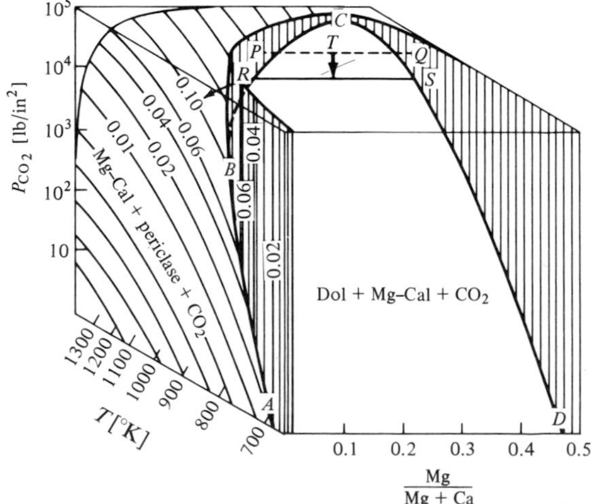

The space curve *ABRCSD* also defines a sloping surface that intersects the steep divariant surface and produces the dolomite decomposition curve. At CO_2 pressures below this surface dolomite is not stable and Mg-calcite cannot be used except to define a minimal temperature. A typical situation of crystallization in the ternary system is illustrated in Figure 2.5: the initial *PTX* coordinates are represented by the point *T*, which we assume falls in the field of coexisting Mg-calcite and dolomite, the equilibrium compositions of these minerals being represented by *P* and *Q*, respectively. If the pressure is now lowered sufficiently at constant temperature, the decomposition of Mg-calcite and dolomite begins when their respective compositions are represented by *R* and *S* on the space curve. If the temperature is held constant, the system is invariant and the phase composition cannot change until all the dolomite is consumed. When this happens, the composition of Mg-calcite begins to change along the divariant surface defined by the coexistence of periclase. This path is shown by the arrow in Figure 2.5.

A qualitative thermodynamic analysis of this system is simple and informative. The pertinent reactions are

$$\underset{\textit{Mg-calcite}}{MgCO_3} \rightleftharpoons \underset{\textit{periclase}}{MgO} + \underset{\textit{gas}}{CO_2} \tag{2.d}$$

$$\underset{\textit{dolomite}}{CaMgC_2O_6} \rightleftharpoons \underset{\textit{Mg-calcite}}{CaCO_3 + MgCO_3} \tag{2.e}$$

$$\underset{\textit{dolomite}}{CaMgC_2O_6} \rightleftharpoons \underset{\textit{Mg-calcite}}{CaCO_3} + \underset{\textit{periclase}}{MgO} + \underset{\textit{gas}}{CO_2} \tag{2.f}$$

We shall assume that dolomite and periclase are stoichiometric compounds and that CO_2 behaves as a perfect gas. The first assumption is justified by Figure 2.5 for the range of metamorphic temperatures below 700°C. The second assumption can only be justified for pure CO_2 at temperatures above 500°C and pressures below 700 bar when z does not exceed 1.17. Thus we may write the following equilibrium constants for the reactions

$$K_{(d)} = \frac{P_{CO_2}}{X_{Mg}^{Cal}\gamma_{Mg}^{Cal}} \tag{2.7}$$

$$K_{(e)} = X_{Ca}^{Cal}X_{Mg}^{Cal}\gamma_{Ca}^{Cal}\gamma_{Mg}^{Cal} \tag{2.8}$$

$$K_{(f)} = X_{Ca}^{Cal}\gamma_{Ca}^{Cal}P_{CO_2} \tag{2.9}$$

in which $X_{Ca}^{Cal} = Ca/(Ca + Mg)$ and $X_{Mg}^{Cal} = Mg/(Ca + Mg)$ for magnesium calcite, and γ_{Ca}^{Cal} and γ_{Mg}^{Cal} are the activity coefficients for the same phase.

As standard states for the components of Mg-calcite it is convenient to adopt pure Ca-calcite and pure magnesite since these components are isostructural. The thermochemical properties of these compounds may be used in calculating values for $K_{(a)}$, $K_{(e)}$ and $K_{(f)}$. However, for the present we shall be concerned only with the qualitative behavior of the system. As previously indicated, Reaction (2.d) is divariant and is represented by a surface in Figure 2.5. This divariance is also apparent from Equation (2.7), which is the only one of the three equations applicable to this surface. For example, if X_{Mg}^{Cal} and P_{CO_2} are fixed, the temperature is determined. This result is most apparent if we consider the case where $X_{Mg}^{Cal} \rightarrow 0$, for then γ_{Mg}^{Cal} approaches a constant. But the same result must hold in general since γ_{Mg}^{Cal} is an implicit function of T, P, and X_{Ca}^{Cal}. As P_{CO_2} decreases, X_{Mg}^{Cal} must also decrease. This conclusion is, of course, in perfect agreement with Figure 2.5.

Equation (2.8) refers to the nearly vertical divariant surface of Figure (2.5). This divariant character is also apparent from Equation (2.8) if we remember that $K_{(e)}$ contains the term $\exp(-P\Delta V_{(e)}/RT)$, where $\Delta V_{(e)}$ is the change in volume. We have already indicated that this term is small. The chief contributing factor to the equilibrium represented by Reaction (2.e) is a positive excess Gibbs free energy term, as in Equation (1.32). This term arises from the mutual repulsion of the Ca and Mg components in the carbonate crystal and is probably attributable chiefly to size differences in these ions. Gordon and Greenwood (1970) have calculated the activity–composition relations in $MgCO_3$-$CaCO_3$ solution from the experimental results of Graf and Goldsmith (1955). Their final equations are

$$\ln \gamma_{MgCO_3}^{Cal} = (1 - X_{MgCO_3}^{Cal})^2(2.1718 - 1.8960X_{MgCO_3}^{Cal})$$

$$0 \le X_{MgCO_3}^{Cal} \le 0.18$$

$$\ln \gamma_{CaCO_3}^{Cal} = (1 - X_{CaCO_3}^{Cal})^2(1.2238 + 1.8960X_{CaCO_3}^{Cal})$$

In these expressions the activity coefficients are not significantly dependent on P and T. Goldsmith and Newton (1969) have recently shown that pressure has only a small effect on the calcite-dolomite solvus.

Exchange equilibria

Generally, rocks are composed of at least three or four complex minerals of variable composition. If equilibrium is attained during crystallization, a considerable variety of reactions and corresponding equilibrium constants can be written for such an assemblage. For metamorphic rocks in general the reactions will involve a fluid or gas phase in addition to several mineral phases. These reactions may be expected to exhibit relations analogous to those just discussed for the system CaO-MgO-CO_2, except that they may be of even greater complexity. Some reactions will, however, be related to each other by linear combinations, so that they will seldom all be independent.

If at least two mineral solid solutions occur, it is generally possible to write exchange reactions that relate two or more variable elements of two or more phases. As a hypothetical example we may consider the species A and B (atoms or ions) which form complex mixed crystals with framework structures indicated by α and β. Examples of such framework structures from mineralogy are SiO_3^{2-} and SiO_4^{4-} in pyroxene and olivine, respectively. The hypothetical formulae are (A, B)α and (A, B)β. We may then write the exchange reaction as

$$A\alpha + B\beta \rightleftharpoons B\alpha + A\beta \tag{2.g}$$

for which the equilibrium constant is

$$K_{(g)} = \frac{X_B^\alpha X_A^\beta \gamma_B^\alpha \gamma_A^\beta}{X_A^\alpha X_B^\beta \gamma_A^\alpha \gamma_B^\beta} \tag{2.10}$$

in terms of the atomic fractions and activity coefficients. If A and B are the only miscible species present, Equation (2.10) becomes

$$K_{(g)} = \frac{(1 - X_A^\alpha) X_B^\beta \gamma_B^\alpha \gamma_A^\beta}{(1 - X_A^\beta) X_A^\alpha \gamma_A^\alpha \gamma_B^\beta} \tag{2.11}$$

Under conditions of constant temperature and pressure the activity coefficients will be functions of the X's, and a solution for one X in terms of another is at least implicit. We shall see that such isothermal-isobaric distributions are of considerable utility in the study of natural assemblages.

Because of the regularity implied by the equilibrium distribution functions such as Equations (2.10) and (2.11), the observed distributions of elements in coexisting phases in rocks should be good criteria for equilibrium. The fact that the distributions reflect fundamental energetic properties of the crystalline solutions gives them a great additional value as chemical data.

The dependence of the distribution constant on the temperature and pressure is of great interest, but unfortunately this dependence is difficult to assess since compositional effects operative in the activity coefficients may easily obscure the comparatively small effects of these variables. As we shall show later for specific systems, ΔV in the solid state exchange reactions such as (2.g) is usually so small that only very large pressure effects could be registered. Of course, this fact enhances these same systems as geothermometers.

For the system in which the variable components form ideal solutions with each other, the distribution relation takes a particularly simple form. In Equations (2.10) and (2.11) this corresponds to all the activity coefficients simultaneously being equal to unity.

We shall now turn to some examples of distribution among the naturally occurring mineral assemblages. The general nature of these distributions was first pointed out by Ramberg (1944a,b, 1952) and by Ramberg and DeVore (1951). However, the first conclusive evidence for many elements was obtained by Kretz (1959) in his now classic study of associated garnets, biotites, and hornblendes from the Quebec gneisses. Kretz also made the first study of a third element on the distributions. We begin by considering distributions from the pre-Cambrian metamorphic iron-rich chemical sediments known as "iron formation." These rocks are widespread in the pre-Cambrian, and in eastern Quebec they have been recrystallized to amphibolite and granulite facies rank. The geologic environment and detailed mineralogy of these metamorphosed iron formations have been discussed by Mueller (1960), Butler (1968) and Kranck (1961). Certain examples are also taken from the work of Howie (1955) on the charnockitic gneisses of India and from the works of Clavan *et al.* (1954) and Norton and Clavan (1959) on rocks from Pennsylvania and Delaware.

The two mineral assemblages of interest from the iron formations are as follows:

Ca-pyroxene	Actinolite
Orthopyroxene	Cummingtonite
Cummingtonite	Ca-pyroxene

Actinolite apparently does not coexist with orthopyroxene, which is an interesting fact we touch upon later. We begin by writing down the exchange reactions for Mg^{2+} and Fe^{2+} among the silicates:

$$\underset{\text{Ca-pyroxene}}{CaFeSi_2O_6} + \tfrac{1}{5}\underset{\text{actinolite}}{Ca_2Mg_5Si_8O_{22}(OH)_2} \rightleftharpoons \underset{\text{Ca-pyroxene}}{CaMgSi_2O_6} + \tfrac{1}{5}\underset{\text{actinolite}}{Ca_2Fe_5Si_8O_{22}(OH)_2} \tag{2.h}$$

$$\underset{\text{Ca-pyroxene}}{CaFeSi_2O_6} + \underset{\text{orthopyroxene}}{MgSiO_3} \rightleftharpoons \underset{\text{Ca-pyroxene}}{CaMgSi_2O_6} + \underset{\text{orthopyroxene}}{FeSiO_3} \tag{2.i}$$

$$\underset{\text{Ca-pyroxene}}{CaFeSi_2O_6} + \tfrac{1}{7}\underset{\text{cummingtonite}}{Mg_7Si_8O_{22}(OH)_2} \rightleftharpoons \underset{\text{Ca-pyroxene}}{CaMgSi_2O_8} + \tfrac{1}{7}\underset{\text{cummingtonite}}{Fe_7Si_8O_{22}(OH)_2} \tag{2.j}$$

$$\tfrac{1}{5}Ca_2Fe_5Si_8O_{22}(OH)_2 + \tfrac{1}{7}Mg_7Si_8O_{22}(OH)_2 \rightleftharpoons$$
actinolite *cummingtonite*

$$\tfrac{1}{5}Ca_2Mg_5Si_8O_{22}(OH)_2 + \tfrac{1}{7}Fe_7Si_8O_{22}(OH)_2 \quad (2.k)$$
actinolite *cummingtonite*

$$FeSiO_3 + \tfrac{1}{7}Mg_7Si_8O_{22}(OH)_2 \rightleftharpoons$$
orthopyroxene *cummingtonite*

$$MgSiO_3 + \tfrac{1}{7}Fe_7Si_8O_{22}(OH)_2 \quad (2.l)$$
orthopyroxene *cummingtonite*

The equilibrium constants corresponding to these reactions are

$$K_{(h)} = \frac{(1 - X_{Mg}^{Act})X_{Mg}^{Ca\text{-}Px}\gamma_{Fe}^{Act}\gamma_{Mg}^{Ca\text{-}Px}}{(1 - X_{Mg}^{Ca\text{-}Px})X_{Mg}^{Act}\gamma_{Fe}^{Ca\text{-}Px}\gamma_{Mg}^{Act}} \quad (2.12)$$

$$K_{(i)} = \frac{(1 - X_{Mg}^{Opx})X_{Mg}^{Ca\text{-}Px}\gamma_{Fe}^{Opx}\gamma_{Mg}^{Ca\text{-}Px}}{(1 - X_{Mg}^{Ca\text{-}Px})X_{Mg}^{Opx}\gamma_{Fe}^{Ca\text{-}Px}\gamma_{Mg}^{Opx}} \quad (2.13)$$

$$K_{(j)} = \frac{(1 - X_{Mg}^{Cum})X_{Mg}^{Ca\text{-}Px}\gamma_{Fe}^{Cum}\gamma_{Mg}^{Ca\text{-}Px}}{(1 - X_{Mg}^{Ca\text{-}Px})X_{Mg}^{Cum}\gamma_{Fe}^{Ca\text{-}Px}\gamma_{Mg}^{Cum}} \quad (2.14)$$

$$K_{(k)} = \frac{(1 - X_{Mg}^{Cum})X_{Mg}^{Act}\gamma_{Fe}^{Cum}\gamma_{Mg}^{Act}}{(1 - X_{Mg}^{Act})X_{Mg}^{Cum}\gamma_{Fe}^{Act}\gamma_{Mg}^{Cum}} \quad (2.15)$$

$$K_{(l)} = \frac{(1 - X_{Mg}^{Cum})X_{Mg}^{Opx}\gamma_{Fe}^{Cum}\gamma_{Mg}^{Opx}}{(1 - X_{Mg}^{Opx})X_{Mg}^{Cum}\gamma_{Fe}^{Opx}\gamma_{Mg}^{Cum}} \quad (2.16)$$

In these equations X refers to the atomic fraction $Mg/(Mg + Fe^{2+})$, whereas the superscripts have the following definitions: Cum is cummingtonite, Act is actinolite, Opx is orthopyroxene, and Ca-Px is calcium pyroxene. The component end members are indicated by the subscripts; as an example, the activity coefficient of $Ca_2Mg_5Si_8O_{22}(OH)_2$ is γ_{Mg}^{Act}. It is important to realize that neither the mole fractions nor the activity coefficients of individual ions are referred to here. However, it is obvious that the ionic ratio $Mg/(Mg + Fe^{2+})$ is equal to the mole fraction of the corresponding end member.

The points plotted in Figures 2.6 to 2.9 are observational data for the coexisting minerals of interest. In these figures each point represents an analyzed pair of the minerals indicated. It is apparent that the distribution for actinolite and Ca-pyroxene may be reasonably well represented by a straight 45° line, which is equivalent to the relation $X_{Mg}^{Act} = X_{Mg}^{Ca\text{-}Px}$. Reference to Equation (2.12) shows that this also implies the relation

$$\frac{\gamma_{Fe}^{Act}\gamma_{Mg}^{Ca\text{-}Px}}{\gamma_{Fe}^{Ca\text{-}Px}\gamma_{Mg}^{Act}} = K_{(h)} \quad (2.17)$$

Keeping this in mind we now consider Figure 2.7, which shows a series of distribution points for coexisting Ca-pyroxenes and orthopyroxenes. It should be noted that the curves which are drawn are symmetrical about a diagonal of the graph. These curves obey the equation

$$K = \frac{(1 - X_{Mg}^{Opx})X_{Mg}^{Ca\text{-}Px}}{(1 - X_{Mg}^{Ca\text{-}Px})X_{Mg}^{Opx}} \quad (2.18)$$

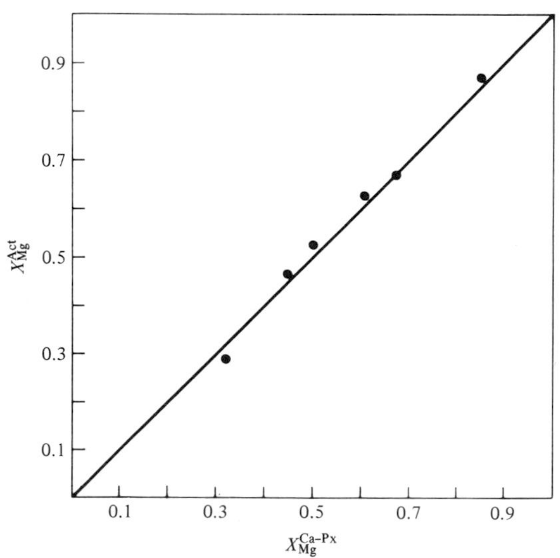

2.6 Distribution of Fe^{2+} and Mg^{2+} between coexisting Ca-pyroxene (Ca-Px) and actinolite in metamorphic rocks.

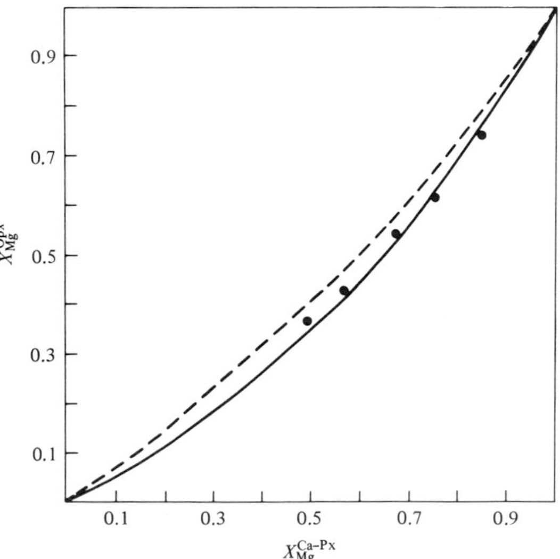

2.7 Distribution of Fe^{2+} and Mg^{2+} between coexisting pyroxenes in charnockites (data from Howie, 1955). Broken line represents magmatic rocks (Kretz, 1963).

where K is as usual the equilibrium constant and is a function of pressure and temperature only. But from Equation (2.13) we then have

$$\frac{\gamma_{Fe}^{Opx}\gamma_{Mg}^{Ca\text{-}Px}}{\gamma_{Fe}^{Ca\text{-}Px}\gamma_{Mg}^{Opx}} = K' \tag{2.19}$$

where K' denotes another constant. Consequently $KK' = K_{(i)}$. Now the activity coefficients of a binary solution are related to each other by a corollary of the Gibbs–Duhem equation (1.8). Thus for constant temperature and pressure we have for Ca-pyroxene:

$$(1 - X_{Mg}^{Ca\text{-}Px})\frac{\partial \ln \gamma_{Fe}^{Ca\text{-}Px}}{\partial X_{Mg}^{Ca\text{-}Px}} + X_{Mg}^{Ca\ Px}\frac{\partial \ln \gamma_{Mg}^{Ca\text{-}Px}}{\partial X_{Mg}^{Ca\text{-}Px}} = 0 \tag{2.20}$$

and similarly for the other binary solid solutions.

From the characteristics of binary activity coefficients it seems unlikely that Equation (2.19) could hold under any condition other than that in which all the activity coefficients are equal to 1 or that K' is some constant. This may be seen by considering ratios such as $\gamma_{Mg}^{Ca\text{-}Px}/\gamma_{Fe}^{Ca\text{-}Px}$ in terms of Equation (2.20). For if $\gamma_{Mg}^{Ca\text{-}Px}$ decreases, $\gamma_{Fe}^{Ca\text{-}Px}$ must increase for the same change in $X_{Mg}^{Ca\text{-}Px}$.

Saxena and Ghose (1971) determined the activity–composition relation in orthopyroxene, using the data on the partitioning of Fe^{2+} and Mg between the crystallographic sites M1 and M2 in the mineral. These data indicate that orthopyroxene is moderately nonideal (positive deviation) at 600°C but becomes ideal with increasing temperature (Saxena, 1973). From the results presented here (Figure 2.7) and elsewhere (Saxena, 1971), it may be noted that Ca-pyroxene would also be somewhat nonideal at 600°C and the function K' would remain nearly constant because of an appropriate adjustment in the activity coefficients. Of course, it must always be kept in mind that these conclusions depend on chemical data containing substantial errors that are frequently difficult to evaluate.

If we now consider the fractionation of Mg and Fe^{2+} between coexisting cummingtonites and actinolites (Figure 2.8), it appears that the distribution of points is somewhat asymmetric. The distribution cannot, therefore, be approximated by a function such as Equation (2.18). If the minerals represent a wide compositional range, it may be possible to find the nonideality parameter such as W by solving the equation

$$\ln K_{(k)} = \ln K'_{(k)} + \frac{W^{Act}}{RT}(1 - 2X_{Mg}^{Act}) - \frac{W^{Cum}}{RT}(1 - 2X_{Mg}^{Cum}) \tag{2.21}$$

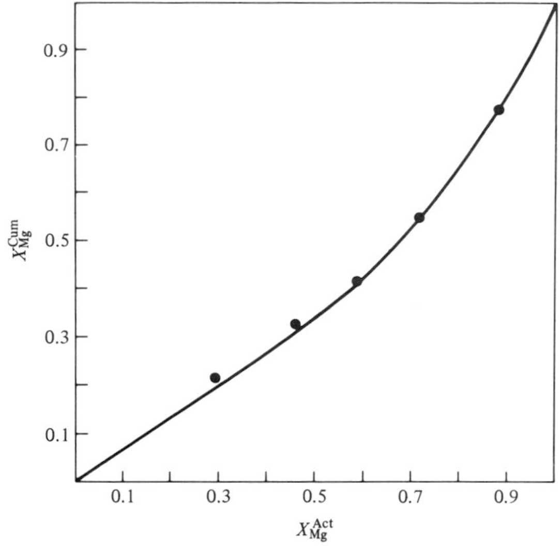

2.8 Fe^{2+}-Mg^{2+} distribution in coexisting cummingtonite and actinolite (data from Mueller, 1960). The curve is drawn free hand to show the asymmetric form of distribution.

where $K'_{(k)}$ equals $K_{(k)}$ when all the activity coefficients are unity. Since the number of data points are few, no attempt to calculate W has been made here.

The distribution of Fe and Mg between coexisting orthopyroxene and Ca-pyroxene in metamorphic rocks has been investigated by Kretz (1963). The distribution data indicate that $K_{(i)}$, when calculated by using Equation (2.18), is close to 1.82 in rocks of high metamorphic grade, some of which contained different mineral assemblages and therefore might be of different bulk chemistry. Thus the effect of bulk chemical differences is not detectable and the distribution is essentially invariant with respect to the total assemblage in which they occur. Saxena (1973) has presented a detailed discussion of these aspects of distribution coefficients.

In the foregoing development we have said little about the effect of temperature and pressure on the distribution equilibria. In the analysis of the observational data, we have assumed isothermal-isobaric conditions. The basis for this assumption was the apparent similarity of the metamorphic grade of all the assemblages considered. We would not expect that this similarity is perfect, however, and temperature differences as great as 100°C may have separated some of the different mineral pairs discussed. A study of the effect of temperature on Reaction (2.i) was made by Kretz (1961, 1963), who compared rocks of widely different origins. He found that the distribution curve for the magmatic rocks fell closer to the 45° (equal distribution) line than that for the metamorphic rocks. The approximate curve for magmatic rocks is shown in Figure 2.7 as a dashed line. This distribution probably represents temperatures around 1000°C if we allow for a small amount of sub-liquidus annealing, whereas the typical metamorphic temperatures are probably around 500–600°C.

The direction of the displacement of the distribution curve by higher temperatures is easily explained in terms of Equation (1.47). For Reaction (2.i) we may write this as

$$\left(\frac{\partial \ln K_{(i)}}{\partial T}\right)_P = \frac{\Delta H^0_{(i)}}{RT^2} \tag{2.23}$$

Now it is apparent that Reaction (2.i) is displaced to the right from the observed distribution; consequently $\Delta G^0_{(i)} < 0$. Since the standard entropy change must be small in this type of reaction involving only similar solids, one would also expect that $\Delta H^0_{(i)} < 0$. Consequently $K_{(i)}$ should decrease with an increase in temperature, as was observed.

The effect of pressure may similarly be evaluated from the general relation (1.48), which may be written as

$$\left(\frac{\partial \ln K_{(i)}}{\partial P}\right)_T = -\frac{\Delta V^0_{(i)}}{RT} \tag{2.24}$$

Under standard conditions $\Delta V^0_{(i)} \simeq -1.9$ cm^3. This is a very small value, but from Equation (2.24) we see that it will have the effect of displacing the curve away from the 45° line, with increasing pressure, an effect opposite to that of increasing temperature.

The distribution of minor elements

The laws that govern the distribution of minor elements are, of course, identical to those we have just discussed for the major elements. However, because the chemical potential rises steeply as a function of composition at low concentrations, minor component distributions are sensitive to variations in the major components. This is essentially true when the distribution of the major element between two phases is very unequal. Such, for example, is the case with the distribution of Mg between the coexisting orthopyroxenes and Ca-pyroxenes (Figure 2.7). On the other hand, when the major elements are equally distributed, as is the case with coexisting Ca-pyroxenes and actinolites (Figure 2.6), the distribution of the minor element may show great regularity. Figure 2.9, shows the observed distribution of manganese.

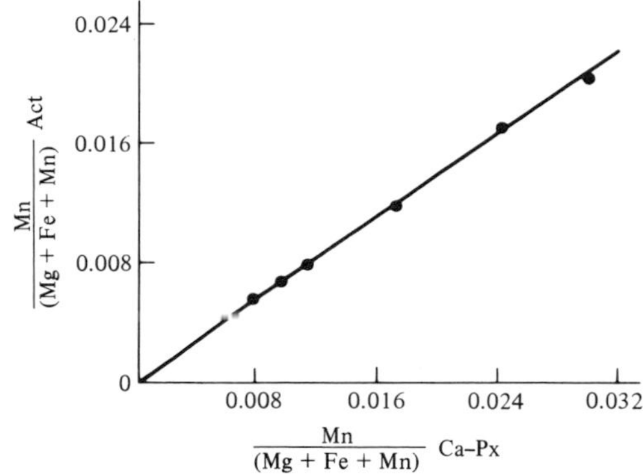

2.9 Distribution of Mn between coexisting Ca-pyroxene and actinolites. (After Mueller, 1960.)

In the case of the distribution of Mn between the other silicates and cummingtonite, further complications arise because of the nonideal behavior of the latter mineral. A full discussion of these points is beyond the scope of this work; for further details the reader is referred to the original paper (Mueller, 1961b).

2.10 V_2O_3 distribution between coexisting biotite and hornblende. (After Kretz, 1959. From the *Journal of Geology*, *67*:371–402, by permission from the University of Chicago Press.)

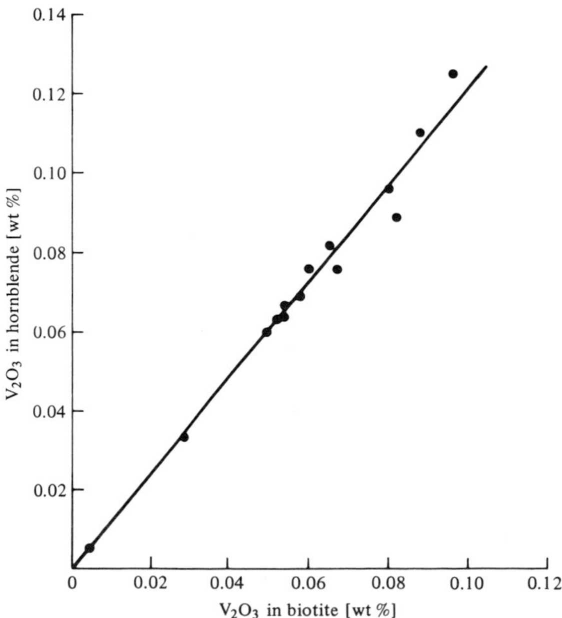

The influence of one element on the distribution of another was first studied in detail by Kretz (1959). A particularly fine illustration is represented by his study of the distribution of vanadium among coexisting biotites, hornblendes, and garnets. The results of this study are shown in Figures 2.10 to 2.12. Figure 2.10 shows the orderly distribution for biotite and hornblende. However, the distribution for biotite and garnet (Figure 2.11) shows an apparent enormous scatter, and on the basis of vanadium alone one could be inclined to doubt that equilibrium was attained here. Figure 2.11 has been divided into sectors representing different ranges of the distribution, and when these sectors are examined it is found that they represent different Ca-contents. The effect of the ratio Ca/(Fe + Mg + Mn + Ca) on the distribution is best shown in Figure 2.12.

2.11 V_2O_3 distribution between coexisting garnet and biotite. (After Kretz, 1959. From the *Journal of Geology*, *67*:371–402, by permission from the University of Chicago Press.)

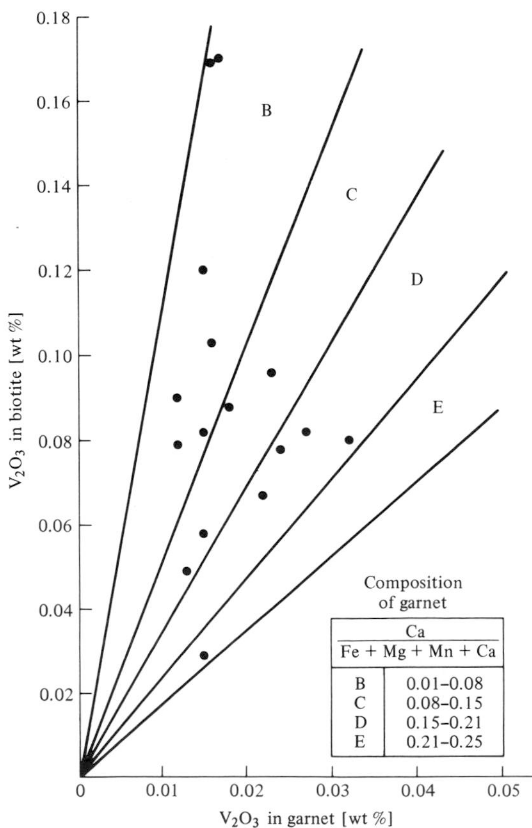

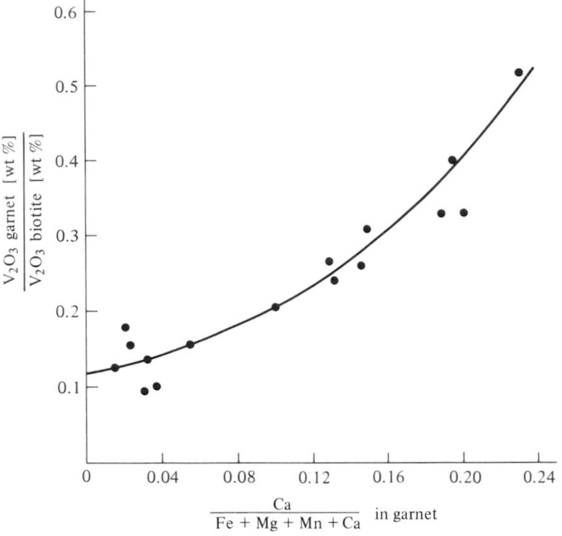

2.12 Effect of Ca concentration on the V_2O_3 distribution between biotite and garnet. (After Kretz, 1959. From the *Journal of Geology*, *67*:371–402, by permission from the University of Chicago Press.)

Coupled equilibria involving solid solutions and gases

The equilibrium of the average metamorphic mineral assemblage requires the simultaneous consideration of all the various types of reactions discussed thus far. As an illustration of such an equilibrium we again use the iron-formation assemblages, since they are of only moderate complexity. Kranck (1961) has found the following association in Quebec:

Ca-pyroxene
orthopyroxene
calcite
quartz
graphite

Magnetite is not reported in this rock but is widespread in other rocks of the region. Following our systematic method of approach we write the possible reactions which involve the end members of the component minerals:

$$\underset{\text{calcite}}{CaCO_3} + \underset{\text{orthopyroxene}}{MgSiO_3} + \underset{\text{quartz}}{SiO_2} \rightleftharpoons \underset{\text{Ca-pyroxene}}{CaMgSi_2O_6} + \underset{\text{graphite}}{C} + \underset{\text{gas}}{O_2} \qquad (2.m)$$

$$\underset{\text{calcite}}{CaCO_3} + \underset{\text{orthopyroxene}}{FeSiO_3} + \underset{\text{quartz}}{SiO_2} \rightleftharpoons \underset{\text{Ca-pyroxene}}{CaFeSi_2O_6} + \underset{\text{graphite}}{C} + \underset{\text{gas}}{O_2} \qquad (2.n)$$

$$\underset{\text{graphite}}{C} + \underset{\text{gas}}{O_2} \rightleftharpoons \underset{\text{gas}}{CO_2} \qquad (2.o)$$

These three reactions are linearly independent. This means that all three are required to define the system but that various combinations of them, obtained by addition or subtraction, may be used to obtain other significant relations. Thus subtracting Reaction (2.n) from Reaction (2.m) yields the exchange reaction (2.i). Similarly if Reaction (2.o) is added to Reactions (2.m) and (2.n) we obtain:

$$\underset{\text{calcite}}{CaCO_3} + \underset{\text{orthopyroxene}}{MgSiO_3} + \underset{\text{quartz}}{SiO_2} \rightleftharpoons \underset{\text{Ca-pyroxene}}{CaMgSi_2O_6} + \underset{\text{gas}}{CO_2} \qquad (2.p)$$

$$\underset{\text{calcite}}{CaCO_3} + \underset{\text{orthopyroxene}}{FeSiO_3} + \underset{\text{quartz}}{SiO_2} \rightleftharpoons \underset{\text{Ca-pyroxene}}{CaFeSi_2O_8} + \underset{\text{gas}}{CO_2} \qquad (2.q)$$

Of the entire set of six reactions, only three are independent.

Of course, there are also additional reactions which need to be considered for special information. Thus if we wish to know f_{CO} we can use the reaction

$$\underset{\text{graphite}}{C} + \underset{\text{gas}}{\tfrac{1}{2}O_2} \rightleftharpoons \underset{\text{gas}}{CO} \qquad (2.r)$$

However, we shall not be concerned with this or additional species here.

We have seen that in attempting to treat the more complex systems we encounter difficulties arising out of nonideality in the phases that are either gaseous, liquid, or solid solutions. In the following discussion we shall treat calcite as pure $CaCO_3$ since the quantities of $MgCO_3$ and $FeCO_3$ dissolved in the calcite at ordinary metamorphic temperatures are small and will have little effect on the chemical potential of $CaCO_3$. Similarly, graphite and quartz may be treated as pure since they dissolve almost no other components in this system.

We have seen in the previous section that both pyroxenes apparently form almost ideal solutions so that no difficulty is encountered from this source. However, no such assumption can be made for the gaseous phase in any natural system. Since little is as yet known of the interaction of such species of H_2O with CO_2, we can only speak of the fugacity of the latter gas. Actually

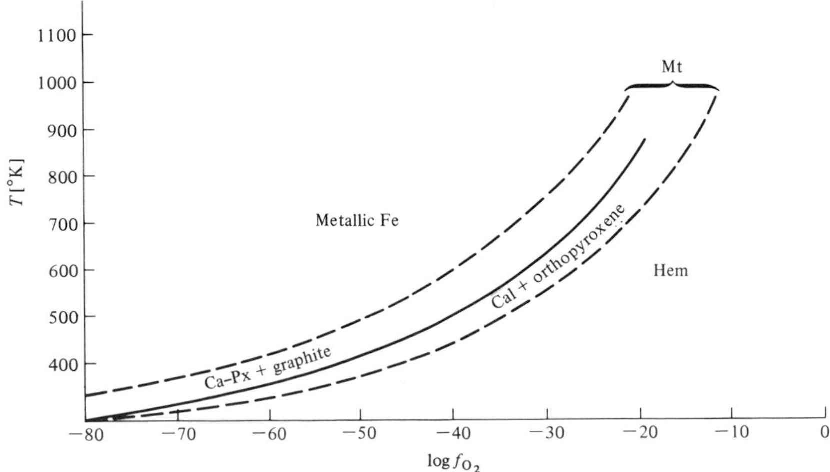

2.13 Fugacity of O_2 as a function of temperature in the assemblage Ca-pyroxene-orthopyroxene-calcite-quartz-graphite where $X_{Mg}^{Opx} = 0.305$, $P = 1$ atm. The fields of iron oxides are indicated by broken lines.

we shall not concern ourselves here with either gas. Instead we consider the fugacity of O_2 and the oxidation state of the system. To do this we begin by writing the equation of equilibrium for Reaction (2.m):

$$\Delta G_{(m)} = 0 = \Delta G^0_{(m)} + P\Delta V^S_{(m)} + RT \ln \frac{X_{Mg}^{Ca\text{-}Px}}{X_{Mg}^{Opx}} + RT \ln f_{O_2} \qquad (2.25)$$

in which f_{O_2} is the fugacity of oxygen, and the X's have the same meaning as in Equation (2.13). The second term on the right contains the volume change of the solids, which amounts to -20 cm^3 in this case. This is a large volume change, and it will have the effect of displacing the reaction strongly to the right at higher pressures. For example, at 600°K and 10^4 atm total pressure $(P\Delta V^s_{(m)}/2.303\ RT) = 1.74$.

If we now consider the third term we see by comparison with Figure 2.7 that the ratio $X_{Mg}^{Ca\text{-}Px}/X_{Mg}^{Opx}$ can assume a maximum value of 1.82, which is the value of the distribution constant. Since log 1.82 is only 0.26 we see that this is only a slight effect and that this effect tends to displace the $\log f_{O_2}$ curve to higher pressures as the fraction $Mg/(Mg + Fe^{2+})$ increases.

Under standard conditions ($P = 1$ atm, $X_{Mg}^{Ca\text{-}Px} = X_{Mg}^{Opx} = 1$) we obtain T as a function of $\log f_{O_2}$. That plot is shown in Figure 2.13, in which the curve for Reaction (2.m) is superimposed on the stability field of the iron-oxygen minerals. It is evident that Reaction (2.m) falls within the field of magnetite, which is to be expected from the presence of this mineral in the region (Kranck, 1961).

We have seen that the second term in Equation (2.25) could cause a considerable shift in the curve of Figure 2.13. However, the absolute magnitude of this shift in terms of oxygen pressure is insignificant. The same is not true, however, for f_{CO_2} of Reaction (2.p) which must also be satisfied. By an approximate calculation f_{CO_2} lies in the range 10^3–10^4 atm at 600°K. In this case we must also consider the third term of Equation (2.25), which may shift the curve by a factor of 0.26. In terms of the high CO_2 pressures, this is an enormous change. We shall see later on that this contrast in the fugacities of the major and minor gases is also closely tied to their mobility and their relationship to structural conditions.

Intrinsic and extrinsic stability

Solid solutions may be classified as either *intrinsically* or *extrinsically* stable. Intrinsic stability arises from the sign and magnitude of the excess free energy terms and is thus a characteristic of the solution *per se*. If the excess free energy of mixing is positive in sign, unmixing occurs as in the case of Mg-calcite, an example of intrinsic instability. Ideal solutions are, of course, always intrinsically stable since they have no excess free energy. Some solutions that have a negative excess free energy have an added intrinsic stability over the ideal model.

Ideal solutions may, however, be extrinsically unstable since this form of instability arises solely from the instability of the component end members relative to some reaction products. Thus we shall see later that ferrosilite ($FeSiO_3$) is extrinsically unstable relative to fayalite (Fe_2SiO_4) and silica. In the experience of the writers these two forms of stability are commonly confused.

Ionic processes

The tendency of mineral phases to dissociate into ions in the aqueous environment gives rise to very important effects over the entire range of petrologic phenomena. The range of conditions for which ionic processes are best understood are those near standard pressure and temperature. At higher temperatures and pressures these processes are less well understood, and more conventional nonionic parameters are usually chosen to describe the system. However, many features of silicate melts cannot be understood without considering their ionic character, and this will undoubtedly form an important branch of petrologic investigation in the future.

The subject of ionic processes in the low-temperature range has received considerable attention in recent years, chiefly in the lucid publications of Garrels, who has discussed in great detail the stability fields of numerous oxides, carbonates, and sulfides in relation to easily measurable parameters such as the oxidation potential (Eh) and the negative logarithm of the hydrogen ion concentration (pH).

The relationship of these variables to the familiar fugacities or partial pressures may be illustrated by considering the stability fields of the iron oxides. The univariant oxidation reaction for magnetite to hematite is

$$\underset{magnetite}{2Fe_3O_4} + \underset{gas}{\tfrac{1}{2}O_2} \rightleftharpoons \underset{hematite}{3Fe_2O_3} \qquad (2.s)$$

which is, of course, invariant at any given temperature and total pressure. If the stability fields are to be considered in an aqueous environment we may conveniently substitute the two variables Eh and pH for f_{O_2} to recover our univariant relationship under standard conditions. To do this we write first the important half cell reaction

$$H_2O \longrightarrow 2H^+ + e \qquad (2.t)$$

If we add this to Reaction (2.s) we obtain

$$\underset{\textit{magnetite}}{2Fe_3O_4} + \underset{\textit{solution}}{H_2O} \longrightarrow \underset{\textit{hematite}}{3Fe_2O_3} + \underset{\textit{solution}}{2H^+} + 2e \qquad (2.u)$$

We have written one-way arrows in these reactions to indicate that these half cells do not represent equilibrium. If we now make use of Definition (1.45) we obtain

$$Eh_{(u)} = E^0_{(u)} + \frac{RT}{F} \ln a_H \qquad (2.26)$$

or

$$Eh_{(u)} = E^0_{(u)} - \frac{2.303\, RT}{F} \text{pH} \qquad (2.27)$$

In a similar way other ions such as Fe^{3+} may be related to the stability fields of the solids. This we may write

$$\underset{\textit{magnetite}}{Fe_3O_4} + \underset{\textit{solution}}{8H^+} \longrightarrow \underset{\textit{solution}}{3Fe^{3+}} + \underset{\textit{solution}}{4H_2O} + e \qquad (2.v)$$

$$\underset{\textit{hematite}}{Fe_2O_3} + \underset{\textit{solution}}{6H^+} \longrightarrow \underset{\textit{solution}}{2Fe^{3+}} + \underset{\textit{solution}}{3H_2O} \qquad (2.w)$$

The corresponding equations are

$$Eh_{(v)} = E^0_{(v)} + 3\, RT \ln a_{Fe^{3+}} + (2.303)(8)RT\, \text{pH} \qquad (2.28)$$

$$\log K_{(w)} = 2 \log a_{Fe^{3+}} - 6\, \text{pH} \qquad (2.29)$$

Equation (2.29) is an ordinary equilibrium equation for the ionic species Fe^{3+} and H^+ and contains the equilibrium constant $K_{(w)}$. Equation (2.28) is not an equilibrium equation since the voltage $Eh_{(v)}$ in general has a finite value as compared with the hydrogen half cell. However, $a_{Fe^{3+}}$ and the pH of Equation (2.28) may be equilibrium values in relation to other phases and species of the system in which they are measured.

Equations of type (2.27), (2.28), and (2.29) may be used to delineate the stability fields of the solid phases by the use of diagrams. Diagrams may also be used to delineate the fields in which a species of ion is dominant. These graphical methods have been extensively employed by Garrels (1960).

We now consider a class of minerals in which ionic processes operate in a truly geologic scale. These are the larger silicates known as clay minerals, to a lesser extent the zeolites, and in the broadest sense all silicates showing ionic dissociation and exchange. Representatives of the latter group are the feldspars, which exhibit exchange properties at moderate temperatures (Orville, 1963). We shall be concerned mainly with the clay minerals at this stage.

The physical chemistry of ionic processes in silicates has received only scant attention from geologists who have been largely occupied with *ad hoc* investigations. However, some notable contributions have been made by Garrels and Christ (1965). The electrochemistry of clays has been subject to intensive investigation by soil scientists and physical chemists. Especially noteworthy is the work of C. E. Marshall and his co-workers at the University of Missouri who have contributed numerous papers dealing with the fundamentals of ionic processes in clays. In a review paper, Marshall (1955) also examined in detail the theoretical basis for the various types of investigation, which may be classed as thermodynamic, quasithermodynamic, and nonthermodynamic.

Before we consider the problem of ionic processes in clays, it seems desirable to discuss the mineralogic environment or "substrate" for these processes. The most important group of clay minerals is the montmorillonoid group. These minerals may be derived from the mica structure by introducing an extra quantity of charge into the tetrahedral or octahedral layers of mica. For example, Si^{4+} may be substituted for Al^{3+} in the sodium mica paragonite to give a montmorillonite:

$$\underset{\textit{paragonite}}{NaAl_2(Si_3Al)O_{10}(OH)_2}$$

$$\underset{\textit{montmorillonite}}{Na_{0.33}Al_2(Si_{3.67}Al_{0.33})O_{10}(OH)_2}$$

Montmorillonoids are characterized by a deficiency of sodium in the interlayers, and the property of water absorption and expandibility are direct consequences of this deficiency. From these relations we see that clays have a twofold aspect in relation to their energetic properties. The osmotic and ionic exchange processes, which have low activation energies and consequently operate on a short time scale, are dependent on the character of the Si-O tetrahedral and Al-O and Mg-O octahedral framework, the properties of which change only on a long time scale because of the high energies of activation involved. As a consequence ionic processes may be treated relatively independently of the recrystallization processes of diagenesis and metamorphism proper. The latter processes tend to eliminate the montmorillonoids by restabilizing the true mica constitution. Of course, this also implies a continuous alteration of the ionic properties with diagenesis and metamorphism.

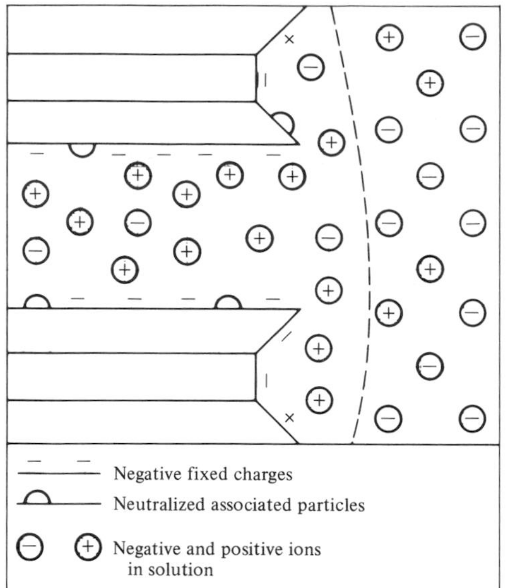

2.14 Effect of a membrane (broken line) on ionic concentrations. Explanation in text.

For a comprehensive understanding of ionic processes in clays we need to be familiar with the general theoretical model of the Donnan equilibrium. The fundamental relations have already been presented in Equations (1.42)–(1.44).[1] These equations give the conditions imposed on two systems or phases separated by a membrane so that at least one charged species is prevented from passing through the membrane. This constraint then gives rise to an unequal distribution of both neutral and ionic species between the two phases. The membrane need not be physical in the ordinary sense, but rather any restriction whatsoever on the mobility of the charged species can set up the effect. It appears that the expandable clay minerals meet this requirement, since their interlayers contain fixed negative charges. Thus the interlayer volume may be regarded as separated by a "membrane" from the external solution in which the clay particles are suspended. The general relations are shown in Figure 2.14. This figure shows diagramatically the difference in ionic concentrations that the membrane effects. Since the fixed internal charge is negative in this case, fewer anions are required on that side of the membrane to neutralize a given quantity of dissociated cations. Thus the cations in the internal solutions are in higher concentration than in the external solution. The hypothetical "membrane" is shown by a broken line. Of course, there is no actual discontinuity at the "membrane" but only a continuous ionic concentration gradient.

[1] A wide variety of applications of this theory are to be found in the book by Bolam (1932).

Figure 2.14 has also been drawn to illustrate that positive fixed charges may occur. Actual clay minerals are undoubtedly far more complicated than this, however.

In the conventional model of membrane equilibria the osmotic pressure plays a prominent role, and if two fluid masses are separated by a membrane in the open air, one of the fluids rises to a higher level than the other because of this pressure. However, it is obvious that our clay model presents a different picture since the external fluid always exerts its hydrostatic pressure on the clay mineral. Consequently, we have the added constraint that the osmotic pressure is zero or that the actual pressures are equal in both internal and external solutions.

The relation corresponding to Equation (1.43) is

$$\text{Osmotic pressure} = P' - P''$$

$$= \frac{RT}{[V_{H_2O}]} \ln \left[\frac{c''_{H_2O}\gamma''_{H_2O}}{c'_{H_2O}\gamma'_{H_2O}} \right] \quad (2.30)$$

with $P' - P'' = 0$.

Here we may represent the internal solution by single primes and the external solution by double primes. Thus we obtain

$$c''_{H_2O}\gamma''_{H_2O} = c'_{H_2O}\gamma'_{H_2O} \quad (2.31)$$

The analogous relation may be obtained for the ions from Equation (1.44):

$$(c'_+)^{\nu+}(c'_-)^{\nu-}(\gamma'_\pm)^{\nu++\nu-} = (c''_+)^{\nu+}(c''_-)^{\nu-}(\gamma''_\pm)^{\nu++\nu-} \quad (2.32)$$

For the distribution of NaCl between internal and external phases we obtain

$$c'_{Na}c'_{Cl}(\gamma_\pm)^2 = c''_{Na}c''_{Cl}(\gamma''_\pm)^2 \quad (2.33)$$

and for $CaCl_2$

$$c'_{Ca}(c'_{Cl})^2(\gamma'_\pm)^3 = c''_{Ca}(c''_{Cl})^2(\gamma''_\pm)^3 \quad (2.34)$$

We may now consider the case of an external aqueous solution of NaCl in equilibrium with an internal solution of the same species in the interlayers of a sodium montmorillonite with a fixed charge of lattice concentration c'_A. The appropriate equation for the ionic equilibrium is Equation (2.33). If c''_{NaCl} is the concentration of NaCl in the external solution, we also have the condition that

$$c''_{Na} = c''_{Cl} = c \quad (2.35)$$

and for the internal phase

$$c'_{Na} = c'_A + c'_{Cl} \tag{2.36}$$

If we substitute Equation (2.36) into (2.34) and make use of (2.35) we get the following equations:

$$c'_{Cl}(c'_{Cl} + c'_A)(\gamma'_\pm)^2 = (c''_{NaCl})^2(\gamma''_\pm)^2 \tag{2.37}$$

$$c'_{Na}(c'_{Na} - c'_A)(\gamma'_\pm)^2 = (c''_{NaCl})^2(\gamma''_\pm)^2 \tag{2.38}$$

Solution of these two quadratic equations for the ionic concentrations of the internal phase gives the following relations:

$$c'_{Cl} = \frac{-c'_A + [(c'_A)^2 + 4(c''_{NaCl})^2(\gamma''_\pm/\gamma'_\pm)^2]^{1/2}}{2} \tag{2.39}$$

$$c'_{Na} = \frac{c'_A + [(c'_A)^2 + 4(c''_{NaCl})^2(\gamma''_\pm/\gamma'_\pm)^2]^{1/2}}{2} \tag{2.40}$$

Equations (2.39) and (2.40) were derived by Bernstein (1960), who also studied this system experimentally. These equations show that at low salt concentrations (ionic strengths) Na^+ should be highly concentrated in the clay interlayer but that Cl^- should be in low concentration relative to the external solution.

Figure 2.15 is a comparison of the theory with the experimental results. From this figure it is apparent that Na^+ behaves essentially as predicted but that Cl^- is more highly concentrated in the interlayer than the theory predicts. This lack of agreement in the case of Cl^- may be due in part to the approximation $\gamma'_\pm = \gamma''_\pm$ employed as explained by Bernstein.

2.15 Experimental (solid lines) and theoretically predicted (broken lines) behavior of Na^+ and Cl^- in solution (Bernstein, 1960).

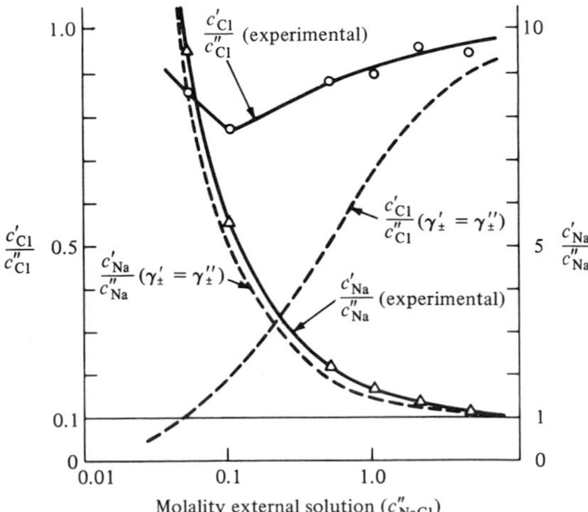

It should be noted that the Donnan effect is only operative at low to moderate salt concentrations and is swamped at the higher concentrations. Figure 2.15 shows that at a molality corresponding to the ionic strength of seawater Na^+ should be concentrated in the clay by a factor of 2 relative to the pore solution.

It has been found through numerous experiments that the cations are readily replaced by H^+ to form hydrogen clays, which then behave as weak acids. This reluctance of hydrogen clays to dissociate is in marked contrast to the Na^+ clays, which behave as salts of strong acids. The Donnan effect is, of course, dependent on this dissociation. It seems probable that the hydrolysis effect of clays following reactions of the type

$$Na^+ + \text{clay} + H_2O \rightleftharpoons H\,\text{clay} + Na^+ + (OH)^- \tag{2.x}$$

plays a very important role in controlling the pH of many natural waters. This should be especially true of seawater, which has had much time to react.

Crystal–melt equilibria

We now consider the equilibria established between crystals and a complex solution of molten oxides or silicates. Later on we shall consider numerous systems of this sort in the physical chemistry of natural magmas. However, at present we shall concern ourselves chiefly with the system albite-anorithite-diopside since this system provides an admirable example of the interrelations of thermodynamic and nonthermodynamic factors.

The problems of silicate melts are in many ways analogous to those of the ionic systems previously discussed. The basis for a successful theoretical treatment of ionic systems was the realization that the statistical unit was the dissociated ion. It has been found that silicate melts are also ionic in character (Tomlinson, 1952); however, many of the ions are really complexes of SiO_4 tetrahedra formed into chain, ring, and group polymers. In a detailed statistical mechanical treatment of such a solution a variety of statistical units must be considered simultaneously. Alternatively, these complex systems may be treated by a method that circumvents some of the complications. This method makes use of *macroscopic* variables such as the pressure, temperature, and the concentrations of hypothetical "molecules" that are adequate to represent all the phase compositions. These molecules are relatively noncommittal from a statistical viewpoint since no assumptions are made regarding their state of dissociation or association. In the system albite-anorthite-diopside these molecules are simply taken as one conventional formula unit of each mineral. The relationship of the statistical mechanical and the macroscopic models have been discussed by one of us (Mueller, 1964).

If we adopt the macroscopic model we may write three reactions of transfer of the components from crystal to melt as

$$\underset{crystal}{\text{Ab}} \rightleftharpoons \underset{liquid}{\text{Ab}} \tag{2.y}$$

$$\underset{crystal}{\text{An}} \rightleftharpoons \underset{liquid}{\text{An}} \tag{2.z}$$

$$\underset{crystal}{\text{Di}} \rightleftharpoons \underset{liquid}{\text{Di}} \tag{2.aa}$$

The equilibrium constants for these reactions are

$$K_{\text{Ab}} = \frac{a_{\text{Ab}}^L}{a_{\text{Ab}}^C} \tag{2.41}$$

$$K_{\text{An}} = \frac{a_{\text{An}}^L}{a_{\text{An}}^C} \tag{2.42}$$

$$K_{\text{Di}} = \frac{a_{\text{Di}}^L}{a_{\text{Di}}^C} \tag{2.43}$$

in which the superscript L refers to the liquid and C to the crystal. It is informative at this point to consider the problem of the statistical units that participate in these reactions. We may, for example, regard the "molecules" of albite and anorthite to be broken up into groups of ions or complexes such as the following:

$$\text{An}: CaAl_2Si_2O_8 = [CaAl]^{5+} + [AlSi_2O_8]^{5-}$$
$$\text{Ab}: NaAlSi_3O_8 = [NaSi]^{5+} + [AlSi_2O_8]^{5-}$$

These complexes are then formally regarded as being the units that form both liquid and solid solutions. The reason for choosing these particular units is that they have the same charge, which enables them to mix without causing charge imbalance locally. We may then substitute expressions for the activities analogous to Definition (1.37) for electrolytes. If we make the additional assumption that the solutions are ideal we obtain for the binary system Ab-An:

$$a_{\text{Ab}} = \frac{N_{\text{NaSi}} N_{\text{AlSi}_2\text{O}_8}}{(N_{\text{NaSi}} + N_{\text{CaAl}} + N_{\text{AlSi}_2\text{O}_8})^2} \tag{2.44}$$

in which N refers to the numbers of moles participating. An analogous expression is also obtained for a_{An}. Also the following relation holds

$$N_{\text{NaSi}} + N_{\text{CaAl}} = N_{\text{AlSi}_2\text{O}_8} \tag{2.45}$$

Substitution of Equation (2.45) into (2.44) gives

$$a_{\text{Ab}} = \frac{N_{\text{NaSi}}}{4(N_{\text{NaSi}} + N_{\text{CaAl}})} \tag{2.46}$$

which holds for both liquid and crystal. We thus see that the activities are directly proportional to the mole fractions of the end members since the ionic groups NaSi and CaAl are equivalent to these. Also we see that the complex $AlSi_3O_8$ representing most of the framework drops out. Thus for all practical purposes we may write for the liquid:

$$a_{\text{Ab}}^L = X_{\text{Ab}}^L \tag{2.47}$$

$$a_{\text{An}}^L = X_{\text{An}}^L \tag{2.48}$$

in which X stands for the mole fractions of the end members. The analogous expressions for the solid solutions are

$$a_{\text{Ab}}^C = X_{\text{Ab}}^C \tag{2.49}$$

$$a_{\text{An}}^C = X_{\text{An}}^C \tag{2.50}$$

These expressions may now be tested by using the experimental data of Bowen (1913) on the system. In order to do this, we must use the general relation (1.47) with (2.41) and (2.42). On integration we obtain

$$\frac{X_{\text{Ab}}^L}{X_{\text{Ab}}^C} = \exp\left[\frac{\Delta H_{\text{Ab}}^f}{R}\left(\frac{1}{T_{\text{Ab}}^0} - \frac{1}{T}\right)\right] \tag{2.51}$$

$$\frac{X_{\text{An}}^L}{X_{\text{An}}^C} = \exp\left[\frac{\Delta H_{\text{An}}^f}{R}\left(\frac{1}{T_{\text{An}}^0} - \frac{1}{T}\right)\right] \tag{2.52}$$

In these expressions ΔH_{Ab}^f and ΔH_{An}^f are the molar heats of fusion of pure albite and anorthite, and T_{Ab}^0 and T_{An}^0 are the temperatures of fusion of these same pure components. The simultaneous solution of Equations (2.51) and (2.52) yields a univariant curve of the type shown by the full curve in Figure 2.16. By adjusting ΔH_{Ab}^f and ΔH_{An}^f, the curves may be made to approximate the experimental points fairly well as was originally pointed out by Bowen. The values of the constants are approximately

$$T_{\text{Ab}}^0 = 1370°\text{K} \qquad \Delta H_{\text{Ab}}^f = 12{,}720 \text{ cal/mole}$$
$$T_{\text{An}}^0 = 1823°\text{K} \qquad \Delta H_{\text{An}}^f = 29{,}000 \text{ cal/mole}$$

It is also possible to evaluate the effect of pressure on this system. To do this we must use Equation (1.48) and integrate between the lower limits of zero pressure

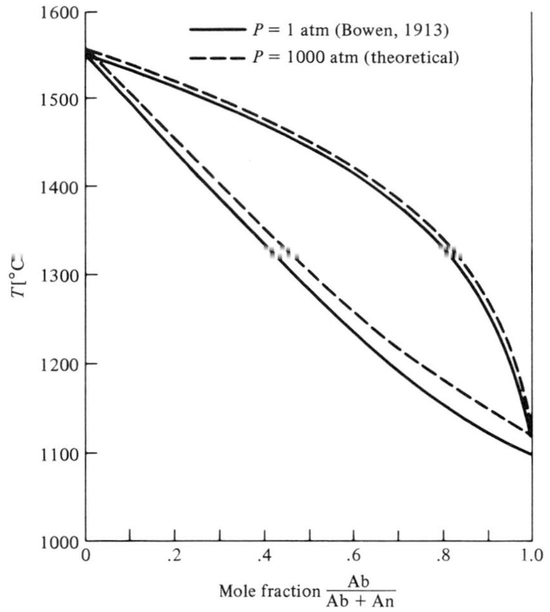

2.16 Theoretically calculated effect of pressure on the plagioclase melting relations using simplifying assumptions as discussed in text.

(regarded the same as 1 atm here) and the variable pressure of interest. We thus obtain

$$\left(\frac{X^L_{Ab}}{X^C_{Ab}}\right)_P = \left(\frac{X^L_{Ab}}{X^C_{Ab}}\right)_0 \exp\left(-\frac{P\Delta V_{Ab}}{RT}\right) \tag{2.53}$$

$$\left(\frac{X^L_{An}}{X^C_{An}}\right)_P = \left(\frac{X^L_{An}}{X^C_{An}}\right)_0 \exp\left(-\frac{P\Delta V_{An}}{RT}\right) \tag{2.54}$$

In these expressions the factors $(X^L_{Ab}/X^C_{Ab})_0$ and $(X^L_{An}/X^C_{An})_0$ are the values obtained in Equations (2.51) and (2.52), and ΔV_{Ab} and ΔV_{An} are the volume changes for Reactions (2.y) and (2.z). The simultaneous solutions of Equations (2.53) and (2.54) yield the dashed curve of Figure 2.16 when $P = 1000$ atm. It is plain that mechanical pressure will have little effect on this system. The values of volume changes used in this calculation are

$$\Delta V_{Ab} = 9.0 \text{ cm}^3/\text{mole} \qquad \Delta V_{An} = 2.5 \text{ cm}^3/\text{mole}$$

which really represents the difference in volume between the crystals and corresponding glasses at room temperature. The calculation is then not strictly correct since we have not taken account of α and κ as defined in Equations (1.5) and(1.6).

In the ternary system Ab-An-Di or in more complex systems, it is usually impossible to identify the statistical units participating, and for such systems the macroscopic approach is the only one possible. We therefore begin by defining the following ternary mole fractions:

$$X^L_{Ab} = \frac{N^L_{Ab}}{N^L_{Ab} + N^L_{Ab} + N^L_{Di}} \tag{2.55}$$

$$X^L_{An} = \frac{N^L_{An}}{N^L_{Ab} + N^L_{An} + N^L_{Di}} \tag{2.56}$$

$$X^L_{Di} = \frac{N^L_{Di}}{N^L_{Ab} + N^L_{An} + N^L_{Di}} \tag{2.57}$$

In these expressions N refers to the numbers of moles of $NaAlSi_3O_8$, $CaAl_2Si_2O_8$, and $CaMgSi_2O_6$ in the liquid. We also assign the macroscopic activity coefficients γ^L_{Ab}, γ^L_{An}, and γ^L_{Di} to these components since we cannot expect the ternary liquid to behave ideally. The activities of the crystalline components will, of course, have the same form as in the binary system since the crystals remain binary solutions. The expressions for the ternary liquid activities become

$$a^L_{Ab} = X^L_{Ab}\gamma^L_{Ab} \tag{2.58}$$

$$a^L_{An} = X^L_{An}\gamma^L_{An} \tag{2.59}$$

$$a^L_{Di} = X^L_{Di}\gamma^L_{Di} \tag{2.60}$$

We shall confine our discussion to the field of crystallization of plagioclase. The equilibrium constants corresponding to the transfer Reactions (2.y) and (2.z) are

$$K_{Ab} = \frac{X^L_{Ab}\gamma^L_{Ab}}{X^C_{Ab}} \tag{2.61}$$

$$K_{An} = \frac{X^L_{An}\gamma^L_{An}}{X^C_{An}} \tag{2.62}$$

It is apparent the K_{Ab} and K_{An} must be the same as in the binary system since they are only functions of T and P.

There are three major types of isobaric space curves which exist on the liquidus: (1) the isotherms, (2) the equilibrium paths of crystallization, and (3) the three-phase boundary. We shall be concerned chiefly with (1) and (2).

On the ternary liquidus the following relation holds:

$$X^L_{Di} = 1 - X^L_{An} - - X^L_{Ab} \tag{2.63}$$

and for the crystal

$$X^C_{An} + X^C_{Ab} = 1 \tag{2.64}$$

If we substitute Equations (2.61) and (2.62) into (2.63) we obtain the general equation of the isotherm for the plagioclase field:

$$X^L_{Di} = 1 + X^C_{An}\left\{\left(\frac{K_{Ab}}{K_{An}}\right) - \left(\frac{K_{An}}{\gamma^L_{An}}\right)\right\} - \frac{K_{Ab}}{\gamma^L_{Ab}} \tag{2.65}$$

If the ternary solution is ideal, this equation assumes a linear form and the isotherms are straight lines.

The effect of pressure on the ternary liquidus may also be computed if the system is ideal. For some arbitrary high pressure, Equation (2.65) becomes

$$X^L_{Di} = 1 + X^C_{An}(K'_{Ab} - K'_{An}) - K'_{Ab} \tag{2.66}$$

in which the primed equilibrium constants are related to their unprimed zero-pressure equivalents by the relations

$$K'_{Ab} = K_{Ab}\exp\left(-\frac{P\Delta V_{Ab}}{RT}\right) \tag{2.67}$$

$$K'_{An} = K_{An}\exp\left(-\frac{P\Delta V_{An}}{RT}\right) \tag{2.68}$$

In order to calculate the path of equilibrium crystallization in the field of plagioclase, it is desirable to combine Equations (2.61) and (2.62) to form the equation for the exchange reaction:

$$\frac{N^L_{An}}{N^L_{Ab}} = \frac{K_{An}\,\gamma^L_{Ab}}{K_{Ab}\,\gamma^L_{An}}\left(\frac{X^C_{An}}{1 - X^C_{An}}\right) \tag{2.69}$$

The initial coordinates at the beginning of crystallization we determined by the molar qualities N^{0L}_{Ab}, N^{0L}_{An}, and N^{0L}_{Di} present. At any stage we have

$$N^{0L}_{An} - N^C_{An} = N^L_{An} \tag{2.70}$$

$$N^{0L}_{Ab} - N^C_{Ab} = N^L_{Ab} \tag{2.71}$$

$$N^{0L}_{Di} = N^L_{Di} \tag{2.72}$$

which give the numbers of moles of each component remaining in the liquid at any time. If Equations (2.69), (2.70), and (2.71) are combined with the definition $X^C_{An} = X^C_{An}/(N^C_{An} + N^C_{Ab})$, the two following relations result:

$$N^L_{An} = \frac{N^{0L}_{An} - X^C_{An}(N^{0L}_{An} + N^{0L}_{Ab})}{\left(1 - \dfrac{K_{Ab}\gamma^L_{An}}{K_{An}\gamma^L_{Ab}}\right)(1 - X^C_{An})} \tag{2.73}$$

$$N^L_{Ab} = \frac{N^{0L}_{Ab} - (1 - X^L_{An})(N^{0L}_{An} + N^{0L}_{Ab})}{\left(1 - \dfrac{K_{An}\gamma^L_{Ab}}{K_{Ab}\gamma^L_{An}}\right)X^C_{An}} \tag{2.74}$$

If these equations are now substituted into Equation (2.57) we obtain the path of equilibrium crystallization for the given initial conditions. Such curves have frequently been discussed in relation to experimental data. The results of this analysis are depicted in Figure 2.17 which is a molar plot of the system.

The various features of the liquidus diagram may be readily deduced from Equations (2.65), (2.73), and (2.74),

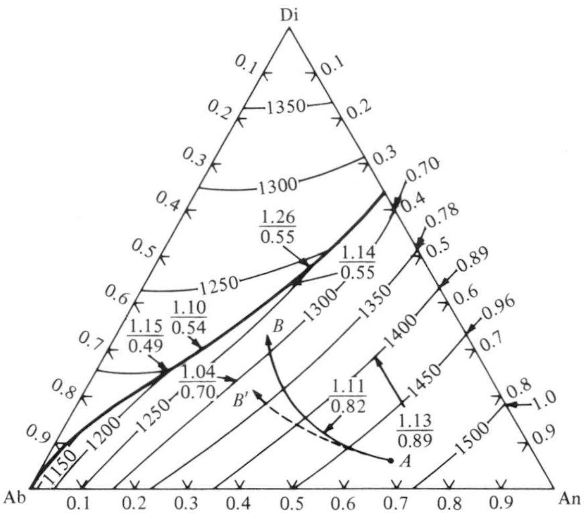

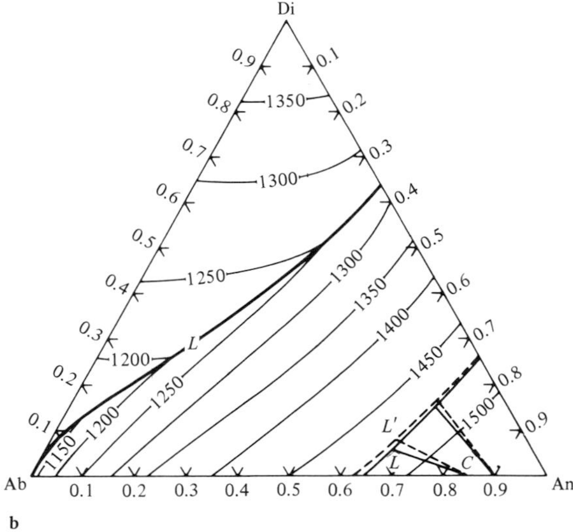

2.17 (a) Molar plot of the system albite-anorthite-diopside (data from Bowen, 1915; after Mueller, 1964). The plotted numerical values are the activity coefficients of albite and anorthite. In the ternary system these values are separated by a bar ($\gamma^L_{Ab}/\gamma^L_{An}$) with the albite value on top. Arrows point to coordinates. Numerical values of γ^L_{An} are also shown on the join anorthite-diopside for various isotherms. The curves *A-B* and *A-B'* are theoretical crystallization paths as discussed in the text. (b) Tie lines *C-L* at 1 atm and *C-L'* at 1000 atm show the shift in the 1450°C isotherm with pressure.

while Relations (2.61) and (2.62) may be used to determine the activity coefficients from the experimental data of Bowen (1913). The results are shown in Figure 2.17, which is a molar plot of the system. Thus we note that the condition $\gamma_{\mathrm{An}}^{L} < 1$ is apparently the chief contributing factor to the convexity of the isotherms toward the anorthite corner. The effect of nonideality can also be seen on the crystallization path (Figure 2.17a). The ratio $K_{\mathrm{An}}/K_{\mathrm{Ab}} < 1$, and since $\gamma_{\mathrm{Ab}}^{L}/\gamma_{\mathrm{An}}^{L} > 1$, there is a tendency for the thermal and compositional An factors to oppose each other. As an illustration, two hypothetical crystallization paths A–B and A–B' are shown in Figure 2.17a. In the path A–B the ratio $K_{\mathrm{An}}\gamma_{\mathrm{Ab}}^{L}/K_{\mathrm{Ab}}\gamma_{\mathrm{An}}^{L}$ has been arbitrarily set equal to 0.265 corresponding to point A. In the curve A–B' the ratio $\gamma_{\mathrm{Ab}}^{L}/\gamma_{\mathrm{An}}^{L}$ has been arbitrarily set equal to one and the ratio $K_{\mathrm{An}}/K_{\mathrm{Ab}}$ varies with the temperature as in the binary system. This curve was obtained by successive approximations by fitting the isotherms from the given initial coordinates at A. The true course of equilibrium crystallization, when both the purely thermal effects and the activity coefficients are taken into account, must lie between the two curves. From Figure 2.17b we see how the 1480°C isotherm is shifted away from the anorthite corner under 10^3 atm pressure and how the increase in pressure steepens the slope of the tie lines corresponding to a higher An-content of the liquid relative to the crystal. The latter effect is in harmony with Le Chatelier's principle, as may be seen from the volume changes on melting. For further discussion see Mueller (1963b, 1964).

Equilibrium and fractional crystallization paths such as curves A–B and A–B' have been frequently discussed in the literature, and usually such curves involve binary solid solutions. However, recently Roedder (1974) provided a geometric discussion of crystallization paths in systems with ternary solid solutions.

The system albite-anorthite-diopside provides an interesting example of ideal and nonideal behavior in silicate melts. The equilibrium of major and trace elements between plagioclase crystals and coexisting melts have been studied by Kudo and Weill (1970), Drake and Weill (1975), and Drake (1976), who have obtained many interesting results. Many more such data involving controlled experiments on ternary and quaternary systems are needed to understand the activity-composition relations, which form the basis of magmatic differentiation.

3 Kinetics of mineral systems

General

In our attempts to understand the origin and evolution of rocks, we are forced at every turn to consider the problems of kinetics. If the answers we seek are values of the thermodynamic variables of the physical and chemical environment of crystallization, we must first establish that the measurements represent equilibrium within the context of the study. The concept of equilibrium is intimately connected with the volumetric and geometric properties of the system. This arises from the universal dependence of chemical and many "physical" processes on particle diffusion. Thus the volume throughout which equilibrium is attained is always of limited size and dependent on the time scales of the events involved. This kinetic restriction on equilibrium was clearly recognized by Harker (1932), whose studies led him to conclude that generally a mineral formed at any given point within a rock depends upon the local composition within a radius of "a small fraction of an inch." This concept of local equilibrium was referred to as *mosaic equilibrium* by Korzhinskii (1957). However, this term seems undesirable because it implies a certain discontinuity when we should really expect a continuous variation in the intensive variables with distance.

The existence of chemical potential gradients is revealed by compositional gradients in a given mineral phase, and this provides a sensitive index of the effectiveness of diffusion (Mueller, 1960). But chemical potential gradients are not revealed by variations in bulk composition or inhomogeneities at phase boundaries. The latter facts do much to ease the petrologist's task since an accurate determination of mineral proportions is usually far more difficult to obtain than the mineral compositions. That this point has been widely misunderstood is shown by the disproportionate attention given to modal analyses in the petrologic literature.

In addition to defining the spatial extent of equilibrium, mineral composition gradients are also in principle capable of yielding information about the kinetic mechanisms during crystallization. Thus the compositional gradient of certain species may arise through restrictions on the diffusion rate of another species. We shall discuss this problem in some detail.

The petrologist is also concerned with the factors that govern mass transport on a geologic scale, that is, the flow and diffusion of chemical species through large rock units or between various levels in a planet's crust. This large-scale mass transport is of two types: (1) mechanical transport along mechanical potential gradients, and (2) chemical transport by diffusion along chemical potential gradients. Of course (1) and (2) become identical as perfect equilibrium is approached. Also it should be noted that perfect mechanical as well as chemical equilibrium can be attained only by the second means of transport. Purely mechanical transport such as the hydrodynamic flow of fluids may completely disregard the chemical requirements, and thus will usually give rise to chemical potential gradients. As an example, we may consider a magma that is intruded into a region in response to increased pressure at depth. The magma immediately finds itself out of chemical equilibrium with the host rocks. In addition, it may also be out of mechanical equilibrium with them, such as when a basic melt is intruded above an acidic layer. These imposed instabilities may give rise to contact metamorphic and metasomatic effects in the wall rock and to endomorphic effects in the crystallizing magma. These effects result from the chemical potential gradients set up by temperature, pressure, and compositional differences and also lead to mechanical disruption due to density inversion.

The processes of consolidation and diagenesis, which are at work in sedimentary deposits, also involve considerable mass transport. Some of this consists of the squeezing out of water and the release of gravitational energy. However, connate waters are usually brines, and therefore there must also be considerable exchange of the dissolved constituents with sedimentary minerals as diagenesis proceeds. Similarly, at higher temperatures additional more tightly bound fluids and gases find their way to the surface as recrystallization continues on the ascending scale of metamorphism. Since macroscopic pores are virtually absent in metamorphic rocks, the volatile constituents must move by intergranular diffusion or by diffusion and hydrodynamic flow in transient fractures and imperfections. Whether or not large-scale transport takes place at any substantial rate then depends upon the rates of the reactions that provide the volatiles as well as the rates of the transport processes themselves.

Although mass transport in rocks may occur by movement along intergranular boundaries and flaws, any substantial metasomatic alteration of the minerals must involve substantial lattice diffusion as well. In general the rate of intergranular diffusion or diffusion along internal lattice imperfections will greatly exceed that of pure lattice diffusion, although the latter need not extend as far to be effective. Much controversy in the petrologic literature has centered around the question of mass transport, and this has been especially true of diffusive transport.

Next to the problem of mass transport rates the most important question is that of the rates of the heterogeneous chemical reactions themselves. For most of these mechanisms can scarcely be imagined, and consequently theoretical equations cannot be directly proposed. An exception is the relatively simple decomposition reaction such as $Mg(OH)_2 \rightarrow MgO + H_2O$, for which we shall construct a highly idealized model. For more complex reactions we may, however, adopt the method of simplest hypotheses. Using this method Greenwood (1963) assumed that the decompositions of talc and anthophyllite were first-order reactions, and the data seem to fit this simple model. However, it would scarcely be expected that this surprising result could be extended to many other systems. The reason for this expectation is that although decomposition reactions themselves may be first order, the rates of the complex products formed should depend heavily on nucleation and growth rates. Helgesson (1971) has concluded that diffusion is the rate-controlling step in many heterogeneous reactions.

Closely related to the problem of diffusion and reaction rates is the matter of the kinetics of rock deformation. We are not concerned here with deformation by fracture or gliding in crystals, which is of a mechanical nature. The close analogy with chemical reactions and diffusion is found in deformation by recrystallization creep, a form of "plastic flow" that characterizes deep-seated deformation in planetary crusts and interiors. This type of deformation depends on a thermal surmounting of the energy barrier, the energy barrier itself having been distorted by the stress field (Glasstone *et al.*, 1941). The thermal diffusion is thus given a preferential direction, which results in the deformation. Such effects require comparatively light stresses operative over long time intervals. It seems obvious that much of the local equilibrium observed in high-grade metamorphic rocks must be the result of such recrystallization creep or "dynamic annealing," which enables a growing crystal to become homogeneous and adjust to its neighbors as well as relieving internal stresses. Since the rate of this type of deformation should contain a factor $\exp(-\Delta E/RT)$, where ΔE is the activation energy, we should expect it to be highly temperature-dependent. It is clear that this is the meeting ground of chemical and structural petrology and that the study of the deformation of rocks is concerned with both. This will be especially apparent when we consider the relation between the structural behavior of rocks and their chemical compositions.

Quasiequilibrium processes in a temperature gradient

The concept of quasiequilibrium provides us with a flexible theoretical model to view a variety of mass-transport phenomena. In this model we compare two regions, spatially separated, in which the thermodynamic variables have values characteristic of the region, each region being regarded as in internal equilibrium. Mass transport then takes place between the two regions as a consequence of the difference in chemical potentials.

The transport may be due to a difference in the temperature of the two regions, in which case both mass and heat transport occur. The classic petrologic example of this is the contact metasomatism around intrusive bodies (Chapter 13). Again the transport may be attributable to a composition gradient of a chemical species that has its fugacity controlled by the presence of certain phases in a local region. Such gradients are common in a variety of petrogenic environments.

The quasiequilibrium model has been applied to vapor transport of solids by Mandel (1962) and by Lever and Mandel (1962). We shall present Mandel's mathematical treatment here.

Let us consider a class of reactions represented by the following:

$$\underset{\textit{cuprite}}{Cu_2O} + \underset{\textit{gas}}{2HCl} \rightleftharpoons \underset{\textit{gas}}{2CuCl} + \underset{\textit{gas}}{H_2O} \tag{3.a}$$

$$\underset{\textit{hematite}}{Fe_2O_3} + \underset{\textit{gas}}{6HCl} \rightleftharpoons \underset{\textit{gas}}{2FeCl_3} + \underset{\textit{gas}}{3H_2O} \tag{3.b}$$

in which the oxide is brought into the vapor phase by reaction with some volatile such as HCl. If the enthalpy change for such reactions as (3.a) and (3.b) is positive, then the oxide will be transported from a region of higher temperature to one of lower temperature and there be deposited again as an oxide, provided the conditions are favorable. The requirement that the enthalpy change be positive for this mode of transfer to occur may be verified from Relation (1.47), which indicates that $K_{(a)}$ and $K_{(b)}$, the equilibrium constants for Reactions (3.a) and (3.b), become smaller with decreasing temperature if $\Delta H_{(a)}$ and $\Delta H_{(b)}$ are greater than zero.

To examine the problem more systematically we may imagine that a linear pore of uniform cross-section A and length L connects two regions (1 and 2) which are at temperatures T_1 and T_2, respectively. Let us consider the generalized reaction that may represent any such reactions as (3.a) and (3.b). This is

$$\underset{\textit{solid}}{S} + \underset{\textit{gas}}{aA} \rightleftharpoons \underset{\textit{gas}}{bB} + \underset{\textit{gas}}{cC} + \underset{\textit{gas}}{dD} + \cdots \tag{3.c}$$

in which a, b, c, and d are the ordinary molar stoichiometric coefficients. Now N_0 moles of the gaseous substance A (corresponding to HCl) are added to M_0 moles of the solids (corresponding to the oxide) at one end of the previously evacuated tube. The equilibrium constant for Reaction (3.c) is

$$K = \frac{P_B^b P_C^c P_D^d}{P_A^a} \tag{3.1}$$

The flux of each component is assumed to be given by Fick's first law [Equation (1.74)], and for the case of the steady state the individual fluxes are related as follows:

$$-\frac{1}{a}\frac{D_A}{RT}\frac{\partial P_A}{\partial x} = \frac{1}{b}\frac{D_B}{RT}\frac{\partial P_B}{\partial x} = \frac{1}{c}\frac{D_C}{RT}\frac{\partial P_C}{\partial x} = \cdots \tag{3.2}$$

in which x is the distance variable, and D is the coefficient of diffusion. In general D will be a function of both the composition and temperature of the gas phase.

If the temperature gradient between regions 1 and 2 is sufficiently small and if the gas properties such as the density and composition do not vary too much, we may regard the diffusion coefficients as substantially constant.[1] Then to a first approximation we may define constants of the type α, γ, and δ such that

$$\frac{D_B}{D_A} = \alpha, \qquad \frac{D_B}{D_C} = \gamma, \qquad \frac{D_B}{D_D} = \delta \tag{3.3}$$

[1] For details of the argument regarding the temperature gradient the reader is referred to the original paper by Mandel (1962).

We may then integrate Equation (3.2) to obtain

$$P_A = K_A - \frac{a}{b}\alpha P_B$$

$$P_C = K_C + \frac{c}{b}\gamma P_B \qquad (3.4)$$

$$P_D = K_D + \frac{d}{b}\delta P_B$$

in which the K's are the integration constants. To evaluate the integration constants the following conservative conditions are used:

$$\frac{1}{b}\int_0^L \frac{P_B}{RT} A dx = \frac{1}{c}\int_0^L \frac{P_C}{RT} A dx = \frac{1}{d}\int_0^L \frac{P_D}{RT} A dx = \cdots \qquad (3.5)$$

$$\frac{1}{a}\int_0^L \frac{P_A}{RT} A dx = \frac{N_0}{a} - \frac{1}{b}\int_0^L \frac{P_B}{RT} A dx = \cdots \qquad (3.6)$$

The initial pressure is defined as

$$P_0 = \frac{N_0 R\overline{T}}{AL} \qquad (3.7)$$

in which $\overline{T}$ is the mean temperature.

If we now use Equations (3.5) and (3.6), Equations (3.4) become

$$P_A = \overline{P}_A + \frac{a}{b}\alpha(\overline{P}_B - P_B)$$

$$P_C = \overline{P}_C + \frac{c}{b}\gamma(\overline{P}_B - P_B) \qquad (3.8)$$

$$P_D = \overline{P}_D + \frac{d}{b}\delta(\overline{P}_B - P_B)$$

in which the barred symbols stand for average values of the partial pressures. These are defined as follows:

$$\overline{P}_B = \frac{R\overline{T}}{L}\int_0^L \frac{P_B}{RT} dx$$

$$\overline{P}_A = \frac{R\overline{T}}{L}\int_0^L \frac{P_A}{RT} dx$$

etc.

When one of the species (here taken as B) is much heavier and larger than any other species and if, in addition, the equilibrium constant is small, we obtain the following approximations to Equations (3.8):

$$P_A = P_0$$

$$P_C = \overline{P}_C = \frac{c}{b}\overline{P}_B \qquad (3.9)$$

$$P_D = \overline{P}_D = \frac{d}{b}\overline{P}_B$$

The equation for the steady state rate of transport of B is

$$R_T = -\frac{1}{b}\frac{\overline{D}_B}{R\overline{T}}\frac{(P_{B_2} - P_{B_1})}{L}A \qquad (3.10)$$

in which we again indicate mean values by the bar.

If we now combine Relations (3.1), (3.9), and (3.10), we obtain the rate of transport in terms of measurable quantities:

$$R_T = \frac{\overline{D}_B}{R\overline{T}}\frac{P_0^{a/M}(K_1^{1/b} - K_2^{1/b})}{b^{b/M}c^{c/M}d^{d/M}\cdots[(K_1^{1/b} + K_2^{1/b})/2]^{[1-(b/M)]}}\frac{A}{L} \qquad (3.11)$$

in which $M = b + c + d \cdots$ and K_1 and K_2 are the equilibrium constants in regions (1) and (2), respectively.

An additional assumption embodied in Equation (3.11) is that the rate of Reaction (3.c) is large compared with the rate of diffusion.

It is also possible to derive transport rates analogous to Equation (3.11) for the case of the large equilibrium constant and for reactions whose specific rate constants must be incorporated in R_T (Mandel, 1962).

Quasiequilibrium processes under isothermal conditions

In the foregoing discussion we were concerned with mass transport between two regions in which a given equilibrium constant had a different value because of temperature differences. However, marked temperature gradients appear to be of only limited occurrence and confined to such local regions as igneous contact zones. The quasiequilibrium concept may also be extended to the isothermal–isobaric case prevalent in regional metamorphism. In such a situation a given equilibrium constant has the same value everywhere, but certain intensive variables vary between adjacent regions because of kinetic barriers to diffusion. The case is especially well illustrated by such species as CO_2, CO, H_2O, O_2, and H_2, which may have quite different mobilities in the earth's crust. To see this we need only consider a solution to Equation (1.76), the equation of Fick's second law. A solution for the case of a constant diffusion coefficient is given by Darken and Gurry (1953). This solution states that the total amount of the diffusing substance that has crossed a unit area of boundary surface at time t is proportional to $(Dt)^{1/2}(c_s - c_0)$, where c_s and c_0 are the highest and lowest initial concentrations of the diffusing species. The same result follows more simply where a steady state prevails, as may be seen from Equation (3.10).

The above principles are best illustrated by considering specific systems, although petrologically significant data are as yet somewhat meager. One such system, which was studied by one of the writers, consists of the mineral assemblages of the Quebec iron formation (Mueller, 1960). In this case gross differences in the assemblages may be attributed to differences in the oxidation state of adjoining rock types. Consequently, at least some of the species H_2O, CO_2, CO, and H_2 in addition to O_2 must have had chemical potential and concentration gradients at the time of crystallization. Two contrasting adjoining assemblages are the following:

Relatively oxidized assemblage	Relatively reduced assemblage
Hematite	Magnetite
Magnetite	Cummingtonite
Actinolite	Actinolite
Talc	Ca-pyroxene
Cummingtonite	Calcite
Calcite	Dolomite
Dolomite	Quartz
Quartz	

Not all of the above minerals of each assemblage are always found together. Also, we shall, at this time, be concerned with only certain minerals of the assemblages.

The oxidation state of a given assemblage is of course immediately apparent from the presence or absence of hematite. In the oxidized assemblage the fugacity of oxygen is fixed by the coexistence of both oxides. Independent evidence also indicates that the fugacities of H_2O and CO_2 were everywhere constant throughout the part of the iron formation studied (Mueller, 1960). This restricts the variation of the species H_2, O_2, and CO to the reduced iron formation, which contains only the oxide phase magnetite. Because of its high kinetic velocity and small size, it might be expected that H_2 would be the controlling species in determining the oxidation state since it could diffuse more rapidly than O_2 and CO. However, as we have already seen, the rates of transfer by diffusion also depend on the absolute concentrations; therefore, it is helpful to see what these concentrations are likely to be.

The equilibrium of hematite and magnetite is governed by Reaction (2.s). If we consider the equilibrium fugacity of oxygen approximately proportional to the concentration, then at 800°K (a reasonable estimate of the temperature of crystallization for these rocks), we find that $f_{O_2} \approx 10^{-17}$ atm. Now the equilibrium constant for the reaction

$$H_2 + \tfrac{1}{2}O_2 \rightleftharpoons H_2O \tag{3.d}$$

is $10^{13.7}$ at this temperature. If the fugacity of water is 1000 atm, we obtain $f_{H_2} = 10^{-1.8}$ atm. If on the other hand $f_{H_2O} = 1$ atm, then $f_{H_2} = 10^{-4.8}$ atm. Since it is unlikely that f_{H_2O} will be many orders of magnitude lower than this over these silicates, we conclude that the concentration of hydrogen will be many orders of magnitude greater than that of oxygen, but that both will be much lower than the concentration of H_2O. A similar analysis of the species CO and CO_2 leads to $f_{CO} = 10^{-2.4}$ atm at $f_{CO_2} = 1000$ atm and $f_{CO} = 10^{-5.4}$ atm at $f_{CO_2} = 1$ atm. The range of $f_{CO_2} = 1000$ atm is most compatible with the occurrence of dolomite and quartz, as may be seen from the data of Weeks (1956). Thus we see that hydrogen is probably the controlling species in the diffusive transfer process.

Available data from the contact of the two iron formation members discussed above are shown in Figure 3.1, which is a plot of the atomic fraction $Mg/(Mg + Fe^{2+})$ and $Mn/(Mg + Fe^{2+} + Mn)$ in the cummingtonite[2] against the footage in a diamond drill core taken roughly perpendicular to the contact. It is apparent that $Mg/(Mg + Fe^{2+})$ falls off markedly as the distance from the oxidized iron formation increases. One possible interpretation of this is as follows.

Consider the reaction that must obtain at equilibrium in both assemblages:

$$\underset{\textit{cummingtonite}}{\tfrac{3}{7}Fe_7Si_8O_{22}(OH)_2} + \underset{\textit{fluid}}{\tfrac{4}{7}H_2O} \rightleftharpoons \underset{\textit{magnetite}}{Fe_3O_4} + \underset{\textit{quartz}}{\tfrac{24}{7}SiO_2} + \underset{\textit{fluid}}{H_2} \tag{3.e}$$

[2] Actually, cummingtonite is not present in the oxidized assemblage at this point, but the composition of the hypothetical cummingtonite that could exist may be determined from the composition of the observed actinolite.

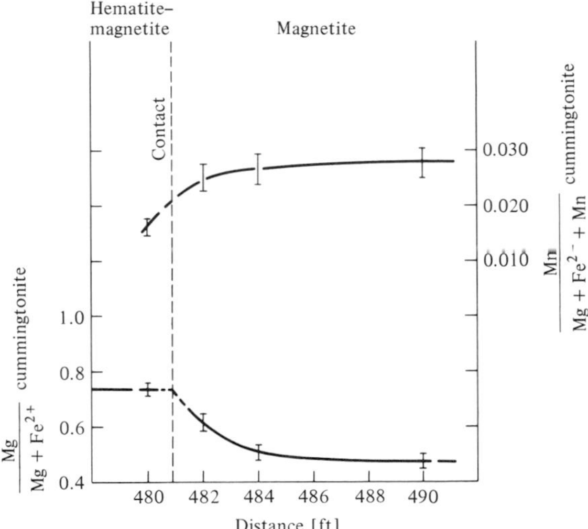

3.1 Variation of the atomic fractions $Mg/(Mg + Fe^{2+})$ and $Mn/(Mg + Fe^{2+} + Mn)$ in cummingtonite with diamond-drill-hole footage in the Quebec iron formation (Mueller, 1960, 1967). The lengths of the bars show the uncertainty in the analytical data.

Assuming that cummingtonite is ideal, the equation of equilibrium for this reaction is

$$K_{(e)} = \frac{f_{H_2}}{(1 - X_{Mg}^{Cum})^3 f_{H_2O}^{4/7} (\gamma_{Mg}^{Cum})^3} \tag{3.12}$$

Since f_{H_2O} is regarded as a constant and $\gamma_{Mg}^{Cum} \sim 1$, we obtain the approximation

$$K'_{(e)} \approx \frac{f_{H_2}}{(1 - X_{Mg}^{Cum})^3} \tag{3.13}$$

which shows the functional relation between the composition of the cummingtonite and the hydrogen fugacity. If we now assume that the fugacity approximates the partial pressure of H_2 and that Fick's first law holds, we may substitute Equation (3.13) into Equation (1.74) to obtain the hydrogen flux as a function of the composition of cummingtonite and the gradient in this composition:

$$J_{H_2} = -\frac{3D_{H_2}K'_{(e)}}{RT}(1 - X_{Mg}^{Cum})^2 \frac{dX_{Mg}^{Cum}}{dx} \tag{3.14}$$

In the foregoing discussion it was assumed that the mobilities of all the other components such as Mg, Fe, Si, and O are negligible as compared with the mobility of hydrogen, but that Reaction (3.e) proceeds rapidly compared with the rate of hydrogen diffusion. Thus within each small local volume such as is represented by one of the sampling points in Figure 3.1, equilibrium obtains and Equations (3.12) and (3.13) apply.

The effect of Mn could presumably be similar although the fugacity of oxygen required to oxidize Mn^{2+} to Mn^{3+} is much higher under standard concentrations of Mn^{2+}. However, since the quantity of Mn^{3+} and Mn^{2+} dissolved in the minerals is not known, this factor cannot be evaluated. Also it is apparent that the consideration of Mn^{3+} introduces an additional degree of freedom into the system. Consequently, the gradient in Mn can be attributed to more factors than the gradient in Mg and Fe^{2+}. Analogous problems have been discussed by Fisher (1974) in terms of nonequilibrium thermodynamic concepts. A comprehensive review of mixed volatile (H_2O-CO_2) equilibria is provided by Kerrick (1974).

Crystal kinetics

We now consider the kinetics of recrystallization and diffusion as they apply to individual crystals. Experimental data on these subjects are very few in number, but evidence of the qualitative sort from natural mineral assemblages is not lacking. This is particularly true of the feldspar family, whose members show prominent optical and x-ray effects of the chemical concentration gradients set up by diffusion (or lack of diffusion) and by recrystallization.

The tendency for plagioclase to form zoned crystals is well known, as is the phenomenon of unmixing in the alkali feldspars. Unmixing phenomena in the plagioclase feldspar series are much less apparent, but provide an interesting subject for future investigations.

The kinetic behavior of the plagioclase and alkali feldspar series is in sharp contrast. The zoning in plagioclase is usually a well-defined concentric structure that reflects the external morphology of the crystal. The anorthite content of the individual zones may be constant or show a slight continuous variation. Normally there is a decline in average anorthite content outward from the crystal center. Normal zoning with a steady decline in anorthite content is readily explained in terms of Figure 2.17 if it is assumed that the rock crystallized from a cooling melt. Under these circumstances the early-formed, anorthite-rich crystals are subsequently enveloped in layers of successively decreasing anorthite content, which are the products of the anorthite-depleted melt. Under static conditions, that is, when the crystals are not subject to shearing forces, no homogenization can take place because of the high activation energies involved in the required diffusion of Si^{4+} and Al^{3+} which must accompany the exchange of Na^+ and Ca^{2+}. Thus equilibrium crystallization probably never occurs under static conditions in this system.

Reversal, or a tendency for the later zones to be more anorthite-rich, is also easily explained in terms of changing conditions. For example, the anorthite content of the melt might be increased by reaction with basic inclusions in the magma, and this excess of anorthite would precipitate as a zone covering the previously formed crystals.

The above-mentioned dependence of the diffusive process in plagioclase on the simultaneous exchange of Si^{4+} and Al^{3+} that must accompany the exchange of Na^+ and Ca^{2+} was pointed out by Goldsmith (1952). Such movements involve the breaking of Si—O and Al—O bonds. The energy of a single Si—O bond is of the order of 90 kcal/mole (Pauling, 1948), and it seems probable that the activation energy for diffusion of the tetrahedrally coordinated Si^{4+} and Al^{3+} ions might be several times this value. Also, we should expect a considerable entropy of activation because of the relatively complex movements involved.

In the alkali feldspars we encounter a far different situation. It is well known that perthites, the two-phase exsolution products of high-temperature single-phase feldspars, form readily in the geologic environment although they have not been formed under laboratory conditions. Perthite type structures are also known from a variety of other mineral systems. It is obvious that in the alkali feldspars the kinetic barrier to diffusion must be much lower than in plagioclase. This is of course readily explained by the fact that only Na^+ and K^+ need be exchanged in the diffusive process, and the activation energy involved is, as we shall see, only of the order of 40 kcal/mole.

The diffusion of Na^+ in the minerals analcite, sodalite, nepheline, orthoclase, albite, and acmite, and in the volcanic glass obsidian, was studied quantitatively by Sippel (1963). The measured species was the radio nuclide sodium 24. Table 3.1 shows the results of Sippel's investigation in terms of the measured diffusion coefficients for various temperatures.

We may represent the coefficient of diffusion by the following equation:

$$\log D = \log D_0 - \frac{\Delta E^{\ddagger}}{2.303RT} \tag{3.15}$$

If this equation is compared with Equations (1.80) and (1.72), we see that D_0 is a frequency factor, and $\Delta E^{\ddagger}$ is the energy of activation. It is apparent that $\log D_0$ will change relatively slowly with the temperature as compared with the last term in Equation (3.15). However, in a Na-K

Table 3.1 Diffusion coefficients

Substance	Temperature [°C]	D [cm²/sec]
Obsidian	357	4.9×10^{-10}
	458	6.2×10^{-9}
Analcite	357	1.31×10^{-9}
Sodalite	580	8.7×10^{-11}
	617	2.5×10^{-10}
	675	1.0×10^{-9}
Nepheline	576	3.8×10^{-11}
	617	4.5×10^{-11}
	700	2.7×10^{-10}
	800	1.4×10^{-9}
Microcline	850	2×10^{-10}
Orthoclase	850	5×10^{-11}
Albite	850	8.0×10^{-11}
	940	2.8×10^{-10}
Acmite	940	$<10^{-11}$

After Sippel (1963).

solid solution such as those treated here, it should be a function of the Na/(Na + K) ratio except for small composition changes.

The results of Table 3.1 are plotted in Figure 3.2. This plot shows that, where the data are sufficient in number, $\log D$ is a linear function of $1/T$, which is the simplest interpretation of Equation (3.15). The activation energy and the frequency factor D_0 may be obtained directly from this plot. For example, for the mineral sodalite, Sippel obtained

$$\log D = \log 6.6 - \frac{42{,}500}{2.303RT} \tag{3.16}$$

There are only two points for albite feldspar, but the corresponding equation is

$$\log D = \log (2.09 \times 10^{-3}) - \frac{38{,}100}{2.303RT} \tag{3.17}$$

similar values have been obtained by Bailey (1971) for high temperature diffusion in albite.

Table 3.1 also shows that in order to get appreciable lattice diffusion on the laboratory time scale it is necessary to anneal the minerals at temperatures that are quite high relative to many petrogenic environments. Sippel compared the minerals studied in terms of their "diffusion temperature," T_D, the temperature at which arbitrarily $D = 2.5 \times 10^{-10}$ cm²/sec, and T_F, the temperature of fusion, both being expressed in °K. He found an approximately linear relation between T_D/T_F and the anion valence electron density (the number of anion valence electrons per cm³). Thus, if we exclude analcite and

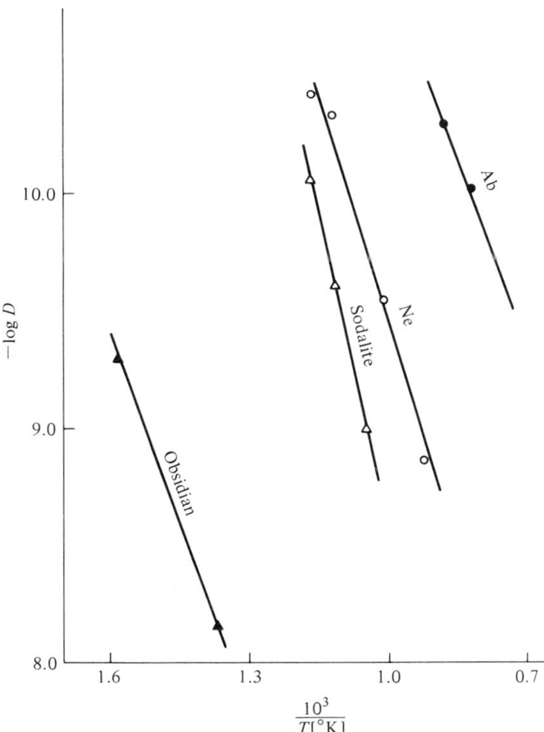

3.2 Logarithm of diffusion coefficient plotted against $1/T$. (Data are from Table 3.1.)

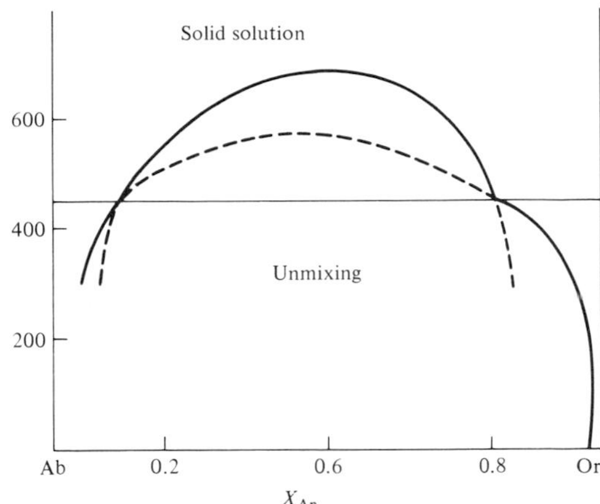

3.3 Proposed equilibrium diagram for the alkali feldspars. (After Laves and Goldsmith, 1961.) See also Fig. 9.9 (Chapter 9).

obsidian, sodium diffuses with greatest ease in sodalite and nepheline, which have the lowest anion valence electron densities, and most reluctantly in acmite pyroxene, which has the highest; the feldspars lie at an intermediate position. The parameter T_D is analogous to the "Tammann temperature," which we shall refer to later.

In order to understand the diffusive processes, we should first know the equilibrium reference states. Unfortunately, it is the kinetic properties themselves that prevent us from knowing these. One of the equilibrium diagrams proposed for the alkali feldspars is depicted in Figure 3.3. This diagram, proposed by Laves and Goldsmith (1961), is thought to approximate the true equilibrium relations attainable only in periods of time that are large compared with the time scales of laboratory experiments. Each point on the unmixing curve also represents a particular degree of ordering of Al and Si, with the lowest degree of order at the highest temperature.

The fact that the unmixing relations exist is evidence that the kinetic properties of these feldspars will be strongly dependent on composition. We can see this by referring back to Equation (1.83), which shows that if the activity coefficient changes negatively with composition, diffusion against the concentration gradient results. These compositional effects should be most marked at the lower temperatures. Thus the diffusion coefficients deduced in the 800–900°C range (Table 3.1) are probably of little use in the study of unmixing or perthite formation in the alkali feldspars.

The lattice mobility of a substance may be crudely estimated by use of the "Tammann temperatures" (Welch, 1955) already mentioned. The Tammann temperature T_T in °K is related to the temperature of fusion T_F also in °K as follows:

$$\frac{T_T}{T_F} = \alpha$$

where $\alpha \sim 0.5$ for many substances. These Tammann temperatures can be regarded as minimum temperatures at which the solid will enter into reactions. If we apply this concept to the data of Sippel, we conclude that since T_F for albite is approximately 1400°K, we should expect some observable lattice diffusion at 700°K, which is, however, far below the temperatures at which significant lattice diffusion was observed. Nevertheless the melting points of minerals provide a good indication of their response to certain processes which depend on lattice diffusion. An example of one such process is the response to deforming stresses as deduced from the geologic evidence. To illustrate, we arrange some common minerals in descending order of their melting points:

forsterite > quartz > enstatite > anorthite
> diopside > fayalite > orthoclase > albite > NaCl

We shall see that the minerals above orthoclase (excluding quartz) are dominant in basic bodies that generally resist plastic deformation and respond to stresses by fracturing, whereas this is less true of the alkali feldspars. Also, NaCl deforms readily in shallow salt domes. The response is, however, greatly complicated by such factors as grain size and accessibility to fluxing solutions. Thus impermeable beds of quartzite may be very competent, but small quartz grains in flaser gneisses generally recrystallize more readily than does feldspar under the same circumstances. Also, some substances such as calcite ($T_F = 1310°C$), which have high temperatures of fusion, show much evidence of deformation by recrystallization creep under deep-seated metamorphic conditions as well as under laboratory conditions (Griggs *et al.*, 1958). These points are discussed in greater detail in Chapter 8.

Kinetics of dissociation

Among the heterogeneous reactions, only dissociation lends itself to a simple mechanistic treatment. We shall begin by setting up a relatively idealized model, derive the pertinent equations, and then compare the system with the results of experiments.

One of the dissociation reactions that has received considerable attention is the dehydration of crystalline $Mg(OH)_2$, the mineral brucite. Our idealized model for the dissociating brucite grain is depicted in Figure 3.4. For simplicity the grain is assumed to be spherical and isotropic (brucite is actually hexagonal), with an initial radius r_0. This grain is assumed to decompose along a reaction interface (indicated by the broken line) that moves inward at a constant rate with the "molecules" of $Mg(OH)_2$ decomposing layer by layer; consequently, the radius at any elapsed time t is r. The water given off is assumed to escape so rapidly that diffusion does not govern the reaction rate. This argument seems plausible enough since it might be expected that the reacted shell of MgO would be quite porous. An additional assumption is that the nucleation of MgO does not govern the reaction rate. This again seems a reasonable assumption since it might be expected that even MgO crystals of near molecular size might be more stable than $Mg(OH)_2$.

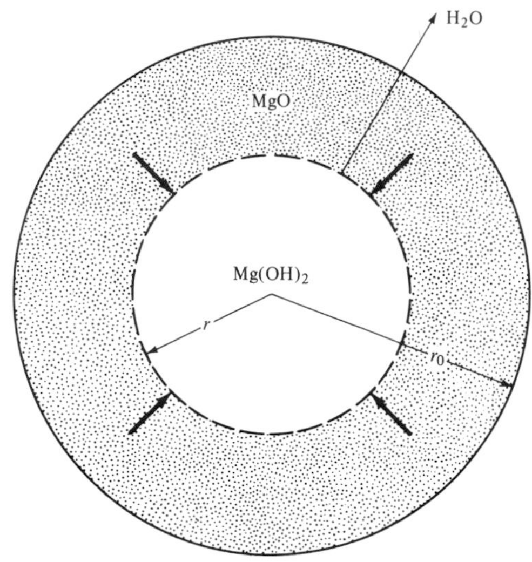

3.4 Idealized model for the dissociation of brucite. This model may be used to derive a simple equation for the dissociation (see text).

Assuming the foregoing we may define the following quantities: the rate constant is K_A the number of molecules of brucite reacting per cm² per second, and A is the area of the reaction interface. Consequently, the total flux of H_2O molecules for the grain is

$$J = K_A A \text{ molecules/sec} \tag{3.18}$$

and the fractional rate of weight loss is

$$-\frac{dq}{dt} = J\,\frac{18/(6.02 \times 10^{23})}{\frac{4}{3}\pi r_0^3 \delta_B} \tag{3.19}$$

where 6.02×10^{23} is Avogadro's number and δ_B is the density of brucite.

Substituting Equation (3.18) we have

$$-\frac{dq}{dt} = \frac{K_A A 1.245 \times 10^{-23}}{\frac{4}{3}\pi r_0^3} \tag{3.20}$$

If we now assume that each brucite "molecule" has a thickness T_M and that c_s is the number of molecules per unit of surface, we obtain the equation for the constant rate of decrease of the radius

$$\text{constant} = \frac{dr}{dt} = \frac{-K_A T_M}{c_s} \tag{3.21}$$

We may integrate this to obtain

$$r - r_0 = \frac{-K_A T_M}{c_s}\, t \tag{3.22}$$

where t is the elapsed time. If we substitute Equation (3.22) into (3.20) we get

$$-\frac{dq}{dt} = \frac{3.74K_A \times 10^{-23}}{r_0^3}\left(-\frac{K_A T_M}{c_s}t + r_0\right)^2 \quad (3.23)$$

On integration we obtain the fractional weight loss as a function of the elapsed time:

$$q = -\frac{3.74 \times 10^{-23}K_A}{r_0^2}\left(\frac{B}{3}t^3 + Br_0t^2 + r_0^3t\right) \quad (3.24)$$

in which we have set $B = -K_A T_M/c_s$.

This system has been investigated in detail by Gregg and Razouk (1949), and these authors have shown that the decomposition curves have three segments (1) the initial or induction period in which the rate of decomposition increases, (2) the major decay period in which the rate decreases, and (3) the final decay segment in which pure MgO is approached asymptotically. The induction period may be due to nucleation (according to Gregg and Razouk) or simply lag in bringing the system to the desired temperature. The segment representing the major decay period may be represented by the equation

$$1 - (1 - \alpha)^{1/3} = Kt + b \quad (3.25)$$

where α is the fraction decomposed, K and b are constants, and t is the elapsed time. This equation is obviously of the same form as Equation (3.24). However, Gregg and Razouk found no simple relation between the particle size and the decomposition rates. Rather, they found an optimum size at which the rate reached a maximum. Over the range in which the rate increases with diminishing particle size the increase appears to be proportional to the edges of brucite fragments rather than the area of the whole grain, as it is in the idealized theory presented above.

For the segment of the curve approaching pure MgO asymptotically, Equation (3.25) obviously does not apply since its curve must intersect the horizontal axis. For this segment an equation based on the unimolecular law (first-order rate equation) is applicable. The equation is

$$\ln(1 - \alpha) = K_1t + \text{constant} \quad (3.26)$$

A further discussion of the theory of dissociation is beyond the scope of this book. However, the subject has been extensively treated by Mample (1940) and others. Special reference is made to the book edited by Garner (1955).

In addition, the dissociation and hydration of brucite have been discussed by Fyfe *et al.* (1958). These authors also discuss the laboratory investigation of the decomposition rates of some other hydrous minerals.

Kinetics of complex heterogeneous reactions

We now consider the kinetics of some reactions of a greater degree of complexity than simple dissociations. For these reactions no detailed mechanism can be suggested, but the data may in some cases suggest the orders of the reactions.

The most detailed quantitative study of a petrologically significant system was presented by Greenwood (1963), who investigated three simultaneous reactions in the laboratory. These are as follows:

$$\underset{\text{talc}}{7Mg_3Si_4O_{10}(OH)_2} \longrightarrow \underset{\text{anthophyllite}}{3Mg_7Si_8O_{22}(OH)_2} + \underset{\text{quartz}}{4SiO_2} + \underset{\text{gas}}{4H_2O} \quad (3.f)$$

$$\underset{\text{anthophyllite}}{Mg_7Si_8O_{22}(OH)_2} \longrightarrow \underset{\text{enstatite}}{7MgSiO_3} + \underset{\text{quartz}}{SiO_2} + \underset{\text{gas}}{H_2O} \quad (3.g)$$

$$\underset{\text{talc}}{Mg_3Si_4O_{10}(OH)_2} \longrightarrow \underset{\text{enstatite}}{3MgSiO_3} + \underset{\text{quartz}}{SiO_2} + \underset{\text{gas}}{H_2O} \quad (3.h)$$

Experimental rate studies were made at a pressure of 1000 bar and at temperatures of 830°C, 815°C, and 805°C. Under these conditions, Reactions (3.f) to (3.h), inclusive, all proceed to the right at observable but different rates, and the system is in the stability field of enstatite and quartz. The reactions may also be written in terms of the volumes of the phases involved:

$$k\ \text{talc} \longrightarrow l\ \text{anthophyllite} + m\ \text{quartz} + H_2O \quad (3.f)$$

$$n\ \text{anthophyllite} \longrightarrow o\ \text{enstatite} + p\ \text{quartz} + H_2O \quad (3.g)$$

$$q\ \text{talc} \longrightarrow r\ \text{enstatite} + s\ \text{quartz} + H_2O \quad (3.h)$$

In these reactions the lower-case letters refer to the stoichiometric coefficients on a volume basis. So that if $k = n = q = 1$, then $l = 0.882$, $m = 0.096$, $o = 0.806$, $p = 0.0826$, $r = 0.711$, and $s = 0.168$.

The assumption is now tentatively made (and later justified by comparison with experimental results) that all the reactions are first order. Let us define the volumes of the various minerals consumed in time t in the three reactions:

$X_{(f)}$ is the volume of talc consumed in Reaction (3.f); $X_{(h)}$ is the volume of talc consumed in Reaction (3.h); $Y_{(f)}$ is the volume of anthopyllite formed in Reaction (3.f); $Y_{(g)}$ is the volume of anthophyllite consumed in Reaction (3.g); $Z_{(f)}$ is the volume of quartz formed in Reaction (3.f); $Z_{(g)}$ is the volume of quartz formed in Reaction (3.g); $Z_{(h)}$ is the volume of quartz formed in Reaction (3.h); A is the total volume of anthophyllite present at time t; T is the total volume of talc present at time t; Q is the total volume of quartz present at time t; and E is the total volume of enstatite present at time t.

Thus we have

$$A = Y_{(f)} - Y_{(g)}$$

and

$$T = (T^0 - X_{(h)} - X_{(f)})$$

where T^0 is the initial volume of talc, which we set equal to unity, giving

$$T = (1 - X_{(h)} - X_{(f)})$$

Consequently

$$-\frac{dT}{dt} = \frac{dX_{(h)}}{dt} + \frac{dX_{(f)}}{dt}$$

Since Reactions (3.f) and (3.h) are assumed to be first order we have

$$\frac{dX_{(f)}}{dt} = K_{(f)}T \tag{3.27}$$

and

$$\frac{dX_{(h)}}{dt} = K_{(h)}T \tag{3.28}$$

where $K_{(f)}$ and $K_{(h)}$ are the specific rate constants. We also define the quantity

$$\kappa = K_{(f)} + K_{(h)}$$

and combine Equations (4.27) and (4.28) to obtain

$$-\frac{dT}{dt} = \kappa T \tag{3.29}$$

which integrates to

$$T = C_1 e^{-\kappa t}$$

where C_1 is the integration constant. At $t = 0$, $T = T^0 = 1$, so that $C_1 = 1$ and

$$T = e^{-\kappa t} \tag{3.30}$$

For the formation of anthophyllite we have

$$Y_{(f)} = \frac{l}{k} X_{(f)} \tag{3.31}$$

So that

$$\frac{dY_{(f)}}{dt} = \frac{l}{k}\frac{dX_{(f)}}{dt} \tag{3.32}$$

If we now combine Equations (3.27), (3.30), and (3.32) we obtain

$$\frac{dY_{(f)}}{dt} = \frac{l}{k} K_{(f)} e^{-\kappa t} \tag{3.33}$$

which integrates to

$$Y_{(f)} = -\frac{l}{k}\frac{K_{(f)}}{\kappa} e^{-\kappa t} + C_2 \tag{3.34}$$

where C_2 is the constant of integration. At $t = 0$, $Y_{(f)} = 0$, so that

$$Y_{(f)} = \frac{l}{k}\frac{K_{(f)}}{\kappa}(1 - e^{-\kappa t}) \tag{3.35}$$

The decomposition rate of anthophyllite according to Reaction (3.g) is given by

$$\frac{dY_{(g)}}{dt} = K_{(g)}(Y_{(f)} - Y_{(g)}) = K_{(g)}\left[\frac{l}{k}\frac{K_{(f)}}{\kappa}(1 - e^{-\kappa t}) - Y_{(g)}\right] \tag{3.36}$$

in which we have used Equation (3.35).

Equation (3.36) may be rewritten as

$$\frac{dY_{(g)}}{dt} + K_{(g)} Y_{(g)} = \frac{l}{k}\frac{K_{(g)}K_{(f)}}{\kappa}(1 - e^{-\kappa t}) \tag{3.37}$$

which is of the general form

$$\frac{dY}{dX} + PY = R$$

where P is a function of X or a constant, but not a function of Y. This equation has the integration factor $e^{\int PdX}$, so that the solution takes the form

$$e^{\int PdX}Y = \int e^{\int PdX}R\,dx + C_3$$

where C_3 is the constant of integration. Consequently the solution to Equation (3.37) is

$$Y_{(g)} = C_3 e^{-K_{(g)}t} + \frac{l}{k}\frac{K_{(f)}}{\kappa} - \frac{l}{k}\frac{K_{(f)}K_{(g)}}{\kappa}\frac{e^{-\kappa t}}{(K_{(g)} - \kappa)} \tag{3.38}$$

Since at $t = 0$, $Y_{(g)} = 0$, Equation (3.38) reduces to

$$Y_{(g)} = \frac{l}{k}\frac{K_{(f)}}{\kappa(K_{(g)} - \kappa)}\left[\kappa e^{-K_{(g)}t} + (K_{(g)} - \kappa) - K_{(g)}e^{-\kappa t}\right] \tag{3.39}$$

The total amount of anthophyllite at any time is

$$A = Y_{(f)} - Y_{(g)}$$

Thus by Equations (3.35) and (3.39) we have

$$A = \frac{l}{k}\frac{K_{(f)}}{(K_{(g)} - \kappa)}(e^{-\kappa t} - e^{-K_{(g)}t}) \tag{3.40}$$

The amount of quartz formed by the three reactions is

$$Q = Z_{(f)} + Z_{(g)} + Z_{(h)} = \frac{m}{l}Y_{(f)} + \frac{p}{h}Y_{(g)} + \frac{s}{q}X_{(h)} \tag{3.41}$$

If we now substitute Equation (3.30) into (3.28) and integrate we get

$$Y_{(h)} = \frac{K_{(h)}}{\kappa}(1 - e^{-\kappa t}) \tag{3.42}$$

We may now substitute the expressions for $Y_{(f)}$, $Y_{(g)}$, and $X_{(h)}$ into Equation (3.41), which yields

$$Q = \frac{1}{\kappa}\left\{(1 - e^{-\kappa t})\left[\frac{s}{q}K_{(h)} + \frac{m}{k}K_{(f)} + \frac{lp}{kn}\frac{K_{(f)}K_{(g)}}{(K_{(g)} - \kappa)}\right] - \frac{lp}{kn}\frac{K_{(f)}\kappa}{(K_{(g)} - \kappa)}(1 - e^{-K_{(g)}t})\right\} \tag{3.43}$$

The quantity of enstatite at any time may be obtained from the expression

$$E = \frac{r}{s}Z_{(h)} + \frac{o}{n}Y_{(g)} \tag{3.44}$$

And substitution of $Z_{(h)}$ and $Y_{(g)}$ gives

$$E = \frac{1}{\kappa}\left\{\left[\frac{r}{q}K_{(h)} + \frac{ol}{nk}\frac{K_{(f)}K_{(g)}}{(K_{(g)} - \kappa)}\right](1 - e^{-\kappa t}) - \left[\frac{ol}{nk}\frac{K_{(f)}\kappa}{(K_{(g)} - \kappa)}\right](1 - e^{-K_{(g)}t})\right\} \tag{3.45}$$

The specific rate constants $K_{(f)}$, $K_{(g)}$, $K_{(h)}$, and κ may be obtained by differentiating the expressions for T, A, Q, and E with respect to t and evaluating the slopes at $t = 0$. These initial slopes may be obtained from the curves fitted to the experimental data of the mineral volumes as a function of time.

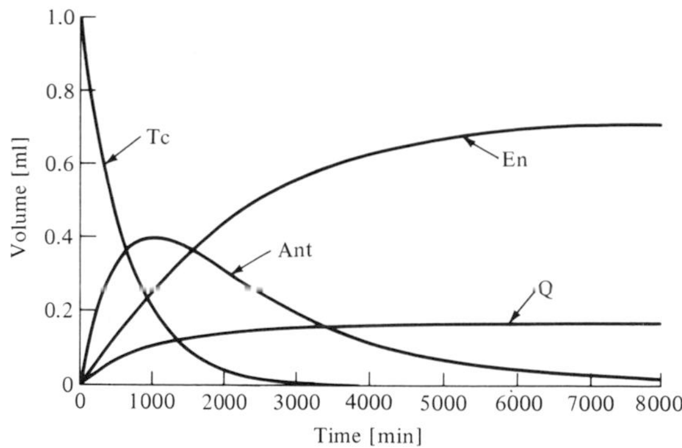

3.5 Volumes of talc, anthophyllite, enstatite, and quartz as functions of time at 830°C and 1000 bar, as calculated from measured rate constants. Note that although the stable assemblage is enstatite plus quartz, anthophyllite is the most abundant phase from 1000 to 1500 min. Figure after H. J. Greenwood, *Journal of Petrology*, *V. 4*, p. 317–351, 1963. Printed by permission from Oxford University Press.

Figure 3.5 shows the theoretical curves determined by the derived expressions and the measured initial slopes for 830°C and 1000 bar. These curves show that the volume of anthophyllite culminates and exceeds the volume of the stable phase enstatite although the latter mineral and quartz are the stable phases. This has important implications in experimental silicate chemistry and in certain rapidly formed mineral assemblages in nature.

Kinetics of rock deformation

One of the most far-reaching applications of kinetic theory is to be found in the study of rock deformation. The phenomena range in scale from individual crystals to the planetary bodies. In fact, the subject is intimately connected with the fundamental time scales of planetary process that produce differentiated crusts and continents. For example, the time scale of crustal processes on a hot planet such as Venus should be much contracted relative to a cooler planet such as Earth. This has important implications relative to the distribution of crustal material on the two planets as well as the distance and time scales of isostatic adjustments. It is also effective in evaluating the theories of sea floor spreading.

It is important to keep in mind that we are primarily concerned here with deformation by recrystallization (recrystallization creep), an essentially chemical process, and not cataclastic deformation, which is primarily physical. This distinction is an important one since most laboratory experiments contain a greater element of the latter. The energetics of recrystallization creep are closely related to those of lattice diffusion since in creep the thermal diffusion is directed along the chemical potential gradient imposed by the stress field. In each case bonds must be broken and diffusive energy barriers surmounted.

We have seen before that the coefficient of diffusion is related to the specific rate constant by the expression $D = \lambda^2\kappa$, where λ is the distance between successive equilibrium positions of the diffusing particle. We now consider two layers of atoms, one of which is in motion relative to the other. We will call the interatomic distance within this layer and perpendicular to the direction of transport λ_2, that parallel to the direction of transport λ_3, and that perpendicular to the shearing layers λ_1. We now define the viscosity of the material as

$$\eta = \frac{f\lambda_1}{\Delta v} \quad \text{(in poise)} \tag{3.46}$$

where f is the differential shearing stress and Δv is the difference in velocity between the two layers. We shall find that generally η is an implicit function of the temperature, the total hydrostatic pressure, and the differential stress f. However, in one very important case, that of Newtonian flow, η is a function only of the temperature and pressure.

We have already stated that the effect of the stress is to deform the energy barriers so that diffusion is greater in one direction than another.[3] Thus the atoms are not "pushed along" by stress but are moved solely by their thermal energy, just as in diffusion under hydrostatic conditions. In terms of potential energy the deformation amounts to a product of three factors: (1) $\lambda_2\lambda_3$, the area of the unit of flow upon which f acts, (2) f the differential stress, and (3) $\frac{1}{2}\lambda$ the distance to the top of the energy barrier. Consequently, one side of the energy barrier is lowered by an amount $\frac{1}{2}f\lambda_2\lambda_3\lambda$ and the other side is increased by the same amount. It may now be shown by a method similar to that used in deriving Equation (1.81) for the net diffusive flux that

$$\Delta v = \lambda K\left[\exp\left(\frac{\frac{1}{2}f\lambda_2\lambda_3\lambda}{kT}\right) - \exp\left(-\frac{\frac{1}{2}f\lambda_2\lambda_3\lambda}{kT}\right)\right] \tag{3.47}$$

[3] This view of flow and diffusion mechanisms is inherent in the theory of absolute reaction rates (Glasstone, *et al.*, 1941).

or

$$\Delta v = 2\lambda K \sinh\left(\frac{f\lambda_2\lambda_3\lambda}{2kT}\right) \tag{3.48}$$

in which K is the specific rate constant for diffusion under hydrostatic conditions. We shall find that in many cases the four quantities λ_1, λ_2, λ_3, and λ are but poorly defined and represent some average values of the interatomic distances in different directions. However, we shall also see that the ultimate expressions we seek are in many cases satisfied by knowing these quantities to a factor of 2 or 3, so that for some purposes we are justified in ignoring the differences between them. Consequently, we assume in the future that

$$\lambda_1 = \lambda_2 = \lambda_3 = \lambda$$

If we divide Equation (3.48) by λ_1, we obtain an expression for the shear strain rate:

$$\dot{\gamma} = \frac{\Delta v}{\lambda_1} = \frac{2K\lambda}{\lambda_1}\sinh\left(\frac{f\lambda_2\lambda_3\lambda}{2kT}\right) \tag{3.49}$$

The shear strain rate may be independently measurable from the geometry and dimensions of a given type of deformation.

Now, if Equation (3.48) is substituted into Equation (3.46), the expression for the viscosity is

$$\eta = \frac{\lambda_1 f}{2\lambda K \sinh(fV/2kT)} \tag{3.50}$$

in which $V = \lambda_2\lambda_3\lambda$, the approximate volume of the unit of flow.

Equation (3.50) is a quite general expression for the viscosity in terms of the absolute reaction rate model. We may also reduce this expression to a much simpler form under certain conditions. If, for example, f, the shear stress, is small it may be that

$$fV \ll 2kT$$

Under these circumstances Equation (3.50) reduces to

$$\eta = \frac{\lambda_1 kT}{\lambda VK} \tag{3.51}$$

which defines Newtonian type flow in which the viscosity is independent of the stress and the strain rate.

The condition $fV \ll 2kT$ is apparently met in some of petrologic systems, and Equation (3.51) is experimentally confirmed for many silicate melts in particular (Chapter 12). Also, surprisingly enough, analogous equations apply to some materials as refractory as polycrystalline aluminum oxide when they are of fine grain size (Folweiler, 1961).

If we wish to express the specific rate constant in constant in terms of the enthalpy and entropy of activation we may write:

$$K = \frac{kT}{h} \exp\left(\frac{-\Delta H^{\ddagger}}{kT}\right) \exp\left(\frac{\Delta S^{\ddagger}}{k}\right) \tag{3.52}$$

where h is Planck's constant. The enthalpy term embodies the effect of hydrostatic pressure. Under the simplest condition this takes the form of a linear $P\Delta V$ term in which $\Delta V^{\ddagger}$ is the volume change of activation. Since we should expect $\Delta V^{\ddagger}$ to be positive, the viscosity should increase with the pressure. As a result pressure and temperature should have opposite effects on the change of viscosity with depth in the Earth's mantle. Also, since the rate of temperature increase with depth is thought to decrease while the pressure increases almost linearly, a viscosity minimum is to be expected at some depth within the upper mantle. Presumably this minimum coincides with the observed zone of low seismic velocity at several hundred kilometers.

Equations (3.49) may be used to obtain the extent of deformation ζ as a function of time if expressions that relate γ, ζ and f can be found.

One of the most interesting applications of the theory is to the postglacial uplift that has occurred in Canada and in Fennoscandia. In this context ζ refers to the "uplift remaining," or actually the uplift that can still occur in the future. Following Post and Griggs (1973) we set

$$\dot{\gamma} = K_1 \dot{\zeta} \tag{3.53}$$

and

$$f = K_2 \zeta \tag{3.54}$$

in which K_1 and K_2 are constants that depend on the geometry of uplift. The equation (Mueller, 1975) is then

$$\dot{\zeta} = C_0' \sinh A'\zeta \tag{3.55}$$

where

$$C_0' = \frac{2\lambda K}{\lambda_1 K_1} \qquad A' = \frac{VK_2}{2kT} \tag{3.56}$$

Integration of Equation (3.55) then gives

$$\log \tanh A\zeta + C = C_0 t \tag{3.57}$$

which is the desired relation between the uplift remaining and the time. In this expression C is a constant of integration that depends on the presently remaining uplift. The other constants are related as follows to those of Equation (3.56):

$$C_0 = \frac{A'C_0'}{2.303} \qquad A = \frac{A'}{2} \tag{3.57}$$

If the presently remaining uplift can be inferred in some way, it is obvious that the constants A and C_0 can be obtained by filling geologic data on the change of ζ with the time. Fortunately, accurate data in the form of elevated and dated sea beaches are available.

Figure 3.6 shows a least-square fit in which ζ is given in meters and the time past in years for the central area of uplift of Fennoscandia. The corresponding rate of uplift in cm/yr from Equation (3.55) is also shown. An equally good fit is obtained from the more extensive data of Lliboutry (Mueller, 1976).

The viscosity of the region of the earth's mantle in which flow occurs in this example is found to fall in the range of 10^{22} to 10^{23} P, which is similar to that first obtained by Haskell (1935) using a fluid dynamics approach. The strain rate is found to fall in the range of 10^{-16} to 10^{-15} sec^{-1}, which is typical of large-scale geologic processes. As first pointed out by Lliboutry, the

3.6 Uplift and rate of uplift for the Fennoscandia postglacial uplift as a function of time (after Mueller, 1975). ζ is the central uplift remaining at various times in the past. Data points are as presented by Post and Griggs (1973; following McConnell, 1968, and are based on original data of Sauramo, 1958). Presently remaining uplift in the central region is assumed to be 180 m. Sizes of the crosses reflect estimated uncertainties. The curves are based on reaction rate theory (see text).

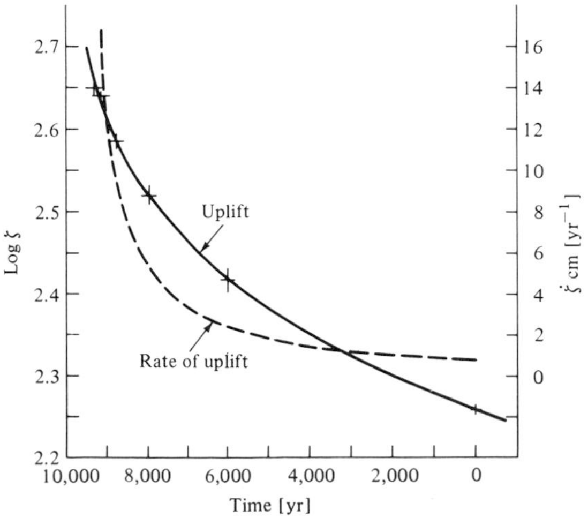

mode of flow in this case is non-Newtonian, as shown by the curvature of log ζ versus the time in Figure 3.6. However, this flow should assume a more Newtonian character as the stresses and strain rates diminish with time.

We shall see that the equations presented here have many applications in comparative planetology and in the interpretation of crustal features on planets such as Venus and Mars.

4 Classifications of rocks

Chemical classification

There is nothing very exciting about rock classifications, yet petrology could scarcely do without them. Classifications take many forms, and so we may justify them in various ways. Some, based on such characteristics as color, are useful chiefly in the field and will not concern us further here. Again, classification based on texture and implicit in such rock names as *diabase* will be of only secondary interest. We shall be concerned chiefly with the chemical and mineralogic classifications that aid us in summarizing much diverse information of importance in understanding rock genesis.

All chemical and mineralogic classifications are genetic in character, otherwise they could scarcely lead to any knowledge of the origin of rocks. Of the two types of classifications—chemical and mineralogic—the former is most widely employed in the study of igneous or igneous-appearing rocks, whereas mineralogic classifications are nearly confined to metamorphic rocks. The characterization of igneous rocks in terms of *normative* minerals, which we discuss later, is not really a mineralogic classification scheme at all but simply makes use of mineralogic formula units instead of elements or oxides employed in ordinary chemical work.

We shall not concern ourselves here with a review of the numerous chemical classifications of igneous rocks that have been proposed throughout the last century. Such a review may be found in the works of Johannsen (1939); also, Barth (1962) has given an excellent summary of the salient points of the most widely used classification schemes.

4 Classifications of rocks

The most useful classifications begin by reducing the rock analyses (always given in weight percents of the oxides) to *molecular numbers*, which equal the weight percents divided by the formula weights of the oxides. These numbers may then be used to form various useful parameters such as the molecular fractions $Na_2O/(Na_2O + K_2O)$ or $MgO/(MgO + FeO)$ as in the system of Osann (1899), for example. In other classifications, such as the first introduced by Cross *et al.* (1902), molecular numbers of the oxides are combined to form normative minerals. However, the end result is presented in terms of the weight percents of these normative minerals. A modification of this procedure was introduced by Niggli (1936), who presented the analysis in terms of equivalent molecular unit percentages. Since this is the system most used today, we give an example of such a calculation in this chapter. However, we shall first need to examine in greater detail the chemical compositions of rocks as they occur in nature.

Table 4.1 shows a series of average analyses of igneous and igneous-appearing rocks taken from Nockolds (1954). Many of these, such as the extrusive lavas, are of unquestioned igneous origin, but a considerable number of the plutonic examples could at best be proved to have been molten only after painstaking field and laboratory investigations, and generally such investigations are very limited. However, since we are for the moment concerned only with chemical characteristics, the question of origin is of secondary importance. The first part of the table (Nos. 1–16) is arranged to give a spectrum of the common compositional types. These rocks, ranging from granites to gabbros, include what is generally believed to be the major calc-alkali series arrived at by fractional crystallization of a basaltic or gabbroic magma under crustal conditions. The second part of the table (Nos. 17–30) contains rocks that are for the most part undersaturated with respect to SiO_2. Their place in the differentiation scheme is much less well

Table 4.1 Average analyses of igneous and igneous-appearing rocks

	Alkali granite	Alkali rhyolite	Peralkaline granite	Peralkaline rhyolite	Calc-alkali granite	Calc-alkali rhyolite	Adamellite	Dellenite	Granodiorite	Rhyodacite	Tonalite	Dacite	Diorite	Andesite	Gabbro
	1	2	3	4	5	6	7	8	9	10	11	12	13	14	15
SiO_2	73.86	74.57	71.08	72.31	72.08	73.66	69.15	70.15	66.88	66.27	66.15	63.58	51.86	54.20	48.36
TiO_2	0.20	0.17	0.40	0.42	0.37	0.22	0.56	0.42	0.57	0.66	0.62	0.64	1.50	1.31	1.32
Al_2O_3	13.75	12.58	11.26	10.88	13.86	13.45	14.63	14.41	15.66	15.39	15.56	16.67	16.40	17.17	16.84
Fe_2O_3	0.78	1.30	4.28	2.92	0.86	1.25	1.22	1.68	1.33	2.14	1.36	2.24	2.73	3.48	2.55
FeO	1.13	1.02	2.19	2.42	1.67	0.75	2.27	1.55	2.59	2.23	3.42	3.00	6.97	5.49	7.92
MnO	0.05	0.05	0.11	0.14	0.06	0.03	0.06	0.06	0.07	0.07	0.08	0.11	0.18	0.15	0.18
MgO	0.26	0.11	0.25	0.16	0.52	0.32	0.99	0.63	1.57	1.57	1.94	2.12	6.12	4.36	8.06
CaO	0.72	0.61	0.84	0.68	1.33	1.13	2.45	2.15	3.56	3.68	4.65	5.53	8.40	7.92	11.07
Na_2O	3.51	4.13	4.92	5.17	3.08	2.99	3.35	3.65	3.84	4.13	3.90	3.98	3.36	3.67	2.26
K_2O	5.13	4.73	4.21	4.42	5.46	5.35	4.58	4.50	3.07	3.01	1.42	1.40	1.33	1.11	0.56
H_2O^+	0.47	0.66	0.39	0.45	0.53	0.78	0.54	0.68	0.65	0.68	0.69	0.56	0.80	0.86	0.64
P_2O_5	0.14	0.07	0.07	0.03	0.18	0.07	0.20	0.12	0.21	0.17	0.21	0.17	0.35	0.28	0.24
No. of analyses	(48)	(21)	(53)	(39)	(72)	(22)	(121)	(58)	(137)	(115)	(58)	(50)	(50)	(49)	(160)

Data are from Nockolds (1954).

established, and they are not nearly so abundant as the table seems to imply. In all cases the odd numbers in the table represent the deep-seated or plutonic members, whereas the even numbers refer to their extrusive equivalents.

In Figure 4.1 the weight percentages of the various oxides are shown plotted against the weight percent of SiO_2 for representative members of the calc-alkali series. Later we shall discuss some variation diagrams for particular rock suites; however Figure 4.1 is only intended at this point to give a graphic picture of the most abundant igneous rock types.

We have said that certain ratios or fractions have an important part in rock classification schemes. An example of such a parameter is the ratio Fe_2O_3/FeO. Although this parameter has been given surprisingly little attention by petrologists in the past, its importance in the interpretation of many types of rocks—igneous, metamorphic and sedimentary—can hardly be overestimated. This point is illustrated in Figure 4.2, in which the same series shown in Figure 4.1 is plotted in terms of Fe_2O_3/FeO. It is immediately apparent that the extrusives are generally more oxidized than their plutonic equivalents but that the difference in degree of oxidation vanishes at the basic end of the series. We shall return to this point later.

Table 4.2 is a simplified representation of the rock types of Table 4.1 and has been derived from a similar table of Nockolds (1954) by excluding certain rocks such as the peridotites, which probably have no equivalent melts under crustal conditions. We do not present a tabulation of the mineral proportions or "modes" of these rocks since such information is readily obtainable from the descriptive petrographies such as that of Johannsen (1939) or Williams *et al.* (1954). However, we shall have many opportunities to consider the igneous minerals when we treat the origin of magmatic rocks.

As previously indicated, a rock analysis may be summarized in terms of the Niggli normative minerals it

	Tholeiitic basalt	**Alkali syenite**	**Alkali trachyte**	**Calc-alkali syenite**	**Calc-alkali trachyte**	**Monzonite**	**Latite**	**Nepheline syenite**	**Phonolite**	**Nepheline monzonite**	**Nepheline latite**	**Essexite**	**Ordanchite**	**Theralite**	**Nepheline tephrite**
	16	17	18	19	20	21	22	23	24	25	26	27	28	29	30
SiO_2	50.83	61.86	61.95	59.41	58.31	55.36	54.02	55.38	56.90	50.38	52.95	46.88	47.34	45.97	44.82
TiO_2	2.03	0.58	0.73	0.83	0.66	1.12	1.18	0.66	0.59	2.49	1.43	2.81	2.53	2.34	2.65
Al_2O_3	14.07	16.91	18.03	17.12	18.05	16.58	17.22	21.03	20.17	19.97	19.14	17.07	16.59	16.64	15.42
Fe_2O_3	2.88	2.32	2.33	2.19	2.54	2.57	3.83	2.42	2.26	2.77	3.25	3.62	4.50	2.94	4.28
FeO	9.00	2.63	1.51	2.83	2.02	4.58	3.98	2.00	1.85	3.96	2.86	5.94	5.72	8.27	6.61
MnO	0.18	0.11	0.13	0.08	0.14	0.13	0.12	0.19	0.19	0.13	0.20	0.16	0.20	0.26	0.16
MgO	6.34	0.96	0.63	2.02	2.07	3.67	3.87	0.57	0.58	2.15	2.02	4.85	5.04	5.73	7.27
CaO	10.42	2.54	1.89	4.06	4.25	6.76	6.76	1.98	1.88	6.01	5.33	9.49	8.65	10.45	10.32
Na_2O	2.23	5.46	6.55	3.92	3.85	3.51	3.32	8.84	8.72	6.35	6.55	5.09	5.33	5.06	5.30
K_2O	0.82	5.91	5.53	6.53	7.38	4.68	4.43	5.34	5.42	3.97	4.37	2.64	2.63	1.37	1.26
H_2O	0.91	0.53	0.54	0.63	0.53	0.60	0.78	0.96	0.96	1.37	1.12	0.97	0.87	0.66	1.56
P_2O_5	0.23	0.19	0.18	0.38	0.20	0.44	0.49	0.19	0.17	0.45	0.37	0.48	0.60	0.31	0.35
CO_2								0.17	—	—	—				
Cl								—	0.23	—	0.09				
SO_3								—	0.13	—	0.34				
No. of analyses	(137)	(25)	(15)	(18)	(24)	(46)	(42)	(80)	(47)	(7)	(16)	(15)	(27)	(6)	(8)

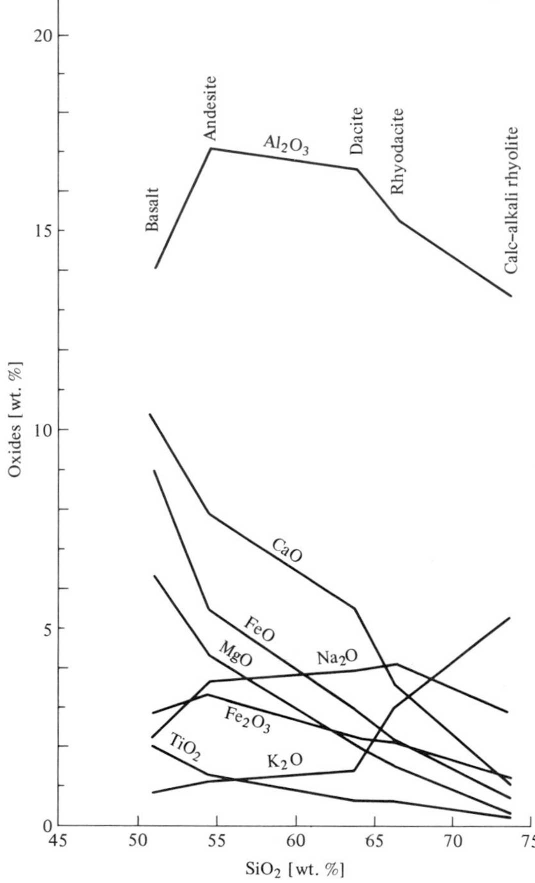

4.1 Variation diagram for rocks of the calc-alkali series.

contains. A sample calculation for the calc-alkali granite [(5) in Table 4.1] is given in Table 4.3. In accord with the rules, the formula unit is normalized to one cation. The following normative minerals result:

Apatite	$\frac{1}{5.33}Ca_{3.33}P_2O_8$
Ilmenite	$\frac{1}{2}FeTiO_3$
Orthoclase	$\frac{1}{10}K_2Al_2S_6O_{16}$
Albite	$\frac{1}{10}Na_2Al_2Si_6O_4$
Anorthite	$\frac{1}{5}CaAl_2Si_2O_8$
Corundum	$\frac{1}{2}Al_2O_3$
Magnetite	$\frac{1}{3}Fe_3O_4$
Enstatite	$\frac{1}{2}HySiO_3$
Ferrosilite	$\frac{1}{2}FeSiO_3$
Quartz	SiO_2

For example we see from Table 4.3 that 0.058 mole of K_2O plus 0.0580 mole of Al_2O_3 plus 0.348 mole of SiO_2 combine to form 0.0580 mole of $K_2Al_2Si_6O_{16}$ or

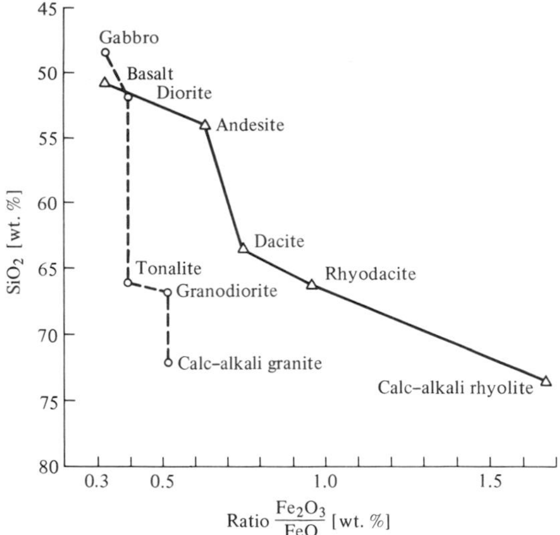

4.2 Plot of the oxidation ratio against weight percent SiO_2 for calc-alkali series.

0.580 mole of the normative mineral. The example presented is a very simple one. For a more complicated calculation, such as that for the certain undersaturated rocks, the reader is referred to Johannsen (1939), who has tabulated the many steps in detail. One of the several types of complications that may arise illustrates the shortcomings of normative calculations in general: thus it is sometimes necessary to calculate both the normative olivine $\frac{1}{3}(Mg, Fe)_2SiO_4$ and pyroxene $\frac{1}{2}(Mg, Fe)SiO_3$. In this case the Mg and Fe must be arbitrarily distributed between the normative minerals, and the convention is to distribute them equally in proportion to their abundances.

Mineral facies classification[1]

The facies concept was first proposed by Eskola (1915) as a consequence of a study of the metamorphic rocks of the Orijarvi region of Finland. The concept is based on the following premise according to Eskola:

1. "In any rock of a metamorphic formation which has arrived at a chemical equilibration through metamorphism at constant temperature and pressure conditions, the mineral composition is controlled only by the chemical composition."

[1] For a detailed discussion of particular facies the reader is referred to Chapter 10.

Table 4.2 Classification of igneous rocks

Other essential minerals	Essential alkali feldspar (Or or Ab or both)	Essential feldspar content			
		$\frac{100\,Or^a}{Or + Pl^b} > 60\%$	$\frac{100\,Or^a}{Or + Pl^b} = 40\text{–}60\%$	$\frac{100\,Or^a}{Or + Pl^b} = 40\text{–}10\%$	$\frac{100\,Or^a}{Or + Pl^b} < 10\%$
Quartz > 10% of rock	Alkali granite (alkali rhyolite) per alkaline granite (per alkaline rhyolite)	Calc-alkali granite (calc-alkali rhyolite)	Adamellite (dellenite)	Granodiorite (rhyodacite)	Tonalite (dacite)
Quartz or feldspathoids < 10%	Alkali syenite (alkali trachyte) per alkaline syenite (per alkaline trachyte)	Calc-alkali syenite (calc-alkali trachyte)	Monzonite (latite)	Mangerite (doreite)	Diorite (andesite) gabbro (basalt)
Feldspathoid > 10% of rock	Nepheline syenite (phonolite)		Nepheline monzonite (neph. latite)	Essexite (nepheline ordanchite) glennmvirite (analcite ordanchite)	Theralite (nepheline tephrite) teschenite (analcite tephrite)

After Nockolds (1954).

[a] Orthoclase or K-feldspar.

[b] Plagioclase.

Table 4.3 Niggli normative mineral composition of the calc-alkali granite

This calc-alkali granite is the same rock as No. (5) in Table 4.1

	Wt. %	Mole No.	Ap	Ilm	Or	Ab	An	C	Mt	En	Fer	Q
SiO_2	72.08	1.20			0.348	0.298	0.0390			0.0129	0.0133	0.489
TiO_2	0.37	0.00463		0.00463								
Al_2O_3	13.86	0.136			0.0580	0.0497	0.0195	0.009				
Fe_2O_3	0.86	0.00539							0.00539			
FeO	1.67	0.0233		0.00463					0.00539		0.0133	
MnO	0.06	trace										
MgO	0.52	0.0129								0.0129		
CaO	1.33	0.0237	0.00423				0.0195					
Na_2O	3.08	0.0497				0.0497						
K_2O	5.46	0.0580			0.0580							
H_2O^+	0.53	0.0294										
P_2O_5	0.18	0.00127	0.00127									
Equivalent molecular units			0.0068	0.0093	0.580	0.497	0.098	0.018	0.0162	0.0258	0.0266	0.489
Percentage normative minerals			0.384	0.523	33.1	28.1	5.5	1.02	0.915	1.46	1.50	27.6

Eskola (1920) subsequently provided the following definition:

2. "A mineral facies comprises all the rocks that originated under temperature and pressure conditions so similar that a definite chemical composition has resulted in the same set of minerals, quite regardless of their mode of crystallization, whether from magma or aqueous solution or gas, and whether by direct crystallization from solution (primary crystallization) or by gradual changes of earlier minerals (metamorphic recrystallization)."

Later, Eskola (1939) more concisely defined a given mineral facies as follows:

3. It consists of those rocks that contain identical minerals if their bulk chemical composition is identical, but whose minerals vary according to definite laws as the bulk chemistry changes.

In the same publication he also stated that

4. The definition of membership in a mineral facies is simply that the mineral composition be the same for the same chemical composition.

The classification of metamorphic rocks according to the facies concept has received much attention in the petrologic literature. The history of the development of the ideas leading up to the facies concept and the various modifications introduced have been discussed by Turner (Fyfe *et al.*, 1958; Fyfe and Turner, 1966, Turner, 1968). It is not our intention to review again all the ideas of what constitutes a metamorphic facies, but we shall be concerned with examining the idea in terms of the physicochemical concepts previously discussed.

It has been said (Ramberg, 1952; Fyfe *et al.*, 1958) that no requirement of equilibrium is implicit in the definitions (3) and (4) above. It is certainly true that it is possible to imagine a mechanically contrived series of rocks of identical bulk composition and composed of minerals of *virtually* the same composition. However, it is exceedingly improbable that such a series could appear in nature under any condition other than thermodynamic equilibrium. Thus it seems to the writers that the requirement of equilibrium is implicit in every one of Eskola's definitions of mineral facies.

Ramberg (1952) has, however, pointed out an inadequacy in the original definitions. This is that these definitions leave the way open for the description of a great number of facies, in fact one corresponding to each point in the P–T field, since at each such point identical bulk compositions give rise to identical mineral compositions that, at least in principle, are unique to that point. To avoid this complication, Ramberg proposed the following definition:

5. "Rocks formed or recrystallized within a certain P, T-field, limited by the stability of certain critical minerals of defined composition, belong to the same mineral facies."

4.3 The mineral facies after Eskola (1939). The diagram has been modified to exclude the "magmatic" facies. For a newer classification see Chapter 10.

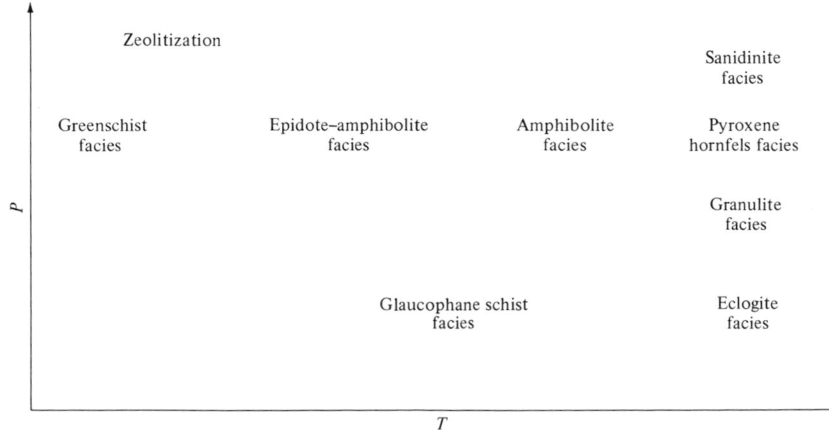

This definition is in harmony with Eskola's original concept of a small number of facies, and recognizes that rocks of identical bulk composition may exhibit coexisting minerals of varying composition and still be assigned to one facies.

Ramberg (1952) also pointed out that some simple rocks, such as a quartzite or a granite gneiss, would maintain the same mineral assemblage over a wide P–T interval. However it is important to define what we mean by the "same" mineral. A quartz-feldspar assemblage formed under granulite facies conditions might appear to be the same as one formed under amphibolite facies conditions. However, if we should examine the coexisting quartz and feldspar carefully we would at the very least find subtle differences. Such differences would, for example, appear in the distribution of certain elements such as germanium between the quartz and feldspar. Also, it is possible that a single mineral like quartz may reflect the P–T field of crystallization through lattice vacancies.

We may represent the mineral facies by a simple diagram (Figure 4.3), which is the modified form of a similar diagram given by Eskola (1939). It should be noted that Eskola did not consider zeolites as representing a distinct facies because he did not think that they represented an equilibrium system. However, Turner (Fyfe *et al.*, 1958) has advanced them to the status of a distinct facies representing the lowest temperatures and pressures of recognized metamorphism. Work by Thompson (1970) and Liou (1971) on P–T conditions of the zeolite facies is discussed in Chapter 10, where we shall also discuss a modern facies classification based upon experimentally determined mineral stability information.

Relation between bulk chemistry and mineral composition

Before we enter into a discussion of the stability of particular mineral assemblages as a function of pressure and temperature, it is well to consider the mutual compositional relations among minerals and the relation to the bulk composition of the rock. If we adopt the definition of a facies as given by Ramberg (1952) [in (5), above], we see that before we can accurately classify a rock as to facies we must first settle questions not only of the bulk composition but also of the compositions of the individual minerals since these may reflect the P–T changes in a continuous way. For example, a series of samples of a certain definite assemblage of minerals of identical composition but of differing proportions (and consequently of different bulk compositions) would, under equilibrium conditions, reflect identical P–T conditions of formation. This is an example of an isothermal–isobaric invariant condition, and in a three-component system the assemblage would fall in a three-phase field whereas in a six-component system it would fall in a six-phase field, and so on. If the P–T conditions now change continuously, the sides of the three-phase triangle or of the six-phase polygon become distorted. Finally, one or more of the minerals become unstable and then a new assemblage forms.

However, this type of disappearance of an assemblage must not be confused with the disappearance due to bulk composition changes. Thus we must be careful in each case to establish that the appearance of a new mineral assemblage is not in some way a response to the latter factor. This applies especially to dispersed or volatile components that may have been subsequently lost. In the past there has been a tendency to attribute the appearance or disappearance of minerals to facies changes without sufficient attention to compositional effects. A good illustration of such errors is the use of manganese content of garnet as a facies indicator without establishing that the manganese content of the rock is constant or that the system is sufficiently restricted as to degrees of freedom. In the average rock, which has approximately twelve major components, the analysis of compositional effects becomes difficult indeed. This is particularly borne out by assemblages showing varying degrees of oxidation. In the past these oxidation effects were largely disregarded by failure to distinguish hematite from magnetite and by the arbitrary inclusion of Fe^{3+} with Fe^{2+} or Al^{3+}.

No suitable analytical method has yet been devised to depict the effects of bulk compositional changes on mineral assemblages. We are therefore largely confined to the diagrammatic or graphical presentations, such as the grossly simplified AKF or ACF diagrams that are in current use. Such diagrams make use of implicit projections from multicomponental space on a three-component triangular diagram. We shall be forced to adopt a similar method; however, this method makes direct use of the tetrahedron, which introduces considerably more rigor and generality.

Figures 4.4 to 4.8 show the four-component systems capable of depicting nearly all the petrographically significant minerals.[2] Figures 4.5, 4.6, and 4.7 are derived from Figure 4.4 by successive replacement of FeO by

[2] Some standard mineral reference book should be used in conjunction with these diagrams to relate mineral compositions to names and abbreviations.

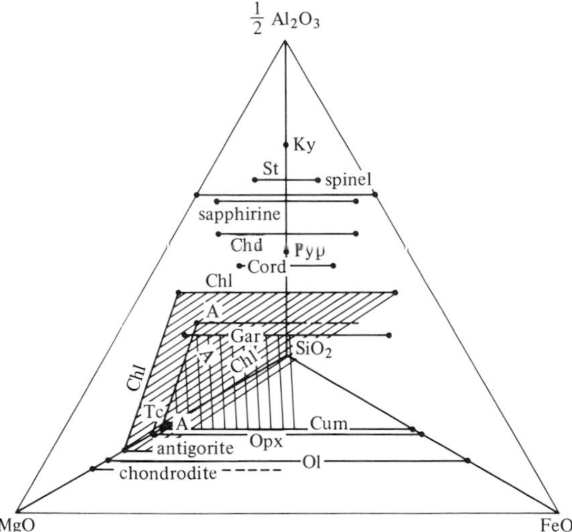

4.4 Mineral compositions in the system MgO-Al_2O_3-SiO_2-FeO. Chlorite and anthophyllite compositions fall on striped planar surface.

CaO, $\frac{1}{2}Na_2O$, and $\frac{1}{2}K_2O$. Since we are not primarily concerned at this stage with genetic relationships, all the common minerals and well-defined mineral solid solution series have been plotted together regardless of their experimental or observationally determined facies positions. However, most of the minerals plotted turn out to be the simply characterized metamorphic and high-temperature phases as distinguished from such comparatively ill-defined low-temperature minerals as the clays, for example. Many of the minerals plotted contain additional components that cannot be shown. In most cases the additional component is H_2O, but some minerals (such as scapolite) also contain CO_2, Cl, etc. It is thus necessary to specify the activities of these extra components before the mineral equilibria can be discussed. Chiefly because of these additional components, the figures contain a number of coincident species. For example, in Figure 4.5 Ca-scapolite is coincident with epidote, and lawsonite is coincident with anorthite.

The reference subsystem MgO-Al_2O_3-SiO_2 occupies the left face of the tetrahedra of Figures 4.4 to 4.7 inclusive. In Figure 4.4, in which this subsystem is combined with FeO, we see a number of linear and planar joins that depict the solid solubility of such minerals as the amphiboles, pyroxenes, chlorites, etc., with respect to Fe^{2+}, Mg^{2+}, Al^{3+}, and Si^{4+}. Where all of these elements are involved, planes such as those that represent chlorite and anthophyllite result. The limits of the joins as drawn are somewhat arbitrary since they are chiefly based on a limited number of natural occurrences from the literature. Since, as we have seen in Chapter 3, solid solutions of Mg^{2+} and Fe^{2+} usually show close to ideal behavior, we should expect that the solution represented by Mg^{2+}-Fe^{2+} joins should be stable in the intrinsic sense. Thus we can attribute limitations in the range of a given Mg^{2+}-Fe^{2+} series only to the extrinsic instability of the end members. For example, very magnesian cummingtonite is not found, whereas the corresponding anthophyllite polymorph is the rule. From this we assume that the reactions

$$\underset{\textit{cummingtonite}}{Mg_7Si_8O_{22}(OH)_2} \rightleftharpoons \underset{\textit{anthophyllite}}{Mg_7Si_8O_{22}(OH)_2} \quad (4.a)$$

$$\underset{\textit{anthophyllite}}{Fe_7Si_8O_{22}(OH)_2} \rightleftharpoons \underset{\textit{cummingtonite}}{Fe_7Si_8O_{22}(OH)_2} \quad (4.b)$$

are both displaced to the right for the generally prevailing P–T condition. The relations involving Al^{3+} and Si^{4+} as applied to the aluminous anthophyllite gedrite are less clear cut, and it is possible that intrinsic characteristics of the solid solutions may also be involved here.

When CaO is substituted for FeO (Figure 4.5), we see that the solid solubility is much more restricted, but that a number of new series involving substitution of Al^{3+} for Mg^{2+} and Si^{4+} arise. The actual amount of substitution shown is again somewhat arbitrary since we have generally terminated a series with a well-known

4.5 Mineral compositions in the system MgO-Al_2O_3-SiO_2-CaO.

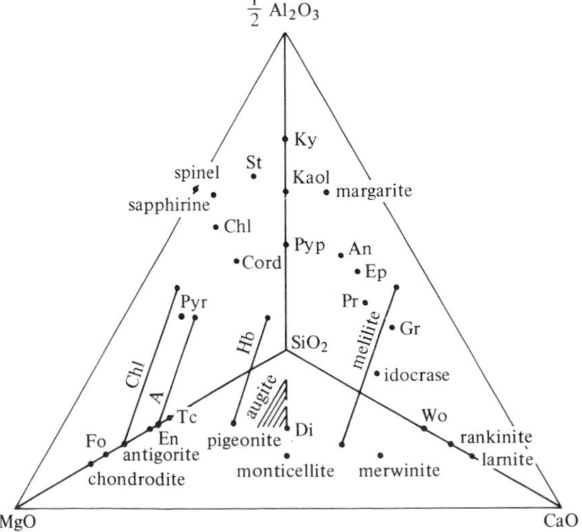

species, although it is possible that the maximum possible substitution covers a wider range. Thus we terminate the hornblende series on the high aluminum end with the quaternary species tschermakite ($Ca_2Mg_3Al_4Si_6O_{22}(OH)_2$) when, according to stoichiometric consideration, the series could bave been extended to $Ca_2Al_5Al_4Si_3O_{22}(OH)_2$. However, no amphiboles approaching this composition are known.

Also it should be clear that Ca^{2+}-Mg^{2+} substitutions become more common at higher temperatures when series such as the pyroxenes expand their range. However, little attempt has been made to indicate this feature in the figures.

In Figures 4.6 and 4.7 we note the lack of naturally occurring compounds in the field MgO-Na_2O-K_2O-SiO_2 a fact which greatly simplifies these diagrams. However, these tetrahedra contain the important feldspar and mica minerals and the quaternary amphibole gluccophane. Similarly, Figure 4.8 shows an absence of minerals within the ternary system Na_2O-CaO-SiO_2, although a considerable number of solid solutions occur within the quaternary system. Only three of the latter have been shown, although many others occur among the zeolites.

As was previously discussed, a given mineral assemblage may be depicted by tie lines that form lines, triangles, or tetrahedra within the major tetrahedron, depending on the number of associated minerals. The construction of such elements turns the compositional diagram into an equilibrium or phase diagram since such lines can have meaning only within the context of equilibrium. Although it is usually difficult to predict the most stable assemblage in the absence of detailed experimental or observational data, a tentative prediction may sometimes be made on the basis of compo-

4.6 Mineral compositions in the system MgO-Al_2O_3-SiO_2-Na_2O.

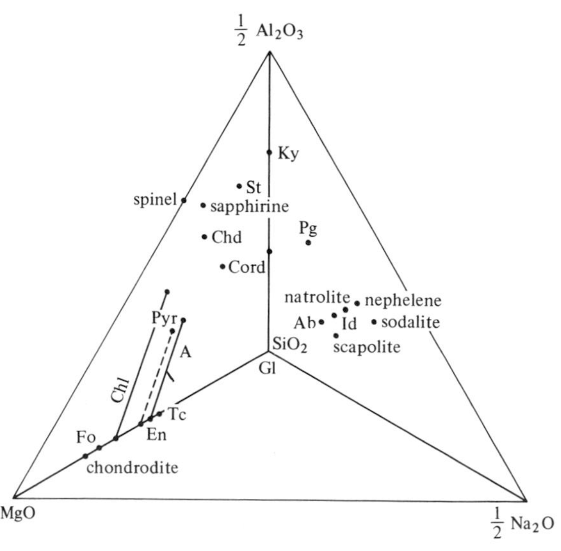

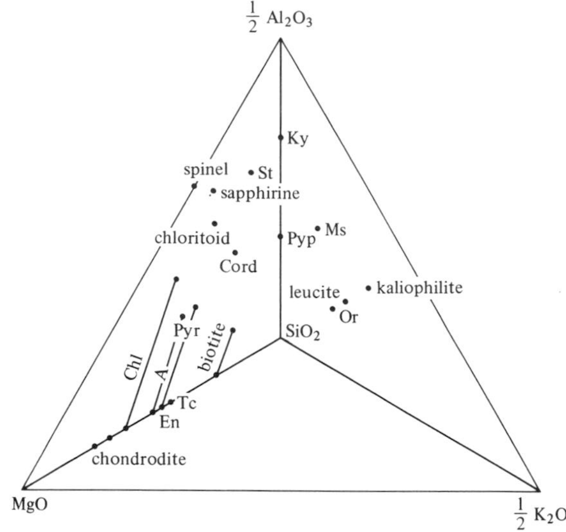

4.7 Mineral compositions in the system MgO-Al_2O_3-SiO_2-K_2O.

4.8 Mineral compositions in the system CaO-Al_2O_3-SiO_2-Na_2O.

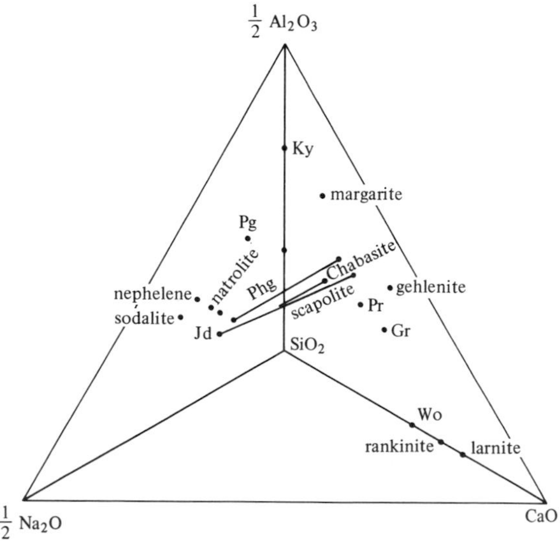

sitional compatibility. It might be expected, for example, that the assemblage talc-enstatite would generally be unstable with respect to anthophyllite since

$$\underset{talc}{Mg_3Si_4O_{10}(OH)_2} + \underset{enstatite}{4MgSiO_3} \rightleftharpoons \underset{anthophyllite}{Mg_7Si_8O_{22}(OH)_2} \quad (4.c)$$

Another example of the same type of relationship is

$$\underset{diopside}{2CaMgSi_2O_6} + \underset{talc}{Mg_3Si_4O_{10}(OH)_2} \rightleftharpoons \underset{tremolite}{Ca_2Mg_5Si_8O_{22}(OH)_2} \quad (4.d)$$

According to Greenwood (1963), Reaction (4.c) is displaced to the right for all pressures less than approximately 10^4 atm at ordinary metamorphic temperatures.

In order to illustrate the relationship of the tetrahedra to the three-component diagrams in conventional use, we have plotted the familiar ACF diagram used to depict the pyroxene hornfels facies within the major tetrahedron (Figure 4.9). The conventional ACF diagram is shown as a small triangle in the upper right of the figures. We see that cordierite and grossularite actually fall off the diagram and must be projected from the SiO_2 corner to form the conventional diagram. Figure 4.9 is rigorous insomuch as no extraneous components are required to form the minerals involved. However, the auxiliary assumption that the phases are nearly stoichiometric is implicit in the figure. The stable assemblages are shown linked by dashed tie lines where the tie lines do not coincide with the triangle. The fact that all the tetrahedra formed have apexes in the corner SiO_2 shows that the extra-phase quartz must be present before we can use the conventional ACF diagram. To illustrate this point further we note that the conventional diagram apparently excludes the association of anorthite and wollastonite. However, in the tetrahedron we see that if quartz is just excluded, the association of anorthite, grossularite, wollastonite, and diopside becomes possible. Similarly, in the absence of quartz, andalusite and enstatite can be associated.

At some lower temperature or higher water pressure certain of the hydrous minerals shown in Figure 4.5 become stabilized. The possible reactions of these phases with other hydrous or anhydrous phases may readily be seen by considering the various elements (points, lines, triangles, or tetrahedra) in the major tetrahedron. For example, we may consider a rock of the amphibolite facies consisting of pure tschermakite, which coincides with the Al-rich termination of the hornblende series (Figure 4.5). At a higher temperature or lower P_{H_2O} this amphibole may break down according to the reaction

$$\underset{tschermakite}{Ca_2Mg_3Al_4Si_6O_{22}(OH)_2} \longrightarrow \underset{enstatite}{MgSiO_3} + \underset{forsterite}{Mg_2SiO_4} + \underset{anorthite}{2CaAl_2Si_2O_8} + H_2O \quad (4.e)$$

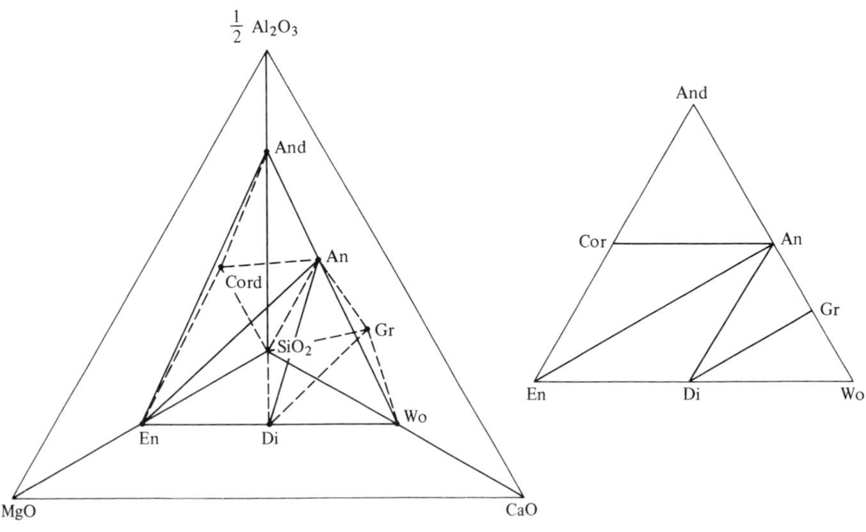

4.9 Relationship between ternary and quaternary representations.

From Figure 4.5 we see that tschermakite lies in the triangle enstatite-forsterite-anorthite. For a more complex example we consider a rock of bulk composition that falls in the tetrahedron staurolite-kyanite-quartz-epidote, which is a magnesium-poor assemblage of the epidote amphibolite facies. Under the higher temperature conditions of the granulite facies, such a rock will give way to an assemblage consisting of plagioclase-sillimanite and garnet according to the following reactions:

$$\underset{\textit{staurolite}}{3Mg_2Al_9Si_4O_{24}H} + \underset{\textit{quartz}}{\tfrac{11}{2}SiO_2} \longrightarrow \underset{\textit{pyrope}}{2Mg_3Al_2Si_3O_{12}} + \underset{\textit{sillimanite}}{\tfrac{23}{2}Al_2SiO_5} + \tfrac{3}{2}H_2O \quad (4.f)$$

$$\underset{\textit{epidote}}{2Ca_2Al_3Si_3O_{12}(OH)} + \underset{\textit{kyanite}}{Al_2SiO_5} + \underset{\textit{quartz}}{SiO_2} \longrightarrow \underset{\textit{anorthite}}{4CaAl_2Si_2O_8} + H_2O \quad (4.g)$$

From Figure 4.5 we see that the reactant tetrahedron has been transformed into the product triangle.

In the foregoing we have been dealing with stoichiometric phases; however, virtually all metamorphic minerals are complex solutions. According to Ramberg's definition of metamorphic facies the compositions of the coexisting minerals tell us where we are in the continuous spectrum of P–T conditions of a single mineral facies, with a definite composition defining the facies boundary. We have already seen in earlier sections that changes in the distribution curves or tie lines may define the P–T changes very precisely if we can separate their effects from those related to composition variation. Where a sufficient number of phases are present to reduce the number of degrees of freedom to one or two, the composition of a single mineral can give the P–T conditions of crystallization. As we have previously indicated, such elements as a tetrahedron defining a phase assemblage in Figure 4.9 change shape slightly with changes in P–T conditions. For example, in the tetrahedron enstatite-diopside-quartz-anorthite a rise in temperature increases the mutual solubility of the pyroxenes, and the enstatite-diopside tie line contracts. However, we must be very careful to separate such effects from the effects of bulk composition changes. As an illustration, petrologists have in the past frequently attempted to relate variation in the Al-content of silicates to P–T changes. Turning again to Figure 4.5, we see that the three-phase assemblage chlorite-anthophyllite-hornblende can, in principle, show a wide variation of Al/Mg ratios under isothermal–isobaric conditions. These relations are shown diagrammatically in Figure 4.10 as various triangular elements. Although the absolute value of this

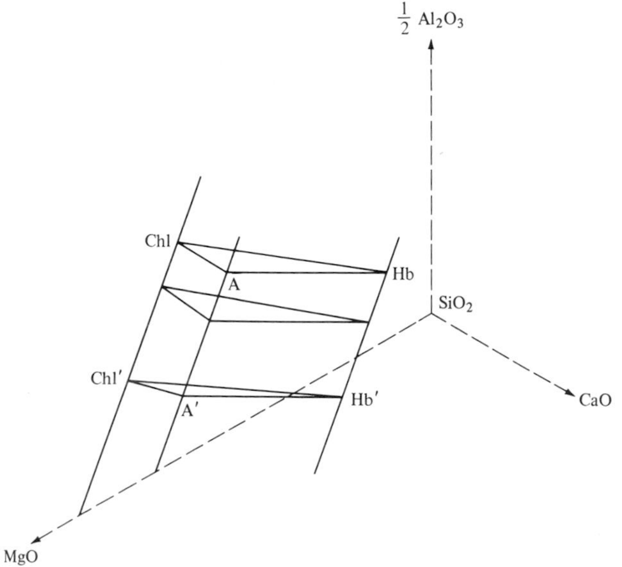

4.10 Compositions of coexisting chlorite (Chl), hornblende (Hb), and anthophyllite (A).

ratio in the minerals thus cannot reflect the P–T condition, the precise form of the triangle does indeed reflect them because each mineral is linked by a very definite distribution relation of the type treated in Chapter 2. The corresponding exchange reaction for coexisting hornblende and anthophyllite may be written as

$$\underset{\textit{hornblende}}{Ca_2Mg_5Si_8O_{22}(OH)_2} + \underset{\textit{anthophyllite}}{Mg_6Al_4Si_6O_{22}(OH)_2} \rightleftharpoons$$

$$\underset{\textit{hornblende}}{Ca_2Mg_3Al_4Si_6O_{22}(OH)_2} + \underset{\textit{anthophyllite}}{Mg_7Si_8O_{22}(OH)_2} \quad (4.g)$$

The equation of equilibrium, which does not concern us here, takes a very complex form owing to the necessity for simultaneous consideration of the distribution of Mg^{2+}, Al^{3+}, and Si^{4+} in the octahedral and tetrahedral sublattices.

We have attempted to illustrate the relationships between bulk compositions of the rock and mineral compositions, which must be understood before a rock can be properly classified as to mineral facies. Of course, if a reasonably complete series of chemically analyzed rocks of identical bulk composition are available, it is possible to assign these to various facies by simply identifying their minerals. This is well illustrated by a grouping of basic metamorphic rocks, according to facies presented by Ernst (1963a). Table 4.4 shows that there is no significant difference between the average glaucophane schists, green schists, and epidote amphibolites except for possibly oxidation state and CO_2 content. In fact, Ernst found that individual members of the group showed more variation among themselves than the groups did. This is an admirable illustration of the reality of the metamorphic facies represented.

Table 4.4 Comparisons of rocks of different metamorphic facies

	10 glaucophane schists	10 green schists	10 epidote amphibolites	137 tholeiitic basalts
SiO_2	47.90	48.36	49.10	50.83
TiO_2	2.11	1.79	1.30	2.03
Al_2O_3	15.05	15.30	15.43	14.07
Fe_2O_3	4.29	5.98	3.58	2.88
FeO	7.06	6.66	7.54	9.06
MnO	0.13(6)[a]	0.25(8)	0.30	0.18
MgO	5.89	6.37	7.06	6.34
CaO	9.50	7.46	9.90	10.42
Na_2O	2.98	3.50	2.67	2.23
K_2O	0.83(9)	0.59	0.54	0.82
P_2O_5	0.13(5)	0.27(7)	0.15	0.23
H_2O^+	3.29	3.42	2.25	0.91
CO_2	1.18(7)	0.16(4)	0.00(3)	—

After Ernst (1963a).

[a] Where fewer than 10 analyses are averaged, the number of analyses is indicated in superior parentheses.

5 Chemical and petrogenic processes in space

Solids in space

It seems safe to infer from astrophysical data that solids in the form of absorbing clouds are very abundant in interstellar space. Such matter with associated gases is thought to comprise a considerable portion of the total mass of the galaxies in which it forms dark bands within the spiral arms. These regions are also the sites of active star formation, as for example in the nebula of Orion, where they are closely associated with luminous gas clouds.

In the study of extraterrestrial solids, petrology merges with astrophysics and the research aims become identical. Whether observed from afar, or more closely as in the case of meteorites, these solids are capable in principle of yielding a great variety of information on the contemporary nature and history of development of the universe. Solids may form only under a restricted range of conditions, low in temperature by most astrophysical standards, although frequently under high pressures. Under conditions of thermodynamic equilibrium, they bear a distinct relation not only to each other but also to the surrounding dispersed medium. This makes it possible in principle to deduce the physicochemical environment of crystallization of meteorite minerals, or, by using the method in reverse, to disclose the mineralogic nature of remote bodies from the spectra of gases associated with them. In addition to thermodynamics there are also more direct methods of studying extraterrestrial solids. The classical method of light absorption, which led to the discovery of interstellar dust, may be extended to include the effect of polarization of light, which may be related to the chemistry of the solids (Dufay, 1957; Stecher and Williams, 1966; Crutcher, 1973).

Because of the large quantities of mass tied up in solids, they play an important part in the chemical balance of the universe and are much involved in the continuous flux of matter in the birth and extinction of stars and planets. From the principle of uniformity we infer that the contemporary observations of dust and gas clouds and the fact of their concentration in regions of active star formation have pertinence to the problem of the origin of the solar system. Conversely, we may also expect that deductions regarding the conditions of origin of our sun and planets are, with proper modifications, of wider applicability to contemporary astrophysical problems.

The problem of the origin of the solar system

This problem has in the past been approached from both the physical and chemical points of view, which of course should complement each other. Our concern here will be primarily with the latter since it bears most immediately upon petrology. We shall not review the history of ideas connected with this subject,[1] depending as it does upon celestial mechanics and other subjects beyond the scope of this book. However we shall draw attention to some of the most pertinent ideas.

The most widely accepted model for the origin of the solar system today holds that the sun and planets condensed out of a nebula of essentially the same composition as the sun. The roots of the nebular hypothesis go back at least to Kant and Laplace, who, however, did not attempt to present details. A number of distinguished astrophysicists have advanced this hypothesis, the most notable contribution being that of von Weizäcker (1944), who emphasized the role of turbulence eddies in the spacing of the protoplanets around the sun.

From our point of view a very interesting hypothesis was put forth by Alfvén (1954), who published a very detailed model of the conditions leading up to the condensation process. His ideas, which are finding renewed acceptance today, are concerned with magnetohydrodynamic processes in the solar nebula. In his book, as well as in his later papers (Alfvén, 1962, 1963), he has postulated the existence of magnetic fields in the already formed central body (sun or planet) which interact with ionized gases that are initially of solar composition. He does not treat the processes leading up to the condensation of the sun itself, but attempts to account for all the other bodies including the asteroids, satellites, and the rings of Saturn.

According to his model, the nebular gases are prevented from falling toward the central body as long as they are ionized. However, when the gases cool they form clouds in sequence of the ionization potentials of the various atomic species and fall toward the central body. For example, He and H atoms, which have the highest ionization potentials, are deionized first, whereas Fe and Mg, which have much lower potentials, become deionized later. In the next step the gases are again ionized when their kinetic energies of free fall reach their energies of ionization. Presumably this process can lead to a marked separation of the elements. This differentiated material, whose fall has been brought to a stop, is then swept to the equatorial plane by the magnetic forces, and forms the planets and satellites. It will be helpful to keep these ideas in mind when we discuss the chemistry of the condensation process. The first detailed treatments of the chemical conditions imposed by the solar composition were given by Eucken (1944), Latimer (1952), and especially by Urey (1952), who has made the most outstanding contribution toward our understanding of these processes. Most of Urey's conclusions about the hypothetical condensation conditions stand today, although some modifications are required to take into account the equilibria of the more complex mineral compounds. Recently, as we shall see, there have been further extensive investigations of condensation processes in the solar nebula.

Condensation from a gas of solar composition

The abundances of the elements in the sun are apparently known well enough that certain conclusions can be made regarding the condensation sequence of a cooling gas of solar composition. However, the precise values of these abundances are still uncertain and subject to constant refinement. This problem is reasonably well defined only as long as the gases do not deviate greatly from this composition, and these conditions hold only during the initial condensation stage for the element of interest. This is the problem as discussed by Urey, and much of our treatment shall differ only in detail from his.

[1] A short elementary review of the major physical hypotheses on the origin of the solar system is to be found in the book by Pickering (1961).

5 Chemical and petrogenic processes in space

Recently, Grossman (1972) has also attempted to trace the condensation process through changing conditions imposed by the process itself.

In approaching the problem it is first necessary to decide which molecular compounds of each element are abundant in the gas phase at a particular stage of the condensation sequence. This is no easy task since the necessary thermochemical data are limited,[2] and it is always possible that some more stable molecules exist in the P–T range of interest. In Table 5.1 we have listed the logarithms of the solar abundances of the major elements required to account for condensates of petrogenic interest. Application is obviously chiefly to the inner planets and to the smaller bodies such as the meteorites, since we have omitted helium which forms an important part of the giant planets. In Table 5.2 are shown what, from the thermochemical data, appear to be the major atomic and molecular species of the most abundant elements for a wide range of P–T conditions below approximately 1000 atm hydrogen pressure and 2300°K. Obviously not all the species listed will be abundant in the same range. We have also indicated the major condensation products to be expected. The latter are not necessarily the most stable condensates but are minerals which by their wide occurrence seem to be the most stable in a variety of environments stemming from the solar gas phase.

Only at temperatures in excess of 2000°K and at low hydrogen pressure need we consider ionization. The conditions favoring equilibrium ionization have been discussed by Alfvén (1954) for the case of a gas with a number density of 10^5 atoms/cm^3 ($P_{H_2} \simeq 10^{-14}$ atm at 273°K). If thermal equilibrium prevails, the degree of ionization of atoms of any given species i or X_i is given by Saha's expression for the ionization equilibrium constant:

$$K = \frac{X_i n_e}{1 - X_i} \tag{5.1}$$

where n_e is the electron density, X_i is a fraction, and K is a function chiefly of the temperature, except at the highest pressures. Alfvén found that above 2000°K a substantial number of Fe and Mg atoms were ionized. However, we shall not discuss this point further since this combination of high temperatures and low pressures does not seem likely for the condensation conditions of most substances. Just as in the case of ionization, the abundances of the molecular species are usually sensitive to the total pressure as well as the temperature. To illustrate this we first consider a system composed of the three most abundant elements in Table 5.1, the system C-O-H. Because these elements are an order of magnitude more abundant than the metals Si and Mg with which oxygen readily combines to form condensates at high temperatures, the latter element will scarcely be diminished in the cooled nebular gases.[3] This makes it possible to treat the system independently of the high temperature processes which may have preceded it. The species that we must consider as possible abodes of oxygen in the gases are H_2O, OH, CO, and CO_2. The hydroxyl molecule is unimportant except at low pressures

[2] The chief sources of the data in this section are the JANAF (Joint Army, Navy, Air Force) tables compiled by Dow Chemical Company. Other data are derived from sources referred to elsewhere in this book.

[3] Uncertainties in the elemental abundances also make such corrections superfluous.

Table 5.1 Abundance of common elements in the solar system

Element	Abundance	log (abundance)	Element	Abundance	log (abundance)
H	2.6×10^{10}	10.415	Cl	1970	3.294
O	2.36×10^7	7.373	Ca	7.36×10^4	4.866
C	1.35×10^7	7.13	F	3630	3.56
N	2.44×10^6	6.388	Ni	4.57×10^4	4.66
Si[a]	1.00×10^6	6.000	Cr	1.24×10^4	4.093
Mg	1.05×10^6	6.021	P	1.27×10^4	4.103
S	5.06×10^5	5.704	Mn	8800	3.945
Fe	8.9×10^5	5.949	K	3240	3.511
Na	6.32×10^4	4.80	Ti	2300	3.362
Al	8.51×10^4	4.93	Co	2300	3.362

Data from Cameron (1968).

[a] Arbitrary value.

and high temperatures as may be shown from the reaction

$$\underset{gas}{H_2O} \rightleftharpoons \underset{gas}{\tfrac{1}{2}H_2} + \underset{gas}{OH} \qquad (5.a)$$

For example, at 800°K, $K_{(a)}$ the equilibrium constant for this reaction is only 10^{-15}, but even at 2200°K, $K_{(a)}$ is still only 10^{-3}, and this would require $P_{H_2} \cong 10^{-6}$ atm in order to have $P_{OH} \cong P_{H_2O}$. Similar reasoning also holds for atomic hydrogen the dominant species in the interstellar gases. However, the hydrogen pressures required for the condensations of most solids at high temperatures are apparently so high that only H_2 is important. The equilibria remaining to be considered are therefore the following:

$$\underset{gas}{CO} + \underset{gas}{3H_2} \rightleftharpoons \underset{gas}{H_2O} + \underset{gas}{CH_4} \qquad (5.b)$$

$$\underset{graphite}{C} + \underset{gas}{2H_2} \rightleftharpoons \underset{gas}{CH_4} \qquad (5.c)$$

$$\underset{graphite}{C} + \underset{gas}{H_2O} \rightleftharpoons \underset{gas}{CO} + \underset{gas}{H_2} \qquad (5.d)$$

$$\underset{gas}{CO_2} + \underset{gas}{H_2} \rightleftharpoons \underset{gas}{H_2O} + \underset{gas}{CO} \qquad (5.e)$$

$$\underset{graphite}{C} + \underset{gas}{2H_2O} \rightleftharpoons \underset{gas}{CO_2} + \underset{gas}{2H_2} \qquad (5.f)$$

$$\underset{gas}{CO_2} + \underset{gas}{4H_2} \rightleftharpoons \underset{gas}{CH_4} + \underset{gas}{2H_2O} \qquad (5.g)$$

It is apparent that of the six reactions, only Reaction (5.e) is essentially independent of the pressure.

In order to determine which species are the most abundant at various temperature and hydrogen pressures, we make use of the condition that the abundance ratios are fixed by Table 5.1, which is based on an arbitrarily set value of the abundance of Si at 10^6. This reduces the possible variance of the system. Thus the hydrogen pressure becomes a univariant function of the temperature at a fixed ratio of P_{CO}/P_{CH_4} and P_{CO_2}/P_{CH_4} and is independent of P_{CO}/P_{CO_2}.[4] We first set $P_{CO} = P_{CH_4}$ and find

$$(P^0)^2 = K_{(b)}^{-1} \frac{P_{H_2O}}{P_{H_2}^0} \qquad (5.2)$$

where the superscript on P_{H_2} indicates the imposed conditions. From Table 5.1 we obtain

$$A_{H_2O} = A_O - \tfrac{1}{2}A_C = 10^{7.225} \qquad (5.3)$$

$$A_{H_2} = \tfrac{1}{2}A_H = 10^{10.114} \qquad (5.4)$$

where A refers to the relative abundances. Then $P_{H_2O} = 10^{-2.89}P_{H_2}^0$ and $P_{H_2}^0 = K_{(b)}^{-1/2}10^{-1.45}$. If we calculate $K_{(b)}$ as a function of T we see that at 2200°K, $P_{H_2}^0 = 10^{2.50}$ atm. At higher pressures than this all oxygen occurs as H_2O and carbon as CH_4, which is of some

[4] All gases assumed to be ideal at high temperatures; P represents the partial pressure of the gas.

Table 5.2 Distribution of the elements in the gas of solar composition and the ultimate condensates from this gas

Element	Generally abundant gas species	Ultimate condensates
H	H_2	H_2O
O	H_2O, CO	Silicates
C	CO, CH_4, CO_2	Graphite, hydrocarbons, carbonates
N	NH_3, N_2	Liquid to solid NH_3, N_2 N_2 in planetary atmospheres
Si	SiO, SiH_4, SiS	Mg-Fe silicates
Mg	Mg	Mg-Fe silicates
S	H_2S, S_2, S, SiS	FeS
Fe	Fe	Metal, silicates, oxides
Na	Na, NaOH	Silicates
Al	Al, $(AlO)_2$, Al_2O, AlOH	Silicates

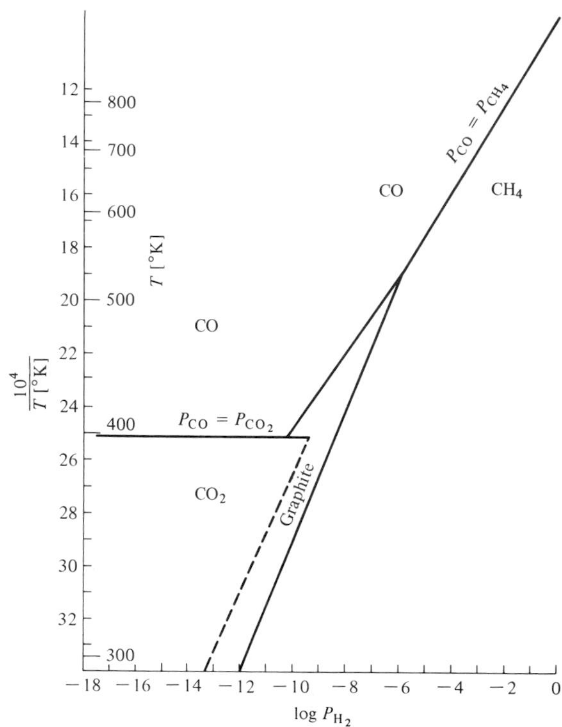

5.1 Plot showing the fields of dominance of the various carbon species in a gas of the inferred composition of the solar nebula. This depicts the situation as condensation of major C-, O-, and H-bearing phases just begins. This figure is unaffected by the high-temperature condensation of the silicates since silica is far less abundant than carbon, oxygen, or hydrogen.

consequence in the analysis of the high temperature condensations. At lower temperatures $\log P^0_{H_2}$ drops rapidly as an almost linear function of T^{-1} (Figure 5.1).

In the lowest temperature range CO_2 becomes an important species. From Reaction (5.e) we obtain the relation

$$\frac{P_{H_2}}{P_{H_2O}} K_{(e)} = 1, \qquad \text{if } P_{CO_2} = P_{CO} \tag{5.5}$$

Since $P_{H_2O} = 10^{-3.11} P_{H_2}$ by taking account of the distribution of oxygen between H_2O and CO, it may be shown that Equation (5.5) corresponds to a temperature of about 400°K.

It is possible now to locate the P_{H_2}–T range in which graphite can begin to precipitate. This may be done by considering Reactions (5.c), (5.d), and (5.f). The second of these yields

$$P_{H_2} = K_{(d)} \frac{P_{H_2O}}{P_{CO}} \tag{5.6}$$

We may assume that in the field in which CO dominates among the carbon species, all carbon is bound in this molecule, while the rest of the oxygen occurs as H_2O. From this we calculate that the ratio P_{H_2O}/P_{CO} is $10^{-0.13}$, and Equation (5.6) becomes

$$P_{H_2} = K_{(d)} 10^{-0.13} \tag{5.7}$$

Also the expression corresponding to Reaction (5.c) becomes

$$P_{H_2} = 10^{-2.98} K_{(c)}^{-1} \tag{5.8}$$

it having been assumed that all the carbon is bound as CH_4 in the field of methane. These relations are shown graphically in Figure 5.1. However, the lower left area of this figure is an indeterminate region. The reason for this may be seen by considering Reaction (5.f), for which we write the equation

$$P_{H_2} = K_{(f)} \left(\frac{P_{H_2O}}{P_{H_2}}\right)\left(\frac{P_{H_2O}}{P_{CO_2}}\right) \tag{5.9}$$

We also have the following connection between the abundances:

$$A_{H_2O} = A_O - 2A_{CO_2} \tag{5.10}$$

Consequently,

$$P_{H_2} = K_{(f)} \left(\frac{A_O - 2A_{CO_2}}{\frac{1}{2}A_H}\right)\left(\frac{A_O - 2A_{CO_2}}{A_{CO_2}}\right) \tag{5.11}$$

which shows that the total pressure depends on the abundance of CO_2 or, conversely, of H_2O. There is not enough oxygen to convert all the carbon to CO_2. Thus we have a somewhat arbitrarily fixed boundary of graphite by the condition that $P_{CO_2} = P_{CH_4}$ along the dashed line in Figure 5.1. Reaction (5.c) was the only one of these reactions considered by Urey (1952), who consequently did not establish the conditions for the beginning of precipitation of graphite. If gas is cooled along an isobar through the field of CO at $\log P_{H_2} = -7$, for example, precipitation of graphite begins at about 500°K. However this changes the CO/H_2O ratio and consequently shifts the boundary toward the lower right.

The next most abundant constituent of the gas is nitrogen, for which it may be shown that the only important species will be N_2 and NH_3, since the conditions are far too reducing for the nitrogen oxides. The critical reaction is therefore

$$\tfrac{1}{2}N_2 + \tfrac{3}{2}H_2 \rightleftharpoons NH_3 \qquad (5.h)$$
$$\textit{gas} \quad \textit{gas} \qquad \textit{gas}$$

From Table 5.1 we obtain the relation

$$\frac{A_N}{A_H} = \frac{2P_{N_2} + P_{NH_3}}{2P_{H_2}} = 10^{-4.02} \qquad (5.12)$$

while the equation of equilibrium for Reaction (5.h) is

$$K_{(h)} = \frac{P_{NH_3}}{(P_{N_2})^{1/2}(P_{H_2})^{3/2}} \qquad (5.13)$$

Let us determine which nitrogen species is the most abundant at 2200°K and $P_{H_2} = 10^3$ atm. For this temperature, $K_{(h)} = 10^{-4.91}$, so that

$$P_{N_2} = P_{NH_3}^2 10^{0.82} \qquad (5.14)$$

If we substitute this into Equation (5.12), we obtain the equation

$$P_{NH_3}^2 + \frac{10^{-0.82}}{2} P_{NH_3} - 10^{-4.84} P_{H_2} = 0 \qquad (5.15)$$

The solution of this equation yields $P_{NH_3} = 10^{-1.05}$ atm and $P_{N_2} = 10^{-1.28}$ atm, showing that nitrogen occurs chiefly as ammonia gas at $T = 2200$°K and $P_{H_2} = 10^3$ atm. By contrast, at 500°K and $P_{H_2} = 10^{-6}$ atm, which falls in Figure 5.1, N_2 is the most important species with $P_{NH_3} = 10^{-4.48} P_{N_2}$.

We next consider the elements Si and Mg, which dominate in the silicate condensations. First we must again decide which are likely to be the most abundant species in the gas so that we can assign abundance values consistent with the elemental abundances. We have already seen that oxygen occurs as H_2O, CO, and CO_2, although the latter species is important only at low temperatures and pressures. We shall first determine the abundances at 2200°K and $P_{H_2} = 10^3$ atm, which will be found to be in the upper range of interest for the silicate condensations. From Reaction (5.b) we find that $P_{CH_4} = 10^{3.38} P_{CO}$, so that virtually no oxygen is tied up as CO. Consequently

$$A_O = A_{H_2O} \qquad (5.16)$$

In the case of silicon we must take into consideration such species as SiO, SiH_4, SiO_2, and Si. We can determine the abundances of SiO_2 and Si relative to SiO from the following reactions:

$$SiO + H_2O \rightleftharpoons SiO_2 + H_2 \qquad (5.i)$$
$$\textit{gas} \quad \textit{gas} \qquad \textit{gas} \quad \textit{gas}$$

$$SiO + H_2 \rightleftharpoons Si + H_2O \qquad (5.j)$$
$$\textit{gas} \quad \textit{gas} \qquad \textit{gas} \quad \textit{gas}$$

which are essentially independent of the total pressure. The equations of equilibrium for the two reactions are:

$$\frac{P_{H_2O}}{P_{H_2}} K_{(i)} = \frac{P_{SiO_2}}{P_{SiO}} \qquad (5.17)$$

and

$$\frac{P_{H_2}}{P_{H_2O}} K_{(j)} = \frac{P_{Si}}{P_{SiO}} \qquad (5.18)$$

Since $(P_{H_2}/P_{H_2O}) = (\tfrac{1}{2}A_H/A_O) = 10^{2.74}$, we obtain

$$\frac{P_{SiO_2}}{P_{SiO}} = 10^{-4.72} \quad \text{and} \quad \frac{P_{Si}}{P_{SiO}} = 10^{-3.92} \qquad (5.19)$$

which eliminates SiO_2 and Si from consideration. To find the relation between SiO and SiH_4, we use the reaction

$$SiO + 3H_2 \rightleftharpoons SiH_4 + H_2O \qquad (5.k)$$
$$\textit{gas} \quad \textit{gas} \qquad \textit{gas} \quad \textit{gas}$$

which has the equation

$$\frac{P_{H_2}^3}{P_{H_2O}} K_{(k)} = \frac{P_{SiH_4}}{P_{SiO}} \qquad (5.20)$$

Since this depends on the pressure, $P_{H_2} = 10^3$ atm at 2200°K yields

$$\frac{P_{SiH_4}}{P_{SiO}} = 10^{-0.67} \qquad (5.21)$$

which tells us that we must distribute silicon between SiH_4 and SiO.

By similar methods we can show that atomic Mg is the only important species of this element present.

In order to correctly treat the condensation of Mg and Si from the solar gas it is necessary to go a step beyond Urey's discussion, which assumes MgO and SiO_2 as the condensates, and treat instead the most stable compounds of this system, which are minerals such as olivine and pyroxene. This was the approach of Wood (1963), who attempted to identify these condensates with certain chondritic meteorites. Wood utilized the reaction

$$2Mg + SiO + 3H_2O \rightleftharpoons Mg_2SiO_4 + 3H_2 \qquad (5.l)$$
$$\textit{gas} \quad \textit{gas} \quad \textit{gas} \qquad \textit{olivine} \quad \textit{gas}$$

We may put the equation of equilibrium of this reaction into the following convenient form[5]:

$$K_{(l)} = \frac{P_{H_2}^3}{P_{Mg}^2 P_{SiO} P_{H_2O}^3} \left(\frac{P_{H_2}}{P_{H_2}}\right)^3 \qquad (5.22)$$

[5] The effects of solid solubility in olivine are disregarded here since all iron will occur as metal under these comparatively reducing conditions.

in which the factor in parentheses has been introduced in order to transform the ratios of partial pressures to abundance ratios. If we now take into account the result (5.21) we have[6]

$$A_{Si} \propto P_{SiO} + P_{SiH_4} \quad (5.23)$$

$$A_{Si} \propto P_{SiO} + K'_{(K)}P_{SiO} = P_{SiO}(1 + K'_{(K)}) \quad (5.24)$$

where

$$K'_{(K)} = K_{(K)}\left(\frac{P^3_{H_2}}{P_{H_2O}}\right) \quad (5.25)$$

Consequently the logarithmic form of Equation (5.22) becomes

$$\log P_{H_2} = \tfrac{1}{3}[-\log K_{(l)} + 6 \log A_H - 6 \log 2 - 2 \log A_{Mg} - 3 \log A_O - \log A_{Si} + \log(1 + K'_K)] \quad (5.26)$$

which may be solved by successive approximation or by other means.

In his work Wood utilized older abundance data than those presented in Table 5.1, specifically the data of Goldberg *et al.* (1960). In particular, these older data, which were based solely on the sun's atmosphere, show a far lower Fe/H ratio than those of Table 5.1, and the same is true to a lesser degree for the ratios Si/H and Mg/H. Using the older value, Equation (5.26) yields $\log P_{H_2} = 3.48$ at $T = 2200°K$, the condensate at this temperature being liquid Mg_2SiO_4. According to Wood's results, condensation occurs over a range of temperatures and pressures, with solid Mg_2SiO_4 appearing in the lower range of values.

These pressures of condensation obtained are very high for most plausible nebular models, and Wood (1963) found it necessary to introduce the *ad hoc* assumption of shock waves from a stellar explosion to account for the simultaneous existence of the high temperatures and pressures. However the pressure drops off fairly rapidly with the temperature as the field of solid silicates is encountered. An additional complication, not mentioned by Wood, is introduced at this point by the increasing quantities of oxygen tied up as CO as the pressure drops below the field of methane. The high pressures are somewhat meliorated by certain deviations from the solar composition that might be brought about, for example, by the effects of magnetic fields as proposed by Alfvén. It may be seen from Equation (5.26) that any local accumulations of the species Mg, SiO, and H_2O would bring about a reduction in the equilibrium P_{H_2}. In addition, the newer and higher values of Fe/H, Si/H, and Mg/H will tend to favor the condensation of the silicates.

[6] A_{Si} here refers to the abundance of total silicon and not to the abundance of gaseous atomic Si.

According to Wood, as the temperature drops, olivine will be converted to pyroxene by reaction with solar gas:

$$\underset{gas}{SiO} + \underset{gas}{H_2O} + \underset{olivine}{Mg_2SiO_4} \rightleftharpoons \underset{pyroxene}{2MgSiO_3} + \underset{gas}{H_2} \quad (5.m)$$

As was previously mentioned, very little iron can be expected to enter the high temperature silicates since this element will occur only in the reduced form as atomic Fe or as particulate metallic condensates. We shall return to this subject when we discuss the system O-Si-Mg-Fe-Ni for more general conditions.

Next to silicon and magnesium, the most important condensate elements are sulfur and iron. In the reducing environment of the solar nebula, sulfur will occur chiefly as S_2, S, and H_2S, with such species as COS playing a minor role. At the higher pressures, H_2S takes precedence according to the reaction

$$H_2 + S \rightleftharpoons H_2S \quad (5.n)$$

with

$$P_{H_2}K_{(n)} = \frac{P_{H_2S}}{P_S} \quad (5.27)$$

If we set $P_{H_2S} = P_S$, we see that for this condition to be fulfilled at 2200°K, P_{H_2} cannot exceed $10^{-1.48}$ atm. Also, if we substitute S_2 for S in the lower temperature range, as for example at 600°K, we see that for the condition $P_{H_2S} = P_{S_2}$, P_{H_2} cannot exceed $10^{-14.86}$ atm, which lies far to the left in Figure 5.1. However, Urey (1952) has shown that in this temperature range metallic iron is converted to FeS by the reaction

$$\underset{metal}{Fe} + \underset{gas}{H_2S} \rightleftharpoons \underset{troilite}{FeS} + \underset{gas}{H_2} \quad (5.o)$$

in a gas of solar composition.

Although iron does not condense as oxides and ferrous silicates, at the higher temperatures it was demonstrated by Latimer (1952) and by Urey (1952) that this becomes possible when the temperature drops somewhat below 400°K. In this range the solar gas will be in equilibrium with magnetite by the reaction

$$\underset{gas}{\tfrac{3}{4}Fe} + \underset{gas}{H_2O} \rightleftharpoons \underset{magnetite}{\tfrac{1}{4}Fe_3O_4} + \underset{gas}{H_2} \quad (5.p)$$

We shall see that this result has important implications for the early history of the terrestrial planets.

In recent years a number of authors have extended Urey's work on the problem of the condensation of the solar nebula. We refer particularly to papers by Larimer (1967, 1973), Blander and Katz (1967), and Grossman (1972). Grossman made the first attempt to trace the condensation process beyond the initial state when the gas composition is given by the solar nebular abundances of the elements. It should be noted that most of Grossman's calculations are for a total pressure of 10^{-3} atm. Some results of his study are as follows:

1. Among the first silicate and oxide minerals to condense are, with falling temperature, corundum, melilite, perovskite, spinel, and diopside, after which olivine and enstatite assume major importance, as was shown by Wood. A typical sequence is shown in Figure 5.2.
2. With lowering temperatures and changing composition of the gas phase, corundum, melilite, and perovskite give way by reaction to diopside, spinel, and anorthite.
3. Iron-nickel alloys have higher condensation temperatures than forsterite and enstatite at all total pressures greater than about 7×10^{-5} atm.

The above condensation sequence is interesting because some of the early condensates have been found in certain meteorites such as carbonaceous chondrites and because the early condensation of metallic iron is a mechanism for building the metallic cores of dense planets such as Mercury, Venus, and Earth.

The effect of nonequilibrium ionization

In the foregoing discussion it was always assumed that at least local thermal equilibrium prevailed. This means that the radiation of the system was internally derived and in equilibrium with the local matter. Such will not be the case when interstellar gas and dust clouds are illuminated by stellar sources. The radiation from the stars will generally represent the higher temperatures characteristic of their photospheres and thus will possess a maximum intensity displaced toward the blue end of the spectrum as compared with the locally derived radiation. The effects of this high-energy radiation will be superimposed on the local thermal equilibria, causing additional

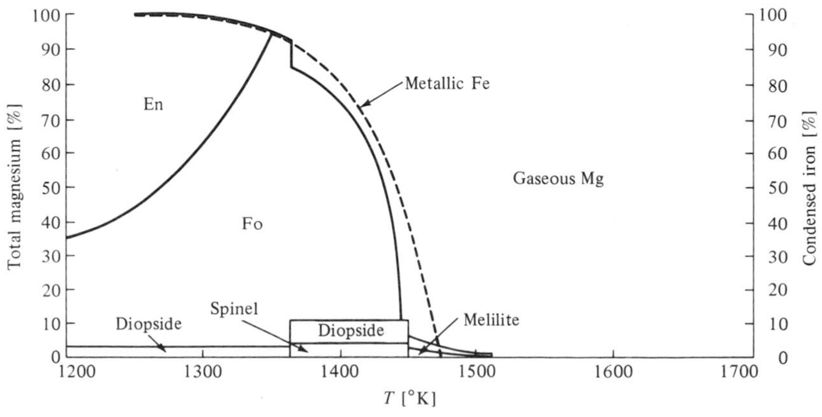

5.2 The distribution of Mg between crystalline phases and vapor. The first Mg-bearing minerals to condense are melilite and spinel, but the fraction of the total Mg contained by them only reaches a few percent by 1450°K, where the melilite is destroyed in the production of diopside and more spinel. These curves are drawn for the ideal melilite solid solution model and are almost indistinguishable on this scale from those for the nonideal case. After forsterite appears at 1444°K, Mg is rapidly condensed with decreasing temperature. At 1362°K diopside reacts with spinel, releasing 8 percent of the total Mg to forsterite. At 1349°K, forsterite begins to react with the gas to form enstatite, and the forsterite-to-enstatite ratio gradually decreases with falling temperature. The condensation curve of metallic Fe is shown for reference. Forty-six percent of the Fe has condensed before the first appearance of forsterite. (After Grossman, 1972).

dissociations of molecules and ionizations of particles. However, since the rates of recombination will be little affected, changes in the concentrations of many species will be very pronounced. The degree of this effect may be estimated by applying a factor $W = \omega/4\pi$ to Saha's equation (5.1) as follows (Struve, 1950):

$$KW = \frac{X_i}{1 - X_i} n_e \tag{5.28}$$

W is the dilution factor and ω is the solid angle subtended by the stellar source at the region of interest. Equation (5.28) gives the increased ionization due to the nonequilibrium radiation, and consequently K must be regarded as a function of the temperature of the source, W being unity in its photosphere. However, Equation (5.28) can only be a crude approximation within a gas and dust cloud since much of the radiation of the source will be filtered out by the cloud itself, preventing illumination by the black-body spectrum as is required in this relation.

Effects similar to the above are observed in the upper atmospheres of the planets due to the ultraviolet solar spectrum, and these effects account in part for Earth's ionosphere. In general when equilibrium between solid condensates and gases is involved, ionization of the latter will operate to displace the temperature of condensation downward and the corresponding pressure upward since the particles must be neutralized before condensation can take place. The effects of dissociation are more difficult to evaluate since some dissociated particles may rapidly combine with the condensates.

System O-Si-Mg-Fe-Ni-P

This system is of great importance in extraterrestrial chemistry and planetology because these elements lie near the top of the abundance scale and are expected to form most of the condensed bodies of the inner solar system. We have already studied some of the stability relations in a hypothetical equilibrium condensation from the solar gas. However, in most of the problems we will be concerned with, the chemistry of the environment will depart markedly from this. We will find this to be true of most meteorites as well as the observable parts of the terrestrial planets.

It is convenient to consider first the system Fe-Ni, which has great importance in meteorites. The phase diagram, based chiefly on the work of Owen and Sully (1939), Owen and Liu (1949), Kaufman and Cohen (1956), and Goldstein and Ogilvie (1965), is shown in Figure 5.3. The effect of phosphorus is shown in Figure 5.4. We shall be interested primarily in the subsolidus below 1200°K. The nature of the crystallization (or in this case of the ex-solution) paths in such diagrams is already familiar from the feldspar fusion relations discussed in Chapter 2. It is useful to think of these diagrams as superimposed on the more complex relations involving the silicates. Because of the general immiscibility and consequent small degree of coupling between the Fe-Ni-P and silicate systems, they may be regarded as essentially independent except as Ni and P affect the activity of iron.

The major part of the system consists of O, Si, Mg, and Fe, whose phase relations for the liquidus temperatures and 1 atm total pressure are best known through the work of Bowen and Schairer (1935) and of Muan and Osborn (1956). Bowen and Schairer's diagram, shown in Figure 5.5, is for the reduced part of the system which is formally regarded as consisting of the three

5.3 Binary phase diagram of the system Fe-Ni.

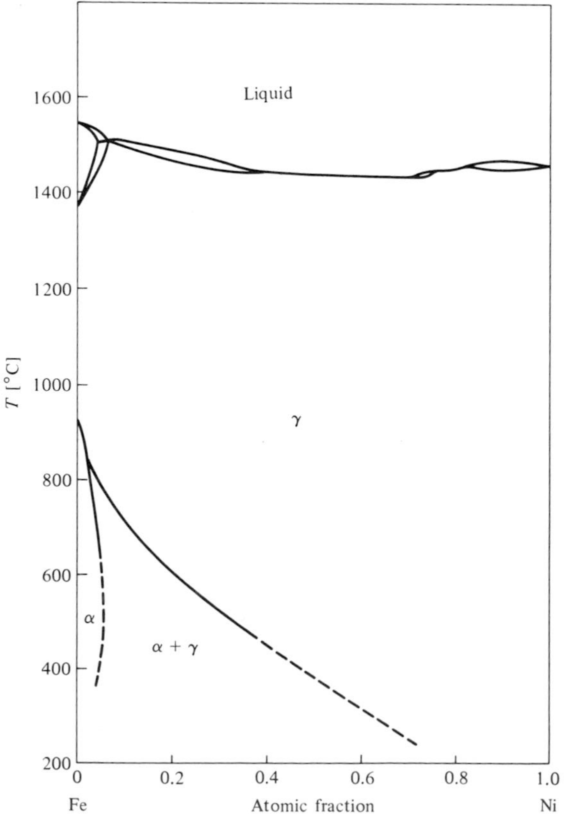

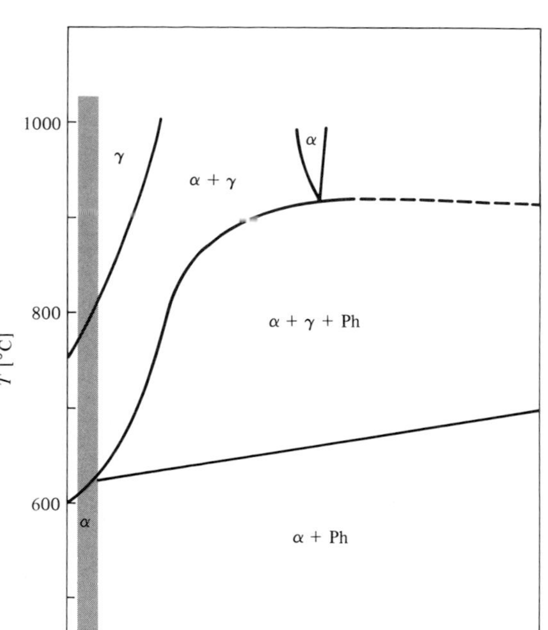

5.4 Section of the system Fe-Ni-P at 5.5 percent Ni. The shaded vertical bar indicates the spread in P in certain iron meteorites (After Goldstein and Doan, 1972).

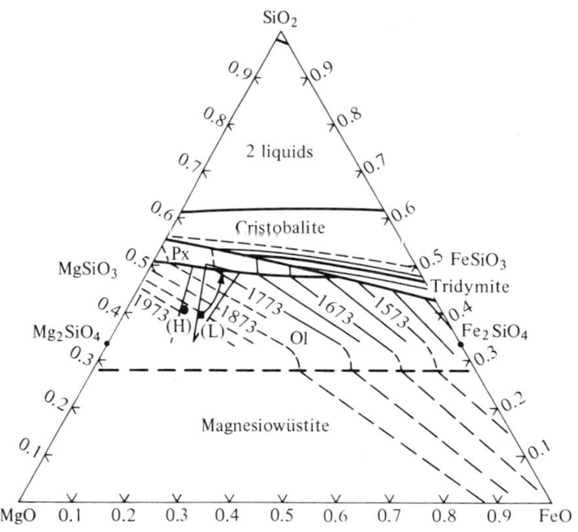

5.5 Crystallization in the ternary $FeO\text{-}MgO\text{-}SiO_2$ system after Bowen and Schairer (1935). Their diagram has been replotted on the basis of mole fractions of the components. Isotherms are in °K. The plotted points refer to average composition of the high (H) and low (L) iron chondrite groups. Olivine-pyroxene tie lines and one equilibrium crystallization path are shown (see text).

components MgO, FeO, and SiO_2. However, solid metallic iron was always present as an extra phase during the experimental work since the containing crucibles were made of this substance. Thus it really makes no difference if we consider the system as one of three or four components since the counting of an extra component also requires the counting of an extra phase. This is analogous to our treatment of the phase relations in the system $MgO\text{-}CaO\text{-}Al_2O_3\text{-}SiO_2$ (Figure 4.9), in which SiO_2 may be disregarded when quartz is present.

The possible paths of crystallization in this system are very diverse, as was demonstrated by Bowen and Schairer. However there are close analogies with the system albite-anorthite-diopside. For example, in the primary field of olivine a series of curved paths of equilibrium crystallization describe the behavior during cooling just as in the field of plagioclase. However, given certain peculiar initial compositions, the path may cross into the field of pyroxene and again return to the pyroxene-olivine boundary somewhat later!

In our more general treatment of the system it is necessary to treat several reactions among the phases olivine, pyroxene, metal, ferrite, and gas. The most useful of these reactions are the following:

$$\underset{olivine}{\tfrac{1}{2}Mg_2SiO_4} + \underset{pyroxene}{FeSiO_3} \rightleftharpoons \underset{pyroxene}{MgSiO_3} + \underset{olivine}{\tfrac{1}{2}Fe_2SiO_4} \tag{5.q}$$

$$\underset{pyroxene}{2MgSiO_3} + \underset{metal}{2Fe} + \underset{gas}{O_2} \rightleftharpoons \underset{olivine}{Mg_2SiO_4} + \underset{olivine}{Fe_2SiO_4} \tag{5.r}$$

$$\underset{pyroxene}{2MgSiO_3} + \underset{ferrite}{\tfrac{2}{3}Fe_3O_4} \rightleftharpoons \underset{olivine}{Mg_2SiO_4} + \underset{olivine}{Fe_2SiO_4} + \underset{gas}{\tfrac{1}{3}O_2} \tag{5.s}$$

$$\underset{metal}{3Fe} + \underset{gas}{2O_2} \rightleftharpoons \underset{ferrite}{Fe_3O_4} \tag{5.t}$$

There are additional reactions in the system involving the components $FeSiO_3$ and $MgFe_2O_4$. No reliable thermochemical data exist for $FeSiO_3$, and the mixing properties of $MgFe_2O_4$ and Fe_3O_4 are not well known. However we shall be able to avoid explicit consideration of these components except in Reaction (5.q). The equations of equilibrium corresponding to the four reactions may then be written as follows:

$$K_{(q)} = \frac{X_{Mg}^{Px}(1 - X_{Mg}^{Ol})}{X_{Mg}^{Ol}(1 - X_{Mg}^{Px})} \tag{5.29}$$

$$K_{(r)} = \frac{(1 - X_{Mg}^{Ol})^2(X_{Mg}^{Ol})^2}{(X_{Mg}^{Px})^2(a_{Fe}^{M})^2 P_{O_2}} \tag{5.30}$$

$$K_{(s)} = \frac{(1 - X_{Mg}^{Ol})^2(X_{Mg}^{Ol})^2 P_{O_2}^{1/3}}{(X_{Mg}^{Px})^2(a_{Fe}^{Fer})^{2/3}} \tag{5.31}$$

$$K_{(t)} = \frac{a_{Fe}^{Fer}}{(a_{Fe}^{M})^3 P_{O_2}^2} \tag{5.32}$$

In these equations all the silicates and ferrite have been treated as binary solutions of their Mg^{2+} and Fe^{2+} end members. In addition, the silicates have been treated as ideal solutions whereas the activities of the ferrite and metal components have been left in the implicit form to allow for possible nonideality.

Adequate thermochemical data exist for the calculation of $K_{(r)}$, $K_{(s)}$, and $K_{(t)}$ as functions of the temperature (Mueller, 1965); $K_{(q)}$ cannot be calculated directly because of $FeSiO_3$. However, we can make some use of the natural distributions of Mg^{2+} and Fe^{2+} among the coexisting olivines and pyroxenes. For our purposes the most useful of these distributions are those observed in the ordinary chondritic meteorites, which, as we shall see, may represent an approximation to equilibrium. For our present purposes we shall not need to know $K_{(q)}$ very precisely, and the average value of 1.13 deduced for this parameter from the optical data of Ringwood (1961b) and Keil and Fredricksson (1964) will suffice.[7] In our present calculation we shall assume that $K_{(q)}$ has this constant value for all temperatures. The error introduced by this assumption will be small compared with that introduced into $K_{(r)}$ and $K_{(s)}$ by the thermochemical data. The results of the calculations are shown in Figure 5.6. We note first that this figure is divided into three major fields distinguished by the occurrence of either metallic iron, ferrite, or hematite. In each case the boundary curve between these fields has been calculated using the pure substances Fe, Fe_3O_4, and Fe_2O_3. In the case of the magnetite–metal curve, which separates approximately the fields of metal and ferrite, the metastable extension into the wüstite (FeO) field is shown as a dashed line. The field of wüstite has not been included because this phase apparently does not appear in systems of such high silica content as those under discussion. To the right of the metal–ferrite boundary lies the domain of Reaction (5.r), which is of great importance in meteorites. If the appropriate equation (5.30) is placed in the logarithmic form and if, for the present, we confine ourselves to a system free of Ni with $a_{Fe}^{M} = 1$, we obtain a family of curves, one for each value of $Mg/(Mg + Fe^{2+})$ in olivine or pyroxene. These curves are arrived at by substitution of Equation (5.29) into Equation (5.30) and by solving for $\log P_{O_2}$. Thus the system is univariant when the composition of one of the silicates is arbitrarily fixed. To see how this checks with the phase rule we note first that the system contains the four phases olivine, pyroxene, metal, and gas and the four components Mg, Fe, Si, and O. Consequently there are two degrees of freedom in the unrestricted system, and the arbitrary fixing of the silicate composition reduces this to one. The curves show how the oxygen pressure decreases with increasing magnesium content of the silicates at any given temperature, which has important consequences in explaining the mineralogy of the sequence of meteoritic types.

[7] Experimental values for $K_{(q)}$ (Larimer, 1968) differ slightly from this value, but the difference need not concern us here.

Unfortunately it is impossible at this time to treat Reaction (5.s) in the same way as Reaction (5.r) since a_{Fe}^{Fer} is not known as a function of composition and we cannot assume that it behaves ideally. However, a qualitative evaluation of the way in which the silicate compositions change with P_{O_2} can still be made. Let us first study how Reaction (5.s) is affected in the limit as $X_{Mg}^{Ol} \to X_{Mg}^{Px} \to 0$. Then also $a_{Fe}^{Fer} \to 1$, since $X_{Mg}^{Fer} \to 0$ by the distribution relation between the silicates and ferrite. Equation (5.31) then reduces to

$$K_{(s)} = \frac{P_{O_2}^{1/3}}{K_{(q)}^2} \tag{5.33}$$

Since $K_{(q)}$ may for all practical purposes be regarded as equal to unity, we obtain

$$K_{(s)} \simeq P_{O_2}^{1/3} \tag{5.34}$$

which is the same as the nonequilibrium equation corresponding to the case when all the solution components are in their standard states.

We next examine the limiting case as $X_{Mg}^{Ol} \to X_{Mg}^{Px} \to X_{Mg}^{Fer} \to 1$, for which Equation (5.31) reduces to

$$K_{(s)} = \left[\frac{(1 - X_{Mg}^{Ol})^2}{(a_{Fe}^{Fer})^{2/3}}\right] P_{O_2}^{1/3} \tag{5.35}$$

The different exponents of the numerator and denominator of the bracketed factor in this expression ensure that this term becomes small in the limit, and consequently P_{O_2} becomes large. This result is in qualitative agreement with the experimental work of Muan and Osborn (1956). Reaction (5.s) and Equation (5.31) have broad applications to the crustal rocks of terrestrial planets, especially wherever basic rocks such as basalts and gabbros occur (Chapter 13).

We have plotted the stability fields of the dominant classes of chondritic meteorites in Figure 5.6. For comparison we have also shown (as a dotted line) the state of oxidation of a gas of composition corresponding to the solar nebula. The latter curve corresponds to the equation

$$\log P_{O_2} = 2 \log\left(\frac{2A_O}{A_H}\right) - 2 \log K,$$

where K is the equilibrium constant for the decomposition of water. The position of the ordinary chondrites (OC) well to the left of this curve at high temperature indicates that they were formed under more oxidizing conditions than prevailed in the initial nebular gas at least. However, that this curve enters the field of the carbonaceous chondrites (CC) at about 400°K indicates that their more oxidized nature corresponds to the nebular gas at this temperature.

Because of the generally great abundances of the elements of this system in the observable universe, the reactions discussed should have wide applicability in extraterrestrial petrogenic processes that occur in the proper temperature and pressure range. Unfortunately, far less is known of processes occurring at lower temperatures or high water vapor pressures. The stable silicates in the latter case appear to be serpentines or chlorites of varying Mg/Fe ratios that coexist with magnetite. We encounter some examples in the carbonaceous chondrites (Chapter 6). On Earth the assemblage is of course familiar from the highly altered basic and ultrabasic rocks.

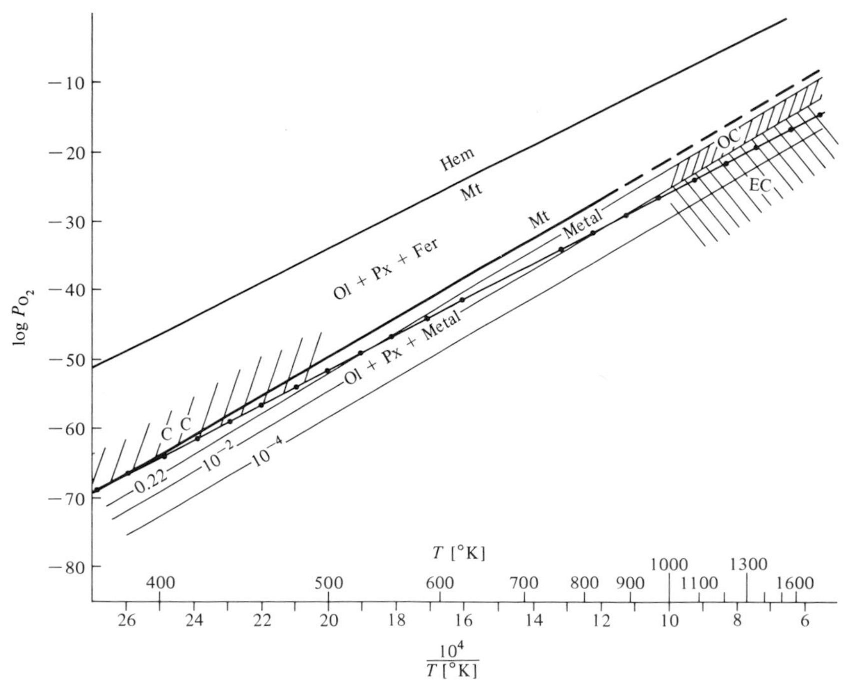

5.6 Stability fields of important minerals in the system O-Si-Mg-Fe. Approximate fields of the three major classes of chondritic meteorites are also shown. Shown are the fields of the ordinary chondrites (OC), enstatite chondrite (EC) and carbonaceous chondrites (CC). The light, numbered curves indicate atomic fractions $Fe^{2+}/(Fe^{2+} + Mg)$ for the hypothetical assemblage olivine-pyroxene-metal gas. However olivine is usually replaced by free silica in the (EC) field. The dotted curve indicates a gas of solar composition with hydrogen pressures high enough to maintain most carbon as methane. Of the three groups, only the field of ordinary chondrites is well defined.

6 Meteorites

General characteristics

There is evidence that our Earth and the other planets and satellites have from the earliest geologic times been bombarded by solid bodies from space. The larger of these bodies, those which reach the surface either whole or as fragments, are called *meteorites*. Recently, Wasson (1974) has described meteorites and reviewed the literature on the subject.

The optical and other atmospheric phenomena accompanying the infall of meteorites and micrometeorites are lumped under the term *meteor*. As is well known, meteors frequently occur in showers of the smaller particles at certain definite times of the year. These showers are known to mark the intersection of Earth's orbit with those of broken-up comets. However, as we shall see, many of the larger meteorites may have quite a different source within our solar system.

In meteorite studies it is important to distinguish between "finds" and "falls." The former are those meteorites that were not observed to fall and thus may have had a long residence time on Earth's surface. There are obvious implications to the chemical and petrologic studies since finds may have been subject to weathering processes that can drastically alter their constitution. Among the meteorite falls, which presumably best reflect the total sample of bodies striking our planet at the present time, about 92 percent are classed as "stones" or stony meteorites since they are composed dominantly of the silicates olivine and pyroxene with a subordinate amount of metal and other compounds. Approximately 6 percent of all falls are "irons" or metallic meteorites consisting dominantly of iron-nickel alloys. The remaining 2 percent consist of "stony irons" that contain approximately equal volumes of metal and silicate (Prior, 1953). However, if one examines instead the statistics of the finds, the irons bulk very large, showing that these are usually more resistant to weathering processes as compared with other meteorites (Mason, 1960). In addition, irons tend to be more noticeable than stone meteorites because of their great density and unusual appearance.

In their flight through the atmosphere meteorites may become so heated by friction that they explode above the ground. The fragments formed by these explosions may reach the surface at low velocities. However the larger meteorites that strike the surface after a direct flight possess so much of their original kinetic energy that they cause explosion craters similar to those produced by bombs. The shock wave generated in the vicinity of such an explosion crater may give rise to transient high temperatures and pressures that favor the synthesis of high-pressure minerals such as coesite and other unusual phases. These have in fact been observed by Chao *et al.* (1960) in the well-known Arizona meteor crater. Such craters are well known from many regions of Earth's surface and are of course very prominently developed on our Moon and on Mars and on Mercury.

During their flight through the atmosphere meteorites develop "fusion crusts" as a result of the heat generated. These crusts, which might more appropriately be termed "reaction crusts," are usually less than a millimeter thick and consist of a variety of oxidation products as well as glass. Suffice it to say that a failure to recognize the fusion crust for what it is could lead to erroneous ideas in the interpretation of the mineral chemistry and origin of these bodies.

Although we shall not be primarily interested here in the interaction of meteorites with Earth's atmosphere, one aspect of this problem is of great importance in the interpretation of their chemistry. This concerns the atmospheric gases that can conceivably be absorbed internally by the meteorites as soon as entry into the atmosphere occurs. That some degree of absorption must occur is more a matter of reason than of experimental evidence, although the latter is of course necessary to establish the quantitative aspects. When a meteorite with pore spaces enters the atmosphere it must in a sense "breathe" in atmospheric gases since the pore spaces are initially in a vacuum. Also since the interior is at the temperature of such matter in space,[1] the condensation of many substances is favored. The importance of such absorption may be critical in the case of certain isotopes and elements that are rare in meteoritic matter.

The dating of events in meteorite history

One of the most important branches of meteoritics is that which is concerned with the "ages" of meteorite materials, or more properly with dating the events in meteorite history that are registered in chemical or mineralogic changes of known time scale. So far most dating has been done by measurement of radioactive isotopic abundances in terms of their ratios. As summarized by Anders (1962), the five principle events in the history of meteorites are:

1. Nucleosynthesis
2. Solidification of certain meteorites from melts in the parent bodies
3. Cooling of parent bodies
4. Break-up of parent bodies
5. Fall of the meteorite

Of these events the time of nucleosynthesis is the most important from the standpoint of cosmologists since it preceded the events of petrogenesis, which are our major concern here. For petrologists the culminating and most interesting period of isotope fractionation occurred when the crystals precipitated from a melt. This event, when evidence is preserved, yields ages which cluster around 4.5 Gyr.

The method of meteorite dating must in each case be suited to the event for which information is sought. In the case of the period of nucleosynthesis the information derived is the time interval Δt between this event and the period of culminating solidification at 4.5 Gyr. The method, which follows an original suggestion by Brown (1947), makes use of the abundance of a stable decay product of a short lived radioisotope such as ^{129}I, ^{205}Pb, etc. In the case of ^{129}I the decay occurs, with a half of 17 Myr, to ^{129}Xe. As determined by Reynolds (1960), Merrihue (1963), and others, the observed accumulation of ^{129}Xe in Bruderheim, Richardton, and other chondritic stones implies that Δt is of the order of

[1] It may be shown that solids in space are at a much lower temperature than that of surrounding gaseous particles.

100 Myr, but with an uncertainty at least as large as this. A large part of this uncertainty stems from the necessity of assuming some rate of synthesis of ^{129}I or other short-lived isotope. These problems appear at present to be insurmountable, but the results obtained thus far are highly suggestive and have stimulated much research.

In order to determine the interval Δt and to date the age of primary solidification as well, it is necessary either that little or no subsequent heating affected the meteorites or that the isotope chosen was little influenced by this heating. It is clear that the $^{129}I/^{129}Xe$ method is sensitive to such heating since Xe isotopes are easily and differentially evolved in the heating range of 500 to 1800°K (Merrihue, 1963). In the case of solidification ages, it is possible to choose methods involving tightly bound metals such as are involved in the rubidium–strontium, rhenium–osmium, or lead–lead methods.

Recently, Herzog *et al.* (1973) have applied the $^{129}I/^{129}Xe$ method to date magnetite from the type-I carbonaceous chondrite, Orgueil. These authors concluded that the I/Xe age of this "primitive" type of chondrite (which incidentally contains no chondrules!) "is similar to that of all other classes of chondrites dated thus far." From this they concluded that the solar nebular stage, involving condensation, chondrule formation, and accretion lasted only 2 Myr or less.

By contrast the cooling history is best reflected by a decay process that has as products the noble gases since these diffuse out of the mineral grains readily and reset the clock to yield younger ages. An example of such a process is the decay of ^{40}K to ^{40}Ar with a half-life of 1.4 Gyr. Other processes involve the decay of uranium and throrium isotopes to lead and helium. These isotopes then yield a spectrum of ages ranging downward from the 4.5 Gyr maximum to less than 0.5 Gyr. Presumably these short ages reflect periods of heating or other disturbances such as collisional shocks (Anders, 1964).

Still more recent events are recorded by the cosmic ray exposure ages. This method dates the time of break-up of the parent bodies to fragments that expose new surface to cosmic ray bombardment. The cosmic rays then generate radioactive and nonradioactive nuclei whose abundances may be used to date the time of break-up. As was first demonstrated by Fireman (1958), this method may also be applied in determining the original size of a meteorite before atmospheric ablation affected it.

Table 6.1 Classification of the meteorites

Groups	Classes and subclasses	
Iron meteorites	Hexahedrites	
	Octahedrites	
	Ni-poor ataxites	
	Ni-rich ataxites	
Stony-iron meteorites	Pallasites	
	Mesosiderites	
Stony meteorites	Enstatite chondrites	
	Ordinary chondrites	olivine-bronzite
		olivine-hypersthene
		pigeonite
	Carbonaceous chondrites	type I
		type II
		type III
	Achondrites	enstatite
		olivine-pigeonite
		hypersthene
		hypersthene-plagioclase
		diopside-olivine
		augite

Classification schemes

As in the case of terrestrial rocks, many schemes of meteorite classification have been devised in the past. Again, as with rocks, most of these schemes made excessive use of a complicated terminology that incorporated place names used to designate individual falls or finds. For example, in the classification of Brezina and others, the Shergotty meteorite (which fell and was recovered at Shergotty, Bihar, India) formed the prototype of the "Shergottite" achondrite group, which is characterized by the presence of augite and maskelynite.[2] To avoid such complications we have followed Mason (1962) and have adopted the classification scheme of Prior (1920) as modified by the suggestions of Wiik (1956) and others. The scheme (Table 6.1) is based chiefly on the following hierachy of characters:

1. Metal/silicate ratio
2. Metal and silicate structure and texture
3. Detailed chemical differences

However we shall find it neccessary in order to accommodate the recent literature of meteorites to superimpose a further minor classification on the broad scheme of Table 6.1. This modification concerns the division of the chondrites as follows:

Degree of compositional uniformity	Total iron content
1. Uniform chondrites	1. High-iron group
2. Nonuniform (unequilibrated) chondrites	2. Low-iron group

The high- and low-iron groups coincide approximately with the bronzite (with 14.7 to 17.2 mol % $FeSiO_3$) and hypersthene (with 17.9 to 21.7 mole % $FeSiO_3$) chondrites, respectively. The uniform (equilibrated?) chondrites, on the other hand, are characterized by the presence of ferromagnesian silicates with constant Mg/Fe ratios throughout the meteorite, and cut across the low- and high-iron groups. The nonuniform (unequilibrated) chondrites, which show a great variability of silicate compositions from grain to grain in any given specimen, also contain representatives from several other groups and are particularly characteristic of the carbonaceous chondrites and the enstatite chondrites.

[2] Maskelynite is a rare feldspar glass that is produced locally within certain meteorites by shock.

It should be mentioned that a more complicated classification code employing letter and number symbols is now in use among meteorite specialists (Keil, 1969). However, we shall not employ this code here since it has few advantages for our purposes and is quite complicated.

Mineralogy

General features

The mineral assemblages and chemical compositions of most meteorites easily serve to distinguish them from terrestrial rocks. Recently the list of meteorite minerals has expanded greatly and this trend seems likely to continue for some time as more sensitive determination methods are developed. However, we shall not attempt here any exhaustive discussion of all these minerals but rather emphasize the major phases and those that particularly characterize the various meteorite classes or are of particular interest as environment indicators. The list presented in Table 6.2 also includes certain phases or mixtures of phases of a noncrystallic nature, and it will be seen that these are of particular significance to our discussion. A more complete list including rarer species has been presented by Mason (1972).

Many of the phases of Table 6.2 are very widespread and occur in most of the meteorite classes. This is the case in particular of graphite, troilite, pyroxene, olivine, silicate glass, and the Fe-Ni alloys kamacite and taenite. Taking the meteorites as a whole, the most abundant and widespread phases are the olivine, pyroxene, Fe-Ni alloys, and troilite, approximately in this order. These minerals comprise the bulk of the most abundant class of the stony meteorites, the chondrites, and in addition are important in the achondrites and stony iron groups. Although olivine is rare in the enstatite and type-I carbonaceous chondrites, it is the major chondrule-forming phase not only in the ordinary chondrites but also in the type-II type-III carbonaceous chondrites. Of the less abundant silicate phases, plagioclase is an almost omnipresent component of the ordinary chondrites where it tends to be quite albitic and of the high temperature variety. However, plagioclase occurs as a major phase only in the mesosiderites and the pyroxene-plagioclase achondrites, where it is usually somewhat calcic in composition.

The status of many of the minor phases in Table 6.2 is uncertain since they are frequently present in such small quantities that they pass unnoticed. Then too, as with the ferromagnesian silicates, their variable compositions may enable them to exist stably over a broad range of conditions with respect to temperature, total pressure, and oxygen fugacities. This appears to be true in particular of the phosphates and spinels. Of particular recent interest are the phases gehlenite, perovskite, spinel, sodalite, and hibonite since they may represent early condensates from the solar nebula, as discussed in Chapter 5.

Iron meteorites

Although a cursory glance at these meteorites gives us an impression of a simple mineralogy, recent work has shown that this is far from true. "Metallic" meteorites frequently possess as many as 8 to 10 phases, but many of these are usually present in such small crystals and are so well concealed that their presence can be detected only after the most careful search.

In the hexahedrites the Fe-Ni alloy is nearly restricted to kamacite since Ni is present to the extent of less than 6 percent, but in the octahedrites the Ni content of the metal is high enough for the characteristic Widmanstätten structure to develop. This structure results when a high temperature homogeneous taenite phase is cooled into the two-phase field of the Fe-Ni phase diagram (Figure 5.3). Under these circumstances plates of kamacite nucleate and grow on the octahedral [111] planes at the expense of the taenite, which at the same time becomes more Ni-rich by diffusion. The geometric character of the Widmanstätten pattern, which is simply a trace of these plates on a polished surface, depends on the orientation of the cut (Figure 6.1). Because the fraction Ni/(Fe + Ni) is almost always less

Table 6.2 Characteristic crystalline and noncrystalline phases of meteorites

Chiefly in chondrites, achondrites, stony irons, and irons		Chiefly in enstatite chondrites, and achondrites	
Kamacite (α iron)	(Fe, Ni, Co) alloy	Enstatite	$MgSiO_3$
Taenite (γ iron)	(Fe, Ni, Co) alloy	Alabandite	MnS
Troilite	FeS	Oldhamite	CaS
Pentlandite	$(Fe, Ni)S_{1-n}$	Sinoite	Si_2N_2O
Graphite	C	Osbornite	TiN
Diamond	C	Tridymite	SiO_2
Orthopyroxene	$(Mg, Fe)SiO_3$	Cristobalite	SiO_2
Clinopyroxene	$(Ca, Mg, Fe, Ti, Al)SiO_3$	Quartz	SiO_2
Olivine	$(Mg, Fe)_2SiO_4$	Sphalerite	ZnS
Plagioclase	$(CaAl, NaSi)AlSi_2O_8$	Niningerite	(Fe, Mg, Mn)S
Alkali feldspar	$(Na, K)AlSi_3O_8$		
Glass	Chiefly rich in Al, K, Na		
Whitlockite	$Ca_3(PO_4)_2$	**Chiefly in carbonaceous chondrites**	
Apatite	$Ca_5(PO_4)_3Cl$		
Schreibersite	$(Fe, Ni)_3P$	Layer-lattice silicate	$(Mg, Fe)_6Si_4O_{10}(OH)_8$
Cohenite	$(Fe, Ni)_3C$	Ferrite	$(Fe, Ni)Fe_2O_3$
Daubreelite	$FeCr_2S_4$	Magnesite	$(Mg, Fe)CO_3$
Ilmenite	$FeTiO_3$	Calcite	$CaCO_3$
Sarcopside	$(Fe, Mn)_3(PO_4)_2$	Dolomite	$CaMgC_2O_6$
Farringtonite	$Mg_3(PO_4)_2$	Gypsum	$CaSO_4 \cdot 2H_2O$
Spinel	$(Mg, Ni)Al_2O_4$	Epsomite	$MgSO_4 \cdot 7H_2O$
Chromite	$FeCr_2O_4$	Organic compounds	H-C-N-S-O
Rutile	TiO_2	Sulfur	S
Carlsbergite	CrN	Akermanite	$Ca_2MgSi_2O_7$
		Gehlenite	$Ca_2AlSiAlO_7$
		Pervoskite	$CaTiO_3$
		Grossular	$Ca_3Al_2Si_3O_{12}$
		Sodalite	$Na_4ClSi_3Al_3O_{12}$
		Hibonite	$Ca_2(Al, Ti)_{24}O_{35}$

than 0.5, the quantity of kamacite generally greatly outweighs that of taenite. However, when this fraction exceeds about 0.25 there is less tendency for unmixing to occur as the temperature is lowered so that the pattern becomes finer grained and discontinuous. Such a fine-grained mixture of taenite and kamacite is called *plessite* and it particularly characterizes the Ni–rich ataxites. The incoherent structure of the Ni–rich ataxites is also crudely mimicked in the Ni–poor ataxites, but since these meteorites usually have a Ni content comparable to the hexahedrites, the origin of their fine granular structure must be quite different.

The phases next in order of abundance to the Fe-Ni alloys are troilite, schreibersite, and graphite. The troilite and graphite commonly occur as rounded nodules, while the schreibersite is found as fine plates (Figure 6.2). Less common minerals are cohenite and daubreelite. Sometimes phosphates and silicates are concentrated in the graphite or troilite nodules as very small grains but in other cases they occur as round blebs which indicate a former molten state. The list of silicate phases found in iron meteorites is now long and includes Ca-pyroxene, orthopyroxene, olivine, plagioclase, alkali feldspar, and glass (Bunch *et al.*, 1970). Some phases, such as schreibersite, have important implications as to the phase relations of the associated metal phases (Goldstein and Doan, 1972).

Stony-iron meteorites

The stony-iron meteorites contain many of the same minerals as the irons. In the pallasites the silicates are nearly entirely olivine and this mineral makes up

6.1 Section of the Rodeo Fine Octahedrite iron meteorite showing the characteristic Widmanstätten pattern. Specimen about 15 cm across. This and the following seven photographs from the Field Museum of Natural History, Chicago, through the courtesey of Dr. E. Olsen.

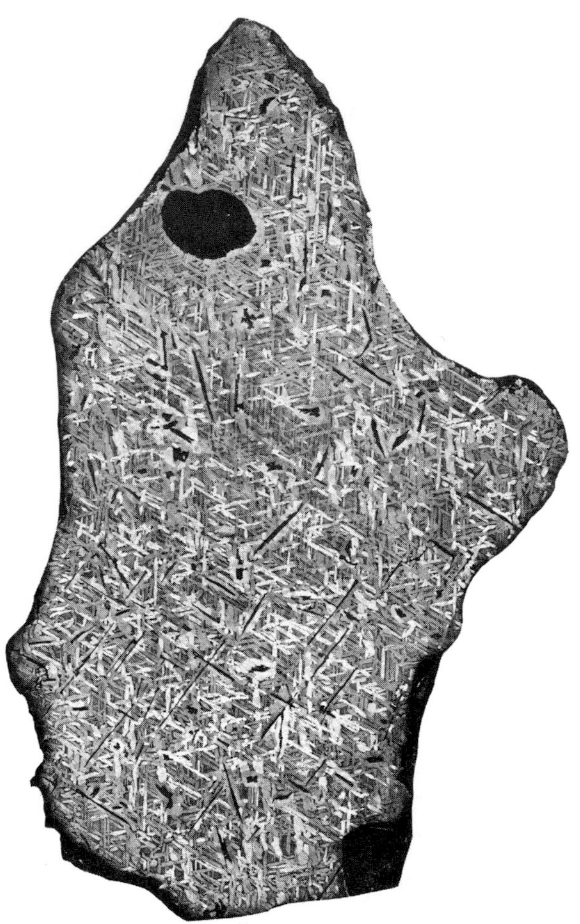

6.2 Section of the Breece Medium Octahedrite iron meteorite showing irregular rounded graphite and troilite nodules and fine plates of schreibersite (dark). Specular plates are part of Widmanstätten pattern of α and γ iron. Specimen about 25 cm across.

approximately 50 percent of the volume (Figure 6.3). However, in the mesosiderites (Figure 6.4) both pyroxene and plagioclase are more common than olivine. Also in the latter the silicates frequently show signs of brecciation; metal grains, however, are of a discontinuous nature and very irregular in form.

Although some of the stony-iron meteorites such as the pallasites contain virtually no other silicates but olivine, the compositions of both the olivine and metal phases do not differ greatly from the same phases in the ordinary chondrites. We shall see that this fact has important implications in terms of the oxidation states of these meteorites.

Stony meteorites

The mineralogy of this quantitatively most important group of meteorites is dominated by the silicates. Although the Fe-Ni alloys are important components of the high to moderately reduced enstatite and ordinary chondrites, they play a very subordinate role in the carbonaceous chondrites and achondrites.

The highly reduced character of the enstatite chondrites and achondrites is reflected in the characteristic sulfide, carbide, and nitride minerals, but the most abundant phases of the class are enstatite, Fe-Ni alloy, and troilite. The occurrence of free silica in these meteorites is due not only to the high degree of reduction but also to the fact that they have a somewhat higher Si/Mg ratio than other chondrites. The high degree of reduction also influences the common minerals, which always have peculiar compositions. Thus as was shown by Ringwood (1961), the Fe-Ni alloys contain several atom percent of silicon in solid solution. Correspondingly also these alloys are relatively low in nickel as a result of the fact that virtually all iron occurs as metal. Consequently the alloy is nearly pure kamacite. Since virtually all the iron is reduced, the pyroxenes are very low in iron. In addition, it was shown by Keil and Andersen (1965) that some of the sulfide phases such as troilite and alabandite are complex solutions of other metals such as Ti and Cr.

The mineralogy of the ordinary chondrites characterizes them as moderately reduced rocks of high temperature origin. The dominant phases in order of abundance are olivine, pyroxene, Fe-Ni alloy, and troilite. The silicates are restricted in composition to values of $Fe^{2+}/(Mg + Fe^{2+})$ between 0.15 and 0.30. In addition there are curious unexplained composition gaps at about 0.20 in olivine and 0.18 in pyroxene (Figure 6.14). Also there is an apparent crude harmony between $Fe^{2+}/(Mg + Fe^{2+})$ in the silicates and Ni/(Fe + Ni) in the bulk alloy (Figure 6.14), although the latter parameter also shows discontinuities. The dominant pyroxenes in chondrites consist mostly of the orthorhombic form, but clinopyroxene is a frequently coexisting and sometimes abundant phase. The olivine, pyroxene, and minor glass and other silicates such as plagioclase form the distinctive spheroidal chondrule structures to be discussed later. By contrast, spheroidal metallic or sulfide particles are very rare. Also in contrast to the metal of the iron and stony-iron groups, the metallic phases of chondrites do not occur in the Widmanstätten structure but rather either as single grains of these phases or as the fine-grained mixture plessite.

Of primary importance to understanding chondrite genesis are the minor compositional characteristics and

6.3 Section of the pallasite meteorite, Springwater. Dark is olivine; light is iron. Specimen about 15 cm across.

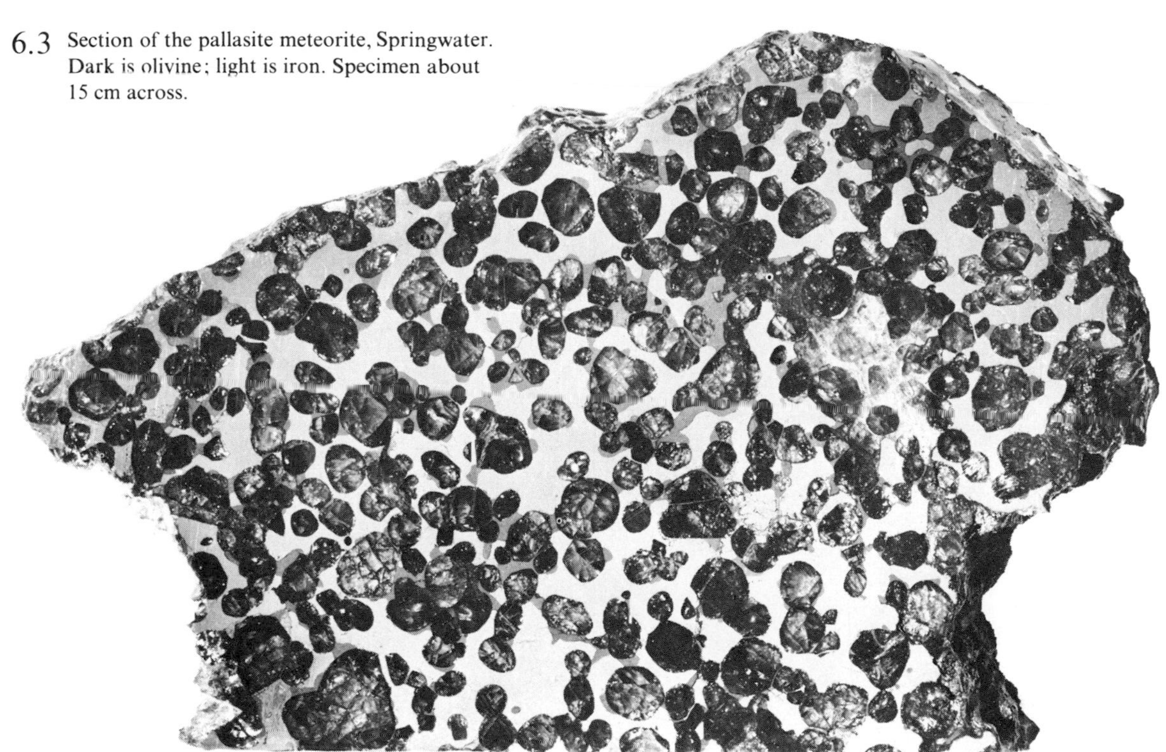

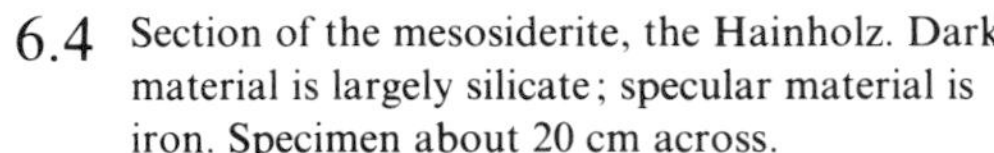

6.4 Section of the mesosiderite, the Hainholz. Dark material is largely silicate; specular material is iron. Specimen about 20 cm across.

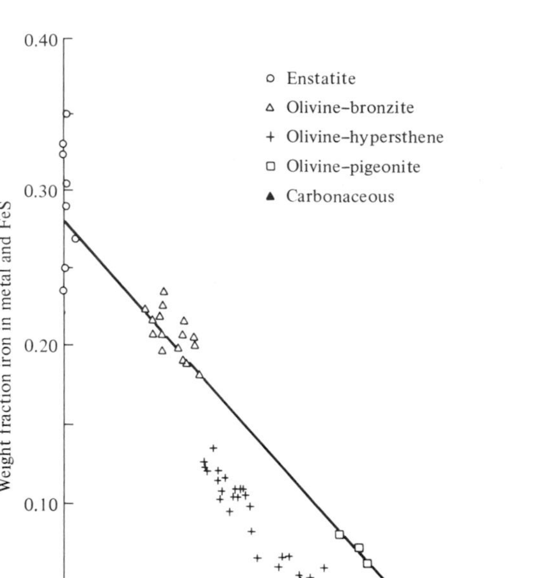

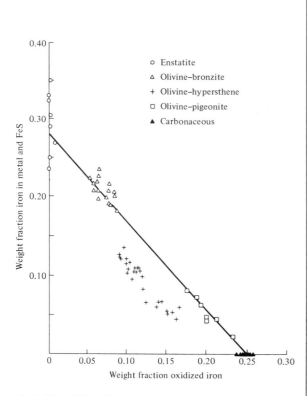

6.5 Plot of the weight fraction of iron in metal and FeS against the weight fraction of oxidized iron in different classes of chondritic meteorites (Mason, 1960). The straight line approximates to Prior's rule (see later in the text).

variations that are the basis of their twofold classification in the high and low iron and uniform and nonuniform groups. The latter division is particularly important because it is a basis for separating the definitely nonequilibrium chondrites with minerals of nonuniform or highly variable compositions from the possible equilibrium or quasiequilibrium uniform chondrites.

The characteristic mineralogy of the carbonaceous chondrites, which indicates generally low temperatures of formation in the presence of a near solar complement of volatile metals, is best developed in types I and II. Type-I meteorites consist almost entirely of a fine-grained mixture of serpentine-like layer-lattice silicate, ferrite, and "organic" material. Sometimes these meteorites also contain small amounts of carbonate, sulfate, and free sulfur phases. Type-II material, on the other hand, consists of a matrix of layer silicates, ferrite, and organic material much like that of type I, but this encloses inclusions and chondrules of olivine and pyroxene that sometimes contain metal, showing that they formed in an

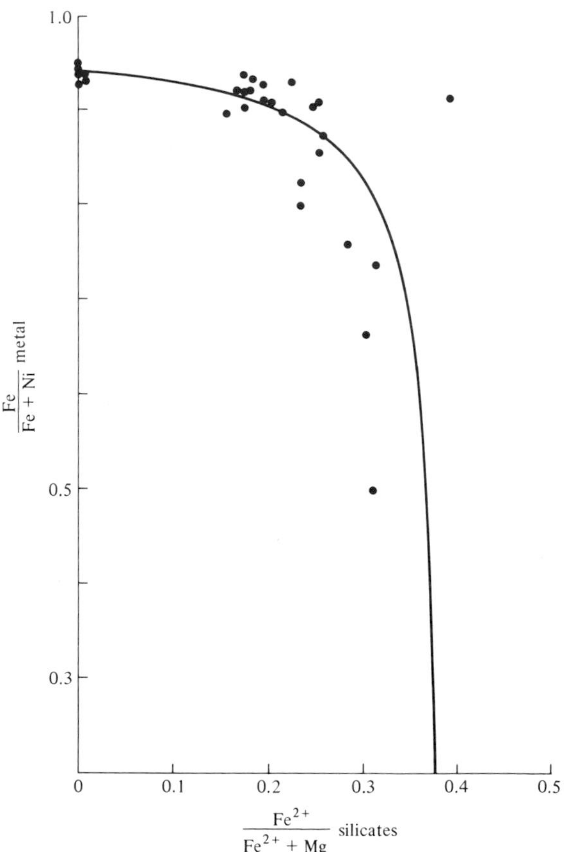

6.6 Plot of the atomic ratios Fe/(Fe + Ni) in the metal against $Fe^{2+}/(Fe^{2+} + Mg)$ in the silicates for various meteorites. The curve is obtained from a strict interpretation of Prior's rule (see later in the text).

environment quite different from that in which the matrix formed.

The type-III carbonaceous chondrites appear to be mineralogically transitional to ordinary chondrites, especially to the olivine-pigeonite group. The silicate present is mostly olivine, and this is accompanied by minor amounts of pigeonite and plagioclase. Generally, ferrite coexists with a small amount of Fe-Ni alloy. The carbonaceous chondrites also contain Ca- and Al-rich inclusions (polymict structures), which are regarded by some authorities as condensates from the solar nebula. Some fragmental or polymict chondrites include mixtures of material that resemble ordinary chondrites and carbonaceous chondrites.

The achondrites contain the same mineral species

as the chondrites but in very different proportions. Also, by definition, they lack the chondrite structures. Their textures are in fact quite prosaic and many of them have the granoblastic textures so common in terrestrial plutonic rocks. Some achondrites are practically monomineralic pyroxenites or dunites consisting of almost pure pyroxene or olivine. However the commonest type of achondrite is the pyroxene-plagioclase variety and some specimens of this closely approach certain terrestrial gabbros both in texture and chemical composition. Few achondrites contain much Fe-Ni alloy and many show a highly brecciated texture. In terms of oxidation state the variation seems to be almost as great as among the chondrites, with specimens ranging from the highly reduced enstatite achondrites to diopside-olivine achondrites in which olivine, with 66 percent of the fayalite component, coexists with magnetite (Mason, 1962a).

Petrography of the chondrites

Although the origin of the crystalline textures of meteoites is not fully understood, the major groups of irons and stony irons and the achondrites present few textural features that are not present in terrestrial rocks. This cannot be said of the meteorites that contain chondrules.[3] In fact, the chondrule structure is so unique and bears so heavily on so many aspects of the origin of meteorites that it deserves special attention.

It has already been stated that the condrules consist overwhelmingly of the silicates olivine and pyroxene, although these may be accompanied by other silicates or glass. The chondrules, which average perhaps 1 mm in diameter, are usually embedded in a matrix consisting either of material similar to themselves or to substances quite different from this, such as the matrix material that has already been described from the carbonaceous chondrites. Although there seems to be an almost infinite variety of chondrules, when one fixes on their details, the vast majority seem to fall into four types or to form composites of these types. These are presented in Table 6.3. Examples of these types are also shown in Figures 6.7 to 6.10.

Table 6.3 Major chondrule types

Chondrule type	Description
Porphyritic olivine	Phenocrysts or porphyroblasts of olivine in a microcrystalline or glassy matrix
Granular olivine	Closely packed equidimensional grains of olivine in a microcrystalline matrix
Barred olivine	Monosomatic or polysomatic skeletal olivine with cross sections that present regularly spaced bars alternating with microcrystalline or glassy material
Radiating pyroxene	Elongate pyroxene crystals radiating from a point usually on the chondrule edge

One of the striking features of the chondrules is their regularity. Generally they seem to deviate little from the spheroidal form although some cases of indentation of one chondrule by another are known. Although they possess very regular forms and are usually undisturbed, the distinctness of the chondrules varies greatly, sometimes even within a single meteorite; some have sharp borders while others merge almost imperceptably with the matrix. This feature of the chondrities has been

6.7 Baretta devitrified glass chondrule fragment indented by porphyritic olivine chondrule.

[3] We note that some members of the chondrite group, such as certain enstatite chrondrites and type-I carbonaceous chondrites, contain *no* chondrules!

6.8 Barred chondrule from the chondrite Tieschitz. Olivine is light material. Interstitial to this is more silica-rich glass and cryptocrystalline silicates.

6.9 Chondrule consisting of radiating pyroxene from the meteorite Dhurmsala.

recognized for many years. Recently it has been interpreted, particularly by Van Schmus and Wood (1967), as pointing to a sequence of progressive metamorphism that results in the eventual destruction of the chondrules.

Another interesting feature of the chondrites is the mode of occurrence of the metal grains. Usually these are very evenly distributed throughout the meteorite and consist of individual grains of kamacite or taenite or plessite. Sometimes they show distinct variations of composition ("diffusion borders") that are related to the boundaries with silicate and other adjacent grains. The metal grains are rarely straight-edged or angular fragments. More frequently they display very irregular forms, showing a distinctly interstitial character with respect to the chondrules and matrix silicates, and sending out delicate apophyses, or being molded closely to the chondrules (Figure 6.10). Because of such features it is possible to infer something of the time sequence of deposition of the several components.

In addition to the above "primary" features of the chondrites, there are also well-defined secondary features in the form of veins of metal and sulfide (Figure 6.11). These veins are found in many chondrites and indicate that in each case the body was solid enough to

6.10 Drawing from a thin section of the chondrite Allegan. Note how the metal (dark) is molded to a chondrule consisting of granular olivine.

6.11 Farmington meteorite. Specimen about 25 cm across.

fracture before the vein material was deposited. Of course it is not implied that this is evidence for fracture filling to the width of the vein but only that the fracture provided a locus and an avenue for the mobilized metal.

Bulk chemical characteristics

The distinguishing features of the major meteorite groups are based directly or indirectly on their bulk chemical differences. The most obvious manifestation of this is the metal/silicate ratio, which separates the iron, stony-iron, and stony-meteorite groups. In the examples of the iron and stony-iron groups the compositions are so extreme that bulk chemical analyses have little obvious meaning except for special studies. The same is true of many achondrites that are such highly differentiated rocks that they are almost monomineralic.[4]

The most intense interest and perhaps the greatest significance is to be found in the chemical compositions and compositional variation within the chondrite group. It has been recognized for some time that the chondrites as a whole are an unusually homogeneous group with respect to the abundances of the nonvolatile metals. It was also recognized at an early date that the different chondrite classes differ markedly from each other in terms of the degree of oxidation of their iron (Nordenskjold, 1878; Wahl, 1910). Subsequently Prior (1916) stated these relations as his now famous rule, to be discussed in the next section. In Table 6.4 are shown the bulk chemical analyses of representatives of the major

[4] Discussion of the bulk chemical characteristic of nonchondritic meteorites and their genetic implications are to be found in Mason (1960) and in Anders (1964).

Table 6.4 Representative analyses of the different groups of chondritic meteorites

	Enstatite (Daniels, Kuil) (Prior, 1916)	Ordinary H (Oakley) (Wiik, 1956)	Ordinary L (Kyushu) (Mason and Wiik, 1961)	Type-III carbonaceous (Warrenton) (Wiik, 1956)	Type-II carbonaceous (Mighei) (Wiik, 1956)	Type-I carbonaceous (Orgueil) (Wiik, 1956)
Fe	23.70	15.15	6.27	4.02	0.00	0.00
Ni	1.78	1.88	1.34	1.43	0.00	0.00
Co	0.12	0.13	0.05	0.09	0.00	0.00
FeS	8.09	6.11	5.89	5.12	3.66(5)[a]	5.65[a]
SiO_2	38.47	36.55	39.93	34.82	27.81	21.74
TiO_2	0.12	0.14	0.14	0.15	0.08	0.07
Al_2O_3	1.78	1.91	1.86	2.18	2.15	1.59
MnO	0.02	0.32	0.33	0.20	0.21	0.18
FeO	0.23	10.21	15.44	24.34	27.34	22.86
MgO	21.63	23.47	24.71	23.57	19.46	15.24
CaO	1.03	2.41	1.70	2.17	1.66	1.18
Na_2O	0.64	0.78	0.74	0.69	0.63	0.71
K_2O	0.16	0.20	0.13	0.23	0.05	0.07
P_2O_5	trace	0.30	0.31	0.20	0.30	0.27
H_2O	0.34	0.21	0.27	0.10	12.86	19.17
Cr_2O_3	0.23	0.52	0.54	0.58	0.36	0.35
NiO	0.11	—	—	0.00	1.53	1.19
CoO	—	—	—	0.00	0.07	0.06
C	0.32	—	0.03	0.19	2.48	2.99
						6.71[b]
Total	99.89	100.29	99.67	100.08	101.00	100.03

After Ringwood (1966) and Mason (1962).

[a] Wiik reported all S as FeS but it is given here as S, and the corresponding Fe is reported as FeO.

[b] Organic matter.

classes of the chondritic meteorites. The same data are shown again in Table 6.5, on a "volatile"-free basis and with only total iron listed. The remarkable constancy of the metal stands out clearly in Table 6.5, whereas the variation of oxidized and reduced iron may be seen from Table 6.4. We note, however, the rather small but significant difference in total iron that defines the high- (H) and low- (L) iron groups, as was first pointed out by Urey and Craig (1953).

The undifferentiated character of the chondrites is again apparent when the nonvolatile metals are compared with their values for the sun (Table 6.6). It is seen that the agreement is quite good.

The great homogeneity that the chondrites as a whole show with respect to the nonvolatile elements does not hold for most minor elements and least of all for the volatile trace metals such as Se, Ag, Ga, Pb, Tl, Bi, Hg, and In.[5] There appears in fact to be a sequence of progressive depletion in these elements extending from the type-I carbonaceous chondrites to the ordinary chondrites. It was first pointed out by Reed *et al.* (1960) that Tl, Bi, Pb, and Hg are several orders of magnitude more abundant in the carbonaceous chondrites than in the ordinary chondrites. Ringwood (1966) has compared the available abundance data from the type-I carbonaceous chondrites with the solar abundances, and found good agreement for more than 30 major and minor elements, with only Y and Yb showing serious deviations.

In summary it may be said that the chondritic meteorites as a whole are remarkably homogeneous with respect to their contents of the major nonvolatile metals. Also, their chemistry corresponds closely to the sun and the solar system as a whole. However there appears to have been considerable fractionation of the minor volatile elements so that all but the type-I carbonaceous chondrites are somewhat depleted with respect to the solar abundances.

[5] Thorough discussions of trace element abundances with references to the original literature are to be found in the review by Anders (1964) and Ringwood (1966).

Prior's rules

As a result of his study of the chemical variations among meteorites Prior (1916, 1920) came to the following important conclusions, which are known as "Prior's rules."

"*The less the amount of nickel-iron in chondritic stones, the richer it is in nickel, and the richer in iron are the magnesium silicates.*"

We observe that one part of the statement refers to the absolute amount of iron, whereas the other part refers

Table 6.5 Representative analyses of the different groups of chondritic meteorites

Data of Table 6.4 recalculated in atom percentages on a water-, carbon-, and sulfur-free bases.

	Enstatite	Ordinary H	Ordinary L	Type-III carbonaceous	Type-II carbonaceous	Type-I carbonaceous
Fe	28.52	26.32	21.85	25.72	26.18	27.34
Ni	1.65	1.74	1.27	1.38	1.41	1.37
Co	0.11	0.12	0.04	0.08	0.06	0.07
Si	34.98	33.17	36.78	32.77	31.85	31.12
Ti	0.08	0.10	0.09	0.14	0.09	0.09
Al	1.91	2.04	2.01	2.42	2.90	2.68
Mn	0.02	0.25	0.26	0.16	0.19	0.22
Mg	29.43	31.72	33.92	33.03	33.19	32.48
Ca	1.67	2.34	1.57	2.19	2.04	1.81
Na	1.13	1.37	1.30	1.25	1.40	1.97
K	0.18	0.23	0.17	0.27	0.07	0.12
P	—	0.23	0.25	0.16	0.29	0.33
Cr	0.32	0.37	0.39	0.43	0.33	0.40
	100.00	100.00	100.00	100.00	100.00	100.00

Data from Ringwood (1966) and Mason (1962a). See Table 6.4 for sources of meteorites.

Table 6.6 Comparison in terms of atom percentage of the major nonvolatile elements of the solar system and the average chondritic meteorite

	Sun[a]	Chondritic meteorites[b]
Si	34.7	36.3
Mg	27.4	33.2
Fe	30.9	21.8
Na	2.19	1.59
Al	1.74	3.45
Ca	1.56	1.78
Ni	0.90	0.995
P	0.24	0.182
Cr	0.17	0.283
Mn	0.087	0.249
Ti	0.053	0.0888
K	0.055	0.115

[a] The values are based on the work of Aller (1961), except for iron which is based on the work of Nusbaumer and Swings (1970, after Cameron, 1966).
[b] After Suess and Urey (1956).

to the composition of the nickel-iron and the silicates. Although the statement is qualitative in nature, it is meaningful only if the major constituents other than oxygen remain constant, and this condition, we have already seen, is actually approached by the chondrites as a whole. As was first indicated by Urey and Craig (1953), one consequence of a strictly quantitative interpretation of Prior's rule is that a plot of the reduced iron against the oxidized iron should yield a straight line with a unit negative slope. The relation may be derived as follows:

Let Fe_T be the total iron (gm); Fe_M be the metallic iron (gm); Fe_S be the Fe^{2+} iron (gm); $\sum'$ be the weight of all constituents other than iron and oxygen; and O_S be the oxygen combined with Fe^{2+} (gm). Then

$$O_S = \frac{16.0}{55.85} Fe_S$$

and for the weight fraction of oxidized and reduced iron we obtain

$$c_{Fe_M} = \frac{Fe_M}{\sum' + 0.28Fe_S + Fe_T}$$

$$c_{Fe_S} = \frac{Fe_S}{\sum' + 0.28Fe_s + Fe_T}$$

Since $Fe_T = Fe_M + Fe_S$, we obtain

$$c_{Fe_M} = \frac{Fe_T}{\sum' + Fe_T + 0.28Fe_S} - c_{Fe_S} \qquad (6.1)$$

which we see has the proper analytical form.

The pertinent observational data required to test Expression (6.1) are plotted in Figure 6.5, taken from Mason (1960). It is clear that while the olivine-bronzite and olivine-pigeonite chondrites plot approximately on a straight line of near unit slope, the olivine-hypersthene chondrites fall below the line. This feature, which is the basis for the high- and low-iron groups, immediately indicates that Prior's rules are obeyed only in a semi-quantitative way.

It is also possible to derive the corresponding relationship between the metal and silicate compositions, which constitutes the other part of Prior's rules (Mueller, 1963b). This at any stage of oxidation the mole fraction of ferrous silicate X^S_{Fe} and the mole fraction of iron in the metal X^M_{Fe} are

$$X^S_{Fe} = \frac{Fe^{2+}}{Fe^{2+} + Mg}$$

$$X^M_{Fe} = \frac{Fe}{Fe + Ni}$$

Then, since the total molar iron content is $Fe_T = Fe^{2+} + Fe$, we obtain the following relation:

$$X^M_{Fe} = \frac{Fe_T - A}{Fe_T - A + Ni} \qquad (6.2)$$

where

$$A = \frac{X^S_{Fe} Mg}{1 - X^S_{Fe}}$$

Now, for any given values of the total iron, magnesium, and nickel (Fe, Mg, and Ni) Equation (6.2) generates a curve in which X^M_{Fe} is a function of X^S_{Fe}. Different values of the parameters Fe, Mg, and Ni yield a family of such curves, which always have the same characteristic form.

Figure 6.6 shows a comparison of one of the family of theoretical curves with some data from the chondritic meteorites. The curve shown in the figure has $Fe_T = 0.40$, $Mg = 0.65$, and $Ni = 0.25$. It is again apparent that the relation is only approximately obeyed, so that both parts of Prior's rules are restricted to a semi-quantitative interpretation.

We will, however, examine one more consequence of Prior's rules, namely that at constant Fe_T, Mg, Ni, and Si, the olivine content must increase regularly at the expense of pyroxene and metallic iron. The general reaction for the process may be written as follows:

$$\underset{pyroxene}{(Mg, Fe)SiO_3} + \underset{metal}{Fe} + \underset{gas}{\tfrac{1}{2}O_2} \rightleftharpoons \underset{olivine}{(Mg, Fe)_2SiO_4} \tag{6.a}$$

The effect of this on the mineral compositions may be seen by rewriting Equation (5.30) as:

$$(\gamma_{Fe}^{M})^2 P_{O_2} K_{(5.r)} = \left(\frac{X_{Mg}^{Ol}}{X_{Mg}^{Px}}\right)^2 \left(\frac{X_{Fe}^{Ol}}{X_{Fe}^{M}}\right)^2 \tag{6.3}$$

In this expression γ_{Fe}^{M} and X_{Fe}^{M} are the activity coefficient and the atomic fraction, respectively, of iron in the metal phase and P_{O_2} is the oxygen pressure assuming it as an ideal gas. Now, since as we have already seen (Chapter 5), $K_{(5.g)}$ the distribution constant is approximately 1.13, so that the first bracketed factor on the right side of Equation (6.3) remains near unity. However X_{Fe}^{Ol}/X_{Fe}^{M} and γ_{Fe}^{M} will change as P_{O_2} increases, and because metal is oxidized in the process, X_{Fe}^{Ol}/X_{Fe}^{M} must increase. Furthermore because of the distribution relation of Equation (5.29), X_{Fe}^{Px} must also increase at the same time. Consequently we may define the ratio

$$R = \frac{X_{Fe}^{S}}{X_{Fe}^{M}} \tag{6.4}$$

which will always increase with increasing P_{O_2}. In this ratio X_{Fe}^{S} is the atomic fraction of Fe^{2+} in the total silicate, both olivine and pyroxene, and so has the same meaning as in Equation (6.2).

Similarly, we may define the mole fraction of the olivine phase as follows:

$$L = \frac{\text{moles } (Mg, Fe)_2SiO_4}{\text{moles } (Mg, Fe)_2SiO_4 + \text{moles } (Mg, Fe)SiO_3 + \text{moles metallic Fe}} \tag{6.5}$$

Then at any stage of the oxidation (or reduction) process there will have been a certain "extent of reaction" in which α mole of $(Mg, Fe)SiO_3$ and Fe reacted to produce α mole of $(Mg, Fe)_2SiO_4$, or vice versa. At the beginning of the oxidative process, when all the iron is in the reduced metallic state, there will be present Px^0 moles of pyroxene of composition $MgSiO_3$, Fe^0 moles of iron (in solution with nickel), and Ol^0 moles of olivine of composition Mg_2SiO_4. We may then write R and L in terms of α:

$$R = \frac{\alpha[(Fe^0 - \alpha) + Ni]}{(Fe^0 - \alpha)(Px^0 + 2Ol^0 + \alpha)} \tag{6.6}$$

$$L = \frac{\alpha + Ol^0}{Px^0 + Ol^0 + Fe^0 - \alpha} \tag{6.7}$$

If α is eliminated from these two equations, we obtain L as a function of R, which we may test against the observational data. In Figure 6.12 are shown plotted two of the family of curves with different values of Ol^0 and Px^0 but with the same values of Fe^0 and Ni. The dashed curve represents the idealized "pure enstatite chondrite" with no initial olivine, whereas the full curve pertains to a realistic amount of initial olivine. It is apparent that although the scatter in the observational data is quite large, the fit is quite good considering that the curve is entirely theoretical. It is important to note that while much of this scatter may be due to errors in the chemical analyses, such scatter is also to be expected from the previously observed deviation from the quantitative version of Prior's rule. This type of scatter of course corresponds to the natural variations in the amounts of Fe, Ni, Mg, and Si that yield different initial values of olivine as well as different Fe^0/Ni ratios.

6.12 Plot of L, the mole fraction olivine in the total moles of olivine, pyroxene, and metallic Fe against R, the ratio of oxidized to reduced iron. The plotted points represent observed values of ordinary chondrites of the low- and high-iron groups, whereas the curves are theoretical oxidation paths. The dashed curve represents the idealized pure enstatite chondrite.

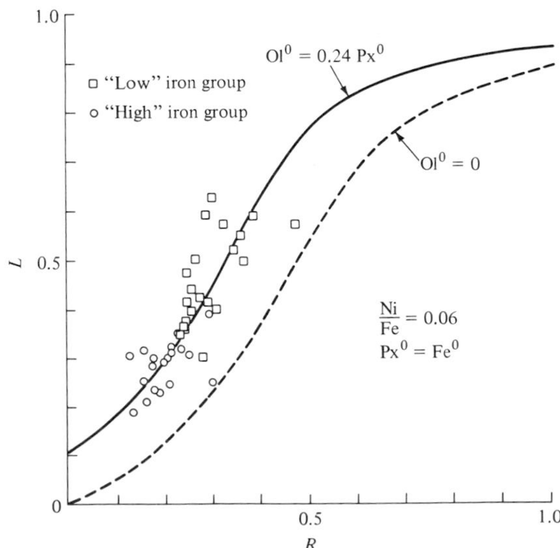

Oxidation and hydration states of meteorites[6]

The ordinary chondrites and related irons and stony-irons

We have seen that although meteorites are on the average quite highly reduced (by terrestrial standards), the range of oxidation is quite large. Because of this and also because oxidation–reduction processes involve such large energy changes, they assume a very critical role in meteorites.

Since approximately 90 percent or more of all meteorite matter lies in the system O-Si-Mg-Fe-Ni, a knowledge of the thermodynamic properties and phase relations of this system can give us a very good idea of at least the gross chemical behavior of this kind of matter. In Chapter 5 we presented the important phase relations of the silica-undersaturated part of the system with particular emphasis on the oxidation reactions that apply to the ordinary chondrites. It was shown that in the typical chondrite assemblage equilibrium demanded that $Mg/(Mg + Fe^{2+})$ increase with decreasing oxygen fugacity at constant temperature or with increasing temperature at constant oxygen fugacity. Consequently, since the composition range of the ferromagnesian silicates of ordinary chondrites is quite restricted and well known, and since, as we shall see, there is ample evidence that these bodies originated at liquidus temperatures, the general field of chondritic matter in the $\log P_{O_2} - T$ diagram is quite well defined, as shown in Figure 5.6.

It is interesting that we can apply the result we have just obtained relative to the oxidation state of the chondrites to other meteorites from the iron and stony-iron groups even though many of these meteorites lack certain of the critical phases of the chondrite assemblage. This may be done as follows: We know for example that although the pallasites generally contain olivine and no pyroxene, the olivine falls in the same compositional range as that of the chondrite olivines. Also because of the known distribution relation between olivine and pyroxene, we know that any hypothetical pyroxene in the pallasite would have a composition similar to that of the ordinary chondrites. Then, since the metallic iron of the pallasites and chondrites is also similar in composition, it follows that for a like temperature of formation the oxygen fugacity in all such meteorites must have been similar.

We may also view the pallasite assemblage in terms of a virtual displacement to the left of Reaction (5.r.). If we begin with olivine and metal of the characteristic pallasite composition and the system remains closed, then P_{O_2} attains the value appropriate to this composition with the formation of a minimal amount of $MgSiO_3$, Fe, and O_2. This conclusion, that many meteorites among the iron and stony-iron groups originated under roughly similar oxidation conditions, has already been arrived at by Olsen and Fredriksson (1966) on the basis of equilibrium involving phosphate and phosphides.

The enstatite chondrites and achondrites

Although these meteorites give an impression of a high degree of reduction, it is desirable if possible to place them in a more quantitative relation to the ordinary chondrites. We have seen that they consist of rather peculiar mineral assemblages in which Ca, Mg, and Mn sulfides, and silicon and titanium nitrides are the characteristic minor phases and in which the Fe-Ni alloys and silicates have unusual compositions. In addition, free silica may occur as tridymite, cristobalite or quartz. Given this situation, we may postulate the following reactions:

$$\underset{silica}{SiO_2} \rightleftharpoons \underset{metal}{Si} + \underset{gas}{O_2} \tag{6.b}$$

$$\underset{pyroxene}{FeSiO_3} \rightleftharpoons \underset{metal}{Fe} + \underset{silica}{SiO_2} + \underset{gas}{\tfrac{1}{2}O_2} \tag{6.c}$$

$$\underset{pyroxene}{MgSiO_3} + \underset{troilite}{FeS} \rightleftharpoons \underset{alabandite}{MgS} + \underset{metal}{Fe} + \underset{silica}{SiO_2} + \underset{gas}{\tfrac{1}{2}O_2} \tag{6.d}$$

$$\underset{pyroxene}{MnSiO_3} + \underset{troilite}{FeS} \rightleftharpoons \underset{alabandite}{MnS} + \underset{metal}{Fe} + \underset{silica}{SiO_2} + \underset{gas}{\tfrac{1}{2}O_2} \tag{6.e}$$

Attention has already been drawn to the conclusion of Ringwood (1961) regarding the solubility of silicon in kamacite of the enstatite chondrites. Since free silica is also present in these meteorites, Reaction (6.b) applies and we may write the following equation of equilibrium:

$$K_{(b)} = X_{Si}^{M}\gamma_{Si}^{M}P_{O_2}, \tag{6.8}$$

[6] Because of the almost undetectable and uncertain amount of water in meteorites other than the carbonaceous chondrites, their precise hydration states are indeterminate.

Table 6.7 Logarithm of the equilibrium constant for Reaction (6.e) as a function of the temperature

T (°K)	$\log K_{(e)}$
298	−49.2
400	−35.5
500	−27.5
600	−22.5
700	−18.7
800	−15.9
900	−13.8
1000	−12.1
1100	−10.7
1200	−9.46
1300	−8.41
1400	−7.62
1500	−6.85

in which $X_{Si}^{M} = Si/(Si + Fe + Ni)$ in the metal and γ_{Si}^{M} is the corresponding activity coefficient. Although adequate thermochemical data exist to determine $\Delta G_{(b)}^{\circ}$ for this reaction, the magnitude and functional form of γ_{Si}^{M} is very difficult to estimate. However approximate calculations show that for the observed silicon contents of the kamacite from enstatite chondrites, P_{O_2} probably lies at least several orders of magnitude lower than the field of ordinary chondrites.[7] Unfortunately insufficient thermal data exist for the evaluation of Reactions (6.c) and (6.d). In the case of Reaction (6.e), adequate data exist for all the components except the pyroxene form of $MnSiO_3$. We may, however, make use of the rhodonite form of this component for which data do exist.[8] We may be reassured that the error involved in this approximation will be small relative to those from other sources. If then we apply our conventional formalism we obtain the equation:

$$\log P_{O_2} = 2 \log K_{(e)} + 2 \log X_{Mn}^{Px} \gamma_{Mn}^{Px} \qquad (6.9)$$

in which $X_{Mn}^{Px} = Mn/(Mg + Fe^{2+} + Mn)$. Also little will be lost by setting $\gamma_{Mn}^{Px} = 1$, since great deviation from ideality is not to be expected in this case. If then we apply the thermochemical data we obtain $\log K_{(e)}$ as a function of temperature (Table 6.7).

According to Mason (1965) the enstatite chondrite Khaipur has $X_{Mn}^{Px} = 3 \times 10^{-4}$. If this value is substituted into Equation (6.9) and γ_{Mn}^{Px} is set equal to unity, then we find that at 1500°K, for example, $P_{O_2} = 10^{-20}$. This lies well to the right of the ordinary chondrites in Figure 5.6 and within the shaded area, by which we designate the approximate range of the enstatite chondrites.[9]

The carbonaceous chondrites

The carbonaceous chondrites contain the most highly oxidized primary meteoritic matter. However, their general field within the T–$\log P_{O_2}$ diagram (Figure 5.6) is but poorly defined as compared with that of the ordinary chondrites. This is particularly true of the upper temperature and oxygen fugacity limits. As we have seen, the type-II and type-III carbonaceous chondrites contain minerals such as olivine, pyroxene, and the Fe-Ni alloys in the form of inclusions and chondrules. These are surrounded by a matrix of the finer grained layer silicates, ferrite, and carbonaceous material. From such a texture, we must infer that the assemblage as a whole is out of equilibrium and was perhaps formed under a variety of temperatures and oxidizing conditions. In addition, these meteorites sometimes contain inclusions of minerals, which have been interpreted as condensates of the solar nebula (Grossman, 1972). It is also possible that the composition of the metals in these meteorites reflect nebular compositions (Grossman and Olsen, 1974).

The occurrence of serpentine-like layer-lattice silicates in these meteorites should be limited by dehydration reactions of the following type:

$$\underset{\textit{serpentine}}{5Mg_3Si_2O_5(OH)_4} \rightleftharpoons \underset{\textit{talc}}{Mg_3Si_4O_{10}(OH)_2} + \underset{\textit{olivine}}{6Mg_2SiO_4} + \underset{\textit{fluid}}{9H_2O} \qquad (6.f)$$

The stability relations of this reaction were studied experimentally by Bowen and Tuttle (1949) and later Olsen (1963) evaluated the effect of Fe_2SiO_4 on the equilibrium. Unfortunately lack of information of the precise nature of the meteorite layer silicates casts some doubt on the applicability of Reaction (6.f); and although talc has never been reported from carbonaceous chondrites, this may be due in part to oversight. If, however, this reaction is even approximately applicable, then it would seem from Olsen's data that the water pressure

[7] Personal communication from I. S. McCallum.

[8] Following a suggestion by I. S. McCallum.

[9] Larimer and Buseck (1974) have estimated crystallization temperatures of 900 to 1100°K and $P_{O_2} = 10^{-28}$ atm for the enstatite chondrites.

should have been quite low (<1 atm) in the range of temperatures below 400°K.

In addition to the hydrated silicates, the type-I carbonaceous chondrites also contain hydrated magnesium sulphates. However, it is somewhat doubtful that the degree of hydration of these minerals can be assigned a preterrestrial origin (DuFresne and Anders, 1962) and it thus seems prudent not to base any conclusion on the presence or absence of these minerals.

Carbon reactions in meteorites

Although carbon is a prominent constituent of only certain meteorites, such as carbonaceous chondrites and certain irons, it is omnipresent and measurable in ordinary chondrites and other meteorites. However, the significance of this element transcends its mere abundance since it can tell us so much about the environment of crystallization of these bodies.

For example, some form of solid carbon such as graphite or its amorphous equivalent is usually present in the chondrites or other meteorites whose field in the T–log P_{O_2} diagram is well defined. If we assume that equilibrium was attained with the gaseous phase, important reactions are some of those already discussed in Chapter 5, such as (5.c), (5.d) and (5.f). In addition, the reactions

$$\underset{graphite}{C} + \underset{gas}{O_2} \rightleftharpoons \underset{gas}{CO_2} \quad (6.g)$$

$$\underset{graphite}{C} + \underset{gas}{\tfrac{1}{2}O_2} \rightleftharpoons \underset{gas}{CO} \quad (6.h)$$

are important since they establish P_{CO} and P_{CO_2} for each temperature and oxidation state. Figure 6.13 shows the values of these partial pressures obtained for the typical assemblage from the ordinary chondrites. It is clear from the figure that in this assemblage CO must dominate over CO_2 and that the partial pressures of these gases reach appreciable values when free carbon is present. It may also be shown (Mueller, 1964) that as P_{O_2} decreases so also does the ratio P_{CO_2}/P_{CO} and the absolute values of P_{CO} and P_{CO_2}.

If it should happen that carbon is scarce, virtually all of the elemental form may occur in solution within the metallic phases so that we must resort to reactions of the following type:

$$\underset{metal}{C} + \underset{gas}{\tfrac{1}{2}O_2} = \underset{gas}{CO} \quad (6.i)$$

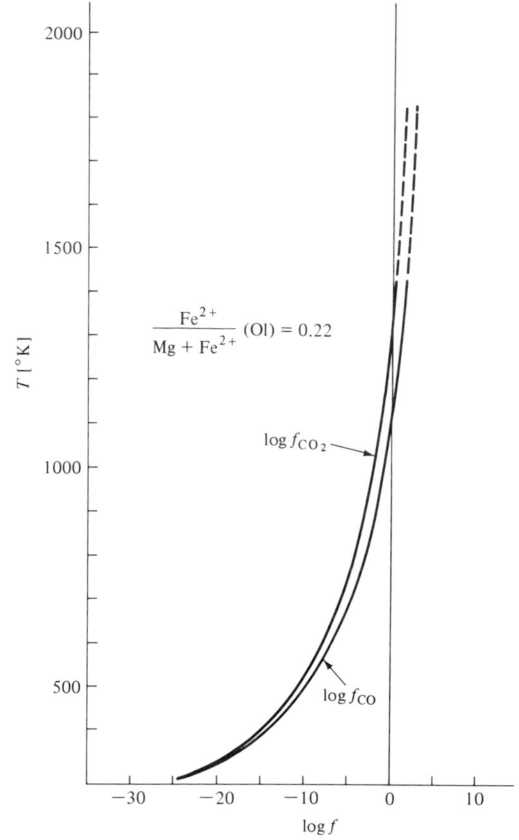

6.13 Theoretical plot of log f_{CO} and log f_{CO_2} against T°K for the assemblage olivine-pyroxene-pure metallic Fe-graphite at the olivine composition shown. This represents a degree of oxidation corresponding to the ordinary chondrites.

It is clear from this reaction that the partial pressures of the gaseous carbon species will be proportional to the activity of carbon in the alloy.

If hydrogen is added to our carbon-bearing system we enter the realm of the complex thermodynamics and kinetics of the so-called "organic compounds." The most prominent examples of this class of compounds are of course the hydrocarbons, and among these the most abundant will be the highly stable paraffin and benzene series. The equilibrium reactions between the paraffin series hydrocarbon, graphite, and hydrogen may be represented as

$$\underset{graphite}{nC} + \underset{gas}{(n+1)H_2} = \underset{gas}{C_nH_{2n+2}} \quad (6.j)$$

The corresponding equilibrium constant is

$$K_{(j)} = \frac{P_{C_nH_{2n+2}}}{P_{H_2}^{n+1}} \qquad (6.10)$$

Thermochemical calculations show that $K_{(j)}$ increases greatly with falling temperature so that in any equilibrium mixture the higher polymers will increase in proportion as the temperature decreases. However, for most temperatures at which equilibrium is readily attainable such light species as CH_4, C_2H_6, etc. will tend to dominate.

Unfortunately space prohibits our further pursuit of this interesting topic. We should, however, mention that the carbon compounds are objects of increasingly active research in meteoritics. Recently, much of this work has been directed toward determining the origin of hydrocarbons of the carbonaceous chondrites (Studier *et al.*, 1972). However it is likely that future investigations will be concerned with carbon as an environment indicator of even wider significance.

Reactions involving sulfur and phosphorus compounds in meteorites

The sulfides, phosphides, and phosphates are among the most abundant of the minor phases in most meteorites. In common with the other mineral assemblages already discussed, these phases have their own contributions to make toward placing limits on the physicochemical environment at various stages of meteorite history.

One of the most obvious reactions is that between metal and iron sulfide:

$$\underset{metal}{Fe} + \underset{gas}{\tfrac{1}{2}S_2} \rightleftharpoons \underset{troilite}{FeS} \qquad (6.k)$$

In terrestrial rocks the corresponding sulfide is pyrrhotite, and it is always deficient in iron by an amount that depends on the fugacity of sulfur.[10] As the work of Toulmin and Barton (1964) clearly shows, at any given temperature the excess sulfur in pyrrhotite (in a thermodynamic sense) is greatest when the sulfide coexists with pyrite and decreases steadily to stoichiometric FeS at the FeS-Fe boundary. For most of the temperature range of interest in meteorites P_{S_2} corresponding to the equilibrium Reaction (6.k) will be less than 10^{-4} atm. However above approximately 1200°K $P_S > P_{S_2}$, so that monatomic sulfur is most abundant in the gas phase.

[10] The phase relations and solution properties of pyrite and pyrrhotite and metal are the subjects of an important paper by Toulmin and Barton (1964).

Sphalerite is a minor phase in some meteorites, and it forms extensive solid solution with pyrrhotite. The system Zn-Fe-S has been studied by Barton and Toulmin (1966) and Scott and Barnes (1971). Using the experimental data of Barton and Toulmin (1966), Schwarcz *et al.* (1975) calculated the following preliminary relation between X_{FeS}^{Sp} (mole fraction of FeS in sphalerite coexisting with troilite), pressure and temperature (kbar and °K):

$$\ln X_{FeS}^{Sp}(P, T) = \ln(1.5 \times 10^{-4}T + 0.394) - \frac{P(5.803 - 4.806 \times 10^{-4}T) - 5.22 \times 10^{-3}P^2}{8.206 \times 10^{-2}T} \qquad (6.11)$$

The FeS mole fraction varies between 0.34 and 0.49 in meteoritic sphalerites. To determine the pressure of equilibrium, it is necessary to estimate temperature at which the final equilibrium values of FeS was attained. A temperature range of 350 to 600°C is possible. Schwarcz *et al.* (1975) assumed the lower temperature and found that the pressure of final equilibrium may vary from 1.7 to 3.4 kbar for most of the meteorites studied. These include Odessa, Campo del Cielo, Sardis, Bogou, and Gladstone.

At temperatures of the order of 1200°K, depending on the presence or absence of additional components, a cotectic relation exists between taenite, troilite, and a Ni-rich melt as determined by the experimental work of Clark and Kullerud (1959). At low Ni content this system quenches to a mixture of troilite and taenite, but if Ni/Fe exceeds 0.2 the quench product consists of an intergrowth of metal, troilite, and pentlandite.

Reactions and phase equilibrium involving phosphorus in its various forms and combinations in meteorites has been studied by Olsen and Fredriksson (1966), Buchwald (1966), Olsen and Fuchs (1967), and Goldstein and Doan (1972). It has been demonstrated experimentally[11] that the phosphorus has the effect of drastically raising the temperature of the transition kamacite $\rightleftharpoons$ taenite in the manner shown in Figure 5.4. In addition, phosphorus also has the effect of lowering the liquidus in the system Fe-Ni-P so that liquid is present at temperatures as low as 1250°K when phosphorus exceeds about 1 percent by weight (Buchwald, 1966). Since reduced phosphorus seldom exceeds 0.5 percent of the metal, conditions for the formation of such liquids are seldom realized. However, it is possible that when the effect of phosphorus is augmented by FeS, melts may at times form at even lower temperatures.

[11] For reference to much of the early experimental work on the system Fe-N-P see the paper by Buchwald (1966).

Although it is unclear what role melting may have played in the history of many iron and stony-iron meteorites, there are certain reactions that relate their phosphide, metal, and phosphate phases, which have relevance quite irrespective of whether melting occurred, since these reactions are formally applicable wherever the critical phases are present. For example, Olsen and Fredriksson (1966) discussed the following reactions between phases they observed in certain octahedrites:

$$\underset{metal}{3Fe} + \underset{metal}{2P} + \underset{gas}{4O_2} \rightleftharpoons \underset{sarcopside}{Fe_3(PO_4)_2} \tag{6.l}$$

$$\underset{olivine}{3Mg_2SiO_4} + \underset{metal}{4P} + \underset{gas}{8O_2} + \underset{metal}{6Fe} \rightleftharpoons \underset{farringtonite}{2Mg_3(PO_4)_2} + \underset{olivine}{3Fe_2SiO_4} \tag{6.m}$$

$$\underset{olivine}{3Mg_2SiO_4} + \underset{schriebersite}{4Fe_3P} + \underset{gas}{8O_2} \rightleftharpoons \underset{farringtonite}{2Mg_3(PO_4)_2} + \underset{olivine}{3Fe_2SiO_4} + \underset{metal}{6Fe} \tag{6.n}$$

We observe first of all that in the case of Reactions (6.l) and (6.m), phosphorus dissolved in the metal phase is involved. If some magnesium is assumed to be dissolved in sarcopside and if Ni is regarded as being present in constant amount, direct application of the phase rule shows Reactions (6.1) and (6.m) to be trivariant. However since Reaction (6.n) involves the additional phase schriebersite, the system it represents is divariant at constant Ni content.[12] Consequently it bears a resemblance to the major assemblage of the ordinary chondrites (Figure 5.6) in that if the composition of olivine is fixed, log P_{O_2} is a univariant function of the temperature. As was previously pointed out, when the compositions of the phases are taken into account in the assemblages studied by Olsen and Frederiksson (1966), it is found that the oxidation states of these iron meteorites are close to that of the ordinary chondrites.

Element distributions in meteorites

We have seen in Chapter 1 and elsewhere that the complex coupled equilibria such as we have been discussing depend in part on the distribution functions of the elements between the coexisting phases participating and that these functions are temperature- and pressure-dependent. However, we have also seen in Chapter 5 that the distributions may have only a small effect as long as the distribution constant approaches unity and the solutions are close to ideal. Under these circumstances such errors as occur by ignoring the temperature effect on the distribution constant are of the same order or smaller than those implicit in the thermochemical data for energetic reactions such as those involving oxidation. In spite of this, however, the element distributions taken by themselves are potentially powerful tools for obtaining the temperatures and pressures of crystallization.

Unforunately most of the important element distribution equilibria that are available in meteorites still have a doubtful status. The best known of these, those which involves the distribution of such elements as Ni, Co, and Fe among the metallic phases, are of course severely limited by kinetics. Since we know that taenite and kamacite unmix at temperatures as low as 700°K, we can scarcely expect them to give us reliable information about any high temperature period. As has been previously indicated (Mueller, 1964), a potential thermometer assemblage for the high temperature stage is likely to be the olivine-pyroxene assemblage. Although there are experimental data on this equilibrium (Larimer, 1968; Medaris, 1969; Nafziger and Muan, 1967; Williams, 1971) the large amount of scatter in these data indicates that the degree of equilibrium attained under laboratory conditions is insufficient to establish the variation of the distribution constant with temperature. Some observational data for the assemblage are summarized in Figure 6.14. It is to be observed that while the precise microprobe data of Keil and Fredriksson (1964) approximate closely the earlier data of Ringwood for which $K_{(5q)} = 1.13$, the former fall into three groups with successively greater displacement from the 45° line as $Fe^{2+}/(Mg + Fe^{2+})$ increases. The linear concentration of these groups approximate the distribution curves for ideal solutions with successively larger values of $K_{(5q)}$, so that if this constant decreases with increasing temperature these data could be interpreted as indicating that the high-iron group represents higher temperatures of crystallization than does the low-iron group. However, we shall see in the next section that such a straightforward interpretation may not be the correct one.

The composition of coexisting Ca-pyroxene and orthopyroxene has been used in estimating temperatures of crystallization of the ordinary chondritic minerals by Bunch *et al.* (1970), Fodor *et al.* (1971), Van Schmus and Koffman (1967), and Bunch and Olsen (1974). The results of the latter authors indicate a recrystallization equilib-

[12] To demonstrate this analytically we would have to take into account several additional exchange reactions involving olivine, schreibersite, and farringtonite.

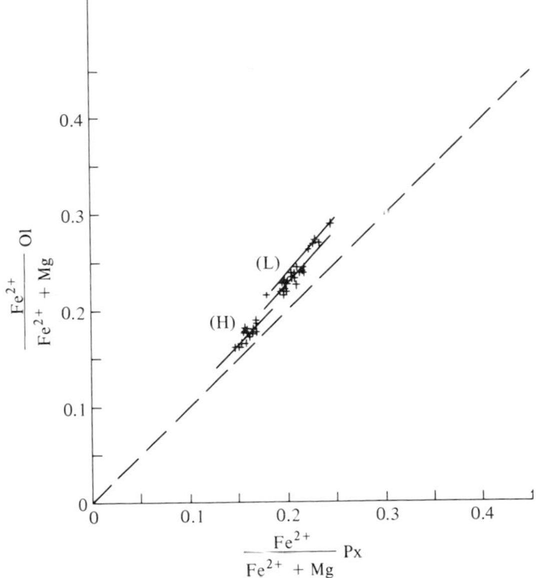

6.14 Plot of the atomic ratio $Fe^{2+}/(Fe^{2+} + Mg^{2+})$ in olivine against the same value for coexisting pyroxenes in high- and low-iron groups of ordinary chondrites. Data from Keil and Fredriksson (1964). The lines that pass through the group points correspond approximately to distribution curves based on the ideal solution model with successively greater values of the distribution coefficient.

rium in the temperature range of 750 to 950°C. Temperature estimates with a two-pyroxene thermometer may be difficult because of the requirement of extreme accuracy in the composition of coexisting minerals and insufficient thermodynamic data on the ternary solution (Saxena, 1976). As pointed out by Bunch and Olsen (1974), it is likely that the temperature estimates by Fodor *et al.* (1971) were higher because of differences in chemical analytical techniques. The results of Onuma *et al.* (1972) on distribution of oxygen isotopes, however, may be interpreted to favor the homogenization of the minerals at igneous temperatures. The problem of temperature estimate using major element fractionation cannot be resolved at the present stage of our knowledge of the thermodynamics of the crystalline solutions.

In addition to the Fe-Mg distribution equilibria, meteorites contain a potentially great number of environment indicators in the form of minor and less known major element distributions. The study by Bunch *et al.* (1970) referred to above indicates that the distribution of Mn is very orderly. It seems likely that the wide variety of responses to kinetic factors to be expected from these distribution functions favors the eventual detailed deciphering of the cooling history of meteorites.

Mechanisms of liquidus crystallization for the chondrites

We have seen that the bulk of the silicate component of meteorites falls in the system MgO-FeO-SiO_2, whose liquidus relations were studied by Bowen and Schairer (1935). It is therefore desirable to examine the silicate component in terms of this system. To do this, however, we must first "normalize" the bulk chemical analyses such as those presented in Table 6.4 by recalculating to eliminate the feldspars and other minor silicate components. We then obtain average values of the subsystems for the high- and low-iron groups of the ordinary chondrites, which plot on Figure 5.5 as two points. Study of the behavior of liquids initially at these two compositions should then tell us, at least to a first approximation, how the ordinary chondrites might have originated by crystallization at liquidus temperatures.

It is clear, first of all, that the paths of equilibrium crystallization should be of the simpler type outlined by Bowen and Schairer (1935). Crystallization should begin with the separation of Mg-rich olivine and the liquid composition should follow a curved path such as that shown in the figure until the pyroxene boundary is attained. At this boundary pyroxene begins to precipitate with olivine and the two minerals enter into certain reactions that involve the SiO_2 component of the melt. During the co-precipitation of olivine and pyroxene the compositions of these two minerals are at each stage defined by a tie line whose precise orientation is a function of the temperature and pressure. This tie line, which forms the base of a triangle whose apex is the liquid composition, moves to the right as crystallization proceeds, and crystallization ceases when the tie line is intersected by the initial bulk composition point. During the entire process there is a continuous exchange of Mg and Fe between the crystals and melt so that when crystallization ceases the mass consists of homogeneous olivine and pyroxene whose compositions as given by the tie line reflect the temperature and pressure at which the last liquid disappeared. Also it should be noted that bulk compositions corresponding to the low-iron (L) group of chondrites should cease crystallization at a slightly lower temperature than those corresponding to the high (H) group.

We may, on the other hand, consider the other mode of crystallization frequently discussed in relation to such systems, that of fractional crystallization, which plays such an important role in the interpretation of terrestrial magmatic rocks. Under this form of crystallization the earlier-formed Mg-rich crystals are constantly removed or isolated from the melt so that the effective bulk composition is continually shifted to the right and the system becomes enriched in FeO. This may go so far that the ternary invariant point is reached and all pyroxene gives way to Fe-rich olivine and free silica as crystallization products. As is well known, fractional crystallization may occur by any of several mechanisms such as crystal settling or armoring through crystal zoning. Detection of either mechanism is evidence for fractional crystallization.

It is evident that, superficially at least, the the uniform chondrites resemble products of equilibrium crystallization whereas the nonuniform chondrites show some evidence of fractional crystallization in the form of zoned crystals and other features. However, we shall have to qualify this conclusion greatly in order to accommodate certain troublesome facts.

First, the observed chondritic textures are entirely contrary to what we should expect from either mode of crystallization described above. The chondrules tell us that, by and large, cooling occurred too rapidly for either equilibrium or fractional crystallization of the classical kinds to occur. There simply was not enough time for exchange to occur between the interiors of macroscopic crystals, nor was there adequate time for zoning to develop except in certain crystals that formed before the ultimate rapid quench, which formed the chondrules. Furthermore, it is well known from numerous experiments and from observations of terrestrial rocks that the continuous adjustment of crystal and liquid compositions required for equilibrium crystallization seldom is possible in silicate systems for kinetic reasons. What results instead is crystal zoning of the type associated with fractional crystallization. However, even this mechanism requires adequate time to change the liquid composition by diffusion over macroscopic dimensions.

As an alternative to the classical equilibrium and fractional modes of crystallization for the chondrites it is interesting to consider a kinetically controlled process which is capable of incorporating the evidence for rapid cooling and the uniform mineral compositions observed in the uniform chondrites. In this model, which is based on a suggestion by Fredriksson (1963), we must consider first an extreme case in which cooling occurs so rapidly that little or no diffusion can occur within the congealing or crystallizing matter. This situation is depicted in Figure 6.15. If, for example, the bulk composition is indicated by P, then the compositions of the coexisting olivines and pyroxenes will be given by the points Ol′ and Px″ which lie on the dashed line which passes through the SiO_2 apex. By contrast the approximate equilibrium compositions of the mineral pairs as taken from Figure 6.14 is given by Ol′–Px′ and Ol″–Px″, as shown by the solid tie lines.

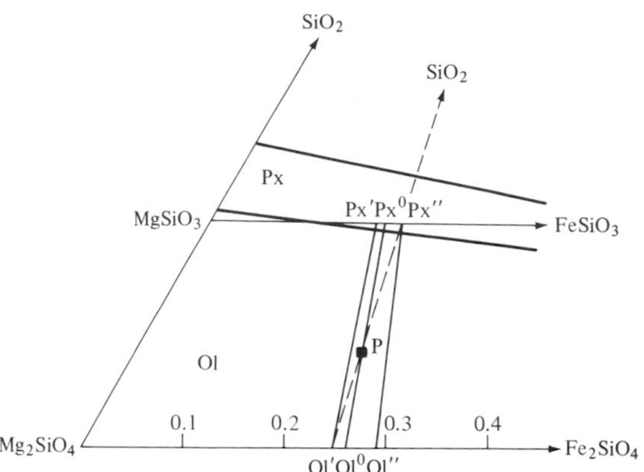

6.15 Diagrammatic enlargement of a portion of the system MgO-FeO-SiO_2 (Figure 5.5) to illustrate the idealized limiting case of rapid cooling and crystallization. The dashed line Ol′–P–Px″, which also passes through the SiO_2 apex, represents cooled mixtures of crystals and glass of varying olivine/pyroxene ratios and constant Mg/Fe^{2+}. The lines Ol′–Px′, Ol–Px^0, and Ol″–Px″ represent near-equilibrium tie lines at temperatures that are probably just below the liquidus.

If indeed the above ideal limiting case was ever realized it must have been obliterated by subsequent annealing or recrystallization. However, it is also easy to argue that the limiting case can never actually occur at all. To see this we fix our attention on the barred chondrules in which the thickness of the "bars" is about 10^{-2} mm and in which they are separated by pyroxene-like glass or microcrystalline material. It is clear then that when the olivine bars formed, Si^{4+} must have diffused away from the olivine site of crystallization and that Mg^{2+} and Fe^{2+} must have diffused toward it. But if this occurred, it also seems likely that Mg^{2+} and Fe^{2+} might at the same time have been able to diffuse sufficiently to affect an exchange approximately commensurate with the equilibrium distribution for the temperature of quenching.[13] In other words the time scales of crystallization

[13] However, it cannot be assumed that this temperature is the "mean" of the liquidus temperatures traversed.

and ion exchange should be similar. Furthermore, it seems likely that the dimensions of the bars in the barred olivine chondrules should be governed to a large extent by the diffusion lengths of the ions involved, particularly of Si^{4+}, which is likely to be the least mobile.

By this means of rapid "quasiequilibrium" crystallization we can then explain the homogeneous crystals and orderly distributions we find in the uniform chondrites. However, we are still required to explain the variations in bulk composition which give rise to chondrules of varying quantities of olivine and pyroxene. It would seem that in order to do this we would have first to be acquainted with the details of formation of the chondrules themselves. Unfortunately we know little or nothing about the chondrule-forming mechanism except that the melt seems to have undergone a fragmentation or disruption which produced the bulk chemical inhomogenieties. We may therefore speak of an "unspecified disruptive process" in which the crystal-liquid suspension is disrupted in a manner which separated olivine and the pyroxene-like liquid. This should occur especially within the field of crystallization of olivine, perhaps simultaneously with the crystallization itself. The general scheme may be outlined as in Figure 6.16. According to this scheme, barred olivine and radiating pyroxene chondrules represent the compositions of possible liquids most closely, whereas the porphyritic chondrules represent suspensions or cumulates of crystals in liquids. However, only the barred type would have a high probability of having the same composition as the bulk meteorite silicates as indicated by the points in Figure 5.5. The radiating pyroxene chondrules of course represent liquids near the pyroxene join so that the disruptive process has brought about marked differentiation in this case. By contrast the porphyritic olivine chondrules should, at least where these are well-developed, indicate a period of relatively slow cooling and crystallization in the field of olivine before the disruption occurred and consequently they should show some enrichment in magnesium.

6.16 General scheme for the derivation of various chondrule types by an unspecified, rapid disruptive process that brings about separation of crystals and liquid.

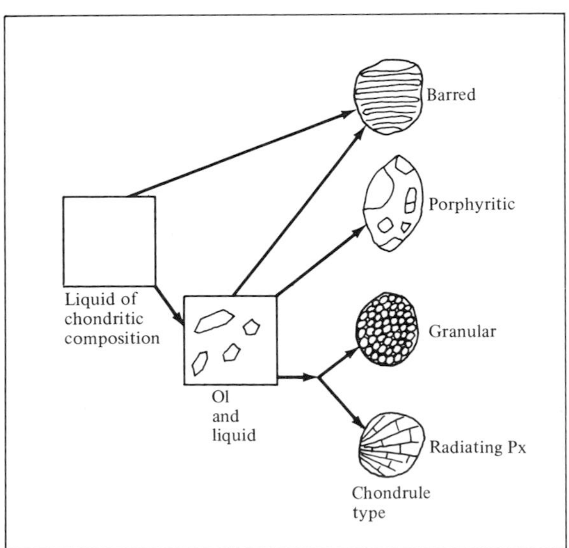

The actual temperature represented by the quasi-equilibrium process and which presumably is reflected in $K_{(5.q)}$ can only be roughly estimated. We may for example assume that there is a lowering of the liquidus temperatures shown on Figure 5.5 by the additional components present in the meteorites. If this amounts to approximately 100°K then the quenching temperatures should be in the range of 1800°K to 1600°K. It may also be that the slightly larger value of $K_{(5.q)}$ in the low-iron group of chondrites actually reflects an extension of the quenching interval to the lower liquidus temperatures implied by the presence of more FeO in these meteorites.

It must be stressed at this point that the above explanation for the uniform chondrites is only one of several. The best known alternative today is that the uniform compositions simply reflect homogenization through recrystallization or "metamorphism" as advocated especially by Wood (1963). Since the recrystallization is thought to have occurred at temperatures substantially below the liquidus, discrimination between the two models would seem to be possible through precise evaluation of $K_{(5.q)}$ and other distribution constants as a function of the temperature. The application of the oxygen geothermometer to the uniform chondrites, as mentioned earlier, has shown that many of these chondrites, notably the Shaw and Burdett meteorites, yield temperatures in the magmatic range and still exhibit highly uniform chemical compositions. In addition, it is especially significant that the Burdett meteorite contains well-defined chondrules. These characteristics indicate that the compositional uniformity of the minerals is probably inherited from the magmatic stage or at very least from just below liquidus temperatures. The simplest explanation of all this seems to be that the chondrules formed during a rapid quench process, as previously explained, and that chemical fractionation occurred as the temperature fell. This explanation thus eliminates the need for any subsequent reheating or "metamorphism" of the material.

Origin of the metallic components of chondrites

Any comprehensive hypothesis of the origin of the chondritic meteorites must account not only for the chondrules and the matrix silicates but also for the metallic grains. In most recent work the origin and even the disposition or mode of occurrence of these grains has been left an open question. However Urey and Mayeda (1959) concluded from their study of a number of chondrites that the grains had been derived from the preexisting "primary object" of Urey's cosmogeny, which had been broken up and reconstituted to form the chondrites. This conclusion was based largely on the presence of some distorted, fractured, and displaced metallic grains, the occurrence of kamacite and taenite in separate crystals, and "diffusion borders" of certain grains.

Before suggesting our own origin for the metallic grains it is desirable to consider first the question of the equilibrium state of the combined system of silicate and metal. Although there is evidence in the form of oxidation and distribution relations previously discussed that there is an approach to chemical equilibrium between the silicates and the metal, we do not know how close the approach is to perfect equilibrium. We have, however, already noted that the metal grains are distributed very regularly among the silicates. This fact tells us immediately that this system could never have been in perfect chemical equilibrium in the presence of a gravitational field. This is true because a state of perfect chemical equilibrium also implies a state of mechanical equilibrium, and the observed regular distribution of grains does not meet this requirement. To see this in terms of the free energy involved we may imagine a 1-cm^3 grain of metal of density δ_M which overlies a 1-cm^3 grain of silicate of density δ_S, where $\delta_M > \delta_S$. If the distance separating the grains is Δh and the acceleration of gravity is g, the energy released by exchanging the grains is

$$\Delta G = \Delta hg(\delta_S - \delta_M) < 0$$

since Δh is taken as positive.[14] Normally of course such mechanical instability also exists in terrestrial metamorphic rocks. Such rocks are too viscous for their constituents to attain their proper levels in the gravity field. However the same does not hold for magmas, so that if either silicate or metal had been molten while in a gravitational field we should expect at least some segregation. Since virtually all hypotheses about the origin of these meteorites assume that a gravitational field was present at the time the metal and silicate particles were brought together, the above considerations appear to rule out the participation of a melt as an agent in the disposition of the grains.

However we have also noted the irregular forms and interstitial character of the metal grains as well as the occurrence of metal as distinct veins (Figures 6.10 and 6.11). It seems clear then that if the metal grains were ever mechanical fragments such as were postulated by Urey and Mayeda (1959) they must have undergone much subsequent alteration. However, the regular distribution of the grains favors a mechanical origin no more than it does a magmatic origin.

Indeed the only mechanism we have left to turn to in explaining most of the observed features of the metallic grains is that of deposition from a vapor phase over a range of temperatures. We do not of course know precisely the form iron and nickel had in such a phase. We know however from thermochemical data that at temperatures as low as 1500°K the vapor pressure of metallic Fe vapor is approximately $10^{-6.41}$ atm and that $P_{\mathrm{Ni}} = 10^{-6.63}$ atm at the same temperature. From kinetic considerations we should therefore expect considerable vapor transport to occur at temperatures immediately below the liquidus so that if the metal vapors had access to hot aggregates of chondrules they should have rapidly deposited the metal grains in the interstitial voids. This conclusion is also supported by certain experiments by Morelock (1962) who observed the growth in 16 hr of whiskers 200 μ in length from a vapor phase at temperatures as low as 1333°K. This form of high-temperature deposition explains the occurrence of plessite, which marks the location of former grains of homogeneous taenite. However to explain the occurrence of separate grains of taenite and kamacite we must assume that some vapor transport also was effective at temperatures below 1100°K where these two phases can grow in equilibrium. In view of the above experimental data this does not seem too severe a requirement.

From the above evidence, then, the following picture emerges: After the rapid quench-formation of the silicate chondrules at silicate liquidus temperatures above 1600°K, the silicate chondrules were incorporated or reincorporated as solids into a parent body with a gravity field. At the same time fairly good chemical communication was established between the minerals and the vapor phase so that there was an approach to equilibrium. Immediately after deposition of the chondrules and other silicate fragments, deposition of metal from a permeating vapor phase began in the interstices

[14] We ignore such effects as the compressibility and the variation of g.

of the silicate grains. The deposition and redeposition or reworking of the metal grains by the vapor phase continued as the temperature declined, and this accounts for the variety of metal grains, including plessite and separate grains of kamacite and taenite. As pointed out by R. E. Marringer (Mason and Wiik, 1961) there appears to have been some subsequent *in situ* movement of the grains, which explains some of the mechanical dislocations observed by Urey and Mayeda (1959). We cannot of course preclude the deposition by mechanical agencies of some preformed grains, nor can we discount very local melting of grains. However the dominant mechanism is regarded as deposition from the interstitial vapor phase, which thus accounts for the regular distribution and interstial character of the grains as well as the presence of later veins.

Pressure of formation of meteorites

In principle all meteorites contain mineral indicators of their temperatures and pressures of crystallization. We have already seen that although evidence of the precise temperature of crystallization of most meteorites is still lacking, certain information on either crystallization or annealing may be obtained through the use of several thermometer assemblages. In the case of possible pressure indicators the outlook is less promising although it should soon be possible to set more severe limits on the pressures of formation of most meteorite bodies than at present.

The question of pressure bears on one of the classical problems of meteoritics—namely, whether these meteorites originated in bodies of planetary size or in bodies of much smaller dimensions, perhaps of the order of asteroids or comets. To early students the broken-up planet hypothesis seemed highly attractive and it has also had its more recent advocates in Brown and Patterson (1948), Zavarisky (1950), Lovering (1957), and Ringwood (1961b). Other investigators since the time of Sorby (1877) have also proposed origins for certain meteorites such as the chondrites which imply a low-pressure origin since they involve processes of a volcanic nature. Certainly plain logic as well as our experience with terrestrial rocks leads us to expect at least some variation in the pressures under which the different classes of meteorites formed.

Unfortunately at present there is a great deal of ambiguity attached even to the limited number of available pressure indicators. Some preliminary results and the discussion of the sphalerite geobarometry have been presented by Schwarcz *et al.* (1975). One of the inherently most simply of the other barometers is the graphite-diamond transition, which we referred to in Chapter 2; it is almost entirely decoupled from the remainder of the system. Since, as we have seen, diamond is a constituent of certain achondrites and irons, it was proposed by Urey (1956) that these implied that the meteorites come from near the center of lunar-sized objects. Although the question is still open in the case of the diamond-bearing achondrites, it was demonstrated fairly convincingly by Lipschutz and Anders (1961) that in the case of at least one iron, Canyon Diablo, the diamonds were probably produced by the shock of impact with Earth in much the same way as coesite is produced in the surrounding crater material.

Soon thereafter it was pointed out by Ringwood (1960) that reactions of the type

$$\underset{\text{metal}}{3\text{Fe}} + \underset{\text{graphite}}{\text{C}} \rightleftharpoons \underset{\text{cohenite}}{\text{Fe}_3\text{C}} \tag{6.o}$$

involved a considerable volume decrease so that cohenite should be favored by pressure. Unfortunately the slope of the boundary curve that separates the metal and graphite assemblage from cohenite is negative, so that if the temperatures are high enough (in excess of 1100°K) cohenite may form under equilibrium conditions even at low pressure. In addition, the equilibrium is complicated by the presence of Ni as has been discussed by Olsen (1964) and by Brett (1966).

If it were not for the unfavorable kinetic properties of the Fe-Ni alloy system, which favors unmixing of taenite and kamacite over a wide range of conditions, this assemblage might be a useful pressure indicator. We know for example that the α-γ transition is pressure-sensitive, and it was pointed out by Uhlig (1954) that the transition is lowered in temperature as the pressure is increased. Uhlig attempted to explain certain features of iron meteorites, such as the occurrence and structure of plessite and the lack of Widmanstätten structure in high Ni meteorites, by calling on high pressures and the sudden release of these pressures in planetary break-up. These conclusions were criticized by Urey and others (Anders, 1964), and consequently the use of the metallic alloy phases as pressure indicators has not found many recent advocates.

When we consider the chondritic meteorites we have even less to go on since the above-mentioned pressure indicators are either absent, difficult to identify, or reflect even a grosser degree of disequilibrium than they do in the irons. However, we do have the chondrules themselves, and in a few cases such phases as tridymite and cristobalite. The difficulty with using the major chemical reactions between chondritic minerals to determine the pressure of crystallization is illustrated by the attempt of Williams (1971) to use the olivine-pyroxene exchange reaction in this way. The result obtained for the pressure of crystallization in this case was 5 ± 5 kbar! Although it was once thought that chondrules might represent a type of porphyroblastic structure, there is nothing like them in any terrestrial metamorphic rock. It is difficult to imagine these fragile structures with all their evidence for rapid cooling as having been produced in an environment of even modest pressure. Indeed their content of glass, microcrystals, and skeletal structures as well as their spheroidal forms and often friable texture militates against their having been exposed to pressures in excess of the kilobar range. It thus seems safe to infer on the basis of the chondrule structures alone that at least the primary features of these meteorites originated under conditions of high temperatures and low pressures.

Origin of meteorites: a critique of theories and hypotheses

General

When we speak of the origin of meteorites it is necessary to be quite specific about what we mean by "origin." That is, we must restrict our statement to a certain time, place, and meteorite type. For example, when we refer to the time of origin we do not ordinarily mean the period of synthesis of the elements which comprise meteoritic matter but rather to the prominent event or series of events that occurred approximately 4.5 Gyr ago. However, it is also possible that we may be interested in the time of occurrence of certain subsequent events such as the differentiation that produced the iron meteorites or gave rise to certain elemental abundance anomalies.

Again, most of what we say regarding the origin of meteorites refers to members of the chondrite classes since as we have seen these are at the same time the most abundant, the least differentiated, and possess the most unique and interesting structures. We are indeed interested in the origin of the irons, stony-irons, and achondrites, but there is almost universal agreement that these meteorites are products of intense differentiation deep within the meteorite parent bodies and so presumably are derivative from chondrite-like matter. Also, as we have seen, the structures and textures of the nonchondritic meteorites, while quite undiagnostic of details of their crystallization history, are nevertheless quite prosaic and have many parallels among terrestrial rocks.

We have also become aware through our mineralogic studies of the chondrites that we must be prepared to consider a certain amount of physical mixing of high and low temperatures and oxidized and reduced components in such meteorites as the carbonaceous chondrites. Then also it has been proposed, in particular by Urey, that meteorites contain evidence of a number of complex events involving the breakup of "primary objects" by collision and the redistribution and reconcentration of this material among other bodies to produce secondary deposits or objects. Such a hypothesis then implies the existence in meteorites of components of different generations.

It is generally conceded that the origin of meteorites in the above sense is connected with the origin of the solar system itself and that a study of meteorites will lead to information about this momentous event. It is therefore necessary to ask first what are the primary features of meteorites and how may they be distinguished from secondary features superimposed at some later date. The answer is that there is quite a good consensus that several types of observational data are of fundamental character and must be accommodated in any satisfactory theory. These are:

1. The correspondence of elemental abundances between the chondrites as a whole, and particularly the type-I carbonaceous chondrites and the composition of the sun
2. The isotopic abundance pattern of chondrites and particularly those of the decay products of short-lived species
3. The chondritic structure

Other types of data which are of varying importance to different students of meteorites are the following:

4. Elemental abundance differences between the different classes of chondrites, the variation of oxidation state reflected in Prior's rule, and the existence of the high- and low-iron groups
5. Character and distribution of the metallic grains in the chondrites
6. The existence of the uniform and nonuniform groups of chondrites
7. Evidence that certain chondrites are mixtures of two or more components
8. Evidence of shock phenomena in meteorites

9. Nature of the distributions of elements between the coexisting minerals and the conclusions that can be drawn from them regarding the attainment of equilibrium and the temperatures and pressures of crystallization

With these points in mind we shall now consider the various ideas that have been proposed for the origin of meteorites.

Direct condensation from the solar nebula or its differentiated equivalent

This involves the classical problem of Urey, which we already discussed in quantitative terms in Chapter 5. The idea that chondrules are the original products of condensation from the primordial nebular gases is an old one. Chamberlin (1928), for example, discussed chondrules in relation to his planetesimal hypothesis and concluded that they resulted from condensation of hot gases thrown off by the early sun. The nebular origin of chondrules has been espoused by Suess (1949), Levin and Slonimskii (1957), and particularly by Wood (1963). Thermodynamic calculations to determine the sequence of condensations from the solar nebula were also performed by Lord (1965), Larimer (1967), and Grossman (1972), whereas kinetic mechanisms were explored by Blander and Katz (1967). As was pointed out by Wood and others, only a few chondritic meteorites exhibit minerals whose compositions are in apparent harmony with direct condensation from a gas of the composition inferred for the solar nebula, since most chondritic silicates are too rich in ferrous iron and indicate a more oxidizing environment (Figure 5.6). Similarly, only certain carbonaceous chondrites appear to contain the early, high temperature Ca-Al-rich condensates such as mellilite, according to Grossman's study.

We have already seen in Chapter 5 that the mineralogic and chemical nature of the condensates depend critically on the relative abundances of hydrogen, the metals, and oxygen as well as the total pressure in the region in which condensation occurs. As emphasized by Grossman, local slight departures from chemical equilibrium due to incomplete reaction of the condensates with the nebular gases could have produced the Ca-Al-rich inclusions. Similarly, the inferred early condensation of metallic iron as compared with the condensation of most of the silicates and the promotion of this differential condensation by an increase in pressure is at least in agreement with the hypothesis that planetary cores may date to the accretion stage. If, however, the bulk of the chondritic meteorites resulted from direct condensation from nebular gases those gases must have undergone substantial differentiation from their initial composition.

Volcanic processes

A volcanic or quasivolcanic origin for chondrules has been advocated by a number of authors, but in recent times chiefly by Ringwood (1961, 1966) and by Fish, Goles, and Anders (1960), although Anders has subsequently changed his mind on this subject.

The greatest unity and self-consistency is shown by Ringwood's theory. According to him the various meteorite groups and classes were derived from primordial matter of the composition of the type-I carbonaceous chondrites. In detail the scheme involves heating, perhaps by short-lived radioisotopes, of parent bodies of yet indeterminate sizes, but no larger than our moon. This heating resulted in complex chemical reactions but especially in the reduction of original ferrous iron by carbon compounds such as free carbon and hydrocarbons. At first, at the inception of the reduction process when temperatures were below 900°K or so, the expelled gases consisted mostly of CO_2 and H_2O, but at higher temperatures these were supplemented by CO and H_2. It is further inferred that H_2O in conjunction with H_2S acted as a powerful solvent and in this way brought about the depletions of many elements that characterize other meteorites relative to the type-I carbonaceous chondrites.

According to this model the degree of reduction attained by a meteorite and consequently the existence of Prior's rules depends on the original carbon content of the material. Thus the most oxidized ordinary chondrites of the low-iron group presumably contained the least carbon originally, whereas in the case of the enstatite chondrites this element was in excess, as is also indicated by presence of graphite in these meteorites.

At the culminating stage of the process, melting is assumed to have occurred in the interior of the parent body or bodies. The resulting eruption of magmas on the surface was then accompanied by a rapid escape of gases consisting mainly of CO and H_2. These phenomena resulted in the violent disruption of the magma and, in combination with rapid cooling, produced the chondrules and typical chondritic structures. The coexistence of metal and silicate particles is explained by assuming that gravitational separation was prevented by turbulence in the magma.

The model of Fish *et al.* (1960) differs in several ways from that of Ringwood. One of the most important differences is that the carbonaceous chondrites are not considered as necessarily primordial. Rather, they are regarded as having formed a near-surface deposit on an asteroid or asteroids whereas the ordinary chondrites, achondrites, stony-irons, and irons comprised the successive layers toward the interior of the same body or bodies. In contrast to Ringwood's model, the volcanic emanations are regarded as having been chiefly S_2 and SiS rather than carbon compounds, water, or hydrogen. Finally, in the model of Fish *et al.* (1960) the observed minor-element fractionations between the various meteorite classes are attributed to the vertical movement of FeS and volatile gases in a complex "sulfur cycle."

Impact and associated processes

This model is largely associated with the name of Urey, who has long advocated various versions of it. Urey (1963, 1964) favors a two-stage process for the formation of the chondrites. In the first stage lunar-sized "primary objects" condensed and agglomerated from the solar nebula. This material is thought to have consisted of low-temperature condensates, much like the carbonaceous chondrites, with an abundance of volatile compounds and iron largely in the ferrous state. As this material heated up, reduction processes, much as in the model of Ringwood, produced metallic iron, and this was differentiated from the silicates by magmatic and gravitational mechanisms. During subsequent cooling of this material the metal differentiated into kamacite and taenite, and diamonds were formed under the high interior pressures.

The second step of the process begins with the collisional break-up of these primary objects. Urey thought that the silicate and troilite fractions would be broken up into such fine dust that they would be lost altogether, but that the metal, because of its greater strength, would be preserved to fall on other primary objects which had not as yet suffered collisions. There the metal fragments would come to be mingled with chondrules produced by some volcanic process.

From our standpoint the most interesting consequence to be drawn from this scheme is that at least a fraction of the metallic particles of the chondrites are from a different source than the silicates themselves. As previously mentioned, Urey and Mayeda (1959) thought that they found evidence for this in their study of the metallic particles of chondrites.

More recently, Ringwood (1966) proposed a version of the impact origin of the chondrites as an alternative to his autoreduction–volcanic model. This model again begins with planetesimals of the composition of the type-I carbonaceous chondrites. However, now the energy required for melting and reducing the bodies is provided by their impact on our moon or some other large body approximately 4.5 Gyr ago. We note that this requires that the moon be a source for meteorites striking Earth, an idea that has been advocated for some time by Urey (1959).

Metamorphism of the chondrites

Students of meteoritics as early as Sorby (1877) saw evidence for recrystallization or metamorphism in the chondrites (see review by Dodd, 1969). However, Wood (1963) has gone so far as to suggest that the major classes of the ordinary chondrites have been derived in this way from a more primitive chondrite, as we have already mentioned. According to him, internal heating of the parent bodies brought about a reaction between the original Mg-rich chondrule silicates and the ferrous-iron-rich matrix so that "Fe^{2+} from the fine Fe_3O_4 grains diffused into the chondrule minerals." At the same time a gradual obliteration of the chondritic texture is thought to have produced the observed range of distinctness of the texture referred to earlier. Much of this recrystallization is thought to have occurred at temperatures as low as 800°K, and Wood has even estimated that at this temperature about 10^7 yr would be required.[15]

It should be mentioned that even in Wood's own opinion much of this scheme "taxes credibility." Although support for Wood's version of metamorphism has not been unanimous, it has been strongly advocated by Anders (1964) and his associates. The proposed process of metamorphism also forms the basis for a classification scheme proposed by van Schmus and Wood (1967).

A critique

When we examine the data pertinent to the origin of the chondritic meteorites there seems to be no compelling evidence that would enable us to choose confidently between the nebular condensation, volcanic, or impact models just discussed. However, we are inclined by the weight of evidence to favor most a form of the volcanic hypothesis, or at least a model involving a condensed environment as distinguished from the dispersed environment implied in the nebular condensation model. However, apparently the volcanic hypothesis has the

[15] This estimate is however based on the erroneous use of intergranular diffusion coefficients when lattice diffusion coefficients should have been used (Mueller and Olsen, 1967).

advantage of accommodating most easily the multiplicity of observations without seriously contradicting any observation. In particular it is true for the nine major characteristics mentioned earlier.

Of the several versions of the volcanic model, that proposed by Ringwood (1966) is worked out in greatest detail. Its first merit is that it begins with the least differentiated type-I carbonaceous chondrites as raw material. As we have seen this material contains no chondrules but does contain, in the form of carbon compounds, substances that are readily capable of affecting the autoreductive process, which is necessary if Prior's rules are to have an evolutionary significance. Also, as we have seen the absolute magnitudes of the carbon monoxide pressures to be expected over the chondritic mineral assemblages seems adequate for the production of highly explosive volcanism at or below liquidus temperatures. In this way Ringwood's model is more realistic than that of Fish *et al.* (1960), which makes use of gases such as S_2 which do not attain such pressures except at unrealistically high temperatures. However, we take this opportunity to suggest that the theory should be modified to account for the regular distribution of the metal grains previously discussed, a feature whose production seems totally beyond the turbulent magma as proposed by Ringwood. Even if such a magma attained the required homogeneity (which is in itself highly uncertain), we have no reason to expect that this state would be maintained when the quenched erupted material accumulated in the gravitational field. This difficulty can be easily surmounted by calling on the equally familiar volcanic mechanism of vapor transport as discussed earlier. In fact, we should expect the deposits of newly formed silicate chondrules to be precisely the environment where transport and deposition by a vapor would be most effective since these deposits would furnish a hot, highly porous medium. Also, the porosity should be quite uniform so that this feature alone could account for the highly uniform distribution of the metal grains.

However, we cannot exclude the possibility that some metal fragments from previously consolidated material might have been reincorporated in the deposit. It seems possible that some such material provides an explanation for the observations of Urey and Mayeda (1959) regarding fractured metal grains.

It seems to us that deposition from a vapor phase can best account for the presence of plessite and individual crystals of kamacite and taenite rather than the coarse unmixed structures (Widmanstätten structures) so common in slowly cooled meteorites. These observations seem to tell us that separate grains of kamacite and taenite can grow directly from a vapor phase at temperatures so low that large-scale unmixing as implied by the Widmanstätten structure is impossible.

As we implied in our discussion of the kinetic model for liquidus crystallization, a rapid quench such as might occur during explosive volcanism is capable of producing the mineral compositional features of the uniform chondrites if the magma is of constant composition on the scale of the specimen which shows uniformity. Furthermore, despite the contrary but unsupported statement by Anders (1964), volcanic processes with their sometimes chaotic regimen have great versatility and are precisely those most capable of producing disequilibrium chondrites such as Renazzo which contain fractions of diverse origin. In addition, they also provide the best environment for the production of zoned crystals, which characterize the nonuniform chondrites. The reason for this lies in the requirement that such crystals have sufficient time for the zones to develop, and this condition is likely to be satisfied only in large bodies of magma.

Turning now to the question of metamorphism we can immediately point to several important inconsistencies in the scheme proposed by Wood (1963). It has already been pointed out by Keil and Fredriksson (1964) that contrary to Wood's hypothesis there appears to be no connection between the iron content of the silicates of chondrites and the distinctness of the chondritic texture. Even a cursory study shows that some of the most iron-rich silicates occur in well-defined chondrules. Also it is very difficult to account for the simultaneous presence of crystals of uniform composition and glass. This is especially true when we consider in detail the process of homogenization outlined by Wood (1963). Although he does not state so explicitly, the process he envisions must involve a reaction of the following type:

$$\underset{\textit{ferrite}}{Fe_3O_4} + \underset{\textit{metal}}{Fe} + \underset{\textit{pyroxene}}{4MgSiO_3} \rightleftharpoons \underset{\textit{olivine}}{2Mg_2SiO_4} + \underset{\textit{olivine}}{2Fe_2SiO_4}, \quad (6.p)$$

otherwise it will involve an oxidation or reduction process of an even more complex nature. Now the very least that is involved in the homogenization is a diffusion of Mg and Fe^{2+}, although Wood considered only the diffusion of Fe^{2+}. However Reaction (6.p) shows that in addition there will also be a reaction involving the breakdown of certain phases and the growth of others. It should be clear that such a process cannot be discussed solely in terms of diffusion. It also seems clear that when such a reaction is pervasive the preservation of glass or even microcrystalline quench products is exceedingly unlikely.

Although we must reject the version of metamorphism proposed by Wood, the hypothesis of recrystallization or annealing deserves our most serious attention. As we have noted, our knowledge of the degree to which the chemical characteristics of the uniform chondrites can be explained by recrystallization can come only through the most careful textural and chemical studies, but particularly from temperature-indicating mineral assemblages. It may turn out that the recrystallization we find evidence of was confined to a large degree to a temperature range immediately below the liquidus rather than to some later separate period of "thermal metamoprhism." In the volcanic model this would merely imply postconsolidation annealing.

In any study of this kind it is best not to require too much of any given model since observed deviations might at a later date be attributable to quite unrelated secondary processes. A particularly important illustration of such a secondary process is that of shock—many meteorites show varying degrees of evidence of it. That shock effects may greatly influence such phenomena as gas absorption has been demonstrated by experiments (Fredriksson and deCarli, 1964). Similarly, shock may at times bring about selective fusion of silicates while leaving the associated metal less affected, as was inferred from evidence found in the iron meteorite Kodiakanal (Olsen and Mueller, 1964). If, as we have reason to believe, the meteorites are fragments of larger bodies, we must expect a wide variety of shock effects depending on the magnitude of the stresses and the states of the colliding bodies. Thus it may be that evidence of early shocks that occurred when the bodies were hot were obliterated by annealing processes or remain only as such features as the metallic veins, while other features such as unhealed fractures or shattered crystals are of a much later origin. Perhaps some of the most useful guides in such matters are the comparisons we can make with analogous tectonic features of terrestrial rocks, but only if these comparisons are tempered with considerable caution.

7 The terrestrial planets

Gross physical and chemical characteristics

The terrestrial planets present us with a sequence of widely varying physical environments that determine their chemical and petrologic characteristics.

In recent years the study of planetary atmospheres, surfaces, and interiors has received great impetus not only from earth-based remote sensors but particularly from the planetary probes, which have added a new dimension to these investigations. Especially important have been the Soviet Venera probes, the U.S. Mariner probes to Venus and to Mars, the U.S. Apollo program of lunar landings and the Viking mission to Mars. Of these only the Mars probes have led to drastic revisions of our ideas of the planets in question; in other cases there have been important confirmations of the older remote sensing discoveries and a filling in of rich detail.

Some of the most pertinent planetary data to our purpose and which we present in part in Tables 7.1 and 7.2 are as follows:

1. Planetary dimensions and form as expressed by the density, radius, ellipticity, and surface topography
2. Properties such as the solar distance and rotation rate
3. Magnetic properties
4. Image properties of the surface and atmosphere, such as the albedo, polarization, and direct visual features

5. Surface temperature ranges and the variation of atmospheric and subsurface temperatures
6. Atmospheric and surface chemistry and mineralogy

The characteristics of categories 1 and 3 are closely related to the bulk chemical composition of the body and the degree of differentiation as revealed by the density distribution within the planet (Urey, 1952). The mean solar distance has broad implications with regard to the rotation rate and the available solar energy. The optical properties of the surface and atmosphere are of course related to the chemical composition, but the relationship is usually very complex except in the case of definitive atomic or molecular spectra. Also it might at first seem that the planetary surface temperatures should be simple functions of the mean solar distance, but a moment's reflection should convince one that they really must depend on nearly every category listed and that the relations are very complicated.

The most important physical data to our study are presented in Table 7.1. These data have been selected largely because of their bearing on the chemical and petrogenic processes such as the escape and retention of atmospheric constituents and chemical reactions between molecules and mineral phases. Of critical importance in this regard are the surface and atmospheric temperatures and the gravitational accelerations. The estimated values of surface temperatures of Mercury and Venus have undergone such drastic revisions in recent years as a result of advances in radiotechnology that we have completely changed our ideas about those planets. In the case of Venus these important results have been confirmed by the planetary probes. The decrease in temperature with height and the resulting temperature minima, which play such an important role in Earth's atmosphere, have also been found to occur on Venus and on Mars although in greatly modified form. In this regard the lowest recorded temperature from infrared radiation of the high atmosphere of Venus is about 220°K, whereas for Mars the corresponding value is about 140°K.

The compositions of the planetary atmospheres in terms of surface partial and total pressures are given in Table 7.2, which reflects our very limited knowledge of planets other than Earth. Many of the tabulated values are upper limits or are accurate, at best, to an order of magnitude. Also in reading the table we must keep in mind the different ways in which the values were determined. For example, it is meaningless to compare the indeterminate low value of the oxygen pressure on the Moon with the corresponding value on Venus since the latter is based on complex deductions of the state of oxidation of the surface rocks, whereas the former value, even if known, probably bears no relation to any equilibrium process. About all we can say is that the oxygen pressure on both bodies is likely to be very low indeed, although for very different reasons. Similarly it is possible that the methane partial pressure on Mars (upper limit determined spectroscopically) is less than it is on Venus (upper limit calculated assuming thermodynamic equilibrium with the surface). In the case of Venus we have sometimes preferred values arrived at from the quasiequilibrium model to those obtained by measurement because the latter are not consistent with other atmospheric molecular abundances and because the measurements seem to be of doubtful validity. Thus the oxygen pressures obtained from the Venera probes appear to be many orders of magnitude too high and are inconsistent with quite reliably measured CO_2/CO ratios.

Origin and evolution

The densities and the observable compositional characteristics of the planets and other extraterrestrial bodies such as the meteorites tend to favor the hypothesis that all these bodies are genetically connected with the sun and were not captured at some later date.

An attempt was made by Urey (1952) to estimate the mean densities of the terrestrial planets under conditions of zero pressure. The method used was to estimate the iron content of the planet, assign to this a density of 7.2 (for liquid iron), and then assume that the rest consists of silicates of density 3.3. Although the uncertainties in the results are very large, they seem to indicate that the densest planet is Mercury followed by Venus, Earth, Mars, and the moon in this order except that the values for Venus and Earth are the same. Thus it would seem that there is a tendency for the innermost planets to contain a preponderance of iron or at least a higher proportion of metals and less admixed oxygen, nitrogen, carbon, and sulfur. We shall see that there is much evidence that the latter elements diffused away from the region of the sun while the metals were retained by various means, and that this loss occurred before the planets assumed their present sizes.

Although the chemistry of the condensation of planetary material from the hypothetical solar nebula is amenable to fairly rigorous treatment, the mode of agglomeration of those condensates into "protoplanets" or "planetesimals" can only be guessed at. The problem of the agglomeration of the condensed grains in Keplerian

Table 7.1 General physical data for the terrestrial planets

	Mean solar distance	Solar radiation intensity	Period of revolution (days)	Period of rotation (days)	Mean radius	Mean density	Surface gravity	Visual albedo	Ellipticity	Velocity of escape (km/sec)	Range of surface temp. (°K)
Mercury	0.387	6.68	88	59	0.38	5.5	0.37	0.058	?	4.3	110–625
Venus	0.723	1.91	225	243	0.965	5.06	0.88	0.760	?	10.4	600–800
Earth	1.0	1.0	365	1.0	1.000	5.52	1.00	0.39	0.00337	11.3	225–315
Moon	1.0	1.0	365	27.3	0.273	3.33	0.16	0.072	—	2.4	75–400
Mars	1.524	0.43	687	1.0	0.532	3.96	0.39	0.198	0.0089	5.1	140–315

Data based in part on similar tabulations presented by Kuiper (1952), Urey (1952), and Öpik (1962) but updated to recent results as documented in text references.

Table 7.2 Atmospheric composition in atmospheres of the terrestrial planets

Values are based on references and calculations discussed in the text

	Ar	N_2	O_2	H_2	CO_2	CO	H_2O	CH_4	NH_3	HCl	HF	Total
Mercury	+[a]	+	~ 0[b]	~ 0	?	?	~ 0	~ 0	~ 0	~ 0	~ 0	?
Venus	+	~ 10	$\sim 10^{-25}$	$< 10^{-4}$	~ 90	$\sim 5 \times 10^{-3}$	$< 10^{-1}$	$< 10^{-10}$	$< 10^{-7}$	$\sim 5 \times 10^{-5}$	$\sim 10^{-6}$	~ 100
Earth	0.00934	0.78	0.21	$\sim 10^{-8}$	3.3×10^{-4}	+	$10^{-4} \sim 10^{-1}$	$\sim 10^{-8}$	+	~ 0	~ 0	1.0
Moon	~ 0	~ 0	~ 0	~ 0	~ 0	~ 0	~ 0	~ 0	~ 0	~ 0	~ 0	~ 0
Mars	1.5×10^{-4}	1.5×10^{-4}	$\sim 10^{-5}$	~ 0	7.3×10^{-3}	$\sim 10^{-5}$	$< 10^{-6}$	$< 10^{-4}$	$< 10^{-5}$	?	?	$\sim 7.3 \times 10^{-3}$

[a] + signifies constituent present.
[b] ~ 0 indicates an indeterminate very low value.

orbits has been discussed in some detail by Alfvén (1954). The rate of agglomeration must of course depend on the number density of grains, but it also depends on such factors as the electrostatic or magnetic properties of these grains, which are very difficult to evaluate. We can, however, assume that the first products of the condensation and agglomeration are low-pressure phases since the pressures attained in bodies from which heavy gases could escape could not exceed 10^3 atm or so over most of their volumes. Because of this, it might be possible to place some limits on the relative amounts of certain constituents which today comprise the deep interiors of the planets. If, for example, carbon were present in the condensates in its full solar complement relative to oxygen, this element could not have condensed entirely as carbonate but must have assumed in part the form of reduced carbon compounds such as graphite or hydrocarbons. However, we have no evidence of great quantities of carbon in the terrestrial planets and we shall in fact take the opposing view that most of the carbon was transported away before the condensation was complete. Given the densities and most plausible compositions of the terrestrial planets, it is necessary to account for the absence of great quantities of N, C, and S as well as the inert gases Ne, A, Kr, and Xe. The deficiency of the inert gases relative to their cosmic or solar abundance are of course known only for Earth's atmosphere and was first pointed out by Suess (1949).

The escape from the vicinity of the sun of volatile constituents can easily be accounted for if these volatiles are not bound by the gravitational field of some massive body. The escape of neutral constituents of planetary atmospheres is a problem in gas kinetics that has been discussed in detail by a number of authors, but most particularly by Spitzer (1952). The discussion is in terms of a "critical level," which is defined as that level of the upper atmosphere beyond which particles with the escape velocity or greater undergo too few collisions to be deflected back into the atmosphere. The total escape rate L for the entire planet is given by the expression:

$$L = 4\pi R_c^2 \frac{n_c}{4} \int_{v_c}^{\infty} v p(v) dv \tag{7.1}$$

in which v_c is the velocity of escape at the critical level, n_c is the number density of the escaping particle at this level, R_c is the distance of the critical level from the planet's center, and $p(v)$ is the Maxwellian velocity distribution function

$$p(v) = \frac{4}{\sqrt{\pi}} \left(\frac{2KT}{m}\right)^{-3/2} v^2 \exp\left(\frac{-mv^2}{2kT}\right) \tag{7.2}$$

We note that in addition to the temperatures and the particle mass m, L is critically dependent on n_c the concentration of the escaping particle. It was shown by Spitzer that on a planet the size of Earth or Venus the major portion of H and He present in the atmosphere would escape during a time corresponding to that geologically available but that atoms as massive as O and N could not escape. However, the latter atoms can escape from Mars and Mercury. It is also necessary to point out that in order to account for the known helium content of our atmosphere, which indicates that this constituent must escape, it is necessary to postulate temperatures of at least 1500°K in the critical layer.

Since the escape times calculated from the gas kinetic theory refer only to the fractional escape of gases already in the atmosphere, they are inapplicable to constituents that must be volatilized before they can escape. Thus, although the greater part of the atmospheric hydrogen should have escaped from Earth, only a very small quantity of hydrogen is ever present in the atmosphere, so that n_c and the absolute escape rate are both very small. Most visible hydrogen on Earth occurs as water and this is obviously escaping at a low rate or is being rapidly replaced from the interior. Because of the small quantity of H and H_2 in the atmosphere, most hydrogen is transferred to the escape layer as H_2O, which is subsequently dissociated by ultraviolet radiation. Thus the rate of transfer of H_2O molecules from the lower atmosphere becomes the controlling factor in the absolute rate of escape of hydrogen. Reviews of this problem have been presented by Kuiper (1952) and by Urey (1959). The conclusion of these authors is that the upward movement of water is greatly retarded by the temperature minimum of the tropopause because the vapor pressure over ice in this cold region is so low. The same situation should hold to varying degrees on other planets with temperature minima, and we may conclude that in general the escape rate of certain constituents will be highly dependent on the temperature profile of the entire atmosphere and not simply on the temperature of the escape level.

As was noted by Suess (1949), the degree of depletion of the inert gases Ne, Ar, Kr, and Xe in Earth's atmosphere is a function of the position in the periodic table. Suess (1949) attempted to relate the depletion to the atomic mass numbers but found that improbably high temperatures were required for the escape of these heavy constituents from a body the size of Earth. More recently, Jokipii (1964) has shown that the degree of depletion of these constitutents may also be related to the first ionization potentials. According to Jokipii, "ambipolar" or differential diffusion of ionized and neutral constituents is a magnetic field, as in the theories of Alfvén (1954) and of Hoyle (1960) could account for the differential depletions. Those constituents that have high ionization potentials, such as H, He, and most other noble gases, would tend to flow away from the region of the sun, while the elements with lower potentials would be retained by the solar magnetic field and later condense to form the terrestrial planets. Thus the escape of many heavy constituents is explained.

From our knowledge of the mean densities of the terrestrial planets we conclude that these bodies retained considerable oxygen. The ionization potential of atomic oxygen, which should be the most abundant oxygen species today in interplanetary space, is 13.5 eV, which is nearly identical to that of the H atom. From this we might at first conclude that oxygen should have diffused away with the other constituents of high ionization potential if ambipolar diffusion had been important. However, oxygen has a great affinity for nearly all the metals, and for carbon as well, so that it appears that much oxygen was condensed as oxides and silicates and thus was retained in the vicinity of the inner planets. However, the high density of Mercury may be accounted for by assuming that the condensates that ultimately formed this planet suffered considerable oxygen ablation at an early date. Also it is possible that the outward diffusion of oxygen in the Mercury region was more effective relative to the metals because the degree of ionization of the latter was higher in this high-temperature region.

The ionization potential of C atoms is 11.2 eV, which is also high relative to the metals. However the outward transport of this element was probably more closely related to its tendency to form light volatiles such as CH_4 and CH, which are stable over a wide range of conditions in a gas of solar composition, as we have seen in Chapter 5. The same argument to a lesser degree should also apply to sulfur, which should have been present as S, HS, and H_2S.

Whether the fractionation was affected by differential diffusion of neutral components or whether ionization played a role depends on the temperature of the sun and the opacity of the nebular gas. If the sun was lower in temperature than it is today, little or no ionization might have occurred and some way of separating the neutral particles would have been required. If, on the other hand, the sun was hotter than it is today and if in addition it possessed an adequate magnetic field, ambipolar diffusion was probably important.

Any of the mechanisms discussed above might have been capable of bringing about a high degree of fractionation in the nebula before or at the time of planet formation. Then the initial processes of consolidation and differentiation should have brought about further changes in the primordial atmospheres of the planets. This would have been all the more true if the newly formed planet passed through a high-temperature stage. If this had occurred there would have been a rapid reaction of the primordial atmospheric gases with the materials comprising the surface and that part of the deep lithosphere accessible to the atmospheric gases so that the atmospheric composition would have been in harmony with these materials. Because the atmospheric mass of a terrestrial planet is always small compared to the mass of the condensed sphere, profound changes in the atmosphere may correspond to only minute emission or absorption of gases by the lithosphere. Knowing this, we can immediately eliminate from discussion such proposals as that of Cameron (1963) that the atmosphere of Venus is a trapped remnant of the solar nebula. Our detailed discussion of this planet will further clarify this point.

The two major sources of the chemical potential gradients which drive the processes of planetary differentiation should be the gravitational field and the temperature gradient. The effect of the gravitational field on the distribution of components of a solution under isothermal conditions was discussed in an elementary way in Chapter 1 where it was shown that complications arising out of nonideal behavior of solutions can greatly modify the distribution with height expected in an ideal solution of heavy and light components. The mechanical aspect of thermodynamic equilibrium in the gravitational field demands that the density of a phase (or phases in a polyphase system) must always decrease upward. However this requirement can still be met if a very dense solution component, which is present in only trace amounts, increases in concentration upward. This may happen if the dense component has a great affinity for a less dense major component. This provides us with an explanation of the high concentration of certain heavy elements such as uranium in the Earth's crust, as was first pointed out by Brewer (1951).

It is obvious that a planetary atmosphere meets the requirement of density stratification since it is the lightest phase[1] we can imagine. Similarly the crustal rocks are composed of dominantly low-density phases such as the feldspars. However there are also elements and phases in the crust that cannot be justified on the basis of gravitational equilibrium. In fact, much of the subject matter of petrology is concerned with these materials. For

[1] Properly the classical definition of a phase looses much of its meaning in a gravitational field; thus we use the term here in a somewhat looser sense.

example, if mechanical equilibrium were attained in the crust, basaltic lavas should not remain on the surface of a continent where they usually overlie lighter rocks. Most of such disequilibrium effects are traceable to the interplay of the temperature gradient and the release of stresses in the comparatively cool crustal rocks. Zones of localized fractures develop in the vicinity of preexisting inhomogenieties where the stresses concentrate, and these regions become the sites of extensive modification of the rocks through metamorphism and magmatic injection. Frequently it appears that the pressures on basic magmas in such regions exceed the hydrostatic head and they are intruded or extruded at higher than normal levels. Also we have seen in Chapter 2 that certain mineral constituents are volatilized in the proper chemical environment and transported down the temperature gradient to be redeposited in a cooler region. Undoubtedly such activity has been very prevalent in crustal rocks of all the terrestrial planets. This is of course in agreement with the classical mechanisms proposed by many field geologists to account for widespread metasomatic alterations in metamorphic terrains.

In general we should expect that processes in planetary bodies are tending toward equilibrium, although many deviations may occur. This tendancy is well illustrated by the theory of isostacy, which treats of bouyant forces operating throughout the vertical extent of the crust. According to this theory, lighter masses of rock float on a denser substratum like icebergs so that the bodies with the greatest vertical dimension stand highest. This is obviously the mechanical response to the same chemical potential gradients which give rise to the density stratification in the gravitational field. However here the response is very imperfect and can be effective only on masses of large dimensions since the crust is too strong to allow small-scale adjustments. Also it is frequently forgotten that the very existence of isostatic conditions depends on the high viscosity or strength of crustal materials. Without this strength the isostatically high masses would flow laterally to form a uniform layer around the planet.

The existence of buoyancy forces on Earth is shown by many ways in addition to isostatic adjustments. The role of these forces in the evolution of major planetary features and in the origin of certain plutonic bodies has been the subject of papers by van Bemmelen (1933), Beloussov (1960), and Ramberg (1963, 1964). According to Ramberg (1964) a sialic crust will develop in either the solid or molten Earth models, and this crust should originally be a uniform shell. The observed present segregation of continental and oceanic areas is derived from this early state by basic volcanism. Some of the early magmas formed below or within the primitive sialic crust are buoyant with respect to the solid sial, but when they are extruded above the latter and crystallize they become too dense and begin to sink. In this process the sial is squeezed out laterally and forms the nuclei of continents, while the subsided region becomes an ocean basin. Thereafter the low-lying basins, which are comprised of solid basic or simatic material, continue to attract basic volcanism because the basic lavas are lighter than their solid counterparts. On the other hand, the high-standing sialic continents largely prevent the extrusion of basic lavas on their surfaces because they are less dense on the average than the basic magmas. By these mechanisms the continents and ocean basins are preserved and continue to grow.

These considerations of buoyancy effects in planetary processes must also be seen in relation to such processes as seafloor spreading, in which the entire seafloor is regarded as having formed by the successive intrusions of basaltic dikes in midoceanic ridges. The whole is thought to represent a gigantic convection cell in which the decending current "subducts" or drags down the continental margin or island arcs. In a recent review, Lowman (1973) has referred to the growth of ocean basins as "oceanization," and has identified it as a major process in the evolution of planetary surfaces.

Chemistry and petrology of Venus

General

Venus is frequently called the sister planet of Earth because of the similarities in density and diameter from which we infer a gross similarity in chemical composition. Venus presents many particularly interesting problems from a petrologic standpoint since its high surface and crustal temperatures should favor chemical reaction times similar to those of laboratory experiments in silicate phase equilibrium. In a sense also Venus is another "Earth experiment" conducted at a higher temperature. The implications for planetory evolutionary theories are obvious and important.

Because of inferred similarities in scale and composition it is natural to assume that Venus evolved in a way similar to Earth and that analogous crustal rock have been derived by differentiation of the mantles of the two planets. It is likely that the analogy stops here, since, as we shall see, the high temperatures on Venus should lead to detailed differences in the lithospheres of the two planets. Also the magnetic field of Venus is very weak, a fact that may be closely related to the observed low rate of rotation.

The high Venus temperatures should favor the establishment of thermochemical equilibrium, at least in local regions, between the atmospheric gases and the minerals of the lithosphere. The postulation of such a "quasi-equilibrium state" enables us to draw far-reaching conclusions about the lower atmosphere and the lithosphere since the major feature of this condition should be widespread and pervasive chemical reactions between the atmospheric molecules and the crustal minerals, so that we may in principle determine the stability ranges of the latter from the atmospheric composition. So far this idea is little more than a hypothesis, referred to as the "interaction model," although favorable evidence is beginning to accumulate.

The critical question of course is whether or not the kinetic properties of the minerals and gases are favorable for the required approximation to equilibrium under Venus conditions. As far as the homogeneous gas reactions are concerned, we seem to be on safe ground, especially in the presence of such common crustal oxide catalysts as Fe_2O_3 and TiO_2, since there is much experimental evidence that reactions involving such species as CO and O_2 proceed rapidly in the temperature range of 400°K to 700°K in the presence of these oxides. However, in the case of the more complex heterogeneous reactions under the large-scale planetary processes involved, equilibrium may be only crudely approximated for the following reasons:

1. A finite time is required for reaction to occur, and this may be long in comparision with perturbing processes in some cases. This is especially true of mass transfer perturbations such as volcanic eruptions.
2. Chemical reaction rates as well as the equilibrium temperatures and pressures should vary throughout the lithospheric and atmospheric reaction zones.
3. Surface temperature gradients, especially in the radial direction, should give rise to chemical mass transport of certain chemical constituents along these gradients.
4. Reactants must be in chemical communication for reaction to occur.

A consideration of possible differentiation mechanisms within the lithosphere of Venus (Mueller, 1969b) shows that there should be a close relation between the chemical compositions of the crust and mantle and the high temperatures. The high temperatures will not only influence the thermodynamic equilibrium and the rates of reactions but also the nature of the material that is available for reaction at each stage of evolution and at each level. These high temperatures should exert a strong influence on the time scale of the differentiation processes themselves as well as the degree to which hydrostatic equilibrium is approached at each level. The differenentiation process, which involves large scale mass transfer, would in part be governed by the kinetics of viscous flow phenomena [as represented by Equations (3.50)–(3.51)], which determine the viscosities of both silicate melts and solid rocks. One consequence, to be discussed later, should be the suppression of topographic relief on Venus relative to that on Earth since high temperatures favor a weak crust and attentuated mountain roots. For mass transport, the high temperatures should favor short relaxation times, which in turn lead to a closer approach to chemical equilibrium of a particular mineral phase in the gravitational field and consequently more complete density stratification than on Earth. However, superimposed on this tendency toward equilibrium, and frequently running counter to it, will be the effect of chemical and physical transport along the temperature gradient. One such effect involving fluid flow, is the upward movement of magmas to form subsurface intrusions and extrusive lava flows. On cooling these magmas form bodies of rock that might be denser than underlying crustal rocks, and gravitational instability could result. Because of the general evenness of the Venus crust this might result in topographic lows at the site of basic volcanism. Also the vapor transport of mineral phases along the radial temperature gradient as described in Chapter 3 could result in the deposition of select dense mineral phases at shallow crustal levels. We shall return to this subject in our discussion of comparative planetology.

Existing data on the stability fields of the most common rock-forming minerals that occur in Earth's lithosphere lead us to believe that many of these same minerals will also be stable in the upper lithosphere of Venus. However they are likely to be concentrated at somewhat different levels because of the high temperatures. We should expect the shallow crustal layers to be composed dominantly of alkali feldspars (densities 2.57–2.61) and quartz (density 2.65).[2] At somewhat lower

[2] Gamma ray spectra obtained from the Venus surface rocks by the Soviet spacecraft *Venera* 8 indicate the weight percents of the following elements: K, 4%; U, 2.2×10^{-4}%; Th, 6.5×10^{-4}% with an estimated error of 30% (Vinogradov *et al.*, 1973). This analysis implies a highly differentiated crust.

levels we can expect such phases as anorthite feldspar (density 2.71), dolomite (density 2.87), and calcite (density 2.70), while at yet greater depths concentrations of ferromagnesian silicates such as pyroxenes, amphiboles, and calc-silicates as wollastonite, for example, with densities greater than 3, should be favored. To the extent that density stratification is realized, there should thus be a tendency to separate mutually reactive minerals of unlike density. Thus, for example, it might happen that the reactive carbonate is in part segregated from free silica in the form of quartz. The result would be a spatial constraint on the achievement of the quasiequilibrium state with respect to the given reaction.

Figure 7.1 shows a diagrammatic cross-section of what we define as the hypothetical "interaction shell" of the Venus lithosphere. Interaction with the atmosphere and the production and absorption of gases are postulated to occur within this shell of indeterminate thickness *S*-*S*′. The interaction shell is pierced by feeder channels of volcanos that periodically contribute gases from deeper, hotter regions of the lithosphere. Presumably the extrusive lavas would rapidly solidify and begin to seek a level compatible with their density. We shall see in this regard that high-standing volcanic shields are practically excluded on Venus and that shallow depressions associated with centers of basic volcanism are to be expected.

7.1 Diagrammatic cross-section of a hypothetical Venus surface region showing details of the interaction shell of thickness *S*-*S*′. The symbols are discussed in the text. The figure is not to scale.

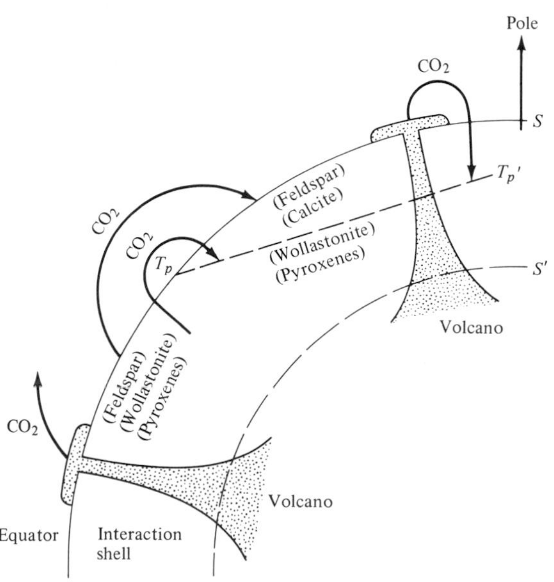

An important feature of the interaction shell is the surface represented by the dashed line T_p-T'_p. Along such a surface a given univariant gas producing or absorbing reaction should be in equilibrium. If we ignore small effects on the equilibrium by pressure acting on the solid reactants and products, this surface defines a virtual isotherm, which is the "effective temperature" of the reaction. At lower latitudes and greater depths than T_p-T'_p the gaseous constituent is emitted, whereas at higher latitudes and shallower depths it is absorbed.

CO_2-producing reactions on Venus

Although the atmosphere of Venus is consistent with the quasiequilibrium interaction model, the precise reactions involved are not yet identified. Since CO_2 is the most abundant atmospheric constituent (Table 7.2), it is particularly interesting to consider how it might be produced and what the time scale of the likely reactions are.

Many of the CO_2-producing reactions of possible importance on Venus are discussed elsewhere in this book in connection with terrestrial metamorphic processes. Broadly these reactions fall into three classes as follows:

1. Reactions between carbonates and quartz
2. Reactions between carbonates and silicates other than quartz
3. Reactions involving water in addition to CO_2

The reactions

$$\underset{\textit{calcite}}{CaCO_3} + \underset{\textit{quartz}}{SiO_2} \rightleftharpoons \underset{\textit{wollastonite}}{CaSiO_3} + \underset{\textit{gas}}{CO_2} \qquad (7.a)$$

and

$$\underset{\textit{orthopyroxene}}{MgSiO_3} + \underset{\textit{calcite}}{CaCO_3} + \underset{\textit{quartz}}{SiO_2} \rightleftharpoons \underset{\textit{Ca-pyroxene}}{CaMgSi_2O_6} + \underset{\textit{gas}}{CO_2} \qquad (7.b)$$

are illustrative of the first class and are particularly interesting because they are important in terrestrial metamorphic processes. Also, certain kinetic and thermodynamic data are available for them.

The second class of reaction is represented by examples such as the following:

$$\underset{\textit{Ca-pyroxene}}{CaMgSi_2O_6} + \underset{\textit{calcite}}{CaCO_3} \rightleftharpoons \underset{\textit{akermanite}}{Ca_2MgSi_2O_7} + \underset{\textit{gas}}{CO_2} \qquad (7.c)$$

This type of reaction may be excluded from further consideration because the equilibrium pressures are too low to be consistent with observed Venus temperatures and the CO_2 pressure.

An example of the third class of reaction is

$$\underset{actinolite}{Ca_2Mg_5Si_8O_{22}(OH)_2} + \underset{calcite}{3CaCO_3} + \underset{quartz}{2SiO_2} \rightleftharpoons \underset{Ca\text{-}pyroxene}{5CaMgSi_2O_6} + \underset{gas}{3CO_2} + \underset{gas}{H_2O} \quad (7.d)$$

Although such reactions may be important on Venus, their consideration at this point would needlessly complicate our analysis.

Figure 7.2 shows the reaction curves for Reactions (7.a) and (7.b) as calculated from thermochemical data and as inferred from terrestrial rocks (Mueller, 1966). The possible range of Venus temperatures and CO_2 pressures are shown as the large cross. The same data for Earth's atmosphere are shown for comparision.

Kinetics of CO_2 production

Complex mineral reactions such as those assumed to be applicable in CO_2 production are likely to involve several distinct steps and mechanisms. In addition the several chemical constituents must diffuse toward and away from the site of reaction. Initially also nucleation must occur. Each of these steps must be characterized by its own rate constant, activation energy, and temperature coefficients. As we have seen in Chapter 1, these parameters may best be evaluated by utilizing the theory of absolute reaction rates.

If we first consider only the elements of chemical decomposition and combination, then for Reaction (7.a) we may infer that there will be one or more steps involving such processes as the decomposition of carbonate and the combination of Ca, O, Si, CaO, SiO_2, etc., as discussed later in a different context in Chapter 8. By logic the overall rate of a stepwise reaction will be governed by the slowest step. However since the overall reaction (7.a) is endothermic, a limit is placed on the enthalphy of activation ΔH^* and consequently on E^*, the experimental activation energy, by the endothermic enthalpy of reaction. At 700°K and $P = 1$ atm, the latter

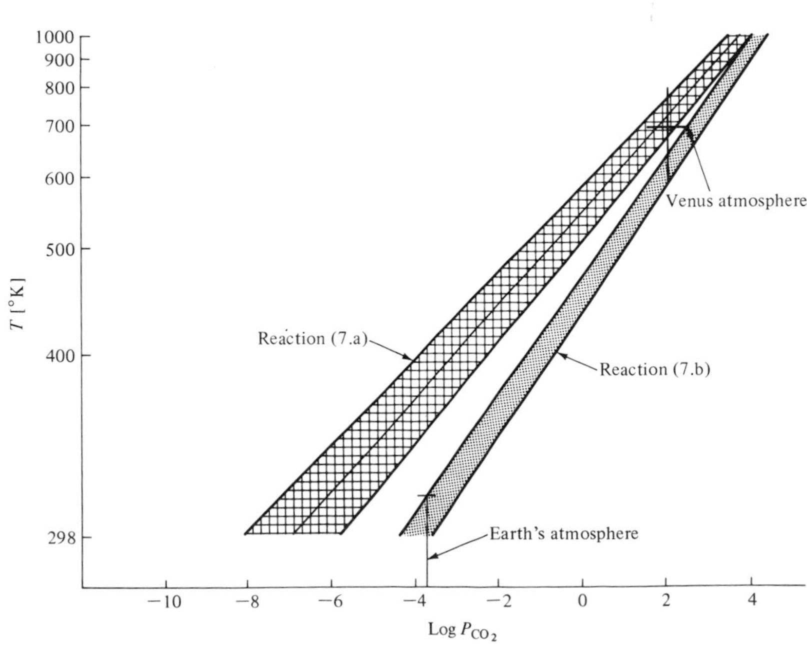

7.2 Comparison of observational data (cross) for the CO_2 pressure in the atmosphere of Venus in the observed temperature range with equilibrium curves for the important CO_2-producing reactions (7.a) and (7.b) discussed in the text. The bandwidths indicate uncertainties in the thermochemical data. The CO_2 pressure for Earth's atmosphere is shown for comparison.

is 21,258 cal/mole, and at 1000°K it is 20,424 cal/mole. It follows from a rather simple relation that $E^*(700°K) \geq$ 21,258 cal/mole and $E^*(1000°K) \geq$ 20,424 cal/mole for the slowest step of the purely chemical part of the wollastonite reaction (7.a). Essentially the same values hold also for the slightly higher Venus pressures (100 atm).

The experimental study of the wollastonite reaction (Kridelbaugh, 1973) provides data for comparison with the values deduced above. The results indicate that under anhydrous conditions the reaction is diffusion-controlled (compare with Chapter 8) with an experimental activation energy of about 22,000 cal/mole when reduced to Venus pressures.

Another important aspect of the kinetics is the role of water, which appears to be present in trace amounts on Venus (Table 7.2). It is likely that water would increase the reaction rates of both the diffusive and chemical steps but it will not decrease the overall rate of the reaction by lowering the activation energy barrier below the limit imposed by the endothermic heat of reaction. Consequently we may infer that Reaction (7.a) probably has an activation energy of at least 20,000 cal/mole or so even under hydrous conditions.

In addition to decarbonation there will also be the reverse or carbonation reactions, and it seems likely that the latter is at least as rapid as the former (Mueller and Kridelbaugh, 1973). The rate of this reaction is important in defining the regions of CO_2 absorption as shown in Figure 7.1.

If, as we have deduced above, the activation energy of Reaction (7.a) is approximately 20,000 cal/mole, then we may compute the time scale of the reaction under Venus conditions. Since under laboratory conditions in the temperature range of 1100°K, substantial reaction occurs in 300 hr or less, we may establish how long it will take under Venus surface conditions ($T \simeq 700°K$) for a similar degree of reaction to occur.

The general relation for the temperature dependence of the rate constant K is

$$\frac{\partial \ln K}{\partial T} = \frac{E^*}{RT^2} \tag{7.1}$$

If this equation is integrated between 1100°K and 700°K we obtain

$$\log \frac{K(700)}{K(1100)} = -2.27$$

Thus under Venus surface conditions the decarbonation reaction to wollastonite proceeds at a rate that is a factor of $10^{-2.27}$ less than at 1100°K. Then a degree of reaction which would require a year at 1100°K would require some hundreds of years under Venus surface conditions. Of course at greater depths within the interaction shell this time scale would be further reduced. We then arrive at the conclusion that the rate of exchange of CO_2 between the Venus atmosphere and lithosphere proceeds at a rapid rate relative to the geologic time scale.

Water on Venus

Although the precise quantity of water on Venus is unknown, the amount present in the atmosphere appears to be very low (Table 7.2). Also the high atmospheric temperatures probably preclude a cold trap as effective as that on Earth. Furthermore we must expect that both the photochemical decomposition of water molecules and the escape of hydrogen from the top of the atmosphere are at least as rapid as these processes on Earth. It cannot however be concluded from this that the low water content of the atmosphere is necessarily due to these processes, since for all we know, the protoplanet of Venus was dryer than Earth's protoplanet to start with.

We must also consider the possibility that the rate-controlling step of the dehydrogenation of Venus is not, as on Earth, the movement of water through the cold trap but rather the escape from the lithospheric layers. It is possible, for example, that the release of water occurs through such reactions as the following:

$$\underset{\textit{anthophyllite}}{Mg_7Si_8O_{22}(OH)_2} \rightleftharpoons \underset{\textit{orthopyroxene}}{7MgSiO_3} + \underset{\textit{quartz}}{SiO_2} + \underset{\textit{gas}}{H_2O} \tag{7.e}$$

$$\underset{\textit{actinolite}}{Ca_2Mg_5Si_8O_{22}(OH)_2} \rightleftharpoons \underset{\textit{Ca-pyroxene}}{2CaMgSi_2O_6} + \underset{\textit{orthopyroxene}}{3MgSiO_3} + \underset{\textit{quartz}}{SiO_2} + \underset{\textit{gas}}{H_2O} \tag{7.f}$$

$$\underset{\textit{biotite}}{KMg_3AlSi_3O_{10}(OH)_2} + \underset{\textit{quartz}}{3SiO_2} \rightleftharpoons \underset{\textit{K-feldspar}}{KAlSi_3O_8} + \underset{\textit{orthopyroxene}}{3MgSiO_3} + \underset{\textit{gas}}{H_2O} \tag{7.g}$$

Although few data exist on the kinetics of such water-producing reactions we have seen in Chapter 3 that the rate of Reaction (7.e) has been studied by Greenwood (1963).

In general we may write for the time rate of change of the volume concentration M of the hydrous mineral

$$-\frac{dM}{dt} = KM - f(T, P_{H_2O}) \tag{7.2}$$

where K is a rate constant of decomposition and f is an undetermined function of temperature and water pressure. While the precise form of f must be determined from experiment, it frequently happens that it is equal to $K'P_{H_2O}$, where K' is another rate constant. However if the temperature is high and P_{H_2O}, the ambient water pressure, is sufficiently low, f will be small relative to KM, which implies there is no back reaction. If this is the experimental situation—as it was in Greenwood's experiments—then E^* may be evaluated. For Reaction (7.e) this was found to be 147,000 calories per mole. As a consequence we find that at the Venusian surface temperature, no significant dehydration of anthophyllite could occur during all geologic time. However at 800°K, $1/K$ equals 10^9 yr and at 900°K it is only 10^4 yr. Since the latter temperature is likely to be encountered at only a few tens of kilometers or less, the kinetics of Venus favor the release of water at moderate depths. If now the escape of hydrogen from the atmosphere is sufficiently slow, f in Equation (7.2), as determined by the return flux of water into the lithosphere, might be only slightly smaller than KM. These conditions would correspond to a quasiequilibrium between the hydrous minerals and the lower atmospheric H_2O molecules but with a constant small leakage of hydrogen through the upper atmosphere to define a steady state.

Oxidation state of Venus

Spectroscopic observations indicate that the atmospheric ratio P_{CO}/P_{CO_2} for Venus is approximately $10^{-4.29}$ (Young, 1972), which corresponds to a very low equilibrium oxygen pressure at the ambient surface temperature (Table 7.2). However it is possible, given the uncertainties in the effective temperature of the quasi-equilibrium and in the thermochemical data, that P_{O_2} may fall nearer the hematite-magnetite boundary. However even these conditions would not lead to oxygen pressures greatly in excess of 10^{-25} atm.

Since the stabilities of many natural compounds are limited by oxygen pressure, it is possible to calculate whether they are likely to occur on Venus. Thus from the reaction

$$\underset{gas}{CO_2} + \underset{gas}{H_2} \rightleftharpoons \underset{gas}{CO} + \underset{gas}{H_2O} \tag{7.h}$$

and a procedure analogous to that used in Chapter 6 we may show that for an upper limit of $P_{H_2O} = 10^{-1}$ atm, $(P_{CO}/P_{CO_2}) = 10^{-4.29}$, and $T = 700$°K, then the partial pressure of hydrogen $P_{H_2} = 10^{-4.34}$ atm. Similarly by utilizing such reactions as (5.b) involving methane or CH_4 we may show that the partial pressure of this constituent must be of the order of 10^{-10} atm or less on the surface of Venus.

By an extension of this method we can determine the upper limits of all the higher paraffin polymers in the gas phase, and this analysis can be extended to other hydrocarbon series as well. For example, once we have the limiting value for the paraffin *n*-octane, we may obtain the related value for ethylcyclohexane from the reaction

$$\underset{gas}{C_8H_{18}} \rightleftharpoons \underset{gas}{C_8H_{16}} + \underset{gas}{H_2} \tag{7.i}$$

The equilibrium constant for this reaction at 700°K is $10^{-0.15}$, which indicate that at low hydrogen pressures $P_{C_8H_{16}} > P_{C_8H_{18}}$. However, such low hydrogen pressures also require that $P_{C_8H_{18}}$ have a very low value and consequently the quantity of ethylcyclohexane will always remain small. In this way we conclude that hydrocarbons are probably present in only trace amounts on Venus.

By means of reactions such as (5.c) and (5.d) which involve graphite it may also be shown that the atmosphere is too oxidizing for this substance to exist in the surface rocks. This result would seem to virtually eliminate the possibility of the occurrence of asphaltic or hydrocarbon substance of any kind, which is of some interest in view of the numerous proposals that these substances should be abundant on Venus. It is easy to show that the same conclusion should also hold for many other hydrogen compounds, as, for example, the important precurser and product of organisms, ammonia. This molecule is related to its components by the reaction

$$\tfrac{1}{2}\underset{gas}{N_2} + \tfrac{3}{2}\underset{gas}{H_2} \rightleftharpoons \underset{gas}{NH_3} \tag{7.j}$$

If we introduce the value $K_{(j)} = 10^{-2.03}$ ($T = 700$°K) for the equilibrium constant for this reaction and if we assume our previously derived value of the hydrogen pressure, we get the limiting relation

$$P_{NH_3} \leq 10^{-2.03}(10^{-4.34})^{3/2}P_{N_2}^{1/2} \tag{7.3}$$

It is clear that with $P_{N_2} \leq 10$ atm (Table 7.2) as inferred for Venus, ammonia can be present only in very minor amounts.

It seems likely that the oxidation state of Venus is controlled primarily by iron compounds of FeO and Fe_2O_3 in the crust since by analogy with Earth these should be very abundant. It seems probable that most of the original oxygen bound as water on Venus has reacted with the rocks long ago and that the hydrogen has escaped. If Venus had started with as much water as Earth, the crust might be highly oxidized to great depths. This may be seen by assuming that the 280 kg/cm² of water estimated to be present on Earth's surface (Goldschmidt, 1933) all reacts with the FeO in basalt as follows:

$$FeO + \frac{1}{2}H_2O \rightleftharpoons \frac{1}{2}Fe_2O_3 + \frac{1}{2}H_2$$

If we assume that each gram of basalt of density 3 gm/cm^3 contains 0.1 gm of FeO, then 280 kg of water is enough to completely oxidize a layer of basalt to a depth of 75 km. If less water than this was originally present on Venus, P_{O_2} may not be high enough for the appearance of hematite as such. If this is the case we must look to reactions of the type (5.s), which yield expressions in which P_{O_2} appears as a multivariant function of complex solid solutions. Thus P_{O_2} might be governed by the movement of FeO into crustal rocks from greater depths. If this movement is restricted, most ferromagnesian minerals will tend to be rich in magnesium, corresponding to a highly oxidized crust.

Although the dominance of Venusian oxidative equilibria by iron compounds relegates all other oxidizable species to a secondary role, the latter will serve by their atmospheric abundances to define the state of oxidation more precisely. In this regard one of the most important elements is sulfur. Sulfur is not abundant in Earth's upper lithosphere, in this respect falling behind titanium, manganese, and phosphorus, although it is very abundant in the cosmos. By analogy with Earth, we should expect sulfur to occur in the Venus lithosphere chiefly as iron sulfide or as sulfates, depending on the local composition. As has been shown elsewhere (Mueller, 1965; Lewis, 1968) the dominant gaseous sulfur species in Venus' atmosphere should be COS and SO_2, although neither have as yet been detected.

Halogen compounds on Venus

One of the most startling discoveries about Venus was that of Connes *et al.* (1967) that the atmosphere contained HCl and HF molecules in the approximate ratios

$$\frac{P_{HCl}}{P_{CO_2}} = 10^{-6.22} \qquad \frac{P_{HF}}{P_{CO_2}} = 10^{-8.30}$$

In order to see whether these observed ratios make sense in terms of the interaction model we may utilize reactions of the type:

$$\underset{halite}{2NaCl} + \underset{andalusite}{Al_2SiO_5} + \underset{quartz}{5SiO_2} + \underset{gas}{H_2O} \rightleftharpoons \underset{plagioclase}{2NaAlSi_3O_8} + \underset{gas}{2HCl} \quad (7.k)$$

$$\underset{fluorite}{CaF_2} + \underset{quartz}{SiO_2} + \underset{gas}{H_2O} \rightleftharpoons \underset{wollastonite}{CaSiO_3} + \underset{gas}{2HF} \quad (7.l)$$

Although we do not know if the necessary mineral phases occur on Venus, conditions do not seem to preclude them. In any case, the reactions provide a starting place for our analysis.

Although there are many uncertainties in the calculations, thermochemical data for Reactions (7.k) and (7.l) yield values that are generally consistent with the observed values (Mueller, 1969d) so that we may infer that the observed molecular abundances of HCl and HF result from the interaction of halogen and silicate phases within the lithosphere. The situation in the hot Venus atmosphere is then something analogous to that in volcanic fumaroles where the same halogen molecules are in quasiequilibrium with the underlying magma (Chapter 15).

The lack of water precludes extensive sedimentary salt deposits on Venus for the same reason that sedimentary carbonates are excluded. Under the prevailing hot, dry conditions, combination of NaCl with silicates is favored not only as in Reaction (7.k) but also with plagioclase to form the mineral sodalite. This is supported by the experimental fact that NaCl and plagioclase readily combine above 1000°K to form this mineral.

Chemical zones of the atmosphere

We have thus far discussed in some detail the consequences of the chemical interaction of the lithosphere and lower atmosphere of Venus. However, many of the expected chemical characteristics, such as the apparent low water content and oxidation state, are in part imposed by conditions in the upper atmosphere. Also most of the atmosphere is of middle height and therefore is too cool for the attainment of internal chemical equilibrium (Howard *et al.* 1974). The upper atmospheric photochemical processes referred to in connection with water affect the stabilities of virtually every species present and need to be considered in any detailed study of the atmospheric abundances. The following are some of the most important decompositions and recombinations that affect the abundances:

$$CO_2 + h\nu \longrightarrow CO + O \quad (7.m)$$

$$CO + O + Z \longrightarrow CO_2 + Z \quad (7.n)$$

$$O_2 + h\nu \longrightarrow 2O \quad (7.o)$$

$$2O + Z \longrightarrow O_2 + Z \quad (7.p)$$

$$H_2O + h\nu \longrightarrow OH + H \quad (7.q)$$

In addition to the neutral particles, many ions would also be produced. In these reactions, the symbol Z represents a third particle of any type. Reactions (7.m) to (7.p) have been studied by Shimizu (1963) who calculated the theoretical abundances of O, CO, and O_2 as a function of height in the upper atmosphere of Venus. Shimizu's work shows that if diffusion is not too important the above constituents will accumulate in certain levels of the upper atmosphere. We must therefore consider the possibility that a spectrogram might register greater amounts of these species than would be expected from the thermochemical equilibria in the lower atmosphere. However, it is also possible that exchange with the latter and subsequent thermochemical reaction with the lithosphere occur so rapidly that the photochemical effect would be only a minor one.

From the foregoing observations and theoretical consideration, we are inevitably led to a chemical model for the vertical structure of the Venus atmosphere. In this model (Figure 7.3) the atmosphere is regarded as consisting of three zones with distinct chemical characteristics. Zone I is the zone of thermochemical equilibrium in which the quasiequilibrium abundances are established by reactions with the lithosphere and by homogeneous gas reactions. In zone II the temperature is considerably lower and the reactions of zone I should be effectively quenched. It is this zone that is largely observed in the absorption spectra, so that these should effectively reflect the equilibria of zone I to the extent that quenching occurs. In zone III the previously discussed photochemical reactions dominate, and H and He diffuse upward and escape from the exosphere. To get a more complete picture of Venus, this zonal structure should be considered in conjunction with Figure 7.1.

The atmosphere and surface of Mercury

Mercury is among the least massive of the planets, but its high density and small radius result in a surface gravitational acceleration similar to that of Mars. The intensity of solar radiation (including the solar wind) is about 15 times that in the Mars orbit and should favor a comparatively high degree of photochemical dissociation and ionization of molecules with the attendant recombination reactions. We have already indicated that

7.3 Chemical model of the vertical structure of the Venus atmosphere.

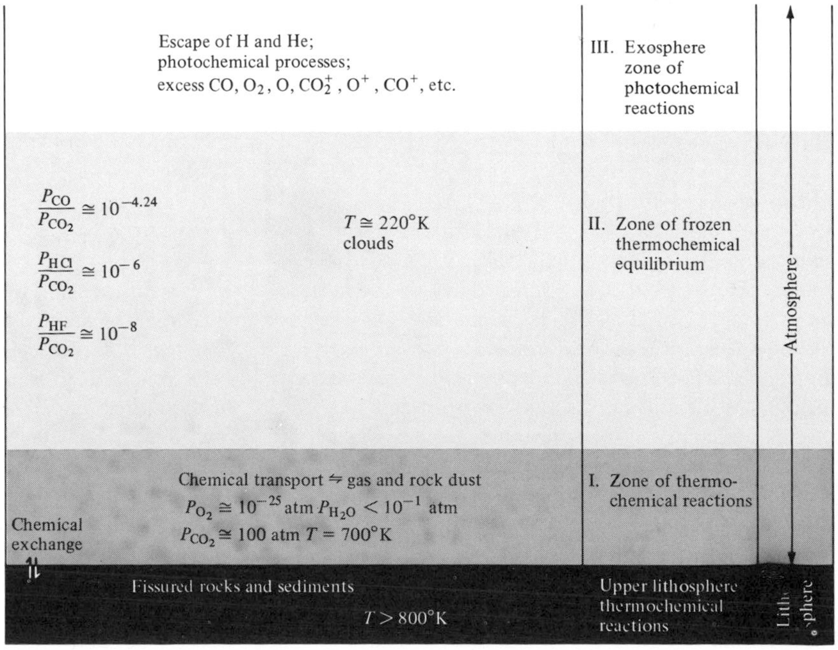

such species as O and N should escape from Mars on the basis of gas kinetics, and we have also referred to the theory of Öpik and Singer (1960), which indicates that heavier species such as CO might escape from that planet should they become ionized. We should then, from every standpoint, expect oxygen to escape from Mercury at a considerably greater rate than from Mars. Thus, if the atmosphere of Mercury is of a density comparable to that of Mars we should expect that the species CO_2, H_2O, and CO would be dissociated throughout much of sunlit atmosphere and that the products of this dissociation would escape at varying rates. This tendency will also be favored by the conclusions to be drawn from the recent observations of Murdock and Ney (1970), which indicate that the dark side of Mercury is at a temperature far above 0°K, presumably because of the transfer of heat to this side by an atmosphere.

If all the oxygen in the atmospheric molecules is ultimately subject to escape, it may be that the surface rocks of Mercury are in state of reduction perhaps even corresponding to meteoritic material. However, by analogy with Venus, it would seem that the upward flux of ferrous iron by endogenic processes would prevent the reduction from progressing much beyond the magnetite-metal boundary. If the degree of reduction does correspond to the metal–magnetite boundary, then at 700°K (which is close to the temperature for rocks a little below the surface on the bright side) $P_{O_2} \sim 10^{-33}$ atm and $P_{CO} \sim P_{CO_2}$, whereas on the metal side of the boundary $P_{CO} > P_{CO_2}$. Also these relations for P_{CO} and P_{CO_2} hold for a wide range of temperatures. If we assume that P_{O_2} has the above value or less we may calculate an upper limit for CO and CO_2. For CO we use the reaction

$$\underset{gas}{CO} \rightleftharpoons \underset{graphite}{C} + \underset{gas}{\tfrac{1}{2}O_2} \qquad (7.r)$$

which has $K_2 = 10^{-13}$ at 700°K. Then if $P_{O_2} \leq 10^{-33}$ atm, we obtain $P_{CO} \leq 10^{-3.5}$ atm. If P_{CO} rises above this value, graphite is depositied. Also, P_{CO_2} will have a comparable or even lower value. Thus, if the surface rocks are as reduced as this, the abundances of gaseous carbon species are severely limited. However, since the atmosphere is probably quite tenuous, they may still be major constituents in relation to other species.

The possibility of other reactive molecules—such as SO_3, SO_2, NO, etc.—being major atmospheric components also appears very remote. Most of these molecules are stable only under rather oxidizing conditions or, in any event, decompose photochemically to lighter species which would escape. However, the N_2 molecule is very resistant to photochemical processes and, in addition, is the thermodynamically stable species over a wide range of conditions. Therefore, it is possible that N_2 might occur with CO in the lower atmosphere where protection of the overlying atmosphere is adequate. Other than these, the most likely constituent of the Mercury atmosphere would appear to be radiogenic argon 40. By analogy with Mars, we might set an upper limit of about 10^{-3} atm for this constituent assuming that the rate of production on the two planets is proportional to their masses and that the rates of outgassing are comparable. Actually we should expect these two factors to counterbalance each other since Mercury probably contains less radiogenic potassium than Mars, but should, because of its higher temperatures, have a higher rate of outgassing.

Thomas (1974) has speculated that H_2O might be the dominant atmospheric constituent on Mercury through its production from solar wind protons. However it is difficult to see how this would be possible in view of its easy photochemical decomposition and escape on this planet.

The paucity of information permits little more speculation about the atmosphere of Mercury. However, the planetary probe Mariner 10 recently returned images (Murry *et al.*, 1974) that show the surface to be cratered in a manner similar to that of the Moon, which is not unexpected in view of the known atmosphere. This information is important however, in evaluating the distribution of planetesimal and meteoritic matter in the early solar system. It is also important relative to the mechanical and chemical properties of the lithosphere since the support of such rough topography requires considerable strength.

Mars

General features

Mars is sufficiently remote from the sun and of sufficiently small mass so that the resultant tenuous atmosphere and remoteness combine to assure a climate characterized by considerable temperature variation (Table 7.1); furthermore, temperatures are, by and large, extremely low. Recent planetary probes culminating in Viking I and II (Owen *et al.*, 1976; Clark *et al.*, 1976; Farmer *et al.*, 1976) indicate that the Martian atmospheric surface pressure is as low as 7 mbar attributable chiefly to CO_2. Constituents such as N_2, Ar, CO, H_2O, O_2, O_3, etc., are present in only trace amounts but are detectable (Table 7.2). The Martian atmosphere, like those of all cooler planets, including Earth, is unquestionably out of chemical equilibrium with the surface rocks. Thus, there is little doubt of a Venus-like relationship in which atmospheric constituents are directly relatable to the lithosphere. However, P_{O_2} is apparently high enough (Table 7.2) to result in oxidation of the surface rocks as indicated by their red coloring (Mutch *et al.*, 1976).

An important characteristic of Mars is its marked observed ellipticity or fractional excess of the equatorial radius over the polar radius. According to some authors, this shape results from convection currents (Runcorn, 1973) or other forms of "dynamic support" which prevent the attainment of hydrostatic equilibrium and which would result in a smaller value of the "dynamic ellipticity" $(C - A)/C$, where C and A are the moments of inertia about the polar and equatorial axes, respectively.

From Table 7.1 we see that the mean density of Mars is approximately 3.96, which corresponds at zero pressure to a density of approximately 3.7 (Urey, 1952). Thus, although there is little evidence for a massive metallic core such as apparently occurs in Mercury, Venus, and Earth, there may be a small core or at least a downward concentration of dense silicates and oxides. In any event, the mean density is sufficiently great to permit an abundance of rocks as dense as terrestrial or lunar basalts.

The prevalence of dust storms on Mars indicates that its atmospheric circulation results in high winds compared with those on Earth; otherwise its tenuous nature would prevent dust from being significantly transported. Analyses of the atmospheric dust, which is swept up by prominent and highly visible storms, have been made by infrared spectrometer (Hanel *et al.*, 1972). They show that the components of this dust, which are thought to consist predominantly of feldspars, have SiO_2 content of about 60 percent. If this dust is surface material it implies that Mars, like the other terrestrial planets, is significantly differentiated.

The surface of Mars is characterized by several classes of features which, thanks to the planetary probes, are now fairly well known. These are:

1. Meteorite impact craters similar to those of the Moon and Mercury
2. Circular basins similar to the lunar mare basins
3. Volcanic shields, domes, and cones similar to terrestrial volcanic structures
4. Structural grabens and other fractures
5. Canyons and "channels"
6. A variety of aeolian deposits including, in particular, the laminated deposits of the polar regions as well as numerous dune fields

In addition to these specific features it has been found possible to divide the Martian surface into the following areas of broad extent (McCauley *et al.*, 1972):

1. Cratered plains
2. Moderate-density cratered terrain
3. Densely cratered terrain
4. Chaotic terrain
5. Grooved terrain
6. Lineated terrain
7. Mountainous terrain
8. Knobby terrain
9. Laminated terrain
10. Pitted terrain

Most of these terms are self-explanatory. The laminated terrain appears to be characteristic of the polar regions. It is interesting, however, that although the laminated terrain appears to be composed of stratified beds, there is no unambiguous evidence of tectonic folding such as is prevalent on Earth. It is likely, however, that some circular structures in the laminated terrain represent either basins or domes truncated by errosion.

Among the most visible features of Mars are the polar ice caps. Although it is not yet known whether this ice consists of CO_2 or H_2O, the polar caps are now thought to consist mostly of water ice (Murry *et al.*, 1972). It is likely that the laminated deposits are related to the mode of deposit of ice in the polar regions.

Surface features as a clue to rock types and chemistry

Although we have referred to the known chemical composition of the atmosphere, the infrared studies of Martian dust clouds, and the presence of CO_2 and H_2O ices, we have very little firm evidence regarding the chemistry, mineralogy, and petrology of the surface rocks and the underlying lithosphere. Yet we may infer certain chemical and petrologic characteristics from the surface features previously listed. For example, from the presence of large shield volcanos such as Nix Olympica (Figure 7.4) it appears basic or at least fluid lavas were at different times extruded onto the surface of Mars (McCauley *et al.*, 1972). This inference is also supported by the numerous arcuate fronts of lava flows, which are clearly visible on the smooth plain surrounding these volcanos and by the many collapsed lava tubes on their flanks. The 600-km diameter of Nix Olympica makes it the largest volcanic edifice known on any planet. The enormous shield exhibits a central crater and a peripheral scarp which faces outward. The latter feature, which is otherwise puzzling, might be explained as follows: It seems likely that the more fluid and denser lavas (fluidity and density generally go together) were extruded at comparatively low elevations or reached these elevations more readily because of their fluidity, whereas the less dense and more viscous lavas formed the shield itself. Perhaps the shields are in part composed of pyroclastics. It is then possible that at least incipient isostatic adjustment led to subsidence of the smooth plains unit relative to the shield. The considerable radius defined by the scarp may indicate the large horizontal scale of isostatic adjustment.

If the Martian shield volcanos are basaltic in composition, this must be reconciled with the fact that they exhibit generally high reliefs—sometimes as great as 15 km—with respect to their surroundings. Although a part of this height can be explained by possible density contrasts with the lower lying lava flows as discussed above, it seems unlikely that these mountains comprise the lightest rocks of the Martian crust. If the crust is

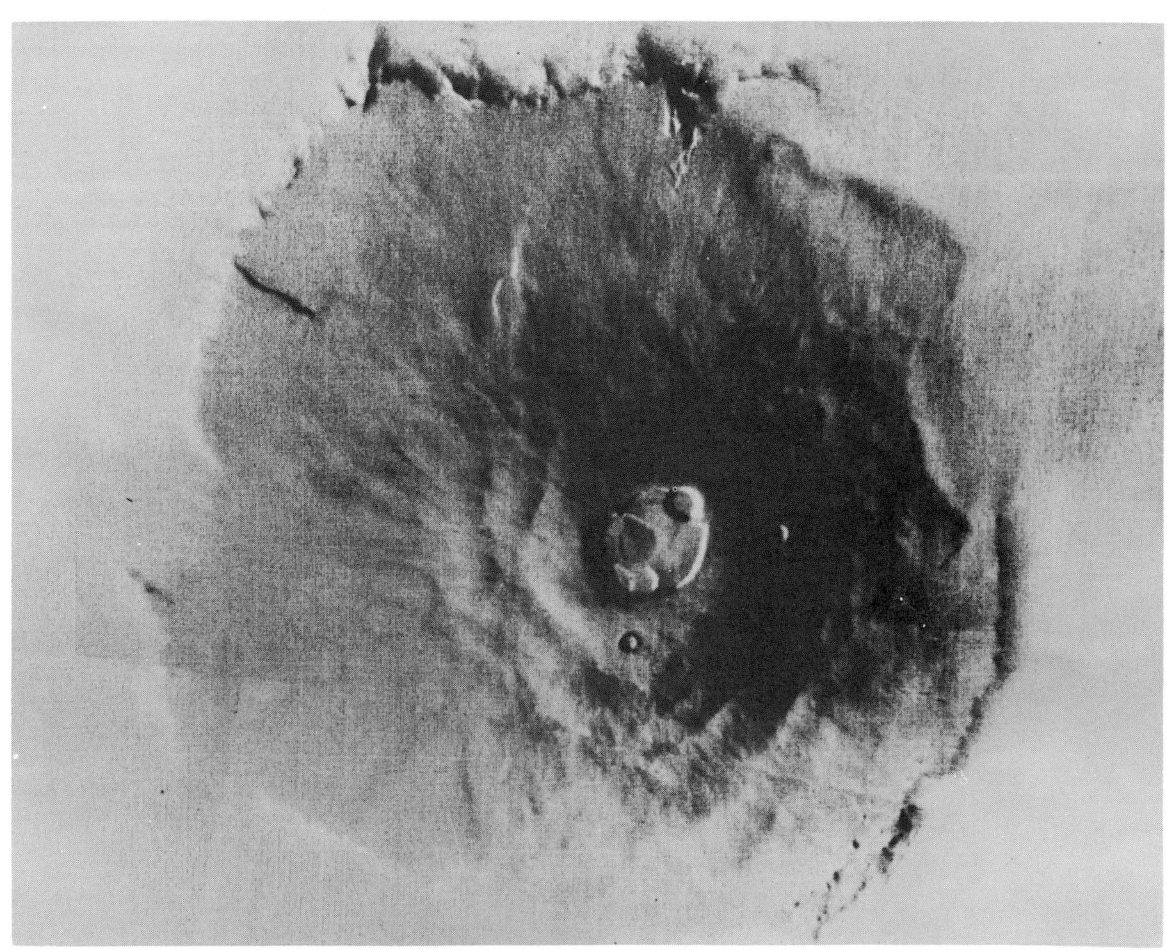

7.4 The Martian shield volcano Nix Olympica. Note the multiple central craters, the radiating lava flows, and the conspicuous outward-facing scarp. The entire structure is approximately 600 km across. Photograph courtesy of the National Space Sciences Data Center.

acidic, as may be inferred from the infrared spectra of dust, and if the shield rocks have a density of 2.9 or more, as we should expect from basalts, then these enormous masses could not be in isostatic equilibrium. Given this situation, the best explanation would seem to be that the Martian crust is far more rigid than that of Earth, perhaps because of lower crustal and subcrustal temperatures. Also, the volcanos on Mars are comparatively young and have yet not had time to subside in the viscous lithosphere.

Although distinctive rock units are still difficult to identify on Mars the laminated terrain of the polar regions appears to consist of eroded sedimentary layers, which reveal at least rudimentary tectonic activity in the form of basin and dome structures. The comparative youth of the laminated terrain and the volcanic regions is indicated by the lack of impact craters. The laminated terrain has been referred to as evidence of a polar sink of windblown sediments and frozen deposits of CO_2 and water. If this is true it seems that the rates of deposition and of their subsequent gentle folding greatly exceed the rates of meteoritic cratering.

Lowman (1973) has recognized on Mars evidence of the process of "oceanization," which is somewhat similar to the process described earlier by Ramberg. However, according to Ramberg's theory the driving force is gravity. In this process ocean basins repeatedly become the seats of basic volcanism because the dense lavas cannot attain continental elevations. We have already proposed a variant of this idea to explain the scarp around Nix Olympica. On Mars recognizable processes of magmatic or tectonic crustal evolution appear to have followed an early history of cratering and impact basin formation analogous to the history of the Moon. Mars also has had an extensive history of erosion which appears to continue unabated to the present. In addition to dune fields and laminated terrain, these erosive forces have given rise to widespread channels and canyons. Although some attribute these features to an early period of water erosion, this seems highly unlikely in view of the ease with which water molecules are dissociated and lost on Mars. More likely they are the result of long-acting wind erosion.

The Moon

General

The beginnings of man's interest in the Moon are lost in antiquity, but an excellent summary of the known history of scientific lunar studies has been provided by Baldwin (1949). Recently these investigations have culminated in the Apollo series of manned lunar landings. The literature describing the Apollo findings are too numerous to be discussed here and the reader is referred for details to such publications as the proceedings of the Lunar Science Conferences.[3]

Of the investigations that predate the lunar probes, those of Baldwin in particular stand forth as prophetic. In 1949 he sketched out in his "Face of the Moon" most of the important findings of the rocket explorers. Those are as follows:

1. The extreme age of the dominant lunar features—that is, that these features are approximately as old or older than Earth's most ancient rocks
2. The dominant role of meteoritic impacts in forming not only the lunar craters but also in forming the great mare basins and many other major surface features
3. The basic (as distinguished from acidic) nature of the maria lava flows and the role of tectonics and volcanism in the production of the maria
4. The sequence of relative ages of most of the major lunar features including the highlands, maria, craters, rays, etc.
5. The essentially anhydrous nature of the lunar surface as related to the weak gravity field
6. The relatively rapid accumulation of meteoritic matter and crater formation in the early history of the Moon and the gradual decline in the frequency of impacts

In addition to these major characteristics, which have been established by the recent studies, there has also emerged a picture of the moon as a fairly well differentiated body as reflected in the chemistry of its rocks. Not surprisingly these recent investigations have also revealed a variety of hitherto unknown or barely suspected minor features such as stratigraphic layering and volcanic doming. However, compared to Earth, and seemingly also Venus and Mars, we must concede that the Moon is a relatively simple and monotonous body.

Gross features as clues to the Moon's interior

We see from Table 7.1 that the mean density of the Moon is only 3.34 gm/cm^3. It is clear that this low density, as compared with other terrestrial planets, places severe constraints on the lunar composition. In fact, many

[3] Published in *Geochimica et Cosmochimica Acta*.

individual rocks from the lunar surface exceed the mean density of the Moon. Since considerations of mechanical and gravitational stability give every reason to believe that the average density of lunar material increases with depth, it is inescapable that these dense surface rocks represent differentiated material that found its way to the surface. As discussed in Chapter 6 and later in Chapters 13 and 14, magmatic differentiation under both meteoritic (relatively reducing) and terrestrial (relatively oxidizing) conditions can give rise to magmas enriched in iron provided that water pressure is low. Ringwood and Essene (1970) have inferred that some lunar basalts were derived from source material of pyroxenitic composition by partial melting. However, the pyroxenites and the general model of the Moon they propose seem to have densities too high to be consistent with the mean lunar density. Partial melting by virtually any mechanism should, however, lead to melts enriched in iron, and such melts could move to the surface because of their density contrast with solid rocks. However, once these melts solidified on or near the surface they would engender gravitational instabilities and anomalies such as have been observed in the flooded mare basins.

Although metallic iron is an ever-present minor constituent of the lunar surface rocks, the mean density of the Moon also rules out any substantial quantity of this metal in the interior and in particular the occurrence of any significant metallic core. A discussion of the nature of the core has been provided by Brett (1973), among others. The density and moment of inertia indicate that the maximum radius of an Fe core may not exceed 400 km. An Fe-FeS core could be as large as 700 km in radius.

Table 7.3 Geological evolution of the moon

Stage	Events	Mechanism/cause/source
I (4.7–4.6 Gyr ago)	(a) Formation of moon	(a) Precipitation, fission, or capture; mechanism unknown
	(b) Strong heating, to temperatures over 1000°C, of outer part of moon	(b) Energy of accretion, fission, and tidal interaction of earth; short-lived isotopes possibly important
II (4.6–3.7 Gyr before present)	(a) First differentiation, forming global crust of aluminum-rich gabbroic rocks and differentiates	(a) Partial melting, forming primary aluminum-rich basaltic magma; magmatic differentiation formed subordinate quantities of anorthosite and felsite; KREEP possibly formed by initial melting
	(b) Heavy cratering	(b) Infall of bodies related to origin of moon, from vicinity of earth-moon system
	(c) Shear faulting	(c) N-S compression caused by slowing of moon's rotation during recession from earth
	(d) Mare basin formation	(d) Infall of large objects from vicinity of earth-moon system, possibly protomoons
	(e) Formation of Archimedian craters	(e) Infall of objects from vicinity of earth-moon system, after mare basin formation but before mare filling
	(f) Formation of highland volcanics such as Cayley formation	(f) Generation and eruption of high-aluminum basaltic magmas; volcanic origin of Cayley formation uncertain; interpreted as deposits related to mare basin formation by some
III (3.7–3.2 Gyr before present)	Second differentiation of moon; mare filling	Generation of basaltic magmas by partial melting of deep interior; maria formed by repeated eruptions, partly localized in circular mare basins
IV (3.1 Gyr ago to present)	Postmare events; concurrent impact cratering by asteroid-belt bodies and comets, minor mare and highland vulcanism, tension faulting, and mass-wasting	

After Lowman (1973, 1976).

Surface features and the Moon's history

Although details of the Moon's evolution are obscure, there is evidence of broad episodes or periods which may be deduced from the surface features. A sequence of events which seems to be quite well substantiated is shown in Table 7.3, which has been taken from Lowman (1973). It is perhaps unusual that a planet reveals so clearly the sequence of episodes from its history. In the case of the Moon this is a direct result of its small size and consequent lack of a significant atmosphere as well as the limited internal activity that also seems to go with small size.

Evidence for the first stage of lunar evolution are not good and may even be lacking. It seems to be restricted largely to isotopic and other compositional data such as the age of the oldest lunar soil component as well as certain theoretical considerations of the energy of planetary fission and reaction, which are hypothetical processes at best. Evidence for the second stage is substantial since it is founded on a great variety of morphologic evidence as well as the variation across the surface of chemical composition. For example, the great age of the lunar highlands (IIb in Table 7.3) is supported not only by their high crater density but also by direct Rb-Sr dating, whereas evidence for early differentiation of this material is found in the x-ray fluorescence scanning of the surface (Adler *et al.*, 1972). The clear evidence for the subsequent development of the mare basins by the infall of large planetesimals as well as the later craters and their rays was pointed out some time ago by Baldwin (1949) and others. However it remained for the teams of lunar investigators, aided by the Apollo landings, to clearly delineate such products of large-scale volcanic eruptions as the Cayley formation. It remains now for us to review some of the detailed chemical and mineralogic evidence from the lunar rocks themselves.

Lunar rocks and minerals

A detailed description of the lunar rocks is beyond the scope of this book, but many such descriptions are to be found in such works as the Proceedings of The Second to Sixth Lunar Science Conferences (Houston, Texas, 1971–1976) and later books and papers.[4] Such studies show the lunar rocks to consist largely of basalts and various crystalline rocks such as norites, breccias, and microbreccias. Acidic rocks such as rhyolite also occur. Among the basic rocks are picritic basalts, ferrobasalts, and feldspathic norites. There is some indirect evidence for anorthosites. Glassy or partly glassy rocks are very abundant, with glass forming a common matrix in the breccias. Superimposed on all these rock types are the lunar "soils," which are composed of comminuted fragments of the previously mentioned rocks. These soils also contain a meteoritic component as identified from the composition of taenite, kamacite, and phosphide particles (Goldstein and Yakowitz, 1971). Textures in both rocks and soil tend to be fine and many of the rocks are profoundly crushed and show evidence of shock transformations. Among the breccias varying degrees of recrystallization and partial melting may be seen. The basalts are generally vesicular or vuggy, indicating that gas evolution occurred just prior to consolidation. Texturally the lunar rocks correspond precisely to what one would expect from materials subjected to repeated catastrophic meteoritic or planetesimal impacts and superimposed vulcanism.

Compared to terrestrial rocks the chemical variation among the lunar rock types is quite restricted since most fall in the basic to ultrabasic categories (Table 7.4). Those acidic rocks that do occur are found largely as interstitial material in basalts and tend to be glassy. Many of the basalts, and especially those obtained from Mare Tranquilitates (Apollo 11) and Mare Serenitatis (Apollo 17) are very high in TiO_2 when compared to terrestrial basalts. Many of the basalts are also high in FeO and CaO and tend to be very low in K_2O and Na_2O. However, certain lunar rocks, designated "KREEP," contain relatively high concentrations of K_2O(K) and rare rare-earth elements (REEP).

The mineralogy of the lunar rocks is predominantly that of the common basic and ultrabasic terrestrial rocks. Among the most abundant minerals are calcic plagioclase, pigeonitic and augitic pyroxenes, olivine, ulvöspinel, chrome spinel, ilmenite, and rutile. Silica-saturated phases such as quartz, tridymite, crystobalite, and alkali feldspars occur interstitially as last-stage products of crystallization-differentiation. In addition to these terrestrially common minerals, the common meteoritic minerals such as metallic iron (kamacite and taenite) and troilite are almost omnipresent, although in only small quantities. In addition there are many minor phases, some of which have not previously been described except from lunar rocks. Examples are baddeleite, pyroxferroite, tranquillityite, and zirconolite ($Fe_8^{2+}Zr_2Ti_3Si_3O_{24}$). Occasionally the plagioclase feldspar is found to be transformed to maskelynite, perhaps due to shock effects as in meteorites.

[4] For example the results of the Apollo 17 preliminary examination team (*Science 182*, 659–672, 1973) or the book by S. R. Taylor (1975).

The low silica content and basic character of the lunar basalts results in magmas of low viscosity (Murase and McBirney, 1970), allowing them to flow great distances. It has been noted that, unlike the case of Martian lava flows, the lunar flows do not show well-defined arcuated fronts. Perhaps this indicates that the lunar lavas are of lower viscosity than those of Mars. If the widespread Cayley formation is truely volcanic it would seem that pyroclastics are also quite common on the Moon. The formation of pyroclastics depends on abundant dissolved gases in the lunar magmas and is consistent with their common vesicular character. Perhaps before these gases were lost the lunar lavas were even less viscous than the experiments of Murase and McBirney indicate.

The precise nature of the gases, which in remote times were released from the lunar magmas, is still unknown; however, from the common occurrence of metallic iron and other reduced phases present in the rocks we may infer that these gases were also reducing in character. By analyses of lunar mineral assemblages consisting of olivine, pyroxene, and metallic iron and by comparison of those assemblages with meteorites (Figure 5.4), it may be inferred that the oxygen fugacity at the time of crystallization of many lunar basalts was of the order of 10^{-13} atm, which is several orders of magnitude less than the corresponding value for terrestrial basalts. At this oxygen fugacity, hydrogen would have been more abundant than water on a molecular basis within the vesicles, although the water pressure also could have been high enough to contribute significantly to the physical characteristics of lava.

Space limitation prohibits further description and discussion of lunar mineralogy and petrology. A review of the Moon has been recently completed by Taylor (1975). Anderson (1973, 1975) has discussed the composition of the lunar interior. Smith (1974) has summarized the essential features of lunar minerals.

Comparative planetology

Thus far there have been few attempts to exploit the very great potential to learn about planetary evolution through a comparative approach to the planets. We close this chapter with a few such comparisons which may also serve as a summary of the sometimes confusing detailed information just presented. Many comparisons have already been made or are implicit in the data contained

Table 7.4 Representative lunar rock compositions

	Apollo 17 basalt 75055,6[a]	Apollo 17 noritic breccia 72275,2[a]	Apollo 12 KREEP 12033 BGI (glass fragment)[b]	Apollo 12 interstitial rhyolitic glass 12040[c]
SiO_2	41.27	45.76	48.4	73.71
TiO_2	10.17	1.54	1.8	0.74
Al_2O_3	9.75	19.23	15.5	12.82
FeO	18.24	8.70	11.0	1.73
MnO	0.29	0.11	—	0.10
MgO	6.84	11.63	7.6	0.10
CaO	12.30	11.72	10.0	1.81
Na_2O	0.44	0.52	0.8	0.75
K_2O	0.09	0.23	1.0	7.74
P_2O_5	0.07	0.27	0.8	—
BaO	—	—	—	0.30
S	0.19	0.08	—	—
Cr_2O_3	0.27	0.20	—	0.07
	99.92	99.99		99.87

[a] Data from Apollo 17 preliminary examination team (1973).

[b] Data from Meyer (1971).

[c] Data from Brown *et al.* (1971).

in Tables 7.1 and 7.2. Table 7.2 in particular illustrates the dramatic differences in atmospheric density and composition that result, on one hand, from the sequence of planetary masses and, on the other, from temperature variations. Small planets such as the Moon, Mars, and Mercury possess tenuous atmospheres, and their rapidly fluctuating temperatures militate against any sort of thermodynamic equilibrium. Even on Earth temperatures are far too low for the attainment of thermodynamic equilibrium; instead, conditions have given rise to the "homeostatic state" of living organisms which, while feeding on the solar energy flux, maintain their own environments in conformity with thermodynamic laws but behind the protection of kinetic barriers that prevent the attainment of equilibrium. The atmosphere of Earth, with its apparently anomalous abundance of O_2 molecules, gives us a hint that we may always be able to recognize planets that support life if the chemistry of that life is similar to our own. Similarly the quasiequilibrium state of the Venus atmosphere seems to provide us with a prototype for the detection of hot planets beyond our solar system if these planets are massive enough to possess atmospheres.

One of the most fertile fields of comparative studies is in global tectonics or the differential workings of

7.5 Hypothetical cross-sections of a basic volcanic edifice on planets of different temperatures and compositions.

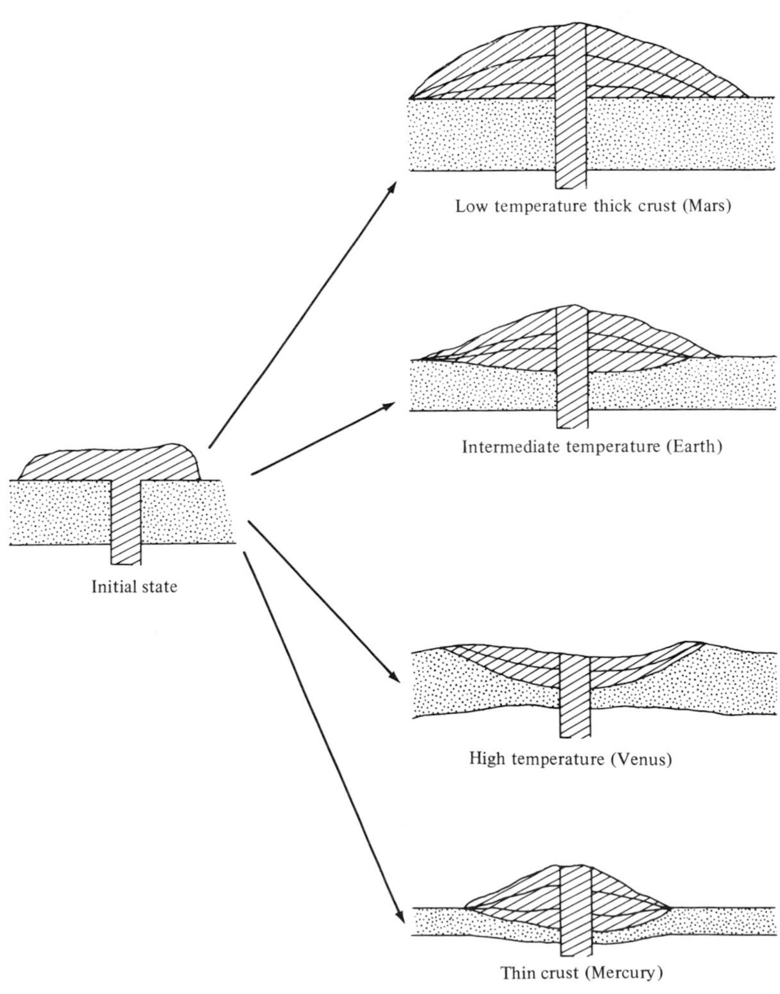

forces that shape planetary surfaces and give rise to the different rock types on the terrestrial planets. Here, too, we are concerned with thermal effects but also with the internal forces and mechanisms that bring about mass movements and chemical changes. Generally useful in such studies is the Earth-born "theory of isostacy," or that degenerate form of mechanical equilibrium that prevails in planetary crusts too rigid for true hydrostatic equilibrium. In the Airy hypothesis of isostatic equilibrium we consider a low-density crust of density δ_c floating on a subcrust or mantle of higher density δ_m. According to this theory any elevation Δh above the standard base level is compensated by a "root" component ΔR, and it may easily be shown that

$$\Delta h = \frac{\Delta R(\delta_m - \delta_c)}{\delta_c}$$

so that the height of an elevation such as a mountain range is a function of the density contrast and the size of the root. Because the root as well as the elevation can exist only by virtue of its rigidity or viscous resistance to flow, and because the rigidity is strongly influenced by the temperature, we should expect that there might be an inverse relation between the magnitude of relief and the temperature. Although the strength of rocks is diminished both with respect to brittle failure and viscous flow when the temperature is increased, only the latter effect lends itself readily to quantitative analysis. The pertinent equations for viscous flow have been discussed in Chapter 3, in particular Equations (3.50) and (3.52) for the viscosity and the flow rate constants, respectively. As indicated, the enthalpy of activation for typical planetary materials is high, of the order of 100,000 to 150,000 cal/mole, which implies a large decrease in viscosity with increasing temperature. Thus, as we have indicated earlier in this chapter, we should expect markedly lower relief on Venus when compared with Earth or Mars. Indeed the latest data seem to indicate that relief on Venus is only a third of that on Earth and perhaps even less.

Although the differences in relief of two planetary surfaces should reflect temperature differences of the subcrusts, there are other factors to be taken into account. Complicating factors result, for example, from the enhanced chemical differentiation which is also a product of high temperatures, as discussed for Venus. Thus on a hot planet there will be a greater tendency for the low melting fraction, such as granite, to migrate to the surface than on a cooler planet. Of course this process must result in the subcrustal source regions of the granite being correspondingly enriched in the higher melting residue consisting of Ca, Mg, Fe, etc. As we have seen, such basic rocks not only have higher melting ranges but also have a greater resistence to viscous deformation than do acidic rocks. Thus high temperatures lead to a "basification," which tends to counteract the expected decrease in viscosity. Quite obviously, also, high temperatures would tend to drive out water, and this would have a similar effect.

A general picture of the effects of temperature and composition on planetary relief is shown in Figure 7.5, which illustrates the response to the extrusion of basic lavas over a less basic crust to form a typical shield volcano. In the case of a cool planet such as Mars, which apparently possesses a thick crust, the result is a well-supported edifice of great relief. Similar behavior is shown by Earth-like intermediate temperatures except that there is a greater tendency for subsidence of the edifice. By contrast, on the hot Venusian-type planet the volcanic edifice never forms at all. Rather, extrusion results in a basin-like center of volcanism that crudely mimics a meteorite impact crater. This may in fact explain the crater-like radar images of Venus. The behavior of a hypothetical basic extrusion on Mercury would be similar to that on Earth. Although Mercury's subcrust might well have a high temperature, it is also likely to be far more basic than Earth's upper mantle. The result is that Mercury can support a relief as great as Earth or the Moon.

8 Metamorphism and metamorphic mechanisms

Introduction

Metamorphism

We made a brief reference to metamorphism of meteorites in Chapter 6. Metamorphic rocks constitute an important part of the Earth's crust. During the last two decades very significant progress has been made in the experimental techniques for the study of rock systems. Particularly notable in this regard is the development of high-pressure techniques, the electron-microprobe analyzer for chemical analysis, and the sophisticated methods of studying solid-gas equilibria with controls of fugacities of O_2, CO_2, and H_2O. Theoretical work has been concerned with the method of graphical representation of mineral assemblages, application of Schreinemaker's method of analysis of phase equilibrium, the nature and composition of crystalline and fluid solutions, and the treatment of open and closed systems during metamorphism. As a result of the various experimental and theoretical studies (see reviews by Hewitt and Wones, 1971; Winkler, 1974), our understanding of terrestrial metamorphism has improved tremendously over the last decade. Recently, Turner (1968), Miyashiro (1973), and Winkler (1974) have described and discussed the petrography and petrogenesis of the metamorphic rocks. Therefore in this and the following three chapters the aim is not to provide the reader with exhaustive information on all aspects of metamorphism but to expose the physicochemical nature of metamorphism through the use of experimental results, chemical thermodynamics, and kinetics.

Metamorphism may be defined as the process of recrystallization of matter in the solid state with varying degrees of involvement of an intergranular fluid phase. In this and the following chapters, we shall be only concerned with rocks recrystallized above the pressure and temperature at which weathering and diagenesis take place.

On a physical basis metamorphism may be described according to pressure, temperature, and spatial extension. Pressure and temperature can be high, medium, or low, and the distribution of rocks can be local or regional. Combination of the three factors can result in a number of physical classes of metamorphic types. However, in nature there is a continuous variation in the three factors, and distinct boundaries do not exist. It is possible to consider some broad divisions such as the regional dynamothermal rocks and the contact metamorphic rocks. The former denote rocks metamorphosed in large orogenic belts extending from a hundred to several hundred kilometers, while the latter are rocks metamorphosed in the vicinity of an igneous body. A detailed treatment of contact metamorphic rocks has been given by Reverdatto and Sharapov (1970). In our treatment of the metamorphic rocks we shall not be greatly concerned with these divisions. In the theoretical and experimental analysis of rock-forming assemblages, such distinctions are not very significant.

Pressure, temperature, and composition are the most important variables in metamorphism. Figure 8.1 shows Miyashiro's (1972b) estimate of geothermal gradients for low-, medium-, and high-pressure metamorphism. The low-temperature limit of metamorphism may be very close to the equilibrium curve for the reaction jadeite + quartz = albite. The high-temperature limit of metamorphism is determined by the melting of the rocks. Initiation and completion of melting for polyphase rocks is a function of the bulk chemistry; therefore, the upper limit is uncertain. Figure 8.1 shows that melting in the assemblage H_2O-quartz-albite-orthoclase starts around 650°C when the water pressure is 3 kbar. If

8.1 Estimate of geothermal gradients for low-, medium-, and high-pressure metamorphism. A more detailed analysis is presented in Figure 10.1.

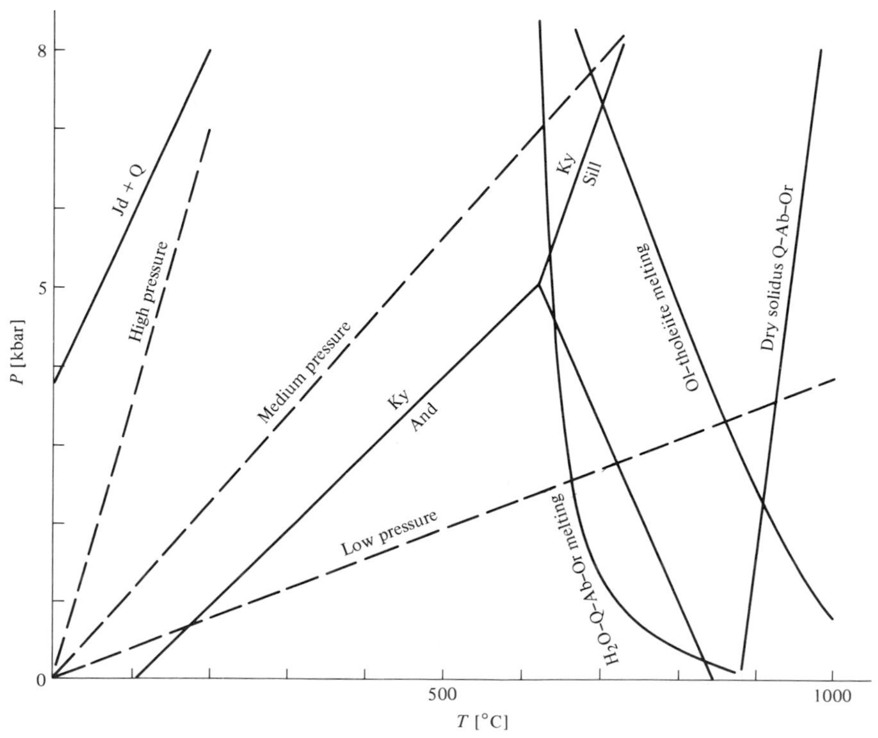

water is not present the initiation of melting is deferred until the temperature reaches over 900°C. This is also true for other systems such as olivine-tholeiite shown in the figure.

One of the important discoveries of the last decade is that of paired metamorphic belts (Miyashiro, 1961). A pair is composed of a low-pressure belt and a high-pressure belt running parallel to an island arc. In Japan three such paired belts have been mapped and are considered to have formed due to underthrusting of an ocean floor along a Benioff zone beneath island arcs and continental margins (Miyashiro, 1961, 1967b, 1973). Figure 8.2 shows a schematic cross-section of an island arc as envisaged by Miyashiro. The figure shows a subdivided eugeosynclinal zone: one of the divisions, on the continental side, contains the belt of low-pressure regional metamorphism, granitic plutonism, and andesitic volcanism; the other division, on the oceanic side, contains the belt of high-pressure regional metamorphism and mafic and ultramafic rocks.

Various types of mineralogic assemblages form by metamorphism, depending on the total pressure, partial pressure of fluid species, temperature, and composition of the solid material being transformed. The stability of many individual minerals has been determined experimentally. Such information provides us with the maximum to minimum range of physical conditions under which a mineral assemblage containing a particular mineral can form. Many minerals have a wide pressure and temperature range of stability and therefore cannot be used in narrowing the estimates of P and T of rock formation. However, there are a number of minerals whose P and T limits of stability are useful in petrogenesis. The polymorphs of Al_2SiO_5—kyanite, sillimanite, and andalusite are examples of such minerals, and Figure 8.1 shows the three curves of polymorphic transitions. The stability of many multicomponent minerals such as amphiboles and biotites is a complex function of P, T, and the concentration of the variable components. The hydroxyl-containing minerals are further critically dependent on the water pressure (P_{H_2O}), the stability generally increasing with increasing P_{H_2O}. Certain other minerals such as staurolite and

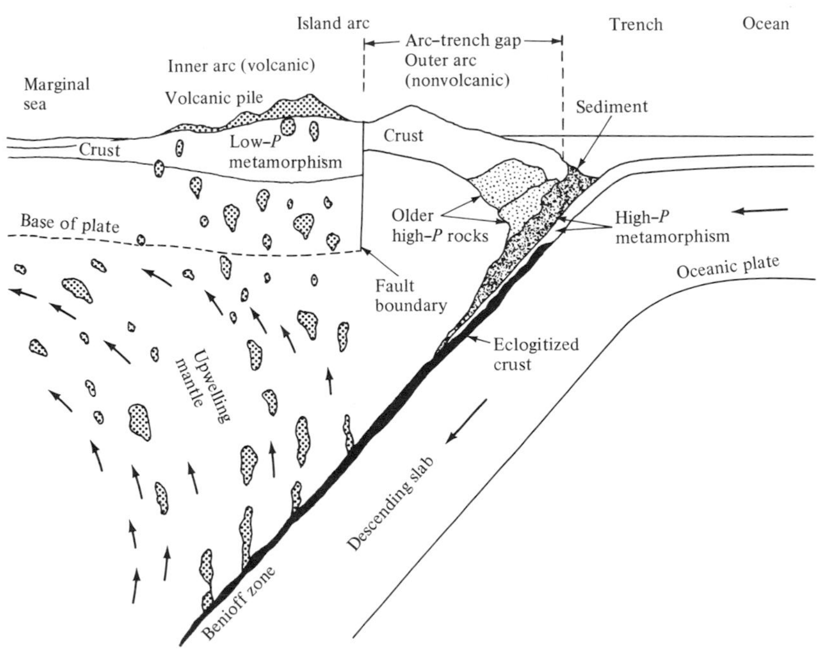

8.2 Schematic cross-section of an island arc with special reference to Northeast Japan Arc. Magmas generated along the Benioff zone are distinguished from magmas produced in the upwelling mantle by different symbols. (After Miyashiro, 1972b) Reprinted by permission of the *American Journal of Science*.

chloritoid require definite ranges of oxygen pressure for crystallization. The evaluation of such functional dependence of the mineral compositions and mineral assemblages is an important part of metamorphic petrogenesis, and will be discussed in Chapters 10 and 11.

The important species in the intergranular fluid phase are CO_2, H_2O, and O_2. The importance of the fugacity of oxygen (f_{O_2}) on the stability of ferromagnesian minerals was discussed by Eugster (1959) and Mueller (1960). Since then, several field, theoretical, and experimental studies have been conducted on assemblages involving variable f_{O_2}. Various types of solid buffers have been used in experiments to control the oxidation. Table 8.1 lists the constants for calculating oxygen fugacity of the commonly used buffers as a function of pressure and temperature, using the equation

$$\log f_{O_2}(\pm\Delta) = \frac{A}{T} + B + \frac{C(P-1)}{T} \qquad (T \text{ in } ^\circ\text{K}, P \text{ in bar})$$

It appears that the most common value of oxygen fugacity in metamorphic rocks may be in the magnetite field. We shall see later that a considerable variation in oxidation state may be present locally in areas such as the metamorphosed iron formations of Quebec and Sweden.

Many metamorphic systems have been investigated under pure water pressure. The thermochemical data for water up to 10,000 bar and 1000°C have been tabulated by Burnham *et al.* (1969). In natural systems, however, water is only rarely a sole constituent of the fluid phase. In the last few years special attention has been given to the carbonate-water systems by Greenwood (1967b), Skippen (1971, 1974) and Metz and co-workers (1968–1971) among others. Greenwood (1969, 1973) determined the solution properties of H_2O-CO_2 between 0 and 500 bar and 450 and 800°C. In the absence of solution properties at other temperatures, an ideal solution model has generally been used (Skippen, 1971, 1974; Gordon and Greenwood 1970).

In Chapter 9 we shall present the results of experimental work on P and T stability of many important rock-forming minerals. The subject of metamorphic mineral facies, which was referred to in Chapter 4, will be treated in detail in Chapter 10. We reserve Chapter 11 for a discussion of important metamorphic reactions and systems under various physical and chemical conditions. In the remaining part of this chapter, we present a discussion of the physical and kinetic factors important in the chemical reorganization and spatial rearrangement of matter during metamorphism.

Metamorphic mechanisms

The mineralogic and structural characteristics of metamorphic rocks[1] are the result of time-dependent mechanisms that have acted to modify the conditions set by thermodynamic equilibrium. They may be categorized as follows:

1. *Microstructures or textural features*: grain shape size, orientation, and distribution, as well as fluid inclusions, crystal zoning, twinning, etc.

[1] We have omitted the effects of such processes as shock metamorphism. For a treatment of this subject the reader is referred to the book edited by French and Short (1968).

Table 8.1 Constants for calculating f_{O_2} of solid buffers

$$\log f_{O_2}(\pm\Delta) - \frac{A}{T} + B + \frac{C(P-1)}{T} \quad (P \text{ in bar}, T \text{ in } ^\circ\text{K})$$

Buffer	Abbreviation	A	B	C	Δ	Original source
Fe_2O_3-Fe_3O_4	HM	25,680	15.953	0.019	±1.98	Robie and Waldbaum (1968); Robie *et al.* (1966)
Mn_3O_4-'MnO'	HMN	25,680	13.38	0.081	±0.05	Huebner and Sato (1970)
Ni-NiO	NNO	24,930	9.36	0.046	±0.02	Huebner and Sato (1970)
SiO_2-Fe_2SiO_4-Fe_3O_4	QFM	25,738	9.00	0.092	±0.10	Wones and Gilbert (1969)
Fe_3O_4-'FeO'	MW	32,730	13.12	0.83	—	Derived by Eugster and Wones (1962) from Darken and Gurry (1945)

After Ganguly (1972).

2. *Macrostructures*: compositional and textural banding, veins, segregations, fractures, folds, and various other linear, planar, and more complex elements of dimension larger than normal grain size
3. *Compositional features*: mineral compositional (including isotopic) variations, compositional relations among coexisting minerals and variations of mineral composition with distance in a rock

The macrostructures of rocks reflect bulk properties such as compressibilities, thermal expansions, rigidity, viscosity, etc., which are in turn directly dependent on the analogous properties of the individual minerals and of the textural relations among the latter. Thus a quartz vein, which is observed to cut a gabbro, may ultimately have resulted not only from the susceptibility of this basic rock to brittle fracture (so that a low-pressure zone was formed) but also to the high solubility and consequent mobility of quartz in the aqueous vapor which very probably penetrated the fracture.

Microstructures and textures most clearly show the interplay of specific rate and equilibrium processes. As we noted in Chapter 1, perfect thermodynamic equilibrium in the gravitational field would result in a form of mineral layering in which the denser phases would always underlie less dense phases—a far simpler state of affairs than we commonly observe. On the other hand, much of the compositional banding we attribute to metamorphic differentiation is at least partly due to equilibrium tendencies. The quartz vein that cuts the gabbro is found at its particular location primarily because the initial fracture provided a low-pressure and hence more stable environment for quartz to grow than in the more restricted environment of the solid rock. Again, although we cannot attribute such features as the spatial distribution of mineral porphyroblasts to equilibrium tendencies, physicists and metallurgists have found that the shapes of minerals and the details of the intercrystalline boundaries can be explained at least partially in this way.

The compositional relations among coexisting minerals in many metamorphic rocks show perhaps the most convincing evidence of chemical equilibrium. However, impressive as these relations are, they can again only represent an approach to the true equilibrium state since they do not satisfy the criteria of simple gravitational stability. The partial and restricted nature of this type of equilibrium is further illustrated by its confinement to small rock volumes.

A metamorphic feature that has given rise to much controversy is the orientation of anisotropic crystals in deformed rocks; it is usually interpreted as resulting from mineral growth under differential stress. We shall see that although this orientation has in the past been treated as a static equilibrium it must result at least in part from kinetic mechanisms since strict static equilibrium is incompatible with differential stress.

We must also remember that the bulk compositions of metamorphic rocks do not have the same significance that they have in many igneous rocks since the former depend, in part, on the details of the sedimentary or igneous processes that deposited the original minerals, and in part on subsequent metasomatic changes they may have undergone. Thus although the complex early history governs the gross character of the ultimate metamorphic assemblages, details will reflect the degree of local metasomatism or diffusional communication, which determines the extent of equilibrium attained. Indeed, metamorphic processes are frequently classified as either "metasomatic" or as "isochemical," but it is clear that such a division can have meaning only in relation to the volume of diffusive communication. For example, all rocks that have recrystallized are metasomatic in the sense that volumes at least as large as the crystal dimensions have undergone compositional change. Equally obvious is the conclusion that virtually all metamorphism results in loss of H_2O, CO_2, etc. Thus virtually all metamorphic rocks are metasomatic in one sense or another.

Origin of microstructures and textures

Equilibrium aspects: crystal form, surface, and interfacial energy

The textural feature of metamorphic rocks that particularly reflects thermodynamic equilibrium is the fundamental crystal form since the development of the faces of a crystal is directly dependent on surface or interfacial energy.[2] By contrast, such features as crystal size and orientation reflect equilibrium in a less well defined manner. In the case of a single crystal growing in an isotropic medium, the equilibrium form is elucidated by Wulff's (1901) theorem, which states that the surface tension of specific surface free energy γ for any face is proportional to the perpendicular distance to that face from an interior point (Wulff point) of the crystal. It

[2] We speak of "surface free energy" when a crystal grows in an isotropic medium such as air, water, a silicate melt or in a vacuum. Where growth occurs in a crystalline aggregate we speak of the "interfacial free energy."

should be stressed that γ as used here refers to a single component system with no adsorption of additional components. Under these conditions γ is defined as

$$\left(\frac{\partial F}{\partial \Omega}\right)_{T,V,n_i} = \gamma \tag{8.1}$$

where F is the Helmholtz free energy[3] of the entire system, Ω is the surface area of the interface, T is the temperature, V is the volume of the system, and n_i is the number of additional components. It is important to realize that γ must always be positive since it represents a "tension" acting inward on a surface contour. If this tension were directed outward, the surface would increase indefinitely (Landau and Lifshitz, 1958). The surface free energy of the entire crystal then is

$$\int \gamma(\bar{n})d\Omega \tag{8.2}$$

where $\bar{n}$ refers to the outward normal, and the integral extends over the entire surface. By Wulff's theorem γ is a function of the outward normal at each surface point. The general relations are illustrated for the two-dimensional case in Figure 8.3. Apparent from this figure are the deeply cusped free-energy minima for the low-index faces (full lines) and the less well defined minima for the higher index faces (broken lines).

Although Wulff's theorem is unambiguous in identifying the low free-energy faces of single crystals in isotropic media, most crystals we encounter in nature occur in crystalline aggregates, so that we must be concerned not only with the variation of free energy from face to face but also with its variation with respect to the orientation of other crystals, which form the different interfaces. The question then arises whether in a simple one-component system a single crystal or a crystalline aggregate of the same mass has the lowest free energy. This problem was examined in some detail by Fullman (1957), who showed that the interfacial free energy of a crystalline aggregate may indeed be so low that the aggregate attains a lower free energy than the single crystal. In particular this appears to be the case for certain spherulitic growths. Such growths are common in both igneous and metamorphic rocks (Misch, 1964). Read and Shockley (1950) have shown that the interfacial free energy of such aggregates can be attributed to grain boundary dislocations.

Although we are interested in the relationships between different grains of the same mineral, more frequently metamorphic rocks consist of more than one type of crystal with the added complication of adsorbed additional species on the interfaces. As was stressed by Smith (1948), the interfacial free energy between different grains largely governs the grain shapes and interfacial angles. Following Johnson (1959) and Kretz (1966) we discuss the relations among F, the Helmholtz free energy of the total system; F_α, the Helmholtz free energy of the phase α; F'_σ, the specific Helmholtz interfacial free energy for the interface σ; $\Gamma_{i,\sigma}$, the interfacial excess of the component i per unit area of interface σ; κ, the number of components in the system; and θ, the dihedral angle of interfaces between the phases α and β (Figure 8.4).

Now if a portion of the two phases α and β with the interface σ has a total Helmholtz free energy F then

$$F'_\sigma = \frac{F - F_\alpha - F_\beta}{\Omega} \tag{8.3}$$

and

$$F'_\sigma + \frac{PV_\sigma}{\Omega_\sigma} = \gamma_\sigma + \sum_{i=1}^{\kappa} \Gamma_{i,\sigma}\mu_i \tag{8.4}$$

where μ_i is as usual the chemical potential of the species i. If absorption is negligible and the PV_σ term is sufficiently small, obviously $F'_\sigma = \gamma_\sigma$.

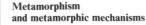

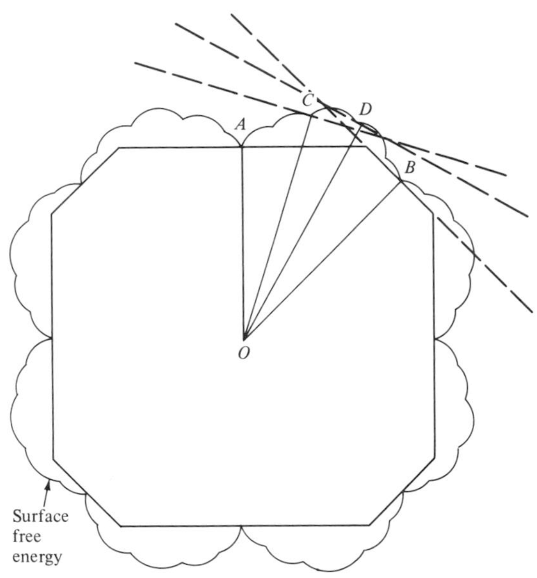

8.3 Illustration of Wulff's theorem. Plot of surface free energy and resultant crystalline polyhedra. The perpendicular OA is proportional to the surface free energy of that face. (After Herring, 1951.)

[3] Helmoltz free energy F is given by the relation $F = U - TS$ where U is the internal energy, T absolute temperature, and S entropy.

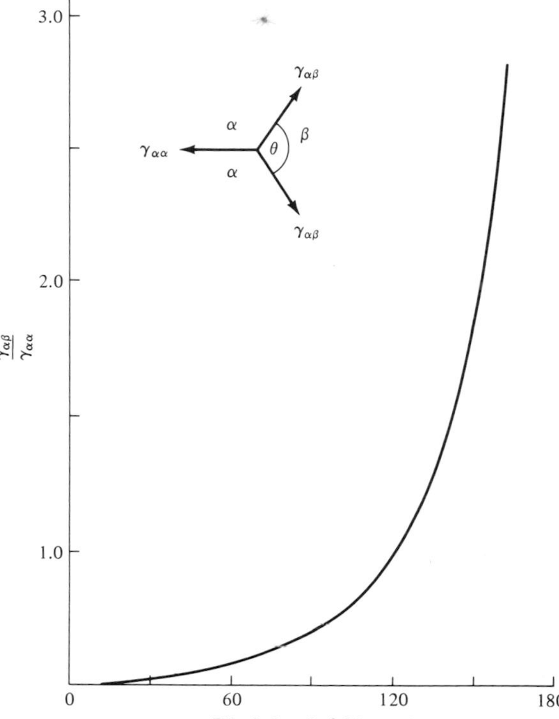

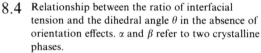

8.4 Relationship between the ratio of interfacial tension and the dihedral angle θ in the absence of orientation effects. α and β refer to two crystalline phases.

If the effect of the crystallographic orientation of the grains at the interface is negligible, it is possible to derive an expression for the ratio of the surface tensions between like and unlike phases by a simple balancing of forces (Figure 8.4). This yields

$$\frac{\gamma_{\alpha\beta}}{\gamma_{\alpha\alpha}} = \frac{1}{2\cos\frac{1}{2}\theta} \tag{8.5}$$

It is apparent that if $\gamma_{\alpha\alpha} = \gamma_{\alpha\beta}$, as when like grains are in contact, $\theta = 120^\circ$. We shall see however that the kinetics of grain growth sometimes modifies this result so that in the case of certain metallic systems the most probable interfacial angle is closer to 111° (Feltham, 1957).

From Figure 8.4 we see that a dihedral angle of 180° is excluded in the simple model, which ignores the dependence of γ on the crystallographic orientations of the adjoining grains. Yet such large angles are very common between different metamorphic grains (Figure 8.7). Large γ values can be explained by deriving a more complicated equation of equilibrium that takes into account the variation of γ with the orientation. According to Herring (Swalin, 1962) the expression is

$$\gamma_{\alpha\alpha} - 2\gamma_{\alpha\beta}\cos\tfrac{1}{2}\theta + 2\frac{\partial\gamma_{\alpha\beta}}{\partial(\frac{1}{2}\theta)}\sin\tfrac{1}{2}\theta = 0. \tag{8.6}$$

Then if $\partial\gamma_{\alpha\beta}/\partial(\frac{1}{2}\theta)$ is not negligibly small, γ will depend on the orientation. If, for example,

$$\frac{\partial\gamma_{\alpha\beta}}{\partial(\frac{1}{2}\theta)} \to -\frac{\gamma_{\alpha\alpha}}{2}$$

then $\theta \to 180^\circ$.

It is worth pointing out here that in the case of porphyroblasts high on the crystalloblastic series $\theta \to 180^\circ$, so that $\gamma_{\alpha\beta}$ should then be quite low.

The temperature dependence of F' is given by

$$\left(\frac{\partial F'}{\partial T}\right)_{V,n_i} = -\frac{S}{\Omega} \tag{8.7}$$

where S/Ω is the entropy per unit of interface. Thus the γ plot (Figure 8.3) will also vary with the temperature, and different crystallographic forms may have minimum values of $\int \gamma d\Omega$ in different temperature ranges.

The interrelation of crystallographic form and other aspects of crystal growth is illustrated by the metamorphic staurolite crystal depicted in Figure 8.5.

8.5 Crystallographic form and crystal growth in staurolite.

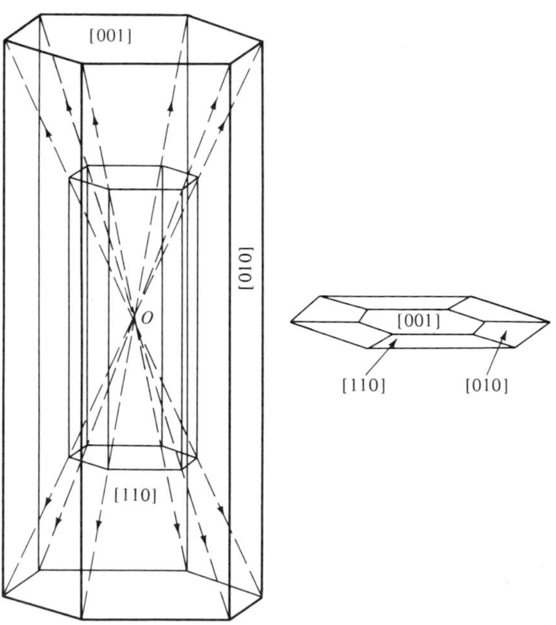

Staurolite is orthorhombic and is high on the crystalloblastic series so that it tends to assume a characteristic euhedral form with pronounced prismatic [110], side pinacoid [010], and basal [001] faces. From the relative prominences of these faces, one would expect that $\gamma_{110} < \gamma_{010} < \gamma_{001}$ if we assume that each face grew in contact with similar material. Staurolite also shows the peculiar but characteristic "chiastolite structure" in which planes of inclusions divide the crystal into sectors that reveal the growth patterns. In Figure 8.5 we have shown the growth vectors along the junction of these sectors. According to Hollister and Bence (1967), the composition of the sectors differ significantly.

Typical metamorphic textures which illustrate grain forms and boundary relations are shown in Figures 8.7 and 8.6. Figure 8.6, which depicts a polymineralic pyroxene-scapolite-sphene rock (Kretz, 1966a) illustrates the relatively simple case of three grains of the same mineral in mutual contact. Also the statistics (Table 8.2) of the angles for the same rock (carefully measured with a universal stage to obtain the true values of θ) show that the angle is 120° with a standard deviation of only 7.5° for scapolite contacts. It is possible that some of the deviation from 120° is attributable to orientation effects or to failure of equilibrium. Yet the results of Table 8.2 are impressive testimony of the tendency toward an equilibrium grain shape.

In Figure 8.7 are shown the mutual contacts of several biotite and hornblende grains (Kretz, 1966a). In this case the dihedral angles are governed almost entirely by the low-index, low-γ faces of one or the other of the grains. As a result the interface a is parallel to [110] of hornblende while the interface b is parallel to [001] of biotite. Vernon (1968, 1970) has carried out an extensive study of grain boundaries in granulites. The results indicate the importance of such studies in petrogenesis.

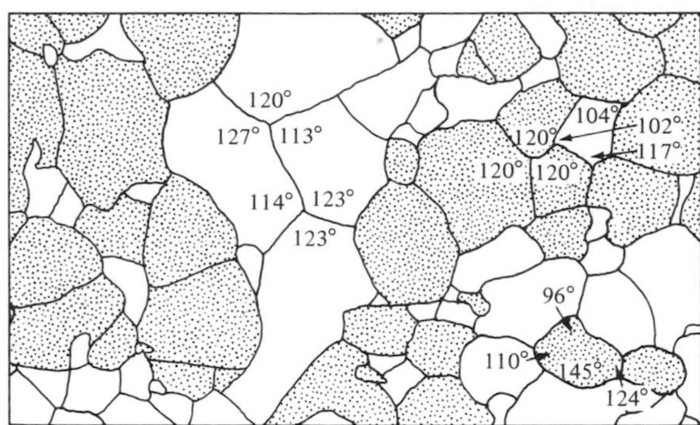

8.6 Three grains of the same mineral in mutual contact. (After Kretz, 1966a. *Journal of Petrology*) Reprinted by permission of the Oxford University Press, Oxford.

The crystalloblastic series

It was first pointed out by Becke (1913) that the metamorphic minerals may be arranged in a series according

Table 8.2 Grain boundary angles in a pyroxene-scapolite-sphene rock

Grain boundary angles at junctions of three scapolite grains (°)			Dihedral angles (°): Pyroxene V Scapolite-scapolite	Dihedral angles (°): Scapolite V Pyroxene-pyroxene
106	120	134	140	124
105	127	128	129	104
114	120	126	122	105
119	120	121	124	104
113	120	127	110	102
114	123	123	145	117
110	123	127		
115	120	125		
112	117	131		
116	119	125		
104	120	138		
116	117	127		
111	122	127		

Data from Kretz (1966a).

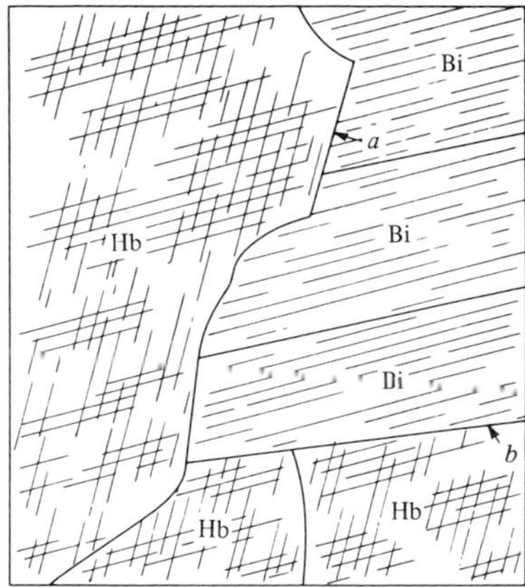

8.7 Contacts between biotite (Bi) and hornblende (Hb) crystals. *a* and *b* denote interfaces. (After Kretz, 1966a, *Journal of Petrology*) Reprinted by permission of the Oxford University Press, Oxford.

to their tendency to display their crystallographic forms against lower members of the series. He arranged them in order of decreasing tendency to form faces as follows:

1. Sphene, rutile, magnetite, hematite, ilmentite, garnet, tourmaline, staurolite, kyanite
2. Epidote, zoisite
3. Pyroxene, hornblende
4. Ferromagnesite, dolomite, albite, mica, chlorite
5. Calcite
6. Quartz, plagioclase
7. Orthoclase, microcline

However it was recognized by Becke that there are many exceptions to this scheme. The complicated nature of the series has already been illustrated for hornblende and biotite in Figure 8.7. Kretz (1966a) proposed a refinement of Becke's concept in which minerals are replaced by particular interfaces with quartz. Then we would have

1. [110] garnet-quartz
2. [001] biotite-quartz
3. [110] hornblende-quartz
4. [110] cordierite-quartz
   etc.

Although several different types of energy have been invoked to explain the crystalloblastic series, it seems clear that interfacial energy is all that is involved. Thus it happens that mineral facies that are high in the series have particularly deep interfacial energy cusps when they contact minerals of faces lower in the series. If we again consider the staurolite crystal previously referred to (Figure 8.6), we note that the growth sector of the basal pinocoid, which normally corresponds to the highest γ value, usually has anhedral boundaries with quartz (Hollister and Bence, 1967). It appears, however, that when a euhedral porphyroblast is developed, the interface with the fine-grained matrix has some average value characteristic of this matrix since there is little evidence of accommodation of the crystal faces to the individual matrix grains. It is likely that in the case of minerals so high in the series the porphyroblast dominates the interfaces and the difference between individual matrix grains is obscured.

Kinetic aspects of crystallization

Grain growth in metals

Rate phenomena are prominent in the processes of nucleation and grain growth. Both processes are highly dependent on the size, shape, and surface character of the nuclei and grains and at the same time determine what these characteristics become. Thus although the interfacial energy determines the ideal equilibrium form, this is approached only in very small crystals under special conditions (Herring, 1951). However, the faces of low interfacial energy assert themselves locally even in coarse-grained metamorphic rocks, as illustrated in Figure 8.7.

In the study of both nucleation and grain growth, one of the most important features is the grain size distribution or the number of grains in each class interval of dimension. These data are particularly useful because grain growth is seldom observed directly but is determined through the observed distribution after a period of nucleation and growth.

The most thorough studies of crystal growth have been made by metallurgists. Burke (1948) in particular showed that the driving force of grain growth in annealed metals is the interfacial free energy. Feltham (1957) demonstrated that the log normal distribution holds for aluminum and tin after a period of isothermal grain growth. He incorporated this distribution in a detailed kinetic model, which is based on the activation energy for intergranular diffusion and on the interfacial free energy.

An interesting observation made by Feltham was that both the grain diameters and the numbers of sides of the metal grains followed the log normal relation so that these variables are related (Figure 8.8).

According to Feltham the time rate of increase of a metal grain diameter D_i of the ith grain is

$$\frac{dD_i}{dt} = \frac{2}{3} a\left(\frac{kT}{h}\right)\left[\exp\left(\frac{-\Delta F_f^*}{kT}\right) - \exp\left(\frac{-\Delta F_b^*}{kT}\right)\right] \quad (8.8)$$

In this equation a is the lattice spacing h is Planck's constant ΔF_f^* is the free energy of activation for an atom to jump from an adjacent grain to the grain in question, and ΔF_b is the free energy of activation for the reverse process. The factor of 2 occurs here from the possibility of the addition of atoms to both ends of D_i. The factor $\frac{1}{3}$, on the other, hand, takes into account the property that only about $\frac{1}{3}$ of the total atoms have their displacements directed toward the grain in question.

Geometry requires that when crystalline grains have a large number of sides, many of the surfaces will be concave outward so that the radius of curvature falls outside the grain. In this case the sign of the radius of curvature ρ_i of the ith grain is regarded as positive. while for smaller grains with convex boundaries it is negative. If the bounding surfaces are regarded as segments of a sphere and dA is an element of area, then as the surface advances a distance $d\rho_i$, the surface area grows by an increment $(2/\rho_i)dAd\rho_i$ while the volume grows by $dAd\rho_i$. Consequently the surface free energy per atom of volume v created by this growth is

$$W_S = \frac{2v\gamma}{\rho_i} \quad (8.9)$$

where γ is the specific interfacial free energy.

If the bounding surfaces are not spherical but are elements of a circular cylinder the energy is

$$W_C = \frac{v\gamma}{\rho_i} \quad (8.10)$$

Feltham used the mean of the two values or

$$W = \frac{3}{2}\frac{v\gamma}{\rho_i} \quad (8.11)$$

If now ΔF_D^* is the free energy of activation for grain boundary diffusion, Feltham assumes that

$$\Delta F_f^* = \Delta F_D^* - W \quad (8.12)$$

and

$$\Delta F_b^* = \Delta F_D^* + W$$

and if $W \ll kT$, equation (8.8) becomes

$$\frac{dD_i}{dt} = a\left(\frac{2v\gamma}{\rho_i h}\right)\exp\left(-\frac{\Delta F_D^*}{kT}\right) \quad (8.13)$$

It is clear first of all that according to Equation (8.13) the metal grains will grow only if ρ_i is positive and will diminish in size when it is negative. The grain diameter for which $(dD_i/dT) = 0$, corresponding to an infinite ρ_i, and which divides the grain size distribution into growing and shrinking grains, is D^*, the most probable diameter. Also according to Equation (8.13), even if γ is regarded as a constant independent of crystal orientation, D_i will not vary directly with the time, since ρ_i varies in magnitude.

Now it has been found that in at least some metals the most probable interfacial angle is 111° rather than 120° as predicted by simple equilibrium theory (Feltham, 1957) so that D_i must be taken to correspond to such grains. We now impose the constraint that the total area of a planar section or the volume of the metal must remain constant. Then if ρ_i is determined by a function of D_i and if the observed log normal distribution and the relation of Figure 8.8 are taken into account, Equation (8.13) can be integrated to yield a relation between the most probable grain diameter and the time. Then if D^* and D_0^* are the most probable instantaneous and initial diameters respectively,

$$(D^*)^2 - (D_0^*)^2 = \left(\lambda va\frac{\gamma}{h}\right)t\exp\left(-\frac{\Delta F_D^*}{kT}\right) \quad (8.14)$$

where λ is a numerical constant of the order of unity.

8.8 Relation between most probable grain diameter and the number of sides (n) of annealed polycrystalline tin. (After Feltham, 1957.)

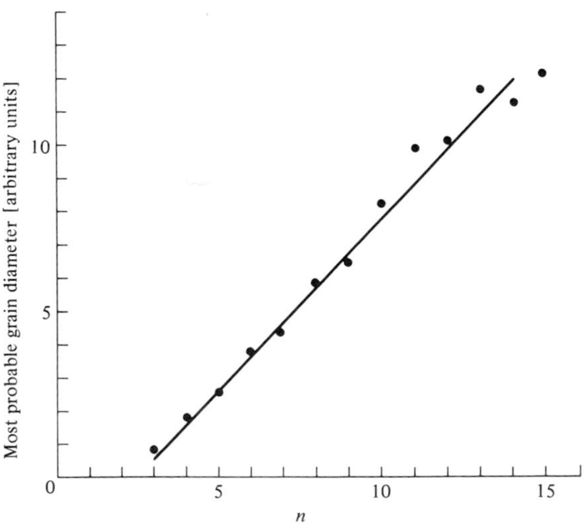

Nucleation and growth of crystals in metamorphic rocks

We have seen that a log normal grain size distribution can be related to interfacial energy effects in the case of isothermal grain growth in single-component metals. It is interesting that in metamorphic rocks very similar or identical size distributions characterize the porphyroblasts or isolated mineral grains of one mineral, which are enclosed by a matrix consisting of other minerals. The nature of these distributions was first recognized by Galwey and Jones (1963) (see also Jones and Galwey, 1964). In their study of garnet porphyroblasts in metamorphic contact aureoles, these authors found that if N equals the sum of the number of grains in a class interval of average grain radius r and the number in all classes of greater radius, then a straight line is produced when $\log [N/(100 - N)]$ is plotted against r.

The discovery of Galwey and Jones was followed by a detailed study by Kretz (1966b) of grain size and spatial distributions of garnets and biotites from metamorphic aureoles and of phlogopite, pyroxene, and sphene in marbles of a regional metamorphic terrain. It is interesting that Kretz as well as Galwey and Jones attributed the observed distributions mainly to nucleation-growth phenomena rather than to interfacial energy effects. To follow the argument of Kretz it is helpful first to review the elements of nucleation kinetics (Jacobs and Tompkins, 1955).

For nucleation that involves only a single step we proceed as follows: if initially there are n_0 potential nucleus-forming sites we may write for the rate of nucleus formation

$$\frac{dn}{dt} = K_1(n_0 - n) \tag{8.15}$$

where n is the concentration at time t, and K_1 is the specific rate constant. If the elimination of nucleus-forming sites through ingestion by growing nuclei is neglected, Equation (8.15) may be integrated to yield

$$n = n_0[1 - \exp(-K_1 t)] \tag{8.16}$$

or

$$\frac{dn}{dt} = K_1 n_0 \exp(-K_1 t) \tag{8.17}$$

which is referred to as the "exponential law" of nucleation. If K_1 is small, because of a large energy of activation, Equation (8.16) may be written as

$$n = n_0 K_1 t \tag{8.18}$$

and Equation (8.17) becomes

$$\frac{dn}{dt} = K_1 n_0 \tag{8.19}$$

which is known as the "linear law" of nucleation.

A stable nucleus may also result from a bimolecular process involving the combination of two active intermediaries, each of which is formed at a constant rate. If the concentration of an active species is A we have

$$\frac{dA}{dt} = K' \tag{8.20}$$

where K' is a rate constant. Then the quantity of A formed in time t is $K't$, if there is no reverse reaction and if their rate of combination is small compared with K'. Thus the rate of stable nucleus formation is

$$\frac{dn}{dt} = K(K't)^2 \tag{8.21}$$

K being another rate constant. Consequently

$$n = \frac{K(K')^2 t^3}{3} \tag{8.22}$$

To generalize, if the reaction of $\beta - 1$ entities are required to form a stable nucleus, we have

$$n = \frac{K(K')^{\beta - 1} t^{\beta}}{\beta} = Dt^{\beta} \tag{8.23}$$

or

$$\frac{dn}{dt} = D\beta t^{\beta - 1} \tag{8.24}$$

which is the "power law" of nucleation.

The result of Kretz's (1966b) investigation of the size distribution among phlogopite grains is shown in Figures 8.9 and 8.10. Since Figure 8.10 is a plot on a log-probability scale, the straight line plot of the large grains indicates a log-normal distribution in this range. The break that divides the large and small grains at y is also a characteristic of biotites, pyroxenes, and sphenes.

Although Kretz considered the possibility of the influence of interfacial energy on the grain-size distributions, he concluded that it could not be operative in the case of isolated porphyroblasts. His conclusion that the distribution results from nucleation phenomena, which continue through a considerable time span of mineral growth, is supported by a number of characteristics of the assemblages and at the same time yields a very simple interpretation. However this simplicity depends to a certain extent on the assumption that the grains grew at a constant linear rate. It is interesting that in Feltham's analysis of one-component metals this assumption holds only for the largest grains in the assemblage [Equation (8.14)].

In the case of phlogopite, Kretz assumed that a grain with a maximum dimension l_i and a linear rate of growth G had grown for a time t_i, so that

$$t_i = \frac{l_i}{2G} \tag{8.25}$$

Now the largest grains are found to fall in the class interval 2.6–2.8 mm so that these grains grew for a time of about 2.7/2G sec. In these terms it is possible to define a time scale of grain growth based on the largest grains for which $t \simeq 2.7/2G$ sec, since this is the time at which all grain growth is completed. To be more precise, we set the time of completion of crystallization at 2.8/2G sec, where 2.8 mm is the top of the class interval occupied by the largest grains. Thus as time passes from $t = 0$ to $t = 2.8/2G$ sec, new nuclei form and the number of mineral grains increases. For example, at $t = 1.4/2G$ sec, n has increased from 0 to 8 grains of each 100 grains that finally appear. Thus there exists a definite relationship between the number of grains n and t 2G, which is a measure of the time of nucleation and simultaneous grain growth. If the data are plotted with log n as a function of t 2G, we obtain Figure 8.11. Comparison of this figure with Figure 8.10 shows that the segments x-y and y-z of the curve seem to represent two different nucleation rates, with the x-y portion essentially linear so that it yields the equation

$$\log n = 1.41t2G - 1.07 \tag{8.26}$$

and

$$\frac{dn}{dt} = (6.49G)0.0854 \exp\,[(6.49G)t] \tag{8.27}$$

which is an example of the exponential law as represented by Equation (8.17). According to Kretz the nucleation rate declined during the latter one-third of the crystallization period represented by the curve segment y-z.

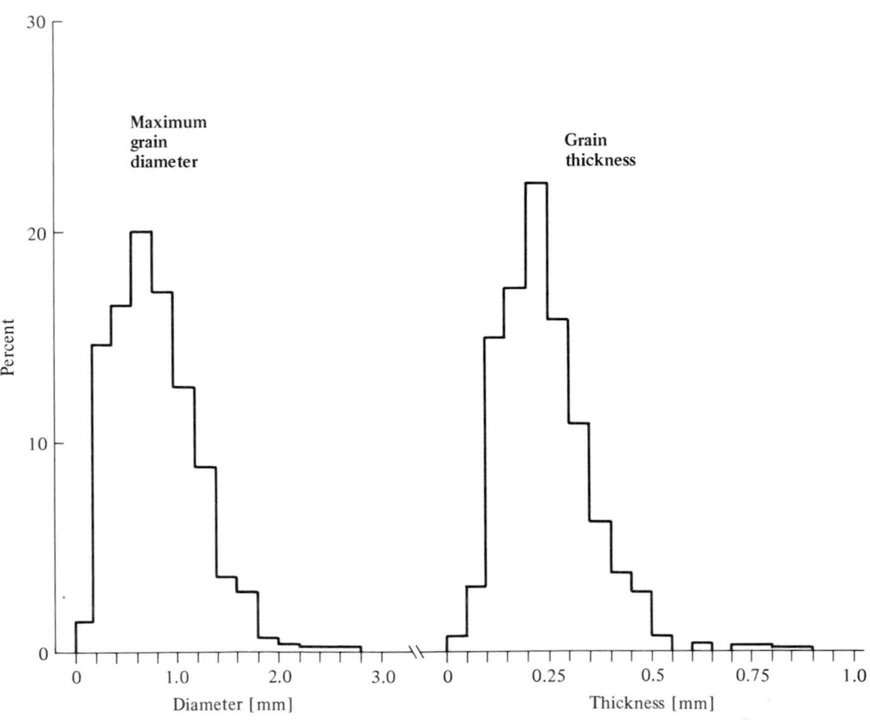

8.9 Histogram (615 grains) of phlogopite grain size in marble for maximum grain diameter and grain thickness. (After Kretz, 1966b, *Journal of Geology*.) Reprinted by permission of The University of Chicago Press.

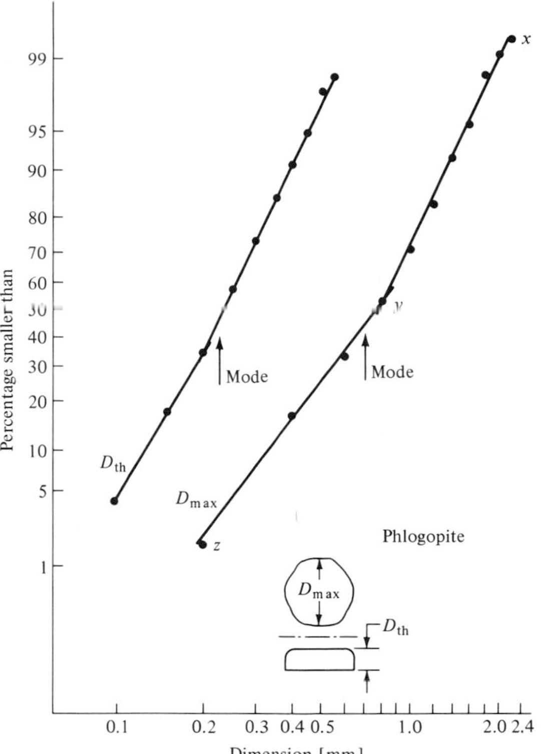

8.10 Cumulative size-distribution curves for maximum diameter (D_{max}) and thickness (D_{th}) of phlogopite grains plotted on log-probability paper. (After Kretz, 1966b, *Journal of Geology*.) Reprinted by permission of The University of Chicago Press.

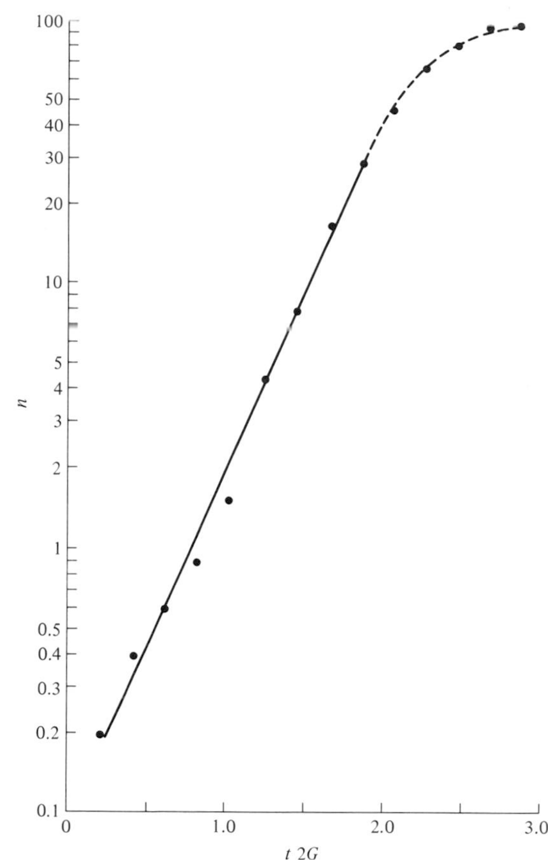

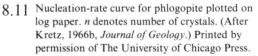

8.11 Nucleation-rate curve for phlogopite plotted on log paper. n denotes number of crystals. (After Kretz, 1966b, *Journal of Geology*.) Printed by permission of The University of Chicago Press.

Kretz also found grain-size distributions similar to that of Figure 8.9 for biotite, pyroxene, and sphene. However the distribution for garnet in one rock did not correspond to a log-normal curve; rather, it was found to lie between the log-normal and arithmetic-normal distributions. In another rock, the distribution was bimodal.

Kretz (1973) considered three models of garnet growth assuming the crystals to be spherical. In the first, second, and third models, the respective rates of increase of radius (r), surface area (a), and volume (v) are assumed to be constant. The second model represents a growth rate intermediate between the first and third. Model 1 refers to diffusion-controlled growth (dr/dt = constant), while model 3 corresponds to surface-reaction-controlled growth (dv/dt = constant, therefore dv/da = constant) at constant temperature.

Figure 8.12 shows compositional profiles from center to edge in four different size garnets in one specimen (No. 285) from the Yellowknife Group, Canada. It is obvious that progressively smaller crystals nucleated and began growth at progressively later points in time. To consider the correlation between the crystal growth rate in terms of either radius, surface, or volume and composition, we write the following relations:

$$X = f(t) \tag{8.27a}$$

$$\frac{dX}{dt} = f_1(X) \tag{8.27b}$$

where $f(t)$ and $f_1(X)$ are unspecified functions of time and composition [X, e.g., Fe/(Fe + Mn)], respectively.

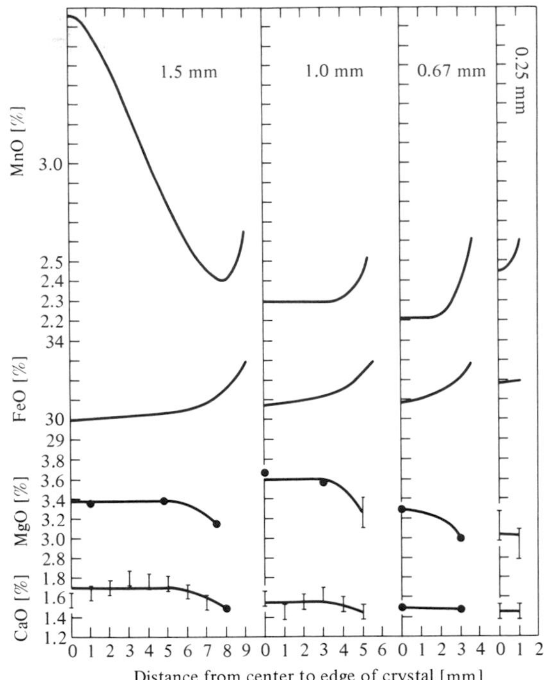

8.12 Chemical profiles from center to edge of garnet crystals of diameter 1.5, 1.0, 0.67, and 0.25 mm. The Mn and Fe data have been averaged with regard to both distance and concentration. In the Mg and Ca profiles, a vertical bar (length = 2 standard deviations) indicates 1 or 2 readings and a circle, the mean of 3 to 8 readings. (After Kretz, 1973.) Reprinted by permission: *Canadian Mineralogist*, *12*, 5.

Now:

$$\frac{dX}{dr} = \frac{dX/dt}{dr/dt} \tag{8.27c}$$

By substituting $dr/dt = K_1$ for the first growth model we have

$$\frac{dX}{dr} = \frac{1}{K_1} f_1(X) \qquad \text{first model}$$

Similarly

$$\frac{dX}{da} = \frac{1}{K_2} f_2(X) \qquad \text{second model}$$

$$\frac{dX}{dv} = \frac{1}{K_3} f_3(X) \qquad \text{third model}$$

Kretz (1973) plotted weight-percent of FeO against volume, area, and distance and found that the compositional data as presented in Figure 8.12 best fits the second growth model for which the relationship between t and n is approximately given by

$$n = 1.35(K_2 t)^{2.07} \tag{8.27d}$$

By differentiating we have

$$dn/dt = 2.79K_2^{2.07}t^{1.07} \approx 2.8K_2^2 t \tag{8.27e}$$

which is an expression for the nucleation rate. Kretz also derived the following expression for the reaction rate

$$\frac{dv(t)}{dt} = \left(\frac{7}{2}\right)(11.2/120^{1/2})K_2^{7/2}t^{5/2} \tag{8.27f}$$

where $v(t)$ is the volume of the product at the time of observation.

Another important result of Kretz's study was the conclusion that the spatial distribution of garnet grains was such that the size of a grain was independent of the distance to its nearest neighbors. The implication here is that the chemical constituents of the grains were not primarily locally derived and that diffusion must have been quite effective through distances of a centimeter or more. This should not be construed to mean, however, that equilibrium was attained through this distance.

Rates of metamorphic reactions

It is necessary to distinguish the rates of metamorphic reactions as such from rates of nucleation and simple grain growth. These reactions may be either homogeneous or heterogeneous. Although the kinetics of many types of homogeneous fluid reactions are quite well known, these are of less interest than the heterogeneous reactions for which not even adequate theory exists.

The simplest of the heterogeneous reactions between solids is the dimorphic inversion. However, experience shows that even this type of reaction may be complex. As an illustration we refer to the work of Ernst and Calvert (1969) on the inversion of the Montery Porcelanite (cristobalite) to quartz. Their study, which involved isothermal heating at 300, 400 and 500°C and at 2000 bar water pressure, indicates that the percent of conversion of porcelanite to quartz obeys the relation

$$\frac{dc_Q}{dt} = K \tag{8.28}$$

where c_Q is the concentration of quartz and K is the rate constant. If this is the true picture, the rate of conversion is independent of the amount of reactant present. Obviously such could not be the case if the inversion occurred uniformly throughout the mass of porcelanite. It could, however, be explained if the reaction occurred on internal surfaces or nucleation centers that remained constant in area or number throughout the reaction period. This might happen if the reaction began along individual grain surfaces and if the number of grains then increased continuously so as to maintain a constant total reaction interface.

The most common type of heterogeneous metamorphic reaction is that which occurs between two or more reactant phases to produce additional quantities of product phases which have already been nucleated. It is conceptually simple to visualize such reactions as occurring between two grains in contact, and indeed there is evidence such contact reactions between solids do proceed at measurable rates. For example Brindley and Hayami (1965) studied the reaction between grains of MgO and SiO_2 in the temperature range of 1100 to 1400°C for times up to 144 hr and under essentially anhydrous conditions. They found that the reaction rate was diffusion controlled and that the end product was the following sequence of contacting phases:

$MgO \quad Mg_2SiO_4 \quad MgSiO_3 \quad SiO_2$

The original grain boundary was identified between MgO and Mg_2SiO_4 and the thickness of the Mg_2SiO_4 layer always greatly exceeded that of $MgSiO_3$. These characteristics point to a greater mobility of MgO than for SiO_2 under these particular conditions.

In Chapter 3 we discussed some of the results of Greenwood's (1963) study of the kinetics of heterogeneous reactions in the system MgO-SiO_2-H_2O. Here it should be mentioned that such complex phases as anthophyllite may not form spontaneously—so that, as was the case here, seed crystals must be introduced.

Greenwood's result that all the reactions studied by him were first order cannot be easily explained. However, the problem is elucidated somewhat if we consider how the vapor of fluid phase may enter into such reactions. For greater simplicity we illustrate this point by setting up a hypothetical reaction scheme for the formation of wollastonite from calcite and quartz. The overall reaction is

$$\underset{\textit{calcite}}{CaCO_3} + \underset{\textit{quartz}}{SiO_2} \longrightarrow \underset{\textit{wollastonite}}{CaSiO_3} + \underset{\textit{fluid}}{CO_2} \tag{8.a}$$

But we know from theoretical considerations that the following reactions also occur at least to some degree:

$$\underset{\textit{wollastonite}}{CaSiO_3} \longrightarrow \underset{\textit{fluid}}{CaO} + \underset{\textit{fluid}}{SiO_2} \tag{8.b}$$

$$\underset{\textit{quartz}}{SiO_2} \longrightarrow \underset{\textit{fluid}}{SiO_2} \tag{8.c}$$

$$\underset{\textit{calcite}}{CaCO_3} \longrightarrow \underset{\textit{fluid}}{CaO} + \underset{\textit{fluid}}{CO_2} \tag{8.d}$$

Now for simplicity it may be assumed that each grain vaporizes at a constant rate K^- per unit volume and reforms with a specific rate constant K^+. We may then write for the net rates of Reactions (8.b), (8.c), and (8.d):

$$J^-_{(b)} = K^-_{(b)} - K^+_{(b)}P_{CaO}P_{SiO_2} \tag{8.29}$$

$$J^-_{(c)} = K^-_{(c)} - K^+_{(c)}P_{SiO_2} \tag{8.30}$$

$$J^-_{(d)} = K^-_{(d)} - K^+_{(d)}P_{CaO}P_{CO_2} \tag{8.31}$$

If we now solve Equations (8.29) and (8.30) for P_{CaO} and P_{SiO_2}, respectively Equation (8.31) becomes

$$J^-_{(d)} = K^-_{(d)} - K^+_{(d)}\left(\frac{K^-_{(d)} - J^-_{(b)}}{K^+_{(b)}}\right)\left(\frac{K^+_{(c)}}{K^-_{(c)} - J^-_{(c)}}\right)P_{CO_2} \tag{8.32}$$

In this way the rate of formation of wollastonite can be related to the differential rates of decomposition into the fluid phase of both reactants and products. Furthermore, although we have arbitrarily chosen to decompose the solids into the species CaO, SiO_2, etc., it may be shown that if the reactions *between* these other fluid phase constituents are rapid relative to the heterogeneous reaction then the choice of constituent is immaterial. Thus, for example, we might have chosen Ca or O_2 as decomposition products instead of CaO. They would of course be linked by the equilibrium equation $P_{CaO} = KP_{Ca}P_{O_2}^{1/2}$. We can also see from Equation (8.32) that if P_{CO_2} is low enough the rate of disappearance of calcite is directly proportional to the volume of calcite present, which corresponds to a first-order reaction as defined by Greenwood. It is apparent that Equations (8.29) to (8.32) could be derived only with a highly idealized model and that the true kinetic behavior may well depend on complex intermediate steps which are not known.[4] The model does illustrate however how the

[4] It has been found experimentally (Chapter 7) that the rate of formation of wollastonite from calcite and quartz is diffusion-controlled under anhydrous conditions (Kridelbaugh, 1973; see also discussion by Helgesson, 1971).

overall reaction such as (8.a) may depend on the participation of a fluid phase and how the restriction of actual contact between grains is in this manner removed.

There is one type of homogeneous reaction that is of great interest to mineralogists and petrologists and which provides considerable insight into the thermal behavior of minerals through the entire temperature range of metamorphism and magmatism. This reaction is the intracrystalline exchange of two or more particles (atoms, ions, etc.) between two or more structurally and energetically nonequivalent lattice sites. The thermodynamics of such reactions—long termed "order–disorder" phenomena—was first given a comprehensive treatment by Bragg and Williams (1934). The extention of the theory to include time-dependent change of order was made by Dienes (1955).

We may represent the general case (Mueller, 1969c) of the time rate of change of the concentration of a given species K on a lattice site i of any crystal as follows:

$$-\frac{dc_i^K}{dt} = \sum_{L=1}^{m-1}\sum_{j=1}^{n-1} K_{ij}^{KL}\phi_{ij}^{KL}c_i^K c_j^L - \sum_{L=1}^{m-1}\sum_{j=1}^{n-1} K_{ji}^{LK}\phi_{ji}^{LK}c_j^K c_i^L \tag{8.33}$$

In this expression L is one of m species other than K and j is one of n sites other than i. The c's are the molar or atomic concentrations (per cm^3) and K^{KL} and K^{LK} represent the specific rate constants for the particular particles and sites indicated. For example K^{KL} is the rate constant for the transport of the species K from the site i and the species L from the site j. These constants are by definition functions solely of the temperature and volume (or pressure). However, the ϕ factors, which are analogous to activity-coefficient products in macrosystems are also compositionally dependent. The ϕ factors, like the rate constants, also contain the thermodynamic properties of the activated states and so are more complicated than ordinary activity coefficient products.

In the case of the distribution of two particles (quasibinary solution) between two nonequivalent sites which are present in equal numbers, Equation (8.33) may be written as

$$-\frac{dX}{dt} = \tfrac{1}{2}c_0[K_{12}\phi_{12}X_1(1-X_2) - K_{21}\phi_{21}X_2(1-X_1)] \tag{8.34}$$

Here c_0 is the total concentration of sites both 1 and 2 and

$$X_1 = \frac{2c_1}{c_0},$$

X being taken as the mole fraction of one of the two species. It is easy to see that at equilibrium when $(dX_1/dt) = 0$, Equation (8.34) reduces to

$$K^0 = \frac{K_{12}}{K_{21}} = \frac{X_2(1-X_1)}{X_1(1-X_2)}\frac{\phi_{21}}{\phi_{12}} \tag{8.35}$$

A well-known metamorphic mineral. which may be represented by Equations (8.34) and (8.35), is orthopyroxene in which Mg^{2+} and Fe^{2+} are distributed between the two nonequivalent M_1 and M_2 sites. The thermal behavior of this mineral has been extensively studied by Virgo and Hafner (1969) and by Saxena and Ghose (1970) and it has been found that the equilibrium distribution differs quite significantly from the "ideal" case in which $(\phi_{21}/\phi_{12}) = 1$, particularly in the metamorphic range of temperatures. However the ideal case is approached at 1000°C. The work of Virgo and Hafner also indicates that the activation energy that enters into the specific rate constants, and which may be obtained from their temperature variation, is only of the order of 20 kcal in the temperature range 500 to 1000°C. Yet the degree of disorder that corresponds to 500°C is retained at lower temperatures and persists even on the geologic time scale. The simplest explanation of this behavior is that the order-disorder process consists of two consecutive steps in which the lower energy step ($\sim$20 kcal) governs the overall reaction above 500°C and that the high-energy step governs it below this temperature. As a consequence the equilibrium degree of order corresponding to temperatures below the critical transition temperature $T_{Tr} \sim 500°C$ can be attained only through complete recrystallization. The critical transition temperature may be different for different minerals. In anthophyllite, for example, Seifert and Virgo (1975) report ordering up to a temperature as low as 280°C.

Mineral orientation

One of the most obvious features of many regionally metamorphosed rocks is a marked orientation of mineral grains. In our treatment of igneous rocks we touched on this subject as it applied to the orientation attained through flowing magmas and in postmagmatic recrystallization. Here, however, we consider it as one of the least-understood features of metamorphic textures.

Rocks whose fabric elements have been oriented by pervasive deformation were termed "tectonites" by Sander (1930), who stressed the relation between orientation and deformation. This orientation may take the form of mere elongation of grains irrespective of crystallographic orientation or, more frequently, of both dimensional and crystallographic orientation. Frequently also there is strong crystallographic orientation but no dimensional orientation. The different forms of orientation for both igneous and metamorphic rocks are shown in Table 8.3.

The simplest and most easily understood orientation effect is the purely dimensional orientation, which arises through pressure solution in a clastic sediment. In this illustration of "Rieke's principle" the upper and lower surfaces of spheroidal grains that are in direct contact, and thus under load stress, undergo dissolution while the dissolved material is redeposited in the intergranular spaces. The result is a fabric in which the grains roughly approximate prolate ellipsiods and which have the crystallographic orientation of the original clastic elements except for effects of compaction. This fabric has been described from diagenetically altered sandstone and interpreted by Weyl (1959). Also, it has been produced in the laboratory by Fairbairn (1950).

An almost equally well known type of orientation is that which occurs when elongate or flat objects such as amphibole needles or mica flakes are immersed in a flowing fluid or a ductile matrix. The most familiar example of this is the lineation of crystals in lava flows. This type of orientation must also occur during dynamic metamorphism when highly competent linear or flat elements are suspended in an incompetent matrix. However, it is likely that oriented metamorphic porphyroblasts grow in size at the same time as the matrix minerals crystallize so that more complicated orientation mechanisms are also involved. The dynamics of orientation of competent objects in a ductile matrix has been reviewed in some detail by Ramsay (1967).

Although crystal orientation through fracturing (cataclastic flow) is poorly documented, it seems reasonable that it could be effective in cases where elongate cleavage fragments are produced. Thus it seems likely that cleaving of feldspar grains during the formation of flaser gneisses would give rise to some coincident crystallographic and dimensional elongation of the typical lens-shaped augen that occur in such rocks.

The orientation effects of gliding mechanisms within individual crystals is familiar from experimental rock deformation studies. Thus Griggs *et al.* (1960) found that crystals of calcite in highly necked cylinders of deformed marbles were characterized by having their c axes at high angles to the axis of extension. In addition, these crystals also exhibited orientation of their a axes. Several different types of gliding—including twin gliding—are thought to contribute to this type of deformation. It is likely that the rate-determining process in gliding is thermally activated flow (Heard, 1963) and that it is in this way closely related to recrystallization creep proper. It has also been suggested (Hahn *et al.*, 1967) that thermally activated gliding in calcite is accompanied by slip along crystal boundaries and that this slip is thermally activated. Such processes are thought to be dependent on dislocations and crystal imperfections such as vacancies and impurity ions.

The most important and to a degree the most interesting lattice orienting mechanism is recrystallization. Typical examples are the strong orientations of the c axis in quartz, calcite, and amphiboles and of micas and other

Table 8.3 Types of mineral orientation

Orientation type	Example	Probable mechanism
Dimensional	Quartz in orthoquartzite	Pressure solution (Rieke's principle)
Crystallographic	Quartz or calcite in metaquartzite	Recrystallization creep
Dimensional-crystallographic	Calcite in marble	Translation and twin gliding
Dimensional-crystallographic	Amphibole in lava flows	Hydrodynamic flow
Dimensional-crystallographic	Feldspar in flaser gneiss	Mechanical orientation by cataclastic flow
Dimensional-crystallographic	Amphibole in amphibole rock	Recrystallization creep in similar medium
Dimensional-crystallographic	Amphibole in feldspathic rock	Recrystallization creep in different medium

platy minerals. That purely crystallographic orientation mechanisms exist is well illustrated by the prevalence of lattice orientation in recrystallized quartz and calcite, which exhibit no dimensional orientation. That these mechanisms are also operative in hornblendes and micas is strongly indicated by the virtual absence of mechanical stress effects either within the oriented grains or in the associated matrix minerals.

The progressive orientation of quartz in a quartzite through recrystallization is shown in Figure 8.13. There is little or no evidence for dimensional orientation of the quartz, but the pertrofabric diagram shows the orientation of the quartz *c* axis in the Harkless quartzite of the Inyo Mountains, which has suffered deformation in the aureole of the Papoose Flats granitic pluton. The sedimentary beds exhibit considerable compressive thinning, deformation of fossils, and boudin formation. From the separation and thinning of the boudinaged beds it was estimated that an overall compressive thinning of at least 80 percent occurred. Presumably the thinning and extension of the sedimentary beds occurred during emplacement of the pluton. The degree of orientation as defined by the *c* axes is directly related to this thinning. The general diagram of Figure 8.14 shows the typical relation between quartz and calcite orientation, the foliation, lineation, and the probable principle stresses of

8.13 The progressive orientation of quartz in a quartzite through recrystallization. (After Sylvester and Christie, 1968.) Reprinted by permission of The University of Chicago Press.

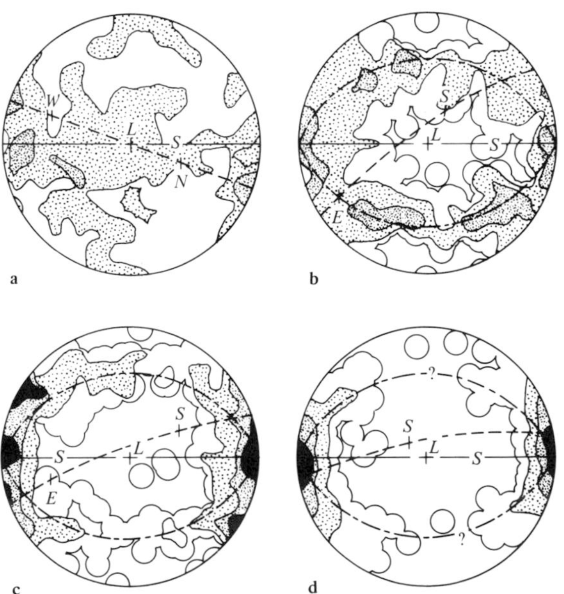

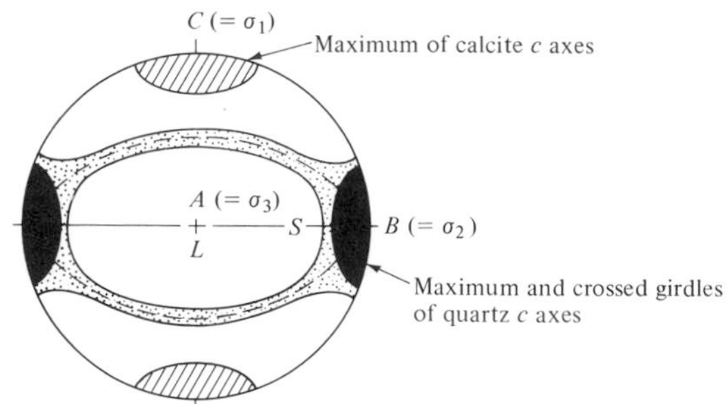

8.14 Relation between quartz and calcite orientation. (After Sylvester and Christie, 1968.) Reprinted by permission of The University of Chicago Press.

any deformation. The orientation of calcite recrystallized during experimental deformation conforms to this pattern as found by Griggs *et al.* (1960) and by Heard (1963). Marked orientation of olivine crystals have also been produced by experimental deformation (Ave'Lallemont and Carter, 1970). As was pointed out by Sylvester and Christie for Harkless quartzite, the correlation between the degree of preferred orientation and the degree of thinning of the beds suggests that the *magnitude of the strain* is the factor most responsible for preferred orientation. We shall see that this conclusion is compatible with attributing the orientation to kinetic rather than equilibrium factors.

It is possible that some degree of orientation of elongate and platy mineral grains in regionally metamorphosed rocks is due to fluid flow of a less competent matrix around these grains. However this mechanism cannot explain the exceedingly common occurrence of clots of almost pure amphibole, mica, etc., in which orientation is highly pronounced. It is clear that in such cases the orientation must be more closely related to the crystal structure than to the grain dimensions.

Lattice orientation of minerals of deformed rocks has been treated by a number of authors notably, Verhoogen (1951a), MacDonald (1957), Kamb (1959), and Ramberg (1959). Without exception these authors have attempted to show that orientation is a response to differential stresses that modify the thermodynamic stability so that a given orientation with respect to the stress field represents the lowest free energy state of the crystal. In all of these papers the oriented phase is treated as a static entity with certain stability relations irrespective of its surroundings. Although the thermodynamic deficiencies of these treatments are in themselves evident, it is clear that they fail even more seriously in that they treat the orientation process as a static balancing of forces. That this cannot be the true picture is indicated by the fact that rocks that contain oriented mineral grains show evidence of considerable strain, as cited above for the Harkless quartzite. But if such is the case, the time rate of strain almost certainly enters into a full description of the system. Indeed, considered in the light of kinetic theory the problem of mineral orientation can be phrased as follows:

"Given a certain initial distribution of mineral phases, composition, and state of stress, what is the lowest free energy path along which a rock can yield to relieve the stress and achieve a state of mechanical and chemical equilibrium?"

This is clearly a modified version of LeChatelier's principle. However we note that we are here concerned with the path of energy change rather than with the energy state of the end products. The relief of stress in an atomistic sense can result only from the motion of strain through gliding or recrystallization creep. That is, the minerals in a sense grow in such a way as to allow the total rock to deform as expiditiously as possible to relieve the stress. In terms of kinetics there will always be a number of competing mechanisms by which a rock can deform, but among competing mechanisms the most rapid one will always take over.

Although little information is as yet available to aid in constructing a kinetic model for mineral orientation, several mechanisms suggest themselves. The most obvious of these are the internal deformation mechanisms such as the various forms of gliding and recrystallization, which we have already seen appear to be thermally activated. According to Buerger and Washken (1947) and Griggs *et al.* (1960) another source of the activation energy is the mechanical strain that is accumulated in some types of deformation. In any case such internal motions are probably accompanied by intergranular motions also involving thermal or strain energy activation. The growth of crystals under stress may also involve elements of the model suggested by Feltham (1957) in which the rate of growth of a given interface depends on the interfacial energy [Equation (8.13)]. If we apply this to amphibole, for example, we should expect that the orientation of the c axis in the direction of extension of a rock would be favored because the highest rate of growth should occur in the direction of (001) rather than in the direction normal to the dominant prism [110] faces with low values of γ. An example of such growth is indeed given by Misch (1969) who describes sodic amphiboles with their c axes oriented in the direction of extension. The rate of extension was initially so great that these crystals were incipiently boudinaged. However the growth of the crystals then kept pace with the rate of separation of the boudins so that the single-crystal structure was maintained. The quasiseparation is marked by segments of different composition.

It is perhaps not meaningful to speculate whether internal plastic deformation or surface properties play the most important role in thermally activated crystal orientation. It is more likely that these are but two aspects of the same phenomenon, which depends more fundamentally on the anisotropic properties that govern both interior and exterior movements.

Origin of macrostructures

Metamorphic rocks contain a variety of structures on scales larger than the normal grain size (1 mm). These structures range from microscopic cracks and veins to large folds, faults, and dikes. However we shall concern ourselves only with those aspects of structures that stem primarily from chemical processes or that are greatly modified thereby. For example, we omit consideration of the formation of open fractures or other features that result from mechanical forces. It is convenient to tabulate the structures of interest as follows:

I. Modified relict structures
  - Deformed and recrystallized primary sedimentary structures such as pebbles, fossils, graded bedding, etc.
  - Pinch and swell structures
  - Boudinaged and microboudinage
  - Folded beds, veins, and like bodies of metamorphic origin
  - Pressure shadow mineralization

II. Newly developed structures
  - Dilatent and nondilatent veins or other bodies, both zoned and unzoned, including pinch and swell secretion veins
  - Gneiss or schist bridges
  - Vein aureoles and selvages

Like those of purely mechanical origin, the dominantly chemical and mineralogic features of rocks are valuable guides to the kinematics and dynamics of deformation. In addition, they also serve as detailed records of the chemical kinetic mechanisms and driving forces of petrogenesis.

It may be observed that in the group-I structures the element of mechanical behavior appears to be dominant in that these features depend on the bulk properties such as yield strength, density, viscosity, etc. On the other hand, in group-II chemical properties seem dominant. However it should be apparent that mechanical deformation and chemical migration are so closely related that no clear distinction can be made between them.

The utilization of deformed relict sedimentary structures in the analysis of strain has been reviewed in depth by Ramsay (1967). However our concern here is mainly with what these structures can tell us about the chemical history of rocks. Consequently we shall be particularly interested in the relationships of the primary group-I structures and the newly developed group-II structures since these relationships bear directly on such important petrogenetic questions as the origin of quartzo-feldspathic veins, pegmatites, and other controversial rocks.

The presence of heterogenieties such as competent beds brings about the localization of fractures to the competent parts. This behavior gives rise to the familiar boudinage structures when the dislocated beds are pulled apart by further plastic yielding of the incompetent material. Of course the relative amounts of fracture and flowage will depend on the magnitude of stress and the contrast in competence between the different beds. Even when the stresses are weak enough for the entire rock mass to yield in a ductile way the varying viscosities of different beds are revealed by several geometric forms. In the case of extension parallel to the beds, "pinch and swell structures" in the least ductile layers represent periodic necking down or thinning which corresponds to boudinage in more competent layers. On the other hand, compression parallel to the bedding can give rise to folds of varying wavelength depending on the thickness t and the relative viscosities of adjacent beds. Thus if a single, relatively competent bed of viscosity η_1 is embedded in a less competent medium of viscosity η_2, the dominant wavelength W_d is given by

$$W_d = 2\pi t\left(\frac{\eta_1}{6\eta_2}\right)^{1/3} \qquad (8.36)$$

This expression, which was derived independently by Biot and by Ramberg, is discussed in detail by Ramsay (1967).

The bulk properties of rocks and their consequent structural behavior depend on the mineralogic constitution as well as the temperature, confining pressure, and magnitude of stress. It is apparent then that a rock may respond quite differently to stress at different periods of its history when it is composed of different minerals. Thus an argillaceous limestone, which yields by brittle fracture under low temperature and near surface conditions, may lose much of its strength at only moderate metamorphic temperatures. But this same rock may again return to a state of high competence after its constituents react to form calc silicates. Ramberg (1956) has emphasized the relation between chemical composition and the manner of deformation under high-grade metamorphic conditions. He has found the following sequence of decreasing tendency for rupture among silicate rocks under tension:

1. Peridotite, pyroxenite, skarn rocks
2. Amphibolite
3. Coarse-grained quartzo-feldspathic rocks
4. Mica schist

The calcite marbles are rather unusual in that their response to stress appears to be largely independent of the presence of water (Griggs *et al.*, 1960). It was also found by Heard (1963) that the deformation of calcite marbles could be represented by the general form of the rate equation for fluid flow [Equation (3.48)], at least at low strain rates. An extension of this theory to higher strain rates by the introduction of separate terms for internal gliding and for slip along crystal boundaries was proposed by Hahn *et al.*, (1967). These problems have been reviewed by Elliot (1973). The relations found and inferred by Heard (1963) between viscosity, strain rate and temperature are shown in Figure 8.15 for extension parallel to the foliation plane in Yule marble. It appears from his results that in the geologic range of strain rates the viscosity is independent of the strain rate as given by Equation (3.51) except in the lowest temperature range. Thus if we can estimate the temperature and pressure at the time of deformation of a given marble, it should be possible to use Equation (8.36) to estimate the viscosity of beds of other composition that have been deformed within the marble. Behavior such as is shown by Figure 8.15 also informs us that the so-called "yield point"—the point at which the transition from elastic to plastic behavior occurs—depends critically on the temperature and strain rate and probably vanishes entirely at low strain rates.

It is interesting and important that the temperature sensitivity of rocks to plastic or fluid deformation is matched by an almost equal sensitivity to higher stress modes which bring about fracture. Thus Griggs *et al.* (1958) found that hard rocks such as granite and pyroxenite lost most of their strength at temperatures considerably below beginning of melting. For example, it was found that granite under a confining pressure of 5 kbar was deformed to the same extent by a differential stress of 5 kbar at 800°C or by a differential stress of 20 kbar 25°C. The mechanism of deformation of such rocks in this high stress mode is largely cataclastic flow

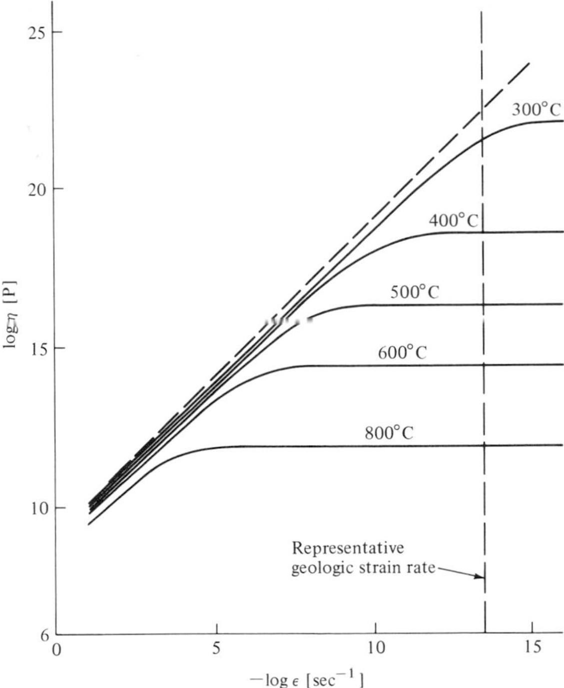

8.15 Relationship between viscosity, strain rate, and temperature in Yule marble. (After Heard, 1963.) Reprinted by permission of The University of Chicago Press.

with some internal gliding and is largely localized in shear zones. The application of such experiments to nature appears to be confined to very local fracturing in planetary lithospheres; however, it may be of considerable importance in the interpretation of earthquakes.

Boudinage structures and their associated mineralization are quite important as sources of information on the relation between mechanical deformation and chemical transport of new materials into the site. In the plastic yielding of incompetent beds surrounding the boudins the rate of creep is governed by the rate of such processes of gliding and recrystallization, and this form of movement is usually adequate to prevent the appearance of open cracks. Filling between the boundins thus occurs simultaneously with dislocation. If conditions are unsuitable for the transport and deposition of mineral matter from outside the system, the gaps between the boudins are filled with the incompetent layers that flow bodily into the low-pressure zone in which a "virtual" incipient opening is forming. If conditions are favorable for the deposition of new mineral matter the gaps may be filled by pegmatitic or other vein minerals. All these forms of filling are evidence for the unity of response of vein-forming processes under both high- and low-grade metamorphism. Ramberg (1961), in his description of quartz-calcite veins in low-grade Caledonian rocks of Norway, has presented particularly good evidence for the relation of such veins to ordinary quartzo-feldspathic pegmatites. Ramberg (1956) also described many examples of interboudinage pegmatite minerals from Greenland. Particularly interesting is his observation that such mineral-forming sites often result in pegmatites with quartz cores. Presumably they are evidence for the continued availablity of transportable silica after the depletion of corresponding feldspar by precipitation in the border zones. It is likely that this merely reflects the greater solubility of silica, as compared with feldspar, at lower temperatures during the waning stages of metamorphism.

Misch (1969) has shown that boudinage also occurs on a microscale and that it can be good evidence for synkinematic crystallization. Lower limits to boudinage are obviously set by the scale of the crystallization or recrystallization.

A frequently observed microstructure is the pressure shadows of quartz associated with garnet and pyrite grains. This phenomenon is clearly a manifestation of the same driving force behind the filling of incipient and real cracks and voids by mineral matter on a macroscopic scale. Geometrically identical structures have been observed around boudinage fragments of competent rocks that have undergone tectonic rotation and in the process have picked up pegmatite minerals in a "snowball" fashion (Ramberg, 1956). Such structures indicate that the pegmatite material was itself quite competent, presumably because it was coarse grained, and could continue to form pressure shadows even after the original boudin was enclosed by it.

A characteristic related to crystal pressure shadows is the spiral or snowball structure of garnet, which apparently reflects rotation due to tectonic movements. This phenomenon has been the subject of detailed studies by Spry (1963) and by Rosenfeld (1970). Rosenfeld in particular has shown that this feature of garnets can be utilized to place certain limits on the properties of rocks and orogenic forces during deformation. Thus he has inferred (Rosenfeld, 1968) that the viscosity of mantling rocks of certain gneiss domes in Vermont must have been less than 3×10^{20} P at the time of crystallization of these garnets. Rosenfeld (1970) has also used these same features to place limits on the differential stress. It is interesting that the viscosity thus obtained is lower than that commonly accepted for the low seismic velocity zone of the mantle which is associated with postglacial rebound (Chapter 3). Thus there appears to be mechanisms for lowering crustal viscosities considerably during orogenic movements.

Veins are frequently late-formed structures in metamorphic rocks although their character and composition may be closely related to the nature of this rock. Following Goodspeed (1940) we categorize a vein as either *dilatent* or as *nondilatent*, which means that either it has formed in a fracture in which the wall rock has been dilated to approximately the thickness of the vein or that the vein material was deposited through replacement with little or no mechanical separation of the vein walls. It is obvious, however, that dilatent veins need never have been more than incipient fractures at any stage of filling. We shall not concern ourselves with the criteria for dilatency and replacement since these have been exhaustively treated by Goodspeed and by others. However, we draw attention to an interesting feature of some dilatent veins of a discontinuous nature in which the vein is interupted by a schist or gneiss "bridge." It appears that such bridges are left when a secretion pegmatite forms in a discontinous incipient fracture (Ramberg, 1956). The bridge is an important aid in the interpretation of the chemical mechanism.

As we have indicated, the chemical and mechanical properties of rocks cannot readily be separated in plastic deformation. Thus the origin of all kinds of veins and secondary structures of those veins can be understood only from the standpoint of thermodynamics and kinetics which govern the atomistic behavior. We shall explore these factors in more detail in the next section.

Origin of compositional variations

The forms of variation

The different forms of compositional variation encountered in metamorphic rocks are as follows:

1. Compositional variations in single crystals
   Continuous variations
   Discontinuous variations
2. Compositional variations from crystal to crystal in the same rock
   Continuous variation
   Discontinuous variation
   Random variation
   Interrelated variation
3. Bulk chemical variations associated with varying proportions of different phases

We have discussed general kinetic aspects of compositional variation in Chapter 3, where we also gave a specific illustration of the relation between the composition of spatially separated but chemically connected crystals and from which we inferred that this connection can be maintained through heterogeneous reactions and the differential mobilities of certain constituents such as H_2O, H_2, CO_2, CO, etc. We pursue this topic further and consider also the growth processes of single crystals.

It has been pointed out (Mueller, 1967a; Misch, 1968) that the compositional variation with distance of a given phase is almost always less pronounced than the bulk chemical variation across the layers of banded metamorphic rocks. The nature of these variations have been illustrated in Figure 3.1. It is apparent that in each case the phase compositional variation is of a quite continuous nature. However we should not be surprised at occasional discontinuities since these could arise through such processes as postcrystallization mechanical displacement.

The compositional variation from grain to grain is sometimes duplicated within single grains of certain metamorphic minerals and it may assume a magnitude almost equal to the compositional zoning exhibited by magmatic minerals. However, metamorphic zoning of this type appears to be much more continuous as a general rule. A typical example of metamorphic mineral zoning, which is frequently found in garnets of the low to intermediate grade is characterized by a concentration of manganese in the early-formed core. A somewhat different type of zoning is shown by staurolite, in which the composition varies from sector to sector (Hollister and Bence, 1967). We shall see that unlike the normal type of compositional variation between grains that is shown by all mineral species, the sharp zoning in single grains is attributable to an almost complete failure of diffusional communication and equilibrium between zones.

It is always difficult to decide in any given case whether the bulk compositional variations in gneisses or schists are attributable to

1. Original differences between sedimentary or igneous bands
2. Metamorphic differentiation, perhaps induced by parallel shears, tension fractures, or even by microfolding as described by Misch (1969)
3. Emplaced foreign material

Then too we encounter the variations of individual minerals and bulk compositions both within well-defined veins and in the rock surrounding these veins and which point to the veins and associated aureoles as distinct entities that have been superimposed on the rock in some way. Systematic variations of tourmaline compositions from the wall to the core of pegmatites have been described in detail by Staatz *et al.* (1955). Variation in wall rock plagioclase associated with certain pegmatites is shown in Figure 8.16. Frequently also this type of zoning is accompanied by concentrations of dark minerals (selvages) in the wall rock next to the vein.

When compositional variations of a phase occur within a vein the variation is generally related to the characteristic zoning in texture or phase proportions. However, Ramberg (1956) found that many of the biotite grains in small veins and segregations in the gneisses of west Greenland showed little or no variation although single crystals sometimes extended completely across the veins.

In many instances also there is a definite reaction relation between vein and wall rock minerals. For example, Ramberg noted that in the Caledonian rocks of Norway, referred to earlier, chlorite, calcite, and quartz of the veins could be related to the epidote and actinolite of the country rock as follows:

$$\underset{\textit{epidote}}{2Ca_2Al_3Si_3O_{12}(OH)} + \underset{\textit{actinolite}}{3Ca_2(Mg, Fe)_5Si_8)_{22}(OH)_2} + \underset{\textit{fluid}}{10CO_2} + \underset{\textit{fluid}}{8H_2O} \rightleftharpoons \underset{\textit{chlorite}}{3(Mg, Fe)_5Al_2Si_3O_{10}(OH)_8} + \underset{\textit{calcite}}{10CaCO_3} + \underset{\textit{quartz}}{21SiO_2} \quad (8.e)$$

It is interesting now to examine the thermodynamic factors in these compositional variations.

Thermodynamic factors in spatial variations in composition

We have seen earlier that in at least some instances polycrystalline aggregates such as spherlites represent lower free-energy states than the corresponding single crystals. A related question is whether the multiphase systems of metamorphic rocks can indeed ever be in internal thermodynamic equilibrium. In the case of the terrestrial environment, at least, the answer is simply that a polyphase system cannot be in equilibrium unless it satisfies the requirement imposed by the gravitational field that the density must increase downward (see Chapter 1). This means that in such a system the individual phases must segregate into large-scale layers in which each layer contains only phases of the same density. It is obvious that this requirement would result in a few large monophase layers. However, this type of density layering is unknown in metamorphic rocks even at the dimensional scale of the mineral grains that comprise them. It follows from this that, despite the general picture of phase compatibility and orderly element distributions, the detailed spatial distribution of phases must be determined by other than equilibrium factors. Yet despite these inhibitions on mechanical equilibrium, the driving forces that determine the gross disposition of phases are thermodynamic in character. This is true, for example, of the observed tendency for the establishment of continuous compositional variation of phases as related to the diffusion of constituents since the tendency in such diffusion is to level chemical potential gradients. Also, it has frequently been suggested that one of the chief driving forces of metamorphic differentiation is the pressure differential that is established between the

8.16 Variation in plagioclase composition at the contact with a pegmatite. (Data from Kretz, 1968.)

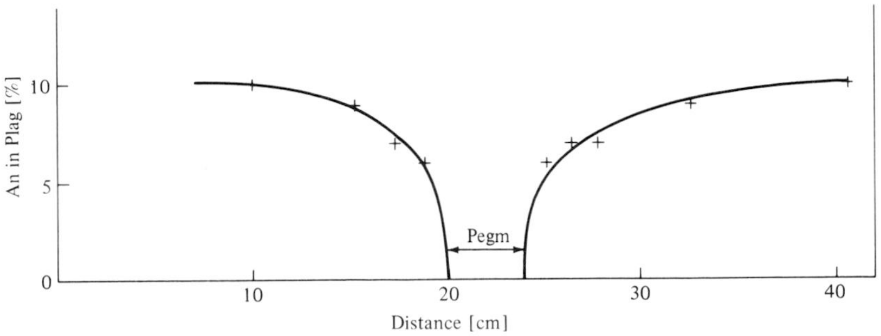

country rock and certain incipient voids or low-pressure areas that occur in cracks, crests of folds, etc. Mineral constituents are thought to migrate from high- to low-pressure regions because an increase in pressure always increases the chemical potential. In some cases, migration is one step in a heterogenous reaction so that the reaction itself depends on the ability of certain of its products to migrate. An example is Reaction (8.e) since it is known that the solids on the right-hand side of this reaction are greater in volume than those on the left. Presumably if the fluid H_2O and CO_2 could escape from the rocks, chlorite, calcite, and quartz would be favored in low-pressure vein sites—as is observed.

Numerous other examples of the action of thermodynamic forces in petrogenesis might be given. However proof of the existence of chemical potential gradients is only one step in an interpretation, and it can cast little or no light on the much more difficult kinetic problem that must be solved to give a complete picture.

Kinetic factors in spatial compositional variations

The rate at which any heterogeneous chemical process occurs in rocks should depend on the following factors:

1. Rates of diffusion
   Intrinsic mobilities of the various chemical species
   Concentrations of the diffusing species in the phase in which diffusion occurs most rapidly
2. Rates of the heterogeneous chemical reactions, including nucleation if new phases are formed

In addition to the diffusion of individual particles, the transport of fluids bearing the mobile constituents by hydrodynamic flow in fractures and in other openings may play a role in some cases, particulary where long-distance transport is required. However, it is unlikely that such flow is rate-determining in most cases since it would be only a step in a sequence that probably involves slower chemical processes under items 1 and 2 above.

By "intrinsic mobility" we mean the mobility attained by virtue of the coefficient of diffusion D_i, which characterizes the species i in a particular environment. In the simple illustration of the diffusion of one gaseous isotope into another isotope of the same element we find that

$$D = \frac{1}{3}\frac{\bar{v}}{c\sigma} \tag{8.37}$$

where $\bar{v}$ is the mean speed of the particle, c is the total concentration of both isotopic particles, and σ is the collision cross section. $\bar{v}$ may be evaluated from the gas kinetic formula

$$\bar{v} = \left(\frac{2.55kt}{m}\right)^{1/2} \tag{8.38}$$

in which k is Boltzmann's constant and m is the particle mass. If we make the assumption that many fluids behave at least qualitatively like a perfect gas, we conclude that the intrinsic mobility is approximately inversely proportional to the square root of the particle mass. On this basis H_2 should be one of the intrinsically most mobile of molecules. However, if the particle in question interacts strongly with other particles or with mineral interfaces, σ may assume large and highly variable values so that D could be greatly diminished by such behavior.

By the effect of concentration on mobility we mean simply that the latter is limited by the quantity of the diffusing constituent in the medium in which diffusion occurs. This concentration will generally be quite independent of its concentration in some other chemical compound as when it is tightly bound in mineral grains of the system. This point is particularly well brought out in the case of the diffusion of molecular constituents such as H_2O which contain the common isotope O^{16} and the much rarer isotope O^{18}. Since the chemical properties and masses of these two isotopes differ only minutely, their intrinsic mobilities and diffusion coefficients should be essentially the same. We may then write

$$\begin{aligned} J_{H_2O^{16}} &= -D\frac{dc_{H_2O^{16}}}{dX} \\ J_{H_2O^{18}} &= -D\frac{dc_{H_2O^{18}}}{dX} \end{aligned} \tag{8.39}$$

or

$$\frac{J_{H_2O^{18}}}{J_{H_2O^{16}}} = \frac{dc_{H_2O^{18}}}{dc_{H_2O^{16}}}$$

We find that in natural systems the isotopic abundances are approximately $c_{H_2O^{16}} = 500c_{H_2O^{18}}$ (Clayton and Epstein, 1958) so that we obtain

$$\frac{J_{H_2O^{18}}}{J_{H_2O^{16}}} = \frac{1}{500}$$

Consequently the rate of exchange of these isotopes by diffusion, as is required in metamorphic recrystallization, should be governed by the concentration of the H_2O^{18} molecule that is in lowest supply. This is simply another example of the slowest step determining the overall rate of a consecutive kinetic process.

On a rigorous theoretical basis the intrinsic mobility

and concentration factors are not easily separated since both depend on interactions between particles. Thus we may say that the O_2 molecule has a low degree of mobility because its concentration is almost always vanishingly low. However, this characteristic depends directly on the tendency of O_2 to interact strongly with its surroundings. In particular, it reacts strongly with H_2 to form the stable and hence highly abundant H_2O molecule.

A petrologically interesting illustration of how the mobility of an element may be increased by reaction with another constituent is afforded by the system Mg-Si-O_2-H_2. Under anhydrous conditions enstatite and quartz can decompose according to the reactions

$$\underset{enstatite}{MgSiO_3} \rightleftharpoons \underset{gas}{Mg} + \underset{gas}{SiO_2} + \underset{gas}{\tfrac{1}{2}O_2} \tag{8.f}$$

$$\underset{quartz}{SiO_2} \rightleftharpoons \underset{gas}{SiO_2} \tag{8.g}$$

which have the equilibrium constants

$$K_{(f)} = P_{Mg}P_{SiO_2}P_{O_2}^{1/2} = 10^{-58.6} \tag{8.40}$$

$$K_{(g)} = P_{SiO_2} = 10^{-24.6} \text{ atm} \tag{8.41}$$

Now if the value $P_{O_2} = 10^{-18}$ atm is chosen as representative for a typical regional metamorphic rock we find that $P_{Mg} = 10^{-25}$ atm. In much the same way we could calculate the equilibrium fugacities for such molecular species as SiO, MgO, etc., and show that virtually all of these would be present only in vanishingly small concentrations. However, the total concentration of magnesium and silicon in the vapor phase can be greatly enhanced if water is present, and especially if it is present as a high-density fluid as was shown by Morey and Hesselgesser (1951) in their experiments. Wasserburg (1958) attributed this enhancement to the reaction

$$\underset{quartz}{SiO_2} + \underset{fluid}{2H_2O} \rightleftharpoons \underset{fluid}{Si(OH)_4} \tag{8.h}$$

From the stoichiometric coefficients of this reaction it is apparent that it is driven to the right by increasing the pressure. Thus at 870°K and a vapor pressure of 1000 bar the partial pressure of SiO_2 in the vapor phase is 0.88 bar. Thus water under high pressure has increased the concentration of vaporized silica many orders of magnitude over that corresponding to simple vaporization as given by reactions such as (8.f) and (8.g).

Morey and Hesselgesser also studied the solubility of the mineral enstatite, and their data provide us with simultaneous information on the concentrations of MgO and SiO_2 in the vapor phase. These were found to be 3.26×10^{-2} and 1.68×10^{-1} bar, respectively, at $T = 873$°K and $P_{H_2O} = 1000$ bar. Thus it is clear that at least under these conditions enstatite dissolves incongruently and that more SiO_2 than MgO goes into solution. It was subsequently pointed out (Mueller, 1967a) that the most likely reaction that occurred in Morey and Hesselgesser's experiments was

$$\underset{enstatite}{2MgSiO_3} + \underset{fluid}{2H_2O} \rightleftharpoons \underset{olivine}{Mg_2SiO_4} + \underset{fluid}{Si(OH)_4} \tag{8.i}$$

Accordingly it seems likely that Mg_2SiO_4 was metastably deposited in their runs although it was not reported.

Solubility relationships such as these, as experimentally determined and interpreted according to solution models, are necessary as a basis for the study of the transport properties of different elements in metamorphism. They are crucial to understanding the origin of large metasomatic complexes (Chapter 14).

The twofold character of diffusive processes in metamorphism requires that they be effective both within single crystals and between crystals that are spatially separated (Mueller, 1967). The vaporization processes just discussed not only supply the material for diffusion between grains but are also closely related to lattice diffusion, since these reactions almost invariably occur during recrystallization when lattice mobility is greatest. It is generally agreed, however, that large-scale mass transport is not possible through crystal lattices. Experiments and theoretical considerations indicate rather that diffusion coefficients for transport along surfaces and through intergranular fluids should be many orders of magnitude greater than corresponding coefficients for lattice diffusion (Glasstone *et al.* 1941). An example of lattice diffusion under static conditions is the unmixing of immiscible crystalline phases in the pyroxenes and the feldspars, which results in perthitic textures. In each case where such unmixing is pronounced it involves relatively low activation energies. Where higher activation energies are involved, as in the formation of plagioclase components of peristerites, the unmixing is on a very fine scale. By contrast in most metamorphic recrystallization even such high activation energies as are entailed in the exchange of $(NaSi)^{5+}$ and $(CaAl)^{5+}$ units are overcome by differential stresses that act to lower the energy barriers so that not only lattice diffusion but heterogeneous chemical reactions are favored. It may be inferred on the basis of experiments with such compounds as Al_2O_3 (Cobel, 1963) that in diffusive deformation processes in which Equations (3.50) and (3.51) apply, the rate-controlling step is the diffusion of oxygen ions since this species has a much greater activation energy of diffusion than does a metal cation such as Al^{3+}. Consequently if recrystallization creep occurs the lower energy barriers for diffusion of such species Ca^{2+}, Na^{+}, Al^{3+}, and Si^{4+} will be overcome simultaneously with that of oxygen so that compositional homogenization which might be impossible under static conditions will be achieved under stress (Mueller, 1967a).

In addition to the homogenization of zoned or potentially zoned minerals under stress we need also to consider the mechanisms of formation of such zones, in the first place, by purely metamorphic crystallization. Perhaps the oldest explanation of compositional zoning in metamorphic crystals is that it arises through progressive changes in grade which are registered in the successive zones. However, this explanation scarcely elucidates the mechanism by which the zoning occurred and probably is not valid in most cases. The most satisfactory explanation for the common type of metamorphic zoning shown in Figure 8.17 was given by Hollister (1966) and Atherton (1968) who suggested that it involves Rayleigh type fractionation. In the latter process a certain weight of rock substance W_R in association with the garnet and containing a weight W_R^E of the element in question reacts to form a weight fraction M_{Gar}^E of this element in the outer shell of the growing garnet so that

$$M_{Gar}^E = \frac{dW_R^E}{dW_R} \tag{8.42}$$

Specifically this equation states that a weight increment dW_R of the rock reacts to contribute an increment dW_R^E of the element to the garnet. Since perfect fractionation is assumed, the quantity of the element added to the garnet is effectively removed from the (rock) system. If D is regarded as either a thermodynamic or steady-state distribution coefficient the following simple distribution relation may be assumed to hold:

$$D = \frac{M_{Gar}^E}{W_R^E/W_R} \tag{8.43}$$

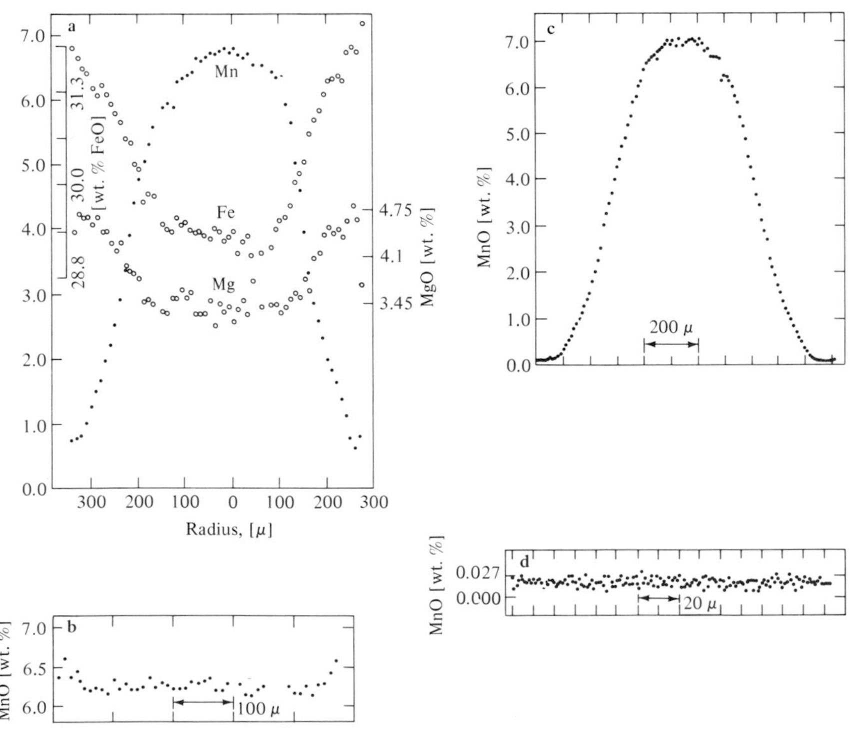

8.17 (a) Mn, Fe, and Mg profiles across the type garnet. (b) Example of Mn profile of nearly homogeneous garnet. (c) Example of Mn profile across a garnet with edge Mn content approaching zero. (d) Mn profile across cleavage of a biotite grain from the same rock as in (a). (After Hollister, 1966.) Reprinted from *Science*, *154*, pp. 1647–1651, Fig. 3. Copyright 1966 by the American Association for the Advancement of Science.

If now Equations (8.42) and (8.43) are combined and integrated we obtain

$$\frac{M_R}{M_R^0} = \left(\frac{W_R}{W_R^0}\right)^{D-1} \qquad (8.44)$$

where $M_R = W_R^E/W_R$ is the weight fraction of the element in the rock exclusive of the garnet. The superscript ^{0} indicates the condition before any garnet was present. if we bear in mind that $W_R = W_R^0 - W_{Gar}$, where W_{Gar} is the total weight of crystallized garnet, Equation (8.44) may be written as

$$M_{Gar}^E = DM_R^0\left(1 - \frac{W_{Gar}}{W_R^0}\right)^{D-1} \qquad (8.45)$$

This equation gives the weight fraction of a given element at the edge of a garnet in terms of, M_R^0, the weight fraction of the element in the rock as a whole and the total weight of crystallized garnet, W_{Gar}. Hollister found it possible to fit Equation (8.45) quite well to the manganese data (Figure 8.17). However, he did not further discuss the heterogeneous reactions by which this form of fractionation might occur. This may be done as follows.

According to Hollister (1966) the rock specimen represented by Figure 8.17 consisted of biotite, chlorite, staurolite, ilmenite, graphite, quartz, and plagioclase. The biotite and chlorite are the chief sources of manganese in the garnet. Since these minerals were the most abundant ferromagnesian silicates, they presumably were also the major sources for Fe^{2+} and Mg in the garnet. If we consider the possible reactions involved we are led to something like the following:

$$\underset{biotite}{K(Fe, Mg, Mn)_3(Si_3Al)O_{10}(OH)_2} + \tfrac{6}{23}\underset{staurolite}{(Fe, Mg, Mn)_2Al_9Si_4O_{23}(OH)} + \tfrac{57}{23}\underset{quartz}{SiO_2} \rightleftharpoons \tfrac{27}{23}\underset{garnet}{(Fe, Mg, Mn)_3Al_2Si_3O_{12}} + \underset{feldspar}{KAlSi_3O_8} + \tfrac{26}{23}\underset{fluid}{H_2O} \qquad (8.j)$$

$$\underset{chlorite}{(Mg, Fe, Mn)_9Al_3(Al_3Si_5)O_{20}(OH)_{16}} + \underset{quartz}{4SiO_2} \rightleftharpoons 3\underset{garnet}{(Fe, Mg, Mn)_3Al_2Si_3O_{12}} + \underset{fluid}{8H_2O} \qquad (8.k)$$

These reactions actually represent three independent sets of reactions, one each for Fe, Mg, and Mn. If we consider only the manganese reactions, we may write the following distribution coefficients corresponding to Reactions (8.j) and (8.k):

$$D_{(j)}^{Mn} = \frac{(X_{Mn}^{Gar})^{81/23}}{(X_{Mn}^{Bi})^3(X_{Mn}^{St})^{12/23}} \qquad (8.46)$$

$$D_{(k)}^{Mn} = \frac{(X_{Mn}^{Gar})^9}{(X_{Mn}^{Chl})^9} \qquad (8.47)$$

We note that if $X_{Mn}^{Bi} \sim X_{Mn}^{St}$, both expressions approximate the form of Equation (8.43). However, Reaction (8.j) and Equation (8.46) also imply K-feldspar as a decomposition product of biotite so that if any substantial quantity of garnet is formed from local biotite the rock should contain either K-feldspar or plagioclase with the equivalent amount of this component in solution. The only alternative is for $KAlSi_3O_8$ to be transported and deposited outside the system.

The zoning in garnets of Yellowknife area was discussed previously. Kretz (1973) considers that garnets have grown principally at the expense of chlorites according to the reaction:

1 chlorite (aluminous) + 4 quartz = 3 garnet + 8 H_2O

Several possibilities of diffusional relationships between garnet and chlorite have been discussed by Kretz. In general the interpretation of Hollister and Atherton seems to be valid. However, Kretz finds that Equation (8.45) is too restrictive. In particular, the factor D is not necessarily constant. Edmunds and Atherton (1971) suggested that the curious increase in the Mn content found in the outer margins of garnets (see Figure 8.12) may be due to an increase in D as a result of decrease in growth rate. Kretz finds that such Mn concentration may be adequately explained by a restriction on diffusion in chlorite during the reaction.

The compositional growth patterns such as have been discussed depend not only on the kinetics of transport and crystal growth but very fundamentally on the thermodynamic properties of the material-supplying reactions such as (8.f) and (8.g). The degree to which such reactions occur at any stage of garnet growth will govern the availability of the different constituents—Fe, Mg, and Mn-and depends on the temperature and pressure (or metamorphic grade). As is well known, reactions involving volatiles are governed in part by the permeability of the rock. To use another illustration, it is apparent that the availability of iron for garnet formation would also depend on such reactions as

$$\underset{magnetite}{Fe_3O_4} + \underset{Al\text{-}silicate}{Al_2SiO_5} + \underset{quartz}{2SiO_2} + \underset{fluid}{H_2} \rightleftharpoons \underset{garnet}{Fe_3Al_2Si_3O_{12}} + \underset{fluid}{H_2O} \qquad (8.l)$$

which involve the relative mobilities of H_2 and H_2O.

Summary

It should be apparent that the many aspects of metamorphic crystallization are highly interdependent. The first requirement for the growth of a crystal in a given specific location is that it represents the lowest free energy state attainable both with respect to the reactant crystals and fluids from which it is derived and in relation to other sites at which nucleation and growth might have occurred. Secondly, it is necessary that the components of the reaction have sufficient mobility to reach the site of growth; and finally that the material-producing reactions, nucleation, and growth must all be rapid enough to occur in the time available. These kinetic factors determine what is the "attainable" thermodynamic state. For the reactants and products to approach a state of local equilibrium it is necessary that each part of every crystal, including the interior, be in diffusional equilibrium with every other crystal of the local volume. It appears that the most effective agent for activating this communication is the recrystallization that occurs under differential stress. When stresses are ineffective, as in the case of strong crystals such as garnet, the exchange process between the interior of the crystal and both its own exterior and surrounding minerals is suspended and the peculiar zoning record of growth is retained. The precise spatial distribution of crystals in metamorphic rocks is a result of kinetic factors such as those that favor certain nucleation sites. These kinetic factors totally dominate and mask the driving forces of the gravitational potential. The same kinetic factors also greatly modify the influence of the equilibrium forms of crystals that result from interfacial free energies. In particular the dynamic processes of flow (gliding recrystallization, etc.) in stressed rocks result in the pervasive crystallographic co-alignment of crystals, which is a response that favors the flow process itself.

9 Metamorphic minerals

Experimental appraisal of the stability of minerals

Although many metamorphic minerals are stable under a variety of pressure and temperature conditions, there are many minerals that are stable only over a narrow range of P and T. Such *index* minerals provide useful petrogenetic information. Index minerals have played important roles in many classical studies in delineating metamorphic zones. However, there are relatively few minerals that are pure anhydrous phases. The stability of most metamorphic minerals, therefore, is a function of not only P and T but also of the fluid pressure and bulk chemistry. In this chapter, we present the P, T and compositional stability relations of many metamorphic minerals.

During the last 20 years a rapid progress has been made in the experimental study of metamorphic reactions. These studies have improved our understanding of the stability of different minerals and have led to a revision in the P and T boundaries of metamorphism and antexis. However, the bulk of the experimental data is for reactions involving pure solid and fluid phases and not solutions. In the second part of this chapter we present a theoretical analysis of the solution effect on the reactions.

The review of the experimental work is not exhaustive, and the stability relations of mineral systems such as carbonates and oxides have not been discussed in detail. The chemical formulae and abbreviations for several metamorphic minerals are listed in Table 9.1.

9 Metamorphic minerals

Table 9.1 Abbreviations of mineral names and chemical formulae

Abbr.	Mineral	Formula	Abbr.	Mineral	Formula
Anth	anthophyllite	$Mg_7Si_8O_{22}(OH)_2$	Kaol	kaolinite	$Al_2Si_2O_5(OH)_4$
Ab	albite	$NaAlSi_3O_8$	K-fs	K-feldspar	$KAlSi_3O_8$
Acm	acmite	$NaFeSi_3O_8$	Ky	kyanite	Al_2SiO_5
Act	actinolite	$Ca_2(Mg, Fe)_5Si_8O_{22}(OH)_2$	Lm	laumontite	$CaAl_2Si_4O_{12}\cdot 4H_2O$
Al-Sil	Al-silicate	Al_2SiO_5	Lw	lawsonite	$CaAl_2Si_2O_7(OH)_2\cdot H_2O$
Alm	almandine	$Fe_3Al_2Si_3O_{12}$	Mag	magnesite	$MgCO_3$
An	anorthite	$CaAl_2Si_2O_8$	Me	melilite	$Ca_2Al_2SiO_7$
And	andalusite	Al_2SiO_5	Ms	muscovite	$KAl_3Si_3O_{10}(OH)_2$
Ann	annite	$KFe_3AlSi_3O_{10}(OH)_2$	Mt	magnetite	Fe_3O_4
Anlc	analcite	$NaAlSi_2O_6\cdot H_2O$	Ne	nepheline	$NaAlSiO_4$
Bi	biotite	variable	Ol	olivine	$(Mg,Fe)_2SiO_4$
Br	brucite	$Mg(OH)_2$	Omp	omphacite	Jd + Di + Hd + Acm + Tschermak.
Calc	calcite	$CaCO_3$			
Chd	chloritoid	$(Fe,Mg)Al_2SiO_5(OH)_2$	Or	orthoclase	$KAlSi_3O_8$
Chl	chlorite	variable	Per	periclase	MgO
Cor	corundum	Al_2O_3	Pg	paragonite	$NaAl_3Si_3O_{10}(OH)_2$
Cord	cordierite	$(Fe,Mg)_2Al_3(Si_5Al)O_{18}\cdot\frac{1}{2}H_2O$	Phl	phlogopite	$KMg_3AlSi_3O_{10}(OH)_2$
Cum	cummingtonite	$(Fe, Mg)_7Si_8O_{22}(OH)_2$	Plag	plagioclase	$NaAlSi_3O_8$—$CaAl_2Si_2O_8$
Cz	clinozoisite	$Ca_2Al_3Si_3O_{12}(OH)$	Pr	prehnite	$Ca_2Al_2Si_3O_{10}(OH)_2$
Di	diopside	$CaMgSi_2O_6$	Pu	pumpellyite	$Ca_2Al_3(SiO_4)(Si_2O_7)(OH)_3$
Dol	dolomite	$(Ca,Mg)CO_3$	Px	pyroxene	variable
En	enstatite	$MgSiO_3$	Pyp	pyrophyllite	$Al_2Si_4O_{10}(OH)_2$
Ep	epidote	$Ca_2(Al,Fe)_3Si_3O_{12}(OH)$	Pyr	pyrope	$Mg_3Al_2Si_3O_{12}$
Fa	fayalite	Fe_2SiO_4	Q	quartz	SiO_2
Fer	ferrite	$(Fe,Mg)Fe_2O_4$	S	serpentine	$Mg_3Si_2O_5(OH)_4$
Fos	forsterite	Mg_2SiO_4	San	sanidine	$KAlSi_3O_8$
Fs	ferrosilite	$FeSiO_3$	Sap	sapphirine	$Mg_2Al_4SiO_{10}$
Fss	feldspar	variable	Sd	siderite	$(Fe,Mn)CO_3$
G	graphite	C	Sill	sillimanite	Al_2SiO_5
Gar	garnet	variable	Sp	spinel	$(Mg, Fe)(Al, Cr)_2O_4$
Ge	gehlenite	$Ca_2Al_2SiO_7$	Sph	sphene	$CaTiOSiO_4$
Gr	grossularite	$Ca_3Al_2Si_3O_{12}$	Sps	spessertine	$Mn_3Al_2Si_3O_{12}$
Gl	glaucophane	$Na_2(Mg,Fe)_3Al_2Si_8O_{22}(OH)_2$	St	staurolite	$(Fe,Mg)Al_{18}Si_8O_{44}(OH)_4$
Hb	hornblende	variable	Stilp	stilpnomelane	variable
Hd	hedenbergite	$CaFeSi_2O_6$	Tc	talc	$Mg_3Si_4O_{15}(OH)_2$
Hem	hematite	Fe_2O_3	Tr	tremolite	$Ca_2Mg_5Si_8O_{22}(OH)_2$
Her	hercynite	$FeAl_2O_4$	V	vapor	variable
Heu	heulandite	variable	Wr	wairakite	$CaAl_2Si_4O_{12}\cdot 2H_2O$
Hy	hypersthene	$(Mg,Fe)SiO_3$	Wo	wollastonite	$CaSiO_3$
Il	ilmenite	$FeTiO_3$	Z	zoisite	$Ca_2Al_3Si_3O_{12}(OH)$
Jd	jadeite	$NaAlSi_2O_6$			

Laumontite, wairakite, lawsonite, and related minerals

The hydrous silicates of Ca and Al forming at low temperatures have been studied by Liou (1971b). These are

laumontite ($CaAl_2Si_4O_{12}4H_2O$),
wairakite ($CaAl_2Si_4O_{12}2H_2O$),
and lawsonite ($CaAl_2Si_2O_7(OH)_2H_2O$),

related by the following reactions:

$$\text{laumontite} = \text{wairakite} + 2H_2O \quad (9.a)$$

$$\text{laumontite} = \text{lawsonite} + 2\text{quartz} + 2H_2O \quad (9.b)$$

$$\text{lawsonite} + 2\text{Quartz} = \text{wairakite} \quad (9.c)$$

The equilibrium conditions are

(9.a)	**(9.b)**	**(9.c)**
230°C at 0.5 kbar	210°C at 3 kbar	305°C at 3.4 kbar
255 ± 5°C at 1 kbar	275°C at 3.2 kbar	390°C at 4.4 kbar
282 ± 5°C at 2 kbar		
297 ± 5°C at 3 kbar		
325 ± 5°C at 6 kbar		

To these data we may add those of Newton and Kennedy (1963) as shown in Figure 9.1. The boundary between wairakite and quartz and anorthite (Liou, 1970) seems to be the upper temperature limit for the zeolites under the condition that P_{total} is equal to P_{fluid}. If P_{fluid} is less than the total pressure, the dehydration curves will shift toward low temperatures. The effect of increasing P_{CO_2} will be similar (see Albee and Zen, 1969).

Stilpnomelane

Eggleton's (1972) structural and chemical data indicate that stilpnomelane may be represented by the formula

$$(Ca, Na, K)_4(Ti_{0.1}Al_{2.3}Fe_{35.5}Mn_{0.8}Mg_{9.3})(Si_{63}Al_9)(O, OH)_{216} \cdot nH_2O.$$

Fe^{3+} is present in large amounts in some stilpnomelanes called ferristilpnomelane by Hutton (1938). Zen (1960) suggests that this may be due to oxidation of Fe^{2+} subsequent to the formation of the mineral. Eggleton (1972), however, considers that from a structural point

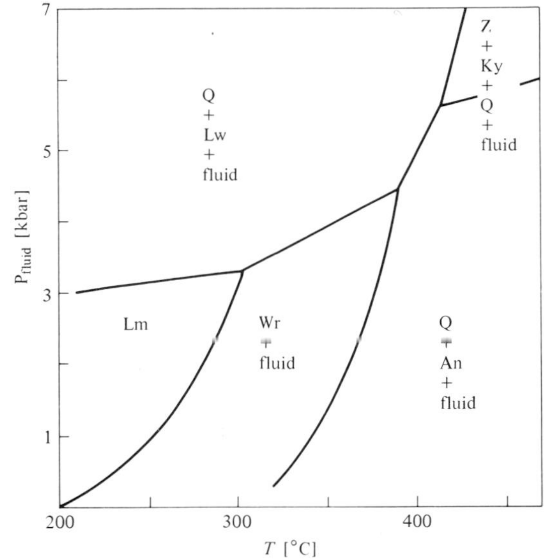

9.1 Minerals in the system $CaAl_2Si_2O_8$-SiO_2-H_2O. Data are from Liou (1971b); boundaries on the right hand above 4 kbar are from Newton and Kennedy (1963).

of view, both ferro- and ferri-stilpnomelane appear equally stable under appropriate oxidation conditions.

There is no experimental data on the synthesis and stability of the mineral. It occurs in rocks of the lowest metamorphic grade as in the schists of western Otago, New Zealand (Hutton, 1940; Brown, 1967) or in the low-temperature and high-pressure glaucophane schists in central Japan (Miyashiro and Seki, 1958). The high-temperature limit of stilpnomelane may lie in the range of temperature in which biotite appears. Generally, stilpnomelane and biotite do not occur in the same rock. However, Brown (1967) has described stably coexisting biotite and stilpnomelane from the Otago schist belt. Actinolite is also commonly associated with this mineral (Turner, 1968, pp. 271–274).

Chlorite

Chlorites form a group of minerals characterized by principal substitutions of Al^{3+} and Mg^{2+} by Fe^{2+}, Fe^{3+}, and Si^{4+}. Those in pelitic schists may be described by the end members

$$Mg_{4.8}Al_{1.2}(Si_{2.8}Al_{1.2}O_{10})(OH)_8\text{-}Fe_{4.8}Al_{1.2}(Si_{2.8}Al_{1.2}O_{10})(OH)_8,$$

with probable variation in Al and Fe^{3+}.

Experimental studies on the stability of chlorite are available mainly for clinochlore

$(Mg_5Al)(Si_3AlO_{10})(OH)_8$.

Clinochlore alone is stable up to temperatures of 670°C to 820°C over the range of P_{H_2O} from 1 to 7 kbar (Fawcett and Yoder, 1966). The breakdown products above 3 kbar are enstatite + forsterite + spinel, and below 3 kbar forsterite + spinel + cordierite.

Clinochlore in combination with other minerals such as quartz, muscovite, and Al_2SiO_5 becomes unstable at much lower temperatures than those of clinochlore alone, which illustrates a general principle of mineral stability. Recent data on the reactions involving chlorite are shown in Figure 9.2. The equilibrium reactions are

chlorite + quartz $\rightleftharpoons$ cordierite + talc + H_2O (9.d)

(Fawcett and Yoder, 1966)

chlorite + muscovite + quartz $\rightleftharpoons$
phlogopite + cordierite + H_2O (9.e)

(Seifert, 1970; Bird and Fawcett, 1973)

chlorite + andalusite + quartz $\rightleftharpoons$
cordierite + H_2O (9.f)

(Seifert and Schreyer, 1970)

chlorite + muscovite $\rightleftharpoons$
phlogopite + kyanite + quartz + H_2O (9.g)

(Bird and Fawcett, 1973)

All chlorites in Reactions (9.d) to (9.g) are magnesium chlorite (clinochlore), and since Mg-silicates tend to be relatively stable, these reactions represent the maximum P and T for chlorite to be stable in metamorphic rocks. Since chlorites in most pelitic rocks are rich in Fe^{2+} and Al^{3+}, the T and P of the disappearance of chlorite in progressive metamorphism will be considerably lower than, for example, the range of 580°C to 620°C at 4 kbar. This is quite evident in Hoschek's (1969) experimental results of the reaction:

(Fe-Mg) chlorite + muscovite $\rightleftharpoons$
staurolite + biotite + quartz + vapor (9.h)

The equilibrium T and P are 565 ± 15°C at 7 kbar and 540 ± 15°C at 4 kbar. Turnock's (1960) result on the synthesis of the ferrous aluminum chlorite (daphnite) suggest a maximum range of 565°C to 595°C at 2 kbar water vapor pressure for the upper stability of this

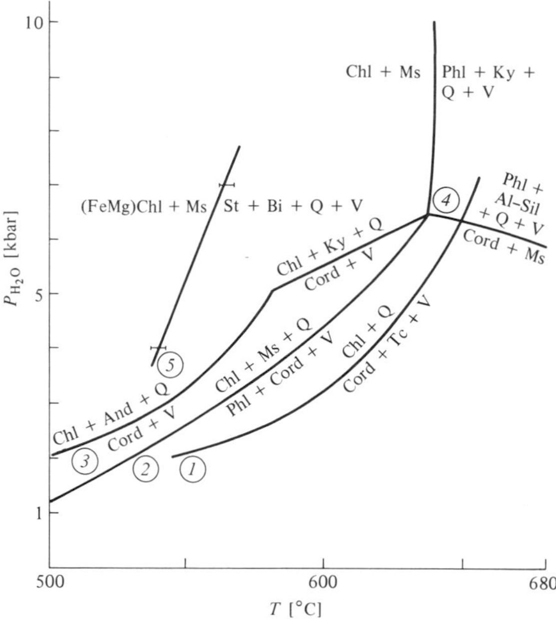

9.2 Experimental data on reactions involving chlorite. The sources are *1*: Fawcett and Yoder (1966); *2*: Siefert (1970); *3*: Siefert and Schreyer (1970); *4*: Bird and Fawcett (1973); *5*: Hoschek (1969).

mineral. It seems, therefore, that with quartz and muscovite, a Fe-Al-chlorite would be eliminated from the assemblage under 500°C. Hsu's (1968) work shows that quartz, Fe-chlorite, and fluid at low f_{O_2} react to form almandine between 500°C and 550°C in the pressure range 500 to 3000 bar. See also Chernosky (1974).

Epidote

The coexistence of albite (An ~ 7 percent) and epidote is regarded by most petrologists as a convenient index for the upper boundary of the greenschist facies. Experimental data on the upper stability limit of epidote is provided by Liou (1973), Nitsch and Winkler (1965), and Holdaway (1967, 1972). Liou (1973) discussed the following two reactions

epidote $\rightleftharpoons$
$grandite_{ss}$ + anorthite + hematite + quartz + fluid (9.i)

epidote + quartz $\rightleftharpoons$
$garnet_{ss}$ + anorthite + $magnetite_{ss}$ + fluid (9.j)

where the subscript *ss* denotes solid solution. The breakdown of epidote alone occurs in the temperature range of 600°C to 750°C over a P_{fluid} range of 1 to 5 kbar. The epidote + quartz reaction takes place at a somewhat lower range of 520°C to 670°C over a P_{fluid} range of 1 to 5 kbar. The fluid phase contains both O_2 and H_2O, and Reaction (9.i) is at f_{O_2} defined by the hematite-magnetite buffer and Reaction (9.j) by the Ni-NiO buffer. Liou's (1973) discussion shows that the stability of epidote may be bounded by the reactions

$$\text{prehnite} + FeO_x + O_2 \rightleftharpoons \text{epidote} + H_2O \quad (9.k)$$

and

$$\text{epidote} + \text{quartz} \rightleftharpoons \text{hedenbergite} + \text{anorthite} + \text{fluid} \quad (9.l)$$

or

$$\text{epidote} \rightleftharpoons \text{anorthite} + \text{grandite} + FeO_x + \text{quartz} + \text{fluid} \quad (9.m)$$

where x represents the nonstoichiometric O in wüstite. The second high-temperature reaction (9.l) takes place at much lower f_{O_2} than Reaction (9.m). At constant fluid pressure, increasing f_{O_2} enlarges the stability field to both higher and lower temperatures, while reducing conditions diminish the stability field.

If albite is present with epidote, the breakdown of epidote would take place at a lower temperature than that of any high-temperature reactions mentioned above. The experiments of Liou *et al.* (1971) indicate the following equilibrium P and T for the reaction

$$\text{epidote} + \text{chlorite} + \text{albite} + \text{quartz} \rightleftharpoons \text{oligoclase} + \text{tschermakite}$$

$$P_{fluid} = 2\text{ kbar}, T = 475°C\ (f_{O_2} \text{ of QFM buffer}) \quad (9.n)$$

The low-temperature stability limit of epidote under hydrous and oxidizing conditions may be estimated to be 220 ± 50°C over the total pressure range of 1 to 6 kbar and 300°C at 7 kbar (Seki, 1972; Tomasson and Kristmannsdottir, 1972).

Pyrophyllite

Experimental determinations on the upper stability limit of pyrophyllite ($Al_2Si_4O_{10}(OH)_2$) have been made by several workers. However, the results where reversibility could be demonstrated are few, and these are as presented below:

$$\text{pyrophyllite} \rightleftharpoons \text{andalusite} + 3\text{quartz} + H_2O \quad (9.o)$$

T[°C]	P_{H_2O} [kbar]	
400 ± 15	1.0	Hemley (1967)
410 ± 15	1.8	Kerrick (1968)
430 ± 15	3.9	Kerrick (1968)
490 ± 5	2.0	Althaus (1966)
525 ± 5	7.0	Althaus (1966)

There is a marked difference of about 80°C in the results of Kerrick and Althaus. The combined results of Kerrick and Hemley explain the transition between kyanite- and andalusite-bearing assemblages (Chinner, 1966).

The low-temperature stability of pyrophyllite may be established by the equilibrium P and T of the following reaction:

$$\underset{\text{kaolinite}}{Al_2Si_2O_5(OH)_4} + 2SiO_2 \rightleftharpoons \underset{\text{pyrophyllite}}{Al_2Si_4O_{10}(OH)_2} + H_2O \quad (9.p)$$

This reaction has been studied by many workers, more recently by Althaus (1966), Velde and Kornprobst (1969), and Thompson (1970). Figure 9.3 shows the possible fields of pyrophyllite stability according to the experimental results of Althaus and Kerrick and Thompson. The two fields differ significantly. The upper and lower temperature limits are somewhere in between, and the field may extend from about 350°C to 450°C at 2 kbar.

Chloritoid

Important information on the stability of chloritoid has been gathered by Hoschek (1969), Ganguly and Newton (1968), and Ganguly (1969). Petrographic studies indicate that iron-rich chloritoid is formed in regionally metamorphed rocks at about the beginning of the greenschist facies probably according to the reactions

$$\text{chlorite} + \text{Al-silicate (kaolinite, pyrophyllite)} = \text{chloritoid} + \text{quartz} + \text{vapor} \quad (9.q)$$

$$\underset{\text{kaolinite}}{3Al_4Si_4O_{10}(OH)_8} + \underset{\text{diaspore}}{6Al_2O_3 \cdot H_2O} + 4Fe_3O_4\ (\text{or } 6Fe_2O_3) \rightleftharpoons \underset{\text{chloritoid}}{12FeAl_2SiO_5(OH)_2} + 6H_2O + 2O_2\ (\text{or } 3O_2) \quad (9.r)$$

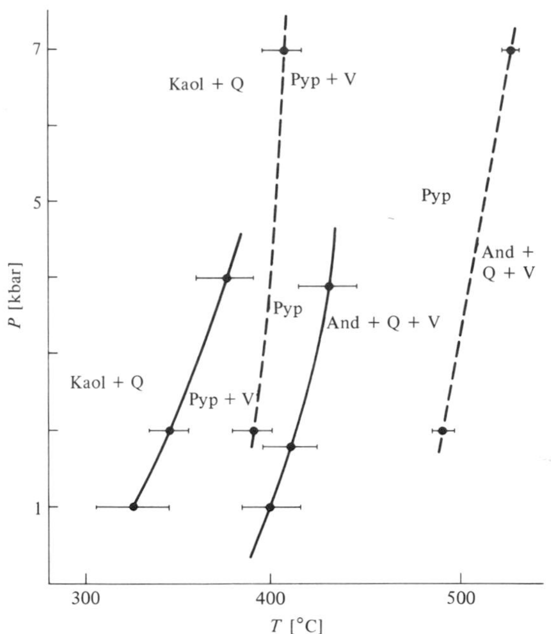

9.3 Stability of pyrophyllite. Two sets of experimental results are shown. The dashed lines represent the results of Althaus (1966), the solid lines of Kerrick (1968) and Thompson (1970).

Chloritoid-bearing assemblages from emery deposits, Samos, Greece (in Halfterdahl, 1961) include diaspore and kaolinite along with muscovite, magnetite, hematite, and rutile. The equilibrium temperature of such assemblages is in the range 375°C to 400°C as inferred from data on the dehydration of diaspore (see Winkler, 1974).

According to Hoschek (1969) the coexistence of chloritoid with albite or with biotite is rare. There is probably a small temperature difference between the reaction

$$\text{chlorite} + \text{muscovite} = \text{chloritoid} + \text{biotite} + \text{quartz} + \text{vapor} \quad (9.s)$$

and the reaction

$$\text{chlorite} + \text{muscovite} = \text{staurolite} + \text{biotite} + \text{quartz} + \text{vapor} \quad (9.t)$$

Therefore these equilibria may be sensitive to small changes in P_{H_2O} or P_{O_2}. Chloritoid and biotite or chloritoid and albite were not found together by Hoschek (1969) in the Central Alps, Austria. Petrologists have also noted the rare occurrence of glaucophane together with chloritoid (see Hoschek, 1969). The stability limits of this association are not well understood. According to theoretical calculations of Hoschek (1969), the equilibrium conditions for the reaction

$$\text{chlorite} + \text{paragonite} + \text{quartz} = \text{chloritoid} + \text{glaucophane} + \text{vapor}$$

may be 450°C at 10 kbar and 500°C at 7.5 kbar. For the reaction

$$\text{chloritoid} + \text{quartz} \rightleftharpoons \text{staurolite} + \text{almandine} + \text{vapor}$$

Hoschek estimated a temperature of ~575°C at 5 kbar, which is not much different from Ganguly's (1969) experimental determination.

Ganguly (1969) has presented a detailed analysis of the stability of chloritoid and its parageneses. Three of these limiting chloritoid breakdown equilibria are presented in Figure 9.4. These equilibria are

$$3\,\text{chloritoid} \rightleftharpoons \text{almandine} + 2\,\text{corundum} + 3\,H_2O \quad (9.u)$$

$$23\,\text{chloritoid} + 8\,\text{quartz} \rightleftharpoons 4\,\text{staurolite} + 5\,\text{almandine} + 21\,H_2O \quad (9.v)$$

$$4\,\text{chloritoid} + 5\,\text{sillimanite} \rightleftharpoons 2\,\text{staurolite} + \text{quartz} + 3\,H_2O \quad (9.w)$$

Unlike many other mineral systems, the stability limit of chloritoid is not significantly affected by the presence of excess quartz. At oxygen fugacities up to those defined by the NNO buffer, excess silica depresses the thermal stability limit of chloritoid by only 15°C. This effect disappears with increasing f_{O_2}. The variation in f_{O_2} within the field of magnetite does not alter the upper thermal stability of chloritoid significantly. It appears (Figure 9.4) that the highest temperature at which magnesian chloritoid can exist in natural assemblages is around 575°C to 600°C. The presence of chlorite and mica introduces a few additional chloritoid equilibria, but all such curves must lie between the limits defined by Reactions (9.v) and (9.w).

Chloritoid in regional metamorphic schists has been generally regarded as one of the source mineral for the formation of staurolite. Ganguly (1969) estimates that, after allowing for the various possible effects such as those of crystalline solubility (mainly Fe and Mg), variable oxygen fugacity, and of lowering of P_{H_2O} relative to total pressure, the temperature at pressures of 4 to 8 kbar for the reaction chloritoid to staurolite is in the range of 500°C to 575°C.

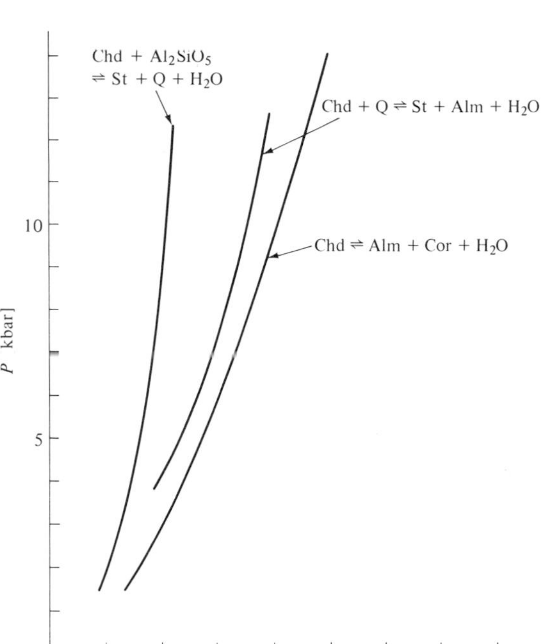

9.4 Section along NNO buffer showing the various stable chloritoid breakdown equilibria. (After Ganguly, 1969.) The experimental data are from Ganguly (1969), Hoschek (1967), and Richardson (1968). Printed from Journal of Petrology (see reference) by permission of the Oxford University Press, Oxford.

Staurolite

The chemical composition and structural formula of staurolite have been discussed by Smith (1968) and Griffen and Ribbe (1973) among others. It appears that staurolite may be represented by the formula $Fe_4Al_{18}Si_8O_{44}(OH)_4$. Most Al is in the Al^{VI} site, and Fe in the Fe^{IV} sites. From Griffen and Ribbe's statistical analysis (as given in their table 4) it appears that Mg may be occupying both Al^{VI} and Fe^{IV} sites. Staurolites may also contain Zn in significant concentration and several other elements such as Ti, Mn, Cr, V, and Co in minor amounts.

Experiments on the stability of staurolite with the composition $4MgO \cdot 9Al_2O_3 \cdot 7.5SiO_2 \cdot 2H_2O$ were conducted by Schreyer and Seifert (1969). Mg-staurolite is stable at fluid pressures greater than 12 kbar within a temperature range that increases with pressure. At 12 kbar it is around 750°C and at 25 kbar it extends from 750°C to 975°C.

Richardson (1968) studied the system Fe-Al-Si-O-H with the oxygen fugacity controlled by the QFM buffer. The reactions are as follows:

$$6\,\text{staurolite} + 11\,\text{quartz} \rightleftharpoons 4\,\text{almandine} + 23\,\text{sillimanite} + 3H_2O \quad (9.x)$$

$$2\,\text{staurolite} + 7\,\text{quartz} \rightleftharpoons 2\,\text{cordierite} + 5\,\text{sillimanite} + H_2O \quad (9.y)$$

$$4\,\text{chloritoid} + 5\,\text{sillimanite} \rightleftharpoons 2\,\text{staurolite} + \text{quartz} + 3H_2O \quad (9.z)$$

Ganguly (1972) has also given experimental data on Reaction (9.x), and his final analysis shows that staurolite plus quartz are stable depending on bulk composition over the temperature interval of 75°C to as much as 150°C at 10 kbar; it has a pressure maximum between 14 to 20 kbar around 600°C as well as a low-pressure termination between 3 to 5 kbar. The association is virtually unstable at oxygen fugacities within the field of hematite. In association with magnetite, the field of staurolite plus quartz (Figure 9.5) becomes restricted to oxygen fugacities between the NNO and HM buffers; the resultant form is a pyramidal volume with the apex around 16 to 20 kbar, 615°C to 625°C, and $\log f_{O_2} = -14$ to -15. The presence of mica and/or chlorite in natural assemblages restricts the field of staurolite further. In the relatively rare assemblages devoid of free quartz, such as reported by Uruno and Kanisawa (1965), the field of staurolite is extended, the major effect being on the upper limit of thermal stability.

In natural assemblages, nearly all minerals are crystalline solutions. The intergranular fluid or vapor phase is also a mixture of various components. Ganguly (1972) has estimated the probable effects of such solution reactions on the stability limits and Figure 9.5 also shows these effects.

Hoschek (1969) investigated the following reactions with the oxygen fugacity at the level of the FMQ buffer:

$$\text{(Fe-Mg)chlorite} + \text{muscovite} \rightleftharpoons \text{(Fe-Mg)staurolite} + \text{(Fe-Mg)biotite} + \text{quartz} + \text{vapor} \quad \text{(Figure 9.2)} \quad (9.aa)$$

$$\text{(Fe-Mg)staurolite} + \text{muscovite} + \text{quartz} \rightleftharpoons Al_2SiO_5 + \text{(Fe-Mg)biotite} + \text{vapor} \quad (9.bb)$$

The equilibrium conditions for the staurolite-forming reaction are 565 ± 15°C at 7 kbar and 540 ± 15°C at 4 kbar. Those for the staurolite breakdown are 675 ± 15°C at 5.5 kbar and 575 ± 15°C at 2 kbar. These ranges of pressure and temperature are close to those of Reactions (9.x), (9.y), and (9.z) represented in Figure 9.5.

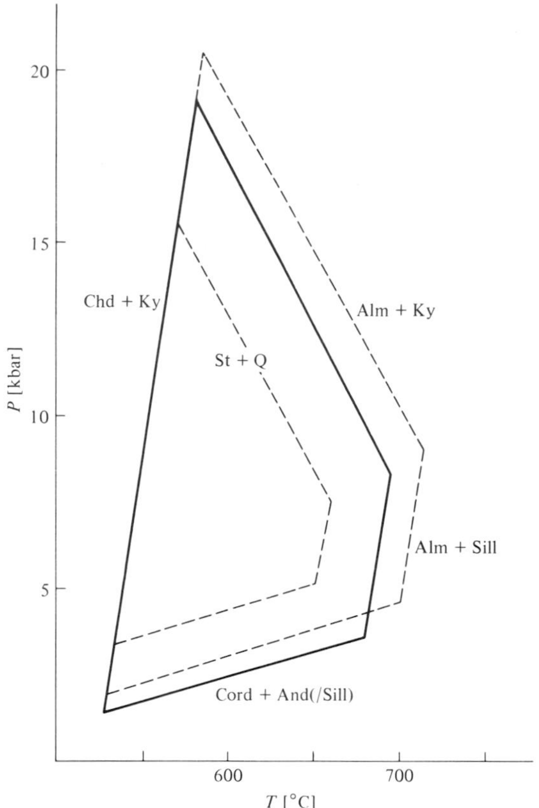

9.5 Stability field of staurolite between NNO and HM buffers. The dashed lines indicate the probable shift as a result of crystalline solution effects. Experimental data involving sillimanite are from Richardson (1968). Reprinted from the *Journal of Petrology* (Ganguly, 1972, see reference) by permission of the Oxford University Press, Oxford.

The Al_2SiO_5 polymorphs

The three Al_2SiO_5 polymorphs—andalusite, sillimanite, and kyanite—are common metamorphic minerals in pelitic rocks. Several experimental determinations of the triple point and the univariant transitions have been made, the more recent being those by Newton (1966b), Althaus (1967), Richardson *et al.* (1968, 1969), and Holdaway (1971). As the difference in the free energy of formation of the polymorphs is small, the experimental accuracy in such determinations has to be very high. Unfortunately, in spite of extensive investigations, the location of the triple point remains uncertain. Zen (1969) has pointed out the possible reasons for the discrepancies in the various experimental results. These reasons include experimental difficulties in pressure determination and variations in the composition and structure of the starting material and the final products. Possible effects of order–disorder in sillimanite have been considered by Greenwood (1972).

Results of three recent studies of the fields of the three polymorphs are shown in Figure 9.6. Richardson *et al.* (1969) give the coordinates of the triple point as 622°C and 5.5 kbar with an uncertainty area. Holdaway (1971) puts the triple point at 501°C and 3.76 kbar. Althaus (1967) determined this point at 595°C and 6.5 kbar. Ganguly (1969) noted that the natural occurrence of chloritoid and sillimanite, although rare does occur in equilibrium in some metamorphic rocks (emery deposits, Naxos, Greece; see Ganguly, 1969). It is possible, therefore, that the low-temperature boundary of the stability field of sillimanite and the high-temperature boundary of the stability field of chloritoid either overlap or they are not significantly separated in the *P–T* field.

9.6 The triple "points" in the Al_2SiO_5 system. Reactions (9.u) and (9.w) are discussed in the text. ⊙: Althaus (1967); ★: Richardson *et al.* (1969); ●: Holdaway (1971); ▲: average possible value.

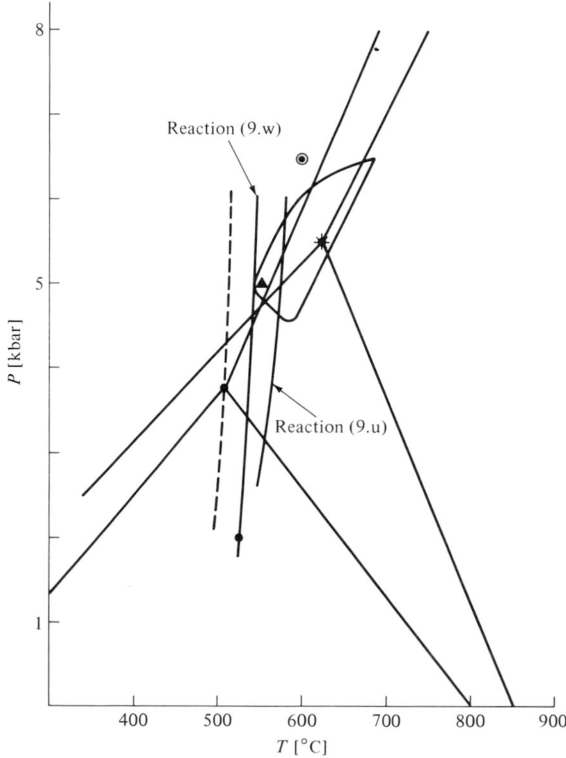

In this respect, Ganguly's (1969) Reactions (9.u) and (9.w) as shown below prescribe the limits within which the triple point may be located. These reactions are

$$\text{chloritoid} \rightleftharpoons \text{almandine} + \text{corundum} + H_2O \quad (9.u)$$

$$\text{chloritoid} + \text{sillimanite} \rightleftharpoons \text{staurolite} + \text{quartz} + H_2O \quad (9.w)$$

Figure 9.6 shows the univariant curves of the two reactions; the dashed curve being Ganguly's estimated shift in curve [Reaction (9.w)] due to possible compositional variations in natural assemblages. It appears that at present we must be satisfied by assuming that the three experimentally determined triple points represent the limits of an area within which the real triple point is located. A reasonable estimate may be 550°C and 5 kbar.

Cordierite

Compositionally, cordierite varies between the end members $Fe_2Al_3(Si_5Al)O_{18}$ and $Mg_2Al_3(Si_5Al)O_{18}$. Most cordierites lie in the $Mg/(Mg + Fe^{2+})$ range of 91 to 46 percent. Most cordierite analyses show a significant concentration of water, ranging from 0.6 to 3.3 wt. percent (Lepezin, 1969). Structurally, cordierite may vary between an orthorhombic "low" structural state (maximum "distortion") and a truly hexagonal "high" form. The variation in structural states found in metamorphic cordierites may not significantly change the energetics, as noted by Newton (1972) who measured the ΔH of solution.

Several reactions that may introduce or remove cordierite under hydrous conditions have been shown above. Figure 9.2 indicates that at 500°C to 520°C cordierite may be formed between 2 to 3 kbar by the reaction:

$$\text{Mg-chlorite} + \text{andalusite} + \text{quartz} \rightleftharpoons \text{cordierite} + \text{vapor} \quad (9.f)$$

The maximum pressure the cordierite stability field reaches under hydrous conditions in presence of micas is about 6.5 kbar at about 640°C. Richardson's (1968) data on the reaction

$$\text{Fe-staurolite} + \text{quartz} \rightleftharpoons \text{Fe-cordierite} + \text{sillimanite} + H_2O \quad (9.cc)$$

are $T = 560°C$, $P = 2$ kbar, and $T = 650°C$, $P = 3$ kbar. The upper stability limit for Fe-cordierite according to the reaction

$$3\,\text{Fe-cordierite} \rightleftharpoons 2\,\text{almandine} + 4\,\text{sillimanite} + 5\,\text{quartz} \quad (9.dd)$$

is shown in Figure 9.7. The stability of Fe-cordierite at about 700°C is limited to a maximum fluid pressure (f_{O_2} · QFM buffer) of about 3.5 kbar.

Figure 9.7 shows the results of Newton (1972) and Hensen and Green (1973). The upper stability limit of Mg-cordierite breaking down to enstatite, sillimanite, and quartz is different, as determined in the dry and wet systems. Newton (1972) suggests that this may be due to the water in the cordierite structure which increases its stability under wet conditions. Green and Vernon (1974), using natural assemblages as starting materials, find that cordierite and amphibole coexist in a divariant band within the temperature range 740°C to 860°C.

9.7 The stability of cordierite. *1*: Hensen and Green (1973); *2*: Newton (1972); *3*: Richardson (1968) (QFM buffer). Shaded areas are divariant bands with amphibole + cordierite + kyanite + quartz (Green and Vernon, 1974). Cordierite + quartz are on the low-pressure side of the band, while amphibole + kyanite + quartz are on the high-pressure side.

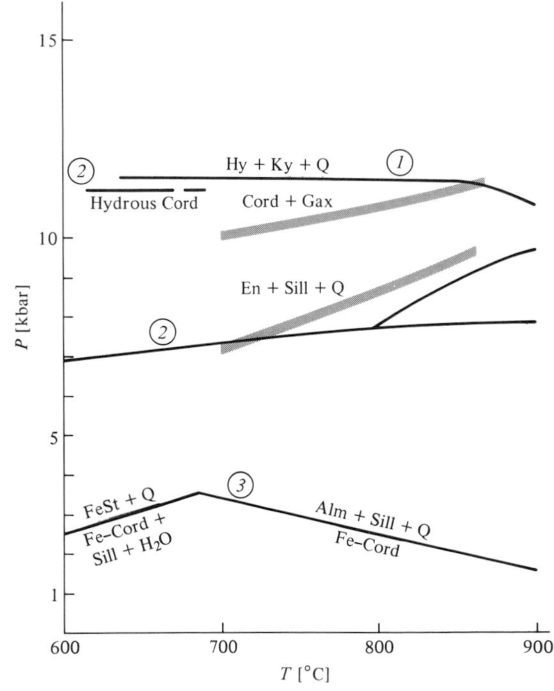

At 740°C, the band is between a pressure of 7.7 and 10.4 kbar at 860°C, between 9.5 and 11.4 kbar. The recrystallized cordierites are in the $Mg/(Mg + Fe^{2+})$ range of 0.89 to 0.94 and the coexisting amphiboles of 0.78 to 0.81. The f_{O_2} is estimated by the authors to be between those of the FMQ and Ni-NiO buffers.

Muscovite-paragonite

The assemblage muscovite and quartz is very common in low- to high-grade metamorphic rocks, whereas the assemblage paragonite-quartz has been noted in low- to medium-grade metamorphic rocks. There is partial crystalline solution between muscovite and paragonite. Recent work on the P–T ($P_{H_2O} = P_{total}$) stability of muscovite or muscovite and quartz was carried out by Velde (1966), Evans (1965), Althaus *et al.* (1970), and Day (1973). The breakdown of paragonite and the assemblage paragonite-quartz have been studied experimentally by Chatterjee (1970, 1972) and Ivanov and Gusynin (1970).

Figure 9.8 shows the univariant curves for the reactions:

$$1\,\text{muscovite} \rightleftharpoons 1\,\text{corundum} + 1\,\text{sanidine} + 1\,H_2O \quad (9.ee)$$

$$1\,\text{paragonite} \rightleftharpoons 1\,\text{corundum} + 1\,\text{albite} + 1\,H_2O \quad (9.ff)$$

$$1\,\text{muscovite} + 1\,\text{quartz} \rightleftharpoons 1\,\text{K-feldspar} + 1\,Al_2SiO_5 + 1\,H_2O \quad (9.ee')$$
Sill/And/Ky

$$1\,\text{paragonite} + 1\,\text{quartz} \rightleftharpoons 1\,\text{albite} + 1\,Al_2SiO_5 + 1\,H_2O \quad (9.ff')$$
And/Ky

Results on the two univariant curves for Reaction (9.ee′) by Day (1973) and Althaus *et al.* (1970) are not significantly different and may be accepted as the upper critical limit in P and T for the reaction. These stability relations are for reactions involving pure water. In rocks the gaseous or fluid phase would contain components other than water. Kerrick (1972) investigated Reaction (9.ee) with the fluid phase as a mixture of CO_2 and H_2O. Kerrick's three experimental results are shown in Figure 9.8. Note that the change in mole fraction by 0.5 changes the temperature stability of muscovite + quartz by over 100°C. Storre and Karotke (1971) and Huang *et al.* (1973) have studied the high-pressure dissociation of muscovite. Their results (up to 30 kbar) indicate that muscovite dissociates or melts at shallow depths. Muscovite in the presence of quartz could be present to a maximum depth of 100 km.

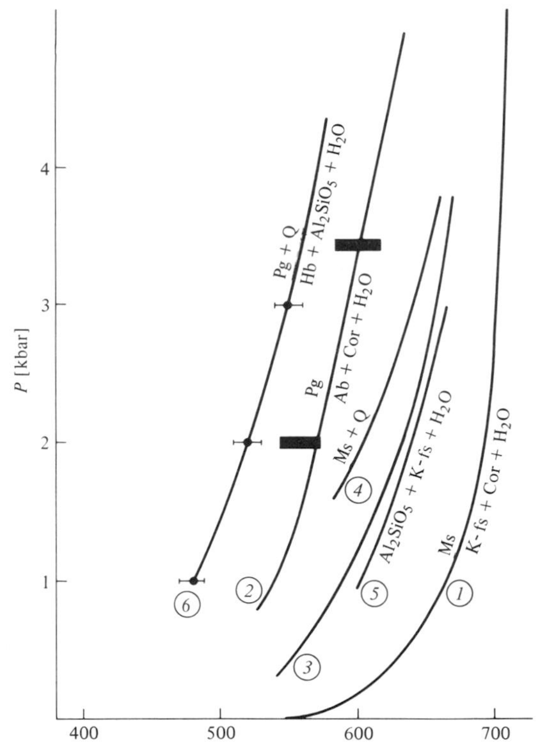

9.8 The mica-quartz reactions. Helgesson (1975) finds that the differences in various curves (e.g. *3*, *4*, *5*) may be due to the different structural states of feldspars in the experiments. *1*: Velde (1966); *2*: Chatterjee (1970); *3*: Althaus *et al.* (1970); *4*: Evans (1965); *5*: Day (1973); *6*: Chatterjee (1972); ■: Kerrick's (1972) muscovite data (50 percent mixture of $H_2O:CO_2$).

Chlorite + muscovite reactions already have been considered. These reactions and the equilibrium P and T are

$$49(\text{Mg-Fe})\,\text{chlorite} + 151\,\text{muscovite} = 44(\text{Mg-Fe})\text{staurolite} + 151(\text{Mg-Fe})\text{biotite} + 360\,\text{quartz} + 696\,\text{vapor} \quad (9.t)$$

$P = 7$ kbar	T°C $= 565 \pm 15$	Hoschek (1969)
$P = 4$ kbar	T°C $= 540 \pm 15$	Hoschek (1969)

$$3\,\text{Mg-chlorite} + 5\,\text{muscovite} = 8\,\text{kyanite} + 5\,\text{phlogopite} + 1\,\text{quartz} + 12\,H_2O \quad (9.g)$$

$P = 7$ kbar	T°C $= 639$–631
$P = 8$ kbar	T°C $= 645$–634

Other reactions involving cordierite have been described in preceding sections.

Feldspars

Before a summary of the results on experimental systems involving feldspars is presented, it is necessary to point out that in most experiments feldspars may be generated metastably. Attainment of chemical equilibrium in the assemblage is particularly difficult because of the high activation energy for intracrystalline diffusion of Al and Si over the nonequivalent tetrahedral sites. This is due to the charge linkage between Ca and Al on the one hand and Na and Si on the other hand (Goldsmith, 1952, Chapter 3). Excellent reviews of the kinetics of order–disorder and exsolution and process of twinning have been presented by Marfunin (1966), Barth (1969), and recently by Smith (1973). A summary of several systems involving feldspars has been presented by Barth (1969) and by Smith (1974). The results of some recent work on feldspar phase diagrams are presented here. Some liquidus relations of the system have been discussed in Chapter 2 as an illustration of crystal-melt equilibria.

$NaAlSi_3O_8$-$KAlSi_3O_8$

The results of experiments at 2.5 and 10 kbar by Orville (1963) and by Luth and Tuttle (1966) were used by Thompson and Waldbaum (1969) to calculate the solvus curve, activity–composition relations, and the effect of pressure on the thermodynamic relations. The change of critical temperature with pressure is given by dT_c/dP = 13.47°C/kbar. The difficulties with the experimental determination of the solvus were outlined by Luth and Tuttle (1966) and were discussed by Thompson and Waldbaum (1969), who suggested that a starting material with excess alkalis may yield experimental results more consistent with natural feldspars than ideal or peraluminous materials.

There is a very significant effect of P_{H_2O} on the sanidine-albite solvus. This was pointed out by Yoder *et al.* (1957), who determined the solvus with a melting minimum of P_{H_2O} at 5 kbar and an estimated critical temperature of 715°C. More recent work on the system alkali feldspar-water was carried out by Morse (1970) and Goldsmith and Newton (1972). Morse (1970) found that the melting at a P_{H_2O} of 5 kbar begins at an isobaric eutectic of 703 ± 2°C with critical composition as $Or_{28.5}$ wt. % (Or = $KAlSi_3O_8$) (Figure 9.9). The liquidus of albite-H_2O at 5 kbar is 758 ± 3°C. Figure 9.9 also shows the K-feldspar-rich water-saturated liquidus

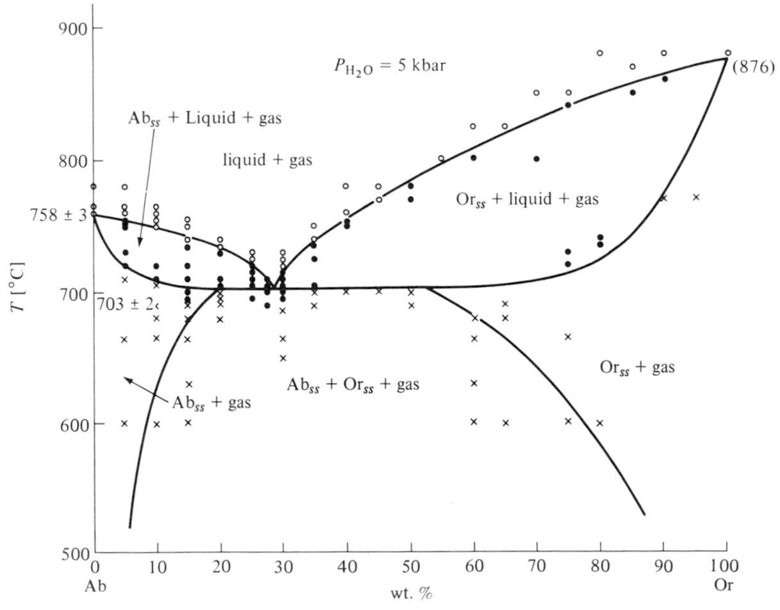

9.9 Equilibrium diagram of the system Ab-Or-H_2O at 5 kbar, projected from H_2O onto the andydrous join. *ss* = solid solution. (After Morse, 1970). Reprinted from the *Journal of Petrology* (see reference) by permission of the Oxford University Press, Oxford.

(Yoder *et al.*, 1957), and the sanidine-H_2O melting temperature as 876°C. Goldsmith and Newton (1972) performed similar experiments on water saturated solvi in the range 10 to 20 kbar P_{H_2O}. The effect of pressure in decreasing the solid solubility on the K-rich limb in 0.60 mole percent/kbar at 450°C; 0.80 mole percent/kbar at 500°C; 1.05 mole percent/kbar at 550°, and greater than 1.65 mole percent/kbar at 600°. These data are found to be in good agreement with the pressure effect calculated by Waldbaum and Thompson (1969). For the anhydrous system Morse (1970) found $T_c = 738°C$ and $X_{Or} = 34$ (wt. %) at 5 kbar. Combining this value of T_c with the value from Orville's 2 kbar results ($T_c = 675$), Morse found dT_c/dP as 18.33°C/kbar, which is significantly different from Thompson and Waldbaum's value.

Waldbaum and Thompson (1969) arrived at the following equation for the molar excess Gibbs energy for the alkali feldspar solid solution:

$$G^E(P, T) = (6326.7 + 0.0925\,P - 4.6321\,T)X_1X_2^2 + (7671.8 + 0.1121\,P - 3.8565\,T)X_2X_1^2 \quad (9.1)$$

where T is temperature (°K), P denotes pressure (bar), and X_1 and X_2 are the mole fractions of the components $NaAlSi_3O_8$ and $KAlSi_3O_8$, respectively. This equation is based on solvus data ranging from 500°C to 700°C and from 2 to 10 kbar. The errors in the computed values for G^E in the P–T range are of the same order of magnitude as the errors in the solvus data. It must also be noted that because of lack of pertinent data in the experimental systems, no account has been taken of Al-Si ordering. The calculated G^E values are, therefore, likely to change significantly as new data on feldspars appear (see Waldbaum and Robie, 1971). We may still use Equation (9.1) for calculating G^E for computing phase equilibria involving alkali feldspars. The results would, of course, be at best semiquantitative.

Albite, anorthite, and plagioclase crystalline solution

As indicated above, phase relations in the plagioclase system are very complex due to the high activation energies involved in intracrystalline diffusion and redistribution of cations. Smith (1974) has presented a possible phase diagram, which he describes as highly uncertain but useful for teaching the complex phenomena to students and for stimulating experimentalists and theorists to make crucial tests. Figure 9.10 shows the phase relations. Starting at the lowest temperature,

9.10 Possible phase diagram for plagioclase feldspars. (After Smith, 1974.) The dashed lines show metastable curves which govern the peristerite, Bøggild and Huttenlocher intergrowths.

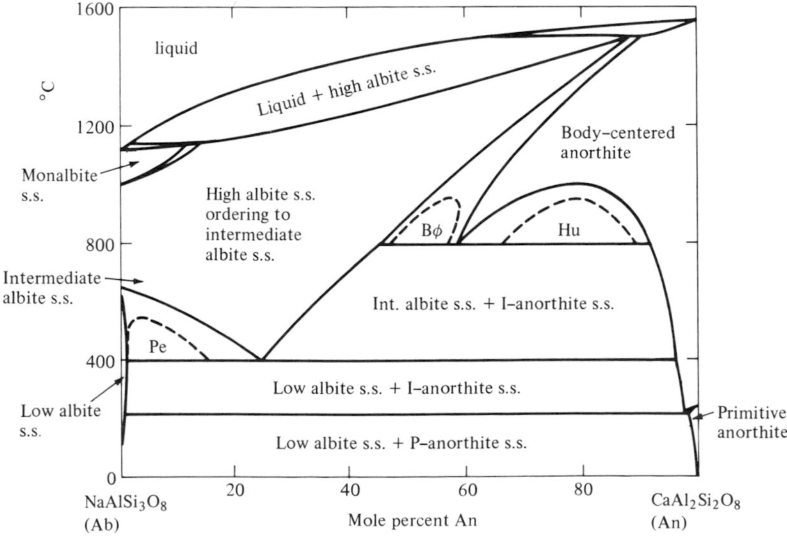

we have a low albite solid solution and a primitive anorthite solid solution. The primitive anorthite inverts to body-centered anorthite near 250°C. The body-centered anorthite becomes weakly disordered with rising temperature and with albite substitution. The disordering with temperature of this anorthite is not significant even up to the melting point. Low albite disorders slightly up to about 550°C where it inverts discontinuously to intermediate albite. The disordering continues without break above 550°C and up to 980°C, where the intermediate albite becomes monoclinic. The central field of the figure belongs to plagioclase solid solution, disordered at the albite end at about 1000°C. This field approaches the ordered anorthite end at about 1500°C. Above the solidus, both binary feldspar liquids and crystals behave essentially as ideal solution as discussed in Chapter 2. With decreasing temperature the high plagioclase crystalline solution continuously orders to intermediate plagioclase solution. The intermediate plagioclase solution eventually gives way to one of the two types of intergrowths depending on composition. In terms of mole percent anorthite, we have, between 2 to 23 peristerite intergrowth, and between 45 to 60 the Bøggild intergrowth. The temperatures at which these intergrowths develop are crucial to phase equilibria, but these are only imperfectly known. There is a third intergrowth, the Huttenlocher (65 to 95 mole percent An) which may continue from low to high temperatures, the binary loop being truncated by the solidus (see Smith, 1974 for discussion of intergrowths).

Ternary feldspar

Seck (1971a, b) presented the ternary solvus data on feldspars at temperatures 650°C, 750°C, 825°C, and 900°C and at pressures ranging from 0.5 kbar to 1 kbar. Seck's results are shown in Figure 9.11 as four isotherms representing the intersection of the ternary solvus surface with the composition plane at the four temperatures. The 900°C isotherm is at 0.5 kbar while the other three isotherms are at 1 kbar. Note that these isotherms are free-hand curves drawn through the distribution data points and are not fitted according to any nonideal ternary solution model. To construct a solution model, binary data are required for all three systems: Ab-Or, Or-An, and Ab-An. For an approximate ternary solution model see Saxena (1973). The ternary miscibility gap expands with increasing pressure (Seck, 1971b) and decreasing temperature. Seck's (1971b) data indicate that a pressure increase of 1 kbar results in the same effect on the solvus as a temperature decrease of 15°C.

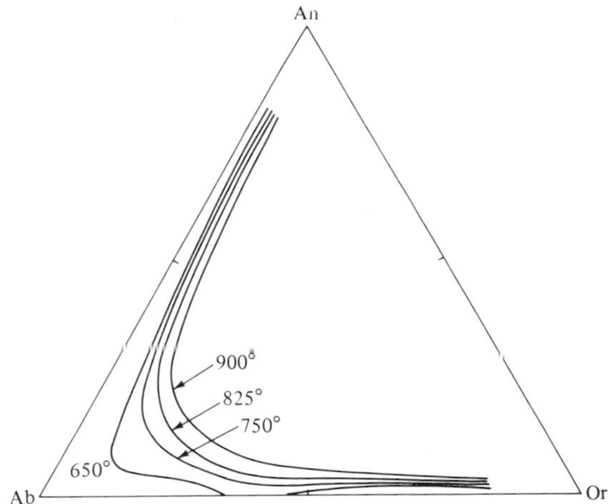

9.11 Ternary feldspar isotherms between 650°C and 900°C at 1 kbar water pressure. (After Seck, 1971.) Reprinted from *Neues Jahrbuch für Mineralogi* by permission.

Biotite

Phlogopite or magnesian biotite is one of the products of the

muscovite + chlorite + quartz

reaction studied by Seifert (1970) and Bird and Fawcett (1973). Figure 9.2 shows the pressure and temperature conditions for various reactions producing phlogopite. The lowest temperature noted by Seifert (1970) is around 495 ± 10°C at 1 kbar water pressure for the reaction

$$1\,\text{chlorite} + 1\,\text{muscovite} + 2\,\text{quartz} \rightleftharpoons 1\,\text{phlogopite} + 1\,\text{cordierite} + 3.5\,H_2O \quad (9.e)$$

In natural systems, however, phlogopite may appear at temperatures lower than 495°C for other reasons (e.g., impure H_2O). The phlogopite breakdown reaction was studied by Yoder and Eugster (1954). This reaction takes place between 1000°C and 1100°C in the pressure range of 1 to 3000 bar.

Much of the experimental work on the stability of annite and the crystalline solution biotite has been done by Eugster and Wones (1962) and Wones and Eugster (1965). Depending on the oxygen fugacity, the equilibrium temperature for the reaction

$$\text{annite} + \text{vapor} \rightleftharpoons \text{sanidine} + \text{magnetite} + \text{vapor}$$

may vary from nearly 400°C to 800°C at 2070 kbar. The breakdown temperature between the Ni-NiO and QFM buffers is between 630°C and 700°C. This temperature is not significantly changed by the addition of quartz. At fixed oxygen fugacity and total pressure, addition of phlogopite to the system would increase the stability field of biotite. With Ni-NiO buffer, annite breaks down at nearly 630°C, but a biotite with Fe/(Fe + Mg) = 0.55 is stable up to 800°C (Wones and Eugster, 1965, their Figure 3).

Garnets

Hsu (1968) studied the system Al-Mn-Fe-Si-O-H with bulk chemistry matching that of almandine ($Fe_3Al_2Si_3O_{12}$) and spessertine ($Mn_3Al_2Si_3O_{12}$). The stability field of almandine is highly dependent on the oxygen fugacity, as shown in Figure 9.12. The curves represent the following equilibria:

$$\text{quartz} + \text{Fe-chlorite} + \text{fluid} \rightleftharpoons \text{almandine} + \text{fluid} \quad (9.gg)$$

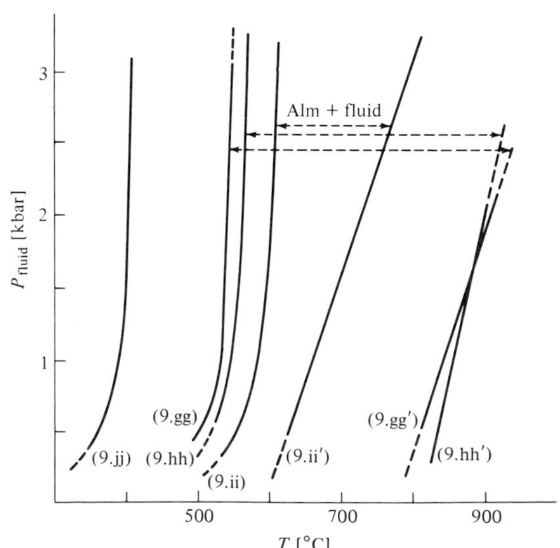

9.12 Stability field of almandine and spessertine. Curves (9.gg–9.gg′) represent the field with f_{O_2} determined by iron – quartz-fayalite buffer. Similarly (9.hh–9.hh′) represent iron-magnetite/iron-wustite and (9.ii–9.ii′) fayalite-magnetite-quartz. Curve (9.jj) is for the reaction quartz + Mn-chlorite + fluid + spessertine. Figure based on the work of Hsu (1968). See text for further explanation.

$$\text{almandine} + \text{fluid} \rightleftharpoons \text{Fe-cordierite} + \text{fayalite} + \text{hercynite} \quad (9.gg')$$

(f_{O_2} determined by iron + quartz + fluid-fayalite buffer)

$$\text{quartz} + \text{Fe-chlorite} + \text{magnetite} + \text{fluid} \rightleftharpoons \text{almandine} + \text{fluid} \quad (9.hh)$$

(9.hh′) as (9.gg′)

(f_{O_2} determined by iron-magnetite and iron-wustite buffers)

(9.ii) as (9.hh)

$$\text{almandine} + \text{fluid} \rightleftharpoons \text{quartz} + \text{hercynite} + \text{magnetite} + \text{fluid} \quad (9.ii')$$

(f_{O_2} determined by fayalite + magnetite + quartz buffer)

It is apparent that with increasing oxygen fugacity the stability field of almandine is reduced. Curve (9.jj) represents the equilibrium

$$\text{Mn-chlorite} + \text{quartz} + \text{fluid} \rightleftharpoons \text{spessertine} + \text{fluid} \quad (9.jj)$$

and is not significantly sensitive to variation in f_{O_2}. Spessertine was found to be stable up to 930°C, the highest temperature used in the study, at 500 bar fluid pressure. Unfortunately the composition of phases such as chlorite are rather uncertain in the experiments and the effect of crystalline solubility cannot be estimated.

Hsu and Burnham (1969) studied phase relationships in the system $Fe_3Al_2Si_3O_{12}$-$Mg_3Al_2Si_3O_{12}$-H_2O at 2.0 kbar fluid pressure, under controlled oxygen fugacities defined by the fayalite + magnetite + quartz buffer. Garnet is stable at temperatures between ~580°C to ~730°C. It is essentially almandine with a maximum of 22 mole percent pyrope found at 648°C.

A study of almandine at high pressure by Keesmann *et al.* (1971) showed that the stability field increases to about 1130°C at 10 kbar, and to 1500°C at 25 kbar, in the presence of metallic iron. At pressures between about 12 and 20 kbar, the products of incongruent melting are hercynite solid solution and liquid, while at higher pressures almandine melts congruently. The high-pressure (15 to 20 kbar) and low-temperature stability of almandine is uncertain, but the breakdown may be from 550°C to 600°C.

Figure 9.13 shows the upper stability limits of the garnet end members. It appears that almandine, andradite, grossularite, and pyrope melt incongruently at high pressures, but at still higher pressures they all melt congruently.

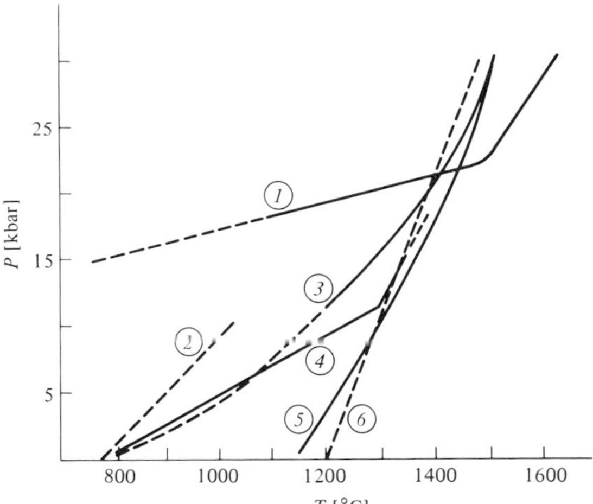

9.13 Upper temperature stability limits of garnet end members (Figure after Keesman *et al.*, 1971). *1*: pyrope (Boyd and England, 1959; Schreyer, 1968); *2*: almandine (Yoder, 1955); *3*: almandine (Keesman *et al.*, 1971); *4*: grossularite (Hays, 1967); *5*: andradite (Huckenholz and Yoder, 1971); *6*: spessertine.

Boettcher (1970) has studied the system

$CaO\text{-}Al_2O_3\text{-}SiO_2\text{-}H_2O$

at high pressures and temperatures. Two of the grossularite-producing reactions with their equilibrium P and T are:

zoisite $\rightleftharpoons$
grossularite + anorthite + corundum + vapor (9.kk)

(P = 3 kbar, T = 640°C; P = 10 kbar, T = 1000°C)

zoisite + quartz $\rightleftharpoons$
anorthite + grossularite + vapor (9.ll)

(P = 4 kbar, T = 580-610°C; P = 5.3 kbar, T = 640–660°C; P = 8 kbar, T = 730°C)

Grossularite, in the presence of excess quartz, disappears according to the following reaction:

grossularite + quartz $\rightleftharpoons$
anorthite + wollastonite (9.mm)

(P = 3 kbar, T = 635°C)

At low pressures (1 kbar) grossularite breaks down to anorthite + wollastonite + gehlenite at 840°C. At high pressure (10 kbar) in the presence of quartz and vapor grossularite melts incongruently to wollastonite + liquid at approximately 800°C.

Gordon and Greenwood (1971) studied the stability of grossularite in the presence of an $H_2O\text{-}CO_2$ mixture. The reactions restricting the grossularite field are, from high to low temperatures,

grossularite + CO_2 $\rightleftharpoons$
calcite + anorthite + wollastonite (9.nn)

grossularite + CO_2 $\rightleftharpoons$
calcite + anorthite + quartz (9.oo)

grossularite + H_2O $\rightleftharpoons$ calcite + zoisite + quartz (9.pp)

The equilibrium boundary for the first reaction at 2000 bar passes approximately through the points; $X_{CO_2} = 0.15$, T = 590°C; $X_{CO_2} = 0.20$, T = 665°C; and $X_{CO_2} = 0.30$, T = 790°C. For the second reaction we have $X_{CO_2} = 0.10$, T = 550°C; and $X_{CO_2} = 0.08$, T = 530°C.

Hydrogarnets

Yoder (1950) studied the hydrogarnets in the system $Ca_3Al_2(SiO_4)_3\text{-}Ca_3Al_2(OH)_{12}$ and found that members of the hydrogrossular solid solution series formed in the presence of water below approximately 850°C and below 2000 atm. Pistorius and Kennedy (1960) showed that the hydration of grossular starts at temperatures below 780°C and is nearly independent of pressure. Roy and Roy (1957) indicated that anhydrous grossularite is more stable than hydrogrossularite at temperatures as low as 400°C. Gordon and Greenwood (1971), in their study described in the preceding section, found only a small hydrogrossular component in their garnets.

Hsu (1968) found that hydrospessertine

$Mn_3Al_2Si_2O_8(SiO_4)_{1-m}(OH)_{4m}$

(m varies between 0 and 1) forms at temperatures below about 600°C, and that more $(OH)_4^-$ substitute for SiO_4^{4-} tetrahedra with decreasing temperature. Matthes (1961) reported a synthesis of hydrospessertine containing as high as 17.6 mole percent of $Mn_3Al_2(OH)_{12}$ corresponding to 3.9 wt. percent H_2O at 500°C and 1000 atm. Hsu (1968) suggested that there may be a relation between the size of the cations and the temperature at which the hydration of the garnet begins. Grossular starts hydrating at 780°C, spessartine at 600°C. Almandine may be hydrating below 450°C, and pyrope probably cannot become hydrous in the laboratory.

Pyroxene and olivine

In the metamorphic rocks, the pyroxenes (enstatite-ferrosillite and diopside-hedenbergite) are usually high in iron. However, orthopyroxene richer in iron beyond 80 mole percent of ferrosillite is rare. This has been explained as due to the instability of orthopyroxene relative to olivine plus quartz. Olsen and Mueller (1966) and Kurepin (1970) presented a thermodynamic analysis of the ferrosillite breakdown reaction. Smith (1971) has restudied Lindsley's (1965) reaction

$$\text{orthopyroxene} \rightleftharpoons \text{olivine} + \text{quartz}$$

and his data together with Kurepin's (1970) theoretical curves are shown in Figure 9.14. Smith (1971) found that at a pressure of 1 kbar or lower the limiting compositions are $Fs_{80}En_{20}$ at 60-700°C, $Fs_{75}En_{25}$ at 800°C, and $Fs_{70}En_{30}$ at 900-1000°C. It is interesting to note that these compositions are fairly close to Kurepin's theoretical values. Kurepin assumed that both olivine and orthopyroxene are ideal solutions. Saxena and Ghose (1971) found that at low temperatures (~600°C) orthopyroxene is somewhat nonideal. It is likely that both orthopyroxene and olivine deviate from ideal solutions to a similar extent and the nonideality effects cancel out in the equation of equilibrium.

It appears, then, that an orthopyroxene with 80 mole percent ferrosillite can be stable at 600°C at a pressure of 1 to 2 kbar. In presence of water, however, Smith (1971) found that at 600°C, 1 kbar pressure orthopyroxene (Fs_{80}) partly broke down to olivine, quartz, and amphibole. Similarly olivine reacted with quartz to yield minor amphibole at 650°C and 1 kbar of P_{H_2O}.

Another experimental result consistent with the discussion in the preceding paragraph is due to Hinrichsen (1966) on the $Mg_7Si_8O_{22}(OH)_2$–$Fe_7Si_8O_{22}(OH)_2$ join at 1 kbar. Hinrichsen found the iron-rich pyroxene (~75 mole percent), olivine, and quartz are stable down to about 625°C, below which they are converted to the assemblage anthophyllite, olivine, and quartz. Fayalite-rich olivine and quartz are shown to be stable at least down to 500°C.

Pyroxenes of the enstatite-ferrosillite and hedenbergite-diopside series are generally considered stable at high metamorphic temperatures (600°C) under anhydrous conditions. However, no actual experimental data are available on reactions involving crystalline solutions at temperatures around 600°C. Reactions involving enstatite and diopside in the presence of a fluid phase (H_2O, CO_2, O_2) will be considered later.

The acmite-jadeite ($NaFeSi_3O_3$–$NaAlSi_3O_8$) pyroxene series has been studied experimentally by Newton and Smith (1967) and Popp and Gilbert (1972) and theoretically by Essene and Fyfe (1967). At 600°C and 4 kbar in the presence of quartz, there is between 4 and 5 mole percent of jadeite component in the acmite-jadeite pyroxene solutions, while at 500°C and 4 kbar it is between 5 and 6 mole percent. Calculations using an ideal solution model for the pyroxene and the breakdown P and T of pure albite (to jadeite plus quartz) of 16.7 kbar at 600°C and 14.5 kbar at 500°C, show that the mole fractions are of the right magnitude. The equilibrium P and T of the reaction

$$\text{albite} \rightleftharpoons \text{jadeite} + \text{quartz}$$

has been determined by Boettcher and Wyllie (1968) and Newton and Smith (1967) among others.

Amphiboles

An excellent review of amphibole chemistry and stability was provided by Ernst (1968). A lot more information has become available on natural amphiboles since then, but few new data on experiments involving amphiboles have been added. The upper stability limit for amphiboles at 2 kbar pressure varies over a wide range of temperature beginning around 500°C for the Fe end members (ferrotremolite, riebeckite) to above 1000°C for the Mg end members (pargasite).

Reactions involving Mg-anthophyllite were studied by Greenwood (1963), Hellner *et al.* (1965), and Hinrichsen (1966). Anthophyllite breaks down to talc and forsterite between 670°C and 680°C at a pressure ranging from 1000 to 4000 bar. The width of the field of stability for Mg-anthophyllite is approximately 100°C. At the upper temperature limit (750°C to 780°C), anthophyllite breaks down to enstatite and quartz according to the reaction

$$\underset{\textit{anthophyllite}}{\tfrac{1}{7}Mg_7Si_8O_{22}(OH)_2} \rightleftharpoons \underset{\textit{enstatite}}{MgSiO_3} + \underset{\textit{quartz}}{\tfrac{1}{7}SiO_2} + \underset{\textit{fluid}}{\tfrac{1}{7}H_2O} \qquad (9.qq)$$

Hinrichsen's result indicates that there may be extensive crystalline solution between magnesian and iron anthophyllite (up to ~60 percent). The field of iron-rich anthophyllite may extend down to 500°C at 1 kbar. At such a temperature the coexisting phases are talc and olivine. The low-temperature stability limit of hypersthene + quartz + anthophyllite, however, is about 620°C.

Gedrite $[(Mg, Fe)_5Al_2Si_6Al_2O_{22}(OH)_2]$ is an aluminous orthoamphibole. The breakdown of gedrite alone

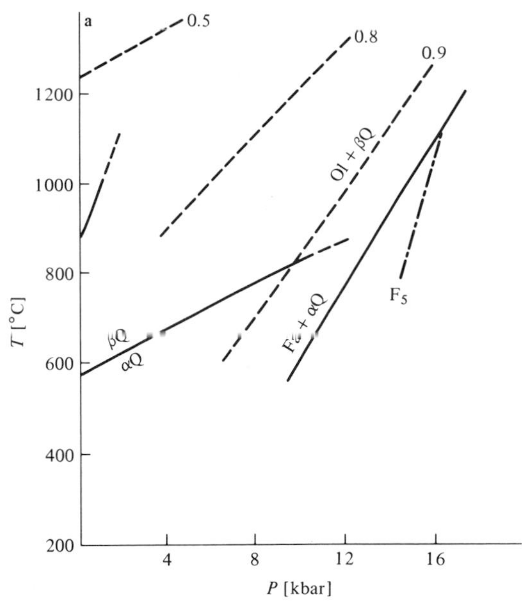

9.14 (a) Pyroxene ⇌ olivine + quartz equilibrium. Numbers refer to mole fraction of Fe^{2+} in orthopyroxene. Dashed curve: Kurepin (1970); solid curve: Smith (1971); dot-dash curve: Lindsley (1965). (b) Proposed approximate phase relations, at constant pressure and water fugacity, between anthophyllite, cummingtonite, orthopyroxene, and talc as a function of the temperature and the atomic fraction Mg/(Mg + Fe). The two-phase field is shown only for the curve (9.ss)–(9.tt) involving cummingtonite and anthophyllite. The temperature of crystallization of the coexisting pairs from Sundius (1933) are not known but have been interpolated. The high-temperature phase boundary of talc and the low-temperature phase boundary of orthopyroxene are shown as heavy dashed lines. The relations are inferred from natural assemblages and phase equilibria studies. Quartz is assumed to be present. The sizes of the symbols do not indicate precision or accuracy. □: Robinson (1963); ○: Sundius (1933); ×: anthophyllite (Klein, 1966); +: cummingtonite (Mueller, 1960). (After Mueller, 1972.)

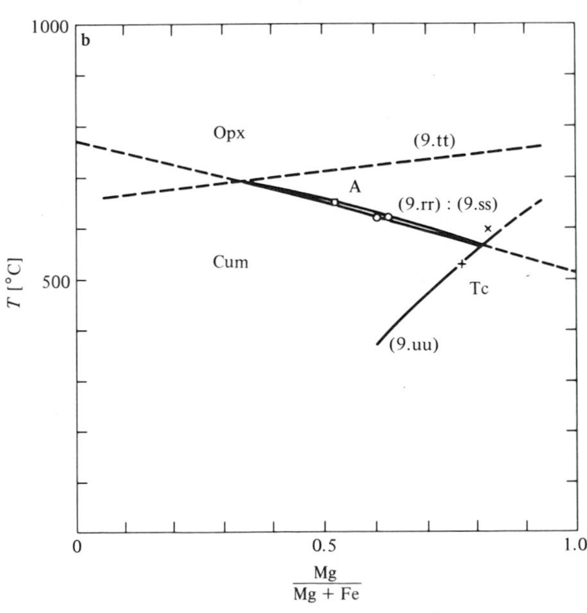

has not been determined experimentally. According to Akella and Winkler (1966), natural gedrite with 2 wt. percent Na_2O breaks down in the presence of quartz and cordierite at about 770°C under 2 kbar water pressure. The solubility of Al would probably increase the stability field of gedrite, and gedrite alone may be stable to higher temperatures than anthophyllite. On the low-temperature side, gedrite + cordierite changes to chlorite + quartz at about 570°C under 2 kbar water pressure.

Little experimental work (however, see Cameron, 1975) has been done on the stability relations of clinoamphiboles, cummingtonite-grunerite,

$$(MgFe)_7Si_8O_{22}(OH)_2.$$

A semiquantitative phase diagram with cummingtonite and anthophyllite has been constructed by Mueller (1972) using data from experiments and natural assemblages. Figure 9.14b shows a constant pressure and water fugacity section of this phase diagram. The curves labelled (9.rr), (9.ss), (9.tt), and (9.uu) represent the following reactions:

$$\underset{\textit{cummingtonite}}{\tfrac{1}{7}Mg_7Si_8O_{22}(OH)_2} \rightleftharpoons \underset{\textit{anthophyllite}}{\tfrac{1}{7}Mg_7Si_8O_{22}(OH)_2} \qquad (9.rr)$$

$$\underset{\textit{cummingtonite}}{\tfrac{1}{7}Fe_7Si_8O_{22}(OH)_2} \rightleftharpoons \underset{\textit{anthophyllite}}{\tfrac{1}{7}Fe_7Si_8O_{22}(OH)_2} \qquad (9.ss)$$

$$\underset{\textit{cummingtonite}}{(FeMg)_7Si_8O_{22}(OH)_2} \rightleftharpoons \underset{\textit{orthopyroxene}}{7(FeMg)SiO_3} + \underset{\textit{quartz}}{SiO_2} + \underset{\textit{fluid}}{H_2O} \qquad (9.tt)$$

$$\underset{\textit{talc}}{7(FeMg)_3Si_4O_{10}(OH)_2} \rightleftharpoons \underset{\textit{cummingtonite}}{3(FeMg)_7Si_8O_{22}(OH)_2} + \underset{\textit{quartz}}{4SiO_2} + \underset{\textit{fluid}}{4H_2O} \qquad (9.uu)$$

The curves (9.rr) and (9.ss) define a narrow two-phase field. Similarly there should be two phase fields for equilibria (9.tt) and (9.uu), which are not shown in the figure. The extension of (9.rr)–(9.ss) in the pyroxene and talc fields is metastable. The position of the curves (9.tt) and (9.uu) are strongly dependent on the water fugacity. The position and slope of the curve (9.rr)–(9.ss) is fixed approximately by the roughly deduced values of the temperatures of crystallization of the naturally coexisting anthophyllite-cummingtonite pairs of Robinson (1963), Robinson and Jaffe (1969a, b), and the separately existing cummingtonite (Mueller, 1960) and anthophyllite (Klein, 1966). The width of the two-phase field and the relation $X_{Fe}^{Cum} > X_{Fe}^{A}$ were derived from the data of Sundius (1933) as based on the Fe^{2+} determination. However, the crystallization temperature is unknown for this pair. The proposed curve is supported by the common occurrence of cummingtonite in low- to intermediate-grade metamorphic iron formations. This deduction applies only to cummingtonite-anthophyllite pairs low in such elements as Al, Ca, etc.

The positive slopes of curves (9.tt) and (9.uu) follow from the general observation that the thermal stability of a ferromagnesian silicate with respect to its high-temperature decomposition products increases with an increase in magnesium content. The extension of curves (9.tt) and (9.uu) to the pure magnesian end members is made to agree with the experimental results of Greenwood (1963, 1971) on the stability of anthophyllite. The curves as extended correspond to water pressures of approximately 5 kbar, a value roughly consistent with the regionally metamorphosed rocks represented.

The upper stability limit of tremolite, as found by Boyd (1959), is shown in Figure 9.15. The reaction is

$$\underset{\textit{tremolite}}{Ca_2Mg_5Si_8O_{22}(OH)_2} \rightleftharpoons \underset{\textit{diopside}}{2CaMgSi_2O_6} + \underset{\textit{enstatite}}{3MgSiO_3} + \underset{\textit{quartz}}{SiO_2} + \underset{\textit{fluid}}{H_2O} \qquad (9.vv)$$

As compared to tremolite, the upper stability limit of ferrotremolite is at least 300°C lower (Ernst, 1966).

The system $NaCa_2Mg_4AlSi_6Al_2O_{22}(OH)_2$ pargasite was studied by Boyd (1959). This amphibole is stable to liquidus temperatures of 1050°C at 1000 bar of fluid pressure.

The high-temperature stability of ferropargasite is several hundred degrees (~450°C) lower than that of pargasite at medium f_{O_2} (fayalite-magnetite-quartz buffer). The stability of ferropargasite as a function of f_{O_2} has been studied by Gilbert (1966) who finds that even at low oxygen fugacity (iron-wustite buffer) there is a difference of about 200°C between the upper stability limit of the two pargasites.

Ernst's (1960b, 1961, 1962) work on sodic amphiboles shows that in presence of fluid, glaucophane $[Na_2Mg_3Al_2Si_8O_{22}(OH)_2]$ and magnesioriebeckite $[Na_2Mg_3Fe_2^{3+}Si_8O_{22}(OH)_2]$ are stable up to the liquidus temperatures of 850°C and 925°C, respectively. The maximum stability for riebeckites

$$[Na_2Fe_3^{2+}Fe_2^{3+}Si_8O_{22}(OH)_2-Na_{2.4}Fe_{4.9}^{2+}Fe_{0.7}^{3+}Si_{7.7}Fe_{0.3}^{3+}O_{22}(OH)_2],$$

on the other hand, is up to about 720°C at f_{O_2} defined by wustite-iron buffer.

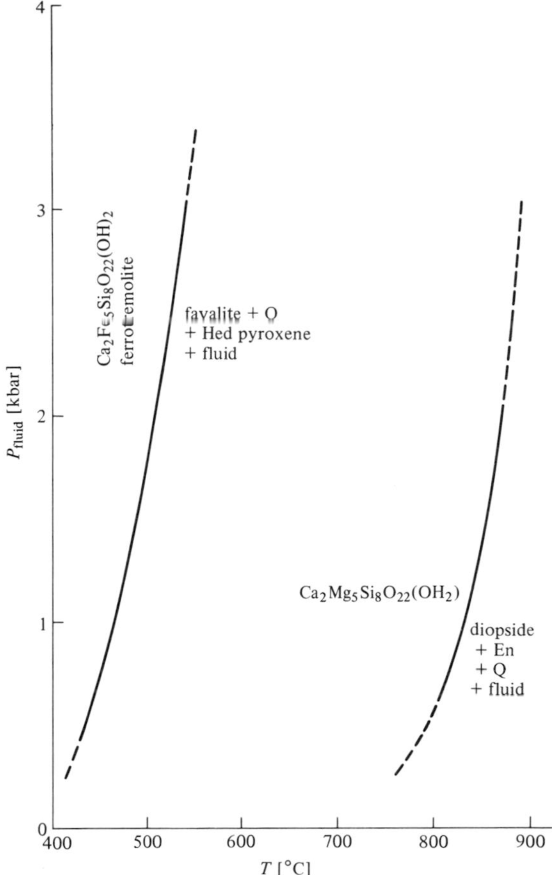

9.15 The thermal stability limits of tremolite (Boyd, 1959) and ferrotremolite (Ernst, 1966). f_{O_2} for ferrotremolite equilibria defined by fayalite-magnetite-quartz buffer.

Richterite $[Na_2CaMg_5Si_8O_{22}(OH)_2]$ and magnesio-richterite $[Na_2Mg_6Si_8O_{22}(OH)_2]$ have been studied by Forbes (1971) and Schreyer and Seifert (1968), respectively. As compared to tremolite, the effect of a substitution of Na for Ca in richterites seems to lend them a greater thermal stability. This is shown by the upper thermal stability temperature of richterites, which, according to Forbes, is between 900°C and 950°C at fluid pressures ranging from 100 to 750 bar.

Charles (1975) has studied richterite and ferro-richterite. Richterite is stable to 1030°C at 1 kbar total pressure and to 1000°C at 200 bar. At higher temperatures, it breaks down to forsterite, diopside, enstatite, melt, and vapor. Ferrorichterite under oxygen fugacity defined by iron-wustite buffer is stable to 715°C at 1 kbar and to 740°C at 5 kbar. Above these temperatures it decomposes to

hedenbergitic pyroxene + fayalite + melt + vapor.

Ferrorichterite on the quartz-fayalite-magnetite buffer is stable to only 525°C at 1 kbar, breaking down to form

$$\text{acmite}_{67}\ \text{hedenbergite}_{33} + \text{fayalite} + \text{magnetite} + \text{quartz} + \text{vapor}$$

at higher temperatures.

The study of natural amphiboles by Ross *et al.* (1969) and by Robinson *et al.* (1971) has yielded useful information on immiscibility in amphiboles. The former authors found that hornblendes, tremolites, and actinolites contain exsolution lamallae of cummingtonite, and vice versa. Sodium- and aluminum-poor anthophyllite and sodium-aluminum-rich gedrite often contain exsolution lamellae of gedrite and anthophyllite, respectively. Several exsolution textures clearly show that there is a miscibility gap between Ca-rich and Ca-poor clino-amphiboles and between gedrite and anthophyllite in the range of temperatures from 500°C to 600°C. Robinson *et al.* (1971) find that the anthophyllite and gedrite are completely miscible in sillimanite grade rocks of south-western New Hampshire and Massachusetts. Such amphiboles on cooling show anthophyllite gedrite intergrowths.

Talc

The lower stability limit of anthophyllite in Greenwood's (1963) experiments is the upper stability limit of talc + forsterite. At a fluid pressure of 2 kbar, talc + forsterite change to anthophyllite at about 680°C. Forbes (1971) studied the system $Mg_3Si_4O_{10}(OH)_2$-$Fe_3Si_4O_{10}(OH)_2$ (talc-minnesotaite). The maximum Fe^{2+} content of talc $[Fe^{2+}/(Fe^{2+} + Mg)]$ is 0.50 at 400°C with f_{O_2} defined by the magnetite-iron, magnetite-wustite buffer. At this f_{O_2}, the mole fraction of Fe^{2+} decreases to 0.19 at 550°C and to 0.05 at 680°C. Extrapolation of the results indicate that the upper stability limit of minnesotaite is about 300°C on the magnetite-iron buffer and 270°C on the fayalite-quartz-magnetite buffer. Attempts to synthesize minnesotaite have been unsuccessful.

Theoretical analysis of stabilities of crystalline solutions

Relatively few systems involving phases of variable composition have been investigated experimentally. Therefore to understand how the P–T stability of a mineral may change as a function of composition, we use theoretical calculations (see Chapters 1 and 2). Although several attempts have been made to refine experimental data on solvi and devise theoretical models, relatively few calculations have been made to show how the effect of crystalline solution may change the measured or calculated pressure and temperature of phase equilibria. One important reason for this lack of calculation may be the incomplete information of the activity–composition relations in many rock-forming silicates. A review of the many studies made on silicate crystalline solutions indicate that wherever Fe-Mg minerals with non-equivalent sites are involved, it is important to consider the intracrystalline partitioning (i.e., the distribution of cations among crystallographic sites). However, many other minerals, such as biotite and garnet, are best treated as ideal solutions until appropriate solution models can be developed (Mueller, 1972). Ganguly's (1973) model for omphacitic pyroxene and Ganguly and Kennedy's (1974) model for garnet are examples of such developments.

Divariant equilibria involving iron-magnesium silicates may be represented by a surface in the P–T–X_{Mg} space. A projection of such a surface on the P–T plane is represented by a band with univariant lines, each with a fixed Fe/Mg ratio. Theoretical calculations using the ideal solution model have been presented for several reactions by Olsen and Mueller (1966), Kurepin (1970), Perchuk (1965), Ganguly (1972), and Evans and Trommsdorff (1974), among others. In all cases, the effect of changing the Fe/Mg ratio on the univariant equilibria has been found to shift the curve systematically toward high or low values of pressure or temperature. In this section some dehydration reactions involving crystalline solutions have been selected to demonstrate the effect of crystalline solubility. In some cases, the effect is the familiar shift in equilibrium P or T with change in X, but in other cases due to the solution effect combined with the P–T effect on water fugacity the divariant surface in the $P_{total}(\simeq P_{H_2O})$–$T$–$X$ space is not of simple form.

Although numerical data have been used in all calculations, the results are at best semiquantitative. This is due to a lack of all the necessary thermodynamic data. Thus the P–T dependence of volume change (ΔV) in the reaction has not been considered. The P–T curves for some pure reactions considered here are also only approximately known. In spite of these difficulties and certain other conditions discussed later, the results are important in metamorphic petrogenesis.

Coexisting cordierite and garnet

Let us consider the reaction

$$\underset{\textit{Fe-cordierite}}{3\,Fe_2Al_4Si_5O_{18}\cdot nH_2O} \rightleftharpoons \underset{\textit{almandine}}{2\,Fe_3Al_2Si_3O_{12}} + \underset{\textit{Al-silicate}}{4\,Al_2SiO_5} + \underset{\textit{quartz}}{5\,SiO_2} + 3n\,H_2O \tag{9.ww}$$

where n represents the possible number of moles of H_2O in cordierite. This reaction has been experimentally studied by Richardson (1968) and more recently by Weisbrod (1973). Weisbrod has also calculated theoretically the number of moles of water in cordierite, which varies as a function of P_{H_2O} at a given temperature.

The derivation of necessary equations for calculating the effect of crystalline solubility (Fe^{2+}-Mg) on phase equilibrium is demonstrated by considering Reaction (9.ww), for which at equilibrium we have (quartz and Al-silicate considered stoichiometric):

$$K_{(ww)}\exp\left(\frac{-P\Delta V_{(ww)}}{RT}\right) = \frac{(a_{Fe}^{Gar})^6}{(a_{Fe}^{Cord})^6}\cdot f_{H_2O}^{3n} \tag{9.2}$$

where $K_{(ww)}$ is the equilibrium constant, $\Delta V_{(ww)}$ is the volume change in Reaction (9.ww) for solids only, P is the pressure in bars, T is the temperature in °K, a is the activity, and f is the fugacity. In this and other calculations the change in ΔV with P and T will not be considered. Besides Reaction (9.ww), we may also consider the following ion-exchange reaction:

$$\text{Fe-cordierite} + \text{Mg-Garnet} \rightleftharpoons \text{Fe-Garnet} + \text{Mg-Cordierite} \tag{9.xx}$$

Considering the ion-exchange on one cation basis, we have

$$K_{(xx)}\exp\left(\frac{-P\Delta V_{(xx)}}{RT}\right) = \frac{a_{Fe}^{Gar}a_{Mg}^{Cord}}{a_{Mg}^{Gar}a_{Fe}^{Cord}} \tag{9.3}$$

Activity–composition relations in the minerals cordierite and garnet are not definitely known. For binary Fe^{2+}-Mg solutions, both minerals may be considered as ideal. Ganguly (1975) has determined approximate activity–composition relations for ternary and quaternary garnets with Ca and Mn, which may be used in the calculations where necessary. Equation (9.2) may be rewritten as

$$\Delta G(P, T) = 0 = \Delta G^0(1, T) + P\Delta V_{(ww)} + 6RT\ln\frac{X_{Fe}^{Gar}}{X_{Fe}^{Cord}} + 3nRT\ln f_{H_2O}(P, T) \tag{9.4}$$

where the a's have been replaced by mole fractions for the ideal solutions. From the experimentally determined equilibrium for the pure phases, we have

$$G^{\circ}(1, T) = -P'\Delta V_{(ww)} - 3nRT \ln f_{H_2O}(P', T) \quad (9.5)$$

From Equation (9.4) and (9.5) we obtain

$$P = P' + \frac{1}{\Delta V_{(ww)}} \left\{ 3nRT \ln f_{H_2O}(P', T) - 3nRT \ln f_{H_2O}(P, T) - 6RT \ln \frac{X_{Fe}^{Gar}}{X_{Fe}^{Cord}} \right\} \quad (9.6)$$

Eliminating $X_{Fe\text{-}Gar}$ from Equation (9.6) with the help of Equation (9.3) we have

$$P = P' + \frac{1}{\Delta V_{(ww)}} \left[3n \left\{ G_{H_2O}(P', T) - G_{H_2O}(P, T) \right\} - 6RT \ln \left\{ \frac{K'_{(xx)} X_{Fe}^{Cord}}{(1 - X_{Fe}^{Cord} + K'_{(xx)} X_{Fe}^{Cord}) X_{Fe}^{Cord}} \right\} \right] \quad (9.7)$$

where $K'_{(xx)} = K_{(xx)} \exp(-P\Delta V_{(xx)}/RT)$.

Compositions of coexisting garnet and cordierite in natural assemblages has been presented by Reinhardt (1968), Dallmeyer and Dodd (1971), and Currie (1971), among others. For demonstrating solid solution effect, we may choose

$K'_{(xx)} = 5.5$ $\quad T°C = 700$ $\quad n = 0.55$

$K'_{(xx)} = 8.5$ $\quad T°C = 600$ $\quad n = 0.5$ or 0.6

These values are consistent with the data from natural assemblages. The value of n_{H_2O} at 700°C has been taken from Weisbrod's (1973) figure and it has been assumed at 600°C. The $\Delta V_{(ww)}$ for the solid phases is -3.95 cal/mole/bar, which is the same as used by Weisbrod. The molar volume of Fe-cordierite is taken to be 237 ± 1 ml/mole as given by Hsu (1968). Figure 9.16 shows the shift in the equilibrium curve as a function of X_{Fe}^{Cord}. The curve for the pure Fe phases is due to Weisbrod (1973) and is somewhat less steep than Richardson's (1968) curve.

As the curves have a small slope, we find that a small variation in the Mg : Fe ratio may require a large change equilibrium temperature. Therefore in natural divariant assemblages from any small area, variation in Fe/Mg in cordierite-garnet assemblage will be comparatively restricted and the range of compositions will be representative of the variation in pressure rather than of the variation in temperature. This is, provided the total pressure and P_{H_2O} are similar and garnets do not contain significant amounts of Mn and Ca. For Ca-rich garnets, the calculations must take into account the nonideal behavior of the Ca component (cf. Ganguly and Kennedy, 1975).

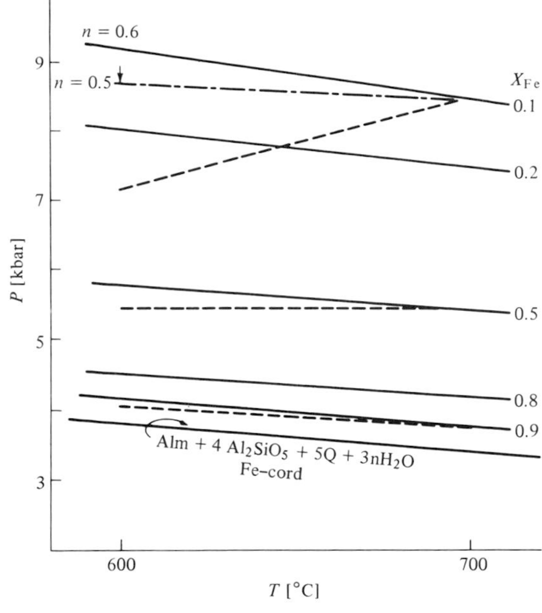

9.16 Solid curves represent the shift in Weisbrod's (1973) experimental curve as a function of X_{Fe} in cordierite. n_{H_2O} in cordierite is 0.6. The dot-dash line indicates the effect of changing n_{H_2O} at 600°C from 0.6 to 0.5. Dashed lines represent a change of $K'_{(xx)}$ from 8.5 to 3.5.

Figure 9.16 also shows the calculated shift at 600°C when n_{H_2O} changes from the arbitrary value of 0.6 to 0.5. At 750°C, Weisbrod's figure shows that the n_{H_2O} versus P curve flattens beyond $n_{H_2O} \sim 0.55$. At 600°C and 3800 bar n_{H_2O}, therefore, may not be very different from 0.55. Note that following Weisbrod (1973) it has been assumed that n_{H_2O} is the same in Fe and Mg end members.

On the basis of our experience with the Fe^{2+}-Mg distribution in coexisting minerals, it was assumed that $K_{(xx)}$ increased with decrease in temperature. Generally the distribution coefficients increase or decrease towards unity with increasing temperature. However, if for some reason (see Currie, 1971), $K_{(xx)}$ decreased to 3.2 at 600°C, the qualitative picture does not change. In Figure 9.16, calculated results corresponding to $K_{(xx)} = 3.2$ are indicated by dashed curves. Wood (1974) suggested that n_{H_2O} may be very different in Fe-cordierite and Mg-cordierite. In the absence of any definite data, this possible difference in n_{H_2O} cannot be considered in these calculations. As the purpose of this section is to demonstrate the extent of crystalline solubility effect, Reactions (9.ww) and (9.xx) could be replaced by similar reactions involving a hydrous phase other than cordierite.

Coexisting muscovite and sanidine

Kretz (1970) has discussed the composition of coexisting muscovite and alkali feldspar at 550°C and 2 kbar. The mole fractions are $X_{Na}^{Ms} = 0.089$ and $X_{Na}^{Fs} = 0.261$. Some tentative information is available on the crystalline solutions muscovite-paragonite and albite-sanidine. Using the expressions of Eugster *et al.* (1971) and Waldbaum and Thompson (1969), it is possible to calculate the Margule's parameters W_{G1} and W_{G2} at 550°C and 2 kbar for both mica and feldspar solutions. For mica these values are $W_{G1} = 3390$ and $W_{G2} = 4751$ cal/mole, and for feldspar $W_{G1} = 2584$ and $W_{G2} = 4624$ cal/mole. Activities in the binary solutions muscovite-paragonite and albite-sanidine can be calculated from the relations

$$RT \ln a_A = X_B^2 W_{G1} + 2X_A(W_{G2} + W_{G1}) \tag{9.8}$$

$$RT \ln a_B = X_A^2 W_{G2} + 2X_B(W_{G1} - W_{G2}) \tag{9.9}$$

We may now consider the reactions

$$\underset{muscovite}{KAl_3Si_3O_{10}(OH)_2} + \underset{quartz}{SiO_2} \rightleftharpoons \underset{feldspar}{KAlSi_3O_8} + \underset{Al\text{-}silicate}{Al_2SiO_5} + \underset{fluid}{H_2O} \tag{9.yy}$$

$$\underset{K\text{-}feldspar}{KAlSi_3O_8} + \underset{muscovite}{NaAl_3Si_3O_{10}(OH)_2} \rightleftharpoons \underset{albite}{NaAlSi_3O_8} + \underset{muscovite}{KAl_3Si_3O_{10}(OH)_2} \tag{9.zz}$$

for which the equilibrium conditions are:

$$K_{(yy)} \exp\left(\frac{-P\Delta V_{(yy)}}{RT}\right) = \frac{a_K^{Fs}}{a_K^{Ms}} f_{H_2O} \tag{9.10}$$

$$K_{(zz)} \exp\left(\frac{-P\Delta V_{(zz)}}{RT}\right) = \frac{a_{Na}^{Fs} a_K^{Ms}}{a_K^{Fs} a_{Na}^{Ms}} = K'_{(zz)} \tag{9.11}$$

Following the earlier approach for the cordierite-garnet equilibrium, the crystalline solubility effect may be calculated by using the expression

$$P = P' + \frac{1}{\Delta V_{(yy)}}\left\{G_{H_2O}(P', T) - G_{H_2O}(P, T) - RT \ln \frac{a_K^{Fs}}{a_K^{Ms}}\right\} \tag{9.12}$$

For the pure phase reaction (9.yy), extrapolated results of Althaus *et al.* (1970), yield $P' = 400$ bar. [The experimental data of Chatterjee and Johannes (1974) yields a pressure as 750 bar at 550°C.] Substituting $P' = 400$ bar, $\Delta V_{(yy)} = 0.0722$, $T = 848$°K, $a_K^{Ms} = 0.92$, $a_K^{Fs} = 0.79$, and the appropriate G_{H_2O} (400 bar, 550°C) from Burnham *et al.* (1969) in equation (9.12), the new equilibrium pressure is calculated to be between 450 and 500 bar. (Note that the composition of the coexisting minerals is strictly at 550°C and 2 kbar, and not 400 bar, as considered here). The shift is rather small. If we change X_K^{Fs} from 0.739 to 0.7, keeping $K'_{(zz)}$ the same as before, the equilibrium pressure again changes only by about 50 bar. The effect is thus not very significant for small changes in composition. We cannot expect large changes in the mole fractions at 550°C since this will put the composition into the immiscibility region of one or both of the crystalline solutions. Note that in this case both mica and feldspar deviate positively from ideal solution. Some of the nonideality effect cancels out, but the muscovite-paragonite solution is more nonideal than albite-sanidine. In the present case the change in equilibrium pressure is less than what it would be if the solutions were ideal.

Coexisting actinolite, ortho- and clinopyroxenes

The three minerals along with quartz and water are related by Reaction (9.aaa):

$$\underset{actinolite}{Ca_2Mg_5Si_8O_{22}(OH)_2} \rightleftharpoons \underset{clinopyroxene}{2CaMgSi_2O_6} + \underset{orthopyroxene}{3MgSiO_3} + \underset{quartz}{SiO_2} + \underset{fluid}{H_2O} \tag{9.aaa}$$

The ion-exchange reactions may be written as

$$\tfrac{1}{5}\text{Mg-actinolite} + \text{Fe-orthopyroxene} \rightleftharpoons \text{Mg-orthopyroxene} + \tfrac{1}{5}\text{Fe-actinolite} \tag{9.bbb}$$

$$\tfrac{1}{5}\text{Mg-actinolite} + \text{Fe-clinopyroxene} \rightleftharpoons \text{Mg-clinopyroxene} + \tfrac{1}{5}\text{Fe-actinolite} \tag{9.ccc}$$

The equilibrium conditions are

$$K_{(aaa)} \exp\left(\frac{-P\Delta V_{(aaa)}}{RT}\right) = \frac{(X_{Mg}^{Cpx})^2 (X_{Mg}^{Opx})^3}{(X_{Mg}^{Act})^5} f_{H_2O} \tag{9.13}$$

$$K_{(bbb)} \exp\left(\frac{-P\Delta V_{(bbb)}}{RT}\right) = \frac{X_{Mg}^{Act}(1 - X_{Mg}^{Opx})}{(1 - X_{Mg}^{Ast})X_{Mg}^{Opx}} = K'_{(bbb)} \tag{9.14}$$

$$K_{(ccc)} \exp\left(\frac{-P\Delta V_{(ccc)}}{RT}\right) = \frac{X_{Mg}^{Act}(1 - X_{Mg}^{Cpx})}{(1 - X_{Mg}^{Act})X_{Mg}^{Cpx}} = K'_{(ccc)} \tag{9.15}$$

Proceeding as before for other reactions, we have

$$P - P' = \frac{1}{\Delta V_{(aaa)}}\Big[G_{H_2O}(P', T) - G_{H_2O}(P, T)$$
$$+ 2RT \ln \{K'_{(ccc)}(1 - X^{Act}_{Mg}) + X^{Act}_{Mg}\}$$
$$+ 3RT \ln \{K'_{(aaa)}(1 - X^{Act}_{Mg}) + X^{Act}_{Mg}\}\Big] \quad (9.16)$$

In Equations (9.13) to (9.15) it has been assumed that quartz is stoichiometric and other minerals are ideal solutions. Orthopyroxene and clinopyroxene (with $M2$ site nearly filled with Ca) are somewhat nonideal (Saxena and Ghose, 1971; Saxena, 1971). The deviations are positive in both cases and will result in part cancellation. The effect of nonideality will be demonstrated with one example later.

Kretz's (1963) study showed that the distribution of Fe^{2+} and Mg between pyroxenes in metamorphic rocks follows the condition:

$$\frac{X^{Cpx}_{Mg}(1 - X^{Opx}_{Mg})}{X^{Opx}_{Mg}(1 - X^{Cpx}_{Mg})} = 1.80 \quad (9.17)$$

Similarly for igneous rocks, the distribution coefficient is 1.30. At 800°C we set the value of the distribution coefficient as 1.67. Since Mueller (1961) found that in the high-grade rocks $X^{Act}_{Mg} = X^{Cpx}_{Mg}$, we have $K'_{(bbb)} = 1.67$. The equilibrium P–T data on Reaction (9.aaa) for Mg end members are given by Boyd (1959) as shown in Figure 9.17. The effect of changing X^{Act}_{Mg} may now be calculated by substituting the appropriate values in Equation (9.16). Since $X^{Act}_{Mg} = X^{Cpx}_{Mg}$, $K'_{(ccc)}$ is unity. $\Delta V_{(aaa)}$ is -0.57 cal/bar (data from Robie and Waldbaum, 1968). Figure 9.17 shows a substantial change in pressure with changing X_{Mg}. The shift in the equilibrium curve is consistent with Ernst's (1966) results on the breakdown of ferrotremolite. However, there are several other points which must be discussed.

In this and other calculations, we neglected the change in ΔV with composition. The change in $\Delta V_{(aaa)}$ will affect the results somewhat. For example, if at X^{Act}_{Mg} the $\Delta V_{(aaa)}$ changes from -0.57 to -0.64 cal/bar/°C, the shift in equilibrium pressure is of the order of 400 bar as shown in Figure 9.17.

As mentioned before, small positive deviations from nonideality are likely in the ferromagnesian minerals. At $X^{Act}_{Mg} = 0.5$, the composition of orthopyroxene is represented by $X^{Opx}_{Mg} = 0.37$. Let us consider that a^{Act}_{Mg} is 0.55 (assumed) and a^{Opx}_{Mg} 0.40 (Saxena, 1973). If $K'_{(bbb)}$ continues to be 1.0, we find that the equilibrium pressure at 800°C will be 2730 bar as against 2150 bar for the ideal case. This shift is also shown in Figure 9.17.

Finally, Figure 9.18 shows the intriguing result of the calculation. At 800°C the equilibrium pressure increases

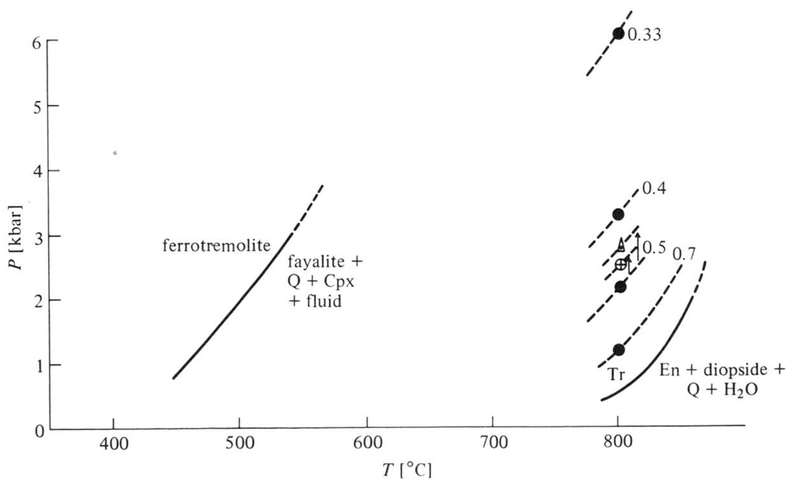

9.17 Equilibrium curves for breakdown of ferrotremolite (Ernst, 1966) and tremolite (Boyd, 1959) and the shift in equilibrium pressure at 800°C as a function of X_{Mg} in actinolite. The data is apparently consistent but see Figure 9.18. Effect of changing ΔV with composition is shown for the curve with $X_{Mg} = 0.5$ by the arrow to a plus-within-a-circle. The arrow to dot-in-a-triangle shows the effect of nonideality.

continuously with decreasing X_{Mg}^{Act}. At 6200 bar X_{Mg}^{Act} is close to 0.3316. If we decrease X_{Mg}^{Act} further, there is no solution to Equation (9.16). The X_{Mg}^{Act}'s for which solutions are obtained at pressures higher than 6200 bar are plotted in Figure 9.18 and show that over the range $X_{Mg}^{Act} = 0.5$ to 0.3316 and probably throughout from $X_{Mg}^{Act} = 1.0$ to 0.3316, there are two equilibrium values of pressure. The high P values may or may not be beyond the stability of the actinolite-pyroxene assemblage under consideration.

To the writers' knowledge there is no compositional data on coexisting actinolite-orthopyroxene and Ca-pyroxene (for all three) occurring in natural assemblages. However, there is no reason why an assemblage with $X_{Mg}^{Act} = 0.33$ should not occur. The effects of probable nonideality and change in ΔV have already been considered, and the pressure shift under such conditions is in the wrong direction. A possible explanation is that the calculated pressure depends critically on the value of the experimentally determined equilibrium pressure for the pure phases.

The system Fe-C-O

The effect of compositional variations on the stability of siderite has been discussed by Mueller (1973) using the experimental results of Weidner (1968, 1972) and French (1971). Phase relations, as presented by French (1971) for an isobaric section with a fluid pressure of 500 bar are shown in Figure 9.19. The section truncates reaction surfaces that separate phase volumes of hematite, magnetite, siderite, and graphite and the two univariant space curves defined by the assemblages siderite-hematite-magnetite-gas (fluid) and siderite-magnetite-graphite-gas (fluid). At oxygen fugacity values below those of the graphite reaction surface (Gr) the total pressure of all gaseous species in the system Fe-C-O are less than 500 bar, so that the system is in a condensed state. However, if an inert gas such as argon is regarded as being present, so that the total pressure is 500 bar, the reaction

$$\underset{siderite}{3FeCO_3} \rightleftharpoons \underset{magnetite}{Fe_3O_4} + \underset{graphite}{3C} + \underset{fluid}{\tfrac{5}{2}O_2} \qquad (9.ddd)$$

becomes univariant and is defined by curve *c* in Figure 9.19. Although we have taken O_2 and the inert gas into account in Reaction (9.ddd), the gas phase will also contain quantities of CO_2, CO, and other species required by equilibrium, as shown by writing the alternative form

$$3FeCO_3 + 5CO \rightleftharpoons Fe_3O_4 + 3C + 5CO_2 \qquad (9.eee)$$

According to French (1971), even at high fluid pressures the phase volume of siderite is limited to temperatures in the vicinity of 500°C or less. However, Weidner (1972) concludes that pure iron siderite is stable to temperatures

9.18 Plot of X_{Mg} in actinolite against equilibrium pressure ($P_{total} = P_{H_2O}$) at 800°C. See text for discussion.

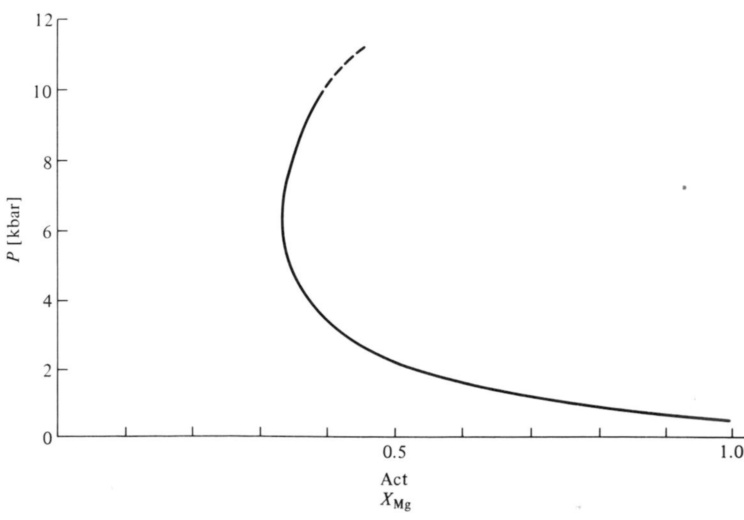

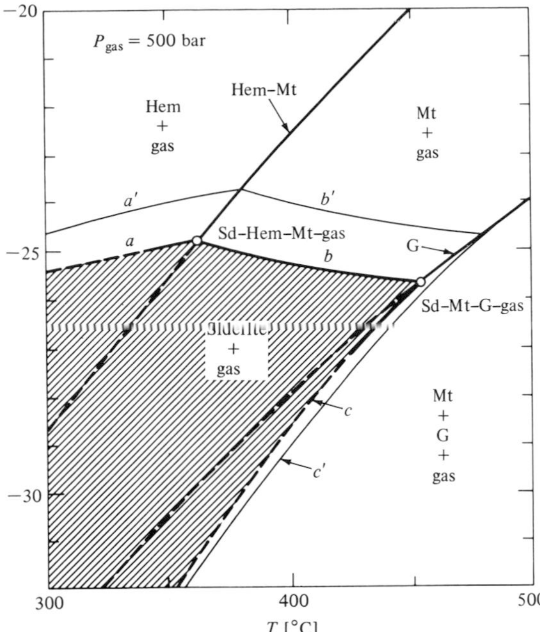

9.19 Phase relations of siderite (Sd), magnetite (Mt), hematite (Hem), graphite (G), and gas at a total pressure of 500 bar as modified after French (1971). Log fugacity of oxygen is plotted against the temperature. The lettered isobaric section curves *a*, *b*, and *c* refer to reactions discussed in the text. The shift in these curves corresponding to a change in Fe/Mg + Fe + Mn + Ca from unity to 0.613 is shown by the curves *a*′, *b*′, and *c*′ assuming siderite behaves as an ideal solution. Although the general aspect of this diagram is in agreement with the data of Weidner (1972) his data indicate that the stability field of siderite is extended to temperatures in excess of 600°C at gas pressures exceeding 5 kbar. (After Mueller, 1973.)

in excess of 600°C at 5 kbar, which is in the range of high-grade regional metamorphism.

Siderite apparently coexists stably with orthopyroxene (Klein, 1966; Butler, 1969) and may represent temperatures higher than 500°C. The additional thermal stability may be due to solid solubility of $MgCO_3$ and $MnCO_3$ as estimated below. Consider Reaction (9.ddd) and the following reactions:

$$\underset{\textit{siderite}}{2FeCO_3} + \underset{\textit{fluid}}{\tfrac{1}{2}O_2} \rightleftharpoons \underset{\textit{hematite}}{Fe_2O_3} + \underset{\textit{fluid}}{2CO_2} \tag{9.fff}$$

$$\underset{\textit{siderite}}{3FeCO_3} + \underset{\textit{fluid}}{\tfrac{1}{2}O_2} \rightleftharpoons \underset{\textit{magnetite}}{Fe_3O_4} + \underset{\textit{fluid}}{3CO_2} \tag{9.ggg}$$

The equations of equilibrium are:

$$\log K_{(ddd)} = \frac{P\Delta V_{(eee)}}{4.576\,T} + \tfrac{5}{2}\log f_{O_2} - 3\log X_{Fe}^{Sd}\gamma_{Fe}^{Sd} \tag{9.18}$$

$$\log K_{(fff)} = \frac{P\Delta V_{(fff)}}{4.576\,T} - \tfrac{1}{2}\log f_{O_2} + 2\log f_{CO_2} - 2\log X_{Fe}^{Sd}\gamma_{Fe}^{Sd} \tag{9.19}$$

$$\log K_{(ggg)} = \frac{P\Delta V_{(ggg)}}{4.576\,T} - \tfrac{1}{2}\log f_{O_2} + 3\log f_{CO_2} - 3\log X_{Fe}^{Sd}\gamma_{Fe}^{Sd} \tag{9.20}$$

Here P is the total pressure, ΔV is for the solid reactants and products, f with subscript O_2 or CO_2 indicates fungacities of the gaseous (fluid) species, X_{Fe}^{Sd} and γ_{Fe}^{Sd} represent the mole fraction and activity coefficient, respectively, of $FeCO_3$ in siderite, and K represents the equilibrium constant and is a function solely of the temperature.

Butler (1969) presented an analysis of a siderite that coexisted in equilibrium with quartz, orthopyroxene, cummingtonite, ankerite, and magnetite. Mole fractions are $FeCO_3$: 0.613; $MnCO_3$: 0.117; $MgCO_3$: 0.243; and $CaCO_3$: 0.027. Although the dependence of γ_{Fe}^{Sd} on composition is not known—the major components being Mg^{2+}, Mn^{2+}, and Fe^{2+}—we shall assume in analogy with silicates that γ_{Fe}^{Sd} does not differ greatly from unity. Accordingly the substitution of X_{Fe}^{Sd} should shift $\log f_{O_2}$ to higher values by a factor of $4\log X_{Fe}^{Sd}$ for Reaction (9.fff) and by $6\log X_{Fe}^{Sd}$ for Reaction (9.ddd). However $\log f_{O_2}$ for Reaction (9.ddd) will be shifted to lower values by a factor of $\tfrac{6}{5}\log X_{Fe}^{Sd}$. The result of substituting $X_{Fe}^{Sd} = 0.613$ is shown in Figure 9.19 as curves *a*′, *b*′, and *c*′. It thus appears that at 500 bar, total pressure siderite of this composition in equilibrium with magnetite would be stabilized at considerably higher temperatures as compared with pure $FeCO_3$. However the maximum thermal stability without precipitation of graphite would not be raised by more than about 25°C. Also since the same shift will occur independently of total pressure, this reasoning also applies to higher pressures as well.

Coexisting biotite, magnetite, and sanidine

Biotite shows a complex substitution of cations. It appears that the Fe^{2+} and Mg^{2+} end members form an ideal solution (Mueller, 1972). In addition, biotite may contain a variable concentration of Al^{3+}, Fe^{3+}, and Ti^{4+}. Rutherford (1973) has studied the stability of biotites along the join annite $K_2Fe_6Al_2Si_6O_{20}(OH)_4$ aluminum biotite $[K_2Al_6Al_2Al_6O_{20}(OH)_4]$. The aluminous biotite is stable to higher temperatures than annite. At 2 kbar fluid pressure, aluminum rich biotite is stable to 555°C on the HM buffer, 763°C on Mt-Hc-Cor, 820°C on NNO, and about 860°C on QFM. We may calculate the effect of crystalline solubility in biotite on the reaction (treating orthoclase as stoichiometric)

$$\underset{\textit{biotite solid solution}}{KFe_3AlSi_3O_{10}(OH)_2} + \underset{\textit{fluid}}{\tfrac{1}{2}O_2} \rightleftharpoons \underset{\textit{magnetite solid solution}}{Fe_3O_4} + \underset{\textit{orthoclase}}{KAlSi_3O_8} + \underset{\textit{fluid}}{H_2O} \qquad (9.hhh)$$

The equilibrium constant is given by

$$K_{(hhh)} \exp\left(\frac{-P\Delta V_{(hhh)}}{RT}\right) = \frac{X_{Fe}^{Mt}\gamma_{Fe}^{Mt} f_{H_2O}}{X_{Fe}^{Bi}\gamma_{Fe}^{Bi} f_{O_2}^{1/2}} \qquad (9.21)$$

or in logarithmic form we have

$$\log K_{(hhh)} = \frac{P\Delta V_{(hhh)}}{2.303\,RT} + \log \frac{X_{Fe}^{Mt}\gamma_{Fe}^{Mt}}{(X_{Fe}^{Bi}\gamma_{Fe}^{Bi})^3} + \log f_{H_2O} - \tfrac{1}{2}\log f_{O_2} \qquad (9.22)$$

where $K_{(hhh)}$, the equilibrium constant, is a function of temperature only. It is obvious that $\log f_{O_2}$ will change by a factor of 2 log $X_{Fe}^{Mt}a_{Fe}^{Mt}/(X_{Fe}^{Bi}a_{Fe}^{Bi})^3$ if we substitute for the value of the latter in Equation (9.22). Let us consider the point $X_{Fe}^{Bi} = 0.91$, $X_{Fe}^{Mt} = 0.96$, the intersection of the curves Ann_{91} and Mt_{96} (Figure 9.20). This point is approximately at 700°C and at a $\log f_{O_2}$ value of about -17. Compared to this point, at 700°C the Ann_{100} point lies at -17.8. If $(\gamma_{Fe}^{Mt}/\gamma_{Fe}^{Bi})^3$ is unity, the difference between the two points according to Equation (9.22) should be much less. This would mean, as expected, that γ_{Fe}^{Mt} and γ_{Fe}^{Bi} are not unity. Fe_3O_4-$FeAl_2O_4$ is non-ideal with positive deviation from ideality. It is likely, then, that a_{Fe}^{Mt} is 0.98 instead of 0.96. This is not enough to explain the shift in $\log f_{O_2}$ and we must consider that a_{Fe}^{Bi} should be of the order of 0.79, which means a considerable negative deviation from ideality in the annite-$KAl_3AlAl_3O_{10}(OH)_2$ solution.

It is apparent from the various examples of phase equilibria considered in this section that the thermodynamic data on solutions is indispensable to the study of petrogenetic processes. As long as necessary data on fluid- and crystalline-solution is unavailable, the pressure and temperature estimates of metamorphic assemblages will remain uncertain.

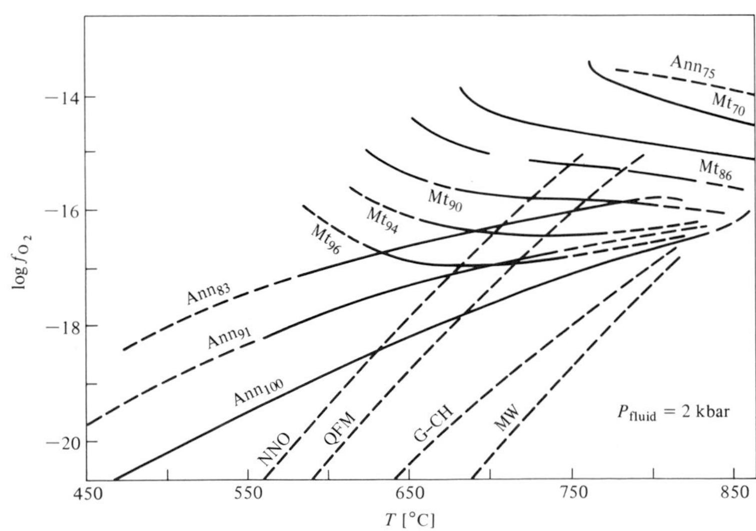

9.20 Contours of aluminous biotite and magnetite crystalline solution. (Data from Rutherford, 1973.) G-CH = graphite-methane (see Table 8.1).

10 Metamorphic mineral facies

Mineral facies classification was briefly discussed in Chapter 4. We begin this chapter by adopting Ramberg's (1952) definition of metamorphic mineral facies: "Rocks formed or recrystallized within a certain *P*, *T*-field, limited by the stability of certain critical minerals of defined composition, belong to the same mineral facies." Thus in Figure 4.3 we labeled various *P–T* areas by different names, such as greenschist facies, granulite facies, etc. The facies concept still provides us with the most convenient framework for the discussion of metamorphic rocks. However, we must not allow its artificiality to bind us rigidly, nor must we lose sight of its highly qualitative and provisional character. At present there is a growing feeling among various workers that the number of facies should be kept to a minimum and the use of subfacies is quite unrealistic. Winkler (1974) has proposed a fourfold classification of metamorphism—very low grade, low grade, medium grade, and high grade—without the use of facies names.

The facies were originally defined on the basis of natural occurrences of mineral assemblages. It is necessary to bring out the correlation between their *P–T* fields and the experimental data as well as the calculations based on thermochemical information. Unfortunately, however, the correlation between experimental systems and the natural assemblages is still not fully developed. Therefore it is preferable to limit the number of facies to the necessary minimum. In order of increasing temperature the facies considered here are zeolite, lawsonite, greenschist, amphibolite, and granulite. The pressure range is from low to medium. The low-pressure equivalents of amphibolite and granulite facies, such as the hornfels and sanidinite facies, are not discussed separately. The high-pressure facies rocks, the glaucophane-schist and eclogites, are discussed in Chapter 11.

Even before the facies concept was introduced by Eskola (Chapter 4), Barrow (1912) recognized zones of metamorphic intensity in Scotland distinguished by the occurrence of particular index minerals. Later Tilley (1924) introduced the term *isograd* for the boundaries of these mineral zones with the implication that each such boundary determined an isothermal (and nearly isobaric) surface extending through the rock. The isograd of a particular mineral marked the first appearance of this mineral in the direction of increasing metamorphic grade. Thus, successive zones were marked by the appearance of chlorite, biotite, almandine, kyanite, and sillimanite in rocks of pelitic composition. Since Tilley emphasized the need for the comparison of rocks of the same bulk composition, the definition of the isograd is in harmony with the concept of Eskola. However, in practice the isograd and metamorphic zone concepts have given rise to many abuses, and investigators have generally not seemed aware of the need to compare isochemical series. Consequently, they have commonly, at least by implication, equated zones observed in the field in particular areas with the universal facies boundaries or subdivisions. Also some authors have entertained the possibility of defining facies in terms of the chemical potentials or fugacities of certain constituents such as H_2O, CO_2, and H_2S (Fyfe *et al.*, 1958). However, since these are compositional variables as strictly defined, their use violates the facies principle and destroys its thermodynamic simplicity.

The inherent difficulty of the isograd-zone concept in metamorphism is that it is very difficult to determine the position of a given rock in the bulk composition field. This can seldom be done accurately with even the best chemical analyses since it is not known whether the equilibrium volume of the mineral assemblages observed in a thin section or represented by analyzed mineral species coincides with the volume of the bulk analysis. Also it is difficult to determine independently the fugacities of the volatile species that are needed to fix the composition with respect to these constituents. Therefore it is far more desirable to make use of the mineral assemblage rather than a single-index mineral. This was of course recognized by Goldschmidt (1911)[1] in his applications of the phase rule to such assemblages.

[1] He was preceded by Van't Hoff and others in application to nonsilicate mineral systems.

However, in order to obtain quantitative results we must investigate the assemblages in more detail than did Goldschmidt, and we must, whenever possible, employ the analytical expressions of thermodynamics and statistical mechanics to treat these multicomponent equilibria. These general methods were already suggested some time ago by Ramberg (1944a, 1952).

One of the early attempts to relate the mineral facies to the temperature and pressure was made by Miyashiro (1961) who superimposed the facies fields on the phase diagram for the Al_2SiO_5 polymorphs. Other general correlations have been made by Hietanen (1967), Turner (1968) and more recently by Winkler (1974). Petrogenetic grids for specific systems have been constructed by Albee (1965a), Hess (1969), Hoschek (1969), Grant (1973), Liou (1971a), and Schreyer and Seifert (1969) among others. Figure 10.1 presents some of the latest experimental data. The figure shows a collection of several important equilibria in the P–T field and a division of regional metamorphic rocks into four mineral facies or four metamorphic grades, namely zeolite/lawsonite facies, greenschist facies, amphibolite facies, and granulite facies. A brief description and the important reactions and theoretical considerations on which these divisions are based are discussed in the following sections.

Diagenesis and metamorphism; the zeolite and lawsonite facies

It seems appropriate to begin our detailed description of metamorphic facies with a discussion of the beginnings of metamorphism. The initial chemical and mineralogic changes that affect rocks at low temperatures and pressures are referred to as "diagenetic." An excellent discussion of the chemistry of diagenesis is given by Berner (1971).

According to Coombs *et al.* (1959) the zeolite facies may be regarded as linking metamorphism with diagenesis. However, because most common sediments are of the nature of shales and graywackes, the most abundant mineral group is that of the clays and their most common representatives are the mica-like "montmorillonoids." The montmorillonoids are essentially dioctahedral or trioctahedral micas with a deficiency of interlayer cations and large and variable amounts of interlayer water. The charge deficiency in the interlayer is compensated by substitution in the octahedral and tetrahedral framework (Grim, 1953). However, we know that when sediments are reconstituted into low-grade metamorphic rocks the montmorillonoids are

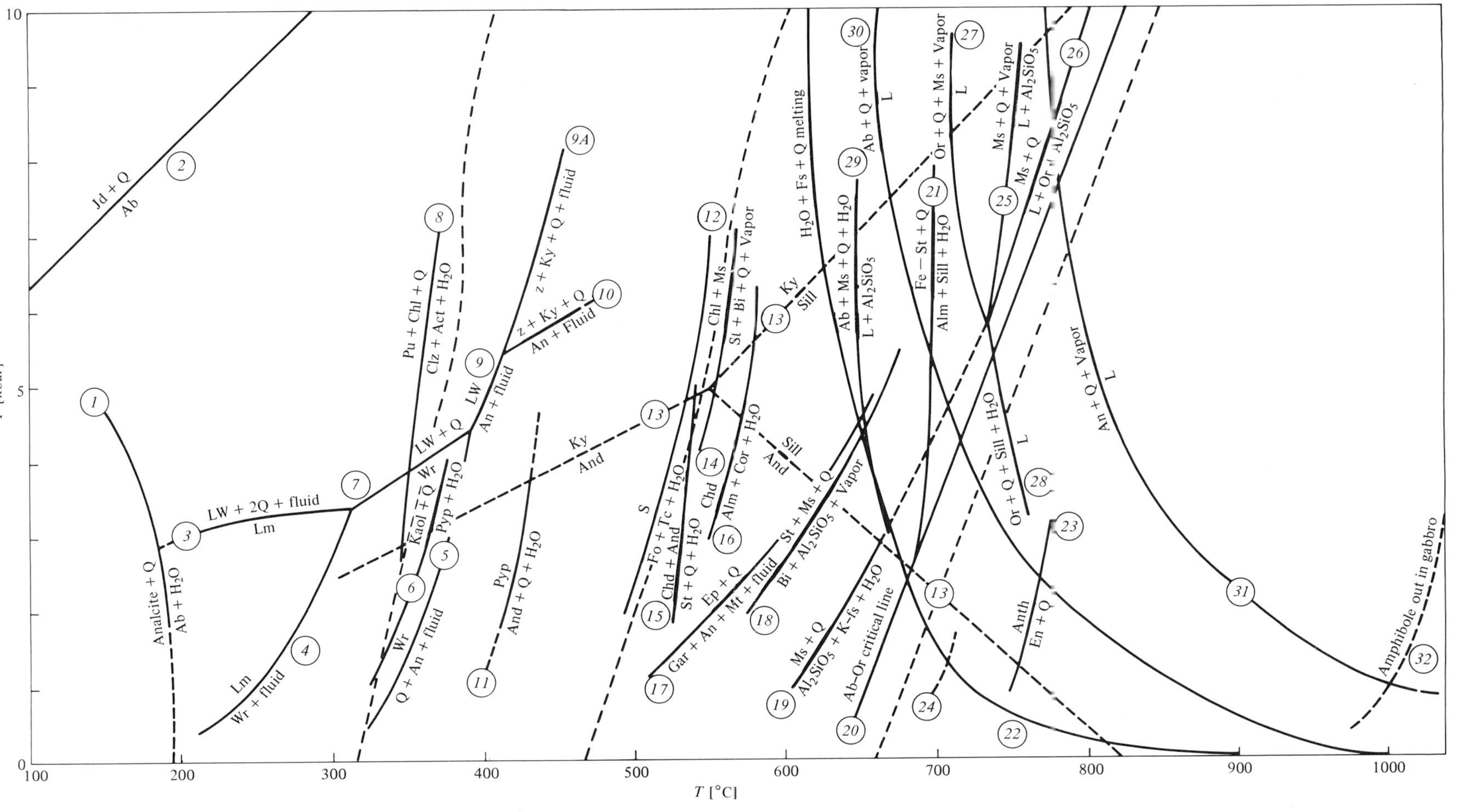

10.1 Collection of some important equilibria. Unless specified all reactions are with $P_{H_2O} = P_{total}$. (*1*) Anlc + Q ⇌ Ab + H_2O (Thompson, 1971); (*2*) Jd + Q ⇌ Ab (anhydrous; Newton and Smith, 1967); (*3*) Lm ⇌ Lw + 2Q + fluid (Liou, 1971); (*4*) Lm ⇌ Wr + fluid (Liou, 1971); (*5*) Wr ⇌ Q + An − fluid (Liou, 1970); (*6*) Kaol + 2Q ⇌ Pyp + H_2O (Thompson, 1970); (*7*) Lw + 2Q ⇌ Wr (Liou, 1971); (*8*) Pu + Chl + Q ⇌ Clz + Act + H_2O (Nitsch, 1971); (*9*) Lw ⇌ An − fluid (Newton and Kennedy, 1963); (*9A*) Lw ⇌ Z + Ky + Q + fluid (Newton and Kennedy, 1963); (*10*) Z + Ky + Q ⇌ An + fluid (Newton and Kennedy, 1963); (*11*) Pyp ⇌ And + 3Q + H_2O (Kerrick, 1968); (*12*) S ⇌ 6Fo + Tc + 9H_2O (Scarfe and Wyllie, 1967); (*13*) And ⇌ Ky ⇌ Sill [inferred from experimental work of Richardson *et al.*, 1968; Holdaway, 1971; and Ganguly's (1969) arguments; see text in Chapters 10 and 11); (*14*) (Fe-Mg)Chl + Ms ⇌ (Fe-Mg)St + (Fe-Mg)Bi + Q + vapor (Hoschek, 1969); (*15*) Chd + Al_2SiO_5 ⇌ St + Q + H_2O (Richardson, 1968); (*16*) Chd ⇌ Alm + Cor + H_2O (Ganguly, 1969); (*17*) Ep ($Ca_2Al_{.75}Fe_{.25})Si_3O_{12}(OH)$ + Q ⇌ Gar + An + Mt + fluid (NNO buffer; Liou, 1973); (*18*) (Fe-Mg)St + Ms + Q ⇌ Al_2SiO_5 + (Fe-Mg)Bi + vapor (Hoschek, 1969); (*19*) Ms + Q ⇌ K-fs + Al_2SiO_5 + H_2O (Day, 1973); (*20*) Ab-Or critical mixing (Morse, 1970); (*21*) Fe-St + Q ⇌ Alm + Sill + H_2O (Richardson, 1968); (*22*) Ab-Or-Q-H_2O system, beginning of melting (Merill *et al.*, 1970); (*23*) Anth ⇌ En + Q (Greenwood, 1963); (*24*) Anth ($X_{Mg} \simeq 0.60$) ⇌ Opx + Q (Hinrichsen, 1966); (*25*) Ms + Q + vapor ⇌ (liquid) + Al_2SiO_5 (Storre and Karotke, 1971); (*26*) Ms + Q ⇌ liquid + Or + Al_2SiO_5 (Storre, 1972); (*27*) Or + Q + Ms + vapor ⇌ L; (*28*) Or + Q + Sill + H_2O ⇌ liquid; (*29*) Ab + Ms + Q + H_2O ⇌ liquid (*27* to *29*; Storre and Karotke, 1971); (*30*) Ab + Q + Vapor ⇌ L; (*31*) An + Q + vapor ⇌ liquid; (*32*) amphibole breakdown in gabbro (*30–32*: Merill *et al.*, 1970). Approximate facies boundaries are shown by large dashed curves. Clz = clinozoisite, Fs = feldspar, Fo = forsterite. Other symbols as in Table 9.1.

usually replaced by true micas with a full complement of alkali ions in the interlayer. Thus if we designate the cation content of a montmorillonoid by a fractional variable n and the water content by the variable m the following reaction is one of the types of transformations that are likely to occur during early metamorphism or advanced diagenesis:

$$\underset{\textit{montmorillonoid}}{(\mathrm{Na, K})_n\mathrm{Al_2(Si_{4-n}Al_n)O_{10}(OH)_2} \cdot m\mathrm{H_2O}} + \underset{\textit{feldspar}}{(1-n)(\mathrm{Na, K})\mathrm{AlSi_3O_8}} \rightleftharpoons$$

$$\underset{\textit{white mica}}{(\mathrm{Na, K})\mathrm{Al_2(Si_3Al)O_{10}(OH)_2}} + \underset{\textit{quartz}}{4(1-n)\mathrm{SiO_2}} + \underset{\textit{fluid}}{m\mathrm{H_2O}} \qquad (10.a)$$

In this illustration the alkali deficiency in the montmorillonoid is balanced entirely by the substitution of Si^{4+} in the tetrahedral framework with the result that silica is released. This reaction is particularly interesting because it does not involve the release or absorption of any water other than that present in the montmorillonoid interlayer.

If the initial material consists of a mixture of kaolinite and feldspar the following reaction should occur:

$$\underset{\textit{kaolinite}}{\mathrm{Al_4Si_4O_{10}(OH)_8}} + \underset{\textit{feldspar}}{2(\mathrm{Na, K})\mathrm{AlSi_3O_8}} \rightleftharpoons$$

$$\underset{\textit{mica}}{2(\mathrm{Na, K})\mathrm{Al_2(Si_3Al)O_{10}(OH)_2}} + \underset{\textit{quartz}}{4\mathrm{SiO_2}} + \underset{\textit{fluid}}{2\mathrm{H_2O}} \qquad (10.b)$$

Reactions of the type of (10.a) and (10.b) probably are always displaced strongly to the right since it would be expected that clay and feldspar mixtures resulting from mechanical sedimentation would be thermodynamically unstable. This is particularly obvious in the case of Reaction (10.b) and has been pointed out by Garrels and Howard (1959).

The reason that reactions such as (10.a) and (10.b) do not occur immediately after sediment deposition is of course a consequence of the high activation energies involved in the reconstitution of the tetrahedral and octahedral frameworks of the clay minerals. The same does not apply to the bulk of the interlayer cations that are readily exchangeable, and such exchanges undoubtedly occur with every alteration in the surrounding fluids. For a discussion of these exchange equilibria see Chapter 2. Also, the free energies of activation for the deposition from solution of certain clays also seem to be low. This is shown by the apparently rapid growth of such clay minerals as glauconite in the surficial layers of some recent marine sediments.

A great variety of types of alterations affect unstable materials under low-temperature conditions. The products of high-temperature processes such as volcanic glasses and igneous and high-grade metamorphic minerals are particularly vulnerable. Among the greatest deposits of montmorillonite clays are those that result from the alteration of volcanic ash beds, probably in an environment of high silica activity engendered by the presence of amorphous silica.[2] Other alterations result in the formation of zeolites and the incipient albitization of plagioclase.

In addition to the reactions between the unstable products of sedimentation and volcanism, progressive alteration of previously stable assemblages also occur within the zeolite facies. Montmorillonoids and other clay minerals formed as stable products in a given diagenetic environment transform to other slightly different assemblages when these conditions are slightly altered. For example, when the temperature and pressure increased somewhat by burial in New Zealand, zeolitic alteration of volcanic deposits took the form of successive replacements of one zeolite by another (Coombs *et al.*, 1959). Thus heulandite-analcite-quartz assemblages give way to laumontite-albite-quartz. Surdam (1973) noted that even very subtle differences in the composition of the fluid phase are significant and can create a large overlap of critical minerals such as laumonite, wairakite, prehnite, pumpellyite, and epidote.

Some of the experimentally investigated equilibria that may occur within the zeolite-lawsonite facies are (Figure 10.1):

$$\text{analcite} + \text{quartz} \rightleftharpoons \text{albite} + \mathrm{H_2O} \qquad (1)$$

$$\text{laumontite} \rightleftharpoons \text{lawsonite} + 2\ \text{quartz} + \text{fluid} \qquad (3)$$

$$\text{laumontite} = \text{wairakite} + \text{fluid} \qquad (4)$$

$$\text{lawsonite} + \text{quartz} = \text{wairakite} \qquad (7)$$

$$\text{pumpellyite} + \text{chlorite} + \text{quartz} = \text{clinozoisite} + \text{actinolite} + \mathrm{H_2O} \qquad (8)$$

The jadeite + quartz = albite curve is also partly in the high-temperature division of this facies. Some other zeolite equilibria are discussed in Chapter 11.

[2] Note that in this case montmorillonite is a *product* of early diagenesis whereas according to Reactions (10.a) and (10.b) it is a *reactant* in this process. Such behavior is entirely compatible and results from a variation chiefly in the composition and temperature of the environment.

Zeolite-lawsonite facies to greenschist facies

The zeolite-lawsonite facies corresponds approximately to the low-temperature metamorphic division of Winkler (1974). As suggested by Winkler, the diagnostic minerals are laumontite, prehnite, pumpellyite, lawsonite, and illite with low crystallinity. In Figure 10.1 the high temperature limit of the zeolite-lawsonite facies may be drawn roughly parallel to the boundary of the reactions

wairakite = quartz + anorthite + fluid (*5*)

lawsonite = anorthite + fluid (*9*)

and

lawsonite = zoisite + kyanite + quartz + fluid (*9A*)

The temperature of the boundary is shown to be lower than the equilibrium temperature of the reactions in the system Al_2O_3-SiO_2-H_2O because the experimental results are modified in the natural system. The addition of Na_2O and Fe_2O_3 would be significant either because of the possibility of new phases and/or because of crystalline solubility in plagioclase and epidote solutions. The latter effect would tend to displace the dehydration curves toward lower temperature. Further, the effect of adding CO_2 to the gas phase would be very significant on most reactions labelled (*1*) to (*8*). As discussed by Albee and Zen (1969) and Coombs *et al.* (1970), increasing the chemical potential of CO_2 may lead to a considerable shift of these equilibria toward lower temperatures.

In addition to the various compositional effects, kinetic considerations make the interpretation of various reactions difficult in this grade of metamorphism. This has been discussed by Zen (1967), Liou (1971), and A. B. Thompson (1970), among others. The nature of the starting material influences the reactions, and laumontite may form metastably in a variety of situations. Due to the slow rate of reactions and the high-activation energies involved in any cation-exchange reactions involving framework silicates, many of the assemblages encountered in this facies may not be in chemical equilibrium.

Although the recent experimental work by Liou (1971) and A. B. Thompson (1970a,b) has given us useful information on the *P–T* conditions of the zeolite facies, the uncertainty due to kinetic and compositional influences in these estimates makes this part of Figure 10.1 generally less meaningful than the division into facies of the region of medium- and high-grade metamorphism.

Greenschist facies

This is the major facies of low-grade regional metamorphism, corresponding to the low-grade division of Winkler (1974). It extends at 5 kbar (P_{H_2O}) from ~350°C to ~550°C. The name comes from the very common green-colored schists that dominate many metamorphic terrains. These schists usually consist of metamorphosed basalt flows, but also contain derivatives of a variety of associated sediments. The equivalent pelitic rocks consisting of slates and phyllites are also very widespread.

The dominant mineral species of the greenschist facies are muscovite, biotite, chlorite, epidote, actinolite, and the carbonates, but the plagioclase and potassium feldspars are also common. Typical representatives of greenschist facies rocks are shown in Tables 10.1, 10.2 (specimens 3 and 4), and Table 10.3 (specimens 11, 14, and 15). The amphiboles of the greenschist facies tend to be nonaluminous or actinolitic except near the high-temperature and pressure limits of the facies. Also, plagioclase tends to be highly sodic in greenschists and is usually referred to as "albite". Apparently this is a consequence of the instability of the anorthite end member relative to epidote-bearing assemblages (Ramberg, 1952). These characteristics of the Ca-amphiboles and plagioclase serve as rough but useful indicators for facies classification in conjunction with the total mineral assemblage.

It is interesting to examine this facies for the type of Al_2SiO_5 polymorph most frequently recorded. Aluminous greenschists should be characterized chiefly by the occurrence of andalusite according to Figure 10.1, although kyanite should also appear in the high-pressure range; sillimanite should be excluded. Some authors have completely ignored the occurrence of andalusite in this facies, preferring to attribute the presence of this mineral to superimposed thermal effects of hidden intrusions, etc. Harker (1932) believed that andalusite indicated rocks "deficient in shearing stress." However, a perusal of the literature shows that it is a fairly common mineral in regionally metamorphosed schists. Harker himself summarizes numerous occurrences of andalusite in the low-grade schists of the British Isles. By way of contrast, kyanite in low-grade schists is very rare indeed.

A typical example of a metamorphic terrain in which greenschist rocks are well developed is the Lake Superior region of the United States and Canada. There basaltic lavas of Precambrian age are associated with slates, graywackes, and iron formations. Many of the lavas show pillow structure and retain ghost textures of

Table 10.1 Chemical analyses of typical metamorphic rocks

	Kyanite-andalusite-sillimanite gneiss	Staurolite-biotite-garnet gneiss	Actinolite-albite-epidote schist	Pelitic schist	Pelitic schist	Ilmenite-magnetite-bearing pelitic gneiss	Ilmenite-magnetite-hematite-bearing pelitic gneiss	Magnetite-hematite-bearing pelitic gneiss	Plagioclase-quartz-biotite-gneiss	Pyroxene gneiss
	1	2	3	4	5	6	7	8	9	10
SiO_2	48.20	60.73	48.60	55.07	39.41	43.36	44.46	44.09	67.92	56.81
TiO_2	0.14	0.74	0.32	0.98	0.66	1.41	1.33	1.69	0.70	1.01
Al_2O_3	32.54	21.18	9.16	21.50	31.21	32.65	31.09	23.64	15.53	17.33
Fe_2O_3	0.23	3.23	4.65	2.72	2.10	0.48	4.28	12.01	0.77	1.87
FeO	2.24	4.16	6.48	6.04	10.36	6.64	6.56	3.66	3.51	5.55
MnO	0.05	0.04	0.25	0.05	0.16	0.03	0.14	0.37	0.05	0.13
MgO	9.30	2.75	18.41	3.00	3.41	3.16	2.85	2.61	2.04	3.46
CaO	0.84	0.79	9.57	0.98	0.14	1.01	0.60	0.85	2.22	6.55
Na_2O	1.66	0.90	0.53	1.43	1.55	1.54	0.87	2.03	3.90	3.54
K_2O	2.32	2.30	0.20	3.01	4.01	6.10	4.69	6.01	2.67	2.24
H_2O^-	0.12			0.02		0.18	0.14	0.12		0.14
H_2O^+	2.26	2.61	1.29	3.53	6.54	3.29	3.32	3.09	0.72	0.85
P_2O_5	0.01	0.09	0.03	0.11		trace	0.16	0.15	0.11	0.26
CO_2	0.22	0.20	0.07	0.12						
F									0.07	
	100.13	99.74	99.88	98.59	99.55	99.85	100.49	100.32	100.21	99.74

1. From Smith Ridge, Boehls Butte Quadrangle, Idaho (Hietanen, 1956).
2. From Agnew Lake area, Sudbury district, Ontario, Canada. Author's No. 2 (Card, 1964).
3. From Agnew Lake area, Sudbury district, Ontario, Canada. Author's No. 6 (Card, 1964).
4. From Agnew Lake area, Sudbury district, Ontario, Canada. Author's No. 1 (Card, 1964).
5. From west-central Vermont. Author's No. 213-1 (Zen, 1960).
6. From Glen Clova, Scotland, Harker Collection No. 83143 (Chinner, 1960).
7. From Glen Clova, Scotland. Harker Collection No. 83121 (Chinner, 1960).
8. From Glen Clova, Scotland. Harker Collection No. 83112 (Chinner, 1960).
9. From West Balmat, New York. Author's No. Bgn 27 (Engel and Engel, 1958).
10. Lützow-Holm Bay, Antarctica. Author's No. JARE 57110802 (Banno *et al.*, 1964).

Table 10.2 Mineral assemblages of the metamorphic rocks of Table 10.1

	Kyanite-andalusite-sillimanite gneiss	Staurolite-biotite-garnet gneiss	Actinolite-albite-epidote schist	Pelitic schist	Pelitic schist	Ilmenite-magnetite-bearing pelitic gneiss	Ilmenite-magnetite-hematite-bearing pelitic gneiss	Magnetite-hematite-bearing pelitic gneiss	Plagioclase-quartz-biotite-gneiss	Pyroxene gneiss
	1	2	3	4	5	6	7	8	9	10
Quartz	3.2	+[a]		+	5	trace	16.4	2.6	32.3	
Plagioclase	15.6	+	+	+		16.4	9.4	13.8	44.4	+
K-feldspar									1.7	+
Muscovite	0.2	+		+	34	26.3	25.2	53.7	0.9	
Paragonite[b]					19					
Andalusite }										
Kyanite }	17.9					9.8	14.9	2.2		
Sillimanite }						12.0	trace	0.5		
Biotite	25.1	+				33.9	26.2	13.3	20.0	
Garnet		+				trace	4.8	2.4		+
Cordierite	37.1									
Staurolite		+					0.3	trace	0.5	
Chlorite			+	+	31					
Epidote			+							
Actinolite			+							
Hornblende										+
Ca-pyroxene										+
Orthopyroxene										+
Chloritoid					9					
Sphene			+	+						
Rutile					1					
Tourmaline						trace	trace	trace		
Calcite	0.5									
Apatite						trace	trace	trace	0.2	+
Zircon									0.1	
Sulfide										+
Opaques			+	+		1.6	2.8	11.2	0.1	+
Plagioclase composition	An_{17}	An_{40}	An_0	An_0						An_{40}

[a] indicates mineral is present.

[b] Paragonite is usually not distinguished from muscovite so that it may comprise some white mica classed as this mineral.

Table 10.3 Mineral assemblages of greenschist, glaucophane schist, amphibolite, and granulite facies rocks

	Metamorphosed graywacke	Pelitic schist	Pelitic schist	Greenstone	Chlorite schist	Sililic schist	Pelitic schist	Aragonite marble	Glaucophane schist	Pelitic schist	Pelitic schist
	11	12	13	14	15	16	17	18	19	20	21
Quartz	+	+	+	+	+	79.0	30.8	+	+	+	+
Plagioclase		+	+	+			30.6			+	
K-feldspar					+[a]						
Muscovite	+					0.6[b]	20.2[b]		+	+	+
Andalusite		+								+	
Sillimanite			+								+
Biotite	+	+	+							+	+
Garnet						6.1	0.2	+		+	+
Staurolite		+									
Chlorite	+			+	+	0.1	7.1				
Stilpnomelane								+			
Epidote				+	+		2.3			+	
Alanite											
Scapolite											
Talc											
Cummingtonite											
Actinolite				+	+			+			
Hornblende											
Na-amphibole						5.4			+		
Ca-pyroxene											
Na-pyroxene						4.0					
Orthopyroxene											
Lawsonite									+		
Magnetite											
Hematite											
Graphite							4.4				
Opaques[c]	+			+		4.7					
Calcite					+		2.3	+			
Argonite								+			
Dolomite											
Pyrite							0.8	+			
Apatite						0.1	0.4				
Sphene							0.9		+		
Plagioclase composition							An_0			An_0	

11. From northern Michigan (James, 1955).
12. Northern Michigan (James, 1955).
13. Northern Michigan (James, 1955).
14. Northern Michigan (James, 1955).
15. East of Mecca, California.
16. Koto-Bizan district, Japan. Author's No. E-7 (Ernst, 1964).
17. Koto-Bizan district Japan. Author's No. E-11 (Ernst, 1964).
18. Cazadero, California (Coleman and Lee, 1962).
19. Cazadero, California (Coleman and Lee, 1962).
20. Elk River-Clarkia area, Idaho (Hietanen, 1963).
21. Elk River-Clarkia area, Idaho (Hietanen, 1963).
22. East central Quebec (specimen No. 11-G).
23. East central Quebec (specimen No. 13-G).
24. East central Quebec (specimen No. 3-M).

Amphibolite	biotite schist	Granite gneiss	Metamorphosed iron formation	Metamorphosed iron formation	Quartz ellipsoid granite gneiss	Feldspathic gneiss	Metamorphosed iron formation	Enderbitic gneiss	Garnet-bearing gneiss
22	23	24	25	26	27	28	29	30	31
10	30	25	+	+	30	+	+	+	+
25	30	30				+		+	+
		35[a]			55[d]	+		+	+
	5	3			5				
					7	+			+
20	30	3			2	+		+	+
3						+		+	+
3	<5	<1			<1				
		<1							
2									
			+						
				+			+		
			+	+					
30								+	
				+			+	+	
							+	+	
			+	+			+		
			+						
5					1	+		+	
			+	+			+		
			+						
<1	<1	<1						+	
An_{30}	An_{30}	An_{30}						An_{40}	

25. East central Quebec. Author's No. DH4-185 (Mueller, 1960).
26. East central Quebec. Author's No. DH 3-172 (Mueller, 1960).
27. Near Fine, New York (specimen No. 4-W).
28. Southwestern Quebec. Author's No. 97-56 (Kretz, 1959).
29. Mt. Reed areas Quebec. Author's No. A-8 (Kranck, 1961).
30. Western Greenland. Author's No. 3618 (Ramberg, 1949).
31. Western Greenland. Author's No. 3286 (Ramberg, 1949).

[a] Microcline, relatively nonperthitic.
[b] White mica.
[c] Undifferentiated iron oxides chiefly.
[d] Microclinic perthite.

igneous plagioclase and pyroxene although their present mineralogy consists largely of chlorite, actinolite, epidote, calcite, and sodic plagioclase. Such is the famous Ely "greenstone" of northern Minnesota (Clements, 1903). Also in northern Wisconsin and Michigan in the Penokee-Gogebic iron range pelitic rocks of this facies are well represented (Irving and Van Hise, 1892). In the latter region we find an apparent increase in metamorphic grade toward the western (Wisconsin) part of the range, with chloritic schists giving way to those richer in biotite. Also in the biotitic slates andalusite appears in the more aluminous beds. Many of the pelitic rocks still show evidence of the original clastic quartz and feldspars. The metamorphic zonal configuration of the eastern part of this region was outlined by James (1955), who traced the low-grade rocks into their amphibolite facies equivalents in central Michigan. Some typical reactions in the greenschist facies have been described by McNamara (1965), Brown (1967), and Mather (1970). McNamara described the characteristic mineral assemblage in the lowest grade Dalradian slates as quartz-albite-phengite-chlorite-calcite-dolomite-anatase. The presence of dolomite is considered to be related to the absence of actinolite and the low-magnesium content of chlorites. A high μ_{CO_2}/μ_{H_2O} may not only inhibit the formation of epidote but may also drive the following reactions to the left:

$$\underset{chlorite}{5(Mg, Fe)_6Si_4O_{10}(OH)_8} + \underset{calcite}{12CaCO_3} + \underset{quartz}{28SiO_2} \rightleftharpoons$$

$$\underset{actinolite}{6Ca_2(Mg, Fe)_5Si_8O_{22}(OH)_2} + 12CO_2 + 14H_2O \quad (10.c)$$

$$\underset{dolomite}{5Ca(Mg, Fe)(CO_3)_2} + \underset{quartz}{8SiO_2} + H_2O \rightleftharpoons$$

$$\underset{actinolite}{Ca_2(Mg, Fe)_5Si_8O_{22}(OH)_2} + \underset{calcite}{3CaCO_3} + 7CO_2 \quad (10.d)$$

Reaction (10.c) is idealized and only shows (Mg, Fe) chlorite component in the (Mg, Fe, Al) chlorite.

The incoming of epidote in Dalradian rocks may be represented by the reaction

$$\underset{muscovite}{15KAl_3Si_3O_{10}(OH)_2} + \underset{chlorite}{9(Mg, Fe)_5Al_2Si_3O_{10}(OH)_8}$$

$$+ \underset{calcite}{32CaCO_3} + \underset{quartz}{21SiO_2} \rightleftharpoons$$

$$\underset{epidote}{16Ca_2Al_3Si_3O_{12}(OH)} + \underset{biotite}{15K(Mg, Fe)_3AlSi_3O_{10}(OH)_2}$$

$$+ 32CO_2 + 28H_2O \quad (10.e)$$

Reaction (10.e) is also a biotite-forming reaction. Chlorite and biotite are crystalline solutions of Fe, Mg, and Al end members. If the carbonate is also a solution of Mg and Ca end members (i.e., dolomite), the stability of the equilibrium assemblage will also change as a function of the activity-composition relation in dolomite.

Mather (1970) described the following mineral assemblages from Aberfoyle area (Scotland):

quartz ± albite ± calcite ± phengite ± chlorite
± epidote ± ferrodolomite ± actinolite

This is a multicomponent system (at least nine) and the molar volumes of such minerals as phengite are incompletely known for a theoretical analysis. It is, however, clear that, besides the effects due to crystalline solutions, the presence or absence of certain minerals in the assemblages at a given P and T must be critically related to μ_{CO_2}/μ_{H_2O}.

Brown (1967) has studied the assemblages in the rocks of greenschist facies in Eastern Otago, New Zealand. Three compositionally different types of rocks with varying proportions of minerals occur in the following assemblages (all contain quartz-albite-epidote-muscovite):

stilpnomelane ± chlorite ± stilpnomelane ± actinolite

chlorite ± biotite ± stilpnomelane ± garnet ± actinolite

biotite ± actinolite ± stilpnomelane ± garnet

A chemical study of the minerals showed the presence of zoning in garnets (77 percent spessertine in the core to 35 percent spessertine 15 μ away on the rim, see Figure 8.17). Epidotes have iron-rich cores and some actinolite crystals have deeply colored hornblende cores. Grain-to-grain compositional variation was also noted. With increasing metamorphic grade, however, these disequilibrium features disappear. The total assemblages do not violate the phase rule and are at least divariant.

One of the most important reactions that occurs within the greenschist facies and at the transition to the amphibolite facies is the following:

$$\underset{calcite}{\tfrac{16}{3}CaCO_3} + \underset{chlorite}{\tfrac{5}{4}Mg_4Al_4Si_2O_{10}(OH)_8} + \underset{quartz}{\tfrac{21}{2}SiO_2} \rightleftharpoons$$

$$\underset{actinolite}{Ca_2Mg_5Si_8O_{22}(OH)_2} + \underset{epidote}{\tfrac{5}{3}Ca_2Al_3Si_3O_{12}(OH)}$$

$$+ \underset{fluid}{\tfrac{19}{2}H_2O} + \underset{fluid}{\tfrac{16}{3}CO_2} \quad (10.f)$$

Ramberg (1952) considered this reaction as defining the boundary between the greenschist and epidote amphibolite facies (which we have incorporated into our greenschist facies). However, it is clear that this reaction has too many degrees of freedom to make it useful except in a very rough way. In fact, if there is ferrous iron present it has three degrees of freedom. The lack of suitable reactions to fix boundaries and of distinctive assemblages within the facies are the main reasons behind our elimination of the epidote amphibolite facies of Eskola.

Schists that contain all the mineral phases appearing on both sides of Reaction (10.f) are very common; an example (specimen 15) is shown in Table 10.3. Some other reactions are discussed in Chapter 11.

Although andalusite is the aluminosilicate of many greenschist pelitic rocks, its range seems restricted to the high-temperature part of the facies. It is possible that this may be attributed to the reactions

$$\underset{\textit{andalusite}}{2Al_2SiO_5} + \underset{\textit{quartz}}{2SiO_2} + \underset{\textit{fluid}}{4H_2O} \rightleftharpoons \underset{\textit{kaolinite}}{Al_4Si_4O_{10}(OH)_8} \quad (10.g)$$

$$\underset{\textit{andalusite}}{Al_2SiO_5} + \underset{\textit{quartz}}{3SiO_2} + \underset{\textit{fluid}}{H_2O} \rightleftharpoons \underset{\textit{pyrophyllite}}{Al_2Si_4O_{10}(OH)_2} \quad (10.h)$$

The standard free-energy change for Reaction (10.g) at 298°K is about −2000 cal, but at 400°K it is about +1500 cal, indicating that kaolinite is stabilized relative to andalusite, quartz, and pure liquid water only in the lower temperature range. The effect of pressure is difficult to evaluate because water will generally not occur as a pure component; however under STP conditions this reaction shows a large volume decrease so that pressure might well act to further stabilize kaolinite. However, if the other fluid constituents lower the activity of water, this will again favor the left side of Reaction (10.g). Reaction (10.h), labeled *11* in Figure 10.1, has been investigated by Kerrick (1968).

In siliceous carbonate rocks (impure limestones) the products of the reaction of dolomite with quartz are tremolite, talc, and calcite, although the CO_2 pressure is frequently too high for this reaction to go to completion in this facies, and for that matter even in the amphibolite facies. Under conditions when the escape of the fluid phase becomes possible and particularly in the high-temperature range of the facies, diopside tends to replace tremolite according to the reaction[3]

$$\underset{\textit{quartz}}{2SiO_2} + \underset{\textit{tremolite}}{Ca_2Mg_5Si_8O_{22}(OH)_2} + \underset{\textit{calcite}}{3CaCO_3} \rightleftharpoons \underset{\textit{diopside}}{5CaMgSi_2O_6} + \underset{\textit{fluid}}{3CO_2} + \underset{\textit{fluid}}{H_2O} \quad (10.h')$$

It is obvious that this reaction also can occur over a range of conditions determined by various independent variables such as P_{H_2O} and P_{CO_2}.

In carbonate rocks containing significant amounts of iron and in the silicate-carbonate iron formations a variety of different iron-magnesium minerals occur. In the carbonate iron formations siderite and ankerite are stable in the presence of chert layers. In the silicate-bearing iron formations the place of carbonates is taken by stilpnomelane and minnesotaite, and in the upper part of the facies by actinolite and cummingtonite, which are frequent products of decarbonation reactions. By contrast, the mineral greenalite is a constituent of many pre-Cambrian iron formations and appears to be a primary chemical precipitate. In Mn-bearing rocks spessertite-rich garnet appears within the greenschist facies, as was pointed out by Ramberg (1949) and by Miyashiro (1953b).

Greenschist facies to amphibolite facies

Although a variety of obvious and well-defined reactions characterize the transition from greenschist to amphibolite facies, none of these appear to be suitable to fix the transition precisely since nearly all involve complex phases which have the effect of smearing out the transition over a broad P–T range. The boundary at high temperature may be suggested to lie slightly below the temperature of the triple point for andalusite-kyanite-sillimanite, so that we may exclude at a glance any rocks from this facies that contain sillimanite. As discussed in Chapter 9, the position of the triple point in the P–T field is somewhat uncertain, Ganguly's (1969) suggestion has been accepted here—namely, that the triple point should lie between the equilibrium curves for the reactions

$$\text{chloritoid} \rightleftharpoons \text{almandine} + \text{corundum} + H_2O \quad (15)$$

[3] Ramberg (1952) suggested that this reaction marked the lower limits of the amphibolite facies.

and

chloritoid + Al_2SiO_5 $\rightleftharpoons$

staurolite + quartz + H_2O (*16*)

We also note that the staurolite-producing reaction

(Fe, Mg) chlorite + muscovite $\rightleftharpoons$

(Fe, Mg) staurolite + (Fe, Mg) biotite + quartz + vapor (*14*)

(*Hoschek, 1969*)

lies within the equilibrium pressure and temperature range of the two reactions. As suggested by Ganguly (1972), the transition from greenschist facies to amphibolite facies should be recognized by the formation of staurolite at the expense of chloritoid or chlorite. The effect of solid solubility in the ferromagnesian minerals and the change in composition in the vapor phase may require that the boundary may be determined by referring to a particular composition of coexisting staurolite and chloritoid or chlorite. This will be compatible with Ramberg's definition of facies. The first appearance of staurolite, without established relationship with chlorite or chloritoid, may not be used as an indication of the beginning of the amphibolite facies. Ganguly (1972) has shown that if the staurolite-producing reaction is governed by oxidation equilibria, the temperature is not necessarily within the range of 525°C to 575°C. For example the reaction

almandine + H_2O + O_2 $\rightleftharpoons$

staurolite + magnetite + quartz

has equilibrium temperatures in the range of 600°C to 675°C at 10 kbar, the temperature varying with oxygen fugacity.

The reaction labeled (*14*) in Figure 10.1:

chlorite + muscovite $\rightleftharpoons$

staurolite + biotite + quartz + vapor (*14*)

involves chlorite and biotite, which are complex crystalline solutions. Unless the effect of solubility on the P–T conditions of this equilibrium is evaluated, the reaction may be used only with caution in marking the beginning of the amphibolite facies.

For a reaction labeled (*12*) involving serpentine, forsterite, and talc,

5 chrysotile = 6 forsterite + talc + 9H_2O

provided that the Fe^{2+} content of talc is not significantly high, the equilibrium P and T would be very close to the suggested boundary between the greenschist and amphibolite facies. The occurrence of the assemblage talc and forsterite may be useful information in mapping the facies.

Amphibolite facies

In rocks of pelitic composition there occurs a group of reactions that particularly characterize the amphibolite facies and that, at the same time, have the interesting feature of involving only crystalline phases. The two simplest of these reactions are

$$\underset{\text{biotite}}{KMg_3(AlSi_3)O_{10}(OH)_2} + \underset{\substack{\text{andalusite}\\ \text{sillimanite}\\ \text{kyanite}}}{2Al_2SiO_5} + \underset{\text{quartz}}{SiO_2} \rightleftharpoons \underset{\text{garnet}}{Mg_3Al_2Si_3O_{12}} + \underset{\text{muscovite}}{KAl_2(AlSi_3)O_{10}(OH)_2} \quad (10.i)$$

$$\underset{\text{biotite}}{KFe_3(AlSi_3)O_{10}(OH)_2} + \underset{\substack{\text{andalusite}\\ \text{sillimanite}\\ \text{kyanite}}}{2Al_2SiO_5} + \underset{\text{quartz}}{SiO_2} \rightleftharpoons \underset{\text{garnet}}{Fe_3Al_2Si_3O_{12}} + \underset{\text{muscovite}}{KAl_2(AlSi_3)O_{10}(OH)_2} \quad (10.j)$$

As these reactions are written, the volume changes under STP conditions are negative in both cases so that the assemblage garnet-muscovite is favored by pressure. The volume changes for the two reactions are as follows when the indicated Al_2SiO_5 polymorphs are involved:

$$\Delta V_{(i)} \begin{cases} \text{andalusite} = -21.59 \text{ ml} \\ \text{sillimanite} = -18.33 \text{ ml} \\ \text{kyanite} \quad = -6.73 \text{ ml} \end{cases}$$

$$\Delta V_{(j)} \begin{cases} \text{andalusite} = -24.26 \text{ ml} \\ \text{sillimanite} = -21.00 \text{ ml} \\ \text{kyanite} \quad = -9.40 \text{ ml} \end{cases}$$

In addition to these reactions involving the biotite end members phlogopite and annite, there are analogous reactions involving the eastonite molecule:

$$\underset{\text{biotite (eastonite)}}{K_2(Mg, Fe)_5Al(Al_3Si_5)O_{20}(OH)_4} + \underset{\substack{\text{andalusite}\\ \text{sillimanite}\\ \text{kyanite}}}{\tfrac{8}{3}Al_2SiO_5} + \underset{\text{quartz}}{\tfrac{10}{3}SiO_2} \rightleftharpoons \underset{\text{garnet}}{\tfrac{5}{3}(Mg, Fe)_3Al_2Si_3O_{12}} + \underset{\text{muscovite}}{K_2Al_4(Al_2Si_6)O_{20}(OH)_4} \quad (10.k)$$

As may be seen from Table 10.2 and 10.3, assemblages containing the phases on both sides of these reactions are very common in amphibolite facies rocks so that a very wide stability range is indicated.

Although the necessary thermochemical data do not exist for a full discussion of the stability relations of this system, we may deduce some of the features if we make certain simplifying assumptions. These are that both biotite and garnet form ideal solutions and that the other minerals are stoichiometric compounds. Under these assumptions the equations of equilibrium for Reactions (10.i) and (10.j) are

$$K_{(i)}\exp\left(\frac{-P\Delta V_{(i)}}{RT}\right) = \frac{(X_{Mg}^{Gar})^3}{(X_{Mg}^{Bi})^3} \tag{10.1}$$

$$K_{(j)}\exp\left(\frac{-P\Delta V_{(j)}}{RT}\right) = \frac{(1 - X_{Mg}^{Gar})^3}{(1 - X_{Mg}^{Bi})^3} \tag{10.2}$$

Direct application of the phase rule to the reactions leads to the conclusion that the system has two degrees of freedom since there are five effective components and five phases. This is also borne out by the equations since these are two in number and relate four variables to each other. The system is geometrically equivalent to a double surface (one for garnet and one for biotite) in P–T–X_{Mg} space which separates the high- and low-pressure assemblages.

If we examine Equations (10.1) and (10.2) for the case of constant temperature, we see that increasing the MgO content of the system corresponds to an increasing equilibrium pressure. This follows from the fact that garnets always contain more iron than the coexisting biotites (Kretz, 1959).

Qualitatively similar conclusions would follow even if biotite and garnet did not form ideal solutions unless the departures from ideality were quite great. This is especially true of the phase-rule deductions since these are totally independent of the character of the solutions.

In discussing the assemblages of the amphibolite facies it is possible to refer in detail to specific occurrences since many of these have been studied. Some of these metamorphic areas embrace a variety of rock types which show great variation in bulk composition but which have been formed under relatively constant P–T conditions. One such area forms part of the extensive regional metamorphic terrain of east-central Quebec (Mueller, 1960). This area, which is located in the southern extension of the Labrador trough of iron-bearing sediments, consists of metamorphosed pelitic sediments, iron formation, and carbonate rocks which were intruded by sills and irregular bodies of diabasic gabbro somewhat before the culminating metamorphism. The diabase was subsequently recrystallized to an amphibolite, which presumably is of the same metamorphic rank as the metasediments. Mineral assemblages from typical representatives of these rocks are shown in Table 10.3.

The type rock of the Quebec suite, the amphibolite (No. 22, Table 10.3) shows the high quartz content which is typical of many amphibolites that are derived from high-temperature gabbros with little or no normative quartz and with high modal feldspar, pyroxene, and olivine. It should also be noted that this specimen contains a large number of mineral phases, which is in part a consequence of the relatively large number of components that attain majority status. By contrast the schist (No. 23) has a simpler mineralogy corresponding to a more extreme chemical composition. Schists of the latter type grade into more feldspathic varieties, and these in turn envelop bands and lenses of granite (No. 24) which contains the same mineral species as the feldspathic schists but in different proportions. It is of interest that these rock types generally contain all the phases that enter into one or the other of the following reactions:

$$\underset{epidote}{2\,Ca_2Al_3Si_3O_{12}(OH)} + \underset{muscovite}{KAl_2(AlSi_3)O_{10}(OH)_2} + \underset{quartz}{2SiO_2} \rightleftharpoons \underset{plagioclase}{4CaAl_2Si_2O_8} + \underset{microcline}{KAlSi_3O_8} + \underset{fluid}{2H_2O} \tag{10.l}$$

$$\underset{epidote}{\tfrac{4}{5}Ca_2Al_3Si_3O_{10}(OH)} + \underset{quartz}{\tfrac{1}{5}SiO_2} \rightleftharpoons \underset{plagioclase}{CaAl_2Si_2O_8} + \underset{garnet}{\tfrac{1}{5}Ca_3Al_2Si_3O_{12}} + \underset{fluid}{\tfrac{2}{5}H_2O} \tag{10.m}$$

Reactions such as these have been proposed by Ramberg (1952) as important in governing the composition of plagioclase. Thus in the amphibolites, Reaction (10.m) would be displaced to the left, and this would account in part for the sodic plagioclase. Reaction (10.l) would serve the same end in the pelitic schists.

In the metamorphosed iron formations of the Quebec suite, the Al-rich minerals of the pelitic schists give way to Ca, Mg, and ferrous iron silicates and carbonates and iron oxides. Here actinolite takes the place of hornblende and Ca-pyroxene finds a stability field. The assemblages differ greatly as a consequence of the considerable variability in the bulk compositional variables including Fe^{3+}/Fe^{2+}. The latter is of course a function of the degree of oxidation. A clear idea of the range in oxygen fugacities involved may be gained by noting the presence or absence of the oxides hematite and magnetite. For example, No. 25 is a relatively oxidized assemblage with hematite and Mg-rich silicates,[4] whereas No. 26 is a more reduced assemblage consisting of ferrous iron-rich silicates and with magnetite as the only oxide phase. In the latter assemblage, cummingtonite, a mineral with a relatively high affinity for ferrous iron, is one of the most abundant phases.[5] It should also be noted that dolomite and quartz coexist in these rocks, which shows that P_{CO_2} probably attained magnitudes of the order of 10^3 atm or more.

Our second example of a suite of amphibolite facies rocks is taken from the classic metamorphic terrain of Glen Clova, Scotland (Chinner, 1960). The rocks are gneisses of pelitic composition and consists of various combinations of the phases of Reaction (10.i) to (10.k). The coexistence of both sillimanite and kyanite, sometimes in the same thin sections, points toward the high pressure part of the facies, perhaps as high as 10^4 bars.

The Glen Clova gneisses exhibit the same general effects of varying oxygen fugacity already discussed in connection with the Quebec iron formation. Some of these effects are clear from the bulk chemical analyses and the corresponding modes of Tables 10.1 and 10.2. Although the three analyses (Nos. 6, 7, and 8) of Table 10.1 are not quite comparable because of the simultaneous variations of several major components, they are enough alike in bulk composition to illustrate the effect of increasing oxidation in the sequence. This is shown by the increase in the ratio Fe^{3+}/Fe^{2+} for the rock, which is accompanied by an increase in iron oxide phases and muscovite and a decrease in ferromagnesian silicates. As in the case of the Quebec iron formation, the ferromagnesian silicates show a systematic increase in $Mg/(Mg + Fe^{2+})$ with increasing oxygen fugacity. Chinner (1960) pointed out that the oxidative process is approximately equivalent to the reaction

$$\underset{\textit{biotite}}{K_2(Fe_5Al)(Al_3Si_5)O_{20}(OH)_4} + \underset{\textit{garnet}}{Fe_3Al_2Si_3O_{12}} + \underset{\textit{fluid}}{2O_2} \rightleftharpoons$$

$$\underset{\textit{muscovite}}{K_2Al_4(Al_2Si_6)O_{20}(OH)_4} + \underset{\textit{hematite}}{4Fe_2O_3} + \underset{\textit{quartz}}{2SiO_2} \quad (10.n)$$

in the field of hematite. In the more reduced region, analogous reactions involving magnetite apply. Another effect closely related to the increase in $Mg/(Mg + Fe^{2+})$ with increasing oxidation is the accompanying increase in manganese content of the same silicates. This is a direct consequence of the resistance of Mn^{2+} to oxidation so that it behaves much as does magnesium, and accumulates in the silicates.

For rocks like those from Glen Clova the presence of an iron oxide in addition to the phases occurring in Reactions (10.j) and (10.k) relates the corresponding Equations (10.1) and (10.2) to reactions like (10.n). As a consequence, the atomic fractions X_{Mg}^{Bi} and X_{Mg}^{Gar} will depend on P_{O_2}. It is evident then that an increase in P_{O_2} corresponds to an increase in the equilibrium total pressure bearing on the solid phases, with the consequence that the P–T range over which the reaction occurs should be greatly influenced by this parameter.

It is easy to see what the above effect of varying the bulk composition through changes in P_{O_2} will have on the identification of "isograds." The appearance or disappearance of the various index minerals such as biotite, sillimanite, and garnet will be a function of P_{O_2} in addition to the temperature and total pressure. Since the effect of this parameter was unsuspected at the time isograds were proposed, it follows that many of these are invalid and may simply represent changes in degree of oxidation. This point has already been made when we discussed the appearance of staurolite in the transition of greenschist facies to amphibolite facies.

An additional interesting feature here is that the assemblage stability relations are almost independent of the water fugacity since no dehydration reactions are involved. Of course a certain minimum P_{H_2O} was necessary in order to stabilize the micas, but any value in excess of this only contributes to the total pressure. However, staurolite also occurs in some of the Glen Clova rocks; and since this mineral reacts with quartz to form garnet, kyanite, and water, the equilibrium P_{H_2O} might be defined in this way.

A third area that shows some interesting contrasts with Glen Clova is the northern Michigan region already referred to in connection with the greenschist facies. Although the green schists give way in north central Michigan to amphibolite facies rocks, the latter appear to have formed under somewhat lower pressures than the rocks of Glen Clova since kyanite is not found. Rather, biotite and quartz coexist with andalusite or sillimanite (James, 1955) so that only the left or high-temperature side of Reactions (10.j) and (10.k) occur (specimen Nos. 12 and 13, Table 10.3).

A fourth area of interest is the regionally metamorphosed envelope of the Idaho batholith. Hietanen (1963a) has described, from the Elk River-Clarkia area,

[4] Actinolite from this assemblage has $Mg/(Mg+Fe^{2+})=0.85$.

[5] See Chapter 2 for a discussion of the distribution of Fe^{2+} and Mg among these minerals and Chapter 11 for a thermodynamic analysis.

the common occurrence of andalusite and sillimanite with biotite, quartz, garnet, and muscovite (Nos. 20 and 21, Table 10.3) so that both reactants and products of Reactions (10.j) and (10.k) again coexist here.

In northern New York State, over a wide region on the flanks of the Adirondack Mountains, amphibolite to granulite facies rocks are in contact with the anorthosite massif. In Table 10.3, the modal phase content of an interesting specimen (No. 27) from this transitional region is presented. This rock is from an extensive body of granite-gneiss of unknown origin which contains many oriented quartz ovoids similar to deformed pebbles of a stretched-pebble conglomerate. The matrix of the ovoids consists of a macrocrystalline assemblage of microcline, quartz, sillimanite, and muscovite, with minor amounts of biotite and low-iron epidote. Thus the entire assemblage represented by reaction labeled (19) in Figure 10.1 is present. Data on coexisting iron-titanium oxides (Buddington and Lindsley, 1964) indicate temperatures of around 600°C to 650°C for the metamorphism of this region, while the P_{H_2O} would be of the order of 3 kbar (see Figure 10.1). However at this pressure, epidote is unstable, as shown by Liou's (1973) experimental results plotted on the figure, and the pressure must be at least 5 kbar. The actual pressure would be between 5 and 7 kbar, the higher limit being defined by the sillimanite-to-kyanite transition. This puts the sample into the middle of the amphibolite facies in the present scheme of facies classification.

Robinson and Jaffe (1969a) described the aluminous enclaves in gedrite-cordierite gneiss from southwestern New Hampshire. The mica schists in the vicinity are characterized by quartz-muscovite-sillimanite-plagioclase-biotite-almandine ± staurolite. Orthoclase may also be present occasionally. Figure 10.1 shows that provided P_{H_2O} is equal to P_{total} and plagioclase is albitic, such an assemblage might form close to the equilibrium P–T of the reaction labeled *(29)*;

$$\text{albite} + \text{muscovite} + \text{quartz} + H_2O = \text{liquid} + Al_2SiO_5 \quad (29)$$

and probably near the kyanite-sillimanite boundary. A good estimate of temperature, therefore, would be 600°C to 650°C at ~6 kbar (Robinson, 1968), which is the middle part of the amphibolite facies.

Evans and Guidotti (1966) described the assemblages

quartz + biotite + plagioclase + muscovite + sillimanite ± garnet

quartz + biotite + plagioclase + muscovite + sillimanite + orthoclase ± garnet

The second assemblage probably represents the transition between amphibolite and granulite facies. Its position in the P–T field should be above 3.5 kbar around 680°C. This position would be on the high-temperature side of the reaction *(29)*. The equilibrium T of the above reaction would increase with increasing mole fraction of anorthite in plagioclase and no phase is lost by melting in these rocks over a range of temperature.

Amphibolite facies to granulite facies

In Figure 10.1 the boundary between amphibolite and granulite facies may be drawn quite close to the univariant curves

$$\text{muscovite} + \text{quartz} \rightleftharpoons \text{K feldspar} + Al_2SiO_5 + H_2O \quad (19)$$

$$\text{muscovite} + \text{quartz} \rightleftharpoons \text{K-feldspar} + \text{sillimanite} + \text{liquid (melt)} \quad (26)$$

Thus any rock containing muscovite + quartz may be excluded from the granulite facies. However, for other reasons the disappearance of this assemblage does not necessarily signify that the boundary of the granulite facies has been approached. In plagioclase-bearing rocks, muscovite + quartz may be eliminated according to the reaction

$$\text{albite} + \text{muscovite} + \text{quartz} + H_2O \rightleftharpoons \text{liquid} + \text{sillimanite} \quad (29)$$

(Storre and Karotke, 1971)

which takes place at high pressures and 50°C to 100°C below the reaction *(26)*: muscovite + quartz. The assemblage muscovite + quartz is also very significantly affected by a change in P_{H_2O}, as shown by Kerrick (1972). The intersection of the curve for reaction *(19)*: muscovite + Q with the granite melting curve, shifts from 650°C and ~3.75 kbar to ~680°C and ~7.75 kbar as a result of a change in X_{H_2O} from 1.0 to 0.5.

As suggested by De Waard (1967), it may be desirable to fix the minimum temperature of the granulite facies by the P–T curve of the reaction *(20)* for albite-K-feldspar critical mixing. This curve has been recently redetermined by Morse (1970) and is given by the equation

$$P(\text{kbar}) = 0.0545(T\,^\circ\text{C}) - 34.82$$

As noted in Figure 10.1, this critical line is rather close to the

$$\text{muscovite} + \text{quartz} \rightleftharpoons Al_2SiO_5 + K\text{-feldspar} + H_2O \text{ (or liquid)}$$

curves with $P_{total} = P_{H_2O}$. We must define this boundary for the anorthite free system because the line shifts toward higher temperature as a function of anorthite content in the system (Morse, 1968). This is also apparent from Seck's (1971a,b) results on the ternary Ab-Or-An system.

The critical line intersects the sillimanite-kyanite boundary at about 850°C at pressures above 11 kbar. The coexistence of kyanite and hypersolvus alkali feldspar should represent very high *P* and *T* for crustal conditions. Similarly, the coexistence of andalusite and hypersolvus alkali feldspar is possible only below about 3 kbar at high temperatures (650°C–800°C). Another reaction likely to be useful in recognising the granulite facies is the conversion of ortho-amphiboles (aluminum-free or low in aluminum) to orthopyroxene. Greenwood (1963) determined the equilibrium curve for the reaction

$$\text{anthophyllite} \rightleftharpoons \text{enstatite} + \text{quartz} + H_2O \quad (23)$$

In natural assemblages, the curve will shift considerably to lower temperatures, and in the high-pressure (> 5 kbar) part of the *P–T* field it may be close to our boundary. In plagioclase-bearing rocks, therefore, the disappearance of aluminum-free orthoamphibole and the appearance of orthopyroxene should be a definite indication that the *P–T* conditions have exceeded those of the amphibolite facies.

Some of the reactions at the highest temperature of the amphibolite facies or at the beginning of the granulite facies may involve partial melting as shown below.

$$\text{plagioclase} + \text{sillimanite} + \text{biotite} \rightleftharpoons \text{sanidine} + \text{cordierite} + \text{garnet} + \text{liquid}$$

$$\text{plagioclase} + \text{sillimanite} + \text{biotite} \rightleftharpoons \text{sanidine} + \text{cordierite} + \text{orthopyroxene} + \text{liquid}$$

$$\text{plagioclase} + \text{biotite} + \text{garnet} \rightleftharpoons \text{sanidine} + \text{cordierite} + \text{orthopyroxene} + \text{liquid}$$

For such assemblages, the removal of the iiquid may enlarge the stability field of the coexisting solid phases.

Granulite facies

This facies includes regionally metamorphosed rocks of an anhydrous character in which feldspars, pyroxenes and garnets are much more abundant than in the lower temperature facies. However, as has been emphasized by Ramberg (1949), this does not imply that the fugacity of water was necessarily lower than that in amphibolite facies rocks since the effects of higher temperatures could offset those of low water content.

The granulite facies is characterized by some common minerals of peculiar aspect. Potassium feldspar tends to be mesoperthitic and is frequently green rather than pink in color, whereas the coexisting quartz is sometimes bluish. Also, the plagioclase that coexists with the potassium feldspar is frequently antiperthite, illustrating the great mutual solubility of the two minerals in this facies. Then, too, some of the mineral associations are unfamiliar in other facies, such as for example the widespread coexistence of orthopyroxenes and potassium feldspars in rocks approaching granites in composition.

There is every reason to believe that the granulite facies represents some of the highest temperatures and pressures of crystallization registered in large bodies of exhumed rock. Some of the mineralogic transformations that mark the transition from the amphibolite facies, or which may take place within the granulite facies, may be described from the Adirondack area of New York. The slight apparent increase in metamorphic grades in the surrounding rocks toward the Adirondack massif is marked by a number of assemblage transformations. One of the most striking of these is the conversion of hornblende in amphibolite into plagioclase-pyroxene assemblages of the granulite facies. This transformation is well shown in Table 10.4, which contains data presented by Engle and Engel (1962). The table shows the chemical compositions and modal phase contents of five average amphibolites which span the distance from the Grenville lowland to the massif. Although Engel and Engel stressed the small differences in K_2O and Fe^{3+}/Fe^{2+} between these rocks, they are remarkably alike chemically and therefore serve well for tracing mineral transformations in progressive metamorphism. The decrease in hornblende and quartz accompanying the increase in plagioclase and pyroxene toward the massif may be illustrated by the reactions

$$\underset{\textit{hornblende}}{2\,NaCa(Mg, Fe)_4(Al_2Si_6)O_{22}(OH)_2} + \underset{\textit{quartz}}{5\,SiO_2} \rightleftharpoons \underset{\textit{orthopyroxene}}{7(Mg, Fe)\,SiO_3} + \underset{\textit{plagioclase}}{2\,NaAlSi_3O_8 + CaAl_2Si_2O_8} + \underset{\textit{Ca-pyroxene}}{Ca(Mg, Fe)Si_2O_6} + \underset{\textit{fluid}}{2\,H_2O} \quad (10.o)$$

$$\underset{\textit{hornblende}}{Ca_2(Mg, Fe)_3Al_2(Al_2Si_6)O_{22}(OH)_2} + \underset{\textit{quartz}}{SiO_2} \rightleftharpoons \underset{\textit{orthopyroxene}}{3(Mg, Fe)SiO_3} + \underset{\textit{plagioclase}}{2\,CaAl_2Si_2O_8} + \underset{\textit{fluid}}{H_2O} \quad (10.p)$$

Previously we have seen in the suite of Quebec rocks (No. 22, Table 10.3) the result of the displacement of these reactions to the left with the consequent development of quartz-rich amphibolities from gabbro.

It is of course impossible to be certain at this stage whether the displacement to the right of Reactions (10.o) and (10.p) represent an increase in temperature since they might just as well reflect a drop in P_{H_2O} toward the Adirondack massif. However, there is no other evidence that this occurred, whereas there seems to be some evidence for an increase in temperature in this direction.

The Grenville rocks of Quebec also appear to be astride the amphibolite-granulite boundary, as is indicated by the assemblages reported by Kretz (1959). This author called attention to the fact that in the Quebec pelitic gneisses studied, Reactions (10.j) and (10.k) are displaced toward the left so that the association garnet-muscovite is not stable within the granulite facies.

The following are some additional important reactions that are obvious from the observed assemblages:

$$\underset{\textit{biotite}}{K(MgFe)_3(AlSi_3)O_{10}(OH)_2} + 3SiO_2 \rightleftharpoons \underset{\textit{orthoclase}}{KAlSi_3O_8} + \underset{\textit{orthopyroxene}}{3(Mg, Fe)SiO_3} + \underset{\textit{fluid}}{H_2O} \quad (10.q)$$

$$\underset{\textit{biotite}}{K(Mg, Fe)_3(AlSi_3)O_{10}(OH)_2} + \underset{\textit{sillimanite}}{Al_2SiO_5} + \underset{\textit{quartz}}{2SiO_2} \rightleftharpoons \underset{\textit{K-feldspar}}{KAlSi_3O_8} + \underset{\textit{garnet}}{(Mg, Fe)_3Al_2Si_3O_{12}} + \underset{\textit{fluid}}{H_2O} \quad (10.r)$$

$$\underset{\textit{sphene}}{CaTiSiO_5} + \underset{\textit{orthopyroxene}}{MgSiO_3} \rightleftharpoons \underset{\textit{Ca-pyroxene}}{CaMgSi_2O_6} + \underset{\textit{rutile}}{TiO_2} \quad (10.s)$$

Table 10.4 Average compositions of progressively metamorphosed amphibolite

	Emeryville area, group 1	Edwards area, group 2	East Edwards area, group 2A	Russell area, group 4	Colton area, group 5
SiO_2	48.20	48.10	47.90	47.69	47.89
TiO_2	1.89	1.72	1.56	1.76	1.56
Al_2O_3	14.45	15.20	14.75	14.52	14.63
Fe_2O_3	3.50	3.37	2.44	2.06	1.85
FeO	10.53	10.59	10.91	11.31	11.20
MnO	0.25	0.23	0.22	0.23	0.25
MgO	6.62	6.87	7.08	7.26	7.41
CaO	10.25	10.01	10.67	11.16	11.54
Na_2O	1.94	1.86	1.97	2.13	2.19
K_2O	0.96	0.98	0.82	0.67	0.58
H_2O^-	0.01	0.03	0.03	0.02	0.03
H_2O^+	1.31	1.26	1.04	0.81	0.72
P_2O_5	0.18	0.17	0.17	0.18	0.14
Quartz	8.6	5.5	2.5	2.2	0.1
Plagioclase	18.5	21.7	29.6	33.5	35.6
Hornblende	68.6	65.8	51.5	35.3	31.3
Ca-pyroxene	0.9	2.8	9.2	16.4	19.2
Orthopyroxene			5.3	7.2	11.3
Opaques	2.1	1.8	1.3	3.2	2.0
Biotite	0.6	1.3	trace	trace	0.2
Sphene	0.4				
Other	0.3	1.1	0.6	2.2	0.3

Data are from the Adirondack Mountains, New York, by Engel and Engel (1962).

Reaction (10.q) may be called the "charnockite reaction" since it produces the characteristic assemblage of these granulites—namely, K-feldspar and orthopyroxene. Although this reaction is a matter of the dehydration of a hydrous assemblage, it is also intimately connected with the high-pressure stabilization of $FeSiO_3$, which is frequently a prominent component of these pyroxenes. We have previously noted that the left side of Reaction (10.r) is the high-temperature equivalent of the assemblage garnet-muscovite and that this association is stable throughout the amphibolite facies. We have also seen that the left side of Reaction (10.r) is stable into the granulite facies. However, under certain conditions when P_{H_2O} falls below a certain value corresponding to the particular bulk composition and temperature, the products of Reaction (10.r) replace this assemblage. Thus over a range of conditions in the granulite facies both product and reactant phases of Reaction (10.r) substitute for the phases represented by Reactions (10.i) to (10.l) in the amphibolite facies.

Reaction (10.s) is one of several involving sphene, which were discussed by Ramberg (1952) to account for the virtual absence of this mineral in granulites. Since $\Delta V^0_{(s)}$ is -2.3 ml and ΔS^0_2 is -0.86 cal/°C, it appears that the left side of this reaction is the high-temperature–low-pressure side, whereas the right side is probably favored by high pressure and lower temperatures.

A number of typical granulite facies mineral assemblages are shown in Tables 10.2 and 10.3. In Tables 10.1 and 10.2 a chemical analysis and the assemblage are tabulated for a pyroxene granulite from Antarctica. It should be noted that in general these specimens exhibit a considerable number of phases including several hydrous minerals. As many as 10 phases are found to coexist in certain specimens. Yet there is nothing contradictory about this if it is kept in mind that variations in independent variables such as P_{O_2} extend the stability range greatly and that the number of major components in these rocks is approximately 12.

The great number and variability of the major components and the effect this has in extending the stability range of hydrous phases makes it practically meaningless to attempt their use as classification tools. It is of little value to classify a rock as a representative of a "hornblende or pyroxene granulite subfacies" if we do not know the effects of such constituents as TiO_2, O_2, and H_2O, which may change in such a way as to stabilize both pyroxene and hornblende-mica rocks at the same temperature and total pressure.

When SiO_2 is a less abundant constituent, a number of interesting assemblages appear in the granulite facies. Some of these assemblages may be read directly from Figures 4.6 to 4.8, although generally several more phases will occur because of the great number of components. Thus olivine commonly occurs in quartz-free gneisses of western Greenland in association with pyroxene, hornblende, biotite, and spinel (Ramberg, 1949). In some silica-poor but MgO- and Al_2O_3-rich rocks, sapphirine- and spinel-bearing assemblages occur (Ramberg, 1948a). Some of the stable assemblages, which may be read from the tetrahedral diagrams of Chapter 4, are

1. hornblende-cordierite-enstatite-sapphirine
2. hornblende-enstatite-sapphirine-spinel
3. hornblende-olivine-enstatite-spinel

In the lime silicate environment the skarns and marbles developed, typically consisting of various combinations of calcite, dolomite, diopside, scapolite, olivine, chondrodite, spinel, amphibole, plagioclase, microcline, biotite, and quartz. Such minerals as olivine and chondrodite are nearly confined to quartz-free rocks, as is usual also for dolomite and spinel. The biotite found here is frequently close to phlogopite, which has such a high thermal stability that it can occur throughout the granulite facies (including the low-pressure part) and even deep into the range of magmatic temperatures.

Magnesium-rich orthopyroxenes apparently do not occur with calcite under attainable geologic conditions, although more iron-rich varieties coexist with this carbonate in the metamorphosed iron formations (No. 29, Table 10.3). The infrequency of this association may be attributed to the reaction

$$\underset{\textit{orthopyroxene}}{3MgSiO_3} + \underset{\textit{calcite}}{CaCO_3} \rightleftharpoons \underset{\textit{Ca-pyroxene}}{CaMgSi_2O_6} + \underset{\textit{olivine}}{Mg_2SiO_4} + \underset{\textit{fluid}}{CO_2} \quad (10.t)$$

as well as Reactions (2.p) and (2.q), which are all displaced strongly to the right. The result of this is that CO_2 fugacities needed to stabilize the left sides of these reactions are so high that they seldom are attained within the crust. However in the assemblages represented by Reactions (2.p) and (2.q) it is found on theoretical considerations that P_{CO_2} is reduced as $Fe^{2+}/(Fe^2 + Mg)$ in the environment increases, and this probably explains the restriction of the assemblage to iron-rich rocks (Mueller, 1966).

11 Experimental and theoretical study of metamorphic systems

In Chapter 9 we considered the formation or breakdown of individual minerals. Information on the stability limits of individual minerals is important in understanding metamorphic systems because no phase assemblage containing a particular phase can be in stable equilibrium outside the stability limits of that phase. However, the stability of a phase may change considerably in the presence of other reacting phases. A study of systems, therefore, is essential to understanding petrogenesis. In what follows, an attempt will be made to bring out the salient features of phase equilibria in the experimental systems. Wherever possible this information will be supplemented by theoretical analysis.

Geometric analysis of phase equilibria

The study of systems is particularly facilitated by geometric analysis of phase equilibria based on Schreinemaker's rules as described by Zen (1966). Before we describe these rules, we may recall the expressions for the phase rule as discussed in Chapter 1. For the condition when total pressure cannot be taken as equal to the sum of the partial pressures, as for example when the pressure of the overburden is in excess of the gas pressure, the phase rule may take the form

$$v = (n + 3) - \phi \qquad (1.56)$$

where as before v is the variance, ϕ the numbers of phases, and n the number of components. While considering Schreinemaker's geometric analysis we shall generally assume the total pressure as equal to the hydrostatic pressure and express the phase rule as

$$v = (n + 2) - \phi \quad (1.49)$$

The number of components and the phases in equilibrium may be counted following the discussion presented in Chapter 1. If $\phi = n + 2$ in an equilibrium system, the coexistence of all phases in the P–T field is represented by an invariant point. With $\phi = n + 1$ there are univariant curves emerging from the invariant point, and with $\phi = n$ there are divariant surfaces intersecting on the univariant curves.

Schreinemaker's rules are the topological expressions of the fact that a given phase cannot take part in an equilibrium reaction outside its own stability field and are as follows:

Rule I: When two divariant assemblages each of n phases meet along a univariant curve of $n + 1$ phases, then on one side of the univariant curve, the divariant assemblage I is relatively less metastable than assemblage II, whereas on the other side of the curve assemblage II is relatively less metastable than assemblage I.

Rule II (Morey–Schreinemaker rule): The P–T region in the neighborhood of each invariant point is divided by univariant curves into $n + 2$ sectors ($n > 1$), each of which is $< 180°$ in angular extent and is occupied by one or more divariant assemblages one of which is unique for the sector.

The rules have been quoted from Zen (1966), where the derivation and original references to Schreinemaker's work between 1915 and 1925 may be found.

In Chapters 1 and 2, we presented examples of univariant, divariant, and trivariant reactions. An assemblage, which belongs to the system K_2O-MgO-Al_2O_3-SiO_2-H_2O, is chlorite, muscovite, phlogopite, cordierite, aluminum silicate, quartz, and water. With five components and seven phases, this assemblage is invariant. If we consider that quartz and water are present everywhere, we shall have five univariant curves radiating from an invariant point. However, in this case, there is one degenerate reaction. (In degenerate reactions certain assemblages of phases, because of compositional coincidences, participate in univariant equilibria which can be described by fewer than n components). Therefore, there are four univariant curves representing the following reactions:

2chlorite + 8Al-silicate + 11quartz $\rightleftharpoons$
5cordierite + 5.5H_2O (Ms, Phl)

2chlorite + 1muscovite + 2quartz $\rightleftharpoons$
1cordierite + 1phlogopite + 3.5H_2O (Al-Sil)

5muscovite + 3chlorite $\rightleftharpoons$
8Al-silicate + 5phlogopite + 3.5H_2O (Cord)

3cordierite + 2muscovite $\rightleftharpoons$
2phlogopite + 8Al-silicate + 7quartz + 1.5H_2O (Chl)

Note that each reaction has been labeled by the phases that are absent. Reaction (Ms, Phl) is degenerate and the curve passes through the invariant point without change in slope (see Figure 11.3, below). The disposition of the univariant curves around the invariant point and the fields of divariant assemblages must be, according to the Schreinemaker's rules, as noted above.

As a result of changing P, T and bulk chemistry (see discussion on bulk chemistry and mineral composition in Chapter 4), we may encounter several different divariant assemblages in the field. These differing assemblages may have formed as a result of changing geothermal gradient, may simply reflect the changing bulk chemistry, or may be due to a combination of both these factors. For a theoretical analysis and the construction of a petrogenetic grid for the entire area of interest, we may count all the phases together that occur in all types of rocks in the area. These phases are not coexisting in the sense of the phase rule (in direct contact or within a defined volume) and the procedure is adopted only to facilitate the counting of the number of possible invariant, univariant, and divariant equilibria present in the multisystems, i.e., rocks of the area. We then say that the number of phases in rocks of all types taken together may exceed $(n + 2)$ with a resulting negative variance. In the example discussed above, such a situation would exist if quartz and water are not present in every assemblage. Korzhinskii (1957) called these the multisystems with negative degrees of freedom. Thus a system with $n + 3$ phases has one negative degree and a system with $n + 4$ phases has two negative degrees of freedom. Zen and Roseboom (1972) and Day (1972) have discussed systems of $n + 3$ phases in ternary systems.

The number of invariant points, univariant lines, and divariant surfaces in a system can be determined by using the combinatorial formula

$$\frac{P!}{\phi!(P-\phi)!}$$

where P is number of all phases to be considered and ϕ is the number of coexisting phases present in equilibrium. ϕ is determined by the phase rule. In a ternary system with $P = 6$, for the number of assemblages with different variance we have

$6!/5!(6-5)! = 6$ invariant assemblages

$6!/4!(6-4)! = 15$ univariant assemblages

$6!/3!(6-3)! = 20$ divariant assemblages

The number of independent geometric parameters necessary for the Schreinemaker's diagrams is given by

$$2 + [\phi - (n+1)](n+2)$$

Thus for a ternary system with five phases, we need to know the coordinates of the invariant point and slopes of the five univariant lines. To construct a diagram with several invariant points, the only information necessary is on the $(1 - v)$ centers. For example, a system of $v = -1$ or a ternary system of $(n + 3)$ phases requires two invariant points and their radiating univariant lines. These would be twelve $\{2 + [6 - (3 + 1)](3 + 2) = 12\}$ in number, i.e., coordinates of the first invariant point, slopes of the five lines on the first invariant point, distance to the second invariant point, and slopes of four univariant lines.

The geometric analysis of phase equilibria is being used extensively for petrologic studies and for constructing models for experimental work. In this chapter that approach will be used wherever possible. It must be noted that it is often useful to consider only some of the phases that are chemically possible in the system. However, in doing so, we may neglect certain phases which individually or together may be more stable than the assemblages being considered. Caution is therefore necessary in the selection of phases. When considering natural assemblages, a schematic grid may be constructed based on chemographic relations as found in the rocks. Such a grid shows the one (of several) possibility most compatible with the natural associations. The grid is constructed also with the assumption of local chemical equilibrium among phases.

One of the important problems in petrology is to determine whether two or more mineral assemblages can be related to one another by changes in P and T of formation, or whether the difference in the mineral assemblages is due to change in bulk chemical composition. Perry (1967), Greenwood (1967a), and Braun and Stout (1975) have studied this problem and developed computer programs to solve mass balance equations for all possible univariant relationships between phases as a direct test for the possible overlap of composition spaces defined by observed mineral assemblages. Although not used in this book, this approach when combined with Schreinemaker's system can illustrate the phase relationships in multicomponent systems in an elegant manner.

Method of interpreting metamorphic assemblages

In many metamorphic areas, some rocks contain a maximum number of phases possible in divariant assemblages. There are, however, many rocks in the same area that lack one or more of the phases. This is generally explained as due to some changes in the bulk chemistry. Similarly, the apparent incompatibility of two or more minerals may be due to a requirement of a rare or unusual bulk chemistry.

Let us consider the following reaction:

$$\underset{\textit{muscovite}}{KAl_3Si_3O_{10}(OH)_2} + \underset{\textit{cordierite}}{3(Fe, Mg)_2Al_4Si_5O_{18}\tfrac{1}{2}H_2O} \rightleftharpoons$$

$$\underset{\textit{orthoclase}}{KAlSi_3O_8} + \underset{\textit{garnet}}{2(Fe, Mg)_3Al_2Si_3O_{12}} + \underset{\textit{sillimanite}}{5Al_2SiO_5}$$

$$+ \underset{\textit{quartz}}{4SiO_2} + \underset{\textit{fluid}}{2.5H_2O} \qquad (11.a)$$

This being a six-component seven-phase reaction, it is univariant and, therefore, all the seven phases may coexist at a certain equilibrium P and T. We may choose any one of the three variables P, T, and Fe/Mg independently. Whether a reaction is univariant or divariant, we may rarely find the mineral phases in the right proportion as required by the equilibrium. If the reactants are not in the right proportion, one of the phases may be used up with the rise in temperature at a given pressure. The temperature at which this happens is greater than the equilibrium temperature for the reaction. In the example of the assemblage above, we may find either muscovite-orthoclase-garnet-sillimanite-quartz, or cordierite-orthoclase-garnet-sillimanite-quartz.

In general it may be possible to divide the system such as Reaction (11.a) into subsystems and consider the intensive variables and bulk chemistry for each subsystem. This is the same as considering the bundle of univariant lines (subsystems) radiating from an invariant point (system) in the P–T field.

Reaction (11.a) can be considered as a combination of the following two reactions:

$$\underset{\textit{cordierite}}{3(Fe, Mg)_2Al_4Si_5O_{18}\cdot\tfrac{1}{2}H_2O} \rightleftharpoons$$

$$\underset{\textit{garnet}}{2(Fe, Mg)_3Al_2Si_3O_{12}} + \underset{\textit{Al-silicate}}{4Al_2SiO_5} + \underset{\textit{quartz}}{5SiO_2} + \underset{\textit{fluid}}{1.5H_2O} \quad (11.b)$$

$$\underset{\textit{muscovite}}{KAl_3Si_3O_{10}(OH)_2} + \underset{\textit{quartz}}{SiO_2} \rightleftharpoons$$

$$\underset{\textit{K-feldspar}}{KAlSi_3O_8} + \underset{\textit{Al-silicate}}{Al_2SiO_5} + \underset{\textit{fluid}}{H_2O} \quad (11.c)$$

Let us consider Reaction (11.a) without Mg^{2+}. A five-component reaction with seven phases is invariant

This may be geometrically represented by the crossing of two univariant curves for Reactions (11.b) and (11.c). The position of the curves in the P–T field may be determined by considering the experimental work. At P_{H_2O} equal to 3 kbar, the equilibrium temperature of Reaction (11.b) may be around 730°C, which is the temperature for the Fe end member reaction (dry) (Richardson, 1968). The equilibrium temperature for Reaction (11.c) is several degrees centigrade lower (630°C to 640°C) (Evans, 1965; Day, 1973). The two univariant curves are schematically shown in Figure 11.1. If we now introduce Mg^{2+} in cordierite and garnet, Reaction (11.b) becomes divariant and may be shown as a surface in Figure 11.1. This figure shows a combination of two P–T curves, one for the iron end members [Reaction (11.b′)] and one for magnesium [Reaction (11.b″)]. The equilibrium reactions are

$$3\,\text{Fe-cordierite} \rightleftharpoons 2\,\text{almandine} + 4\,\text{sillimanite} + 5\,\text{quartz} + 1.5H_2O \quad (11.b')$$

$$3\,\text{Mg-cordierite} \rightleftharpoons 2\,\text{pyrope} + 4\,\text{sillimanite} + 5\,\text{quartz} + 1.5H_2O \quad (11.b'')$$

The two equilibrium curves meet the equilibrium curve for Reaction (11.c) at points $a(P_{Fe}, T_{Fe})$ and $b(P_{Mg}, T_{Mg})$.

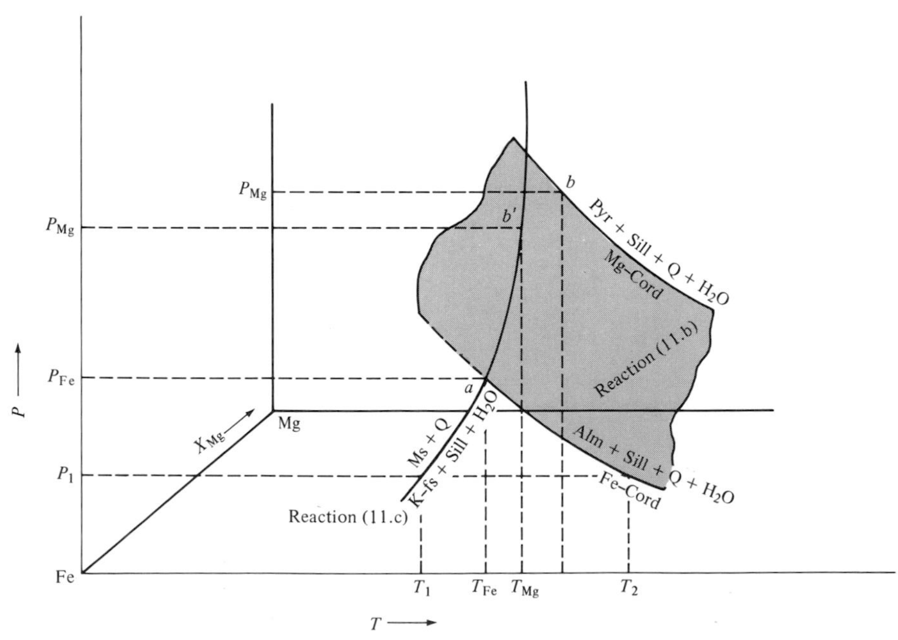

11.1 Schematic representation of the isodimorphic Reaction (11.b) and Reaction (11.c). The curves for Fe and Mg end-members form a surface in the P–T–X_{Mg} space. b' is the projection of the Mg end-member reaction on the P–T plane. See text for discussion.

These points, therefore, represent the invariant equilibrium assemblage given by the combined Reaction (11.a). Note that for a particular Fe^{2+}-Mg^{2+} distribution the assemblage (11.a) continues to be invariant, or at most the assemblage may be regarded as univariant between a and b' or in the temperature range of T_{Fe}–T_{Mg} and the pressure range of P_{Fe}–P_{Mg}. We may increase the variance in Reaction (11.c) by introducing other components in the solid solutions but the assemblage containing muscovite-cordierite-garnet will always have the lowest variance. This in turn would mean an absence or rarity of the three minerals coexisting in any natural assemblage. Table 11.1 shows a list of mineral assemblages from many areas and none of the assemblages contain coexisting muscovite-garnet-cordierite.

With the help of Figure 11.1 we are able to consider the effects of changing bulk chemistry, pressure, and temperature. Since we are concerned only with the six-component seven-phase system, the change in bulk chemistry may be simply considered as a change in the proportion of the minerals in any assemblage.

At the intersection of curve of Reaction (11.b) for the Fe end members and curve of Reaction (11.c)—i.e., at point a—all phases of assemblage (11.a) must coexist irrespective of their proportion in the rock. This would be true even if the Fe/Mg ratio in the rock changes. The changing Fe/Mg ratio would shift the equilibrium pressure and temperature of Reaction (11.a) along the curve of Reaction (11.c) from point a to point b'.

At a pressure P_1, lower than the equilibrium pressure P_{Fe}, and at a temperature of equilibrium (T_1) for Reaction (11.c), we should have muscovite, quartz, sanidine, and sillimanite in equilibrium. Cordierite may not directly participate in the reaction. At the same pressure P_1 and T_{Fe} or any temperature higher than T_1, we may have either muscovite, K-feldspar, sillimanite, and cordierite; or quartz, K-feldspar, sillimanite, and cordierite, depending on whether muscovite or quartz are in excess.

At the same pressure P_1 and the temperature for equilibrium of Reaction (11.b) (T_2), we may have the assemblage quartz, K-feldspar, sillimanite, cordierite, and garnet, where the minerals cordierite, garnet, sillimanite, and quartz are related by equilibrium Reaction (11.b), and K-feldspar does not directly participate. This assemblage would correspond to the situation where quartz is in excess. Supposing the assemblage at temperatures less than T_2 is muscovite, K-feldspar, sillimanite, and cordierite. In this case Reaction (11.b)

$$3\,\text{cordierite} \rightleftharpoons 2\,\text{garnet} + 4\,\text{sillimanite} + 5\,\text{quartz} + 2.5\,H_2O$$

produces quartz, which should react with muscovite according to Reaction (11.c). If cordierite is in excess, we shall finally consume all muscovite and obtain the equilibrium assemblage cordierite, garnet, sillimanite, K-feldspar, and quartz. If there is a large excess of muscovite, cordierite, and quartz will all be used up and we shall have the assemblage muscovite, garnet, sillimanite, and K-feldspar. These results have been summarized in Table 11.2, where the assemblages at a pressure higher than P_{Fe} are also listed. We note that the resulting assemblages are of the same character as occur in many metamorphic areas. However, this analysis shows that to obtain such assemblages, the variation in proportion of minerals (bulk chemistry) alone is not enough, and some variation in pressure and temperature is necessary. However variance in the natural system could also be affected by certain components in solution not considered in the simplified system.

Table 11.1 Coexisting phases in pelitic rocks

Quartz	Biotite	Plagioclase	Muscovite	K-feldspar	Garnet	Sillimanite	Cordierite	References[a]
+	+	+	−	+	+	+	−	1, 2, 3
+	+	+	+	−	+	−	−	4
+	+	+	+	+	+	+	−	2, 5, 6
+	+	+	−	+	+	+	+	3, 6
+	+	+	+	+	+	−	−	5
+	+	+	+	−	+	+	−	7
+	+	+	−	−	+	+	+	8

[a] Data are from (1): Kretz, 1964; (2): Evans and Guidotti, 1966; (3): Dallmeyer and Dodd, 1971; (4): Butler, 1967; (5): Guidotti *et al.*, 1973; (6): Hess 1971; (7): Guidotti, 1970; (8): Lal and Moorehouse, 1969.

11 Experimental and theoretical study of metamorphic systems

Petrogenetic grids in the greenschist facies

A petrogenetic grid for reactions producing biotite and other Al-Fe-Mg silicates in the greenschist facies was described by Brown (1975). The elements considered important are Si, Al, Ti, Fe^{3+}, Fe^{2+}, Mg, Ca, Na, K, H, and O. The phases are muscovite, biotite, chlorite, stilpnomelane, actinolite, K-feldspar, quartz, albite, epidote, magnetite, sphene, and H_2O (fluid). Brown studied the composition of the coexisting minerals in natural assemblages and fixed the composition of the minerals in the grid as follows:

muscovite $K_{1.85}Na_{0.03}Mg_{0.65}Fe^{2+}_{0.35}Fe^{3+}_{0.02}Ti^{4+}_{0.02}Al_{4.24}Si_{6.74}(OH)_4$

chlorite $Mg_{1.97}Fe^{2+}_{2.53}Fe^{3+}_{0.16}Al_{2.30}Si_{2.30}(OH)_8$

biotite $K_{1.84}Na_{0.01}Ca_{0.01}Mg_{2.0}Fe^{2+}_{2.90}Fe^{3+}_{0.22}Ti_{0.16}Al_{2.78}Si_{5.82}(OH)_4$

stilpnomelane $K_{0.35}Na_{0.02}Mg_{0.99}Fe^{2+}_{1.77}Fe^{3+}_{0.09}Al_{0.66}Si_{3.98}(OH)_{2.91}$

actinolite $K_{0.02}Na_{0.44}Ca_{1.66}Mg_{2.47}Fe^{2+}_{1.82}Fe^{3+}_{0.42}Ti_{0.08}Al_{0.67}Si_{7.54}(OH)_2$

Epidote is assumed to contain 20 percent of the iron end member. Other phases are pure species with ideal chemical formulae. The subassemblage muscovite, quartz, albite, epidote, sphene, magnetite, and H_2O is of common occurrence in the greenschist facies. Brown assumes that this subassemblage is always present and considers the representation of the following five "univariant" reactions:

0.85 chlorite + 1.75 K-feldspar + 1.54 quartz + 0.08 albite + 0.05 epidote + 0.02 sphene + 0.02 magnetite $\rightleftharpoons$ 0.76 muscovite + 0.08 actinolite + 0.00 stilpnomelane + 0.39 H_2O (Bi)

1.37 stipnomelane + 0.15 actinolite + 0.61 K-feldspar + 0.40 muscovite + 0.14 sphene + 0.05 magnetite $\rightleftharpoons$ 1.00 biotite + 4.64 quartz + 0.10 albite + 0.19 epidote + 0.86 H_2O (Chl)

1.72 stilpnomelane + 0.18 actinolite + 0.66 muscovite + 0.13 sphene + 0.04 magnetite $\rightleftharpoons$ 0.30 chlorite + 1.00 biotite + 5.18 quartz + 0.12 albite + 0.21 epidote + 0.73 H_2O (K-fs)

1.62 chlorite + 3.93 K-feldspar + 0.05 albite + 0.18 sphene + 0.09 magnetite $\rightleftharpoons$ 1.03 muscovite + 0.52 stilpnomelane + 1.00 biotite + 1.71 quartz + 0.09 epidote + 1.61 H_2O (Act)

1.17 chlorite + 0.04 actinolite + 3.02 K-feldspar + 0.01 albite + 0.17 sphene + 0.08 magnetite $\rightleftharpoons$ 0.64 muscovite + 1.00 biotite + 2.52 quartz + 0.11 epidote + 1.40 H_2O (Stilp)

It must be noted here that the practice of writing reactions using chemical formulae for crystalline solutions with defined compositions is not recommended by the authors for several reasons. The phases are too complex to be defined as end members. The validity of such diagrams is restricted to a definite bulk chemistry. The stoichiometric coefficients may become either too large or too small. However, the literature contains comparatively few examples of the application of

Table 11.2 Scheme of *P* and *T* for various mineral combinations

Phases in parentheses do not participate in the reaction

quartz	muscovite	cordierite	K-feldspar	garnet	sillimanite	$P_{Fe}T_{Fe}$	
quartz	muscovite	(cordierite)	K-feldspar	—	sillimanite	P_1T_1	
—	muscovite	(cordierite)	K-feldspar	—	sillimanite	P_1T_{Fe}	excess muscovite
quartz	—	(cordierite)	K-feldspar	—	sillimanite	P_1T_{Fe}	excess quartz
quartz	—	cordierite	(K-feldspar)	garnet	sillimanite	P_1T_2	excess cordierite
—	muscovite	—	(K-feldspar)	garnet	sillimanite	P_1T_2	excess muscovite
quartz	—	cordierite	(K-feldspar)	garnet	sillimanite	P_1T_2	excess quartz
quartz	(muscovite)	cordierite	—	garnet	sillimanite	$P > P_{Fe}T_1$	
quartz	(muscovite)	—	—	garnet	sillimanite	$P > P_{Fe}T_{Fe}$	
quartz	muscovite	—	K-feldspar	(garnet)	sillimanite	$P > P_{Fe}T_2$	
quartz	—	—	K-feldspar	(garnet)	sillimanite	$P > P_{Fe}T_3$	excess quartz
—	muscovite	—	K-feldspar	(garnet)	sillimanite	$P > P_1T_3$	excess muscovite

Schreinemaker's technique to complex assemblages. Therefore the description of some of these examples is necessary to facilitate the understanding of the geometric approach.

Considering the field occurrence of the assemblages, a petrogenetic grid may be drawn as shown in Figure 11.2. The location of the reaction lines in the P–T space takes into consideration the following observations.

1. Biotite-free rocks exist at low temperatures and over a broad pressure range extending up into the blueschist facies.
2. Biotite with chlorite and muscovite is stable at higher temperatures and also over a relatively broad pressure range.

11.2 A petrogenetic grid for the greenschist facies (Brown, 1975). It is applicable to assemblages containing all the phases of the subassemblage muscovite, quartz, albite, epidote, magnetite, sphene, and H_2O. The compositions of the solid solutions are fixed. (Reprinted from the *Journal of Petrology* by permission.) Note that K-feldspar plots outside the limits of the diagram ($-\infty$ A). It is schematically plotted for convenience to show correct reaction relations (Brown, 1976, personal communication).

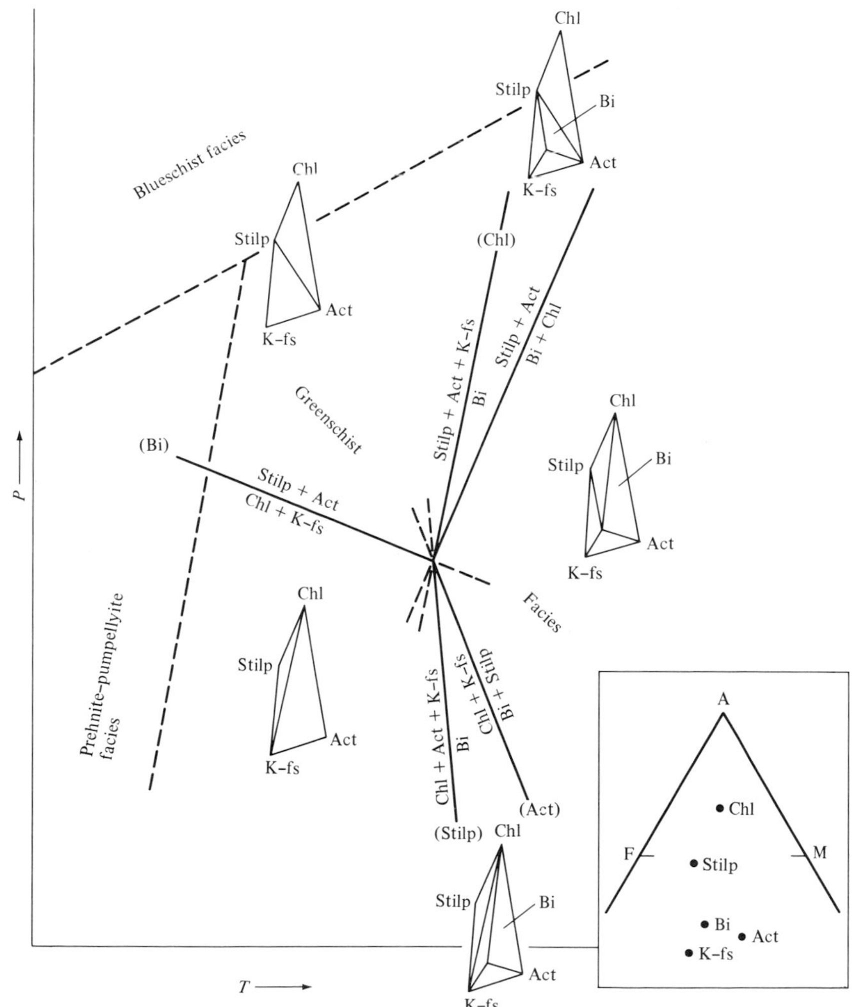

3. A comparison of the greenschist facies rocks from western Vermont and Dalradian and from Otago shows that chlorite + K-feldspar may be confined to the low-pressure rocks (western Vermont, Dalradian) and muscovite + actinolite + stilpnomelane in the relatively high-pressure rocks (Otago).

Brown has listed some other field occurrences which support the grid in Figure 11.2.

The grid shown in Figure 11.2 is applicable only to assemblages containing all the phases of the subassemblage muscovite, quartz, albite, epidote, magnetite, sphene, and H_2O. Further continuous reactions involving changes in composition of the solid solutions cannot be represented on such a grid. However the grid does provide us with the information about the P–T area, where a continuous reaction may be operative. For example, the reaction

chlorite + K-feldspar $\rightleftharpoons$ biotite + muscovite

may operate only between lines (Stilp) and (Act) as a result of a changing Mg/Fe ratio in the coexisting phases. A continuous reaction may operate only in the P–T fields where an assemblage containing both reactants and products of the reaction is stable.

The system K_2O-MgO-Al_2O_3-SiO_2-H_2O

While studying a multicomponent system such as this one, we shall restrict ourselves to the study of a limited number of phases in a limited range of pressure and temperature. This will reduce the work to certain manageable details which will be more relevant to phase equilibria of interest. In the present system, for example, Schreyer and Seifert (1969) find that several important reactions take place over the range of pressure of approximately 4 to 8 kbar at temperatures between 600°C and 720°C, which corresponds to the P–T field of the amphibolite to granulite facies transition. The following analysis is based on Schreyer and Seifert's work.

In the P–T range mentioned above, the crystalline phases that are stable are

quartz (SiO_2),

muscovite [$KAl_3Si_3O_{10}(OH)_2$],

cordierite ($Mg_2Al_4Si_5O_{18}0.5H_2O$),

chlorite [$Mg_5Al_2Si_3O_{10}(OH)_8$],

phlogopite [$KMg_3AlSi_3O_{10}(OH)_2$],

K-feldspar ($KAlSi_3O_8$), and

Al-silicate (Al_2SiO_5).

If we consider as before that quartz and water vapor are present in all assemblages ($P_{total} = P_{H_2O}$), we have a ternary system of $(n + 3)$ phases. Consequently there are six invariant points and 15 univariant lines. Two of the invariant points falling within the P–T range considered here are shown in Figure 11.3. The assemblage of the first point and four univariant reactions were described in Chapter 9. The second invariant point and the associated equilibria are

IP_2: *Phases*

muscovite + aluminum silicate + phlogopite + cordierite + K-feldspar + quartz

Reactions

3cordierite + 2muscovite $\rightleftharpoons$ 2phlogopite + 8Al-silicate + 7quartz + 1.5H_2O (K-fs)

1muscovite + 1quartz $\rightleftharpoons$ 1K-feldspar + 1Al-silicate + 1H_2O (Phl, Cord)

2phlogopite + 6Al-silicate + 9quartz $\rightleftharpoons$ 3cordierite + 2K-feldspar + 0.5H_2O (Ms)

4muscovite + 4phlogopite + 3quartz $\rightleftharpoons$ 3cordierite + 4K-feldspar + 6.5H_2O (Al-Sil)

One of the reactions (Phl, Cord) is degenerate. It does not involve MgO and the univariant curve passes through IP_2 in Figure 11.3.

As discussed in Chapter 9, Reactions (Ms, Phl) and (Al-Sil) at IP_1 mark the upper pressure stability limits of cordierite; the latter reaction is important in forming cordierite porphyroblasts at low pressures. IP_1 at 645°C and 6.5 kbar is fixed by the intersection of the two curves (Seifert, 1970). Reaction (Ms, Phl) marks the upper pressure stability limit of cordierite. However, above Reaction (Chl), cordierite may not be found in potassium-rich rocks. Reaction (Chl) represents the upper P–T limit of a stable coexistence of cordierite and muscovite. Reactions (Ms, Phl) and (Al-Sil) also mark the upper stability limit of chlorite. Not indicated in the figure is the reaction

chlorite + quartz $\rightleftharpoons$ cordierite + talc + H_2O

for which the equilibrium P and T are somewhat higher than those of Reaction (Al-Sil) as considered before.

The position of IP_2 (695°C, 5 kbar) is given by the intersection of the univariant lines (Al-Sil) and (Phl,

Cord). The former reaction (Schreyer and Seifert, 1967) marks the end of the coexistence of muscovite and phlogopite in the presence of quartz and the formation of the cordierite–K-feldspar assemblage typical of the low-pressure contact aureoles. The probable position of the univariant curve (Ms)

$$\text{phlogopite} + \text{sillimanite} + \text{quartz} \rightleftharpoons \text{cordierite} + \text{K-feldspar}$$

is also shown at IP_2. This reaction also leads to the formation of cordierite–K-feldspar paragenesis but at a higher pressure than that of Reaction (Al-Sil).

Assemblages with staurolite

Important mineralogic assemblages with staurolite have been listed in Table 11.3. Although the list does not cover exhaustively the different metamorphic areas, it is obvious that staurolite could occur in a variety of pelitic rocks. We also note that the assemblage: quartz, muscovite, garnet, biotite, chlorite, and staurolite is very common. Magnetite (and/or ilmenite) and Al-silicates are also common. Cordierite and chloritoid may be found occasionally, and in some aluminous rocks gedrite may also be associated (Robinson and Jaffe, 1969a).

Staurolite, cordierite, and andalusite

We may consider the following reactions in the system SiO_2-Al_2O_3-K_2O-FeO-MgO-H_2O as discussed by Hess (1969):

$$1.18\,\text{staurolite} + 1.23\,\text{chlorite} + 9.09\,\text{quartz} \rightleftharpoons 2.77\,\text{cordierite} + 1.00\,\text{garnet} + 4.13\,H_2O \quad (11.d)$$

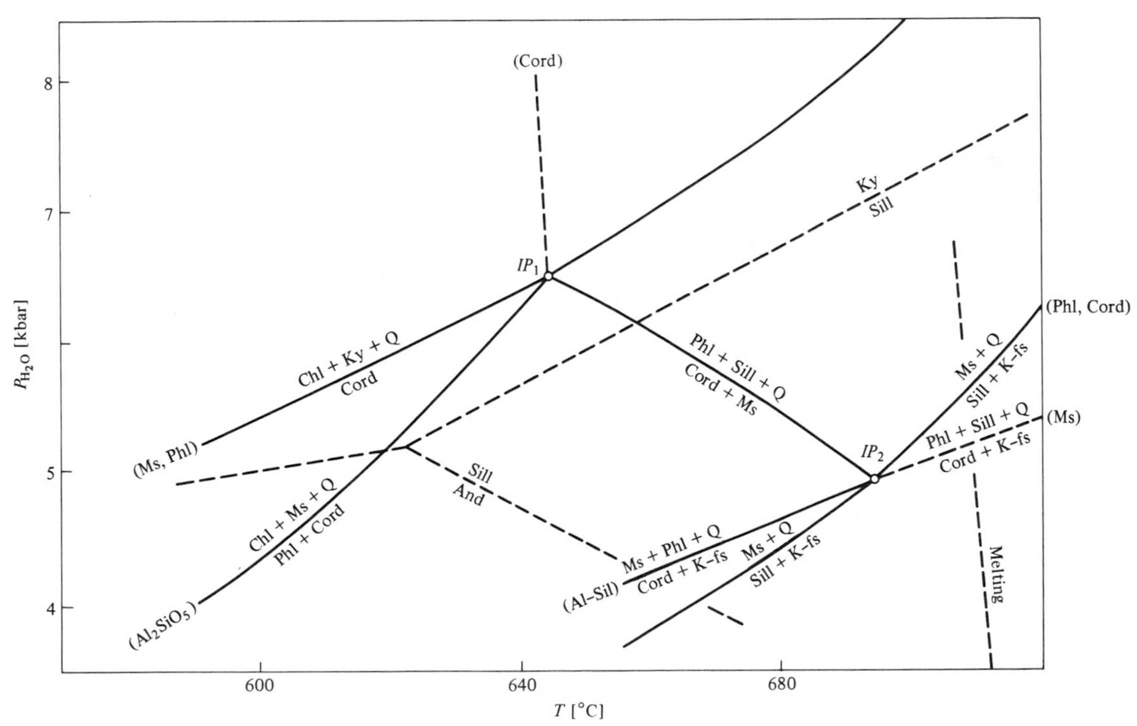

11.3 P–T grid showing some equilibria in the system K_2O-MgO-Al_2O_3-SiO_2-H_2O studied by Schreyer and Seifert (1969). The melting curve is from Schreyer and Seifert's unpublished data. Al_2SiO_5 results are from Richardson *et al.* (1968) and are different from those adopted by us. Quartz and H_2O are present in all assemblages. (Reprinted from the *American Journal of Science* by permission.)

1.00 staurolite + 2.25 quartz $\rightleftharpoons$
0.50 garnet + 0.25 cordierite + 3.5 andalusite + 0.40 H_2O (11.e)

1.00 staurolite + 1.75 quartz $\rightleftharpoons$
0.50 garnet + 0.12 chlorite + 3.89 andalusite + 0.02 H_2O (11.f)

5.65 andalusite + 1.00 chlorite + 3.19 quartz $\rightleftharpoons$
0.67 staurolite + 1.83 cordierite + 2.75 H_2O (11.g)

0.50 staurolite + 1.00 cordierite + 1.00 muscovite $\rightleftharpoons$
5.25 andalusite + 1.00 biotite + 1.75 quartz + 0.25 H_2O (11.h)

0.93 chlorite + 1.00 muscovite + 1.22 quartz $\rightleftharpoons$
1.00 biotite + 0.12 staurolite + 0.70 cordierite + 3.38 H_2O (11.i)

0.15 staurolite + 0.54 chlorite + 1.00 muscovite $\rightleftharpoons$
1.00 biotite + 2.22 andalusite + 2.24 H_2O (11.j)

1.00 staurolite + 1.00 biotite + 5.15 quartz $\rightleftharpoons$
0.75 garnet + 1.38 cordierite + 1.00 muscovite + 0.19 H_2O (11.k)

4.50 staurolite + 1.00 muscovite + 7.30 quartz $\rightleftharpoons$
1.00 biotite + 19.30 andalusite/kyanite/sillimanite + 2.00 garnet + 2.20 H_2O (11.l)

0.25 garnet + 0.62 chlorite + 1.00 muscovite $\rightleftharpoons$
0.42 staurolite + 1.00 biotite + 0.93 quartz + 2.27 H_2O (11.m)

Figure 11.4 shows a petrogenetic grid constructed by Hess, after Albee's (1965a) method, using the following compositions:

cordierite ($Mg_{1.2}Fe_{0.8}Al_4Si_5O_{10}\frac{1}{2}H_2O$);
chlorite [$Mg_{2.6}Fe_{2.4}Al_2Si_3O_{10}(OH)_8$];
biotite [$KMg_{1.5}Fe_{1.5}AlSi_3O_{10}(OH)_2$];
garnet ($Mg_{0.6}Fe_{2.4}Al_2Si_3O_{12}$);
staurolite ($Mg_{0.6}Fe_{1.4}Al_9Si_4O_{24}H$);
muscovite [$KAl_3Si_3O_{10}(OH)_2$];
kyanite/andalusite (Al_2SiO_5); and
quartz (SiO_2).

Although the *P* and *T* shown in Figure 11.4 are only approximate, the chemographic relations may be used in the same way as was done in the previous section. In the figure there are four invariant points with the assemblages

(Chl) staurolite, cordierite, muscovite, andalusite, biotite, quartz, garnet
(Gar) staurolite, cordierite, muscovite, andalusite, biotite, quartz, garnet
(And) staurolite, cordierite, muscovite, biotite, quartz, garnet, chlorite
(Ms, Bi) staurolite, cordierite, andalusite, quartz, garnet, chlorite

Table 11.3 Important assemblages with staurolite

Quartz	Muscovite	Garnet	Biotite	Plagioclase	Magnetite	Al_2SiO_5	Chlorite	Cordierite	Chloritoid	References[a]
+	+	−	−	−	+	−	+	−	+	1
+	+	−	+	+	−	−	+	−	−	2, 3
+	+	−	−	−	−	−	+	−	+	2
+	+	+	+	+	−	−	+	−	−	3
+	+	+	+	+	−	+	+	−	−	3
+	+	+	+	+	−	−	+	−	−	4
+	+	−	+	+	−	−	+	+	−	3
+	+	+	+	+	+	+	−	−	−	4, 5
+	+	+	+	+	+	−	−	−	−	1, 2, 3, 6, 7, 8
+	+	+	+	+	+	+	−	−	−	1, 6, 9

[a] Data are from (1): Schreyer and Chinner, 1966; (2): Chinner, 1967; (3): Guidotti, 1974; (4): Guidotti, 1968; (5): Robinson and Jaffe, 1969a; (6): Hietanen, 1969; (7): Guidotti; 1970; (8): Uruno and Kanisawa, 1965; (9): Hounslow and Moore, 1967 (also contain hematite).

The inferred P–T gradient in the eastern Pyrenees (Guitard, 1965) is represented by the broken line, which shows that the important reactions with increasing temperatures are

garnet + chlorite + muscovite ⇌ staurolite + biotite + quartz

andalusite + chlorite + quartz ⇌ cordierite + staurolite

chlorite + muscovite + quartz ⇌ biotite + staurolite + cordierite

staurolite + cordierite + muscovite ⇌ andalusite + biotite + quartz

staurolite + muscovite + quartz ⇌ biotite + andalusite + garnet

Staurolite coexists stably with cordierite and andalusite in the Pyrenees and parts of Maine and New Hampshire. In the latter area, staurolite persists to the sillimanite zone as described by Green (1963) and Drake (1968). The assemblages are

quartz, muscovite, biotite, garnet, chlorite

quartz, muscovite, andalusite, chlorite, biotite, ±garnet

quartz, muscovite, biotite, chlorite, staurolite, ±garnet

quartz, muscovite, biotite, sillimanite, garnet, ±staurolite

If the rocks of the New Hampshire area are similar in bulk chemistry to the Maine rocks, the absence of cordierite in these rocks may be due to an increase in the pressure gradient toward the kyanite-bearing rocks in Vermont described by Thompson and Norton (1968). The pressure of formation of the assemblages must be above the invariant point (Gar) and on the high-pressure side of Reactions (11.g) and (11.i). The surface formed by the line (Ms, Bi)–(Gar) and the high-pressure side of Reaction (11.i) may represent the upper pressure limit of cordierite stability in these compositions. This surface would be similar to the boundary of the univariant reaction

chlorite + andalusite + quartz ⇌ cordierite + vapor

11.4 Staurolite-cordierite assemblages after Hess (1969). The P–T gradient in the eastern Pyrenees is shown by the broken line.

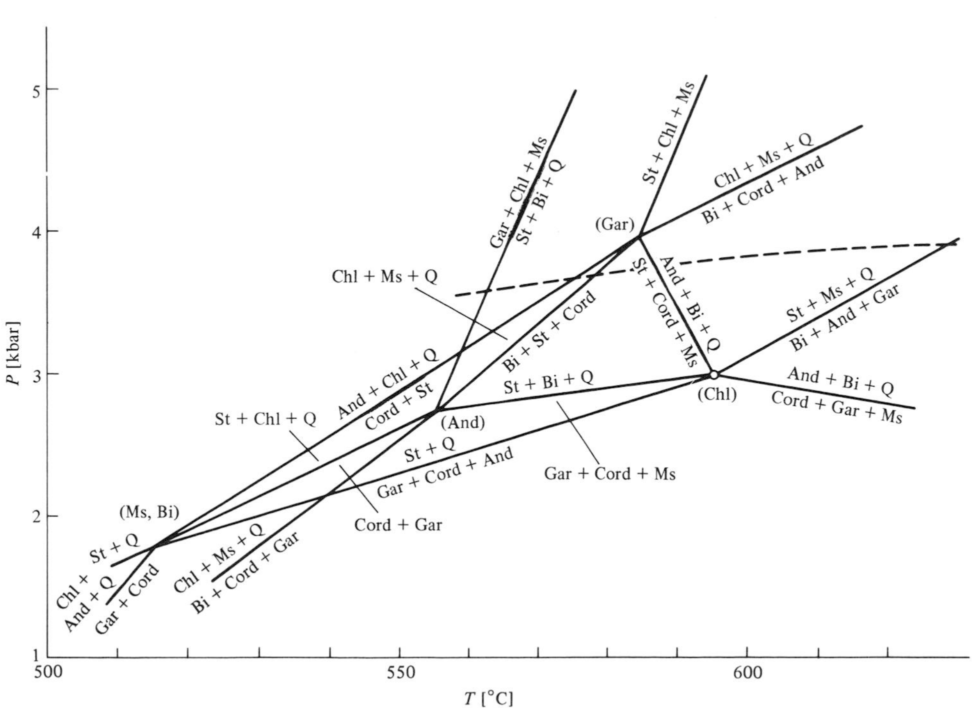

as determined by Seifert and Schreyer (1970) (see Figure 11.3).

Staurolite-Al_2SiO_5

At pressures higher than 5 kbar, andalusite in Reactions (11.j) and (11.m) is replaced by kyanite. Hounslow and Moore (1967) have described the following assemblages from Grenville schists near Fernleigh, Ontario (sample numbers are in parentheses):

quartz, muscovite, biotite, staurolite, plagioclase (*1*)
quartz, muscovite, biotite, garnet, staurolite (*2A*)(*A*)
quartz, muscovite, biotite, plagioclase, kyanite, chlorite (*2B*)
quartz, muscovite, kyanite, hematite, chlorite (*6*)
quartz, muscovite, biotite, garnet, staurolite, plagioclase, kyanite (*7A*, *7B*, *10*)
quartz, muscovite, biotite, garnet, staurolite, plagioclase (*7C*, *7D*, *8*, *9B*)

The above samples are from a band of pelitic schists extending from the southwest [sample (*1*)] to the northeast [sample (*10*)], a distance of 12 miles. Rocks west of sample (*1*) belong to the greenschist facies, and rocks east of sample (*10*) contain sillimanite. If we do not consider plagioclase and the iron ores (ilmenite, magnetite, and hematite), Reactions (11.j) and (11.m) may be applied to the interpretation of these assemblages by referring to Figure 11.4. Acocrding to Reaction (11.m), if the temperature is higher than the equilibrium temperature ($\sim 580°C$) at 5 kbar and if chlorite is used up, we may obtain

muscovite, staurolite, biotite, quartz

or

garnet, muscovite, staurolite, biotite, quartz

i.e., samples (*1*) and (*2A*). In such assemblages without kyanite, garnet contains only a little Mn (0.4 percent spessertine). In kyanite-bearing assemblages, garnet is rich in Mn (12.7 to 16.4 percent spessertine). The assemblage

garnet-kyanite-staurolite-muscovite-quartz-biotite

may have resulted if the temperature exceeded the equilibrium temperature of Reaction (11.j) with staurolite and muscovite being in excess and garnet being stable because of the concentration of Mn. The above analysis shows that the thermal gradient increased toward the northeast. However, the simple scheme above is disturbed by an increased compositional variance not only due to Mn in garnet but also due to a variable concentration of Ca. Along with Reactions (11.j) and (11.m), the following reaction involving plagioclase should be considered:

$$\underset{\text{in garnet solution}}{Ca_3Al_2Si_3O_{12}} + \underset{\text{kyanite}}{2Al_2SiO_5} + \underset{\text{quartz}}{SiO_2} \rightleftharpoons \underset{\text{in plagioclase solution}}{3CaAl_2Si_2O_8} \quad (11.n)$$

Samples (*7A*), (*7B*), and (*10*)—where all of the three minerals garnet, plagioclase, and kyanite coexist—the grossularite content in garnet varies within a narrow limit (8.6 to 11.6 percent). The grossularite content, however, varies between 9.8 and 23.7 percent in garnets not coexisting with kyanite. Hounslow and Moore's data are too few and incomplete (no plagioclase compositions) for a detailed analysis of the problem. However the indications are that these rocks probably formed at ~ 5 kbar and in the range of temperatures 550°C to 600°C.

Staurolite persists up to the sillimanite zone and is unstable within this zone. Guidotti (1968, 1974) described staurolite in the lower sillimanite zone in Maine, it is absent in the upper sillimanite zone.

Staurolite-chloritoid and variation of oxygen fugacity; The system $FeO\text{-}Al_2O_3\text{-}SiO_2\text{-}H_2O\text{-}O_2$

The effect of variable oxygen fugacity has not always been given due consideration in metamorphic petrogenesis. A system where this effect is very significantly demonstrated has been studied by Richardson (1968), Ganguly (1968, 1969, 1972), and Hoschek (1969). We shall adopt Ganguly's treatment in the following theoretical analysis.

The minerals that are of particular importance are

chloritoid ($H_2FeAl_2SiO_7$),
staurolite ($HFe_2Al_9Si_4O_{24}$),
almandine ($Fe_3Al_2Si_3O_{12}$),
Fe-cordierite ($Fe_2Al_4Si_5O_{18}$),
Al-silicate (Al_2SiO_5),
magnetite (Fe_3O_4),
hematite (Fe_2O_3),
hercynite ($FeAl_2O_4$),
quartz (SiO_2), and
corundum (Al_2O_3).

In a system such as this with three important variables—P, T, and fugacities of H_2O and O_2 in the fluid phase—

we shall be concerned generally with divariant equilibria of which the following are considered here:

In presence of excess quartz:

(*1*) 6 staurolite + 11 quartz $\rightleftharpoons$ 4 almandine + 23 Al_2SiO_5 + 3 H_2O

(*2*) 2 staurolite + 7 quartz $\rightleftharpoons$ 2 cordierite + 5 Al_2SiO_5 + H_2O

(*3*) 4 chloritoid + 5 Al_2SiO_5 $\rightleftharpoons$ 2 staurolite + quartz + 3 H_2O

(*4*) 6 staurolite + 3 quartz + 2 O_2 $\rightleftharpoons$ 27 Al_2SiO_5 + 4 magnetite + 3 H_2O

(*5*) 54 chloritoid + 5 O_2 $\rightleftharpoons$ 12 staurolite + 10 magnetite + 6 quartz + 48 H_2O

(*6*) 54 almandine + 6 H_2O + 23 O_2 $\rightleftharpoons$ 12 staurolite + 46 magnetite + 114 quartz

(*7*) 2 almandine + O_2 $\rightleftharpoons$ 2 Al_2SiO_5 + 4 magnetite + 4 quartz

(*8*) 3 almandine + O_2 + 3 H_2O $\rightleftharpoons$ 3 chloritoid + 2 magnetite + 6 quartz

(*9*) 54 almandine + 18 O_2 $\rightleftharpoons$ 27 cordierite + 36 magnetite + 27 quartz

(*10*) 27 cordierite + 6 H_2O + 5 O_2 $\rightleftharpoons$ 12 staurolite + 10 magnetite + 87 quartz

(*11*) 6 cordierite + 2 O_2 $\rightleftharpoons$ 12 andalusite + 4 magnetite + 18 quartz

(*12*) 12 chloritoid + 2 O_2 (or 3 O_2) $\rightleftharpoons$ 12 Al_2SiO_5 + 4 magnetite (or 6 hematite) + 12 H_2O

(*13*) 23 chloritoid + 8 quartz $\rightleftharpoons$ 5 almandine + 4 staurolite + 21 H_2O

(*14*) 3 chloritoid + 2 quartz $\rightleftharpoons$ almandine + 2 kyanite + 3 H_2O

In absence of excess silica:

(*15*) 6 staurolite $\rightleftharpoons$ 4 almandine + 12 Al_2SiO_5 + 11 corundum + 3 H_2O

(*16*) 4 chloritoid + 4 Al_2SiO_5 + corundum $\rightleftharpoons$ 2 staurolite + 3 H_2O

(*17*) 6 staurolite + 2 O_2 (or 3 O_2) $\rightleftharpoons$ 24 Al_2SiO_5 + 3 corundum + 4 magnetite (or 6 hematite) + 3 H_2O

(*18*) 8 almandine + 19 corundum + 3 H_2O + 2 O_2 $\rightleftharpoons$ 6 staurolite + 4 magnetite

(*19*) 2 almandine + 4 corundum + O_2 $\rightleftharpoons$ 6 kyanite + 4 magnetite

(*20*) 3 chloritoid $\rightleftharpoons$ almandine + 2 corundum + 3 H_2O

(*21*) 57 chloritoid + 4 O_2 $\rightleftharpoons$ 12 staurolite + 3 almandine + 8 magnetite + 51 H_2O

(*22*) 48 chloritoid + 6 corundum + 4 O_2 $\rightleftharpoons$ 12 staurolite + 8 magnetite + 42 H_2O

(*23*) 10 staurolite $\rightleftharpoons$ 3 almandine + 31 serpentine + 11 hercynite + 5 H_2O

(*24*) 10 staurolite $\rightleftharpoons$ 3 cordierite + 25 serpentine + 14 hercynite + 5 H_2O

(*25*) 3 chloritoid + 5 kyanite + hercynite $\rightleftharpoons$ 2 staurolite + 2 H_2O

In analyzing the relative dispositions of the different equilibria, a constant $P_t(P_{total})$ section in the P_{O_2}-P_t-T volume is used. A univariant equilibrium curve intersects such a section as a point and the divariant equilibrium plane as a curve. Not all the possible equilibria will be stable at every P_t; therefore a convenient choice of a P_t must be made. Figure 11.5 shows a collection of several equilibria that have been experimentally determined or theoretically inferred on an isobaric section at 10 ± 0.5 kbar. The abbreviation to oxygen buffers are HM, HMN, NNO, QFM, and MW for Fe_2O_3-Fe_3O_4, MnO_4-"MnO," Ni-NiO, SiO_2-Fe_2SiO_4-Fe_3O_4, and Fe_3O_4-"FeO," respectively.

Ganguly (1972) calculated the possible effect of crystalline solubility in some equilibria involving staurolite. This has been discussed before. Another interesting aspect of this work is the calculation of the pressure dependence of the oxidation equilibria and the oxidation limits of staurolite, almandine, and Fe-cordierite. For this purpose a general equilibrium reaction is

$$\sum_i nR_i + n_{O_2}O_2 + n_{H_2O}H_2O = \sum_i p_j P_j \qquad (11.1)$$

where R_i and P_j refer to solid reactants and products respectively and n_i stands for the number of moles of the species i. At equilibrium, we have

$$\Delta G(P, T, f_{O_2}) = 0 = \Delta G^0(1, T) + P\Delta V^S + 2.303\, RT(\mp n_{H_2O} \log f_{H_2O} - n_{O_2} \log f_{O_2}) \qquad (11.2)$$

which at constant temperature (and therefore with cancellation of ΔG^0) transforms to

$$\log \frac{f_{O_2}(P_2)}{f_{O_2}(P_1)} = \frac{1}{2.303 n_{O_2} RT} \times \left[\Delta V^S(P_2 - P_1) \mp n_{H_2O} \int_{P_1}^{P_2} \Delta \overline{V}_{H_2O}\, dP\right] \qquad (11.3)$$

where $\overline{V}_{H_2O}$ is the partial molar volume of H_2O. For $P_{H_2O} = P_{total}$, $\overline{V}_{H_2O} = V_{H_2O}$, in which case the integral may be evaluated using the data for pure water. Figure 11.6 shows an isothermal section (620°C) of the field of stability of staurolite + quartz and staurolite + magnetite + quartz in the P–T–log f_{O_2} space. The diagram also shows the field of stability for the assemblage Fe-cordierite + magnetite + quartz defined by the equilibria

$$(9)\quad 54\,\text{almandine} + 18\,O_2 \rightleftharpoons 27\,\text{cordierite} + 36\,\text{magnetite} + 27\,\text{quartz}$$

$$(10)\quad 27\,\text{cordierite} + 6\,H_2O + 5\,O_2 \rightleftharpoons 12\,\text{staurolite} + 10\,\text{magnetite} + 87\,\text{quartz}$$

$$(11)\quad 6\,\text{cordierite} + 2\,O_2 \rightleftharpoons 12\,Al_2SiO_5(\text{andalusite}) + 4\,\text{magnetite} + 18\,\text{quartz}$$

Figure 11.6 also shows the oxidation limits of almandine ± quartz [equilibria (6), (7), and (8)]. The breakdown

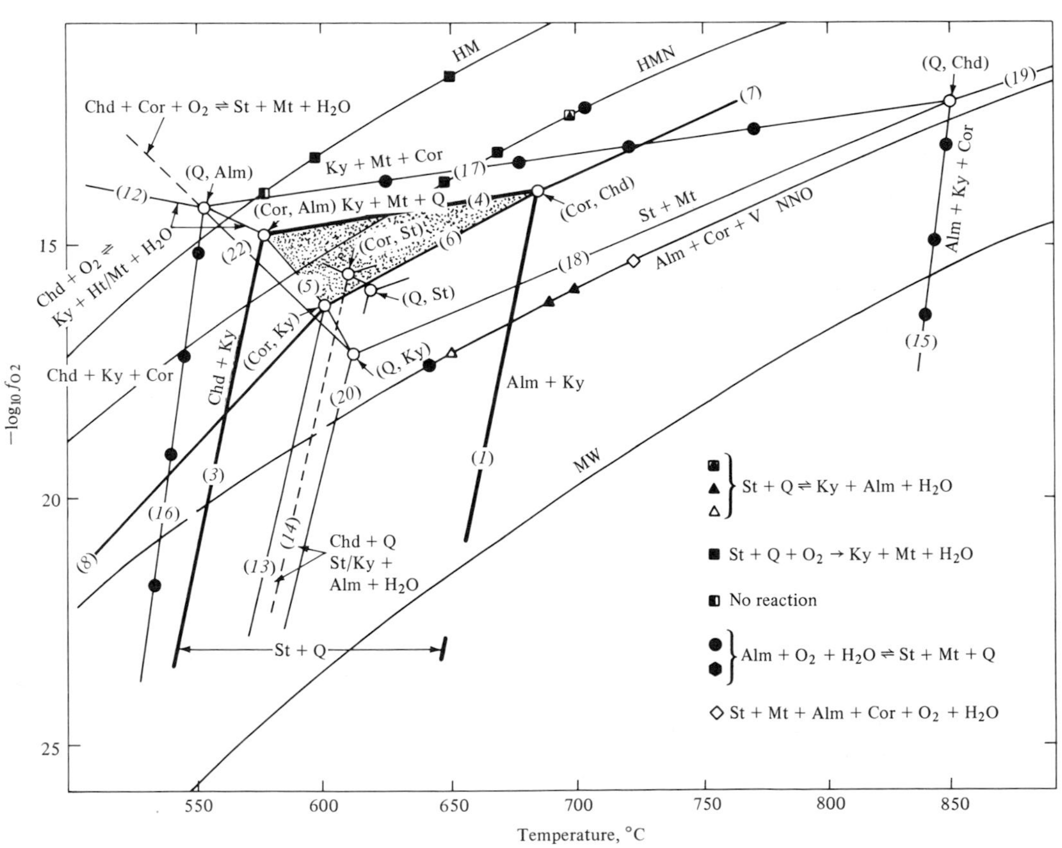

11.5 Complete collection of all equilibria, experimentally determined and theoretically deduced, on an isobaric section at 10 ± 0.5 kbar. The locations of the invariant points (Cor, Alm), (Cor, Ky), and (Q, Ky) and the related equilibria involving chloritoid have been taken, with minor readjustments, from Ganguly (1969), which have been based on the buffer data from Eugster and Wones, 1962). The fields of staurolite + quartz, staurolite, and almandine (± quartz) are shown, respectively, by very heavy, stippled, and medium heavy lines. The stippled area shows the field of staurolite + magnetite + quartz. The field of staurolite + magnetite is confined to the triangle defined by (Q, Alm), (Q, Ky), and (Q, Chd). The numbers represent equilibria discussed in the text. (After Ganguly, 1972.) (Reprinted from the *Journal of Petrology* by permission.)

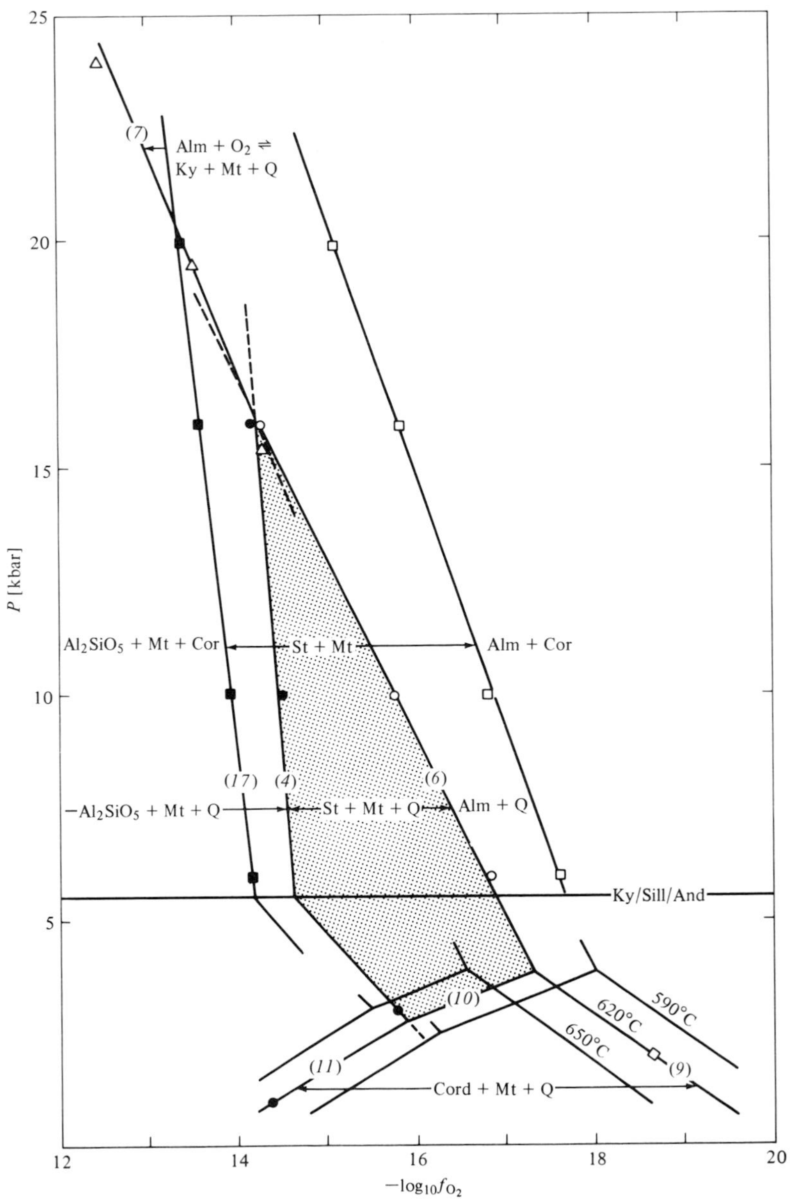

11.6 Isothermal section at 620°C showing the fields of staurolite + magnetite ± quartz and cordierite + magnetite + quartz; sections along 650°C and 590°C are also shown in the low-pressure region to convey the three-dimensional form. Numbers represent equilibria discussed in the text. (After Ganguly, 1972.) (Reprinted from the *Journal of Petrology* by permission.)

reaction of almandine

almandine + O_2 $\rightleftharpoons$ corundum + magnetite + quartz

according to these results takes place at pressures below 4 kbar and at temperatures around 600°C to 675°C. In this P–T region, however, according to Hsu (1968) almandine should break down to hercynite, magnetite, and quartz.

In natural assemblages, because of the reaction between staurolite and mica/or chlorite, the high-temperature stability of staurolite is lower (see Figure 11.4). Two important reactions involving chloritoid and staurolite are

23 chloritoid + 8 quartz $\rightleftharpoons$ 4 staurolite + 5 almandine + 21 H_2O

4 chloritoid + 5 Al_2SiO_5 $\rightleftharpoons$ 2 staurolite + quartz + 3 H_2O

At oxygen fugacity defined by the Ni-NiO buffer, the equilibrium temperatures for the two reactions at 5 kbar are ~540°C and ~570°C, respectively (Ganguly, 1969). Chinner (1967) has described pelitic rocks with the following common assemblages:

chloritoid-muscovite-chlorite
chloritoid-muscovite-garnet
chloritoid-muscovite-chlorite-staurolite-biotite
muscovite-(chlorite)?-staurolite-biotite
muscovite-garnet-staurolite-biotite

All assemblages contain quartz and albitic feldspar in addition. Following Hoschek (1969), we may draw a schematic P–T diagram with the following two invariant and seven univariant reactions (unbalanced):

chloritoid + biotite + muscovite + quartz $\rightleftharpoons$ garnet + staurolite (Chl)

chloritoid + muscovite $\rightleftharpoons$ staurolite + biotite + quartz (Chl, Gar)

staurolite + biotite + quartz $\rightleftharpoons$ garnet + muscovite (Chl, Chd)

chloritoid + quartz $\rightleftharpoons$ staurolite + garnet (Chl, Ms)

chloritoid + biotite + quartz $\rightleftharpoons$ garnet + muscovite (Chl, St)

chloritoid + quartz + chlorite + muscovite $\rightleftharpoons$ staurolite + biotite (Gar)

chlorite + muscovite $\rightleftharpoons$ staurolite + biotite + quartz (Gar, Chd)

chlorite + muscovite + quartz $\rightleftharpoons$ chloritoid + biotite (Gar, St)

chloritoid + quartz $\rightleftharpoons$ staurolite + chlorite (Gar, Ms, Bi)

Figure 11.7 shows the invariant and univariant assemblages, which would become univariant and divariant, respectively, with the introduction of Mg. The dashed line shows a possible P–T gradient in Chinner's area. For a further discussion of a possible univariant assemblage chloritoid-muscovite-chlorite-staurolite-biotite-quartz, more chemical data on the minerals are necessary.

The chloritoid-staurolite assemblage has been discussed by Albee (1972), who studied the chemical composition of coexisting minerals in samples from different areas. On the basis of the mineral-chemical data, the effect of changing Fe/Mg in coexisting minerals on Albee's (1965a) petrogenetic grid could be evaluated. Figure 11.8 shows parts of the grids with mineral phases, biotite, chlorite, chloritoid, cordierite, staurolite, garnet, kyanite, muscovite, and quartz. Figure 11.8a shows the condition when $P_{H_2O} = P_{total}$ or pressure on the solids, and Figure 11.8b shows the condition when $P_{H_2O} <$

11.7 Schematic representation of some equilibria (Hoschek, 1969). Broken line shows a possible P–T gradient in Chinner's (1967) area.

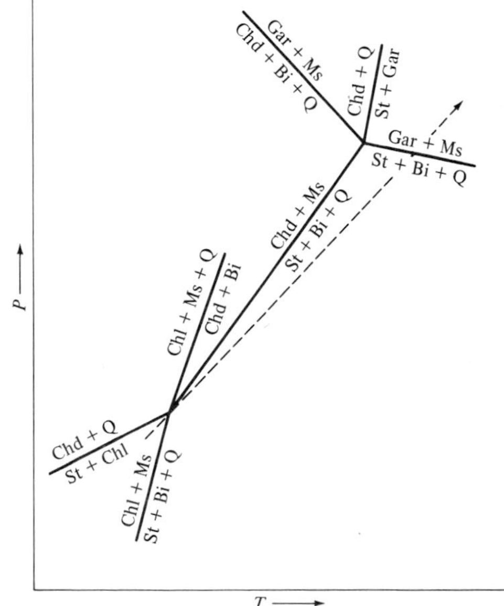

P_{total}. It is obvious that the slope and relative positions of the reaction lines may be quite different in different sections of P_S–T–P_{H_2O} space and, as discussed by Albee (1965), there is probably a continuous transition from one to the other. If the garnet-chlorite join breaks early, the sequence of reactions with excess quartz and muscovite is

chloritoid + kyanite ⇌ staurolite

garnet + chlorite ⇌ chloritoid + biotite

chloritoid ⇌ garnet + staurolite + Mg-rich chloritoid

chloritoid + kyanite ⇌ staurolite + chlorite

chloritoid + chlorite ⇌ staurolite + biotite

chloritoid ⇌ garnet + staurolite + biotite

In the other situation, when garnet-chlorite join breaks down later than the breakdown of chloritoid, the reaction sequence is

chloritoid + Al-silicate ⇌ staurolite

chloritoid ⇌ garnet + staurolite + Mg-rich chloritoid

chloritoid + Al-silicate ⇌ staurolite + chlorite

chloritoid ⇌ garnet + staurolite + chlorite

garnet + chlorite ⇌ staurolite + biotite

Figure 11.9 shows the pseudobinary temperature-composition sections as plotted by Albee (1972) for the garnet-chlorite "join," illustrating the two sequencies of reactions. Similar diagrams for the chloritoid-cordierite join have also been drawn by Albee (1972). Such diagrams together with the Thompson (1957) projections are good illustrations of the facies types. Both the sequences of reactions discussed above seem to be present in the natural assemblages, and a preference of one sequence over the other cannot be made without some information on the P_{total}-μ_{H_2O} relationship.

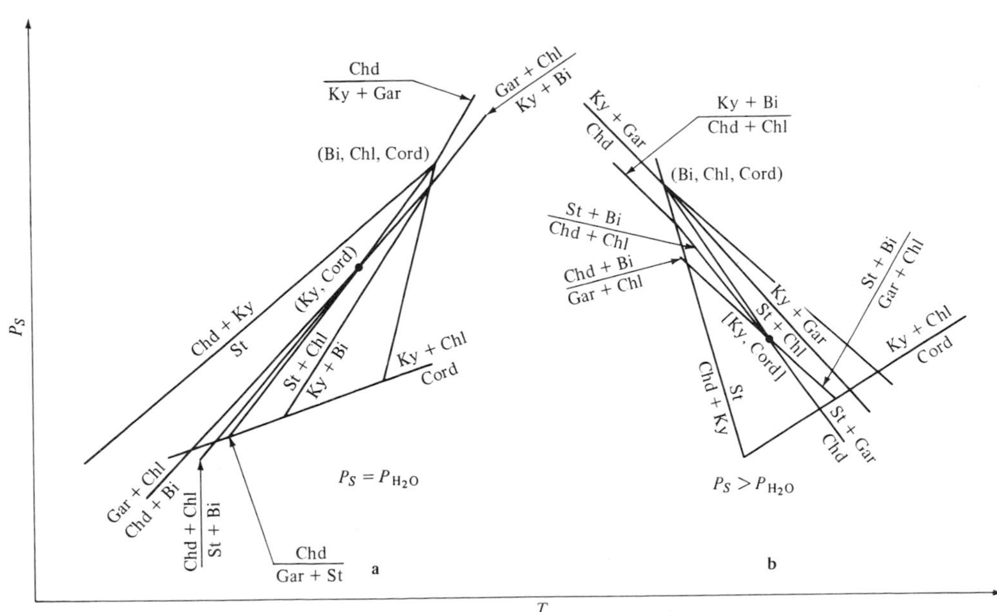

11.8 Albee's (1972) petrogenetic grid for the multisystem biotite, chlorite, chloritoid, cordierite staurolite, garnet, kyanite, muscovite, and quartz. Effect of variation in P_{H_2O} is shown in Figure parts (a) and (b). $P_S = P_{\text{solid}}$. (Reprinted from the *Bulletin of Geological Society of America* by permission.)

The system SiO_2-K_2O-Al_2O_3-FeO-MgO-H_2O in transition from amphibolite to granulite facies

Four or more of the following phases occur in various combinations in metamorphic rocks (see Table 11.1): quartz, biotite, plagioclase, muscovite, K-feldspar, garnet, sillimanite, and cordierite. Except plagioclase, all other minerals can be considered to belong to the system SiO_2-K_2O-Al_2O_3-FeO-MgO–H_2O in which P_{H_2O} is constant everywhere. We shall first consider the possible reactions in the six-component, nine-phase system. Later we shall study the increase of variance in this system as a result of additional components. Quartz, as well as H_2O, may be considered as phases that are always present. This reduces our system to a seven-phase, four-component system. With FeO and MgO completely replacing each other in the ferromagnesian phases, we have the following univariant [ ] and divariant ( ) reactions:

1 muscovite + 3 cordierite $\rightleftharpoons$ 1 K-feldspar + 2 garnet + 4 quartz + 5 sillimanite + 2.5 H_2O [Bi]

1 K-feldspar + 3 cordierite $\rightleftharpoons$ 1 muscovite + 2 garnet + 3 sillimanite + 6 quartz + 0.5 H_2O [Bi]

3 cordierite $\rightleftharpoons$ 2 garnet + 4 sillimanite + 5 quartz + 1.5 H_2O (Bi, Ms, K-fs)

1 muscovite + 1 quartz $\rightleftharpoons$ 1 K-feldspar + 1 sillimanite + 1 H_2O (Bi, Cord, Gar)

4 muscovite + 2 garnet + 9 quartz $\rightleftharpoons$ 3 cordierite + 4 K-feldspar + 2.5 H_2O (Bi, Sill)

1 biotite + 3 cordierite $\rightleftharpoons$ 1 K-feldspar + 3 garnet + 3 sillimanite + 3Q + 2.5 H_2O [Ms]

1 K-feldspar + 3 cordierite $\rightleftharpoons$ 1 garnet + 1 biotite + 5 sillimanite + 7 quartz + 0.5 H_2O [Ms]

2 biotite + 6 sillimanite + 9 quartz $\rightleftharpoons$ 2 K-feldspar + 3 cordierite + 0.5 H_2O (Ms, Gar)

4 biotite + 3 cordierite + 3 quartz $\rightleftharpoons$ 4 K-feldspar + 6 garnet + 5.5 H_2O (Ms, Sill)

1 biotite + 2 quartz + 1 sillimanite $\rightleftharpoons$ 1 garnet + 1 K-feldspar + 1 H_2O (Ms, Cord)

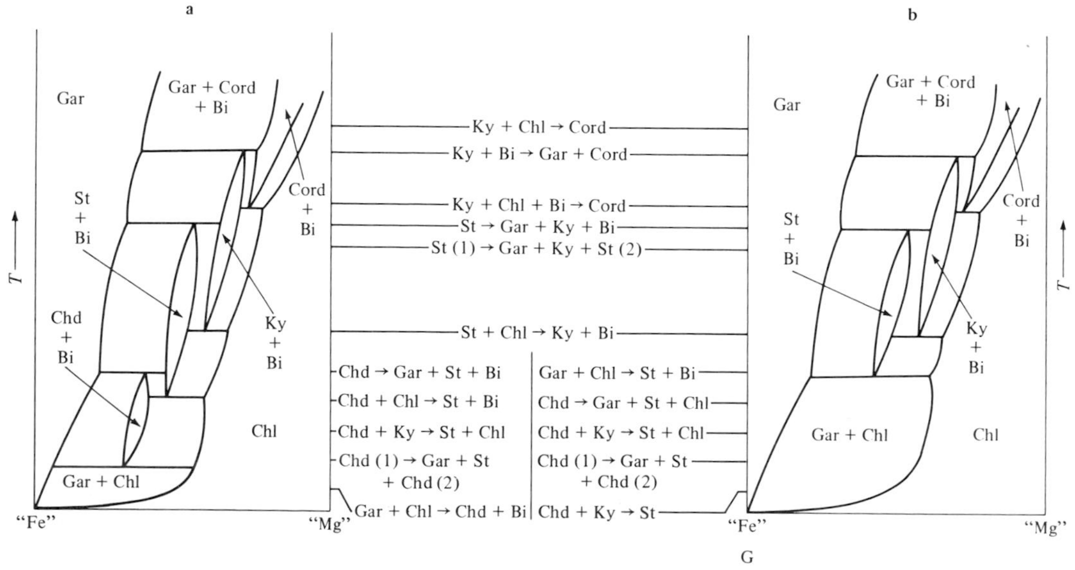

11.9 Comparison of sequences of reactions. (a) Garnet-chlorite breaks early; (b) Garnet-chlorite break late. Figures, as in Albee (1972), are pseudobinary T–X sections. The assemblages in the three-phase field (not labelled) consists of all phases in the adjacent one- or two-phase fields at the same temperature. (Reprinted from the *Bulletin of Geological Society of America* by permission.)

1 muscovite + 3 cordierite $\rightleftharpoons$ 1 biotite + 1 garnet + 6 sillimante + 6 quartz + 1.5 H_2O [K-fs]

3 biotite + 3 cordierite + 2 sillimanite $\rightleftharpoons$ 3 muscovite + 5 garnet + 2 quartz + 1.5 H_2O [K-fs]

2 muscovite + 3 cordierite $\rightleftharpoons$ 2 biotite + 8 sillimanite + 7 quartz + 1.5 H_2O (K-fs, Gar)

2 biotite + 3 cordierite $\rightleftharpoons$ 2 K-feldspar + 4 garnet + 3 quartz + 1.5 H_2O (K-fs, Sill)

1 biotite + 2 sillimanite + 1 quartz $\rightleftharpoons$ 1 muscovite + 1 quartz (K-fs, Cord, H_2O)

2 muscovite + 3 cordierite $\rightleftharpoons$ 2 biotite + 8 sillimanite + 7 quartz + 1.5 H_2O [Gar]

6 muscovite + 2 biotite + 15 quartz $\rightleftharpoons$ 8 K-feldspar + 3 cordierite + 6.5 H_2O (Gar, Sill)

3 biotite + 7 muscovite + 18 quartz $\rightleftharpoons$ 1 garnet + 3 cordierite + 10 K-feldspar + 8.5 H_2O [Sill]

1 biotite + 1 muscovite + 3 quartz $\rightleftharpoons$ 2 K-feldspar + 1 garnet + 2 H_2O (Sill, Cord)

1 biotite + 2 muscovite + 4 quartz $\rightleftharpoons$ 3 K-feldspar + 1 garnet + 1 sillimanite + 3 H_2O [Cord]

3 muscovite + 2 garnet $\rightleftharpoons$ 2 biotite + 1 K-feldspar + 5 sillimanite + 1 quartz + 1 H_2O [Cord]

Dallmeyer and Dodd (1971) have listed a maximum of seven phases—quartz, K-feldspar, plagioclase, sillimanite, cordierite, biotite, and garnet—in rocks of the Hudson Highlands, New York. These phases (without plagioclase) may be related by Reactions [Ms] and the subsystem Reactions (Ms, Gar), (Ms, Sill), and (Ms, Cord), which are univariant and divariant, respectively. The variance could be increased by the additional components in crystalline solution in garnet, biotite, cordierite, and K-feldspar. In the seven-phase assemblages, large-scale mineralogic equilibrium may have been established according to [Ms], or according to a third [Ms] reaction, which can be written by adding (Ms, Gar) and (Ms, Cord), depending on the water fugacity. There are, however, other samples where one or two of the phases are missing. The assemblages are (plagioclase and accessories neglected)

quartz	K-feldspar	sillimanite	cordierite	biotite	garnet
quartz	K-feldspar	sillimanite	—	biotite	garnet
quartz	K-feldspar	sillimanite	cordierite	biotite	—
quartz	K-feldspar	sillimanite	—	biotite	—
quartz	—	sillimanite	cordierite	biotite	garnet

Quartz, biotite, sillimanite and H_2O are always present. If we consider the system SiO_2-K_2O-Al_2O_3-FeO-H_2O, Reaction [Ms] is invariant and Reactions (Bi, Ms, Kfs), (Ms, Gar), (Ms, Sill), and (Ms, Cord) are univariant. One possible configuration of the univariant lines is shown in Figure 11.10. This figure may be used to find the possible assemblages in the system as a function of changing pressure and temperature. At a pressure P_1 lower than P_i (equilibrium P at the invariant point), we have the following assemblages as a function of increasing temperature

T_1	biotite, sillimanite, quartz, K-feldspar cordierite	T_1 equilibrium T for (Ms, Gar)
T_i (1)	biotite, sillimanite, —, K-feldspar, cordierite	Due to changing proportion of minerals in (Ms, Gar). Also 3-phase assemblages possible.
(2)	biotite, —, quartz, K-feldspar, cordierite	
(3)	—, sillimanite, quartz, K-feldspar, cordierite	
T_2	biotite, —, quartz, K-feldspar, cordierite, garnet	From T_i (2). T_2 is equilibrium T for (Ms, Sill). No change in T_i (1) and T_i (3).
T_3(1)	biotite, —, —, K-feldspar, cordierite, garnet	Due to changing proportion of minerals in (Ms, Sill). No change in T_i(1) and T_i(3).
(2)	biotite, —, quartz, K-feldspar, —, garnet	
(3)	—, —, quartz, K-feldspar, cordierite, garnet	

As we reach T_4, the equilibrium temperature for Reaction (Bi, Ms, K-fs)

cordierite $\rightleftharpoons$ garnet + sillimanite + quartz + H_2O

we have the assemblages T_i(1), T_i(3), T_3(1), T_3(2), and T_3(3). The reactions will be as discussed below.

The breakdown of cordierite results in the production of quartz, garnet, sillimanite and H_2O. Therefore, biotite and sillimanite in the assemblage T_i(1) (biotite, sillimanite, K-feldspar, cordierite) will react according to Reaction (Ms, Gar), and biotite cordierite and quartz according to (Ms, Sill) until all biotite is consumed, giving the assemblage sillimanite, cordierite, garnet, quartz and K-feldspar. If biotite is in excess, we may have biotite, garnet, sillimanite, and K-feldspar or biotite, garnet, quartz, and K-feldspar. Assemblage T_i(3) will

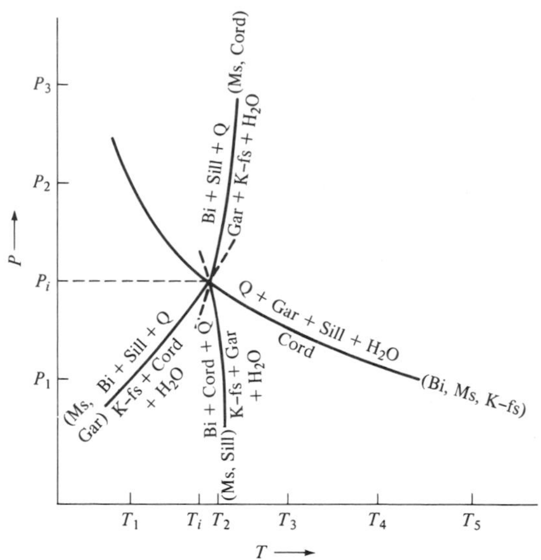

11.10 Schematic representation of possible equilibria in the SiO_2-K_2O-Al_2O_3-FeO-MgO-H_2O system. See text for discussion.

change to sillimanite, cordierite, garnet, quartz, and K-feldspar by a simple addition of garnet. For $T_3(1)$ (biotite, K-feldspar, cordierite, garnet), we have similarly if there is excess biotite; biotite, garnet, sillimanite or quartz and K-feldspar; or if there is excess cordierite; cordierite, sillimanite, garnet, quartz, K-feldspar. For $T_3(2)$ there is no change (because cordierite is absent), and for $T_3(3)$ we have the same general assemblage cordierite, sillimanite, almandine, quartz and K-feldspar. In summary we have

T_1 biotite, sillimanite, quartz, K-feldspar, cordierite
T_i biotite, sillimanite, —, K-feldspar, cordierite
biotite, —, quartz, K-feldspar, cordierite
—, sillimanite, quartz, K-feldspar, cordierite
T_2 biotite —, quartz, K-feldspar, cordierite, garnet
T_3 biotite, —, —, K-feldspar, cordierite, garnet
biotite, —, quartz, K-feldspar, —, garnet
—, —, quartz, K-feldspar, cordierite, garnet
T_4 —, sillimanite, quartz, K-feldspar, cordierite, garnet
biotite, sillimanite, —, K-feldspar, —, garnet
biotite, —, quartz, K-feldspar, —, garnet

It was obvious from the start that since all assemblages included quartz, biotite, and sillimanite, the variation of temperature at a constant pressure ($<P_1$) could give us only the assemblage at T_1(quartz, biotite, sillimanite, K-feldspar, cordierite) which is present in Dallmeyer and Dodd's area. Let us consider the variation of pressure at a constant temperature. Using similar arguments as before, we have the following assemblages at $T_1(<T_i)$ at increasing pressures:

P_1 biotite, sillimanite, quartz, K-feldspar, cordierite
P_i biotite, sillimanite, quartz, K-feldspar, —
biotite, sillimanite, quartz, —, cordierite
P_2 biotite, sillimanite, quartz, —, cordierite, garnet
($T_1 P_i$ continues unchanged).
P_3 biotite, sillimanite, quartz, —, —, garnet

From the assemblages listed above in the two cases, we have the following results regarding the P and T of the rocks in the Hudson Highlands:

biotite, sillimanite, quartz, K-feldspar, cordierite, garnet	$P_i T_i$
biotite, sillimanite, quartz, K-feldspar, cordierite, —	$T_1(<T_i)P_1(<P_i)$
biotite, sillimanite, quartz, K-feldspar, —, —,	$T_1 P_i$ (or any $P > P_1$)
biotite, sillimanite, quartz, —, cordierite, garnet	$T_1 P_2$ [equilibrium P for Reaction (11.b)]
biotite, sillimanite, quartz, K-feldspar, —, garnet	$T > T_i$, $P > P_i$ [T and P, equilibrium values for the reaction (Ms, Cord)]

The variance of the natural assemblages is greater than that of the model assemblages. MgO, CaO, and Na_2O are three components and plagioclase a phase that must be included in the system. There is some interaction between plagioclase and garnet in the form of Ca distribution and some between plagioclase and orthoclase in the form of Na and Ca distribution. Addition of MgO does not change our results, and if Na_2O and CaO are neglected the qualitative results may not change significantly. Dallmeyer and Dodd suggest that the ranges of temperature and pressure of metamorphism may be 700°C to 750°C and 3.0 to 5.5 kbar, respectively. Such changes are enough to produce the kind of changes in the assemblages described above.

Hess (1971) has studied pelitic rocks from south-central Massachusetts. The largest number of specimens contain the assemblage quartz, sillimanite, garnet, cordierite, K-feldspar, and biotite (Reaction [Ms]). Cordierite-free samples are next in number (Ms, Cord), followed by orthoclase free samples (Bi, Ms, K-fs). On the basis of the mineralogy of the rocks in the area and in other adjacent areas, Hess (1971) suggests that large-scale mineralogic and chemical equilibrium was established around 700°C at P = 5 to 6 kbar. He also suggests that temperatures may have varied 50°C to 100°C within the same facies, a conclusion that is supported by the present analysis.

The assemblage quartz, sillimanite, biotite, garnet, and K-feldspar is common in many metamorphic terranes. Kretz (1964) presented an equilibrium analysis of this assemblage from southwestern Quebec. Addition of muscovite to this assemblage leads to the common assemblage

quartz-sillimanite-muscovite-biotite-garnet-K-feldspar

described by Evans and Guidotti (1966) from western Maine. The probable reactions relating the mineral phases are

$$\text{biotite} + 2\text{muscovite} + 4\text{quartz} \rightleftharpoons 3\text{K-feldspar} + \text{garnet} + \text{sillimanite} + 3H_2O$$

$$3\text{muscovite} + 2\text{garnet} \rightleftharpoons 2\text{biotite} + \text{K-feldspar} + 5\text{sillimanite} + \text{quartz} + H_2O$$

In addition to the six phases, plagioclase is present in all assemblages. The following reactions are pertinent to the six-phase system:

$$\text{muscovite} + \text{quartz} \rightleftharpoons \text{K-feldspar} + \text{sillimanite} + H_2O \quad \text{(Bi, Cord, Gar)}$$

$$\text{biotite} + 2\text{quartz} + \text{sillimanite} \rightleftharpoons \text{garnet} + \text{K-feldspar} + H_2O \quad \text{(Ms, Cord)}$$

$$\text{biotite} + 2\text{sillimanite} + \text{quartz} \rightleftharpoons \text{muscovite} + \text{garnet} \quad (\text{K-fs, Cord}, H_2O)$$

$$\text{biotite} + \text{muscovite} \rightleftharpoons 2\text{K-feldspar} + \text{garnet} + 2H_2O \quad \text{(Sill, Cord)}$$

A possible configuration of the univariant lines as suggested by Ganguly (1974) is shown in Figure 11.11. The mineral assemblages resulting from changes in T at a constant pressure P and bulk composition are

$P_1 T_1$	muscovite, quartz, sillimanite, K-feldspar, (biotite)	T_1 is equilibrium T for Reaction (Bi, Cord, Gar). Biotite may not be directly involved.
$T_2(1)$	muscovite, — sillimanite, K-feldspar (biotite)	
(2)	—, quartz, sillimanite, K-feldspar, (biotite)	
T_3	—, quartz, sillimanite, K-feldspar, biotite, garnet	T_3 is equilibrium T for Reaction (Ms, Cord). $T_2(1)$ assemblage unchanged.
T_4	—, —, sillimanite, K-feldspar, biotite, garnet —, quartz, —, K-feldspar, biotite, garnet —, quartz, sillimanite, —, —, garnet	

At pressure P_2 and at increasing temperatures, we have

$P_2 T_1(1)$	muscovite, quartz, (biotite), sillimanite, —, —,	
(2)	muscovite, quartz, (biotite), —, K-feldspar, —	
T_5	muscovite, quartz, biotite, sillimanite, —, garnet	From $T_1(1)$. T_5 is equilibrium T for Reaction (K-fs, Cord, H_2O).
$T_6(1)$	muscovite, quartz, biotite, —, —, garnet	
(2)	muscovite, quartz, —, sillimanite, —, garnet	
T_7	muscovite, quartz, biotite, —, K-feldspar, garnet	From $T_1(2)$ and $T_6(1)$. T_7 is the equilibrium T for Reaction (Sill, Cord).

The natural assemblages in western Maine, with their inferred P and T by comparison with the above model, are

quartz, biotite, muscovite, K-feldspar, sillimanite, garnet	$P_i T_i$
quartz, biotite, muscovite, K-feldspar, sillimanite, —	$P_1 T_1$
quartz, biotite, muscovite, —, sillimanite, garnet	$P_2 T_5$
quartz, biotite, muscovite, —, sillimanite, —	$P_2 T_1$
quartz, biotite, muscovite, K-feldspar, —, —,	$P_2 T_1$
quartz, biotite, muscovite, —, —, garnet	$P_2 T_6$

It is obvious that both P and T must have varied over a range to yield the mineral assemblages found in the area. Therefore the "invariant" assemblage (which becomes univariant by introducing MgO) may have crystallized above 650°C at about 4 kbar. The interaction of plagioclase with Ms and K-fs has been discussed by Evans and Guidotti, who have also drawn attention to a possible existence of the univariant assemblage sillimanite K-feldspar (orthoclase)-muscovite-plagioclase-quartz in a large part of the area. The basic isograd

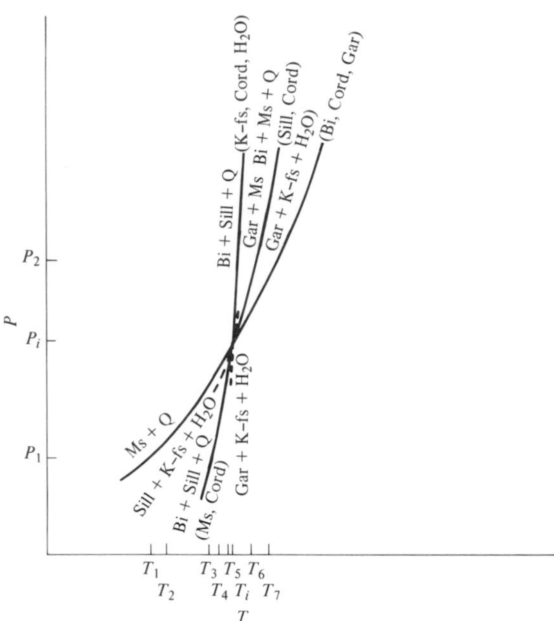

11.11 Schematic representation of reactions in the assemblage muscovite, quartz, K-feldspar, sillimanite, biotite, garnet, H_2O.

reaction is

$$K_{0.94}Na_{0.06}Al_2AlSi_3O_{10}(OH)_2 + SiO_2 + 0.1\,NaAlSi_3O_8 \rightleftharpoons \text{in plagioclase ss}$$
$$1.1(K_{0.86}Na_{0.14}AlSi_3O_8) + Al_2SiO_5 + H_2O$$

It may be that P_{H_2O} and T varied together along the univariant curve for the above reaction or, as suggested by Evans and Guidotti, P_{H_2O} was also a variable "due to a combination of rapid dehydration and low permeability" and was buffered by the above assemblage and controlled by local values of P and T.

Engel and Engel (1960) have described paragneisses from northwest Adirondack Mountains, New York. In a 30-mile distance, the assemblage quartz, biotite, plagioclase, muscovite, and K-feldspar (microcline) changes to quartz, biotite, plagioclase, garnet, and K-feldspar according to the simplified reaction (Sill, Cord)

$$\text{biotite} + \text{muscovite} + 3\,\text{quartz} \rightleftharpoons 2\,\text{K-feldspar} + \text{garnet} + 2H_2O$$

Interaction of plagioclase with other minerals is also evident because of a redistribution of anorthite and albite between feldspars. Variance in Reaction (Sill, Cord) may be increased by considering MgO and MnO in garnet and MgO and other elements in biotite. From Figure 11.11, where if P_i and T_i are approximately 4 kbar and 600°C, respectively, the T of these gneisses may be in the range of 650°C to 700°C at about a pressure of 5 kbar.

Another area, where the assemblage quartz, biotite, plagioclase, garnet, muscovite, and K-feldspar occurs, has been described by Butler (1967). Unlike the northwest Adirondacks, where the temperature must have exceeded the equilibrium T for Reaction (Sill, Cord) to give the assemblage quartz, biotite, plagioclase, garnet and K-feldspar, the temperature here must have decreased during the metamorphism. This retrograde effect would produce the quartz, biotite, garnet, muscovite, and plagioclase assemblage, with all K-feldspar being consumed in the reverse reaction. Note that the K-feldspar-free assemblage has higher Al in muscovites, which is due to bulk chemical changes as discussed by Guidotti (1969) and not due to increasing temperature.

Lal and Moorehouse (1969) have described cordierite-bearing assemblages from Fishtail Lake, Ontario. The assemblages belonging to the system here are

(*1*) garnet, cordierite, biotite, sillimanite, quartz
(*2*) garnet, cordierite, biotite, —, quartz
(*3*) garnet, cordierite, —, —, quartz
(*4*) —, cordierite, biotite, —, quartz
(*5*) —, cordierite, biotite, sillimanite, quartz
(*6*) —, cordierite, —, sillimanite, quartz
(*7*) garnet, —, biotite, sillimanite, quartz

Plagioclase may be present in addition. Referring to Figure 11.10 indicates that if we consider biotite as a phase originally present, garnet, cordierite, sillimanite, quartz may be related by Reaction (Bi, Ms, K-fs):

$$\text{cordierite} \rightleftharpoons \text{almandine} + \text{sillimanite} + \text{quartz} + H_2O.$$

The pressure must be higher than P_i ($\simeq$4 kbar), otherwise the reaction

$$\text{biotite} + \text{sillimanite} + \text{quartz} \rightleftharpoons \text{K-feldspar} + \text{cordierite} + H_2O$$

should have taken place. Assemblages (*2*), (*3*), (*5*), and (*6*) may be interpreted as due to retrograde decrease of temperature. Assemblage (*4*) may or may not be due to reverse reaction, while Assemblage (*7*) may represent any temperature between the equilibrium temperatures of assemblages (*2*) and (*6*).

Grant (1973) has considered the model system Al_2O_3-

K_2O-FeO-MgO-SiO_2-H_2O, with the phases quartz, K-feldspar, muscovite, Al-silicate, biotite, cordierite, garnet, anthophyllite (or orthopyroxene), staurolite, and fluid. A schematic *P–T* grid is shown in Figure 11.12. The phases K-feldspar and muscovite are pure. In the ferromagnesian minerals, Fe^{2+} and Mg^{2+} are distributed such that for Mg/Fe, cordierite > biotite > anthophyllite > garnet. The following divariant assemblages are possible:

I. K-feldspar-muscovite-biotite-cordierite, K-feldspar-muscovite-biotite-garnet, K-feldspar-biotite-garnet-anthophyllite, muscovite-andalusite-cordierite-garnet, muscovite-biotite-cordierite-garnet, biotite-cordierite-garnet-anthophyllite

The system SiO_2-K_2O-Al_2O_3-FeO-MgO-H_2O in transition from amphibolite to granulite facies

II. K-feldspar-muscovite-cordierite-garnet, K-feldspar-biotite-cordierite-garnet, K-feldspar-biotite-garnet-anthophyllite/orthopyroxene, muscovite-andalusite-cordierite-garnet, biotite-cordierite-garnet-anthophyllite/orthopyroxene

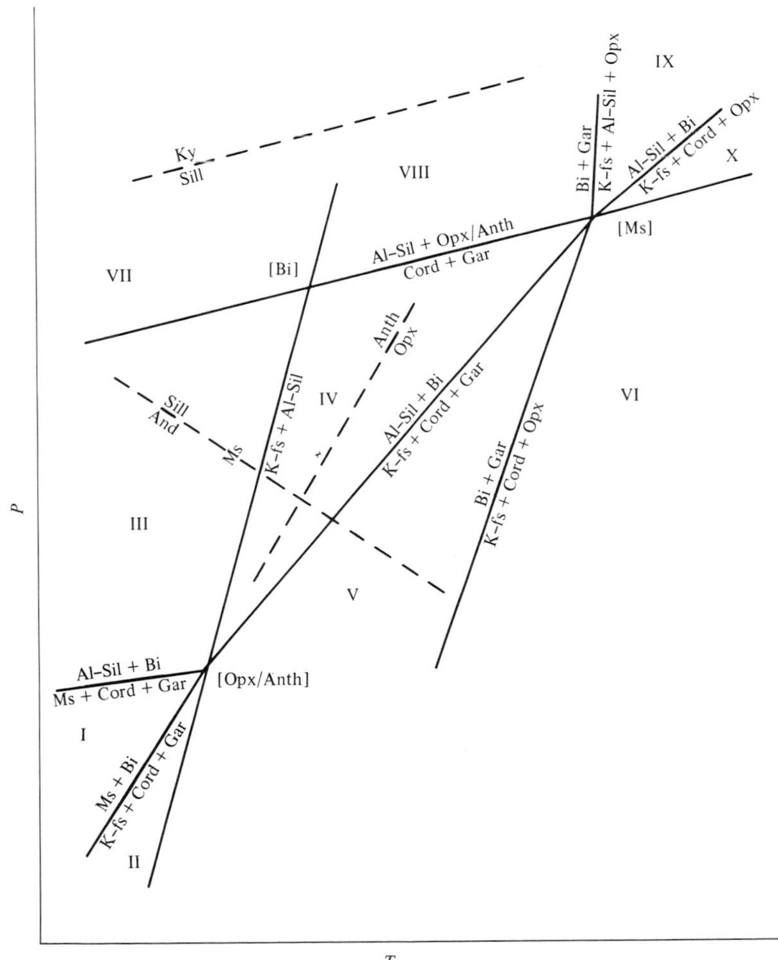

11.12 Phase relations in the system K_2O-Al_2O_3-FeO-MgO-SiO_2-H_2O. Quartz and water are present in all assemblages. (After Grant, 1973. Reprinted from the *American Journal of Science* by permission.)

III. K-feldspar-muscovite-biotite-cordierite, K-feldspar-muscovite-biotite-garnet, K-feldspar-biotite-garnet-anthophyllite, muscovite-andalusite-biotite-cordierite, muscovite-andalusite-biotite-garnet, andalusite-biotite-cordierite-garnet, biotite-cordierite-garnet-anthophyllite (andalusite may be replaced by sillimanite)

IV. K-feldspar-sillimanite-biotite-cordierite, K-feldspar-sillimanite-biotite-garnet, K-feldspar-biotite-garnet-anthophyllite/orthopyroxene, sillimanite-biotite-cordierite-garnet, biotite-cordierite-garnet-anthophyllite/orthopyroxene

V. K-feldspar-andalusite/sillimanite-cordierite-garnet, K-feldspar-biotite-cordierite-garnet, K-feldspar-biotite-garnet-orthopyroxene, biotite-cordierite-garnet-orthopyroxene

VI. K-feldspar-sillimanite/andalusite-cordierite garnet, K-feldspar-biotite-cordierite-orthopyroxene, K-feldspar-cordierite-garnet-orthopyroxene

VII. K-feldspar-muscovite-biotite-cordierite, K-feldspar-muscovite-biotite-garnet, K-feldspar-biotite-garnet-anthophyllite, muscovite-sillimanite/kyanite-biotite-cordierite, muscovite-sillimanite/kyanite-biotite-garnet, sillimanite/kyanite-biotite-cordierite-anthophyllite, sillimanite/kyanite-biotite-garnet-anthophyllite

VIII. K-feldspar-sillimanite/kyanite-biotite-cordierite, K-feldspar-sillimanite/kyanite-biotite-garnet, K-feldspar-biotite-garnet-orthopyroxene, kyanite/sillimanite-biotite-cordierite-orthopyroxene/anthophyllite, sillimanite/kyanite-biotite-garnet-orthopyroxene/anthophyllite

IX. K-feldspar-sillimanite/kyanite-biotite-cordierite, K-feldspar-sillimanite/kyanite-biotite-orthopyroxene, K-feldspar-sillimanite/kyanite-garnet-orthopyroxene, sillimanite/kyanite-biotite-cordierite-orthopyroxene

X. K-feldspar-sillimanite/kyanite-cordierite-orthopyroxene, K-feldspar-sillimanite/kyanite-garnet-orthopyroxene, K-feldspar-biotite-cordierite-orthopyroxene.

Quartz and fluid are possible in all assemblages.

Grant (1973) has provided an extensive list of assemblages from the literature that belong to the various facies types noted above. The curve for andalusite-sillimanite traverses fields III, IV, and V, whereas the transition between orthoamphibole- and orthopyroxene-bearing assemblages is inferred to traverse field IV. In natural assemblages cordierite + garnet are not found with kyanite. Similarly a muscovite-orthopyroxene assemblage, and an orthoamphibole facies V assemblage have not been found. Assemblages recorded under III, IV, and V are quite common in pelitic rocks (Reinhardt, 1968; Wynne-Edwards, 1967; Gable and Sims; 1970). Type VIII assemblages have also been recorded often (Hietanen, 1959; Tilley, 1937).

Anthophyllite-gedrite pelitic rocks

Anthophyllite-bearing pelitic rocks have been described from a number of localities. Natural assemblages in the system FeO-MgO-Al_2O_3-SiO_2-H_2O have been discussed by Robinson and Jaffe (1969b). Using Albee's (1965b) method these authors have calculated a P–T grid shown in Figure 11.13. The phases are: aluminous anthophyllite [$Mg_3Fe_{2.5}Al^{VI}_{1.5}Al^{IV}_{1.5}Si_{6.5}O_{22}(OH)_2$], cummingtonite [$Mg_{2.5}Fe_{4.5}Si_8O_{22}(OH)_2$], cordierite [$Mg_{1.5}Fe_{0.5}Al^{IV}_4Si_5O_{18}$], staurolite [$Mg_{0.5}Fe_{1.5}Al_9Si_{3.75}O_{22}(OH)_2$], garnet ($Mg_{0.5}Fe_{2.5}Al_2Si_3O_{12}$), kyanite ($Al_2SiO_5$), and quartz ($SiO_2$). The "univariant" reactions based on these formulae are

$$36\,\text{staurolite} + 34\,\text{cummingtonite} + 35\,\text{quartz} \rightleftharpoons 44\,\text{cordierite} + 74\,\text{garnet} + 70\,H_2O$$

$$444\,\text{anthophyllite} + 100\,\text{staurolite} + 1273\,\text{quartz} \rightleftharpoons 558\,\text{cordierite} + 218\,\text{cummingtonite} + 326\,H_2O$$

$$32\,\text{anthophyllite} + 218\,\text{kyanite} + 4\,H_2O \rightleftharpoons 52\,\text{cordierite} + 36\,\text{staurolite} + 31\,\text{quartz}$$

$$386\,\text{cummingtonite} + 212\,\text{staurolite} \rightleftharpoons 264\,\text{anthophyllite} + 558\,\text{garnet} + 493\,\text{quartz} + 334\,H_2O$$

$$350\,\text{staurolite} + 766.5\,\text{quartz} \rightleftharpoons 28\,\text{anthophyllite} + 182\,\text{garnet} + 1351\,\text{kyanite} + 322\,H_2O$$

$$14\,\text{anthophyllite} + 38\,\text{kyanite} + 23\,\text{quartz} \rightleftharpoons 25\,\text{cordierite} + 9\,\text{garnet} + 14\,H_2O$$

$$54\,\text{anthophyllite} + 25\,\text{garnet} + 143\,\text{quartz} \rightleftharpoons 53\,\text{cordierite} + 38\,\text{cummingtonite} + 16\,H_2O$$

28 staurolite + 65 quartz $\rightleftharpoons$
4 cordierite + 16 garnet + 102 kyanite + 28 H_2O

102 anthophyllite + 76 staurolite + 334 quartz $\rightleftharpoons$
193 cordierite + 109 garnet + 178 H_2O

The assemblages encountered in the rocks are

quartz-anthophyllite-kyanite-staurolite
(Hietanen, 1959; Robinson and Jaffe, 1969b)

quartz-anthophyllite-cordierite-staurolite
(Tilley, 1937)

quartz-andalusite-cordierite-staurolite

quartz-anthophyllite-staurolite-cummingtonite
(Rabbitt, 1948)

In the absence of garnet, these assemblages would all occur on the low temperature side of Figure 11.13.

11.13 Phase relations in the system $FeO-MgO-Al_2O_3-SiO_2-H_2O$. (After Robinson and Jaffe, 1969b. Reprinted from special paper of the *Mineralogical Society of America* by permission.)

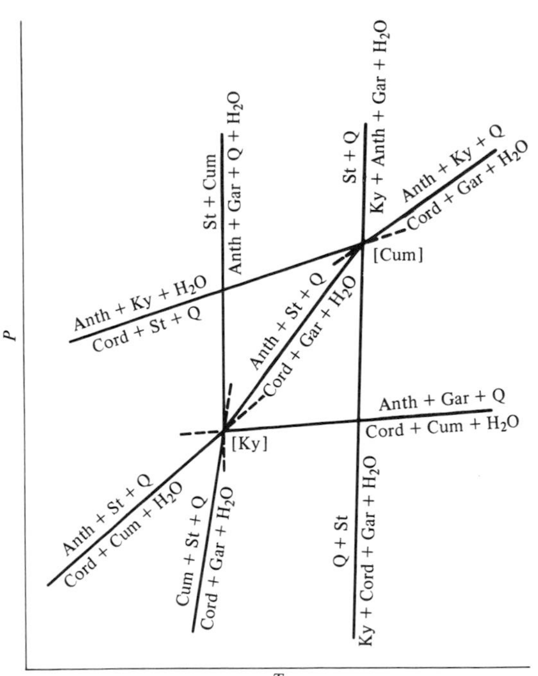

The system $CaO-Al_2O_3-SiO_2-H_2O$

Boettcher (1970) has determined the equilibrium *P* and *T* of several reactions in this four-component system (Figure 11.14). The mineralogic phases with which we are concerned are: anorthite ($CaAl_2Si_2O_8$), corundum (Al_2O_3), gehlenite ($Ca_2Al_2SiO_7$), grossularite ($Ca_3Al_2Si_3O_{12}$), kyanite or sillimanite (Al_2SiO_5), quartz (SiO_2), zoisite [$Ca_2Al_3Si_3O_{12}(OH)$], and liquid or vapor. This system is, therefore, a four-component, nine-phase system with several possible invariant and univariant reactions. Although Boettcher (1970) has given a list of 46 reactions, at metamorphic temperatures and pressures it may suffice to consider the following important reactions:

(*1*) zoisite $\rightleftharpoons$
anorthite + grossularite + corundum + vapor

(*2*) zoisite + quartz $\rightleftharpoons$
anorthite + grossularite + vapor

(*3*) grossularite + quartz $\rightleftharpoons$
anorthite + wollastonite

(*4*) zoisite + kyanite + quartz $\rightleftharpoons$
anorthite + vapor

(*5*) grossularite + corundum $\rightleftharpoons$
anorthite + gehlenite

(*6*) grossularite $\rightleftharpoons$
anorthite + gehlenite + wollastonite

(*8*) grossularite + kyanite + quartz $\rightleftharpoons$
anorthite

For low-temperature reactions see Liou (1970) and Thompson (1070a, b) and Chapter 10.

Three of these reactions involve the disappearance of zoisite from the assemblages. At 4 kbar, zoisite alone breaks down to grossularite, anorthite, corundum, and vapor at about 700°C. With quartz, the reaction temperature at 4 kbar is reduced to 595°C; finally, with kyanite and quartz it may further be reduced to below 500°C. If some Fe^{3+} is present in zoisite, the reaction temperature may be lower by as much as 50°C. Boettcher's results also show that anorthite, wollastonite, and gehlenite are stable throughout the *P*–*T* field up to the liquidus, but grossularite breaks down to anorthite, wollastonite, and gehlenite at rather high temperatures. At 3 kbar this temperature is above 900°C. In the presence of corundum, the reaction temperature at 3 kbar is about 850°C; in the presence of quartz it is reduced to 635°C. At 3 kbar, grossularite forms by reaction (*2*) at 550°C. In presence of quartz, the width of the stability field of grossularite in the system is about 100°C at 3 kbar.

Assemblages with cordierite in the granulite facies

The stability field of cordierite extends beyond the possible high-temperature limit of the division of granulite facies. On the high-pressure side, the stability field is bounded by the reaction

$$\text{cordierite } (\pm \text{ garnet}) \rightleftharpoons \text{orthopyroxene} + \text{kyanite} + \text{quartz}$$

This breakdown has not been established for hydrous cordierite end members. According to Newton (1972) anhydrous Mg-cordierite breaks down to enstatite + sillimanite + quartz between 6 and 7 kbar at ~600°C (Figure 11.15). Hydrous cordierite may be stable up to ~12 kbar. Figure 11.15 shows some other univariant equilibria involving cordierite. The low-temperature equilibria are plotted from Richardson (1968) and refer to Fe-cordierite. The high-temperature equilibria have been investigated by Hensen and Green (1973) at low water fugacity. Their results indicate that the assemblage garnet, cordierite, and sillimanite may be stable up to very high temperatures (~1050°C). This temperature, however, is the maximum limit, and the granulite facies

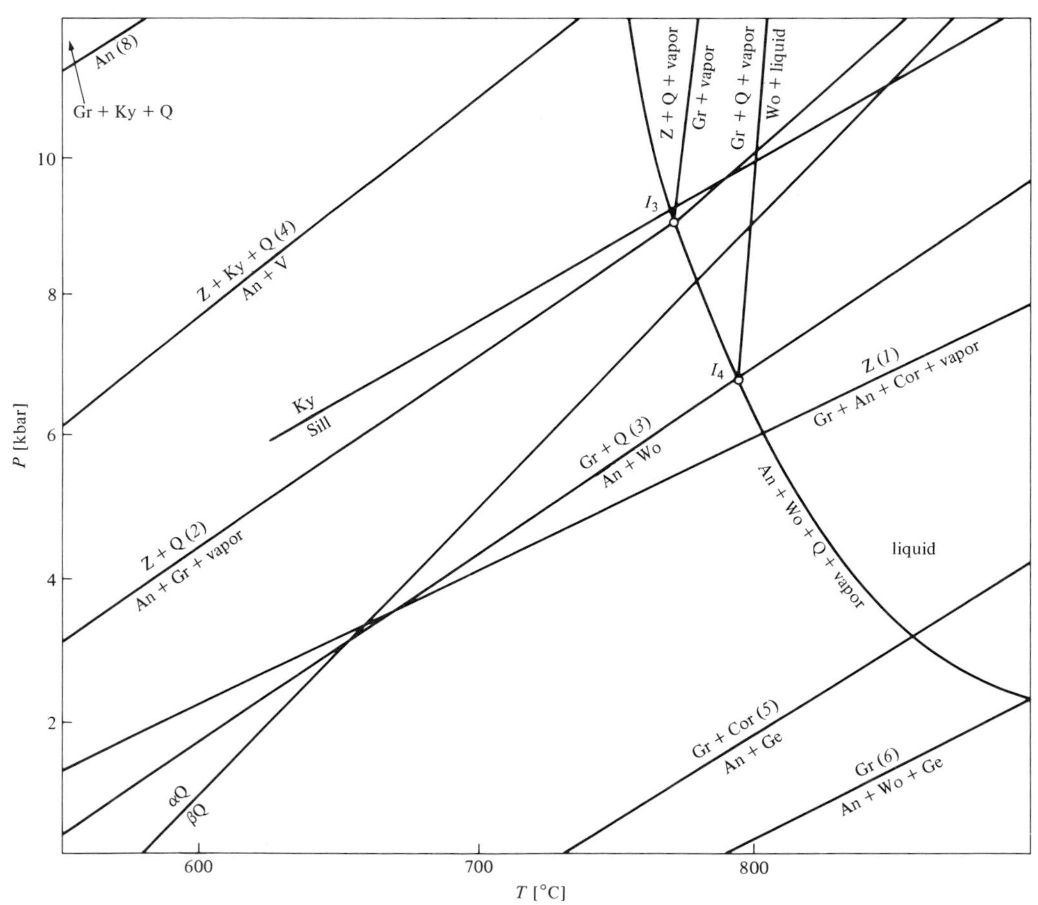

11.14 Some of the reactions in the system CaO-Al_2O_3-SiO_2-H_2O after Boettcher (1970). The reaction numbers (see text) are the same as given by Boettcher except that Reaction (8) is from Hays (1967) and Hariya and Kennedy (1968). I_3 and I_4 are invariant points.

rocks in nature probably never attain this high temperature. The reactions studied by Hensen and Green (1973) in sequence from high pressure and low temperature to high temperature and low pressure are

$$\text{cordierite} + \text{garnet} \rightleftharpoons \text{hypersthene} + \text{sillimanite} + \text{quartz}$$

(see also Grant's figure for this reaction, Figure 11.12)

$$\text{cordierite} + \text{garnet} \rightleftharpoons \text{hypersthene} + \text{sapphirine} + \text{quartz}$$

$$\text{cordierite} + \text{garnet} \rightleftharpoons \text{hypersthene} + \text{spinel} + \text{quartz}$$

$$\text{cordierite} + \text{garnet} \rightleftharpoons \text{olivine} + \text{spinel} + \text{quartz}$$

The stability field for cordierite + garnet, therefore, is bound by these reactions. The highest temperature for this field is ~1140°C, which is probably too high. The experiments involving cordierites suffer from the lack of information on the content of water in the mineral. The water content in cordierite is not fixed and varies with P_{H_2O}. An approximate calculation of water in cordierite at various P_{H_2O} has been carried out by Weisbrod (1973).

Other problems involving the use of the garnet-cordierite assemblage in petrogenesis are the uncertain thermochemical quantities and the slope of the reaction

$$\text{cordierite} \rightleftharpoons \text{almandine} + \text{quartz} + \text{sillimanite}$$

(Froese, 1973)

Until more thermochemical and experimental information are available, we should use this mineral with caution in petrogenetic studies.

The charnockite assemblage

The "charnockite" system may consist of one or two pyroxenes and some or all of the following minerals: hornblende, biotite, orthoclase, garnet, plagioclase, quartz, and other accessory minerals such as ilmenite,

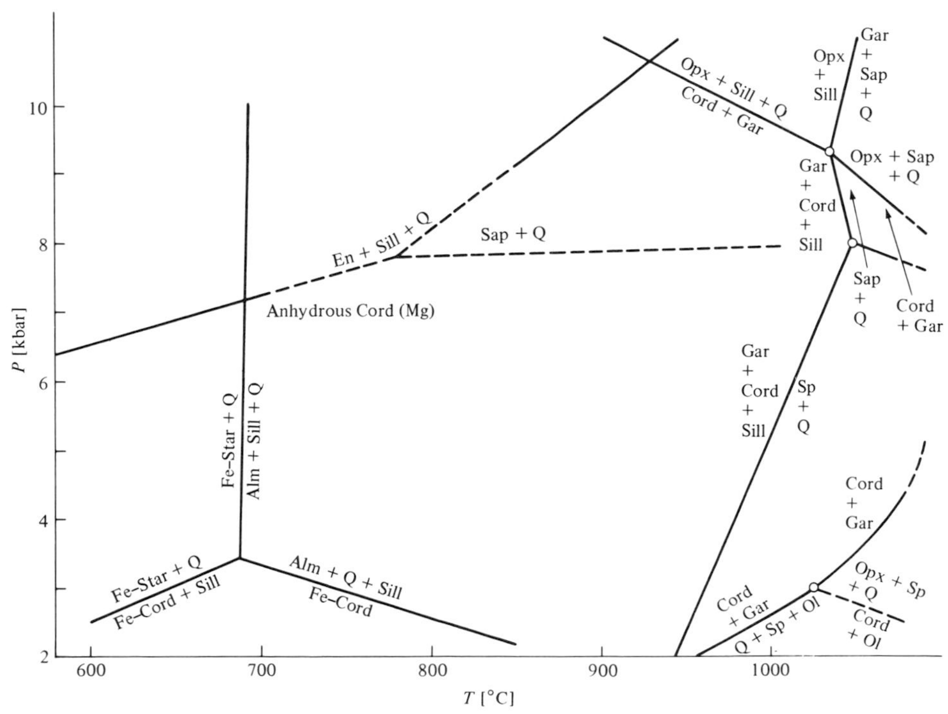

11.15 Some equilibria involving cordierite. Fe system (Richardson, 1968; $P_{total} = P_{H_2O}$); anhydrous cordierite (Newton, 1972); $P_{H_2O} < P_{total}$ system above 900°C (Hensen and Green, 1973).

magnetite, and zircon. The components of this system are

SiO_2-Al_2O_3-FeO-MgO-CaO-K_2O-Na_2O-TiO_2-H_2O.

TiO_2 is significantly high in biotite and is considered to be essential for the stability of the mineral (Kretz, 1964). Na_2O is present in hornblende and in plagioclase. In the former its concentration does not vary significantly. The mineral-chemical data from Varberg (Saxena, 1968) indicated that we may simplify the system by assuming that Na and Ca in hornblende, Ti in biotite, and Ca in garnet are more or less fixed in concentration and the major variation is in the Fe^{2+}-Mg concentration in the minerals.

The scheme of distribution of Fe and Mg in minerals of Varberg samples is

$$X_{Fe}^{Gar} > X_{Fe}^{Hb} > X_{Fe}^{Bi} > X_{Fe}^{Opx} > X_{Fe}^{Cpx}$$

where X is Fe/(Fe + Mg). Assuming that the minerals are ideal solutions, equilibrium constants for mineral pairs lie in the range

		Average
$K_{(Bi-Hb)}$	0.789–0.996	0.90
$K_{(Bi-Cpx)}$	1.324–1.707	1.49
$K_{(Bi-Opx)}$	0.759–1.055	0.90
$K_{(Bi-Gar)}$	0.230–0.338	0.29

The ion-exchange reaction is written as

$$\text{Mg Bi} + \text{Fe}\,\alpha \rightleftharpoons \text{Fe Bi} + \text{Mg}\,\alpha$$

where α is the coexisting phase. K for other pairs may be found by suitable divisions, for example

$$K_{(Bi-Opx)}/K_{(Bi-Cpx)} = K_{(Cpx-Opx)} = 0.60$$

Other compositional averages in Varberg minerals that may be specified are

$$X_{Ti}^{Bi} = 0.5\left(\frac{\text{ions}}{22\text{O}}\right)$$

$$X_{Ca}^{Gar}\left(\frac{\text{Ca}}{\text{Ca} + \text{Fe} + \text{Mg} + \text{Mn}}\right) = 0.28$$

$$X_{An} = 0.26$$

Simplifying the system to eight minerals ($+H_2O$) and seven components (K_2O-FeO-MgO-Al_2O_3-CaO-SiO_2-H_2O), we may consider an invariant assemblage and the several univariant reactions. One could write balanced reactions using the specified variation in Fe:Mg and Ca:Na and consider the slopes of the reactions. Since the molar volume and entropy data are not available for some of the minerals, no attempt will be made here to study the system quantitatively. We may consider the following reactions:

$$\text{biotite} + \text{plagioclase} + \text{orthopyroxene} + 2\text{quartz} \rightleftharpoons \text{clinopyroxene} + \text{garnet} + \text{K-feldspar} + H_2O \quad \text{(Hb)}$$

$$\text{biotite} + \text{garnet} + \text{hornblende} + 4\text{quartz} \rightleftharpoons \text{K-feldspar} + 2\text{plagioclase} + 10\text{orthopyroxene} + 2H_2O \quad \text{(Cpx)}$$

$$3\text{hornblende} + \text{garnet} + \text{quartz} \rightleftharpoons 4\text{plagioclase} + 2\text{clinopyroxene} + 13\text{orthopyroxene} + 3H_2O \quad \text{(Bi, K-fs)}$$

$$11\,\text{biotite} + 9\text{plagioclase} + 37\text{quartz} \rightleftharpoons 3\text{hornblende} + 11\,\text{K-feldspar} + 6\text{garnet} + 3\text{clinopyroxene} + 8H_2O \quad \text{(Opx)}$$

$$\text{biotite} + \text{hornblende} + 3\text{quartz} \rightleftharpoons \text{K-feldspar} + \text{clinopyroxene} + 6\text{orthopyroxene} + \text{plagioclase} + 2H_2O \quad \text{(Gar)}$$

where reactions are balanced with the compositions:

biotite-K(Fe, Mg)$_3$AlSi$_3$O$_{10}$(OH)$_2$

garnet-(Fe, Mg)$_3$Al$_2$Si$_3$O$_{12}$

clinopyroxene-(Ca, Fe, Mg)Si$_2$O$_6$

orthopyroxene-(Fe, Mg)SiO$_3$

plagioclase-CaAl$_2$Si$_2$O$_8$

K-feldspar-KAlSi$_3$O$_8$

hornblende-Ca$_2$(Fe, Mg)$_4$Al$_2$Si$_7$O$_{22}$(OH)$_2$

A qualitative scheme of the reactions is shown in Figure 11.16 using the constraint that water is released on the high-temperature side and at pressures above the invariant point garnet is on the high-pressure side as against plagioclase. This arrangement of univariant equilibria is also consistent with the fact that we have a large P–T field on the low-temperature side where the very commonly occurring assemblage biotite-garnet-hornblende-plagioclase and quartz is stable. Similarly we have a large P–T field where the anhydrous assemblage garnet-orthoclase-clinopyroxene-orthopyroxene-plagioclase may be stable. The position of the "invariant" points is uncertain in the P–T field and is a function of P, T, μ_{H_2O}, and the composition of the various phases. At 5 kbar pressure the point may lie above 750°C above the Ab-Or critical line (see Figure 10.1).

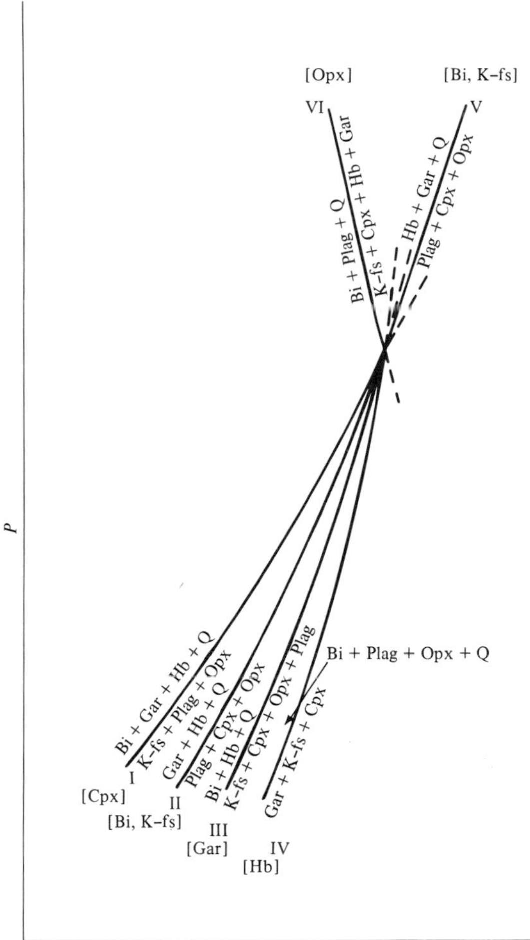

11.16 Schematic representation of the charnockite phase equilibria. Roman numerals represent assemblages discussed in the text.

The divariant assemblages possible in the basic to intermediate compositions are as follows (see Figure 11.16):

I. a. biotite, garnet, hornblende, K-feldspar, plagioclase, orthopyroxene
   b. biotite, quartz, hornblende, K-feldspar, plagioclase, orthopyroxene
   c. biotite, garnet, quartz, K-feldspar, plagioclase, orthopyroxene
   d. hornblende, garnet, quartz, K-feldspar, plagioclase, orthopyroxene

II. a. garnet, hornblende, plagioclase, clinopyroxene, orthopyroxene
   b. hornblende, quartz, plagioclase, clinopyroxene, orthopyroxene, biotite
   c. garnet, quartz, plagioclase, clinopyroxene, orthopyroxene, biotite
   d. plagioclase, clinopyroxene, orthopyroxene, biotite, hornblende, quartz

III. a. biotite, hornblende, K-feldspar, clinopyroxene, orthopyroxene, plagioclase
   b. quartz, hornblende, K-feldspar, clinopyroxene, orthopyroxene, plagioclase
   c. biotite, quartz, K-feldspar, clinopyroxene, orthopyroxene, plagioclase

IV. a. biotite, plagioclase, orthopyroxene, garnet, K-feldspar, clinopyroxene
   b. biotite, orthopyroxene, quartz, garnet, K-feldspar, clinopyroxene
   c. plagioclase, orthopyroxene, quartz, garnet, K-feldspar, clinopyroxene
   d. plagioclase, clinopyroxene, orthopyroxene, garnet, K-feldspar, quartz
   e. hornblende, garnet, plagioclase, clinopyroxene, orthopyroxene, biotite
   f. hornblende, quartz, plagioclase, clinopyroxene, orthopyroxene
   g. garnet, quartz, plagioclase, clinopyroxene, orthopyroxene, K-feldspar

V. a. biotite, plagioclase, K-feldspar, clinopyroxene, hornblende, garnet
   b. biotite, quartz, K-feldspar, clinopyroxene, hornblende, garnet
   c. plagioclase, quartz, K-feldspar, clinopyroxene, hornblende, garnet
   d. quartz, K-feldspar, clinopyroxene, hornblende, garnet

VI. a. biotite, garnet, hornblende, quartz, plagioclase

The seven-, eight-, or nine-phase assemblages found in charnockites of Varberg and elsewhere (Leelanandam, 1967) are obviously the result of an increase in variance in the system due to solid solution of components such as Ti, Na, and Ca.

Magnesian, calcic, and carbonate rocks

This section deals with phase equilibria where CO_2 is involved as a fluid phase and the carbonates appear in the assemblages. We have already referred to the system

Fe-C-O experimentally investigated by French (1971) and Weidner (1973) in Chapter 9.

The system CaO-Al_2O_3-SiO_2-CO_2-H_2O

The carbonate-free system $CaAl_2Si_2O_8$-SiO_2-H_2O as described by Liou (1971) was discussed in Chapters 9 and 10. Rocks of the zeolite facies generally appear between the greenschist facies and the unmetamorphosed sediments. However, in some carbonate-rich sediments, the zeolites may be entirely eliminated by reactions of the following type:

$$\underset{\textit{laumontite}}{CaAl_2Si_4O_{12}\cdot 4H_2O} + \underset{\textit{fluid}}{CO_2} \rightleftharpoons$$

$$\underset{\textit{calcite}}{CaCO_3} + \underset{\textit{kaolinite}}{Al_2Si_2O_5(OH)_4} + \underset{\textit{quartz}}{2SiO_2} + \underset{\textit{fluid}}{2H_2O} \quad (11.o)$$

It is, therefore, important to study the relation between zeolite and carbonate assemblages. We shall follow the theoretical analysis of Thompson (1971b) and present the results of experiments and calculation on the following reaction of interest:

(*1*) 1 laumontite + 1 CO_2 $\rightleftharpoons$ 1 calcite + 1 kaolinite + 2 quartz + 2 H_2O

(*2*) 2 laumontite $\rightleftharpoons$ 1 prehnite + kaolinite + 3 quartz + 5 H_2O

(*3*) 1 laumontite + 1 calcite $\rightleftharpoons$ 1 prehnite + 1 quartz + 3 H_2O + 1 CO_2

(*4*) 1 prehnite + 1 kaolinite + 1 quartz $\rightleftharpoons$ 1 laumontite + 1 H_2O

(*5*) 1 calcite + 1 quartz $\rightleftharpoons$ 1 wollastonite + 1 CO_2
(Harker and Tuttle, 1956; Greenwood, 1967b)

(*6*) 1 prehnite $\rightleftharpoons$ 1 anorthite + 1 wollastonite + 1 H_2O
(Coombs *et al.*, 1959)

(*7*) 1 calcite + 1 quartz + 1 anorthite + H_2O $\rightleftharpoons$ 1 prehnite + 1 CO_2

(*8*) 5 prehnite $\rightleftharpoons$ 2 zoisite + 2 grossularite + 3 quartz + 4 H_2O

(*9*) 1 grossularite + 1 quartz $\rightleftharpoons$ 2 wollastonite + 1 anorthite
(Newton, 1966a; Storre, 1970)

(*10*) 1 anorthite + 2 calcite + quartz $\rightleftharpoons$ 1 grossularite + 1 CO_2

(*11*) 4 zoisite + quartz $\rightleftharpoons$ 1 grossularite + 5 anorthite + 2 H_2O
(Newton, 1966a)

(*12*) 2 zoisite + 5 calcite + 3 quartz $\rightleftharpoons$ 3 grossularite + 5 CO_2 + 1 H_2O

(*13*) 2 zoisite + 2 calcite + 3 quartz + 2 H_2O $\rightleftharpoons$ 3 prehnite + 2 CO_2

(*14*) 2 zoisite + 1 quartz $\rightleftharpoons$ 1 prehnite + 2 anorthite

(*15*) 1 laumontite $\rightleftharpoons$ 1 anorthite + 2 quartz + 4 H_2O
(Thompson, 1970b)

(*16*) 1 wairakite $\rightleftharpoons$ 1 anorthite + 2 quartz + 2 H_2O
(Thompson, 1970b)

(*17*) 1 wairakite + 1 CO_2 $\rightleftharpoons$ 1 calcite + 1 kaolinite + 1 quartz

(*18*) 1 heulandite $\rightleftharpoons$ 1 laumontite + (2 or 3) quartz + 2 H_2O
(Thompson, 1971b)

(*19*) 1 calcite + 1 andalusite + 1 quartz $\rightleftharpoons$ 1 anorthite + 1 CO_2

(*20*) 1 pyrophyllite $\rightleftharpoons$ 1 andalusite + 3 quartz + 1 H_2O
(Kerrick, 1968)

(*21*) 1 kaolinite + 2 quartz $\rightleftharpoons$ 1 pyrophyllite + 1 H_2O
(Thompson, 1970a)

(*22*) 1 pyrophyllite + 1 calcite $\rightleftharpoons$ 1 anorthite + 2 quartz + 1 H_2O + 1 CO_2

(*23*) 2 calcite + 1 pyrophyllite $\rightleftharpoons$ 1 prehnite + 1 quartz + 2 CO_2

(*24*) 2 calcite + 1 andalusite + 2 quartz + 1 H_2O $\rightleftharpoons$ 1 prehnite + 2 CO_2

(*25*) 1 laumontite $\rightleftharpoons$ 1 lawsonite + 2 quartz + 2 H_2O

(*26*) 1 lawsonite $\rightleftharpoons$ 1 anorthite + 2 H_2O

(*27*) 1 calcite + 1 andalusite + 1 quartz + 2 H_2O $\rightleftharpoons$ 1 lawsonite + 1 CO_2

(*28*) 1 pyrophillite + 1 calcite + 1 H_2O $\rightleftharpoons$ 1 lawsonite + 2 quartz + 1 CO_2

(*29*) 1 lawsonite + 1 aragonite + 1 quartz $\rightleftharpoons$ 1 prehnite + 1 H_2O + 1 CO_2

(*30*) 3 lawsonite + 1 calcite $\rightleftharpoons$ 2 zoisite + 5 H_2O + CO_2

(*31*) 3 anorthite + 1 calcite + 1 H_2O $\rightleftharpoons$ 2 zoisite + 1 CO_2

(*32*) 1 lawsonite + 5 anorthite + 2 calcite $\rightleftharpoons$ 4 zoisite + 2 CO_2

(*33*) 12 lawsonite $\rightleftharpoons$ 6 zoisite + 3 kyanite (Sill) + 3 quartz + 21 H_2O

(Newton and Kennedy, 1963)

The references are for the experimental data. Many of these equilibria are shown in Figures 11.17 and 11.18. The zeolite minerals are restricted to fluids extremely low in CO_2 and do not appear in Figure 11.18.

Thompson's calculations show that zeolites, in reactions such as (*1*) and (*3*), are in equilibrium with an H_2O-CO_2 fluid phase having $X_{CO_2} \simeq 0.0075$ for $P_{fluid} = P_{total} = 2000$ bar. In an equilibrium assemblage of laumontite, calcite, kaolinite, and quartz to change the condition to a maximum production of laumontite, X_{CO_2} must be much lower than 0.007. At shallow to moderate depths, the fluid pressure would be generally lower than the experimental fluid pressure of 2000 bar. This will reduce the equilibrium X_{CO_2} to even lower values than 0.007. Wairakite may be more stable under moderate P–T conditions because Reaction (*16*)

wairakite $\rightleftharpoons$ anorthite + quartz + vapor

(see Figures 10.1 and 11.18) occurs at higher temperatures than the laumontite Reaction (*15*) in pure water, and in H_2O-CO_2 fluid the equilibrium X_{CO_2} at 2 kbar is

11.17 P_{H_2O}–T diagram showing experimentally determined univariant curves of equilibria in the system CaO-Al_2O_3-SiO_2-H_2O. The numbered curves refer to the reactions listed in the text. Schematic heavy broken line X [Reaction (4)] and band Y [Reaction (2)] indicate the possible limiting reactions for prehnite relative to laumontite in this system. (After A. B. Thompson, 1971b.)

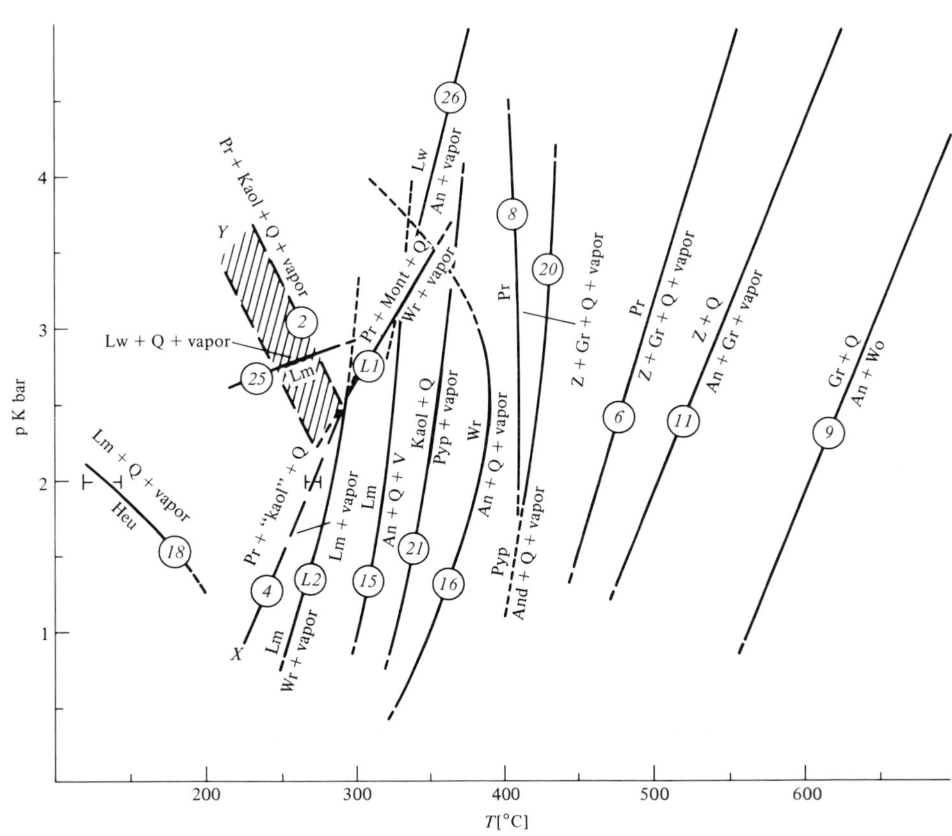

11.18 Calculated (solid) and schematic (dashed) curves of reactions in CaO-Al_2O_3-SiO_2-CO_2-H_2O, for $P_{total} = P_{fluid} = 2000$ bar. The numbered curves refer to reactions listed in the text. Since calcite is considered to be the stable polymorph of $CaCO_3$ in this region, aragonite reactions are not shown. Some univariant curves at schematic "invariant" points have been omitted for clarity. The following "invariant" points are shown: *D*: Gr-An-Wo-Q-Calc-(CO_2-H_2O); *E*: Gr-An-Z-Q-Calc-(CO_2-H_2O); *H*: Pyp-An-And-Q-Calc-(CO_2-H_2O); *J*: Pyp-And-Pr-Q-Calc-(CO_2-H_2O); *K*: Lw-An-Z-Q-Calc-(CO_2-H_2O); *L*; Pyo-An-Lw-Q-Calc-(CO_2-H_2O); *M*: Lw-An-And-Q-Calc-(CO_2-H_2O). The shape and position of curve (*23*) is uncertain. All of the reactions shown are considered to occur in the CO_2-H_2O single-phase region. (After A. B. Thompson, 1971b.)

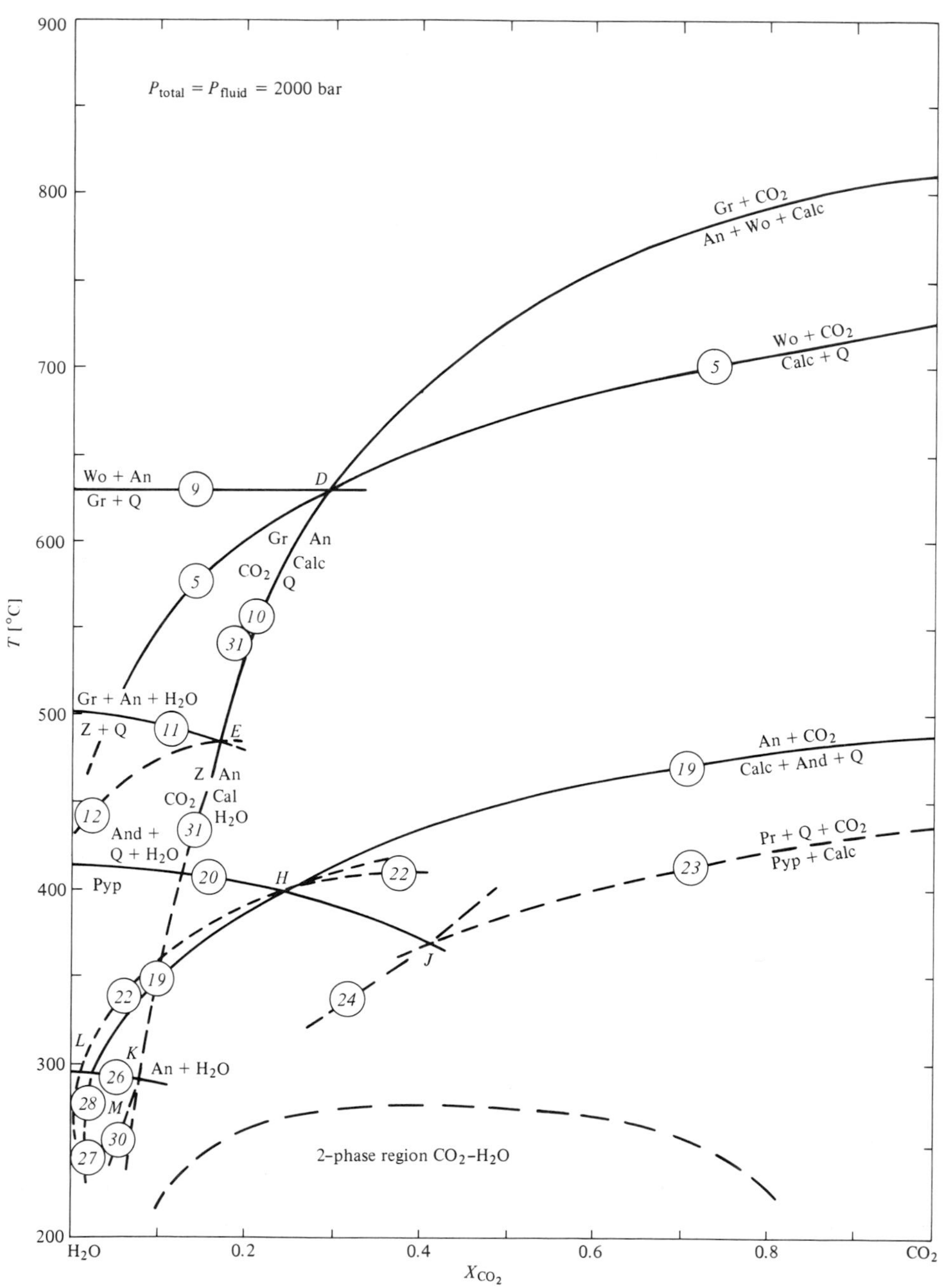

somewhat higher (0.009). Reaction (*17*) will similarly be at higher X_{CO_2} than that of Reaction (*1*). Heulandite may occur at temperatures as low as 120°C at $P_{H_2O} = 2$ kbar [Figure 11.17, Reaction (*18*)].

It is obvious from the conditions of the various zeolite reactions that the situation would be different in open and closed systems. All phases laumontite, calcite, prehnite, and quartz of Reaction (*3*) could occur only as a univariant equilibrium but if CO_2 continues to escape we may have the assemblages laumontite-prehnite-quartz or quartz-prehnite-calcite, which are commonly found. Similarly in volcanic rocks or derived sediments, zeolite-calcite may crystallize if X_{CO_2} is low. However in hyrothermal and vein deposits, where the minerals are precipitated from solutions, calcite and zeolite do not form an equilibrium assemblage.

The effect of an increasing CO_2 content of the fluid on some reactions involving only water (Figures 11.17 and 11.18) is to decrease the equilibrium temperature by the pressure effect on the solids. These reactions are

(*11*) 4 zoisite + 1 quartz $\rightleftharpoons$ 1 grossularite + 5 anorthite + $2H_2O$

(*20*) 1 pyrophyllite $\rightleftharpoons$ 1 andalusite + 3 quartz + $1H_2O$

(*26*) 1 lawsonite $\rightleftharpoons$ anorthite + $2H_2O$

Similarly equilibrium temperatures of reactions which proceed with evolution of CO_2 may decrease by several degrees with increasing X_{H_2O}. For example, the equilibrium temperature at 2000 bar for the reaction

calcite + quartz $\rightleftharpoons$ wollastonite + CO_2

decreases from 730°C at $X_{H_2O} = 0$, to 630°C at $X_{H_2O} = 0.7$. This effect was discussed in Chapter 2.

Gordon and Greenwood (1971) have considered this system at somewhat higher temperatures. They investigated the reactions:

calcite + anorthite + wollastonite $\rightleftharpoons$ grossularite + CO_2

calcite + anorthite + quartz $\rightleftharpoons$ grossularite + CO_2

Their experimental results and calculated equilibria are shown in Figure 11.19. Note that the "invariant" point D in Figure 11.19 and the equilibrium

wollastonite + anorthite $\rightleftharpoons$ grossularite + quartz

lies below 600°C in this figure. The figure also does not include such phases as portlandite, spurrite, and calciochondrodite, which occur on the water-rich side (Wyllie and Haas, 1966). The reactions delineating the field of grossularite are, from high to low temperatures:

grossularite + CO_2 $\rightleftharpoons$ calcite + anorthite + wollastonite

grossularite + CO_2 $\rightleftharpoons$ calcite + anorthite + quartz

grossularite + CO_2 $\rightleftharpoons$ calcite + zoisite + quartz

All these equilibria are highly dependent on X_{CO_2}. Zoisite is restricted to temperatures below 500°C at 2000 bar in fluid with more than 10 mole percent CO_2.

The system $MgO\text{-}SiO_2\text{-}H_2O\text{-}CO_2$

Greenwood (1967b) and Johannes (1969) have studied the reactions in this system. The following phases are of interest. Anthophyllite $[Mg_7Si_8O_{22}(OH)_2]$, brucite $[Mg(OH)_2]$, enstatite $(Mg_2Si_2O_6)$, forsterite (Mg_2SiO_4), magnesite $(MgCO_3)$, periclase (MgO), quartz (SiO_2), serpentine $[Mg_3Si_2O_5(OH)_4]$ and talc $[Mg_3Si_4O_{10}(OH)_2]$. The reactions to be considered are

(*1*) 2 forsterite + $2H_2O$ + $1CO_2$ $\rightleftharpoons$ 1 serpentine + 1 magnesite

(*2*) 4 forsterite + $1H_2O$ + $5CO_2$ $\rightleftharpoons$ 1 talc + 5 magnesite

(*3*) 2 serpentine + $3CO_2$ $\rightleftharpoons$ 1 talc + 3 magnesite + $3H_2O$

(*4*) 1 talc + $3CO_2$ $\rightleftharpoons$ 4 quartz + 3 magnesite + $1H_2O$

(*5*) 1 serpentine + $3CO_2$ $\rightleftharpoons$ 2 quartz + 3 magnesite + $2H_2O$

(*6*) 1 serpentine + 1 brucite $\rightleftharpoons$ 2 forsterite + $3H_2O$

(*7*) 5 serpentine $\rightleftharpoons$ 6 forsterite + 1 talc + $9H_2O$

(*8*) 1 serpentine + 2 quartz $\rightleftharpoons$ 1 talc + H_2O

(*9*) 1 anthophyllite + $4H_2O$ $\rightleftharpoons$ 9 talc + 4 forsterite

(*10*) 2 anthophyllite + 2 forsterite $\rightleftharpoons$ 9 enstatite + $2H_2O$

(*11*) 7 talc $\rightleftharpoons$ 3 anthophyllite + 4 quartz + $4H_2O$

(*12*) 2 anthophyllite $\rightleftharpoons$ 7 enstatite + 2 quartz + $2H_2O$

(*13*) 1 anthophyllite + 9 magnesite $\rightleftharpoons$ 8 forsterite + $1H_2O$ + $9CO_2$

11.19 Experimental results and calculated equilibria for a total pressure of 2000 bar. (After Gordon and Greenwood, 1971.)

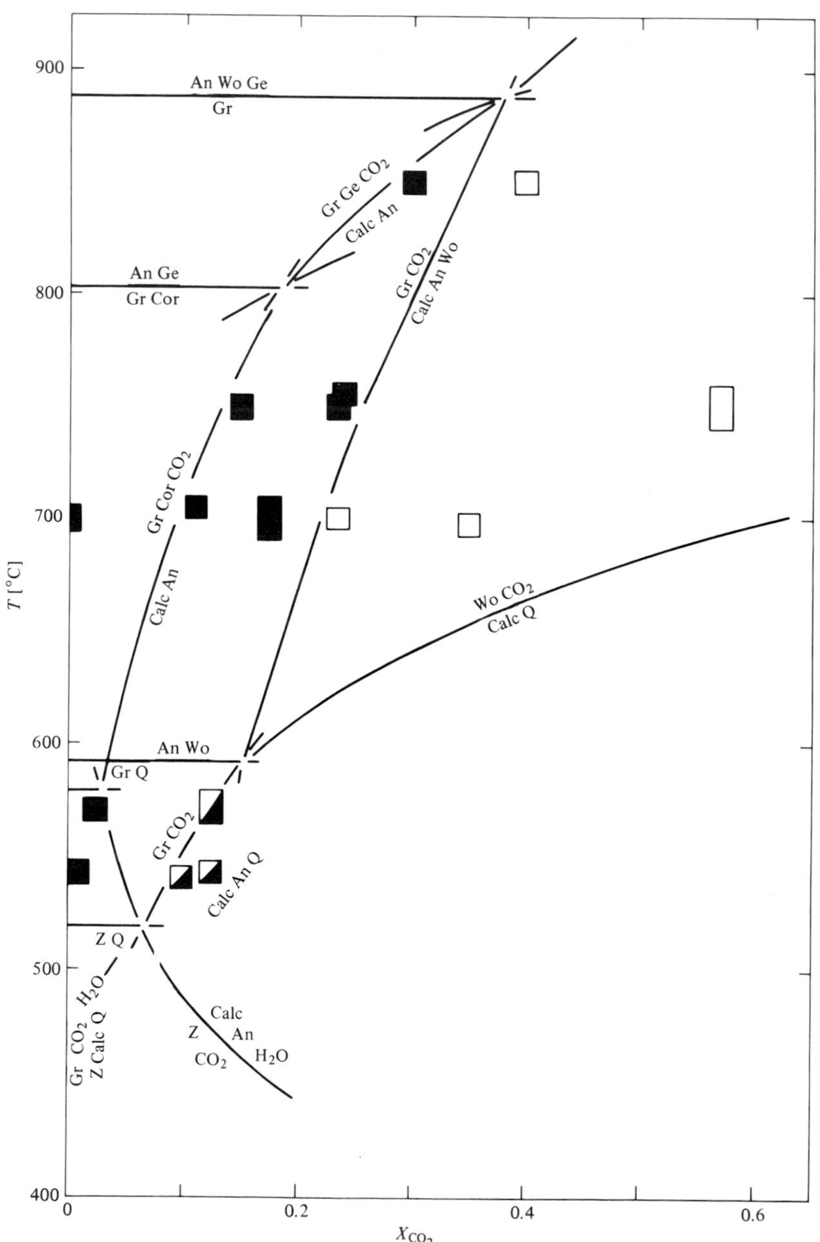

(*14*) 1 anthophyllite + 1 H_2O + 1 CO_2 $\rightleftharpoons$ 2 talc + 1 magnesite

(*15*) 1 anthophyllite + 7 CO_2 $\rightleftharpoons$ 8 quartz + 7 magnesite + 1 H_2O

(*16*) 4 enstatite + 1 H_2O + 1 CO_2 $\rightleftharpoons$ 1 anthophyllite + 1 magnesite

(*17*) 2 forsterite + 2 CO_2 $\rightleftharpoons$ 1 enstatite + 2 magnesite

(*18*) 2 quartz + 2 magnesite $\rightleftharpoons$ 1 enstatite + 2 CO_2

Figures 11.20 and 11.21 show the equilibrium relations in these equilibria as described by Johannes (1969).

11.20 Position of equilibrium curves (1) to (18) in the T–X_{CO_2} field at P_{fluid} = 2000 bar. The solid lines have been experimentally determined; the broken lines have been derived from intersection-point relationships of the equilibrium curves. Those mineral parageneses have been presented that were produced in the individual regions of the T–X_{CO_2} diagram from an MgO:SiO_2 ratio = 2:1 (forsterite composition). The names of typical and especially well-known magnesite occurrences have also been indicated. At pressures greater than 2000 bar the equilibrium curves move to higher temperatures; at lower pressures they move to lower temperatures. (After Johannes, 1969. Reprinted from the *American Journal of Science* by permission.) A = anthophyllite.

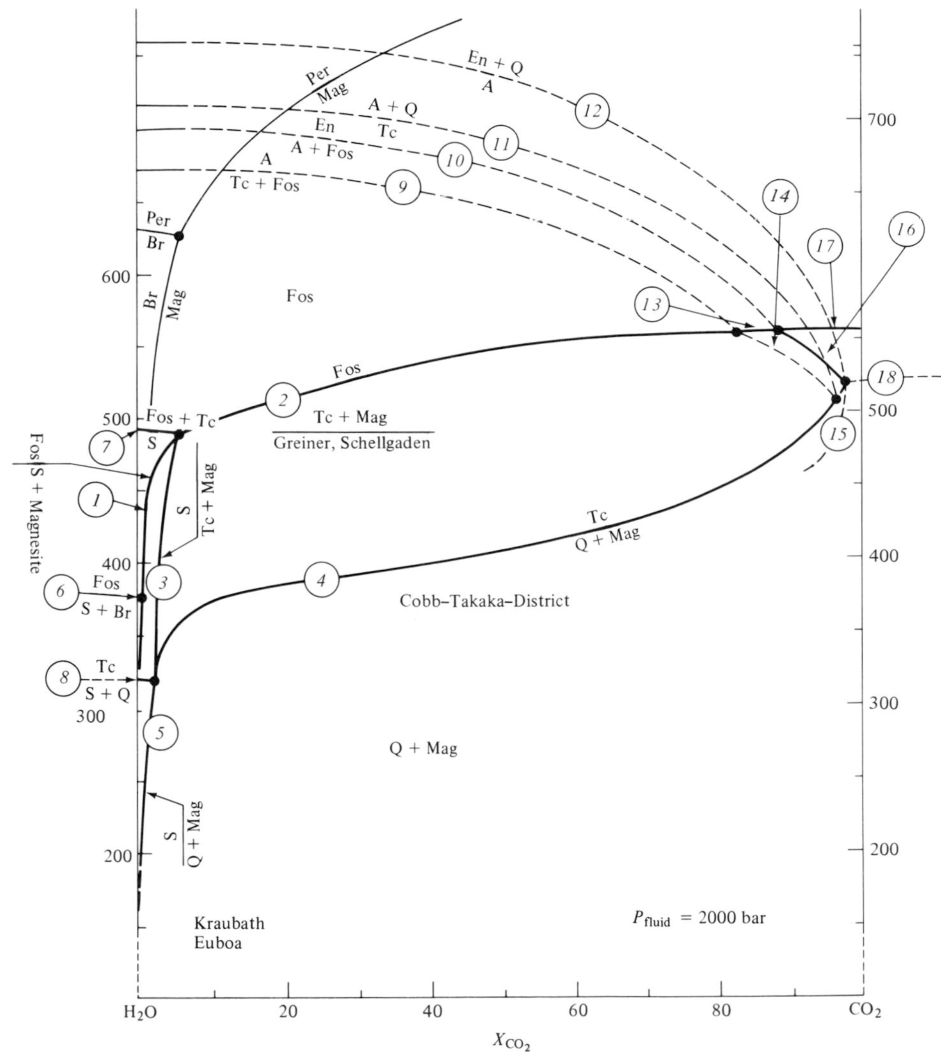

Curves (*3*), (*5*), and (*7*) are the boundaries of the stability field of serpentine. Serpentine and magnesite can form at 2000 bar of fluid pressure between 380°C to 490°C. Serpentinization of olivine may take place at lower temperature according to Reaction (*6*):

$$2\,\text{forsterite} + 3\,H_2O \rightleftharpoons 1\,\text{serpentine} + 1\,\text{brucite}$$

Curves (*3*) and (*5*) show that serpentine can coexist only with a CO_2-poor fluid phase. This has important implications for the origin of certain meteorites, which contain serpentine like phases (Chapter 6).

11.21 Phase relationships in the T–X_{CO_2} field in the region of high-CO_2 contents of the fluid phase at $P_{\text{fluid}} = 2000$ bar. The heavily delineated curves (*2*), (*4*), (*16*), and (*17*) have been experimentally determined; the position of the remaining curves have been estimated. Horizontally ruled field: stability region of the assemblage enstatite + magnesite; perpendicularly ruled field: stability region of the assemblage Mg-anthophyllite + magnesite. (After Johannes, 1969. Reprinted from the *American Journal of Science* by permission.) In this figure A = anthophyllite, Mt = magnesite.

At low temperature quartz + magnesite have their upper stability limit defined by equilibrium curves (*5*), (*4*), (*15*), and (*18*). Above this limit talc + magnesite have a broad field of coexistence. Their upper stability limit mostly matches with the lower stability limit of forsterite which is defined by equilibrium Reactions (*1*), (*2*), (*6*), (*13*), and (*17*). Curves (*8*) and (*4*) specify the lowest temperatures (350°C to ~500°C) at which talc may occur in the system. The equilibrium pressure for Reaction (*8*) may increase somewhat with some iron in talc. Curve (*4*) may be used to estimate the temperature at which quartz and magnesite will react to produce talc. The stability fields of the assemblage anthophyllite + magnesite and enstatite + magnesite are bounded by equilibrium curves (*13*), (*14*), (*15*), and (*16*), and (*16*), (*17*), and (*18*), respectively. These equilibria lie near the CO_2 end of the H_2O-CO_2 join and, therefore, such assemblages indicate an extremely high CO_2 in the fluid phase.

Evans and Trommsdorff (1974) have described metaperidotites from Val d'Efra, Lepontine Alps, consisting of magnesite-bearing chlorite-talc-forsterite schist and a variety of sagvandite with the assemblage chlorite-forsterite-magnesite-talc and enstatite. The textural study, an orderly partitioning of Fe-Mg in coexisting

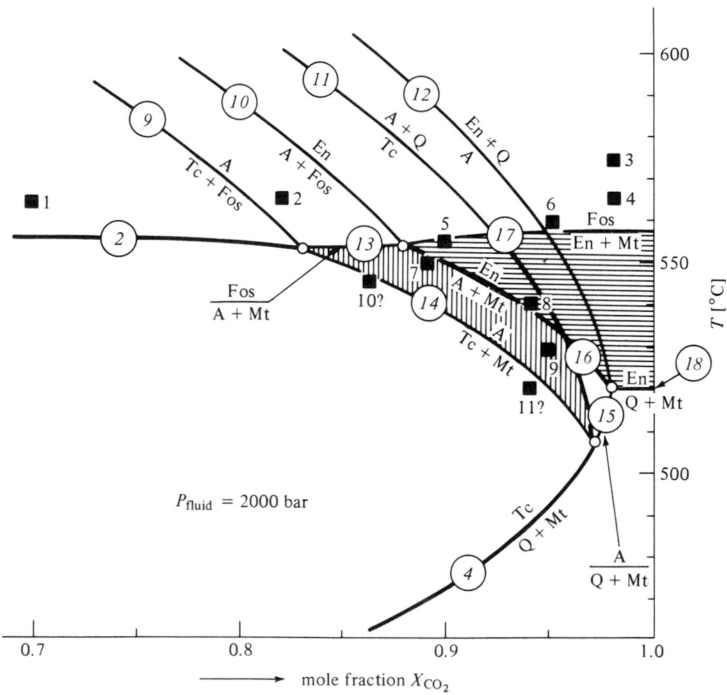

enstatite and talc, and a Nernst-type distribution of Ni in forsterite, enstatite, talc, and chlorite indicate a close approach to chemical equilibrium. The thermochemical data are at present not sufficiently accurate to define the equilibrium P–T for the reaction

$$\text{anthophyllite} \rightleftharpoons \text{talc} + \text{enstatite}$$

On the basis of the associated pelitic assemblages and the general conditions of alpine metamorphism, Evans and Trommsdorff postulate that the equilibrium P and T for

$$\text{enstatite } (X_{Mg} = 0.9) + \text{talc } (X_{Mg} = 0.973) \rightleftharpoons \text{anthophyllite } (X_{Mg} = 0.88)$$

may be ~600°C and ~7 kbar, respectively. The pure Mg-phase equilibrium is estimated to lie at a temperature 200°C (±95°C) higher than the temperature for the equilibrium involving solid solutions.

The system CaO-MgO-SiO_2-CO_2-H_2O

This system, which is important in understanding reactions in siliceous carbonate rocks, has been studied by Skippen (1971) and Metz (1970) and Metz and co-workers (Metz and Trommsdorff, 1968; Metz and Puhan 1970). Skippen (1971) determined the following equilibrium relations for five independent reactions in the system (T°K, P total fluid pressure in bars):

$$\underset{\textit{talc}}{Mg_3Si_4O_{10}(OH)_2} \rightleftharpoons \underset{\textit{orthoenstatite}}{3MgSiO_3} + \underset{\textit{quartz}}{SiO_2} + \underset{\textit{fluid}}{H_2O} \quad (11.p)$$

$$\log_{10}K_{(p)} = \frac{-7422}{T} + 10.54 + \frac{0.104(P\text{-}2000)}{T} \quad (11.4)$$

$$\underset{\textit{talc}}{Mg_3Si_4O_{10}(OH)_2} + \underset{\textit{calcite}}{3CaCO_3} + \underset{\textit{quartz}}{2SiO_2} \rightleftharpoons \underset{\textit{diopside}}{3CaMgSi_2O_6} + \underset{\textit{fluid}}{3CO_2} + \underset{\textit{fluid}}{H_2O} \quad (11.q)$$

$$\log_{10}K_b = \frac{-12930}{T} + 29.16 + \frac{0.494\,(P\text{-}2000)}{T} \quad (11.5)$$

$$\underset{\textit{tremolite}}{Ca_2Mg_5Si_8O_{22}(OH)_2} \rightleftharpoons \underset{\textit{diopside}}{2CaMgSi_2O_6} + \underset{\textit{orthoenstatite}}{3MgSiO_3} + \underset{\textit{quartz}}{SiO_2} + \underset{\textit{fluid}}{H_2O} \quad (11.r)$$

$$\log_{10}K_{(r)} = \frac{-6977}{T} + 9.33 + \frac{0.125\,(P\text{-}2000)}{T} \quad (11.6)$$

$$\underset{\textit{dolomite}}{3CaMg(CO_3)_2} + \underset{\textit{quartz}}{4SiO_2} + \underset{\textit{fluid}}{H_2O} \rightleftharpoons \underset{\textit{talc}}{Mg_3Si_4O_{10}(OH)_2} + \underset{\textit{calcite}}{3CaCO_3} + \underset{\textit{fluid}}{3CO_2} \quad (11.s)$$

$$\log_{10}K_{(s)} = \frac{-14330}{T} + 25.68 + \frac{0.190\,(P\text{-}2000)}{T} \quad (11.7)$$

$$\underset{\textit{talc}}{4Mg_3Si_4O_{10}(OH)_2} + \underset{\textit{dolomite}}{5CaMg(CO_3)_2} \rightleftharpoons \underset{\textit{diopside}}{5CaMgSi_2O_6} + \underset{\textit{forsterite}}{6Mg_2SiO_4} + \underset{\textit{fluid}}{10CO_2} + \underset{\textit{fluid}}{4H_2O} \quad (11.t)$$

$$\log_{10}K_{(t)} = \frac{-75400}{T} + 133.7 + \frac{1.348\,(P\text{-}2000)}{T} \quad (11.8)$$

(Note that in all the equations, the last term involves ΔV of solids only.)

Using these data, it is possible to calculate equilibrium data for any of the 49 reactions (see Skippen, 1971 for details) in the system (with H_2O and CO_2 as the two fluid components). As an example, let us consider the following reaction studied by Metz (1970)

$$\underset{\textit{tremolite}}{Ca_2Mg_5Si_8O_{22}(OH)_2} + \underset{\textit{calcite}}{3CaCO_3} + \underset{\textit{quartz}}{2SiO_2} \rightleftharpoons \underset{\textit{diopside}}{5CaMgSi_2O_6} + 3CO_2 + \underset{\textit{fluid}}{H_2O} \quad (11.u)$$

This reaction is the same as Reaction (11.r) + (11.q)–(11.p) and, therefore, the equilibrium condition is

$$\log K_{(u)} = \log K_{(r)} + \log K_{(q)} - \log K_{(p)}$$

$$= \frac{-12885}{T} + 27.95 + \frac{0.515\,(P\text{-}2000)}{T} \quad (11.9)$$

Expressions such as Equation (11.9) can be used advantageously either to draw fugacity diagrams—for example, $\log f_{H_2O}$ versus $\log f_{CO_2}$—or the T–X diagrams. Since $\log K_{(u)} = \log(f^3_{CO_2} \cdot f_{H_2O})$, Equation (11.9) can be solved for a specified P and T. For T–X diagrams we must solve Equation (11.9) simultaneously with the following equation:

$$\frac{f_{CO_2}}{\gamma_{CO_2}} + \frac{f_{H_2O}}{\gamma_{H_2O}} = P_{(u)} \qquad [= 2000 \text{ bar in Equation (11.9)}] \quad (11.10)$$

This may be done graphically, and the fugacities at the intersections of the two curves [Equations (11.9) and (11.10)] can be converted to partial pressures by using the fugacity coefficients from standard references (see

Skippen, 1971, Table 1). The corresponding mole fractions of CO_2 or H_2O may then be plotted against temperature giving the T–X representation of the equilibrium (11.u).

Metz (1970) has experimentally determined Reaction (11.u), and at 1000 bar the results are somewhat different from those obtained by calculations using Skippen's data. Metz discussed the following four reactions radiating from the invariant point in the T–X diagram:

$$5\,\text{dolomite} + 8\,\text{quartz} + 1\,H_2O \rightleftharpoons 1\,\text{tremolite} + 3\,\text{calcite} + 7\,CO_2 \quad \text{(Di)}$$

$$1\,\text{tremolite} + 3\,\text{calcite} + 2\,\text{quartz} \rightleftharpoons 5\,\text{diopside} + 3\,CO_2 + 1\,H_2O \quad \text{(Dol)} \quad (11.u)$$

$$1\,\text{tremolite} + 3\,\text{calcite} \rightleftharpoons 4\,\text{diopside} + 1\,\text{dolomite} + 1\,CO_2 + 1\,H_2O \quad \text{(Q)}$$

$$1\,\text{dolomite} + 2\,\text{quartz} \rightleftharpoons 1\,\text{diopside} + 2\,CO_2 \quad \text{(Calc, Tr)}$$

The (Dol) reaction was determined experimentally and (Calc, Tr) was calculated thermodynamically. The two curves intersect at the invariant point (Figure 11.22). A thermodynamically calculated curve for the (Q) reaction also meets at this point. This figure shows the lower stability limit of diopside in the presence of CO_2, which lies at 495°C at 1000 bar fluid pressure. If the fluid is very poor in CO_2 ($X_{CO_2} < 0.05$) the temperature of formation of diopside according to the (Dol) reaction may be even less and as low as 440°C at 1000 bar. At moderate pressures (3000 bar) and with $X_{CO_2} \simeq 0.1$, this temperature lies between 550 and 600°C.

Reaction (11.s)

$$\underset{\textit{dolomite}}{3\,CaMg(CO_3)_2} + \underset{\textit{quartz}}{4\,SiO_2} + \underset{\textit{fluid}}{H_2O} \rightleftharpoons \underset{\textit{talc}}{Mg_3Si_4O_{10}(OH)_2} + \underset{\textit{calcite}}{3\,CaCO_3} + \underset{\textit{fluid}}{3\,CO_2}$$

has been investigated by Gordon and Greenwood (1970) and Metz and Puhan (1970). This reaction is considered to initiate metamorphism of siliceous dolomites. Gordon and Greenwood (1970) obtained the following equation:

$$\log_{10} K_{(s)} = \frac{-14490}{T} + 27.00 \qquad \text{(at 2000 bar)} \qquad (11.11)$$

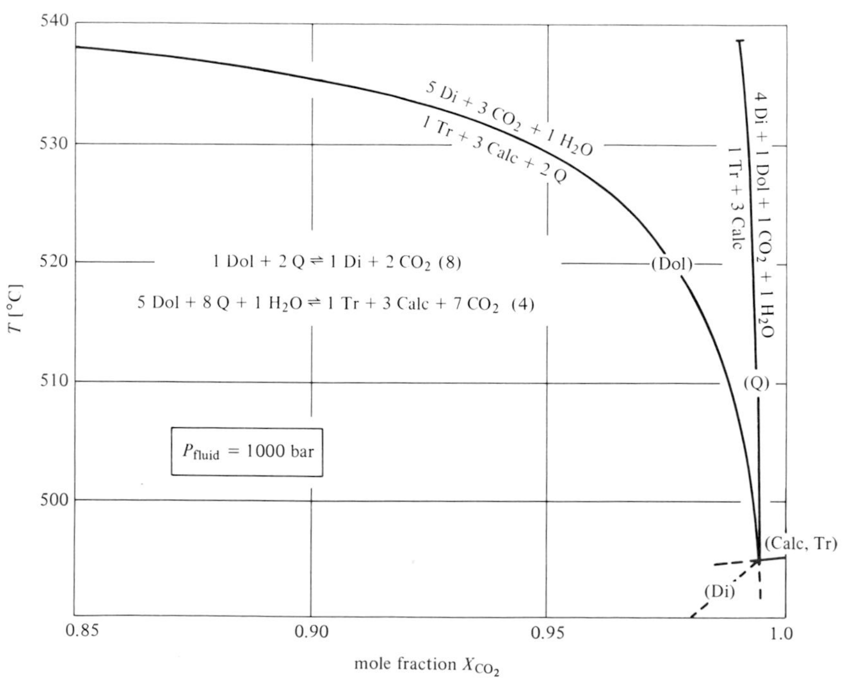

11.22 Isobaric univariant equilibrium curves for the (Di), (Dol), (Q), and (Calc, Tr). (After Metz, 1970.)

which is closely similar to that of Skippen's. In this reaction, however, it is important to consider that both dolomite and calcite are crystalline solutions and the activities of $CaCO_3$ and $MgCO_3$ should appear in the expression for the equilibrium constant. Gordon and Greenwood have determined the activity-composition relation in the $CaCO_3$-$MgCO_3$ solution. The equation for log K using these activities is given by

$$\log_{10} K_{(s)} = \frac{-14145}{T} + 26.07 \quad \text{(at 2000 bar)} \quad (11.12)$$

Although in this case the variations among the three expressions (11.7), (11.11), and (11.12) are within the experimental errors, the effect of crystalline solubility should generally be considered. Metz and Puhan's (1970) results are similar to the others and are shown in Figure 11.23.

Gordon and Greenwood (1970) arrived at the following equation for the activity coefficient for $CaCO_3$ in calcite:

$$\ln \gamma_{CaCo_3} = (1 - X_{CaCO_3})^2(1.2238 + 1.8960 X_{CaCo_3}) \quad (11.13)$$

$0 > X_{MgCO_3} < 0.18$

Skippen (1974) combined the above equation with the solvus data of Graf and Goldsmith (1958) and Harker and Tuttle (1955) to produce the equation:

$$\log a_{CaCO_3} = \frac{68.38}{T} - 0.110 \quad (11.14)$$

Gordon and Greenwood's (1970) equation for a_{MgCO_3} is

$$\log a_{MgCO_3} = \frac{-730.9}{T} + 0.4854 \quad (11.15)$$

Using these activity-composition expressions, Skippen (1974) has calculated the following equilibria between 500 and 3000 bar:

(*4*) 3dolomite + 4quartz + H_2O
= talc + $3CO_2$ + 3calcite (11.s)

(*8*) tremolite + 3calcite + 2quartz
= 5diopside + $3CO_2$ + H_2O (11.u)

(*12*) 5talc + 6calcite + 4quartz
= 3tremolite + $6CO_2$ + $2H_2O$

(*16*) dolomite + calcite + 2quartz
= diopside + $2CO_2$

11.23 Experimentally determined curves for the reaction 3 dolomite + 4 quartz + H_2O = talc + 3 calcite + $3CO_2$ in the T–X field (b) P–T curve for the same reaction at X_{CO_2} = 0.5. (After Metz and Puhan, 1971.)

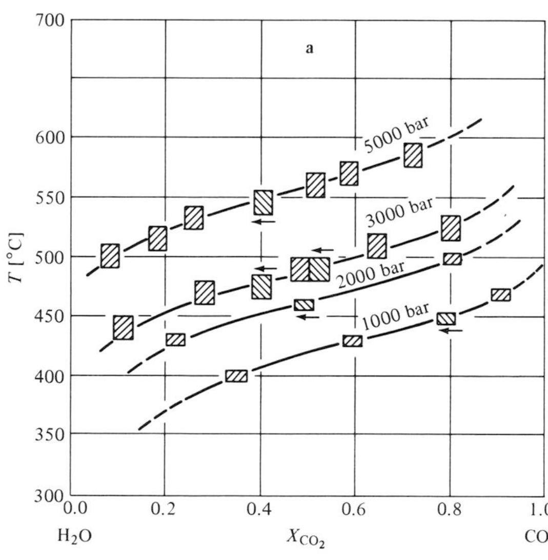

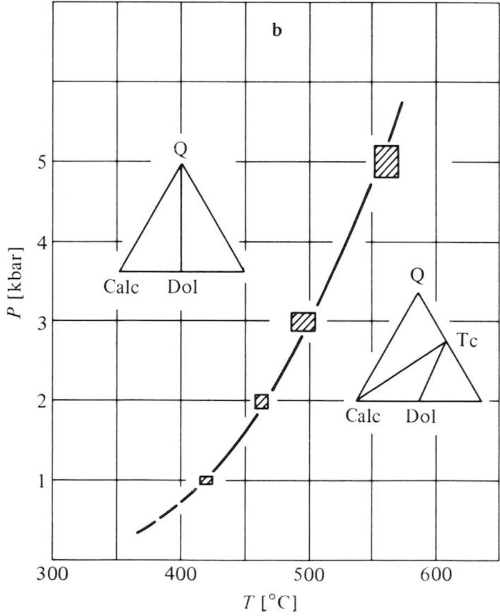

(*17*) 2 talc + 3 calcite
= tremolite + dolomite + CO_2 + H_2O

(*19*) 8 quartz + 5 dolomite + H_2O
= tremolite + 7 CO_2 + 3 calcite

(*20*) tremolite + 3 calcite
= 4 diopside + dolomite + CO_2 + H_2O

(*27*) diopside + 3 dolomite
= 2 forsterite + 4 calcite + 2 CO_2

11.24 Equilibria affecting bulk compositions in the triangle quartz-calcite-dolomite with the activities of $MgCO_3$ and $CaCO_3$ determined by the calcite-dolomite miscibility gap and $P_{CO_2} + P_{H_2O} = P_{total}$, at 500 bar total pressure and at 1000 bar total pressure. (After Skippen, 1974. Figures 11.24 and 11.25 reprinted from the *American Journal of Science* by permission.)

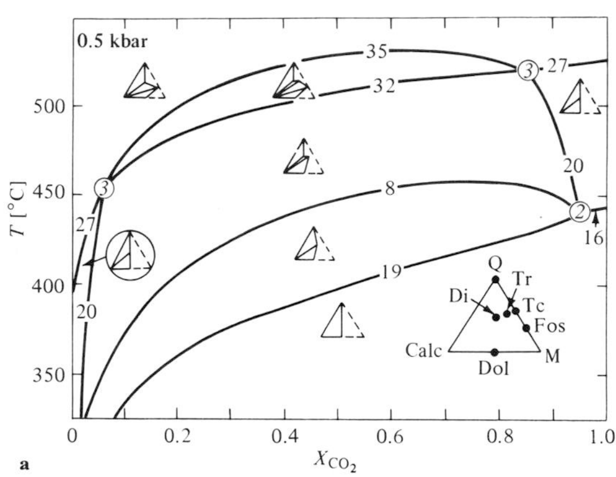

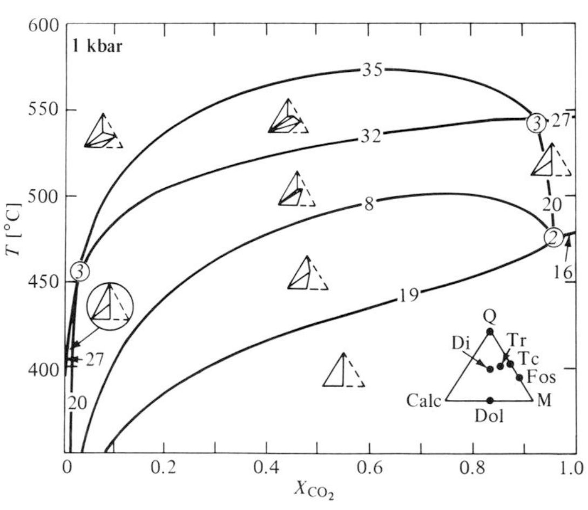

(*32*) tremolite + 11 dolomite
= 8 forsterite + 2 calcite + 9 CO_2 + H_2O

(*35*) 3 tremolite + 5 calcite
= 11 diopside + 2 forsterite + 5 CO_2 + 3 H_2O

Figures 11.24 and 11.25 show these equilibria (numbers in parentheses refer to numbers on curves) with the

11.25 Equilibria affecting bulk compositions in the triangle quartz-calcite-dolomite with the activities of $MgCO_3$ and $CaCO_3$ determined by the calcite-dolomite miscibility gap and $P_{CO_2} + P_{H_2O} = P_{total}$. Equilibria (*12*) and (*17*) are dashed because of large limits of uncertainty. Figures are for 2000 bar total pressure and 3000 bar total pressure.

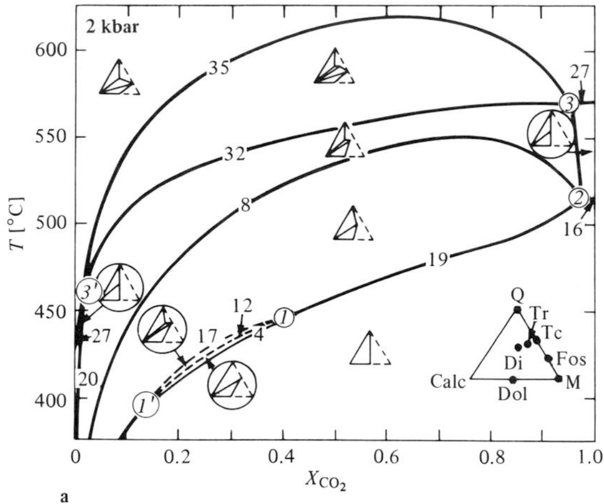

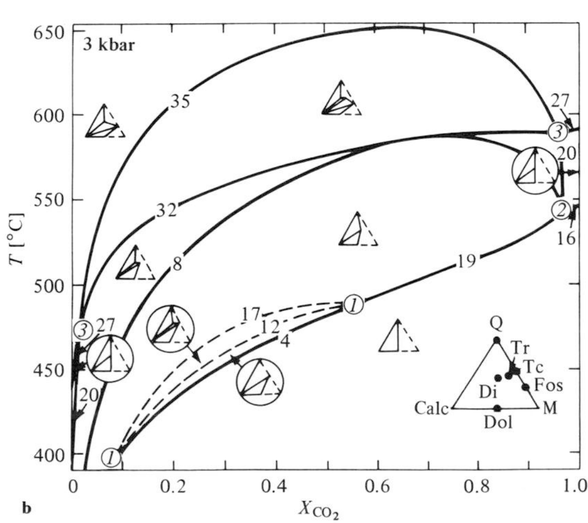

condition that $P_{H_2O} + P_{CO_2} = P_{total}$. If $P_{H_2O} + P_{CO_2} < P_{total}$ the curves are displaced to lower temperatures. Assemblages of five minerals are shown as points numbered as below:

(1) and *(1')*: talc-tremolite-quartz-calcite-dolomite

(2): tremolite-diopside-quartz-calcite-dolomite

(3) and *(3')*: forsterite-diopside-tremolite-calcite-dolomite

There are significant uncertainties associated with these points. Skippen's (1974) discussion, however, shows that the points are consistent with data from natural assemblages. For petrogenesis the points that are important in the field are the five-mineral assemblage points, the maxima in the equilibrium curves, intersections with the $X = 1$ axis *(16, 27)* and the indifferent crossing of equilibrium curves *(8–32)*. To bracket a point such as *(2)*, we may proceed as follows: The lowest temperature at which the five-mineral assemblage tremolite-diopside-quartz-calcite-dolomite exists is located by the final disappearance of tremolite-calcite-quartz-dolomite [Equilibrium *(19)*] in moving from lower to higher grade. The maximum temperature for point *(2)* is given by the appearance of diopside-dolomite-calcite in CO_2-rich fluids according to Equilibrium *(16)*.

Puhan and Hoffer (1973) have described the paragenesis of four mineral assemblages talc-calcite-dolomite-quartz and tremolite + talc + calcite + quartz from the southern part of the Damara Belt in southwest Africa. Even five-phase assemblages tremolite + talc + calcite + dolomite + quartz have been encountered in the area. Associated pelitic rocks contain kyanite, which indicates a minimum pressure of about 5.5 kbar at a temperature of ~600°C. These results would be consistent with a probable extrapolation of Skippen's (1974) results on the invariant point *1*.

Metamorphism of ultramafic rocks

Trommsdorff and Evans (1972) have described the mineral paragenesis in the serpentinite body of Val Malenco, between the Bergell Alps, Italy and the Bernina Mountains, Switzerland. The Malenco body was first metamorphosed into antigorite schist with the assemblage antigorite + olivine + diopside + chlorite + magnetite. Later, due to contact metamorphism, the following zones of progressive metamorphism were developed:

- *A*: antigorite-olivine-diopside zone
- *B*: antigorite-olivine-tremolite zone
- *C*: talc-olivine-tremolite zone
- *D*: anthophyllite-olivine-tremolite zone

The univariant equilibria connecting these assemblages are:

$$5\,\text{antigorite} + 2\,\text{diopside} = \text{tremolite} + 6\,\text{olivine} + 9\,H_2O \quad (11.v)$$

$$5\,\text{antigorite} = 6\,\text{olivine} + \text{talc} + 9\,H_2O \quad (11.w)$$

$$9\,\text{talc} + 4\,\text{olivine} = 5\,\text{anthophyllite} + 4\,H_2O \quad (11.x)$$

$$\text{anthophyllite} + \text{olivine} = 9\,\text{enstatite} + H_2O \quad (11.y)$$

Geologic data indicate a pressure of 2 kbar ($P_{total} \simeq P_{H_2O}$). At this pressure the experimental and calculated results suggests the temperature of transition from zone *A* to *B* [Reaction 11.v)] to be ~420°C (Evans and Trommsdorff, 1970), from *B* to *C* [Reaction (11.w)] ~460°C (Scarfe and Wyllie, 1967), and from *C* to *D* ~675°C [Reaction (11.x)] (Greenwood, 1963). The *B* and *C* zones are 500 + 200 m and 700 ± 300 m wide, respectively. The temperature gradient seems to have suddenly steepened in the latter zone (~210°C). The Fe-Mg solid solution in the phases is small and does not affect the temperature of the equilibria significantly. The thermal gradient is calculated to be ~60°C/km and is inadequate to explain the relative positions of the equilibria in space. Trommsdorff and Evans consider that this may be due to a greater thermal stability of antigorite than that of chrysotile (serpentine), which is the polymorph in experimental work.

Evans and Trommsdorff (1970) have also described a sequence of several critical mineral assemblages in the rocks of the Lepontine and Rhetic Alps. The reactions and the experimentally determined or theoretically calculated temperatures at 2 kbar are

$$\text{serpentine} + 2\,\text{quartz} = \text{talc} + H_2O \quad (310°C)$$

$$\text{serpentine} + \text{brucite} = 2\,\text{forsterite} + 3\,H_2O \quad (380°C)$$

$$5\,\text{serpentine} + 2\,\text{diopside} = \text{tremolite} + 6\,\text{forsterite} + 9\,H_2O \quad (\sim 415°C)$$

$$5\,\text{serpentine} = 6\,\text{forsterite} + \text{talc} + 9\,H_2O \quad (\sim 460°C)$$

$$9\,\text{talc} + 4\,\text{forsterite} = 5\,\text{anthophyllite} + 4\,H_2O \quad (675°C)$$

$$\text{anthophyllite} + \text{forsterite} = 9\,\text{enstatite} + H_2O \quad (700°C)$$

$$7\text{talc} = 3\text{anthophyllite} + 4\text{quartz} + 4H_2O \quad (705°C)$$

$$\text{anthophyllite} = 7\text{enstatite} + \text{quartz} + H_2O \quad (760°C)$$

$$\text{tremolite} + \text{forsterite} = 5\text{enstatite} + 2\text{diopside} + H_2O \quad (\sim 775°C)$$

$$\text{tremolite} = 3\text{enstatite} + 2\text{diopside} + \text{quartz} + H_2O \quad (860°C)$$

The experimental determinations are from Johannes (1968a), Greenwood (1963), and Boyd (1959). The temperatures have been calculated by Evans and Trommsdorff (1970).

The temperature estimates of the various equilibria in ultramafic rocks are subject to the same uncertainty with regard to P_{H_2O} as those of pelitic rocks. Although the water pressure may be generally high and equal to total pressure as in the case of the Malenco serpentinites, the interesting occurrence of the assemblage enstatite-forsterite-chlorite in stable equilibrium (Trommsdorff and Evans, 1969) may best be explained by metamorphism in the amphibolite facies under conditions of low water fugacity. The medium-grade metamorphic P and T is indicated for these rocks because of the associated parageneses of kyanite-sillimanite-muscovite in pelitic rocks and diopside-calcite-anorthite in marbles.

The iron formations

Metamorphic rocks rich in iron have been described from many parts of the world. These rocks involve hydrous and anhydrous silicates, iron oxides, and carbonates and provide an opportunity to study the effect of the fugacity of fluid components on the composition of Fe^{2+}-Mg silicates. A reference to Wones and Eugster's (1965) experimental work on the iron-biotite-annite has already been made. The system FeO-Al_2O_3-SiO_2-H_2O-O_2 investigated by Ganguly (1969, 1972) was also discussed in this chapter and in Chapter 9. All these studies have emphasized the control of oxygen fugacity on the stability of minerals. In this section, before we discuss the paragenesis of natural assemblages from iron formations, we may consider the effect of crystalline solubility in the system ferrotremolite [$Ca_2Fe_5Si_8O_{22}(OH)_2$]-tremolite.

The breakdown curve for ferrotremolite was presented before (Figure 9.15). Here we consider, following Mueller (1973), the effect of Mg^{2+} on the reaction

$$\underset{\text{actinolite solution}}{Ca_2Fe_5Si_8O_{22}(OH)_2} + \tfrac{1}{2}O_2$$

$$= \underset{\text{Ca-pyroxene solution}}{2CaFeSi_2O_6} + \underset{\text{magnetite}}{Fe_3O_4} + \underset{\text{quartz}}{4SiO_2} + \underset{\text{fluid}}{H_2O} \quad (11.z)$$

Ernst's (1966) results on the pure iron system are shown in Figure 11.26. By assuming actinolite and Capyroxene as ideal, we may write the equation of equilibrium for Reaction (11.z) as follows:

$$\log K_{(z)} = \frac{P\Delta V_{(z)}}{4.576T} - 5 \log X_{Fe}^{Act} + 2 \log X_{Fe}^{Cpx} + \log f_{H_2O} - \tfrac{1}{2} \log f_{O_2} \quad (11.16)$$

$\Delta V_{(z)}$, as before, is the volume difference for the solids.

11.26 Phase relations of actinolite and associated minerals at a total fluid pressure of 3000 bar, modified after Ernst (1966). Curve (z′) shows the shift in curve (z) corresponding to a change in Fe/(Fe + Mg) in actinolite from unity to 0.15 [Reaction (11.z)] (After Mueller, 1973.)

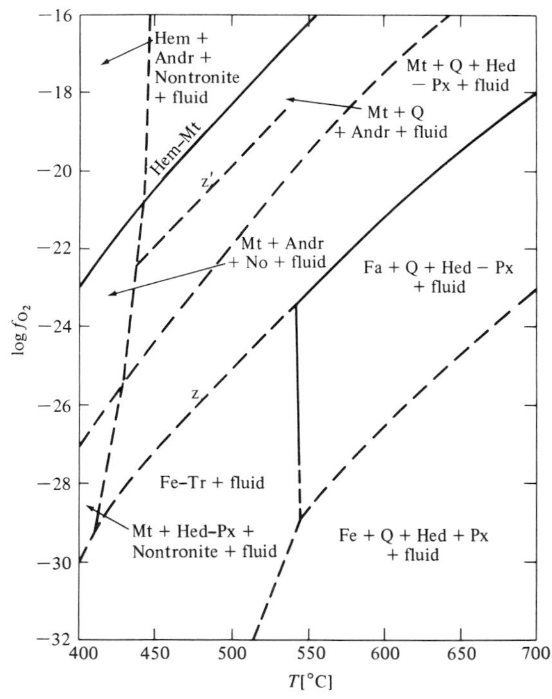

From Mueller's (1961b) data $X_{\mathrm{Fe}}^{\mathrm{Cpx}} = X_{\mathrm{Fe}}^{\mathrm{Act}}$, which reduces Equation (11.16) to

$$\log K_{(z)} = \frac{P\Delta V_{(z)}}{4.576T} - 3 \log X_{\mathrm{Fe}}^{\mathrm{Act}} + \log f_{\mathrm{H_2O}} - \tfrac{1}{2} \log f_{\mathrm{O_2}} \qquad (11.17)$$

We compare an assemblage with arbitrarily chosen values of $X_{\mathrm{Fe}}^{\mathrm{Act}}, f_{\mathrm{H_2O}}, f_{\mathrm{O_2}}$, and P with a reference state in which $X_{\mathrm{Fe}}^{\mathrm{Act}} = X_{\mathrm{Fe}}^{\prime\mathrm{Act}} = 1, f_{\mathrm{H_2O}} = f'_{\mathrm{H_2O}}$ and $P = P'$, where the primes indicate the reference values. We then obtain the following from Equation (11.17):

$$\log f_{\mathrm{O_2}} - \log f'_{\mathrm{O_2}} = 6 \log X_{\mathrm{Fe}}^{\mathrm{Act}} - \frac{2(P' - P)\Delta V}{4.576T} + 2(\log f_{\mathrm{H_2O}} - \log f'_{\mathrm{H_2O}}) \qquad (11.18)$$

ΔV is -18 cm^3 (Robie *et al.*, 1966) under STP conditions.

We now consider the shift in $\log f_{\mathrm{O_2}}$ at a temperature of 500°C. From Figure 11.26, $P' = 3000$ bar and $f'_{\mathrm{O_2}} = 10^{-24.8}$ bar. Note that under the experimental conditions the reference system is reduced to the extent that considerable hydrogen must be present in the fluid phase. From the thermodynamic data on the decomposition of water (Kelley, 1960; Robie and Waldbaum, 1968), we find that at this temperature and oxygen fugacity

$$\frac{f_{\mathrm{H_2O}}}{f_{\mathrm{O_2}}} = 21.4$$

From the experimental data on systems consisting of "pure" hydrogen (Shaw and Wones, 1964) and of "pure" water (Anderson, 1964; Burnham *et al.*, 1969) we find

$$f^0_{\mathrm{H_2O}} = 0.35P^0_{\mathrm{H_2O}} \qquad f^0_{\mathrm{H_2}} = 2P^0_{\mathrm{H_2}}$$

$(T = 500°\mathrm{C}, P = 3000$ bar)

where the superscript refers to the "pure" phase. If we now assume that the fugacities of H_2O and H_2 in the mixture are proportional to their mole fractions, we obtain the following expression for $T = 500°\mathrm{C}$:

$$\frac{f'_{\mathrm{H_2O}}}{0.35} + \frac{f'_{\mathrm{H_2}}}{2} = 3000 \text{ bar}$$

Consequently $f'_{\mathrm{H_2O}} = 1043$ bar under the experimental reference conditions. If we substitute these values into Equation (11.17), we obtain $K_{(z)}$ and $\Delta G^0_{(z)}$, the standard Gibbs free energy charge for Reaction (11.z) at 500°C:

$$K_{(z)} = 10^{15} \quad \text{and} \quad \Delta G^0_{(z)} = -53{,}059 \text{ cal}$$

If we utilize the deduced value of $f'_{\mathrm{H_2O}} = 1043$ bar and the experimental value of $P' = 3000$ bar, we can determine that the shift in $\log f_{\mathrm{O_2}}$ corresponding to arbitrary values of $f_{\mathrm{H_2O}}, P$, and $X_{\mathrm{Fe}}^{\mathrm{Act}}$. To illustrate, we may assume a fluid composition and pressure such that $P = 6000$ bar, $f_{\mathrm{H_2O}} = 3000$ bar, and $X_{\mathrm{Fe}}^{\mathrm{Act}} = 0.15$, a mineral composition that corresponds to high-pressure regional metamorphism on the hematite-magnetite boundary (Mueller, 1960). We then find that

$$\log f_{\mathrm{O_2}} - \log f'_{\mathrm{O_2}} = 5.15$$

It should be noted that this calculation is not particularly sensitive to P or $f_{\mathrm{H_2O}}$ but is dominated by the term $-6 \log X_{\mathrm{Fe}}^{\mathrm{Act}}$, which alone amounts to 4.94. We see then that for values of the total pressure and water fugacity in the range of high-grade regional metamorphism, the enrichment of actinolite in magnesium to $X_{\mathrm{Fe}}^{\mathrm{Act}} = 0.15$ corresponds to an increase of between 5 and 6 orders of magnitude in $f_{\mathrm{O_2}}$. This calculated shift in curve (z) is shown as (z′) in Figure 11.26.

The estimated shift in $\log f_{\mathrm{O_2}}$ calculated from Equation (11.18) is in quite good agreement with the observations from the Quebec iron-formation assemblages, if all the uncertainties in the correlation are considered. This in turn implies a good correspondence between the experimental results of Ernst and the natural assemblages. The presence of both hematite and magnetite in the actinolite–Ca-pyroxene assemblage with $X_{\mathrm{Fe}}^{\mathrm{Act}} = 0.15$ in the metamorphosed Quebec iron formation (Mueller, 1960) places curve (z′) directly on the HM curve. However the values of $P, f_{\mathrm{H_2O}}$, and T are not known or only uncertainly known for the iron formation. Examination of Equation (11.18) shows that decreasing P would have the effect of increasing the shift of $\log f_{\mathrm{O_2}}$ to higher values, whereas decreasing $f_{\mathrm{H_2O}}$ would have the opposite effect. Also, it must be kept in mind that there are experimental uncertainties in curve (z), the precise location of which is only inferred according to Ernst (1966). What is important here is that the compositional variation of the ferromagnesian silicates is identified as the major effect on the shift of $\log f_{\mathrm{O_2}}$ and that the shift (z) to (z′) is of the expected order of magnitude and corresponds to the theoretically expected sensitivity of $f_{\mathrm{O_2}}$ to changes in $X_{\mathrm{Fe}}^{\mathrm{Act}}$. It appears that perturbations arising from uncertainties in the theoretical model and the experimental data are of the same order as those involved in not knowing $P, f_{\mathrm{H_2O}}$, and T precisely.

Kranck (1961) has described the assemblage

calcite + quartz + orthopyroxene + Ca-pyroxene ± cummingtonite ± graphite

from the metamorphic iron formations in the Mount Reed area of northern Quebec. The various phases may be related by the reactions

$$\underset{orthopyroxene}{MgSiO_3} + \underset{calcite}{CaCO_3} + \underset{quartz}{SiO_2} = \underset{clinopyroxene}{CaMgSi_2O_6} + \underset{fluid}{CO_2} \quad (11.aa)$$

$$\underset{orthopyroxene}{MgSiO_3} + \underset{clinopyroxene}{CaFeSi_2O_6} = \underset{orthopyroxene}{FeSiO_3} + \underset{clinopyroxene}{CaMgSi_2O_6} \quad (11.bb)$$

$$\underset{fluid}{CO_2} = \underset{graphite}{C} + \underset{fluid}{O_2} \quad (11.cc)$$

At equilibrium, we have

$$K_{(aa)} \exp\left(\frac{-P\Delta V_{(aa)}}{RT}\right) = \frac{X_{Mg}^{Cpx}}{X_{Mg}^{Opx}} f_{CO_2} \quad (11.19)$$

$$K_{(bb)} \exp\left(\frac{-P\Delta V_{(bb)}}{RT}\right) = \frac{X_{Mg}^{Cpx}(1 - X_{Mg}^{Opx})}{X_{Mg}^{Opx}(1 - X_{Mg}^{Cpx})} \quad (11.20)$$

In these equations both pyroxenes are ideal solutions and calcite is stoichiometric $CaCO_3$. Calcite in the natural assemblages may contain (Mg, Fe) CO_3, although in the absence of dolomite the concentration would be low. By neglecting $\Delta V_{(bb)}$ for the ion exchange, which is small, we may write from Equations (11.19) and (11.20)

$$f_{CO_2} = \frac{K_{(aa)}}{K_{(bb)}} \exp\left(\frac{-P\Delta V_{(aa)}}{RT}\right) [(K_{(bb)} - 1)X_{Mg}^{Opx} + 1]. \quad (11.21)$$

Equation (11.21) shows that the fugacity of CO_2 is a linear function of the composition of orthopyroxene at constant temperature and total pressure.

From the general assemblage with two pyroxenes and the Al-free amphibole cummingtonite and the occurrence of kyanite in the adjacent area, we may infer the temperature and pressure of formation of the assemblage to be around 700°C and 5 kbar, respectively. Extrapolation of the fugacity data in Skippen (1971) shows that γ_{CO_2} may be of the order of 4.5 under these conditions. If total pressure is the same as P_{CO_2}, the fugacity (f_{CO_2}) is 22.5 kbar. This may be reduced by the presence of other fluid components. ΔV (solids) is approximately -25 cm^3 under STP conditions, and $K_{(bb)}$ from the composition of coexisting pyroxenes is 1.82. We may substitute these values in equation (11.21) to obtain $K_{(aa)}$ as 9020.61 or log $K_{(aa)}$ as 3.96.

$K_{(aa)}$ may also be calculated from Skippen's (1974) data on the reactions

$$\text{talc} = 3\,\text{enstatite} + \text{quartz} + H_2O \quad (11.p)$$

$$\text{talc} + 3\,\text{calcite} + 2\,\text{quartz} = 3\,\text{diopside} + 3CO_2 + H_2O \quad (11.q)$$

by subtracting log $K_{(p)}$ from log $K_{(q)}$ at 2 kbar. At 1027°K the calculated log $K_{(aa)}$ is 4.42. At 5 kbar, this K would be even higher and is not consistent with the possible pressure and temperature conditions of the assemblage.

At constant P and T, the relation between f_{CO_2} and the composition of the pyroxenes is shown in Figure 11.27. The ordinate represents $f_{CO_2}/f^0_{CO_2}$, where $f^0_{CO_2}$ represents the standard fugacity of 22.5 kbar. The effect of f_{CO_2} on pyroxenes may be demonstrated with the help of Figure 11.27. Consider a system at 700°C consisting of orthopyroxene, calcite, and quartz with $X_{Mg}^{Cpx} = 0.6$ and $f_{CO_2}/f^0_{CO_2} = 1.6$. If the latter is dropped to about 1.5, Reaction (11.aa) runs to the right and Ca-pyroxene with $X_{Mg}^{Cpx} = 0.75$ is first produced in conjunction with the ion-exchange (11.bb). If f_{CO_2} continues to drop, the reaction is further displaced to the right until the last of the orthopyroxene with $X_{Mg}^{Opx} = 0.45$ is consumed. At lower f_{CO_2} we have Ca-pyroxene ($X_{Mg}^{Cpx} = 0.6$), calcite and quartz. The stability of the pyroxenes is thus a function of the fugacity of CO_2 and their composition. This con-

11.27 Plot of the relative activity or fugacity ratio of CO_2 as a function of the pyroxene composition. (After Mueller, 1966.)

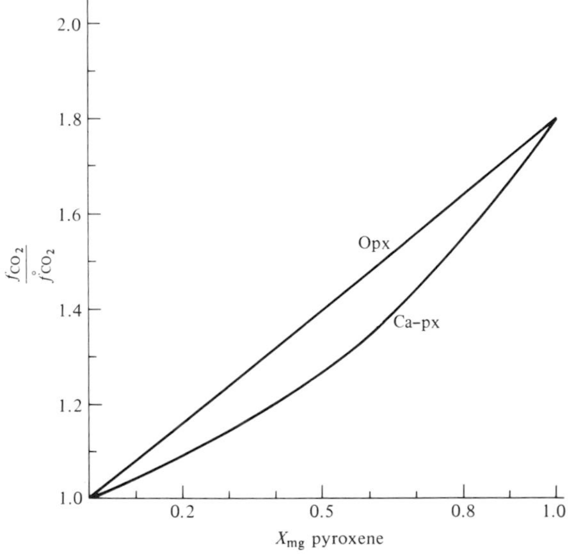

clusion is similar to those arrived at elsewhere in relation to the effect of water fugacity and composition of solid solutions.

In the rocks described by Kranck, the association quartz and dolomite does not occur. Skippen's results [see Figure 11.25; Reaction (*16*)] show that at high f_{CO_2} and 3 kbar, this reaction occurs above 540°C. At 5 kbar the equilibrium temperature may be still higher. The association dolomite-quartz is quite common in other Quebec iron-formation rocks, where orthopyroxene is absent and hydrous minerals such as actinolite and talc are common (Mueller, 1960). These rocks probably represent lower temperatures than the rocks studied by Kranck.

A particularly interesting aspect of the iron formation is the possible correlation between the silicate compositions and the chemical potentials or fugacities of fluid components such as CO_2 and H_2O. We have already considered the oxidation of actinolite. Mueller (1960) discussed the assemblage actinolite-talc-magnetite-hematite-calcite-quartz using the reactions

$$\underset{tremolite}{\tfrac{1}{5}Ca_2Mg_5Si_8O_{22}(OH)_2} + \tfrac{2}{15}H_2O + \tfrac{2}{5}CO_2 \rightleftharpoons \underset{talc}{\tfrac{1}{3}Mg_3Si_4O_{10}(OH)_2} + \underset{calcite}{\tfrac{2}{5}CaCO_3} + \underset{quartz}{\tfrac{4}{15}SiO_2} \quad (11.dd)$$

$$\underset{ferrotremolite}{\tfrac{1}{5}Ca_2Fe_5Si_8O_{22}(OH)_2} + \tfrac{2}{5}CO_2 + \underset{hematite}{Fe_2O_3} \rightleftharpoons \underset{calcite}{\tfrac{2}{5}CaCO_3} + \underset{quartz}{\tfrac{8}{5}SiO_2} + \underset{magnetite}{Fe_3O_4} + \tfrac{1}{5}H_2O \quad (11.ee)$$

At constant temperature under equilibrium assuming ideal solutions we have

$$\frac{1}{K_{(dd)}} = X_{Mg}^{Act} P_{CO_2}^{2/5} P_{H_2O}^{2/15} \quad (11.22)$$

and

$$\frac{1}{K_{(ee)}} = \frac{(1 - X_{Mg}^{Act}) P_{CO_2}^{2/5}}{P_{H_2O}^{1/5}} \quad (11.23)$$

where P is the partial pressure. Therefore

$$P_{H_2O} = \left(\frac{K_{(ee)}}{K_{(dd)}}\right)^3 \left(\frac{1 - X_{Mg}^{Act}}{X_{Mg}^{Act}}\right)^3 \quad (11.24)$$

and

$$P_{CO2} = \frac{K_{(dd)}^{-3/2} K_{(ee)}^{-1}}{(1 - X_{Mg}^{Act})(X_{Mg}^{Act})^{3/2}} \quad (11.25)$$

There is a positive correlation between the enrichment of iron in actinolite and water pressure. Fe-rich actinolites of this assemblage require relatively high gas pressures over a considerable range of composition as compared to Mg-rich actinolites. The CO_2 pressure is really immaterial to the stability of actinolite in the sense that this mineral may be stable in the absence of CO_2. In assemblages that contain dolomite in addition to the other minerals as discussed above, Mueller (1960) found that actinolites are uniform in composition. According to the phase rule, such a system is univariant. Then the compositions of all minerals, as well as the partial pressures of all the gases, are determined at each temperature.

Butler (1969) has discussed changes in gradient of the fluid components in the rocks of the same general area as studied by Mueller and Kranck.

Iron formation with

biotite + amphibole ± magnetite ± hematite
+ plagioclase + quartz + accessory minerals

has been described by Annersten (1968) from northern Sweden. He finds a similar control of oxygen fugacity on the Fe/Mg ratio in silicate minerals as described by Mueller and Butler. An additional result of interest in these minerals is a decrease in Ti content of silicates with increasing oxygen pressure.

Rocks of high-pressure and controversial origin

In this and previous chapters we confined ourselves to rocks of regional and contact metamorphism, generally formed below 10 kbar. There are other rocks that are only occasionally exposed on the surface, and which may have formed over a wide range of temperature and pressure. Some of these rocks, which will now be discussed are glaucophane schists, garnet-pyroxenites, various types of eclogites, grospydites, and other inclusions of mafic types occurring in rocks of high-pressure origin such as kimberlites. Examples of the possible assemblages in these rocks are as follows:

Glaucophane schists: quartz, muscovite, glaucophane, lawsonite, stilpnomelane; or quartz, omphacite, glaucophane, garnet, epidote, aragonite

Garnet-pyroxenites: garnet, clinopyroxene, amphibole, iron ores

Eclogites: omphacite, almandine-pyrope-grossular garnet

Grospydite: (xenoliths in kimberlites) garnet, clinopyroxene, kyanite

We have already discussed equilibrium relations in systems involving such minerals as lawsonite in Chapters

11.28 Some high-pressure facies boundaries. Eclogite-gabbro transition shown by the stippled area (from D. H. Green and Ringwood, 1972.) The area takes into account the experimental results of Ito and Kennedy (1971). The horizontally and vertically striped areas represent high- and medium-pressure granulite facies according to Green and Ringwood (1967). *BD*, *AB*, and *CD* represent the dry solidus, the appearance of garnet, and the disappearance of plagioclase, respectively, in a quartz-tholeiite composition (D. H. Green and Ringwood, 1967.) Other solidus curves for tholeiite, andesite, and dacite compositions are from T. H. Green and Ringwood (1968). Light solid lines represent phase relations in a 1:1 mixture of forsterite: anorthite as determined by Kushiro and Yoder (1966). Lines with filled circles represent phase equilibrium relations in high alumina basalt in the subsolidus field as determined by T. H. Green (1967). *G–G′*: first appearance of garnet; *P–P′*: disappearance of plagioclase. Curves (*1*), (*2*), and (*3*) are from Kushiro (1969) for the albite ⇌ jadeite + quartz equilibria as a function of diopside content. Curve (*4*) is for the equilibrium 4 enstatite + 1 spinel ⇌ 1 forsterite + 1 pyrope from MacGregor (1974). Ec = eclogite.

9 and 10. The mineral glaucophane is stable up to 850°C under hydrous conditions ($P_{H_2O} = P_{total}$) (Ernst, 1961). From Ernst's (1963a) work it appears to be stable up to pressures of about 30 kbar. Recent experimental work has been directed mainly toward a study of recrystallization of natural assemblages under high pressure and temperature. Results of some of these experiments are plotted in Figure 11.28 and discussed below.

The gabbro-eclogite transformation

A study of the gabbro-eclogite transformation was initiated by Kushiro and Yoder (1964). They were followed by Ringwood and D. H. Green (1966a), and D. H. Green and Ringwood (1967, 1972). The latter authors compared the results of Ito and Kennedy (1971) with their own work. Figure 11.28 shows that the transformation of the gabbroic mineral assemblage (plagioclase + pyroxene ± olivine ± spinel) to an eclogite assemblage (garnet + pyroxene ± quartz) takes place through the intermediate zone of a granulite assemblage, which is represented by the coexistence of garnet and plagioclase. The width of the *P–T* region of the granulite facies is a function

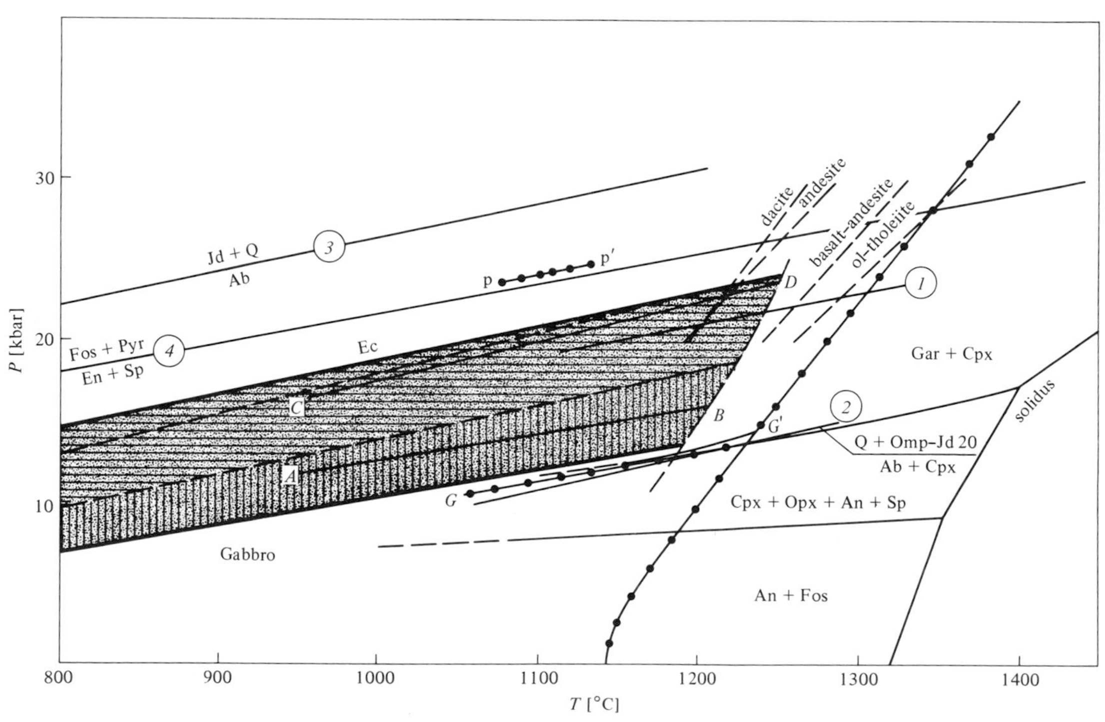

of the bulk chemistry. Generally the more aluminous an assemblage, the wider the range of P over which garnet and plagioclase coexist. In Figure 11.28 the low-pressure boundary of the garnet-plagioclase zone (stippled) marks the appearance of garnet, and the high-pressure boundary, the disappearance of plagioclase. The horizontally and vertically striped areas represent the high- and medium-pressure divisions of the granulite facies as suggested by T. H. Green and Ringwood (1967). Figure 11.28 also shows the solidus curves for various rock compositions as determined by T. H. Green and Ringwood (1968) and T. H. Green (1967). Note that the slopes and positions of the various solidus curves change regularly as a function of composition from dacite to olivine tholeiite. For comparison with the results from rock compositions, results of a compositionally simpler system (forsterite : anorthite = 1 : 1) as determined by Kushiro and Yoder (1966) are shown in the figure. Note that in the absence of any solid solution in plagioclase, the field of coexisting garnet and plagioclase is eliminated and the pressure of formation of garnet-pyroxene in the sodium-free system is considerably lower. Similarly the solidus curve is about 200°C higher than the solidus curves for rock compositions. The reactions that are probable in the experimental system are

$$\underset{\textit{plagioclase}}{\{NaAlSi_3O_8\}} = \underset{\textit{clinopyroxene}}{\{NaAlSi_2O_6\}} + \underset{\textit{quartz}}{SiO_2} \qquad \Delta V_1 < 0 \tag{11.ff}$$

$$\underset{\textit{plagioclase}}{\{CaAl_2Si_2O_6\}} = \underset{\textit{clinopyroxene}}{\{CaAl_2SiO_6\}} + \underset{\textit{quartz}}{SiO_2} \qquad \Delta V_3 < 0 \tag{11.gg}$$

where the { } denotes component of a solution. Equilibrium conditions for Reactions (11.ff) and (11.gg) are

$$K_{(ff)} \exp \frac{(-P\Delta V_{(ff)})}{RT} = \frac{X_{Jd}^{Cpx}\,\gamma_{Jd}^{Cpx}}{X_{Ab}\,\gamma_{Ab}} \tag{11.26}$$

$$K_{(gg)} \exp \frac{(-P\Delta V_{(gg)})}{RT} = \frac{X_{Ts}^{Cpx}\gamma_{Ts}^{Cpx}}{(1 - X_{Ab}^{Plag})\gamma_{An}^{Plag}} \tag{11.27}$$

where Ts stands for Ca-Tschermak's molecule. Three results of Kushiro's (1969) experiments have been plotted in Figure 11.28. We note that with increasing concentration of jadeite the equilibrium

$$\text{albite} \rightleftharpoons \text{jadeite} + \text{quartz}$$

shifts to higher pressures. Other reactions probable in experiments of D. H. Green and Ringwood (1967) and T. H. Green (1967) may be expressed as

$$\underset{\textit{olivine}}{2(Mg, Fe)_2SiO_4} + \underset{\textit{plagioclase}}{\{2CaAl_2Si_2O_8\}}$$
$$= \underset{\textit{clinopyroxene}}{Ca(Mg, Fe)Si_2O_6CaAl_2SiO_6}$$
$$+ \underset{\textit{orthopyroxene}}{\{2(Mg, Fe)SiO_3(Mg, Fe)Al_2SiO_6\}}$$
$$(\Delta V \simeq -37\ cm^3) \tag{11.hh}$$

$$\underset{\textit{Al-orthopyroxine}}{\{4(Mg, Fe)SiO_3\}} + \underset{\textit{plagioclase}}{\{CaAl_2Si_2O_8\}}$$
$$= \underset{\textit{garnet}}{\{(Mg, Fe)_3Al_2Si_3O_{12}\}} + \underset{\textit{Al-clinopyroxene}}{\{Ca(Mg, Fe)Si_2O_6\}} + \underset{\textit{quartz}}{SiO_2} \tag{11.ii}$$

where { } denotes a component of a crystalline solution. Note that orthopyroxene and clinopyroxene are components in the aluminous pyroxenes

$n(Mg, Fe)SiO_3 \cdot MgAl_2SiO_6$

and

$n[Ca(Mg, Fe)Si_2O_6 \cdot CaAl_2SiO_6]$,

respectively. The components $MgAl_2SiO_6$ and $CaAl_2SiO_6$ are Mg- and Ca-Tschermak's molecules, respectively. In pyroxenes, the Al^{3+} is partly in tetrahedral coordination and partly in octahedral. The equilibrium condition for Reaction (11.ii) is

$$K_{(ii)} \exp\left(\frac{-P\Delta V_{(ii)}}{RT}\right) = \frac{(a_{(Mg, Fe)}^{Gar})^3 a_{(Mg, Fe)}^{Cpx} a_{SiO_2}}{(a_{(Mg, Fe)}^{Opx})^4 a_{An}^{Plag}} \tag{11.28}$$

It is obvious that this is a complex relationship, since we have three ternary (Mg, Fe, Ca-garnet; Al, Mg, Fe-orthopyroxene; Al, Mg, Fe-clinopyroxene) and one binary (Na-plagioclase, Ca) solutions involved. However at the lower pressure boundary where pyroxenes may be binary solutions with a fixed Mg/Fe ratio, with other conditions equal, the rock rich in anorthite will yield garnet at a lower pressure. A further complication would be the reaction between the albite component and diopside [Reaction (11.ff)], which would change the pyroxene into omphacite. Although from Kushiro's (1969) work it appears that the diopside + anorthite reaction, to yield a pyroxene with 20 mole percent Ca-Tschermak, takes place at a lower pressure than the formation of omphacite, with 20 mole percent jadeite in the P–T range of the transitional granulite area, both reactions would compete. In the absence of suitable compositional data and the activity-composition relations in the various minerals, it is difficult to evaluate the quantitative nature of the boundaries between pyroxene-granulite and garnet-granulite, and garnet-granulite and eclogite.

Recently MacGregor (1974) has experimentally studied the crystalline solubility of Al_2O_3 in orthopyroxene, using the reactions

$$Mg_3Al_2Si_3O_{12} + (1 - X)Mg_2SiO_4 = (4 - X)MgSiO_3XAl_2O_3 + (1 - X)MgAl_2O_4 \quad (11.jj)$$

$$XMgAl_2O_4 + (1 + X)MgSiO_3 = MgSiO_3XAl_2O_3 + XMg_2SiO_4 \quad (11.kk)$$

$$XMg_3Al_2Si_3O_{12} + 3(1 - X)MgSiO_3 = 3MgSiO_3XAl_2O_3 \quad (11.ll)$$

Note that the composition $n(Mg, Fe)SiO_3 \cdot MgAl_2SiO_6$ may as well be represented by $(n + 1)MgSiO_3 \cdot Al_2O_3$, as done in Reactions (11.jj) to (11.ll). The equilibrium curve for Reaction (11.jj) and the results on the Al_2O_3 content in enstatite in equilibrium with pyrope or spinel are shown in Figure 11.29. The use of the figure in estimating the pressure and temperature of a natural assemblage requires the knowledge of the effect of other components and an independent assignment of either temperature or pressure. From Boyd's (1970) work it would appear that at 30 kbar and 1200°C, the Al_2O_3 content of enstatite in equilibrium with garnet varies from 5.9 to 4 weight percent from the calcium-undersaturated to calcium-saturated assemblage. It is, however, important to note that Figure 11.29 is applicable to all such assemblages that contain all the minerals of equilibrium Reactions (11.jj) and (11.ll) and the presence of an additional phase such as a Ca-pyroxene should significantly influence the Al_2O_3 content of enstatite only if the chemical potential of Al_2O_3 changes in enstatite as a result of this addition. Such a change may be reflected in the change in composition of orthopyroxene such as an increased concentration of Ca. It should be useful to study the distribution of Al_2O_3 between coexisting pyroxenes as a function of P, T, and composition.

11.29 Al_2O_3 content of enstatite in equilibrium with forsterite and pyrope or spinel (heavy bold line defines reaction boundary for Reaction (11.jj); less heavy lines represent Al_2O_3 isopleths whose values (given in circles) represent contents of Al_2O_3 in enstatite ranging from 1 to 15 wt. percent). (After MacGregor, 1974. Reprinted from the *American Mineralogist* by permission.)

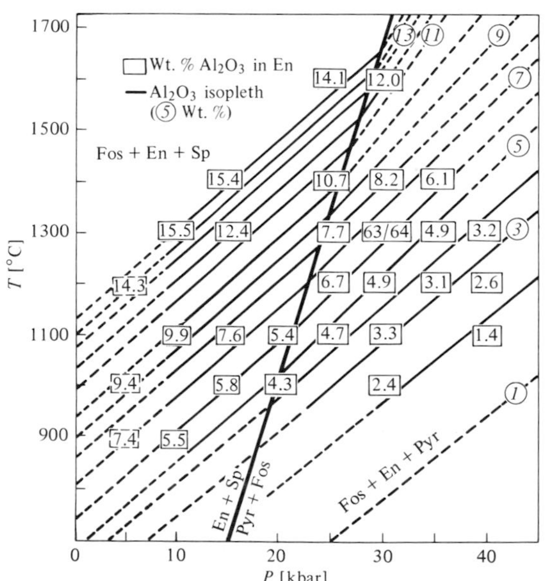

Rocks of apparently high-pressure and low-temperature origin

There are rocks of the so-called glaucophane schist or blue schist facies which occur in the mobile, feebly metamorphosed portions of late and post-Paleozoic eugeosynclines. They are generally concentrated around the margins of the Pacific Ocean, in the Caribbean, and the Alpine–Himalayan region. A summary on occurrence and mineralogic evolution of the blue schists has been given by Ernst (1972a). References to most occurrences can be found there. The minerals of these rocks include glaucophane, lawsonite, aragonite, jadeitic pyroxene, and some other minerals common in the greenschist facies, such as chlorite, actinolite, albite, stilpnomelane, muscovite, pumpellyite, and epidote. Glaucophane is stable over a wide range of physical conditions, but lawsonite, aragonite, and jadeite are stable only at moderate to high pressures. The peculiar mineralogy of this facies of rocks has been explained variously. Some favor the idea that the phase assemblages reflect low-temperature metasomatism accompanying the emplacement and serpentinization of alpine-type peridotites, common in most glaucophane schist terranes, or they represent metastable recrystallization. Others favor an origin due to the equilibrium recrystallization of sedimentary and basaltic rocks at high pressure (either high lithostatic pressure or tectonic overpressure).

The metasomatic production of observed assemblages has been considered by Brothers (1954), Bloxam (1966), and Gresens (1969), among others. Glaucophane-lawsonite schists commonly, but not invariably, occur closely associated with bodies of serpentinites. Essene *et al.* (1965) proposed that deep-seated sedimentary rocks at 200°C to 300°C and 10 kbar were intruded by ultramafic rocks, which were eventually serpentinized, and during this process gave rise to local deformation and fluids rich in Na_2O, CaO, and Al_2O_3. These fluids formed the vein assemblages (diopside, jadeite, lawsonite, etc.) commonly developed in glaucophane schists and metagraywackes at serpentinite contacts. While the formation of contact facies may have formed by some metasomatic introduction of material, the bulk of the glaucophane schists in the Fransiscan region appear to have been metamorphosed isochemically (Ernst, 1959, 1963a,b,c, 1972a; Coleman and Lee, 1963) and formed from preexisting metavolcanics and metasedimentary rocks as is indicated by their compositions (Table 4.4).

Ernst (1972a) has discussed in detail the other hypotheses of origin, which include metastable recrystallization and recrystallization due to tectonic overpressures. Metastable recrystallization of a high albite may result in the assemblage jadeite + quartz at moderate pressures. Ernst (1972a) indicates that such disordered feldspars are not present in major amounts to account for the common presence of jadeite. The influence of tectonic overpressures—an increase of fluid pressure due to stress—has been invoked by Blake *et al.* (1969) and Bailey and Blake (1969). The main difficulty with this hypothesis appears to be the amount of overpressure that can be generated. Under appropriate metamorphic conditions (T = 150°C to 300°C; P_{fluid} = 2 to 4 kbar; strain rates 10^{-13} or 10^{-14}), the possibilities of generating tectonic overpressures exceeding about 1 kbar on a regional scale are remote. This is particularly true for Fransiscan metagraywackes that are also intercalated with weak incompetent metashales.

Although none of the hypotheses need be rejected at the present stage of discussion and argument (see Gresens, 1972, for example), an equilibrium recrystallization of metavolcanics and metasediments at 200°C to 400°C and 5 to 10 kbar seems to be the favored mode of origin of rocks of the blue schist facies. We have already discussed several relevant experimental investigations—lawsonite: Newton and Kennedy (1963), Crawford and Fyfe (1965), Liou (1971); calcite-aragonite: Jamieson (1953), Clark (1957), Boettcher and Wyllie (1968), Newton *et al.* (1969); albite-jadeite: Newton and Smith (1967), Boettcher and Wyllie (1968).

Rocks of high pressure and high temperature

Garnet pyroxenites and some eclogites may have originated at high pressure and high temperature and under completely dry to very low P_{H_2O} conditions.

Garnet pyroxenites. These rocks may be distinguished from the eclogites by the composition of clinopyroxenes. The eclogites contain omphacite (with high ratios of jadeite and Tschermak's component) and lack plagioclase. Garnet pyroxenites contain clinopyroxene of the diopside-hedenbergite type with possible high concentrations of Al_2O_3. They do not necessarily lack plagioclase and may have formed under conditions of the granulite facies or at higher pressure and temperature. Forbes and Swainbank (1974) have described garnet pyroxenite masses from the Red Mountain Pluton, Alaska, where the garnets are rich in grossular (24 to 28 mole percent) and the clinopyroxenes are rich in Al_2O_3 (7 to 8.29 weight percent). Using the experimental data in Figure 11.29, these rocks may be inferred to have recrystallized between 1100°C and 1200°C, and at pressures of 15 to 20 kbar.

Grospydites and other inclusions in igneous rocks. Grospydite is the name for garnet + pyroxene ± corundum ± kyanite xenoliths found in the kimberlite pipes of Yakutia and Zagadochnaya in Siberia (Sobolev *et al.*, 1968). The garnets in the kyanite eclogite and grospydites vary in composition from nearly 80 percent grossularite ($Pyr_{12}Alm_{8.0}Gr_{80}$) to 30 percent ($Pyr_{34}Alm_{36}Gr_{30}$) and are fine examples of the crystalline solution series between grossular and pyrope/almandine. Pyroxenes are jadeite rich and are similar to omphacites in other eclogites. Their inclusion in the diamond-bearing kimberlite, which is an ultrabasic rock with olivine, pyroxene, amphibole, and micas, indicate their origin under high-pressure conditions. T. H. Green's (1967) experimental results show that kyanite-eclogite assemblages are attained at 1100°C at pressures between 20 and 25 kbar.

Several other types of eclogitic and noneclogitic inclusions have been described from basalts and peridotites. Eclogite inclusions with fassaite (clinopyroxene high in Tschermak's molecule), and inclusions with such assemblages as garnet + plagioclase + pyroxene and spinel + garnet + two pyroxenes + olivine or scapolite and plagioclase occur in basanite pipes in Delegate, Australia (Lovering and White, 1969). Dallmeyer (1974) has described eclogite inclusions in periodotite sills in Georgia and North Carolina. The mineral assemblage is garnet and omphacite, with secondary symplectites of clinopyroxene, plagioclase, and amphiboles. The garnets are zoned. The core is generally rich in almandine and poor in pyrope (for example, core: $Alm_{64}Gr_{19}Pyr_{16}$, rim: $Alm_{55}Gr_{17}Pyr_{27}$). Clinopyroxenes are omphacites with interior regions containing an average jadeite content of 28 mole percent. The rims of these pyroxenes are depleted in jadeite component and enriched in the Ca-Tschermak's molecule. The petrography and mineralogy of these inclusions suggest the following possible history:

1. Formation of an igneous peridotite
2. Formation of the subsolidus assemblage garnet $[(Alm + Sp)_{53-65}(Pyr)_{13-26}(Gr + Andr)_{19-27}]$, omphacite ($Jd_{24-30}$, $Ca\text{-}Ts_{4-10}$), and plagioclase (An_{9-15}) at an inferred pressure and temperature of 15 to 20 kbar and 1000°C to 1200°C, respectively
3. Emplacement of the periodotite series within the crust with concomittant intercrystalline reactions, recorded as zoning in garnet and omphacite, as a partial reequilibration at lower pressures
4. Shortly after emplacement, middle Paleozoic metamorphism effected a retrograde breakdown of the primary omphacite to a secondary symplectic composed of clinopyroxene (Jd_{4-9}, $Ca\text{-}Ts_{7-12}$), plagioclase (An_{28-39}), and amphibole.

Eclogites: rocks crystallized under various physical conditions

A lot of confusion can be avoided if we define eclogites as rocks with the definite assemblage of garnet (grossular-pyrope-almandine) and omphacite (a solution of jadeite, diopside/hedenbergite, acmite, and Ca-Tschermak components). When defined this way, the problem of the formation of eclogite reduces to the problem of the stability of the garnet-omphacite assemblage. The stability range of these minerals is by no means established. The task is difficult because both minerals are multicomponent solutions, and although information on many end members is available, little is known about the crystalline solutions. In any case, these minerals are stable over a wide pressure and temperature range and, hence, eclogites may also form under a variety of physical conditions.

Figure 11.30 shows the region of the basalt-eclogite transition of D. H. Green and Ringwood (1967) and some other experimental equilibria. Note that in the dry systems, experimental results are available only in the high-temperature range. Thus there is experimental confirmation for the formation of eclogites in the P–T field marked Z. Such eclogites generally occur as inclusions in basaltic and ultrabasic rocks. Any inferences of other physical conditions, such as those of area Y must be made on the basis of extrapolation of the experimental results. As to the extent of extrapolation, opinions vary and temperatures and pressures as low as 200°C and 7 kbar have been suggested.

Under hydrous conditions of metamorphism, almandine garnet enters the assemblage in the amphibolite facies. Hsu's (1968) results indicate a temperature of about 550°C at low pressures and with f_{O_2} determined by the fayalite-magnetite-quartz buffer. In dry systems, the hydrous phases would break down probably at a lower temperature. However, if we consider that garnet in eclogites contains a substantial amount of grossularite (13.2 to 34.2 mole percent in eclogites in glaucophane schists; Coleman *et al.*, 1965) and pyrope (6 to 13 mole percent), we should not expect a significantly lower temperature than 400°C for the stable recrystallization of garnets. It is doubtful that grossular-pyrope-almandine form an ideal solution, and the formation of a thermodynamically stable garnet with 35 mole percent of grossular and 13 mole percent pyrope at 200°C should be considered unlikely. Even at 300°C and 400°C, the recrystallization of a rock under dry conditions may indicate evidence of partial disequilibrium. It is possible, therefore, that in such eclogites grain-to-grain compositional variation may be very common due to a lack of extensive diffusion.

Area Y in Figure 11.30 represents eclogites that may form in some high-pressure granulite facies or amphibolite facies terranes. The experiments of Essene *et al.* (1970) indicate that under the condition $P_{H_2O} = P_{total}$, amphibole at 700°C is stable up to a pressure of 21 ± 3 kbar. This means that only under dry or partially dry conditions can an eclogite form in the amphibolite facies. This would also be true about the eclogites found in blue schist facies terranes, as for example in California with the physical conditions defined by the area X. The eclogites in many of such areas are not strictly bimineralic (garnet + omphacite) but also contain some hydrous minerals such as amphibole, zoisite, and sometimes micas. Sillimanite has not been reported, but kyanite is occasionally encountered in these assemblages. Therefore, from natural occurrences it would appear that the assemblage garnet + omphacite may be stable in the amphibolite facies with the low-pressure limit defined by the kyanite-sillimanite transition. The formation of eclogites and not of amphibolites or the formation of a mixed assemblage of garnet, omphacite, and hydrous minerals must depend on some critical compositional variables, particularly the fugacity of water.

A particularly interesting example of an eclogite

assemblage with hydrous minerals has been described by Morgan (1970) from Puerto Cabello, Venezuela. Rocks of basaltic composition have been transformed to omphacite-garnet eclogites, "eclogite-amphibolite" with garnet-clinopyroxene-zoisite-subcalcic amphiboles, and to some other amphibolites. The associated country rocks have attained a metamorphic mineral assemblage of middle amphibolite facies (see area with crosses in Figure 11.30; $T = 525 \pm 50°C$, $P = 7$ kbar). The basaltic rocks were only partly open to fluid components, and the formation of the rocks in the sequence

zoisite-chlorite amphibolite → garnet amphibolite → eclogite amphibolite → eclogite

took place with a decreasing chemical potential of H_2O. We should note here that the inference regarding the *in situ* metamorphism of eclogites has to be based on structural conformability of the rock units and the petrographic and chemical evidence of equilibrium among the anhydrous and hydrous phases of the assemblages. If the evidence is not sufficient, an alternate mode of origin—namely, the formation of eclogite at higher pressure and temperature in the mantle and its subsequent transportation to the crust—should not be ruled out (O'Hara and Mercy, 1963; Lappin, 1966; Carswell, 1968). However, generally the tectonic evidence of transportation may be more obscure than the phase-petrologic evidence of equilibrium, particularly because the deformation properties of a dry basic rock enclosed in wet metasediments are not well known.

Many eclogite bodies in western Norway occur in migmatized areas. This led Bryhni *et al.* (1970) to consider the relationship between partial melting of the crustal rocks and the formation of eclogites. An old basement that has been involved in repeated partial melting may lose much of its water, thus creating the condition $P_{H_2O} < P_S$. If such a basement were intruded by basic dykes, conditions could be unfavorable to amphibolite formation but may be favorable to the formation of eclogites under high pressure. In layers of mixed basaltic and granitic composition, at a certain

11.30 Pressure and temperature of formation of various eclogites and transitional facies. See text for explanation. Ec = eclogite.

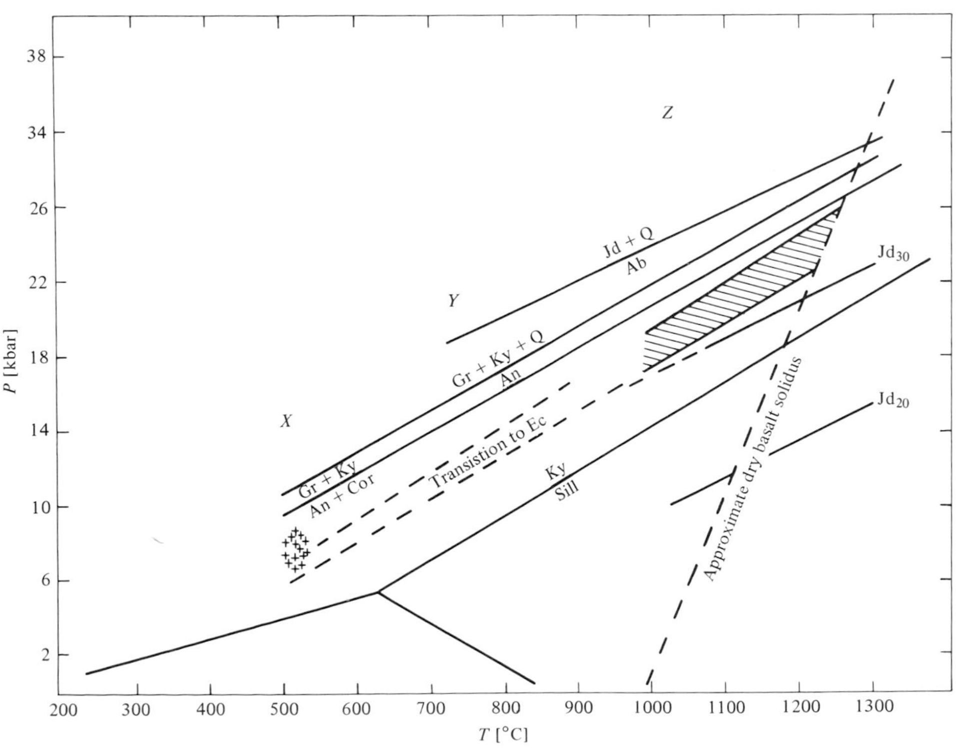

pressure, melting occurs at lower temperature in the granitic part, and under conditions of $P_{H_2O} < P_{total}$ amphibole may break down before the basic part melts. The acidic part is enriched in water and the basic part becomes more suitable for crystallization of eclogite assemblages. In western Norway eclogite lenses are closely associated with pegmatites. Such eclogites also contain poikiloblastic hornblende, whose form suggests replacement of garnet and pyroxene. It is likely that this replacement occurred during the pegmatitic stage.

In summary, we have evidence that eclogites crystallize under various physical conditions. Conclusions regarding the origin of a particular eclogite body should not, however, be based on the nature of the terrane in which they occur unless it can be shown that the phase assemblage is chemically and physically compatible with the surrounding rocks.

Ultramafic rocks at high pressures

Significant progress has recently been made in estimating the pressure and temperature of rocks of the upper mantle (Boyd, 1973; MacGregor and Basu, 1974). The temperature estimate is based on the enstatite-solvus data, and the pressure estimate is based on MacGregor's (1974) data on the concentration of Al_2O_3 in orthopyroxene. Although there is insufficient information on activity-composition relations in pyroxenes, the P–T estimates may be accepted as relatively accurate. MacGregor and Basu (1974) have compiled P and T data on several groups of ultramafic rocks and xenoliths (Figure 11.31). It is interesting to note that a set of samples from a single locality defines a linear trend, which may be interpreted as the geothermal gradient existing at the time the xenoliths were included in the intruding host magma. Another interesting point of this figure is the sudden change in the interpreted geothermal gradients, which occurs at successively greater depths for the Louwrencia, Lesotho, and Kimberley areas. MacGregor (1974) and Boyd (1973) have interpreted this change to coincide with the lithosphere-asthenosphere boundary.

11.31 Summary of P–T plots for ultramafic rock suites from different tectonic environments. (After MacGregor and Basu, 1974. Reprinted from *Science* by permission.)

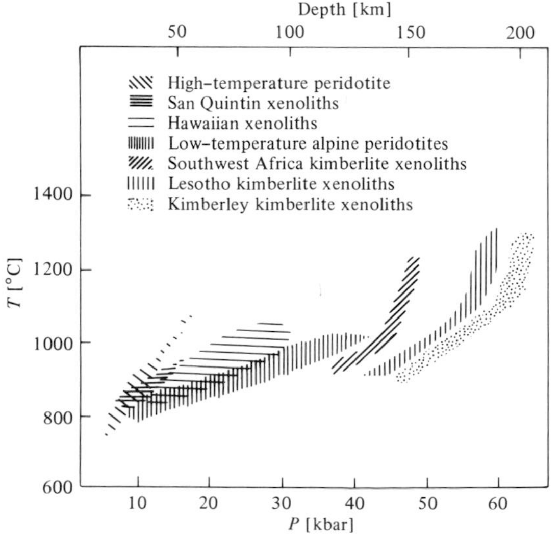

12 Properties, Origin and Environment of Magmas

Structures of silicate melts

Magmas are natural melts which are usually complex solutions of oxides or silicates. However, occasionally magmas composed of sulfides or metals may also occur. The complexity of the physicochemical behavior exhibited by these melts arises from the interactions of the solution components over a broad range of temperatures and pressures. A part of this complexity is revealed by the chemical analyses of rocks such as those presented in Table 4.1 and elsewhere, which frequently show 10 or more major components. However a large part of the behavior of melts is attributable to constituents which either do not regularly appear in the analyses or which appear as only traces in them. This is true for water and such other volatile constituents as HCl, HF, etc., the greater part of which may have been lost from the magma before solidification occurred. Then too the effect of a component is usually more closely related to its molar concentration than to its weight percentage so that light constituents such as H_2O and HF may have a proportionately larger effect (Buerger, 1948).

It is instructive to consider melts in terms of a structural or atomistic model since this approach not only provides us with some qualitative insights but also forms the basis for a more quantitative treatment of certain important properties such as volatile solubilities and viscous behavior. There is considerable evidence that silicate melts—like the crystals from which they are derived—have as their fundamental structural units the SiO_4^{4-} tetrahedral group and the associated larger cations such as Ca^{2+} and Na^{+} (Tomlinson, 1953). Also, as in the crystalline structure, the SiO_4^{4-} groups will in the melt be linked or polymerized to varying degrees into chains, rings, sheets, and three-dimensional frameworks consonant with the composition and the physical environment. In fact we are to regard melts simply as highly disordered versions of the crystalline solids. This is particularly well illustrated in Riebling's (1966) work. He obtained a series of viscosity isotherms for melts of constant SiO_2 content in the system Na_2O-Al_2O_3-SiO_2. A pronounced maximum in viscosity occurs when Al/Na is unity. For more Al-rich compositions, the viscosity of the melt decreases. It is possible that at Al/Na = 1, the melt is closely similar in structure to albite, but with excess Al or Na, the linkages depolymerize into other units. Change in the activation energy for viscous flow is similar to the change in viscosity, reaching a maximum at Al/Na = 1 and decreasing at other compositions probably because of depolymerization. If we pursue this idea to its ultimate extreme an orthosilicate or olivine-like melt should consist of the units Mg^{2+}, Fe^{2+}, and SiO_4^{4-}, a metasilicate or pyroxene-like melt of Mg^{2+}, Fe^{2+}, and Si_nO_{3n} chains or rings, and feldspar melts of Ca^{2+}, Na^{+}, and K^{+} ions in a fairly continuous but disordered $(Si, Al)O_2$ framework. Presumably melts with compositions falling between these unique stoichiometries would be composed of a mixture of polymeric units. Actually this picture is far too simple, and rather elementary energetic considerations tell us that we would expect even in the ranges of unique stoichiometry a distribution of silica and alumina among a variety of groups. It might happen for example that the energy content of the system requires a metasilicate stoichiometry to be expressed as follows:

$$\begin{aligned} MgSiO_3 &\rightleftharpoons Mg^{2+} + (SiO_3)^{2-} \\ 2MgSiO_3 &\rightleftharpoons 2Mg^{2+} + (SiO_4)^{4-} + SiO_2 \end{aligned} \tag{12.a}$$

with both dissociation schemes represented to a certain degree. We should of course expect the simplest units to be favored at higher temperatures even if their formation involves increasing ionization. In spite, however, of the relative chaos introduced by this type of disociation, we may still expect considerable local order to prevail with the familiar associations and linkages being maintained for short distances.

Hess (1971) calculated the degrees of random polymerization of SiO_4 structural units using the statistical methods of Flory (1953). These calculations yield the size distribution of silicate species in a melt as a function of the degree of polymerization. Using the Temkin model of ionic liquids, the degree of polymerization can be correlated to the composition of a binary silicate melt. Activity composition curves for the systems PbO-SiO_2 and Na_2O-SiO_2 and for some other systems calculated from such models are in good accord with the experimental data. The distribution of the silicate species in the melt is a function of temperature and SiO_2 content. The number of SiO_4 monomers at a given temperature decreases with increasing SiO_2 parallel to the polymerization of oxygen in the series olivine-pyroxene-amphiboles-micas-feldspars and quartz.

This structural model must be modified further when volatiles are introduced. Certain of these—water, for example—might be expected to react energetically with both the larger cations and with the SiO and Al-O chains and frameworks present. The latter reaction is particularly important because it leads to rupture of the bridging oxygen bonds. This problem has been examined in detail by Wasserburg (1957) in his important statistical treatment of water solubility in feldspar melts. According to his analysis water may occur.

1. As H_2O molecules segregated in holes or as water of hydration of the large cations
2. As OH replacement of the bridging oxygen atoms
3. As OH in replacing nonbridging or unshared oxygen or as OH in combination with the large cations

In the type 2 occurrence, the splitting of the Si-O chains or framework structures may be represented as follows:

$$\left[-\overset{|}{\underset{|}{Si}}-O-\overset{|}{\underset{|}{Si}}- \right] + H_2O \longrightarrow 2\left[-\overset{|}{\underset{|}{Si}}-OH \right] \tag{12.b}$$

Here the bracketed symbols represent arbitrary polymerized chains or groups. Reaction (12.b) is particularly important because it alters radically the continuity and mean coherence energy of the melt, and its ultimate effect is the production of simple neutral species such as $Si(OH)_4$, which interact—but weakly—with each other. We shall see that this has a profound effect on such properties as the viscosity.

The foregoing picture of the silicate melt is depicted in Figure 12.1. In this figure an irregularly branched Si_nO_{3n} chain crosses the field in association with a six-membered ring, various independent charged or neutral tetrahedra, and cations either singly occurring or in association with neutral particles. In an ordinary magma the small neutral particles would consist dominantly of H_2O molecules, whereas the neutral SiO_4 groups would be represented by $Si(OH)_4$. The mode of occur-

12.1 Diagrammatic picture of silicate melt structures. Cations are indicated by (+) symbol, anions by (−) symbol. SiO_4^{4-} tetrahedra are shown with (−) charges on corners. Neutral species are shown as plain circles. Several polymerized ring and branched-chain structures of these tetrahedra are also shown.

rence of local order is clear in the figure, as is the lack of any long-range order.

It is evident that the inferred ionic melt structure should be reflected in certain physical properties such as, in particular, the electrical conductivity. That this is indeed the case has been shown by numerous experiments (Tomlinson, 1953). However as yet there have been few investigations of either the electrical properties of rock-forming magmas (see Lebedev and Khitarov, 1964 and the recent work by Murase and McBirney 1973) or of the petrogenic consequences thereof. It is hoped, however, that this will form a fruitful field of research for the future.

Solubilities of volatile components

The solubilities of H_2O, HCl, etc., are of critical importance in understanding both the physical and chemical behavior of magmas. In this chapter we shall be concerned largely with the effects on physical properties; we shall not discuss in any detail such phenomena as the depression of liquidus temperatures by volatile components.

The first extensive work on the solubility of water in silicate melts was undertaken by Goranson (1936, 1938), who studied the system $NaAlSi_3O_8$-H_2O and $KAlSi_3O_8$-H_2O and natural granites as well. Burnham and Davis (1971) have determined the P–V–T relationships in the system $NaAlSi_3O_8$-H_2O. Wyllie and his collaborators have carried out an extensive study of water–silicate-melt relations (e.g., Koster van Groos and Wyllie, 1968a,b, 1969; Merill, *et al.*, 1970; Boettcher and Wyllie, 1969). Figures 12.2 and 12.3 show examples of the water–melt relations as a function of pressure and temperature.

It may easily be shown from elementary thermodynamic considerations (Chapter 1) that under isothermal conditions the following equation relates a saturated pure vapor phase to the water dissolved in a silicate melt:

$$V^{v}_{H_2O}dP = V^{M}_{H_2O}dP + RTd \ln a^{M}_{H_2O} \tag{12.1}$$

Here $V^{v}_{H_2O}$ and $V^{M}_{H_2O}$ are the partial molar volumes of water in the vapor and melt, respectively, P is the water pressure, and $a^{M}_{H_2O}$ is the activity of water in the melt and is regarded as independent of P (Wasserburg, 1958). If we now assume that the vapor is a perfect gas we have $PV^{v}_{H_2O} = RT$. Also if $V^{M}_{H_2O}$ remains essentially constant,

12.2 Weight percentage of dissolved water in Stone Mountain granite as a function of saturation water pressure. T = 1173°K. (After Goransen, 1931.)

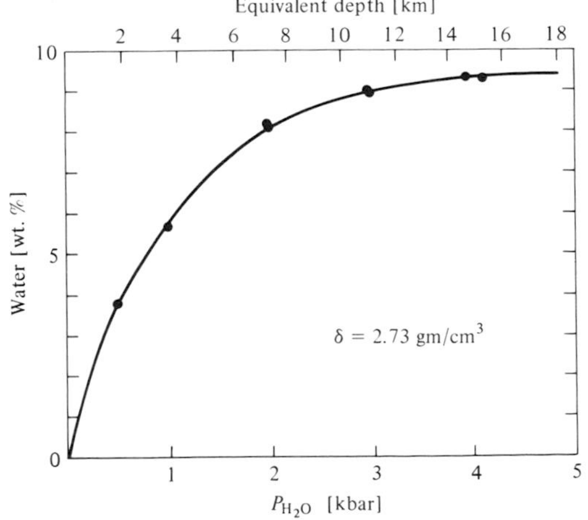

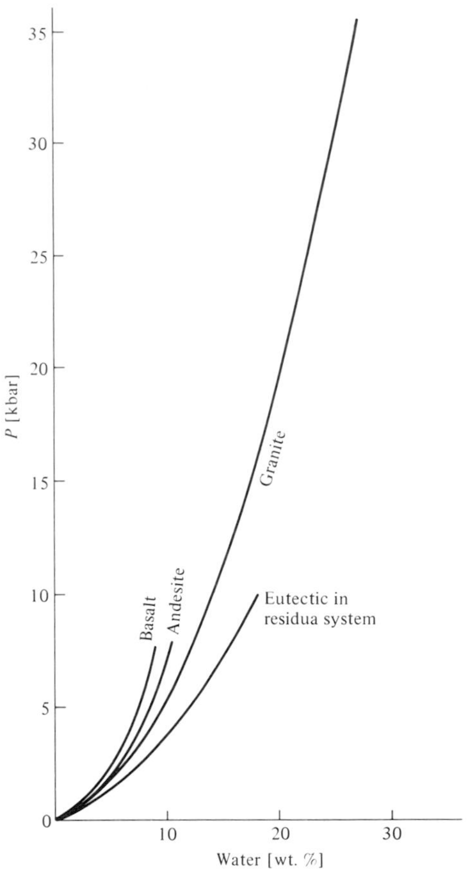

12.3 Solubility of water in rock melts at water saturated liquidus temperatures. (After Stern and Wyllie, 1973.)

Equation (12.1) integrates to

$$\ln\left(\frac{P}{P_0}\right) - \frac{V^{M}_{H_2O}(P - P_0)}{RT} = \ln \frac{a^{M}_{H_2O}}{a^{M}_{0H_2O}} \tag{12.2}$$

in which the subscript 0 refers to the parameters at some reference equilibrium condition which serves as the lower limit of integration. If in addition we assume that the solution of water in the silicate melt is ideal, we obtain the exponential form

$$X^{M}_{H_2O} = \frac{X^{M}_{0H_2O}}{P_0} P \exp\left[\frac{-V^{M}_{H_2O}(P - P_0)}{RT}\right] \tag{12.3}$$

In this expression X refers, as usual, to the mole fractions of water dissolved in the melt.

By an expansion of the exponential term Equation (12.3) may be converted into the following form:

$$X^{M}_{H_2O} = \frac{P}{A + BP} \tag{12.4}$$

in which A and B are constants. This same general form also holds for the weight fraction of water. Obviously Equation (12.4) has validity over an even more restricted range of temperature and pressure than Equation (12.3). In spite of this, however, it was found by Goranson that the general form of Equation (12.4) was a good approximation for the data represented by Figure 12.2.

We have already referred to Wasserburg's application of statistical mechanics to the system $NaAlSi_3O_8$-H_2O and the melt structure model, which forms the basis for this approach. In this model the entropy of mixing S_m is derived by enumerating the total number of states, Ω, accessible to the system. These are then related by the well-known expression

$$S_m = R \ln \Omega. \tag{12.5}$$

These states correspond to the number of ways N_{H_2O} molecules of H_2O can be distributed either as replacements for bridging oxygens [the type 2 structural positions] or as nonbridging water [the types 1 and 3 structural positions]. If we let r represent the number of bridging oxygens per silicate formula weight (eight for $NaAlSi_3O_8$) and N_{Sil} the number of silicate "molecules", then there will be $N_{H_2O} + rN_{Sil}$ ways of placing the first water molecule, $N_{H_2O} + rN_{Sil} - 1$ ways for the second, etc. The number of distinguishable configurations will then be

$$\Omega = \frac{(N^{M}_{H_2O} + rN^{M}_{Sil})(N^{M}_{H_2O} + rN^{M}_{Sil} - 1)\ldots(rN^{M}_{Sil} + 1)}{N^{M}_{H_2O}!} \tag{12.6}$$

which reduces to

$$\Omega = \frac{(N^{M}_{H_2O} + rN^{M}_{Sil})!}{N^{M}_{H_2O}!(rN^{M}_{Sil})!}. \tag{12.7}$$

If we apply Stirling's approximation to Equation (12.7) we find that

$$S_m = -RN^{M}_{H_2O} \ln\left[\frac{X^{M}_{H_2O}}{r - (r - 1)X^{M}_{H_2O}}\right] - RrN^{M}_{Sil} \ln\left[\frac{rX^{M}_{Sil}}{1 + (r - 1)X^{M}_{Sil}}\right] \tag{12.8}$$

since

$$X^{M}_{H_2O} = \frac{N^{M}_{H_2O}}{N^{M}_{H_2O} + N^{M}_{Sil}} \qquad \text{etc.}$$

The model is equivalent to an ideal mixture of $N^M_{H_2O}$ molecules and rN^M_{Sil} "molecular segments." The activities for the water and silicate components are then

$$a^M_{H_2O} = \frac{X^M_{H_2O}}{r - (r - 1)X^M_{H_2O}} \quad (12.9)$$

$$a^M_{Sil} = \left[\frac{rX^M_{Sil}}{1 + (r - 1)X^M_{Sil}}\right]^r \quad (12.10)$$

If we substitute $a^M_{H_2O}$ from Equation (12.9) for $X^M_{H_2O}$ in Equation (12.3), we obtain an alternative expression for the solubility of water in a silicate melt. Wasserburg found that this expression provided a good approximation for the solubility of water in an albite melt in the region $T = 1273°K$ and $P = 2.3 \times 10^3$ bar.

In addition to the data of Goranson, further recent experiments on the solubility of water in complex silicate melts has been obtained by a number of investigators. Much of this work[1] was discussed by Shaw (1964), who also attempted a further refinement of the statistical approach to the system $CaAlSi_2O_8$-$KAlSi_3O_8$-$NaAlSi_3O_8$-SiO_2-H_2O. The model employed was that of the regular solution and it was found to provide a fair approximation for most of the data. The chief deviation occurred in the binary join SiO_2-H_2O, which appears to require an even more complex solution model. It was generally found that a negative heat of solution was required to fit the data. This is instructive because it indicates that these melts have a greater affinity for water than would be the case if the solution were ideal. Also, as was pointed out by Shaw, this behavior practically eliminates the possibility of liquid immiscibility occurring in melts of these compositions.

Burnham and Davis (1971) have determined the P–V–T data on the $NaAlSi_3O_8$-H_2O system. Their results indicate that the pressure dependence of the partial molar volume of water in the melt is significant. The ΔV of solution of H_2O in $NaAlSi_3O_8$ melts at high temperatures and moderate pressures is very large and negative. This is probably due to interaction between H_2O and the silicate melt. Burnham (1967) and Carron (1969) consider the water melt reaction as follows:

$$H_2O + O^{2-}(M) = 2OH^-(M)$$

where (M) stands for melt. Assuming that (OH) ions are predominant in the $NaAlSi_3O_8$-H_2O melt, the calculated value of $\overline{V}_{OH^-}$ from the experimental P–V–T data was found by Burnham and Davis (1971) to be 9.8 cm^3/mole at 20°C. Further assuming that the OH^- and O^{2-} ions in the hydrous glass at 20°C and 1 bar are essentially rigid spheres in a close-packed arrangement, a $\overline{V}_{OH^-}$ of 9.8 cm^3/mole led these authors to a claculated radius for the OH^- ion of 1.38 Å. It is likely that OH^- ions are predominant in the melts.

Information on the solubility of volatile constituents other than water is being sought actively by Wyllie and his collaborators. Some information comes to us second hand from such effects as their influence on the melting behavior. For example, it was found by Wyllie and Tuttle (1959) that the liquidus temperatures of granites and feldspar are *increased* when their melts are subjected to high CO_2 pressures. This certainly indicates a very small solubility of CO_2 in melts of this composition. As Wyllie and Tuttle suggested, there should be a relation between the CO_2 solubility and the amount of cations not bound to SiO_4 groups in the liquid; this is also indicated by the work of Morey and Fleischer (1940). Examples of such liquids are the alkaline magmas with high values of the ratio

$$\frac{K_2O + N_2O + CaO + MgO + FeO}{Al_2O_3 + SiO_2}$$

In such *miaskitic* or *agpaitic*[2] liquids, homogeneous reactions of the following type are pertinent:

$$\underset{melt}{Na_2O} + \underset{melt}{CO_2} \rightleftharpoons \underset{melt}{Na_2CO_3} \quad (12.c)$$

$$\underset{melt}{CaO} + \underset{melt}{CO_2} \rightleftharpoons \underset{melt}{CaCO_3} \quad (12.d)$$

$$\underset{melt}{MgO} + \underset{melt}{CO_2} \rightleftharpoons \underset{melt}{MgCO_3} \quad (12.e)$$

$$\underset{melt}{FeO} + \underset{melt}{CO_2} \rightleftharpoons \underset{melt}{FeCO_3} \quad (12.f)$$

so that the solubility of CO_2 is thereby promoted. Of course the CO_2 partial pressures required to maintain a given solubility will depend on the stabilities of the carbonate complexes. Since $NaCO_3$ is less stable than $MgCO_3$, $CaCO_3$, and $FeCO_3$, the latter constituents should require the highest pressures.

Wyllie and Tuttle, examined, in addition to CO_2, the effects of H_2SO_4, H_3PO_4, HCl, HF, and Li_2O on the melting points of granite and feldspars (Wyllie and Tuttle, 1961, 1964). Some of the volatile components, such as F or CO_2, may be added as salts. As an example, we may consider the system albite-NaF-H_2O at 1 kbar pressure, studied by Koster van Groos and Wyllie

[1] See, especially, the work of Stewart (1958), Tuttle and Bowen (1958), Burnham and Jahns (1962), and Spengler and Burnham (1962).

[2] The reader is referred to the discussion on alkalic rocks in Chapter 14.

(1968a). Addition of NaF to the system albite-H_2O lowers the temperature of initial melting from 910°C to 688°C at 1 kbar. Similarly NaCl also reacts with the melt and doubles the solubility of H_2O in the silicate liquid (albite-H_2O) (Koster van Groos and Wyllie, 1969). It appears from these studies that, as might be expected, certain of the constituents react strongly with the melt and bring about substantial lowering of the liquidus temperatures in some cases. However, the precise solution relations and mechanisms remain unexplored as yet.

Densities of magmas

During the intrusion or extrusion of magmas there is a close relation between the height of the magma column and the pressure which obtains at the source of the column. This may be expressed by the following relation:

$$P_1 \geq \int_{h_1}^{h_2} g\delta_m dh. \qquad (12.11)$$

In this expression P_1 is the hydrostatic pressure at the magma source, which is located at a height h_1 and is extruding magma at a height h_2 in the gravitational field; g is the gravitational acceleration, which is for all practical purposes a constant in crustal regions; and δ_m is the density, which is of course a function of the temperature, pressure, and composition (i.e., $\delta_m = \delta_m(P, T, X)$, where P, T, and X are indications of these parameters at any point in the column). Obviously if P_1 exceeds the integral the magma will rise and be extruded or intruded. Such a condition could result if the sum of the weights of the overburdening rock layers exceeded the weight of the magma column penetrating them. This may be expressed as

$$\sum g\delta_i \Delta h_i \geq \int_{h_1}^{h_2} g\delta_m dh, \qquad (12.12)$$

in which the index i refers to the ith layer of thickness Δh_i and density δ_i; summation is between h_1 and h_2. Inequality could of course also result if the overpressure were developed by organic stresses.

The relation between the density of a melt and its composition is very complex since it depends on the partial molar volume of mixing of each constituent.[3]

[3] Density may be calculated approximately by using the partial molar volumes of oxides as discussed by Bottinga and Weill (1970).

However, for most anhydrous basic magmas the density will be closely related to the iron content. Unfortunately most of the existing density determinations refer to quenched glasses near STP conditions, although some high-temperature data are available. From these data it appears that glasses under STP conditions range from about 2.3 gm/cm^3 for granitic compositions with high-water contents to about 2.8 gm/cm^3 for anhydrous basalts. These values correspond to approximately 10- and 5-percent volume changes on melting for granites and basalts, respectively. The effect of dissolved water on glasses of granitic composition are shown in Figure 12.4.

The effect of temperature and pressure on the densities of melts may be evaluated from the isobaric thermal expansion α and the coefficient of isothermal compressibility κ (see Chapter 1). For magmatic glasses, α falls in the range of 20×10^{-6}/°K and κ in the range 3×10^{-6}/bar. One consequence of the scale relation between α and κ is that the volume increase occasioned by a 1000°K increase in temperature is counteracted approximately by a pressure increase of 10^4 bar. As a result of this, it seems likely that the densities of many magmas at great depth might not be far different than the densities of corresponding glasses under STP conditions. Thus a plateau basalt glass under STP conditions was found to have a density of 2.84 gm/cm^3, whereas at 1673°K and 17,000 bar the density of this same glass was 2.8 gm/cm^3 (Daly, 1933). Although the density changes brought about by thermal expansion and compressibility effects may not seem large, they could play a critical role in a magma with a density distribution such that the two sides of Expression (12.12) are nearly equal.

12.4 Effect of dissolved water on the densities of three rhyolite glasses. Data taken from Tilley (1922).

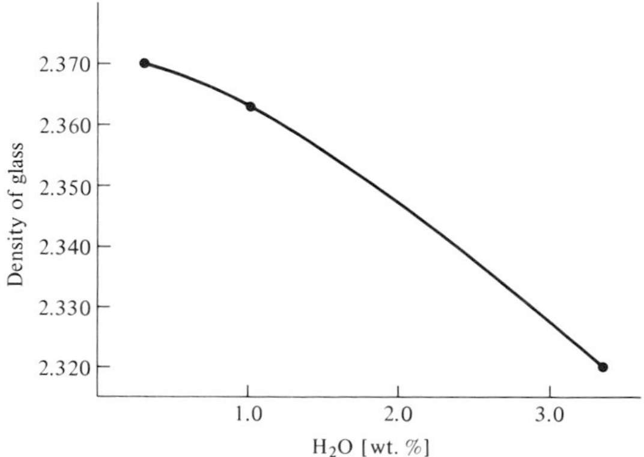

Viscosities of magmas

General

The gross effects of the variation of viscosity among the different magma types are known to every student of elementary geology through their influence on volcanic land forms. The impressive steep-sided stratocones, which are the products of acid and alkaline volcanism, contrast sharply with the lava plains and unimpressive shield volcanoes of basaltic volcanism. But beyond these surficial effects, viscosity plays a more subtle but equally impressive part in many deep-seated magmatic processes. It governs not only the rates of intrusion but also the forms the resulting igneous bodies take in relation to the country rock, and it enters into the details of magmatic differentiation through influences on such processes as crystal settling, filter pressing, and the segregation and flow orientation of solid elements in the magma.

We have already seen (Chapter 3) that viscous flow phenomena are essentially chemical in nature since they involve the thermal diffusion and interaction of atomistic particles. We have also seen that according to the theory of absolute reaction rates a simple expression for the logarithm of the viscosity results when the system reduces to the Newtonian fluid model. This may be written as follows;

$$\log \eta = \log\left(\frac{h}{V}\right) - \frac{\Delta S^{\ddagger}}{2.303R} + \frac{\Delta H^{\ddagger}}{2.303RT} + \frac{P\Delta V^{\ddagger}}{2.303RT} \tag{12.13}$$

The first three terms on the right-hand side of this expression can be combined to form the semiempirical expression

$$\log \eta = \log A + \frac{\Delta E^{\ddagger}}{2.303RT} + \frac{P\Delta V^{\ddagger}}{2.303RT} \tag{12.14}$$

in which the first two terms on the right-hand side represent the conventional Arrhenius equation in which A is the frequency factor or preexponential constant and $\Delta E^{\ddagger}$ is the empirical "energy of activation." The third term on the right-hand side containing the hydrostatic pressure P and the volume change of activation $\Delta V^{\ddagger}$ takes account of the variation of the viscosity with pressure, which may be an important consideration in some geophysical problems.

Effect of temperature and silica content

Expression (12.14) is very useful in the analysis of viscosity data that represent melts of different temperatures and silica contents. Bottinga and Weill (1972) have presented a critical review of the measured viscosity data. If a melt is truely Newtonian then at constant pressure and composition it should yield a linear plot of $\log \eta$ against $1/T$. In Figure 12.5 several such plots are given for natural anhydrous lavas ranging in composition from basalt to alkaline dacite. However, one set of data shown in this figure, that of Minakami (1951), does not refer to measurements of anhydrous lavas at constant composition since the measurements were made on a naturally flowing basalt that was losing its volatiles. It must also be kept in mind that most of the data of this figure were obtained at temperatures far above those possessed by natural magmas. It is only in the low-temperature range that they attain these values. It is apparent, however, that most of the measurements adhere closely to the Newtonian model. It is also clear that the effects of increasing silica content in the basalt-dacite series is very large. Although the effects of silica is apparent largely from the parallel upward displacement of the curves, which is equivalent to an increase in A, careful measurement of the slopes reveal that these

12.5 Variation of the viscosity η (in poises) as a function of temperature. Data taken from Volarovic *et al.* (1906), Kani (1934), and Minakami (1951). The broken vertical lines indicate crystallization.

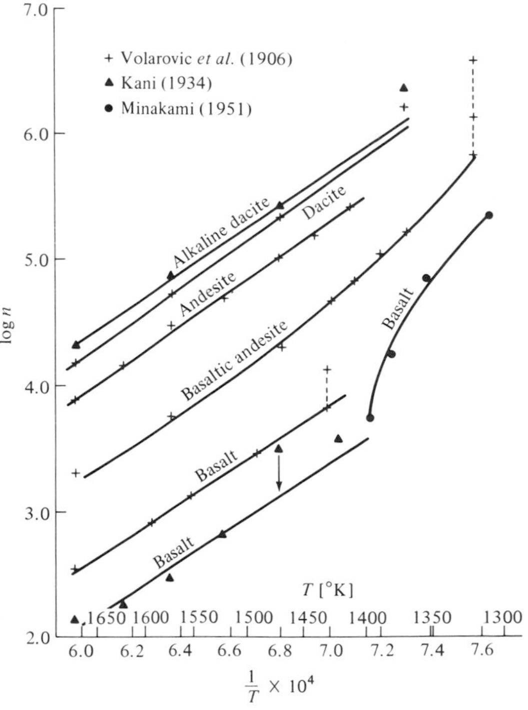

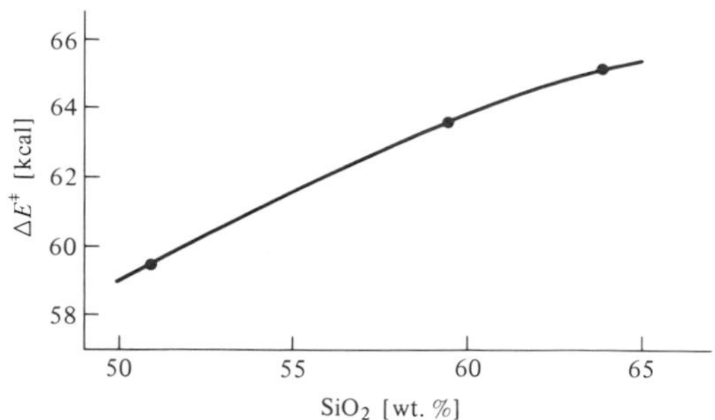

12.6 Variation of the activation energy $\Delta E^{\ddagger}$ as a function of SiO_2 content from three rock melts. $\Delta E^{\ddagger}$ may be obtained directly from plots such as those in Figure 12.5.

also increase with silica content so that this increase implies a substantial increase in $\Delta E^{\ddagger}$. This change is depicted in Figure 12.6.

Effect of water content

The effect of water, referred to previously in connection with the atomistic model of melts and solubility relations, is to break up the high polymer chains and groups to simpler units; as might be expected the viscosity is thereby greatly reduced. In the case of basalts this effect is apparent in the curve obtained by Minakami (1951) (Figure 12.5), which represents the change of viscosity in a natural lava flow in which cooling was accompanied by loss of volatiles, presumably mostly water, and perhaps also by certain simultaneous reactions such as the oxidation of FeO. The bulk of the volatiles were lost in the initial stages when the lava was at a temperature of nearly 1400°K so that on further cooling the viscosity increased at a decreasing rate and eventually the slope of the curve approached asymptotically those of the anhydrous experiments.

As might have been anticipated the effect of water on viscosity is greatest in melts of granitic composition since these have the highest degree of polymerization in the anhydrous state. As in the case of the more basic melts, there is only a small amount of data for melts of granitic composition and most of this has been summarized by Shaw (1965), who presented the results in the form shown in Figure 12.7. Shaw (1963) and Friedman *et al.* (1963) also demonstrated that natural obsidian and rhyolite melts of fixed water contents yield linear $\log \eta - 1/T$ plots similar to those of Figure 12.5. Also, it was found that there were very marked decreases in the experimental activation energies with increasing water content. For rhyolite and obsidian the anhydrous values of more than 10^5 cal are reduced to approximately 4×10^4 cal when more than 6 wt.-percent water is added.

Scarfe (1973) has determined viscosity of some basic magmas. The effect of dissolving 4 ± 0.5 weight-percent H_2O in the melt of basic andesite at 1150°C was to lower the viscosity from 10,000 to 2,500 P. A tholeiitic melt was lowered from 1600 to 600 P. A melt of olivine basalt, however, showed no measurable reduction in viscosity. It seems that ultrabasic melts are already depolymerized to the extent that water is ineffective as a depolymerizer.

It should be apparent from the foregoing that it is very difficult to determine, even approximately, from rock analyses what the viscosity of a magma might have been at the time of intrusion or extrusion. However it is clear that certain highly useful limits may still be specified. For example, it might be determined from the mineral assemblage of a granite that crystallization took place in the neighborhood of 1000°K and that the total pressure cannot have exceeded 1 kbar. Under such conditions the granite melt can, from Figure 12.2, contain a maximum of approximately 5 to 7 weight percent H_2O. Then comparison of these data with Figure 12.7 tells us that it is unlikely that the viscosity could be much less than 10^5 P. By such means it is possible to place close restrictions on the interpretation of certain structural evidence which is common to many plutonic bodies.

12.7 Effect of water content on the viscosity of obsidian melt (Shaw, 1965). Numbers indicate weight percentage of H_2O.

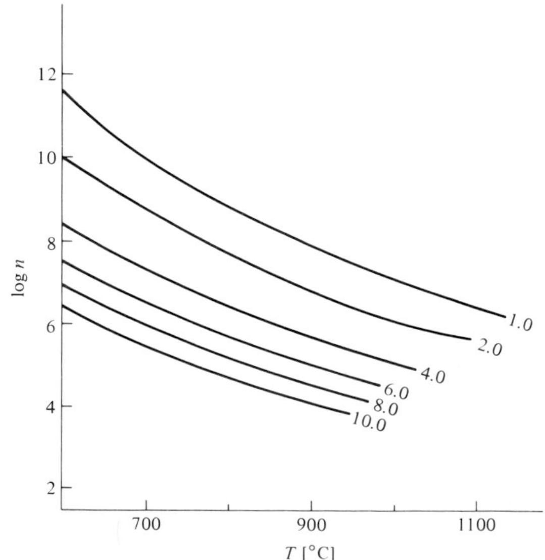

Recently there has been a significant advancement in our understanding of the interrelationship of viscosity and the composition of the magmas, largely due to the work of Burnham (1967), Bottinga and Weill (1970, 1972), and Shaw (1965, 1969, 1972). A method of calculating viscosity in a multicomponent silicate liquid from chemical composition has been presented by Shaw (1972).

Effect of pressure

It has sometimes been suggested that a considerable amount of magma might exist within Earth's upper mantle and that these magmas might be so viscous that they would behave essentially as solids in the transmission of seismic waves. Presumably this high viscosity would result from the high pressures which would in this region overcome the effect of tempeature.

This problem was examined by Birch and Bancroft (1942), who applied a modified version of Equation (12.14) to the problem. These authors assumed that to a first approximation the viscosity is a function of the specific volume alone, so that we have

$$\eta = f(V) \tag{12.15}$$

It is therefore possible to relate the viscosity, the isothermal compressibility κ, and the isobaric coefficient of thermal expansion α as follows:

$$\left(\frac{\partial\eta}{\partial T}\right)_P = \frac{\partial f(V)}{\partial V}\left(\frac{\partial V}{\partial T}\right)_P = \frac{\partial f(V)}{\partial V}\alpha V$$

$$\left(\frac{\partial\eta}{\partial P}\right)_T = \frac{\partial f(V)}{\partial V}\left(\frac{\partial V}{\partial P}\right)_T = \frac{\partial f(V)}{\partial V}\kappa V$$

from which it follows that

$$\left(\frac{\partial\eta}{\partial P}\right)_T = \xi\left(\frac{\partial\eta}{\partial T}\right)_P \tag{12.16}$$

where $\xi = \kappa/\alpha$.

It is now assumed that for a limited range of pressure and at constant temperature

$$\eta = A' \exp\left(\frac{B}{T}\right) \tag{12.17}$$

where A' is a constant and B is another constant independent of T but which is a function of P. Similarly it is assumed that for a limited range of the temperature and at constant pressure we have

$$\eta = A'' \exp(CP) \tag{12.18}$$

where A'' is a constant and C is another constant independent of the pressure but dependent on the temperature. If we now further assume that these separate effects can be combined, we may write the following expression for η as a function of T and P:

$$\eta = A \exp\left(\frac{B}{T} + CP\right) \tag{12.19}$$

in which obviously $A = A'A''$. This expression is now differentiated with respect to the pressure at constant temperature and vice versa, and these results are substituted into Equation (12.16), which then yields the following expression for the pressure coefficient C:

$$C = \frac{\xi B}{T^2} - \frac{\xi P \partial C}{\partial T} - \frac{1}{T}\frac{\partial B}{\partial P} \tag{12.20}$$

which reduces approximately to

$$C = \frac{\xi B}{T^2} \tag{12.21}$$

Now the values of C may also be derived directly from Equation (12.19) if at the same time we have experimentally determined values of η, A, and B for various values of T and P. In this way Birch and Bancroft compared values of C for certain glasses with those obtained from Equation (12.21) and found good agreement. They then applied Equation (12.21) to basalt glass for which B was available and found the small value $C = 0.0005$. This corresponds to an increase in η by only about a factor of 10^4 for a pressure increase of 17,000 bar. It is consequently very improbable that a basaltic magma of exceptionally high viscosity could exist within the upper mantle.

Viscosities of suspensions

Thus far we have limited our discussion to completely liquid magmas. However it frequently happens that crystallization is already in progress at the time of intrusion or extrusion. As we see from Figure 12.5 when isothermal crystallization occurs the viscosity of the suspension increases rapidly. There are two main reasons for this:

1. The melt becomes more viscous by virtue of an increase of silica. This effect may be offset in a closed magma chamber by the simultaneous increase in the concentration of volatiles.
2. A melt with suspended particles is more viscous than a melt of identical composition that does not contain these particles.

It may be shown that the viscosity η_s of a suspension of spheres is related to the viscosity η_0 of the pure suspending liquid as follows:

$$\eta_s = (1 \pm a\phi)^{\pm n}\eta_0 \tag{12.22}$$

Here ϕ is the volume fraction occupied by the spheres and a and n are positive constants. In the case of dilute suspensions, Equation (12.22) reduces to a form first given by Einstein (1906, 1911) in which $a = 2.5$ and $n = 1$. In an analogous formula given by Roscoe (1952) for more concentrated suspensions, $a = -1.35$ and $n = -2.5$. A review of these and other pertinent experimental data regarding suspensions is to be found in the previously quoted paper by Shaw (1965). Shaw (1969) compared experimental data for various concentrations of crystals with the viscosities for rigid spheres in a Newtonian liquid under isothermal conditions using relations such as Equation (12.22). He found that suspended crystals in basaltic liquids have a different effect from that predicted by the equation.

Liquid immiscibility

Under certain conditions a magma may spontaneously split into two immiscible fractions, and this may have important implications not only in magmatic evolution but also in the interpretation of certain structural and textural features of igneous rocks. As yet we know very little about the frequency of occurrence of this phenomenon in silicate magmas, although we know that it must occur when very unlike substances such as silicates, sulfides, and metals are melted together.

Immiscibility in binary oxide systems has been investigated by Kracek (1930) and others. This work has shown that miscibility gaps occur in the systems MO-SiO_2 where MO refers to certain common rock-forming monoxides. The tendency in order of decreasing immiscibility is as follows: MgO > FeO > CaO > SrO. By contrast the oxides BaO, Li_2O, Na_2O, and K_2O exhibit no gaps at all with silica.

The first evidence of an immiscibility field in a multicomponental melt of major petrologic interest was provided by the experiments of Roedder (1951) in the system K_2O-FeO-Al_2O_3-SiO_2. Later Holgate (1954) presented some convincing evidence that two immiscible liquids may at times arise in reactions between basic magmas and siliceous xenoliths.

Figure 12.8 depicts the immiscibility and associated liquid-crystal phase equilibria in the plane $(KAlO_2)_{1/2}$-FeO-SiO_2 of the quaternary system studied by Roedder

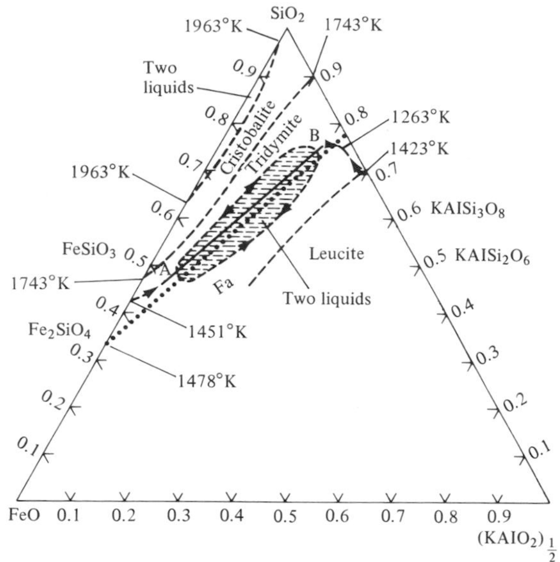

12.8 Molar plot showing liquid immiscibility and associated crystal phase equilibria in the plane $(KAlO_2)_{1/2}$-FeO-SiO_2 of the system K_2O-FeO-Al_2O_3-SiO_2. The letters A and B refer to two coexisting liquids. For explanation see text. (After Roedder, 1951.)

(1951). The component $(KAlO_2)_{1/2}$ was utilized because figures scaled on such a "per cation" basis frequently give convenient plots that do not differ too much from those based on weight percent. The small region of immiscibility near the FeO-SiO_2 join is of little interest since the liquidus temperatures in this region are far too high for nearly all natural conditions. However, the larger region is of great interest since a number of rock types do fall in the analogous natural system $[(K, Na)AlO_2]_{1/2}$-(Fe, Mg, Ca)O-SiO_2.

As an illustration of the equilibrium relations between two typical immiscible liquids in the system represented by Figure 12.8, we may consider the coexisting liquids A and B for $T = 1423$°K. The tie line joining these liquids is of particular interest because it also joins the phase boundaries for coexisting tridymite and fayalite. In this case and in general the activities of the fayalite components of the liquids may be expressed as follows:

$$\begin{aligned} a_{\mathrm{Fa}}^{\mathrm{A}} &= (X_{\mathrm{FeO}}^{\mathrm{A}})^2 X_{\mathrm{SiO_2}}^{\mathrm{A}} (\gamma_{\mathrm{Fa}}^{\mathrm{A}})^3 \\ a_{\mathrm{Fa}}^{\mathrm{B}} &= (X_{\mathrm{FeO}}^{\mathrm{B}})^2 X_{\mathrm{SiO_2}}^{\mathrm{B}} (\gamma_{\mathrm{Fa}}^{\mathrm{B}})^3 \end{aligned} \tag{12.23}$$

Here X is the mole fraction of the oxide and γ is the mean activity coefficient appropriate to the X chosen.[4] Since

[4] See Chapter 2 for a discussion of the choice of concentration variables in complex melts.

at equilibrium we have $a_{Fa}^{A} = a_{Fa}^{B}$, we obtain the relation

$$\left[\frac{(X_{FeO}^{A})^2 X_{SiO_2}^{A}}{(X_{FeO}^{B})^2 X_{SiO_2}^{B}}\right]^{1/3} = \frac{\gamma_{Fa}^{B}}{\gamma_{Fa}^{A}} \tag{12.24}$$

The mole fractions, which may be read directly from Figure 12.8, then yield $(\gamma_{Fa}^{B}/\gamma_{Fa}^{A}) = 3.04$. The relatively large value of the activity coefficient of Fe_2SiO_4 in the B liquid is of course consistent with its low concentration in this phase and expresses its positive excess free energy relative to the A phase.

The detection of liquid immiscibility in the system K_2O-FeO-Al_2O_3-SiO_2 leads us to enquire whether or not an immiscibility region might not also occur in the more complex systems which contain the additional major components MgO, CaO, Na_2O, Fe_2O_3, H_2O, etc. As Holgate (1954) and others have emphasized, no substantial immiscibility fields have been detected in silicate melts of other combinations of these oxides than those of Figure 12.8. However Holgate drew attention to the shape of the liquidus in certain of these systems that seemed to hint that an immiscibility field might be encountered if the liquidus were somewhat depressed by additional components. Such, in fact, may be the case for certain basic magmas of Uganda, Africa, and elsewhere, which were studied by Holgate. In these rocks, quartzite xenoliths show rims of acid glass corresponding to the more complex analogue of liquid B, whereas the basic magma in which the xenolith was immersed is the analogue of liquid A. However, liquid B also resembles certain simple rhyolitic or granitic magmas so that the possibility must be considered that such magmas may at times arise in the differentiation sequence. However there appear to be certain cogent arguments already raised by Bowen (1928) that counter the idea that any substantial body of magma could be generated in this way. We shall have occasion to return to this problem again.

Another case of liquid immiscibility in natural systems has come to light in certain iron meteorites that contain nodules or inclusions of silicates (Olsen and Mueller, 1964). In this case the basic component of the melt is completely devitrified and consists of a mixture of orthopyroxene Ca-pyroxene, and feldspar, whereas the acidic glass is only slightly devitrified and appears to be rich in silica and alkali feldspar components. The two types of glass are separated from each other by an almost planar surface, which presumably represents the original meniscus. It is again informative to examine this occurrence in relation to Figure 12.8. If the melt is too rich in divalent oxide or alkali and aluminum oxide components, such as, for example, the compositions near the fayalite-feldspar join, then the immiscibility region is not intersected. However if, as is the case with the meteorite inclusions, the melt lies near the pyroxene-feldspar join, then such intersection seems highly probable even within the more complex system.

We cannot discuss here all the possible occurrences of liquid immiscibility in magmatic systems. We should, however, mention that another instance of this phenomenon occurs within the compositional range occupied by the rare but interesting carbonatites (Koster van Gross and Wyllie, 1966, 1968), which are usually found in association with alkaline rocks. This problem will be examined in more detail in later chapters.

Origin of magmas

Our knowledge of the field distributions of rock types, as well as elementary logic, lead us to consider two broad categories of magmas:

1. Those which are formed by fusion of preexisting rock approximately in the locality in which their solidified equivalents are found
2. Those which have been introduced from some other, perhaps geologically inaccessible, region

Most igneous bodies encountered were probably derived from magmas of the second type so that we are forced to consider the possible ultimate sources of these magmas.

In the past, petrologists have been greatly concerned with *primary magmas*, which presumably are derived from some source at great depth and which subsequently give rise to *derivative magmas* by various mechanisms of differentiation or contamination. However the only igneous rock that comes close to representing a primary magma is the basaltic or gabbroic compositional type which alone occurs on a circumplanetary scale as basalt flows, diabase dikes and sills and as plutonic bodies such as lopoliths. The composition of these basaltic rocks is remarkably constant both in space and time since they are known from the early Precambrian as well as from contemporary volcanic deposits. These characteristics point to a source within a world-circling shell subject to few of the vagaries of crustal events and inhomogenities. Almost certainly then the source of basaltic magma must lie within Earth's mantle at depths of the order of 50 km or greater. Also, intrusion or extrusion from these depths must frequently occur in such a way that little reaction with crustal rocks takes place, otherwise the constancy of composition in space and time would not be possible. One factor in this lack of reaction with crustal rocks appears to be the speed of intrusion, especially in the case of the larger feeder dikes.[5]

What then of the mode of origin of the primary basaltic magma? Is it derived from a layer in the mantle which is permanently above its melting temperature? We have already seen in our discussion of the effect of pressure on the viscosity that we cannot expect such a melt to exhibit any strength comparable to that of solid rocks. Then, since the mantle clearly transmits shear waves, it seems clear that no substantial bodies of magma can be present at these depths. This then leaves us with the inference that basaltic magmas are derived by partial fusion within the mantle of crystalline material which exists there on a global scale. While the precise nature of this material is still a matter of conjecture, it is possible from theoretical considerations, analogies with certain crustal minerals, and high-pressure experiments to attempt the deduction of some of the mineral assemblages involved. We have also seen (Chapter 11) that there is some evidence from the inclusions found within certain basaltic explosion breccias that this source rock may resemble an eclogite. It is best, however, to postpone our discussion of this interesting topic until we consider the physico-chemical basis of high-pressure differentiation. Here we are concerned more with the possible mechanism of melting. We can, of course, conceive of several different ones. The most obvious of these is a simple rise in temperature, perhaps brought on by mass transfer (convection) or radioactive or chemical heating. Another mechanism is a release in pressure such as might follow unloading or fracturing. A third mechanism is the influx of certain volatile components, which would lower the liquidus temperatures. We tend to perfer the second alternative because it may occur with great rapidity and would be confined to linear or slightly curved belts, as are most regions of volcanism. Also it is known that earthquakes occur at depths of several hundred kilometers along these belts. It thus seems likely that when stresses exceed the values normally associated with plastic or fluid deformations, fractures or shear zones form within the upper mantle. The life time of the actual openings or low-pressure regions in these zones must be short and it is in fact to be expected that the generated magma itself would play an important part in healing them. Although the existence of magma would be very transient, it might at times be transported for significant distances before the healing process would be finished. With a further building up of stresses, the whole process might repeat itself and in this way magma might well be transported from great depths to a level in the crust where openings could be sustained until extrusion could occur.

While release of pressure is a definite mechanism of melting, it is likely that melting may occur independently due to other mechanisms, or the release of pressure due to fracturing may be combined with other factors in producing the melt. To explain the rise of temperature, Green and Ringwood (1967) proposed a model that involves diapiric uprise of solid mantle. Wyllie (1971a) has proposed a further modification of this model by introducing the possible effect of trace amounts of water in the mantle. According to Wyllie (1971b), uprise of the mantle material may begin at the base of the low-velocity zone—at depths of the order of 300 km—and that uprise may be triggered by the outward migration of water from within the deep mantle. Whether any hydrous minerals occur in the deep mantle is uncertain, but it is important to note that even the presence of 0.1 percent water in the mantle can influence the melting relationships very significantly (Wyllie, 1971a). Introduction of water into the mantle beneath island arcs is considered possible according to the new global tectonics (Isacks *et al.*, 1968). Allen *et al.* (1972) have shown experimentally that phlogopite is stable at 1325°C at pressures up to 35 kbar. Similarly amphiboles persist to a maximum temperature of 1090°C and a maximum pressure of 25 kbar.

Some petrologists (Turner and Verhoogen, 1960) also favor the existence of primary granitic magma to account for the widespread occurrence of granitic rocks in the Precambrian shield areas of continents. However there seems to be little evidence in this case for a uniform source rock. Many of the so-called Precambrian "batholiths" have had exceedingly complex histories and do not possess either the petrographic or structural unity of the Mesozoic batholiths, which are in any case not granites. We shall in fact see that the widespread occurrence of granitic rocks on the continents may be more easily related to a more generalized and complex process of continent formation and cannot be attributed to either magmatism or metamorphism acting alone. According to Yoder (1973), however, it is probable that two contrasting types of magmas such as rhyolite and andesite may be generated from the same parental source at shallow depths ($P \approx 20$ kbar, H_2O) through fractional melting.

To summarize, there appears to be a single major type of primary magma, that of basaltic composition, and this is derived from a world-circling layer at some distance below the crust and exceeding 50 km in depth. The source of magma is a layer of solid crystalline material which is

[5] It may be shown (Shaw, 1965) that in the case of instantaneous fracture of the country rock, large masses of basaltic magma of viscosity of 10^6 P and under differential pressures of 1000 bar can be introduced on a time scale of hours or days.

occasionally partially melted. The precise mechanism by which melting occurs is unknown but it may well be through rapid pressure release in deep fracturing. By contrast, there is little evidence for any primary granitic magma source, but the widespread occurrence of granitic rocks on continents appears to be related rather to the general process of continent building, which includes both magmatic and metamorphic elements.

General features of the occurrence of magmatic rocks

Our evidence for the origin and sources of primary magmas such as the basaltic type is necessarily indirect. By way of contrast, we have a wealth of field and laboratory evidence that bears on the origin and history of derivative magmas. In particular, we have accumulated field data for many years on the modes of occurrence of igneous rocks and their structural relations to the country rocks into which they have been intruded.

As might be expected, the modes of occurrence of igneous bodies of different kinds are closely related not only to the physico-chemical properties of the fluid magma but also to the nature of the country rock and the properties of the igneous body that formed on cooling. That is, the multiplicity of features we observe in igneous bodies reflect the total history, which may have been only partly magmatic. We can best illustrate this broad view of igneous rocks by examining certain types of occurrences of basic and acidic igneous bodies within the crustal environment. These occurrences are illustrated diagrammatically in Figure 12.9, which is supposed to represent a more or less typical section through the crust.

12.9 Diagrammatic cross section of Earth's crust to illustrate the forms of igneous rock masses of different composition. All the illustrated structures do not ordinarily occur together as shown here.

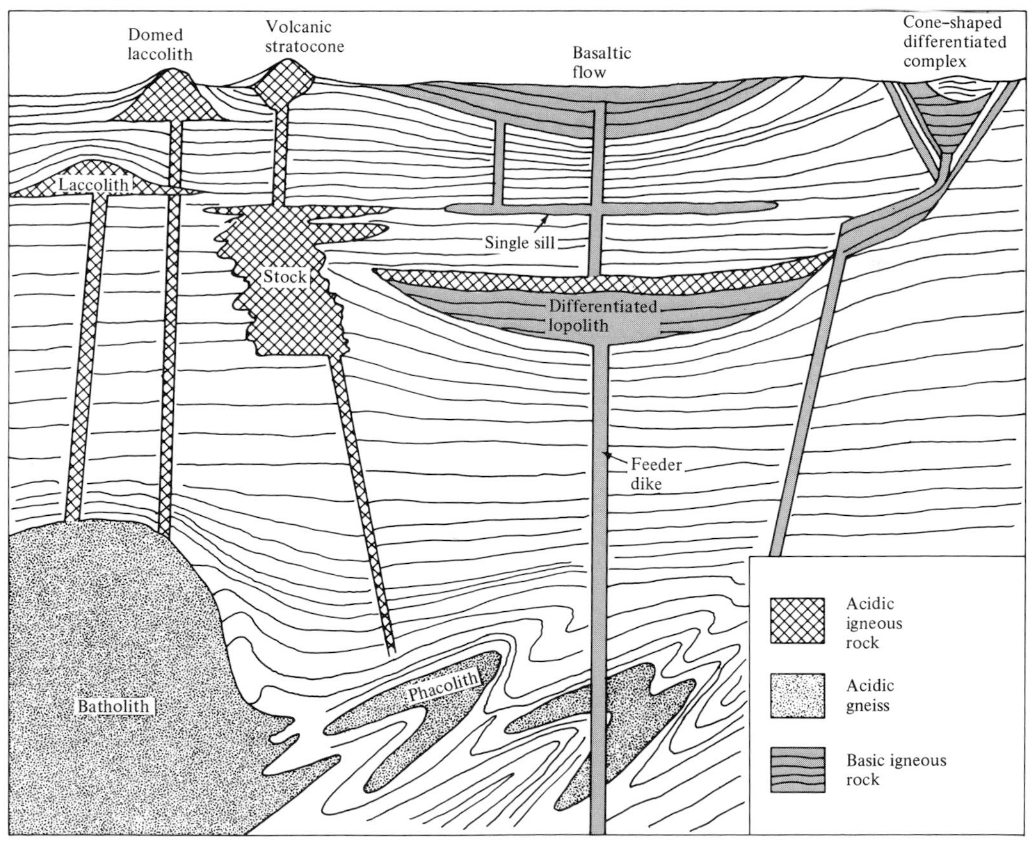

In the region so labeled, we note a small concordant body of granite that occupies the crest of the fold and is therefore known as a *phacolith* (Daly, 1933). As will become clear later, the true igneous nature of such bodies is usually far from clear since there appear to be metamorphic processes that can give rise to similar forms. Then, too, because they are usually formed under deep-seated environments in which regional metamorphism has also been prominent, the original character of these granitic rocks is frequently obscured by postmagmatic events. When such bodies coalesce to form larger bodies, they are known as *batholiths*. These larger bodies frequently extend laterally over thousands of square kilometers and vertically for many thousands of meters. In their upper reaches, where they come in contact with relatively unmetamorphosed sediments, they may show sharp unambiguous intrusive relations, but in their lower reaches they exhibit all the ambiguous characteristics of deep-seated igneous bodies. Because of their great vertical dimensions, the "roots" of batholiths may be grounded within the more basic *sialic* material or even within the *simatic* material, which presumably underlies this. Consequently, the batholith as a whole is affected by strong buoyancy forces which assume the scale of isostatic adjustments. The resulting upward movements, which usually occur some time after the body has formed (whether by magmatic or metamorphic processes), bring about great distortions in the surrounding layers, and these may erroneously be attributed to forceful intrusion of magma. Of course the same type of structural complications can also arise in the vicinity of the smaller bodies if they are subject to orogenic forces. But in general these smaller bodies have densities not greatly different from the mean value of the enveloping rocks so that there will be only weak buoyancy forces involved.

The small concordant granitic lenses and phacoliths frequently grade into the country rock, which then form the *migmatite* complexes which in particular characterize the pre-Cambrian shield areas. We shall see that there is evidence that at least such smaller bodies can form not only by partial melting or *anatexis* of the country rock but also very likely by some means of differential vapor transport or metasomatic process. In neither type of origin is there any great problem of the disposal of the material displaced by the granite, since in any event the intrusion or metasomatism is usually accompanied by profound plastic or fluid deformation which could easily accommodate either increase or decrease of volume. However, in the case of the larger bodies we encounter the "room problem" in the disposition of the replaced country rock. In order to form large homogeneous bodies of granite or granodiorite such as we find in the Sierra Nevada and Idaho batholiths, for example, there must occur either a massive metasomatic transformation or redistribution or an assimilation by magma on a scale scarcely deemed possible under liquidus conditions. The alternative is the improbable forceful shouldering aside of the country rock by the magma. The same problem is encountered to a lesser degree in the case of the smaller stocks such as shown in the figure, which sometimes appear to occupy the former position of unmetamorphosed country rock with little structural disturbance of the surroundings.[6]

As the surface is approached, there is an increasing tendency for the intruding acid or alkaline magma to lift up the roof rocks, and the typical laccolith structure illustrated results. If the viscosity of the magma is high—and this is more probable since volatiles are lost more rapidly in the near-surface region—then an accentuated doming of the laccolith such as is shown will occur (Paige, 1913) and the magma may actually break through to the surface. The extrusion of such viscous magma is one of the commonest features of acid volcanism, and domes and spines of partially solidified magma are frequently extruded from the cones of stratovolcanoes in association with pyroclastics and more fluid lavas as shown in Figure 12.9, labeled "volcanic stratocone."

If we now compare the products of basic magmatism with those just discussed we see a rather marked contrast. This contrast arises largely from the density and viscosity differences of the magmas but is also at least partly related to the density differences between the crystallization products. We have seen that unless a basaltic magma contains considerable water, the density is likely to exceed 2.7 gm/cm^3. This will of course be augmented to the extent that suspended crystals are present. Thus it will frequently happen that even the basaltic magma is more dense than the rocks it intrudes. On the other hand the viscosity will be considerably less than acidic magma under the same conditions. The effect of the low viscosity is the formation of intrusive and extrusive bodies of great lateral dimension as compared with those resulting from acidic magmas. This feature is also illustrated in Figure 12.9, where a massive sequence of relatively thin basalt flows is depicted. Also shown is the basining effect, which is attributable largely to the high density of solidified basalt and usually occurs sometime after deposition. Somewhat analogous processes occur underground when sills are intruded between sedimentary layers. If these are of minor magnitude, as in the "single sill" illustrated, there is no tendency for

[6] An example of such a stock, beautifully exposed, is the Notch Peak intrusive in Utah (Gehman, 1958).

basining to occur although the lateral dimensions may be nearly as impressive as those of the corresponding flow. When, on the other hand, the intrusion involves a large volume of magma, the difference in density between the partly crystalline magma and the country rock coupled with the loss of strength attending heating of the latter may initiate basining early in the history of the intrusion. This would of course be continued after solidification, and the end result is the development of the typical "differentiated lopolith" structure as shown. An added distinction of large sills and lopoliths is the differentiation brought about by the gravitational settling of crystals. In the larger bodies this can result in the formation of a substantial amount of acidic or intermediate rock in the upper layers, as we have also indicated in the figure. Prime examples of such strongly differentiated lopoliths are the well-known Stillwater complex of Montana and the Bushveld complex of South Africa.

Since basaltic magmas have their sources in the mantle, most intrusive bodies of this rock must be connected with it by extensive feeder dikes. These dikes, which are usually straight walled and may be hundreds of miles in length, result from tension fractures that characterize non-orogenic zones of the crust. The relatively quiescent character of most basic igneous activity is also reflected in the structures of lopoliths, which show little evidence of deformation beyond that associated with intrusion or adjustment to the gravitational field.

Sometimes, also, basic igneous activity is associated with block faulting of a type that results in "ring dikes." When this is carried to its ultimate result, cone-shaped blocks of country rock may be pushed upward toward the surface to make way for the intrusion. If the cone-shaped mass is large enough, differentiation may also occur and a layered complex similar to the lopoliths can result, as we have shown in the figure. A well-known igneous body that has been thus interpreted is the Skaergaard complex of eastern Greenland.

13 Differentiation and crystallization of magmas

The existence of rock series and their meaning

The conclusion that the common igneous rocks form continuous series with respect to bulk compositional variation was an outgrowth of the detailed investigations of the nineteenth century petrographers. However, an understanding of the full evolutionary meaning of these series, in particular with respect to physical chemistry, did not come until the turn of the century. The groundwork for this interpretation was laid by such investigators as Michel-Lévy and J. H. L. Vogt and was carried forward by the careful experimentation and solid deductions of N. L. Bowen and his associates. The ground broken by these early investigators is still being explored today, and recent work has revealed some important facets which were virtually ignored by these and even most comparatively recent investigators as well. One of the most important of recent developments is a growing appreciation of the effect of oxidation, while another is concern with the effect of high total pressures on crystallization, partial melting, and differentiation, with particular reference to the origin of basaltic (or gabbroic) and andesitic magmas. Today these two topics are among the most active areas of petrologic research.

Among the evolutionary rock series referred to above, one of the most important is that which ranges in composition from "tholeiitic" basalt to rhyolite. This, the classical "calc-alkali series," bulks very large in the continental batholiths and volcanic belts of orogenic regions and is of some importance in oceanic areas as well. Equally interesting and important from a genetic viewpoint, although far less frequent in occurrence, is the gabbroic series of layered basic complexes represented by such bodies as those of the Bushveld in South Africa and of the Skaergaard in Greenland. Although it was until recently suggested that these gabbroic complexes differ from the calc-alkali series largely in terms of degree of oxidation, this has now been found to be untrue. Although some of these differences may be attributable to differences of structural setting and water content, the factors that lead in one case to the calc-alkali series and in another to the gabbroic series are only beginning to emerge.

The gabbroic complexes present us with the opportunity of observing, virtually in its entirety, the differentiation history of a basaltic magma, chiefly it is thought through the comparatively well-understood mechanism of fractional crystallization. However, it is the very condition of crystallization—that of a static, tectonically stable environment—that makes it possible for the process to go to completion and for its products to be preserved in the original sequence of formation through deposition in orderly gravitationally differentiated layers. The same, unfortunately, does not apply to the calc-alkali series. These never form well-defined, neatly integrated complexes of layered rocks varying uniformly from bottom to top. By contrast, the members of the series are made manifest through multiple intrusions, extrusions, or pyroclastic members which are usually found in chaotic relationships in large volcanic or plutonic provinces. Frequently there is no clear-cut time sequence extending from basalt to rhyolite (or gabbro to granite), but even complete reversal of this sequence. Furthermore, this already chaotic state of igneous affairs is further complicated by evidence of at least some contamination from intruded subcrustal and crustal rocks or compositional modifications attributable to late or post-magmatic fluid-transport processes.

Despite the above difficulties, the calc-alkali series has been widely studied and there is a wealth of useful data already available. However, these studies have revealed significant chemical variations not only within the series from province to province but also apparent continuous transitions toward other far less abundant and well-recognized series of more radical chemical composition. Thus it seems possible that variants of the basalt-rhyolite series might result from the fractional crystallization of basaltic magmas, with considerable but strictly bounded variations in original chemical composition and in particular the degree of silica saturation or supersaturation and volatile content. Then, when this variation corresponds to the degree of undersaturation exhibited by the alkaline basalts, quite different, highly undersaturated alkaline series leading to such rocks as nephelene syenite or phonolites result. In less extreme variations of this nature a trachyte series may develop, but even more bizarre tendencies leading to such rocks as carbonatites must be considered. Given this state of affairs it is well to keep such questions as the following continuously in mind:

1. Do the calc-alkali and other rock series represent a true liquid line of descent formed under fractional crystallization of basaltic magmas in a crustal environment or do they somehow arise by some form of high-pressure subcrustal differentiation?
2. Are the differences both within the calc-alkali series and between other series attributable to original chemical differences in a parent basaltic magma or do they result from other factors such as the physical environment?
3. To what extent are the series products of differentiation mechanisms other than fractional crystallization or melting?
4. To what extent do contamination and other external modifying processes contribute to the formation of rock series?

In this chapter we shall attempt to establish the physicochemical basis for answering these and other questions, and in the following chapters we shall propose some tentative answers.

Homogeneous differentiation

There are two general modes of differentiation by which a magma may undergo compositional change. These are (1) homogeneous differentiation, in which only the magma itself is involved, and (2) heterogeneous differentiation, in which some other phase, either fluid or crystalline, separates from the magma. In the first mode of differentiation, which we shall consider first, diffusion is the primary mechanism, although the chemical potential gradients that drive this diffusion may arise in different ways.

In a completely isolated magma at superliquidus temperatures, differentiation may be induced by gravitational potential differences such as result when a column of matter initially uniform (vertically constant) in composition is subject to a gravitational field, or it may be induced by a temperature gradient that exists in the column. In the case of the gravitational field we can write a phenomenological expression for the vertical flux of a constituent i if we make certain simplifying assumptions. Thus if we assume that we are dealing with an ideal binary solution in which the partial molar volumes of both components are identical, we obtain the following expression for the flux of i:

$$J_i = -D\left(\frac{dc_i}{dh} - c_i\frac{M_i g}{RT}\right) \tag{13.1}$$

In this expression D is as usual the coefficient of diffusion, c_i is the concentration of i, M_i its molecular weight, and g the acceleration due to gravity. It is obvious that the second term in parentheses corresponds to the driving force of the gravitational field whereas the first term corresponds to the ordinary concentration gradient. It may be shown by substitution of reasonable values of c_i, M_i, and T that the second term in Equation (13.1) is of negligible magnitude for most situations and thus may usually be disregarded in computing the flux; however, it is also obvious that it is this term that must be responsible for the adjustment of the system to chemical—and hence perfect mechanical—equilibrium in a gravitational field. In fact this result gives us a hint of the unlikelihood of such adjustments in nature.

At the present time there does not exist, to our knowledge at least, any information on the possible significance of diffusion induced purely by a temperature gradient (Soret effect) in systems of petrologic importance.[1] In any event, we shall see that the thermal gradients necessary for the effect to be operative will generally have a life time that is small relative to the gradients that drive ordinary particle diffusion.

It is interesting now to examine in some detail the case of particle diffusion along a concentration gradient in a magma, since this process must have essentially the same activation energy as other more complicated processes such as are involved in the Soret effect and movement in a gravitational field, and so will have a direct bearing on them also.

The case we shall discuss is diffusion in a homogeneous basaltic magma since this has some important bearings on features of certain gabbroic complexes and other problems. However, the system we shall treat is as a whole not a completely homogeneous one since we assume that it is crystallization occurring at the boundary that sets up the concentration gradients. Yet the diffusion itself within the magma should be directly comparable with that in purely homogeneous systems. We shall assume that D is independent of the composition, varies with T, and is derivable from the bulk viscosity as discussed in Chapter 3. Implicit in the latter assumption is the notion that the same statistical unit that is involved in the fluid flow of basaltic magma is also that which is involved in the diffusion under discussion, or at least that the activation energies are the same. Although this assumption will certainly be invalid when the details of any real situation are considered, we shall present evidence that it can at times yield results that are reliable to within a few orders of magnitude in D.

Under the assumption that the diffusion mechanism is similar to that of fluid flow and that $\lambda_1 = \lambda_2 = \lambda_3 = \lambda$, we may make use of Equations (1.80) and (3.51), which give us the relation

$$D = \frac{kT}{\eta\lambda} \tag{13.2}$$

In this expression k is Boltzmann's constant and λ is the characteristic interatomic distance which we take to be 10^{-8} cm. To obtain approximately correct values of η for a basaltic magma we may consult our viscosity data as given in Figure 12.5 or use Shaw's (1972) method of calculation. We then see that for a magma near its liquidus at 1400°K, $\eta \cong 10^4$ P, although this may vary by a factor of 10 depending on the anhydrous compositional variation and on the volatile content of the magma. If we substitute these values of T, λ and η into Equation (13.2), we obtain the result that $D \cong 10^{-9}$ cm^2/sec. As a rough check on this result we may compare it with experimentally determined values of D presented by Bowen (1921) in his classical paper on diffusion in the system albite-anorthite-diopside. This system, in which the diffusion of feldspar components into diopside was studied, should be at least somewhat comparable to a basaltic system. However, all Bowen's data are concerned with temperatures far above the basalt liquidus. A typical value obtained by him for $T = 1773$°K is $D = 3.5 \times 10^{-6}$ cm^2/sec. If we again consult Figure 12.5 we see that in this general temperature range $\eta \cong 10^2$ P, so that by substitution into Equation (13.2) we obtain $D \cong 10^{-7}$ cm^2/sec for a basaltic magma. Certainly this value is as close to Bowen's value as we have any right to expect and indicates that our assumptions yield values of D that are in the right range.

Having obtained a value of D that is probably correct to within several orders of magnitude, we may consider diffusion in our model of boundary-localized crystallization. We may visualize the system as an extensive plutonic intrusive such as a sill or lopolith which is

[1] This statement of course excludes transport induced in heterogeneous systems by the effect of temperature gradients on equilibrium constants as discussed under the heading of "Quasiequilibrium processes" in Chapter 3.

cooling slowly and in which crystallization is confined to a layer on the wall or floor (border zone). It might happen, for example, that a single phase such as plagioclase or olivine is being deposited there. If we consider a period some time after the intrusion, then we should expect that all but the smallest temperature gradients will have been levelled within the magma.[2] Also we should expect that while the magma in contact with the border should be just saturated, varying degrees of supersaturation should occur within the interior of the magma and that, because of diffusion, the degree of supersaturation should increase uniformly from the border to some maximum value in the interior. We may therefore designate two constants, the saturation concentration c_s and the maximum supersaturation concentration c_0. We shall be concerned with a variable supersaturation concentration c which falls between c_s and c_0 at a distance x from the border. Under the assumption of a constant D, a simple solution for the concentration c as a function of x and t the time may be obtained by integration of Equation (1.76), which is the statement of Fick's second law of diffusion. The boundary conditions for this integration are as follows:

$$c = c_0 \quad \text{at } t = 0, \qquad 0 < x < \infty$$
$$c = c_s \quad \text{at } x = 0, \qquad 0 < t < \infty$$

[2] As we show later, this is justified in terms of the heat conductivity.

13.1 Fractional degree of supersaturation $(c - c_s)/(c_0 - c_s)$ of diffusing constituent of an anhydrous basaltic magma as a function of distance from the point of saturation (crystallization). c_s is saturation concentration, c_0 the maximum supersaturation concentration, and c the variable supersaturation. The system is treated as binary for simplicity, and the diffusing substance may be regarded as either a feldspar or ferromagnesian component. For further explanation see text.

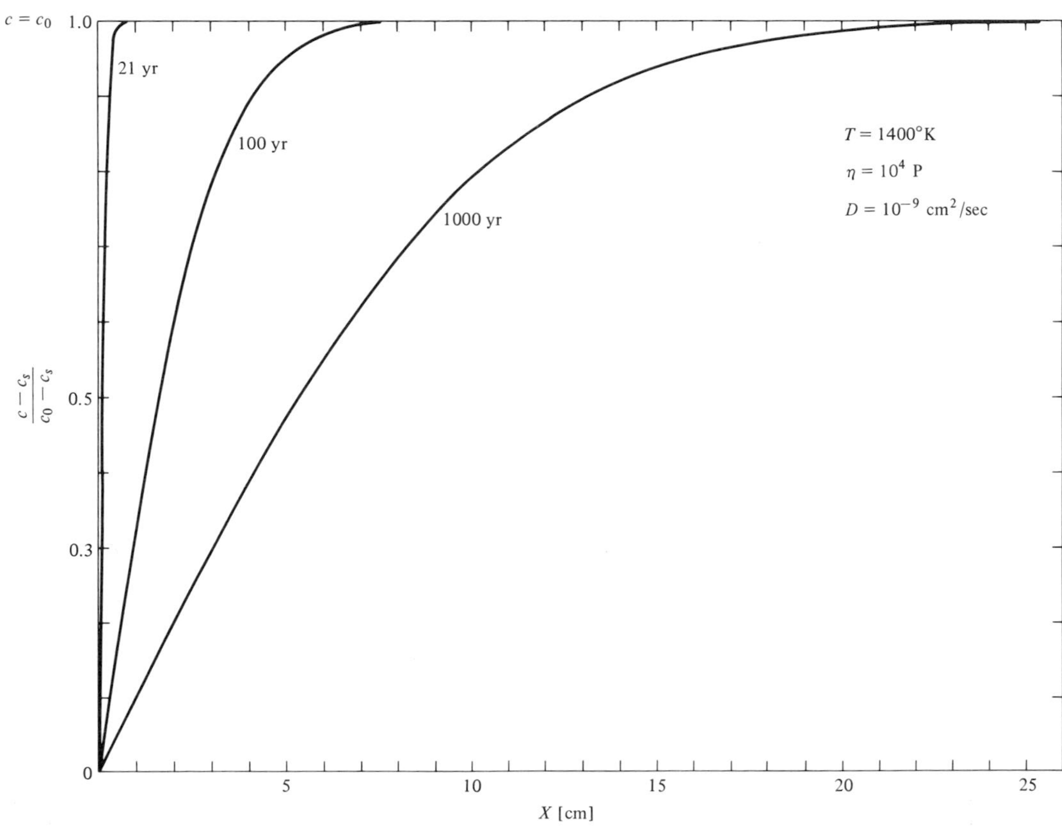

The integration, which is straightforward (Darken and Gurry, 1953), then yields the following expression for the fractional degree of supersaturation:

$$\frac{c - c_s}{c_0 - c_s} = \frac{2}{\sqrt{\pi}} \int_0^{x/2(Dt)^{1/2}} e^{-\beta^2} d\beta \tag{13.3}$$

The right-hand side of Equation (13.3) is the well-known probability integral given in terms of the integration β. Tables of values of this integral for each value of $x/(Dt)^{1/2}$ are readily available. If we make use of such tables and our previously determined value of D, we may obtain the fractional degree of supersaturation as a function of x for any time interval we choose to substitute into $x/(Dt)^{1/2}$. The results for three geologically interesting time intervals are shown in Figure 13.1. It is clear from this figure that although little change in the degree of supersaturation is to be expected beyond 1 cm in time intervals of the order of 1 yr, an interval greater than 1000 yr is adequate for substantial change to occur at distances of 10 cm or more, which is large compared to the scale of banding encountered in gabbroic complexes. Also as was pointed out by Bowen (1921), the area bounded by each curve and the horizontal line corresponding to maximum supersaturation ($c = c_0$) is proportional to the quantity of crystals deposited in the time interval.

In Bowen's treatment of the analogous problem, he assumed that the maximum concentration c_0 was maintained by higher temperatures in the interior region rather than by supersaturation, as we have assumed. However, as Bowen himself pointed out, great temperature differences between the interior and border regions of a plutonic magmatic body are not to be expected because of the high value of the heat diffusivity as compared with that of matter. This may readily be demonstrated by substituting the temperature T for c and $K/C\delta$ for D in Equation (13.3). Here K is the heat conductivity, C the heat capacity, and δ the density of the magma. Now reasonable values of these constants are $K = 2 \times 10^5$ erg/cm/sec, $C = 1.2 \times 10^7$ erg/gm/°K and $\delta = 2.8$ gm/cm^3, so that $K/C\delta = 6.2 \times 10^{-3}$ °K cm^2/sec. Thus it is clear that for similar time intervals the curves of fractional temperature decrease $(T - T_s)/(T_0 - T_s)$ versus x would be much displaced to the right relative to the curves of Figure 13.1. This of course means that the temperature gradients would disappear long before comparable composition gradients.

The above results are in general agreement with the early conclusions of Bowen that it is unlikely that significant changes in composition could affect any large body of magma while it is dominantly liquid. In virtually all cases the higher rates of cooling as compared with those of mass transport by diffusion would be expected to bring about crystallization before large-scale homogeneous differentiation could occur. This conclusion does not however depreciate the important role that diffusion might have over distances up to a meter or more in transporting crystallizing constituents to the site of crystallization. It is also probable that the effects of diffusion working in conjunction with other mechanisms such as convection have been underestimated. We shall return to these questions in our discussion of differentiation of certain gabbroic complexes.

Heterogeneous differentiation by separation of a fluid phase

Effect of the gravitational field

We have already seen that under certain circumstances a second liquid phase may separate from a magma as a result of immiscibility. We saw, however, that a silicate melt can split into two immiscible parts only in certain restricted composition ranges. Also, since such unmixing can occur only by a diffusive mechanism limited by the same conditions just discussed, such a magmatic pair would begin as a highly dispersed suspension which would still have to be segregated by gravitation to form large separate bodies. Evidence for such suspensions of one immiscible silicate melt within another have occasionally been presented but none has been entirely convincing. However, another form of fluid immiscibility, that which involves the formation of a volatile-rich vapor phase, is of very common occurrence in magmatism, and evidence in the form of vesicular rocks is widespread in volcanics of all types.

The separation of a vapor phase from a magma is highly pressure-dependent because of the large change in molar volume involved, and consequently certain complications are present that do not occur in ordinary liquid immiscibility. In order to treat the problem, we shall discuss several idealized situations.

We may begin by considering the condition of separation in an isothermal column of completely liquid magma which is subject to the gravitational field. For simplicity we shall assume that the vapor phase is pure water and the water in the melt forms an ideal binary solution with the silicate component. We shall further assume that the magma has the same solubility with respect to water as the granite depicted in Figure 12.2. The appropriate equation for a binary ideal solution in the gravitational field may be obtained from Equation (1.69) by setting $W = 0$:

$$\frac{X^{\beta}_{H_2O}}{1 - X^{\beta}_{H_2O}} = \frac{X^{\alpha}_{H_2O}}{1 - X^{\alpha}_{H_2O}} \exp\left[\frac{M_s - M_{H_2O}}{RT} g\Delta h\right] \quad (13.4)$$

In this equation M_{H_2O} (= 18 gm/mole) is the molecular weight of water, and since we are treating the solution as binary, M_s may be regarded as the "mean" molecular weight of the silicate components.[3] Then for a hypothetical "granite melt" a value of $M_s = 50$ gm/mole seems appropriate since this is close to the weight of $\frac{1}{5}(KAlSi_3O_8)$. Thus if we choose $T = 1173°K$ from Figure 12.2 and if Δh is taken as positive upward, $X^{\beta}_{H_2O}$ in Equation (13.4) may be regarded as the upper limit of concentration of water as defined by the maximum solubility at the given depth within the magma column. Similarly $X^{\alpha}_{H_2O}$ may be regarded as the variable downward concentration in the column. Let us choose the upper limit as 5 weight percent, which according to Figure 12.2 corresponds to a depth of 3 km. Then by calculation $X^{\beta}_{H_2O} = 0.128$ or 12.8 mole percent H_2O. Figure 13.2 then shows $X^{\alpha}_{H_2O}$ as a function of the depth within the magma column (curve 1). It also shows the curve of maximum solubility (curve 2) of water with depth at 1173°K. It is obvious from this figure that under conditions of solubility approximating those shown, the variation with height of water in the gravitational field is quite small. The intersection of the two curves gives the height at which vesiculation begins, which is in this case 3 km. If diffusive equilibrium were maintained throughout the column during progressive water loss above this point, curve 1 and its intersection with curve 2 would progressively migrate to the left, and vesiculation would be confined to higher and higher levels.

The system just discussed is certainly too idealized for any natural situation. In the first place, such a column of magma would have both vertical and horizontal temperature gradients, which would ensure corresponding chemical and mechanical potential gradients in addition to those induced by the gravitational field. These would result in mass and heat transfer by both diffusion and convection, with the latter process playing the dominant role. As a result the simple gravitational profile of water concentration would be greatly modified. Secondly, any actual magma column would most probably be partially crystalline so that not only the density (and hence the convection) but also the thermal properties would be further modified. Nevertheless,

[3] Also implicit in the derivation of Equation (1.69) is the assumption that the partial molar volumes $\overline{V}_{H_2O}$ and $\overline{V}_s$ are identical.

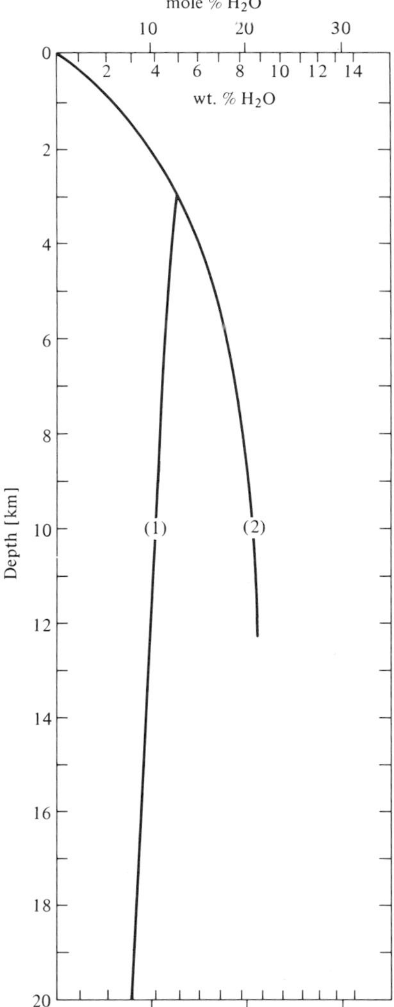

13.2 Highly simplified and idealized diagram to illustrate the effect of the gravitational field in bringing about saturation of water near the top of a magma column. It is assumed that water forms an ideal solution with the silicates and that the temperature of the column is a uniform 1173°K. In addition, the partial molar volumes of water and silicate are taken to be the same. Variation of the mole fraction of water in the melt is given by curve 1 while the variation of the saturation mole fraction is given by curve 2. For further explanation see text.

Figure 13.2 does provide us with a useful reference model in that it is the simplest possible equilibrium state that water could attain in a magma column.

Separation of a fluid phase in a crystallizing magma

Most frequently a fluid phase begins to separate from a magma only after crystallization has already begun. Under these circumstances there may be a strong interaction between the two types of phase separation, and this will have an important influence on when the fluid phase begins to separate. The situation may be illustrated by an analysis of a two-component system consisting of the three phases—liquid, vapor, and crystal (Morey, 1922). This system may be approached through the Gibbs–Duhem relation [Equation (1.7)], which may be written as follows for each phase:

$$\begin{aligned} -V^{V}dP + S^{V}dT + X^{V}_{H_2O}d\mu^{V}_{H_2O} + X^{V}_{s}d\mu^{V}_{s} &= 0 \\ -V^{L}dP + S^{L}dT + X^{L}_{H_2O}d\mu^{L}_{H_2O} + X^{L}_{s}d\mu^{L}_{s} &= 0 \\ -V^{C}dP + S^{C}dT + X^{C}_{H_2O}d\mu^{C}_{H_2O} + X^{C}_{s}d\mu^{C}_{s} &= 0 \end{aligned} \quad (13.5)$$

In these equations the superscripts V, L, and C refer to vapor, liquid and crystal phases, respectively.

We shall now make several assumptions which are required to simplify the problem. These are that the vapor phase consists entirely of water and that the crystal phase is pure silicate so that $X^{V}_{s} = X^{C}_{H_2O} = 0$, while $X^{V}_{H_2O} = X^{C}_{s} = 1$. Consequently at equilibrium $P = P_{H_2O}$, where P_{H_2O} is the vapor pressure of water. Also at equilibrium[4] $\mu^{V}_{H_2O} = \mu^{C}_{H_2O} = \mu^{C}_{H_2O}$ and $\mu^{V}_{s} = \mu^{L}_{s} = \mu^{C}_{s}$. Then by eliminating the chemical potentials we obtain:

$$\frac{dP}{dT} = \frac{S^{V}X^{L}_{H_2O} + S^{C}X^{L}_{s} - S^{L}}{V^{V}X^{L}_{H_2O} + V^{C}X^{L}_{s} - V^{L}} \quad (13.6)$$

Furthermore if we let $\bar{S}^{L}_{H_2O}$, $\bar{V}^{L}_{H_2O}$, $\bar{S}^{L}_{s}$, and $\bar{V}^{L}_{s}$ represent the partial molar entropies and volumes of H_2O and the silicate in the liquid we may write Equation (13.6) as

$$\frac{dP}{dT} = \frac{S^{V}X^{L}_{H_2O} - \bar{S}^{L}_{H_2O}X^{L}_{H_2O} - (X^{L}_{s}\bar{S}^{L}_{s} - S^{C}X^{L}_{s})}{V^{V}X^{L}_{H_2O} - \bar{V}^{L}_{H_2O}X^{L}_{H_2O} - (X^{L}_{s}\bar{V}^{L}_{s} - V^{C}X^{L}_{s})} \quad (13.7)$$

which may in turn be rewritten as follows:

$$\frac{dP}{dT} = \frac{X^{L}_{H_2O}\Delta h_{H_2O} - X^{L}_{s}\Delta h_{s}}{T(X^{L}_{H_2O}\Delta V_{H_2O} - X^{L}_{s}\Delta V_{s})} \quad (13.8)$$

[4] The relations $\mu^{V}_{H_2O} = \mu^{C}_{H_2O}$, $\mu^{V}_{s} = \mu^{C}_{s}$ hold even though we have assumed that $X^{V}_{s} = X^{C}_{H_2O} = 0$. This is true because stoichiometric phases are only a limiting case of a slight solubility which is always present.

where Δh_{H_2O} and ΔV_{H_2O} are the molar heat and volume of evaporation of H_2O and Δh_s and ΔV_s are the molar heat and volume of solution of the silicate at the concentration given by $X^{L}_{H_2O}$ and X^{L}_{s}.

It is clear that the sign of dP/dT will depend on the relative heat and volume changes of solution as compared with those of vaporization. It is to be expected that generally Δh_{H_2O}, Δh_s, ΔV_{H_2O}, and ΔV_s will all be positive, and if $X^{L}_{s}\Delta h_s$ exceeds $X^{L}_{H_2O}\Delta h_{H_2O}$ while the opposite holds for $X^{L}_{H_2O}\Delta V_{H_2O}$ and $X^{L}_{s}\Delta V_s$, the vapor pressure will increase with falling temperature, which is opposite to the case for liquid-vapor equilibrium in the absence of crystals. This is the explanation of the "second boiling point phenomena" so frequently discussed in petrologic literature.

The system $NaAlSi_3O_8$-$KAlSi_3O_8$-H_2O which is of great petrologic interest and is analogous to that just discussed was treated in detail by Tuttle and Bowen (1958). However since it is a three-component system it possesses an additional degree of freedom. At any given total pressure and temperature, crystallization of feldspar may begin before the system is saturated with respect to water. With further cooling P_{H_2O} rises toward the saturation level as the quantity of liquid diminishes. However when the saturation isobar is attained, further cooling is along this isobar, and boiling occurs with only slightly falling temperatures. As boiling continues the liquid becomes slightly enriched in water but greatly diminishes in absolute amount until it is all used up and crystallization is complete. A highly detailed analysis of the system $NaAlSi_3O_8$-H_2O has recently been given by Burnham and Davis (1974).

Although we have so far confined our discussion to water, other volatile constituents also are subject to differentiation processes in magmas. Thus we have reason to believe that the major volatiles of a basaltic magma consist of almost equal parts of H_2O, CO_2, and SO_2, with perhaps a slight preponderance of the latter two constituents (Nordlie, 1967, 1971). However, CO_2 and SO_2 seem to diminish relative to H_2O as differentiation progresses. This effect is proabably partly attributable to vapor-phase solubility and partly to the precipitation of crystalline phases containing sulfur and carbon.

Although it is sometimes difficult to distinguish "volatiles" from "refractories," the solubility of the latter in the vapor of fluid phase is usually quite restricted.[5] However, if we consider any real magma, the solubilities of such constituents as SiO_2 and MgO assume considerable importance in any vapor-transport

[5] As we have already learned in our study of mass transport in metamorphism.

phenomenon. Although any real magma will have many components and hence many degrees of freedom, the problem is somewhat simplified when saturation accompanies crystallization. Let us consider a magma from which olivine and pyroxene have begun to precipitate. If a vapor phase separates from this magma, equilibrium demands that a certain quantity of MgO, SiO_2, etc. will go into the vapor and we may represent this as follows:

$$\underset{melt}{MgO} \rightleftharpoons \underset{vapor}{MgO} \tag{13.a}$$

$$\underset{melt}{SiO_2} \rightleftharpoons \underset{vapor}{SiO_2} \tag{13.b}$$

We also know that such reactions will be promoted by high vapor pressure (Chapter 8) so that the quantity dissolved will depend on the water content as well as the quantity of MgO and SiO_2 in the magma. However, since crystals are present, the following equilibrium must also hold:

$$\underset{olivine}{Mg_2SiO_4} \rightleftharpoons \underset{vapor}{2MgO} + \underset{vapor}{SiO_2} \tag{13.c}$$

$$\underset{pyroxene}{MgSiO_3} \rightleftharpoons \underset{vapor}{MgO} + \underset{vapor}{SiO_2} \tag{13.d}$$

If for the moment we regard these crystals as stoichiometric, the equilibrium constants for these reactions are:

$$K_{(c)} = P^2_{MgO}P_{SiO_2} \tag{13.9}$$

$$K_{(d)} = P_{MgO}P_{SiO_2} \tag{13.10}$$

so that $K^2_{(d)}/K_{(c)} = P_{SiO_2}$ and $K_{(c)}/K_{(d)} = P_{MgO}$, which is the same as the case when no magma is present. However, because of the higher magmatic temperatures, the solubilities will be greater for any given water pressure.

By the same type of reasoning and based on such experiments as those of Morey and Hesselgesser (1951), we should expect the magmatic vapor phase to be charged with a variety of refractory constituents,[6] but especially with silicates such as the feldspars and certain metal oxides. The solubility of the latter will of course be greatly enhanced by the presence of HCl. Now we have already seen that the reactions of H_2O and HCl vapor with refractory oxides are usually endothermic and at the same time may show a decrease in the number of gaseous molecules. Thus if a bubble of such gas is formed at depth in a magma column where the temperature and pressure are both relatively high, and if this bubble is then transported upward to a region of lower temperature and pressure there will be a tendency for the volatilization reaction to be reversed so the oxide is deposited. Thus the concept of vapor transfer or "gaseous transfer" as advocated in particular by Fenner (1926) fits our knowledge of the physical chemistry of magmas quite well. However it remains to be proven in one given case just how effective the mechanism can be.[7]

[6] The heavy-metal content of such a vapor phase based on thermochemical data has been discussed in detail by Krauskopf (1959).

Crystallization and heterogeneous differentiation in magmas

General features

Since the time of Bowen, crystallization, partial melting, and reaction between crystals and liquids in general have come to be widely accepted as the chief mechanisms by which the differentiation of magmas occurs. Although this conclusion is to a great extent the outgrowth of laboratory experiments and physico-chemical arguments, there is also a wide variety of field evidence that supports this idea. In fact, much of this field evidence was also adduced by Bowen (1928). Yet even the wealth of evidence now available is still inadequate to reveal unambiguously the detailed course of differentiation in most common magmatic series even when it is unperturbed by other factors. However, now the problem is at least well defined in most cases.

One of the classical problems of igneous petrology is that concerned with the differentiation of basaltic magma to yield such rock series as the gabbroic complexes, the calc-alkali series of orogenic regions, or other less well known series. Although it is recognized that fractional crystallization alone is probably inadequate to account for this diversity or of even certain characteristics of individual series, we shall present certain evidence that points to this mode of differentiation as dominant in at least some important magmatic provinces. At any rate, the fractional crystallization of a basaltic magma is a highly useful reference process and starting point for our discussion of crystallization-differentiation in general.

We may begin by outlining a number of points of critical information in Table 13.1, which we must have in order to trace the course of evolution in any system undergoing crystallization-differentiation, whether the

[7] Such doubt exists only for vapor transport *within* a magma. There is a wealth of evidence that vapor transport is important in postconsolidation phenomena and at igneous contacts.

Table 13.1 Critical information required to trace the course of differentiation by crystallization

1. Composition of the original system including the volatile content of a primary magma; it is usually useful to summarize this information in terms of normative minerals
2. Composition and proportion of each crystal separating at each stage
3. Stage of appearance and dissappearance of each crystal type
4. Identification and characterization—including the volatiles—of possible pure liquids in the differentiation sequence (liquid line of descent) whether these be represented by quenched glasses or holocrystalline rocks
5. Temperature and total pressure at each stage of crystallization either as recorded in experiments or as inferred from the phase assemblages

system is a simple artificial one or the most complex of natural magmas.

The importance of knowing the composition of the original (primary) magma arises from the control details of this composition exert on the whole course of differentiation, both in terms of compositional trends of the derivative magmas and in terms of the separating crystals. For example, we can account for the dominance of hypersthene in the lower gabbroic complex at Stillwater, Montana, by the abundance of this mineral in the norm of the original magma. Similarly, a knowledge of the compositional details and crystallographic character of the separating crystals is important because the instantaneous change in composition of the magma is always away from the resultant sum of the crystal compositions separating or being resorbed at that moment. For stoichiometric crystals in the primary field of crystallization, this change takes the form of a straight line in composition space; but if the primary crystal is a solid solution the line (crystallization path) will be a curved one. For equilibrium crystallization the crystallization path depends only on the temperature, the total pressure, and the compositional details of the system (expressed as concentrations and activity coefficients). However, in the case of fractional crystallization, this path also depends on the degree of fractionation. In only one case, that of perfect fractional crystallization, do the disequilibrium paths assume a simple form. In the case of perfect equilibrium crystallization involving solid solutions, crystallization in principle ceases and the liquid is consumed after it has undergone only a limited change in composition. The reason for this lies in the implied complete reaction and exchange of materials between liquid and crystals. However, in fractional crystallization where this exchange is less complete, or in the case of perfect fractional crystallization where it is nonexistent, the liquid may undergo such great compositional change that it attains the lowest accessible temperature of the system.

In considering the compositions of crystals and liquids we must concern ourselves not only with the familiar metal oxides that appear in rock analyses but also with the volatiles that may have escaped in the consolidation process. We must concern ourselves not only with the total iron but also with the different oxidation states of this metal.

Of course, if for any given differentiation series, we had all the information listed in Table 13.1 for each stage of the process we should know nearly all there is to know. However, such a wealth of information is never available even in the most exhaustively studied artificial systems. Rather, we invariably can only hope to learn the crystal and liquid compositions, the temperatures, and the pressures at a limited number of points so that at other points we must infer their values.

Flowage differentiation and Muskox Intrusion

Bhattacharji and Smith (1964) conducted experiments to demonstrate the process of flowage differentiation and explain the vertical zoning of rocks in the feeder dike of Muskox Intrusion situated in the Northwest Territories of Canada. In a flowing magma with crystalline and fluid phases, the crystals can segregate to form cumulates because of the inherent flow properties of the mixture. The cumulates form away from the walls and toward the central axis conduit. This process of segregation is called flowage or flow differentiation and has been studied theoretically and experimentally by Bhattacharji and Smith (1964), Bhattacharji (1967), Bhattacharji and Nehru (1972), and Komar (1972a,b).

The feeder dike of Muskox Intrusion is vertically zoned, with norite forming the margins and picrite the central part. Olivine increases in grain size, abundance, and Mg/Fe ratio from the dike margins toward the center while plagioclase decreases in abundance but increases in grain size, anorthite content, and degree of order. Bhattacharji and Smith (1964) showed the following with their model experiment:

1. In laminar flow the solid particles separate from the walls and gradually increase in concentration toward the center as the solid-fluid mixtures are pushed upward in the conduit.

2. Spherical and rod-shaped particles rotate as they move from the walls toward the center.
3. The rate of concentration toward the center increases with increased velocity or shear gradient. Thus constrictions in the conduit accelerate the process.
4. For equivalent shapes, the rate of inward movement of solids increases with the particle size.

These experimental observations suggest that the early crystallized Mg-rich olivine in the flowing magma at Muskox must have migrated away from the walls, leaving the residual fluid depleted in Mg relative to Fe. Succeeding crystals would become more Fe-rich as the process—a flow analogue of gravitative differentiation—continued. Bhattacharji's (1967) other model experiments have clearly established flowage differentiation to be a quantitatively important factor in causing a fractionation of rock types, mineralogic zoning, and chemical differentiation in nature.

System MgO-FeO-Fe_2O_3-SiO_2

We have already encountered this system in our treatment of planetary and meteorite chemistry (Chapters 5 and 6). In terrestrial rocks its importance arises largely from its status as one of the major subsystems of basaltic magmas. Muan and Osborn (1956) and Osborn (1959), who worked out the major liquidus relations at 1 atm total pressure, found it convenient to describe the system in terms of the four following idealized modes of crystallization:

1. Equilibrium crystallization at constant total composition of the condensed system
2. Fractional crystallization at constant total composition of the condensed system
3. Equilibrium crystallization at constant oxygen fugacity
4. Fractional crystallization at constant oxygen fugacity

Of these four, the first two are the simplest and, as we shall see, the most pertinent to the crystallization of natural magmas.

In Figure 13.3 the system is depicted within a tetrahedron, the base of which forms the previously discussed system MgO-FeO-SiO_2 (Chapters 5 and 6).[8] The primary phase volumes or volumes of the tetrahedron, in which crystallization of a given mineral begins, appear in this figure as irregular polyhedra which subdivide the entire tetrahedron. In each of these volumes one crystalline phase, one liquid phase, and a gas phase consisting of a mixture of CO, CO_2, H_2 and H_2O is present. In the experimental system these gases are controlled mixtures that govern the oxidation state and may be omitted in treating the condensed phase relations. Within the volumes there are of course three degrees of freedom.

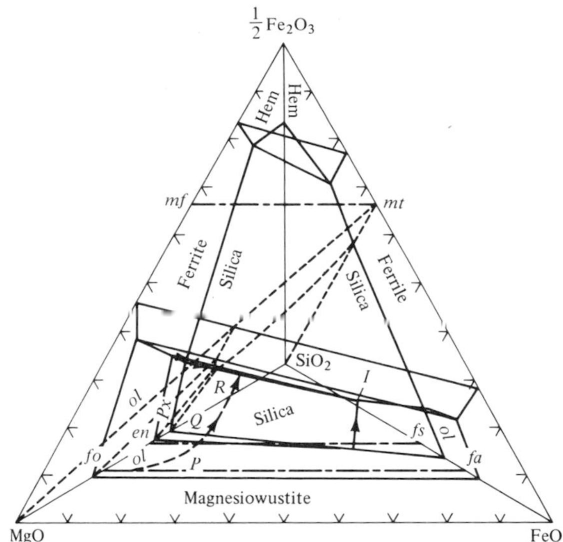

13.3 Molar plot of system MgO-FeO-Fe_2O_3-SiO_2 modified after Muan and Osborn (1956). Primary phase volumes solid solution joins and crystallization paths are shown. The plane MgO-SiO_2-Fe_3O_4 is also shown as light, broken lines. See text for explanation.

The primary phase volumes are separated by surfaces, corresponding to two degrees of freedom, along which two crystalline phases coexist with the liquid and gas. These surfaces in turn intersect to form a number of univariant space curves, the most important of which extend through the lower central part of the tetrahedron. These in particular define the junction of the primary phase volumes of olivine, pyroxene, silica, and ferrite. In the right-hand part of the diagram three of these curves intersect to form the quaternary invariant point *I* at which all four crystalline phases coexist with the liquid and gas.

Also shown in Figure 13.3, as light, dashed lines, is the plane MgO-SiO_2-Fe_3O_4, which is a cross-section of the entire tetrahedron. This is one of an infinite number of

[8] For a summary of the data sources for the systems which form the other faces of the tetrahedron see Muan and Osborn (1956).

such planes that hinge on the tetrahedron edge MgO-SiO_2 and whose apexes vary from $FeO_{3/2}$ to FeO. These planes roughly approximate surfaces of constant oxygen fugacity which decreases steadily toward the highly reduced base system MgO-FeO-SiO_2.

We shall discuss first the case of equilibrium crystallization at constant total composition. In terms of the experimental system, which includes a separate gas phase, this implies conditions under which mass exchange between the gas phase and the condensed system is small. However, it is obvious that in natural magmas a separate gas phase may never appear, so that in this case the condition of constant total composition corresponds to little or no reaction with dissolved constituents such as H_2O and CO. If these conditions hold in each case, the course of crystallization will be similar to that in the classical systems such as albite-anorthite-diopside as discussed in Chapter 3. Let us follow the course of a melt of initial composition *P* (Figure 13.3) which lies in the primary volume of olivine just above the base of the tetrahedron. Then on cooling the homogeneous melt, crystallization begins with the separation of Mg-rich olivine of composition *ol* on the olivine join, which is shown connected to *P* by its tieline. As cooling continues, separation of olivine continues and the liquid changes composition along the equilibrium path *P–Q*. Also because of the free exchange between the liquid and crystal, the composition of the latter moves along the olivine join toward the end member *fa*. Just as in the case of the three-component systems, the tieline connecting liquid and crystal must always pass through *P*. Also because the olivine compositions all lie on the join *fo–fa*, the curved path *P–Q* must lie in a plane that includes this join and *P*, and that intersects the primary phase volume of pyroxene at *Q*. When *Q* is attained, pyroxene begins to crystallize directly from the melt but also forms by reaction of dissolved silica with olivine. For simplicity, the pyroxene join is not shown, but here the behavior is precisely analogous to the olivine join. At this point the crystallization path undergoes a discontinuous change and thereafter follows the curved path *Q–R* to *R*, where it intersects the univarient line just as ferrite begins to precipitate. From this point the liquid coexists with olivine, pyroxene, and ferrite, whose composition tielines form an irregular tetrahedron which is not shown but which moves continuously toward the right along the joins shown separately in Figure 13.4. The last liquid is consumed, and crystallization ceases when the triangular base of this tetrahedron intersects *P*.

In the case of fractional crystallization at constant total composition, a portion of the early-formed crystals

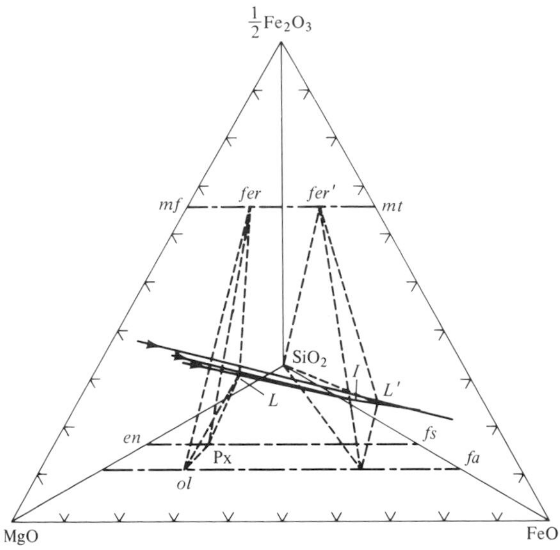

13.4 Molar plot of crystal joins certain phase volume intersections and tie lines of the system MgO-FeO-Fe_2O_3-SiO_2 as shown in Figure 13.3. The irregular tetrahedra which define coexisting phases are discussed in the text.

are prevented from reacting with the melt, so that the effective composition of the total system is diminished by these crystals. For example, if crystallization begins at *P* and the Mg-rich olivines of composition *ol* are isolated from the melt, the *effective* composition of the system is no longer *P* but some composition to the right and upward from this point. As a result, not only does the liquid follow a different path than in equilibrium crystallization but also the composition of the last liquid is shifted toward the iron-rich side of the diagram along the univariant line. In fact, the liquid may move as far as the invariant point *I*, where pyroxene must disappear, or even beyond to the ternary eutectic on the face FeO-$FeO_{3/2}$-SiO_2. The resulting liquid is shown as *L′* in Figure 13.4. It is obvious that such extreme fractional crystallization would result in an enormous enrichment of the liquid in FeO relative to MgO.

During the first stage of crystallization along any such path as *P–Q–R* there is an enrichment of the ratio $FeO_{3/2}/FeO$ in the melt, but later when the univariant line is attained on the beginning of crystallization of ferrite, this trend is reversed. During the latter course of crystallization there is in fact a steady drop in the oxygen fugacity as is necessitated by the falling temperature. This is true because for any given oxygen fugacity the tendency for oxidation increases greatly as the temperature declines. This may readily be seen by examination of such diagrams as Figure 5.4.

We shall discuss only briefly the cases of crystallization under conditions of constant oxygen fugacity. In this mode of crystallization the liquid composition is constrained to a plane (more properly a surface) of the type of MgO-SiO_2-Fe_3O_4, as previously mentioned. However, the point corresponding to *P*, which represents the total composition, must move off this plane during crystallization since oxygen is either added or subtracted. The movement of the bulk composition point is along "isocompositional lines," which are directed from the oxygen corner of the Fe-Mg-Si-O tetrahedron of which the system MgO-FeO-$FeO_{3/2}$-SiO_2 forms a subtetrahedron. Obviously such liquids cannot move toward the FeO-$FeO_{3/2}$-SiO_2 face and so cannot become enriched in FeO. However when crystallization begins as previously in the field of olivine, or indeed in any composition range of significance to basaltic rocks, the liquid will eventually move toward the SiO_2 corner and experience silica enrichment. The fundamental reactions controlling this behavior are of the following types:

$$\underset{melt}{3FeO} + \underset{gas}{\tfrac{1}{2}O_2} \rightleftharpoons \underset{ferrite}{Fe_3O_4} \tag{13.e}$$

$$\underset{melt}{2FeO} + \underset{melt}{MgO} + \underset{gas}{\tfrac{1}{2}O_2} \rightleftharpoons \underset{ferrite}{MgFe_2O_4} \tag{13.f}$$

which shows that as oxygen is added to the system—as is required with falling temperatures at constant oxygen fugacity—ferrite is withdrawn from the liquid and silica accumulates.

Certain details on composition of minerals have been added to Muan and Osborn's work by Speidel and Osborn (1967) who determined the composition of coexisting phases in the system MgO-FeO-Fe_2O_3-SiO_2 as a function of temperature and oxygen fugacity under equilibrium conditions at a pressure of 1 atm. Their results indicate that the distribution of Mg and Fe^{2+} in coexisting olivine, orthopyroxene, and magnetite may be significantly different at lower temperatures than at original crystallization temperatures.

In the artificial system under discussion here the source of this oxygen is largely H_2O and CO_2 of the equilibrated gas phase, which is of controllable composition. In nature the idealization of constant oxygen fugacity is probably never realized since the source of oxygen would frequently be confined to dissolved water and other substances within the magma itself. We shall see, however, that even magmas with large quantities of such volatiles are usually not very oxidizing but are more frequently somewhat reducing.

System $NaAlSi_3O_8$-$CaAl_2Si_2O_8$-$KAlSi_3O_8$-SiO_2-H_2O

The system albite-anorthite-orthoclase-silica-water is not only the major subsystem of basaltic magmas but also assumes increasingly greater importance as differentiation proceeds. When we discussed this system in relation to quartzofeldspathic metamorphic rocks, we introduced the ternary feldspar subsolidus diagram (Figure 9.11) which depicts the important immiscibility relation among these minerals. We shall be obliged to refer to this diagram here also and consider it in conjunction with what is known of the liquidus.

Our chief approach will be through the tetrahedron albite-anorthite-orthoclase-silica considered at some fixed water pressure. Figure 13.5 shows the tetrahedron that corresponds to a saturation water pressure of 5 kbar

13.5 Molar plot of the system $NaAlSi_3O_8$-$CaAl_2Si_2O_8$-$KAlSi_3O_8$-SiO_2-H_2O at 5 kbar water pressure. Hypothetical phase volumes are based on a similar diagram presented by Bateman *et al.* (1963). Possible equilibrium and fractional crystallization paths are also shown and are discussed in the text.

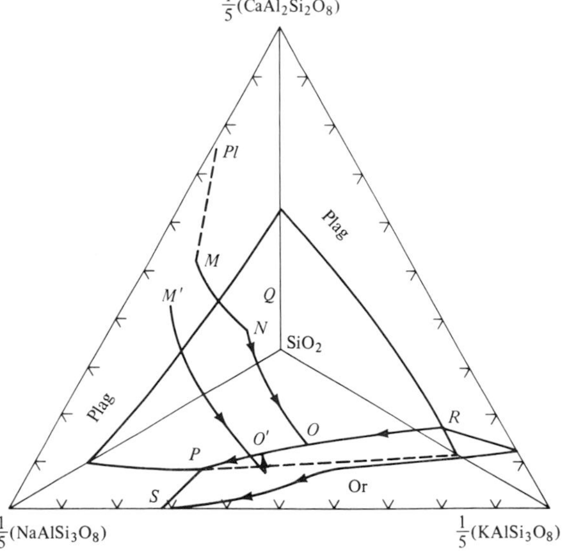

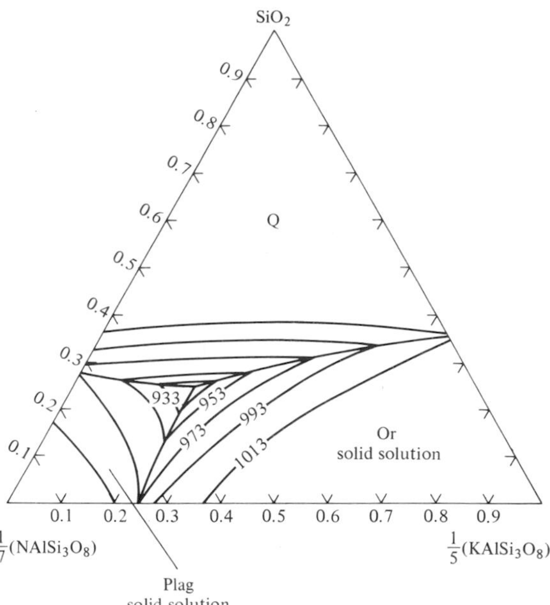

13.6 Molar plot of system $NaAlSi_3O_8$-$KAlSi_3O_8$-SiO_2-H_2O at 5 kbar water pressure. Based on data of Luth and Tuttle (1964). Temperature in °K.

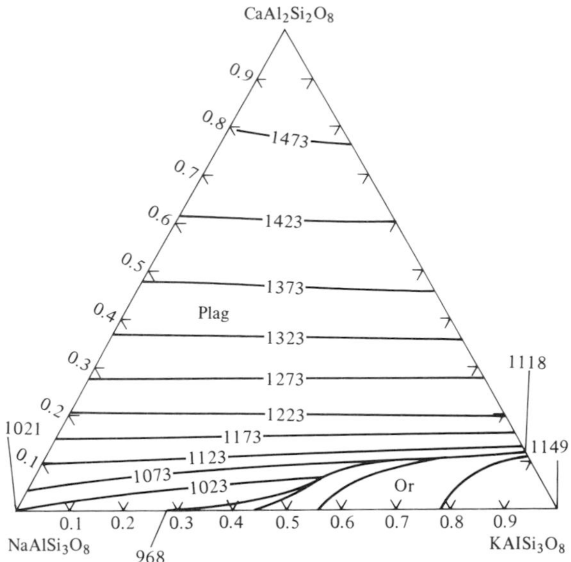

13.7 Molar plot of system $NaAlSi_3O_8$-$KAlSi_3O_8$-$CaAl_2Si_2O_8$-H_2O at 5 kbar water pressure. Based on data of Yoder *et al.* (1957).

as inferred from the limited experimental data.[9] The best known parts of this tetrahedron, the basal and frontal faces, are shown separately in Figures 13.6 and 13.7.[10] These diagrams should also be studied in conjunction with those presented by Tuttle and Bowen (1958) on the system albite-orthoclase-silica-water. Attention is specifically directed to the shift toward the silica-rich corner of the feldspar-silica boundary and the appearance of a leucite field at lower water pressures. The positions of the feldspar-silica phase boundary at several different water pressures are shown in Figure 13.8.

Many of the general qualitative phase relations, as well as their geometric expressions, are similar to those previously discussed for the systems albite-anorthite-diopside and MgO-FeO-$FeO_{3/2}$-SiO_2. This is true despite the fact that much water is present, because all the crystalline phases are anhydrous in character. Also

[9] Data for the tetrahedron base are from Luth *et al.* (1964), for the system albite-anorthite-orthoclase from Yoder *et al.* (1957) and for anorthite-silica join from Stewart (1958). Certain phase boundaries are as those previously inferred by Bateman *et al.* (1963) in a similar diagram.

[10] Such diagrams representing anhydrous projections in systems that are water saturated are referred to as "saturation surfaces" by Tuttle and Bowen (1958).

13.8 Molar plot showing shift of feldspar-silica phase boundary with changes in water pressure in the system $NaAlSi_3O_8$-$KAlSi_3O_8$-SiO_2-H_2O. After data of Tuttle and Bowen (1958). The shaded area shows the compositions of natural granitic rocks.

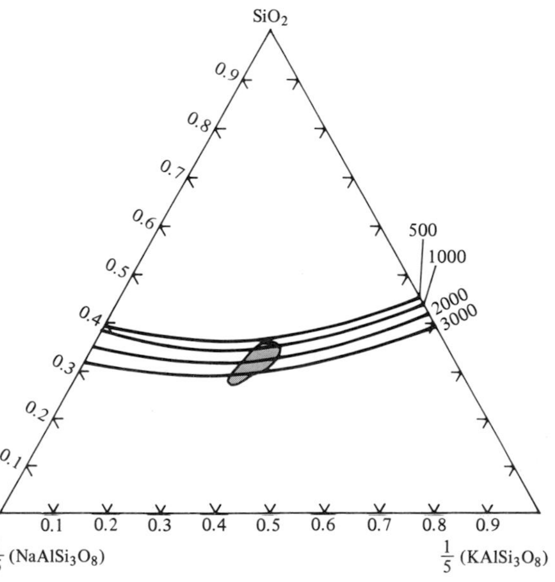

the diagrams at constant water pressure should correspond to reasonably constant water contents of the melts, especially near the tetrahedral base (Tuttle and Bowen, 1958).

The tetrahedron of Figure 13.5 is divided into a relatively small number of phase volumes. A characteristic of some significance is the large size of the plagioclase field as compared with that of orthoclase, a feature which should hold for lower water pressures as well. In no case has any attempt been made to differentiate the various polymorphs such as tridymite or α and β quartz; however, some of these may be located approximately by reference to the isotherms of Figures 13.6 and 13.7.

We shall illustrate differentiation of a melt in this system by considering two different paths of isobaric crystallization, one beginning at M well within the tetrahedron and one beginning at M' somewhat nearer the albite-anorthite-orthoclase face (Figure 13.5). Since both points fall within the field of plagioclase, the first part of the path of crystallization will be somewhat similar to that already discussed for the system albite-anorthite-diopside (Chapter 2). Then if crystallization begins at M, the first plagioclase will be anorthite-rich, corresponding to the point *pl*, which is shown joined to M by the dashed tieline. As crystallization proceeds, this crystal becomes enriched in both orthoclase and albite, but mainly in the latter. Consequently the direction of change of the liquid will have a strong component away from the anorthite corner along the curved path M–N. At N the path intersects the field of quartz, suffers a discontinuity, and thereafter curves downward to O on the univariant space curve along which orthoclase also begins to separate.

So far we have not stated whether the crystallization path refers to equilibrium or fractional crystallization. If equilibrium is maintained and if the projection of M on the albite-anorthite-orthoclase face does not fall within the two-feldspar field at the lowest attainable liquid temperature (Figure 13.5), then crystallization ceases when the last liquid lies somewhere along the path segment corresponding to N–O since no separate orthoclase can crystallize. If, however, the projection of M falls within the two-feldspar field on the albite-anorthite-orthoclase face, crystallization can cease only along the univariant space curve between O and P since in this case two feldspars must appear in the crystalline product along with quartz.

In the case of fractional crystallization, the paths will not only differ in location and curvature but they will also be greatly extended just as in the other systems already discussed. If M–N–O is such a path, the liquid may in fact move all the way to the invariant point P before crystallization ceases; or in the case of lower water pressure, the path of such extreme fractionation will lead to the minima shown in Figure 13.8.

The crystallization path beginning at M' will at first be similar to the segment M–N. However, because the system in this case has less SiO_2, the field of orthoclase is intersected at N' before that of quartz at O'. Other features distinguishing the equilibrium and fractional crystallization paths are similar to those already described for the system M and may be easily deduced by the reader.

Although we have not referred explicitly to the crystal compositions and their tielines, their general character will be similar to those already discussed for the system MgO-FeO-Fe_2O_3-SiO_2. For example, along the path segment N–O these will form a triangle, which intersects the point M under equilibrium conditions. Similarly along the path segment O–P the tielines form a tetrahedron. The precise configurations of these geometric elements is of course critically dependent on the temperature, the pressure, and effective bulk composition of the system as it is expressed in the phase compositions.

Because we have stipulated that the crystallization just discussed is isobaric, it is necessary that water be continuously withdrawn from the system, otherwise P_{H_2O} would build up rapidly as the quantity of liquid diminished. Of course we use this idealization of constant P_{H_2O} only as an illustration and recognize that in nature there would probably be a constant rise in water pressure as differentiation proceeded. Eventually, however, a maximum value of P_{H_2O} consistent with the confining pressure or permeability of the surroundings would be attained.

There is one feature of this subsystem that leads to some rather important petrologic conclusions. This is that the minimum temperatures at which liquids can exist within the system probably all lie within the base of the tetrahedron of Figure 13.5 or its analogs at different water pressures. These are of course the quaternary eutectic points like that of Figure 13.6 and the minima of Figure 13.8. As a consequence, liquids that are products of differentiation in the plagioclase field can approach the base system albite-orthoclase-silica (anorthite = 0) only in this low-temperature region. Conversely, it is unlikely that liquids that have very low anorthite content and that also lie toward the orthoclase-rich part of the diagram are products of differentiation in the field of plagioclase. It appears, however, that the compositions of the majority of rocks of general granitic character do project near the region of the ternary minima as was demonstrated by Tuttle and Bowen (1958) (Figure 13.8).

Isothermal and adiabatic crystallization

In many magmatic systems that contain substantial quantities of water, the possibility exists for crystallization to occur under isothermal conditions when heat is continually lost from the system. Furthermore in some systems crystallization may occur even when the heat of crystallization cannot escape and actually raises the temperature. These modes of crystallization are of course accompanied by substantial loss of water from the melt with a consequent drop in P_{H_2O} so that the liquidus temperatures are increased faster than the actual temperature of the system. Such behavior is especially important in the system albite-orthoclase-SiO_2-H_2O, for which Tuttle and Bowen (1958) have provided an interesting discussion. It is in all probability also a characteristic of the system albite-anorthite-orthoclase-SiO_2-H_2O and the even more complex systems of the calc-alkali and other magma series.

Effect of water on liquidus relations of magmas

In our discussion of the experimental data on the quartzofeldspathic systems we have seen that water has a rather profound effect. The major way in which water acts is of course to generally lower the liquidus tempertures, which is equivalent to an increase in the solubilities of the feldspars and quartz. This is, of course, to be expected in view of what we already know of the reaction between water and the highly polymerized silica structures of such melts (Chapter 12). It is hardly surprizing then that melts corresponding to the com-

13.9 Projection of natural olivine tholeiite-water system. The low-temperature limits of pyroxene and olivine and the high-temperature limit of sphene have been omitted because of uncertainties. (After Yoder and Tilley, 1962.)

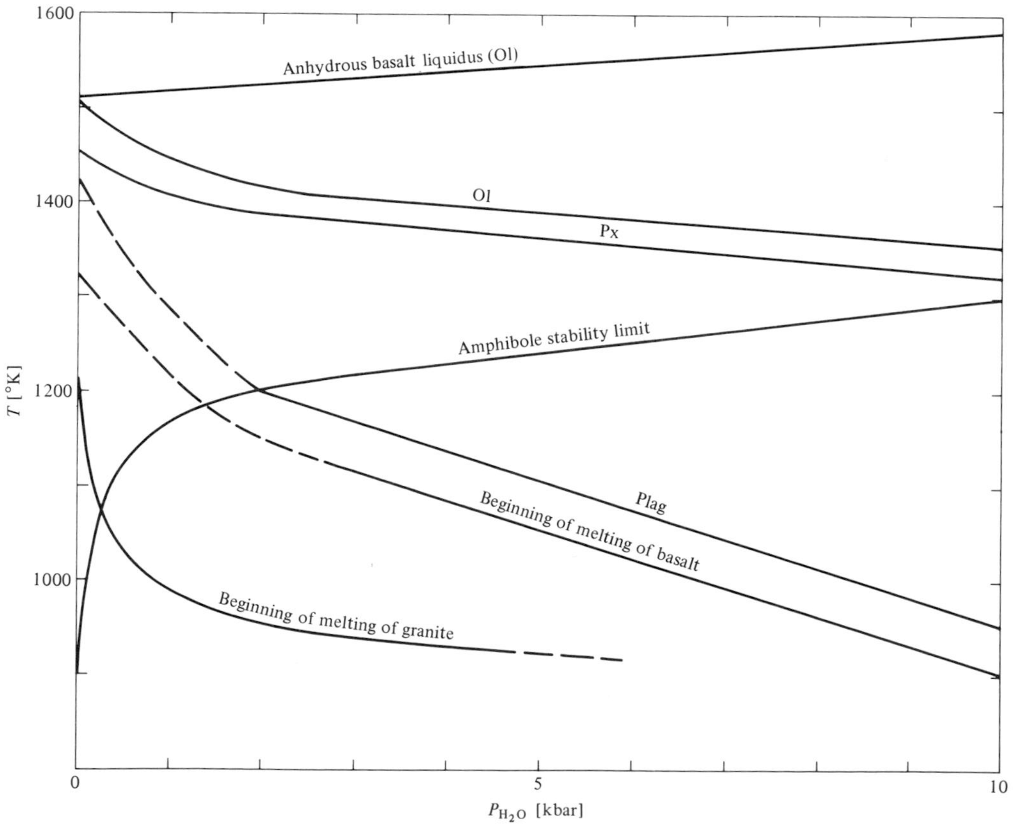

position of natural rocks show a similar effect. Especially interesting and important to petrologic interpretation is the behavior of basalt under varying water pressures, as was determined in the work of Yoder and Tilley (1962). If we examine their experimental results on a typical tholeiitic olivine basalt (Figure 13.9) several features stand out:

1. There is a considerable lowering of the liquidus temperature as defined by the beginning of crystallization of olivine.
2. Olivine and pyroxene are the first minerals to crystallize at all water pressures in the range investigated for this specimen.
3. The temperatures of beginning of crystallization of plagioclase is greatly depressed, corresponding to increased solubility as the water pressure is increased. This has the effect of greatly extending the temperature interval of crystallization or melting.
4. At high water pressures, amphibole crystallizes directly from the melt and forms by reaction between plagioclase and the ferromagnesium minerals.
5. The region of anhydrous basaltic or gabbroic mineral assemblages is virtually confined to water pressures less than 2000 bar.

It is worth noting that most of the depression of the crystallization temperatures of plagioclase occurs within the first 2000 bar where amphibole is not stable. Also, although in the particular example chosen the order of beginning of crystallization of olivine, pyroxene, and plagioclase is the same at all water pressures, the same is not true generally. Specifically, Yoder and Tilley (1962) found that in some high alumina basalts, for example, plagioclase is the first mineral to crystallize at low water pressures, but at pressures in excess of 2000 bars it already lags behind olivine and pyroxene.

Figure 13.9 also shows the curve of the beginning of melting of a typical granite as found by Tuttle and Bowen (1958). The minimum melting curve also coincides well with the values of T and P_{H_2O} of the ternary minima in the system albite-orthoclase-SiO_2-H_2O (Figure 13.8). However, the range of melting of granite is quite restricted as compared with basalt when refractory phases such as pyroxene and hornblende do not occur on the liquidus.

The general melting behavior of a typical calc-alkaline sequence of rocks is shown in Figure 13.10, which is taken from Piwinskii (1973). It is apparent from the figure that the response to increased water pressure

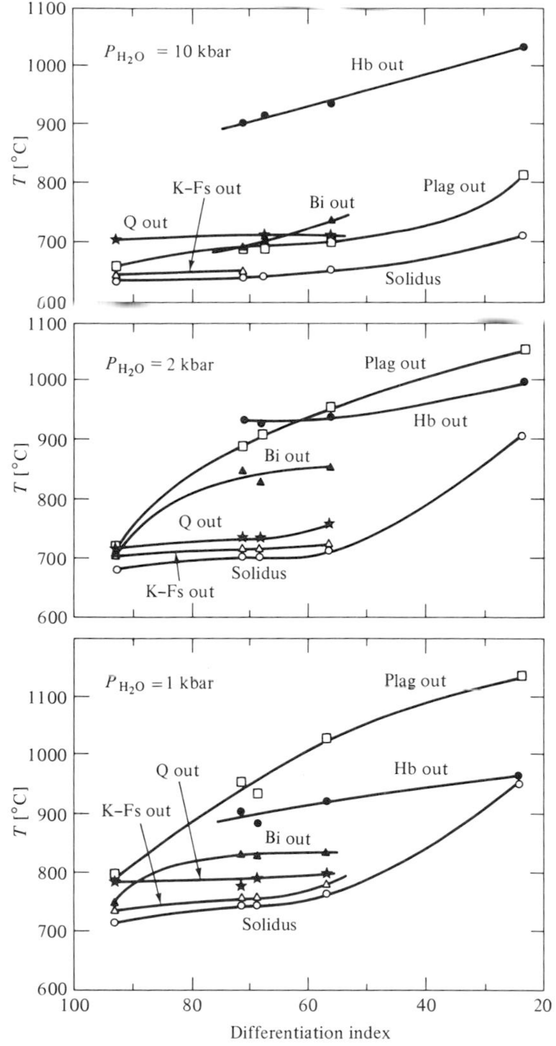

13.10 Phase boundaries of typical members of the calc-alkaline sequence at three different water pressures. The members of the sequence are expressed in terms of the differentiation index (D.I.), which is defined as the sum of the weight percentage of normative quartz + orthoclase + albite. (After Piwinskii, 1973.)

throughout the sequence is similar to that of basalt, with the quartzofeldspathic components showing a marked increase in solubility relative to the ferromagnesians with increased water pressure.

In addition to the direct effect that water and other volatiles have on the solubilities of minerals, there is a more subtle influence that these volatiles exert on the gamut of mineral stability relations. One of the most important of these effects is the complex relation between the composition of the dissolved volatiles and the oxidation and hydration states of the system, including both the melt and the coexisting crystals. It should be stressed that these states are strongly coupled in magmas. However, detailed consideration of these effects is best delayed until we consider certain specific rock types in which some evidence has come to light.

Crystallization of amphiboles and micas

Some of the effects of the crystallization of these minerals on differentiation trends were already well appreciated by Bowen (1928). In considering what these effects will indeed be, the guiding principle is, as in all previous cases, the utilization of compositional information on these minerals since it is these compositions that govern the kinds and degrees of element depletions suffered by the magma. For example, it frequently happens that amphiboles that separate from plutonic calc-alkali magmas are enriched in normative anorthite relative to albite and contain less silica than either feldspar or pyroxene. This stoichiometric relation, which is expressed by the reactions already discussed in connection with progressive metamorphism of amphibolites (Chapter 10), leads to the relative enrichment of magmas in albite and silica. Similarly, when biotite crystalizes early in the differentiation sequence it has the effect, through certain other reactions, of depleting the melt in the orthoclase molecule so that the residual differentiate is of trondhjemitic rather than granitic composition (Barth, 1962).

The obvious hydration effects that stabilize the biotites and hornblendes form one aspect of the more subtle interactions with the magmatic volatile compliment alluded to in the last section. For example, it is apparent that the oxidation state of a system may be defined in terms of the reaction $H_2O \rightleftharpoons H_2 + \frac{1}{2}O_2$ and the activity ratio a_{H_2O}/a_{H_2} so that the system becomes more reduced as this ratio diminishes. We shall see that there is evidence that reduction accompanies the differentiation of the common magmas. Although the precise reason for this effect is still not experimentally established, it seems reasonable that it is connected with the affinity of silicate melts for water and with the crystallization of amphiboles and micas that characterize these magmas.

One noteworthy effect shown in Figure 13.10 is the apparent decline in thermal stability of biotite and hornblende from basic to acid members of the calc-alkaline sequence. A part of this effect may be attributable to the increased Fe/Mg ratio of these minerals since iron-rich ferromagnesians tend to be less stable than their magnesian counterparts. However, the effect may also be explained as follows: if biotite is in equilibrium with the melt, we have for the phlogopite component:

$$\underset{\textit{biotite}}{KMg_3AlSi_3O_{10}(OH)_2} \rightleftharpoons \underset{\textit{melt}}{KO_{1/2} + 3MgO + AlO_{3/2} + 3SiO_2 + H_2O}$$

Then the corresponding equation of equilibrium may be written as

$$(X_{Mg}^{Bi})^3 K = X_{KO_{1/2}}^{L} (X_{MgO}^{L})^3 X_{AlO_{3/2}}^{L} (X_{SiO_2}^{L})^3 P_{H_2O} \phi(P, T, X)$$

where X represents the mole fractions of the constituents indicated and $\phi(P, T, X)$ represents the product of the activity coefficients. Similar relations also apply to the Fe members. It is apparent from the chemical analyses of the rocks (Piwinskii, 1968) that although $X_{KO_{1/2}}^{L}$, $X_{AlO_{3/2}}^{L}$, and $X_{SiO_2}^{L}$ increase, their *percentage increase*, which is significant here, is much less that the *percentage decrease* of X_{MgO}^{L} in the range D.I. (differentiation index) = 60 to D.I. = 95. The same argument also applies to the CaO component, which influences the stability of hornblende. This may be seen from the following tabulation of ratios for the Cartridge Pass tonalite (T) to the Dinkey Creek biotite granite (G) (Piwinskii, 1968):

Oxide	***T/G***
MgO	2.5/0.10
FeO	3.62/0.64
CaO	5.92/1.0
SiO_2	59.14/75.4
$AlO_{1/2}$	18.23/13.5
$KO_{3/2}$	2.19/4.6

Also it should be noted that X_{Mg}^{M} and X_{Fe}^{M}, which show the greatest, change appear raised to the third power in the equation of equilibrium. Thus it seems likely that the decreased thermal stability of the ferromagnesian minerals in the sequence from basic to acid magmas is at least in part due to the decreasing concentrations of their most characteristic components in the melt.

If we compare this effect with the effect of the differential increase of solubility of quartzofeldspathic and ferromagnesian constituents as a function of water pressure (Chapter 14), we note that although water pressure stabilizes ferromagnesians relative to feldspars on the liquidus, the ferromagnesians are subject to increased crystal fractionation from the magma as a consequence. This fractionation in turn gives rise to the low concentrations of MgO, FeO, CaO, etc. in the granitic melts and results in the eventual thermal destabilization of the ferromagnesians that contain these oxides.

Partial melting (anatexis)

In addition to the various crystallization mechanisms, differentiation can also occur through partial or differential fusion or dissolution of existing rocks. After such melting or *anatexis* has occurred, the liquid formed may be separated from the crystalline residue, be transported for varying distances, and intruded into the country rock. Only occasionally will this liquid remain precisely where it formed since it is always subject to at least the body force of gravity.

Although anatexis probably plays an important role in the generation of basaltic magmas under deep-seated conditions, it is discussed most frequently in connection with the formation of granitic melts, since only these can form at temperatures that are readily attainable within the crust. The question then usually boils down to whether or not a given rock—whether it be of sedimentary or of igneous origin—can give rise to a granitic melt.

As in crystallization, it is necessary to distinguish whether melting occurs under equilibrium or disequilibrium conditions. Under equilibrium melting—an idealization that is probably as rare in occurrence as equilibrium crystallization—phases appear and disappear precisely in the reverse order as they do under equilibrium crystallization in the same system. Consequently the first liquid to form is identical in composition and forms at the same temperature and pressure as the last to disappear with crystallization. Also as in crystallization, the composition, temperature, and pressure of this initial anatectic melt is limited to a considerable degree by the bulk composition of the system. For example, in the system albite-orthoclase-SiO_2-H_2O (Figure 13.6) the first liquid to appear as the temperature is raised will coincide with the quaternary minimum only if its bulk composition falls within one of the three triangles whose liquid apex also coincides with this minimum. In any other composition range farther from the minimum, equilibrium melting can only begin at some higher temperature. This point is also illustrated by the difference in beginning of melting of granite and basalt (Figure 13.9). These conclusions are equally applicable to virtually every system of petrologic interest since they depend on the properties of the ubiquitous solid solutions.[11]

In disequilibrium melting the affected mass is in some structural or compositional state not appropriate to the temperature and pressure at which fusion occurs. This is the case, for example, when a zoned plagioclase or a low-temperature microcline crystal is melted before the proper homogeniety or structural state can be attained by diffusion. In the case where a disequilibrium structural state occurs it is difficult to predict whether melting will begin at higher or lower temperatures than under precise equilibrium conditions.

Where compositional factors are the overriding ones the question regarding the possible appearance of a given anatectic liquid can sometimes be decided through knowledge of the mineral assemblages and compositions. For example, if a normal gabbro consists of essentially homogeneous plagioclase and ferromagnesian minerals, the small quantity of normative orthoclase will almost certainly be in solution with far greater quantities of albite and anorthite. It is inescapable that, in such a rock, melting could begin only at temperatures substantially above those of the normal granitic quartzo-feldspathic liquidus (Figure 13.9). However, if another gabbro, identical in bulk composition to the first, is a product of disequilibrium crystallization which has retained a segregated granite component, either through crystal zoning or as interstitial symplectites, it is obvious that melting could well begin at temperatures corresponding to the granitic liquidus. However, if after the initial liquid were formed, there should be a reaction tending toward the homogeneous equilibrium state, this liquid might again disappear.

Tuttle and Bowen (1958) have discussed in some detail the production of anatectic granitic melts and some of the conditions affecting their transportation. Also Winkler (1974) has devoted considerable attention to anatexis. One of the most important factors in their production is the available water. If this is restricted to that present in the rock as hydrous minerals, then there is a definite relation between this water, the solubility of water in the melt, and the quantity of melt produced. For example, a mass Q of granite may contain a fraction of water X_{H_2O} bound only as entirely unstable hydrous minerals. Then, if with anatectic metling, a quantity of melt q is produced, and if under the prevailing temperature and pressure conditions this melt contains a fraction $X^M_{H_2O}$ of water, then obviously

$$QX_{H_2O} = qX^M_{H_2O}$$

[11] See also Presnall's (1969) geometric analysis of partial fusion.

Now at the higher water pressure $X^M_{H_2O} \cong 0.10$, so that if $X_{H_2O} = 0.001$ (as in many dry rocks), then $q/Q = 10^{-2}$. Thus under these circumstances only 1 percent of the granite could melt. This quantity may of course be increased by increasing the quantity of unstably bound water in the original rock. However, it will be decreased to the extent stable hydrous minerals can coexist with the melt.

An additional compositional factor to be considered is the possible formation of a water-rich fluid, which occupies the continuum between magmas and dilute aqueous (so-called "hydrothermal") solutions. Presumably such fluids could form at temperatures considerably below those of the liquidus (saturation surface) in the system albite-orthoclase-SiO_2-H_2O. However, the work of Tuttle and Bowen (1958) indicates that these fluids could only form in a system with excess alkalis so that $[(K + Na)/Al] > 1$. Consideration of the chemical compositions of normal sedimentary and igneous rocks shows that such compositions are very unusual and that consequently we should not expect such fluids to arise very often in nature.

The relatively small quantity of anatectic melt that will ordinarily be produced, the lack of significant superheat, and the decrease of water solubility with decreasing confining pressure all militate against extensive upward transport of such melts within the crust. Consequently although anatexis may play an important role in the formation of certain plutonic migmatites, we shall apparently have to look elsewhere for a source of the more massive outpourings and intrusions of volcanic and plutonic granitic magmas.

It may indeed happen that a number of common and even rare magma types are regularly produced by anatexis within the Earth's upper mantle as suggested particularly for andesites. It is virtually certain that basaltic magmas are produced in this way as is discussed in the next section.

Differentiation under high pressures

Until recent years magmatic differentiation has been discussed largely in terms of minerals that regularly appear on the liquidus under crustal conditions. Specifically the discussion has centered around the feldspars, pyroxenes, quartz, and olivine, with lesser attention paid to the spinelloids, hornblende, and the micas. To be sure, there are modifications, some of which are far from obvious, that are imposed on these minerals by pressure effects within the crustal range. However, we are here concerned with the more dramatic transformations that usher in highly unfamiliar assemblages that only infrequently reach the Earth's surface.

A clue to these transformations is provided by certain familiar high-density minerals of regional metamorphism, which should, by simple analogy, have their counterparts in deep-seated magmatism. Evidence of the latter in fact do occur as eclogitic inclusions in certain basalts and other basic rocks, which are assumed to have their sources within the mantle. Since their minerals assemblages, consisting of garnet and pyroxenes, differ considerably in composition from the feldspar, pyroxenes, and olivines of crustal magmatic rocks, we infer that their appearance on the liquidus should directly influence the course of igneous differentiation at depth. To the extent that we can deduce the compositions of the high-pressure magmatic minerals of the mantle, we can also predict the course of differentiation brought about by either crystallization or partial melting. However we first face the difficulty of not knowing, except in the grossest terms, what the bulk chemical composition of the subcrustal material is. All that we can infer from the comparatively crude geophysical data is that this material probably consists of the constituents MgO, FeO, CaO, SiO_2, and Al_2O_3, with minor amounts of Na_2O, K_2O, and other elements and that it is more basic than any basaltic magma that reaches the surface. In addition we may also infer that basalt is a component of this material since it appears to be derived from it on a world-wide extent by partial melting. Ringwood (1965) was one of the first to attempt a detailed deduction of mantle composition, and for it he proposed the term "pyrolite" which he defined as a mixture of three parts peredotite to one part basalt. Thus pyrolite should yield basalt on partial melting if for no other reason than it was put there for this purpose. Subsequently many other mantle compositions have been discussed but all remain conjectures.

Although we can only roughly surmise the bulk chemistry of the subcrust, it is possible that our knowledge of high-pressure minerals as obtained by direct observation and through carefully controlled experiments may enable us to arrive at certain significant conclusions on at least conceivable differentiation trends. In recent years there have indeed been many high-pressure experiments on rocks and synthetic mixtures that are inferred to resemble mantle material and its differentiates, while other experiments have extended the classical phase equilibrium studies to hydrous magmas and phases that resemble calc-alkaline or other rock suites that correspond to depths of the order of 30 to 60 km. In particular, the latter type of experimentation is exemplified by the work of Yoder and

Tilley (1962), Piwinskii and Wyllie (1968), and Piwinskii (1973) on basalts and the calc-alkaline suite. In this range of experimentation the quartzofeldspathic constituents show the increased solubility with water pressure, which appears to be crucial to the development of the calc-alkaline series under crustal conditions (Chapter 14), while the hydrous minerals biotite and hornblende are stabilized. However, at higher pressures and temperatures the latter again become unstable. It is in the region beyond 20 kbar where this occurs. Here we conceive the primary basaltic and some say andesitic—magmas to be generated. Commencing in this range, the chief minerals found on both the dry and water-saturated liquidus in compositions ranging from periodotite to andesite consist chiefly of the pyroxenes (ortho and clino), olivine, and garnet (Figure 13.11). This has now been confirmed by numerous experiments such as those of Green and Ringwood (1968), Green (1972), Millhollen and Wyllie (1974), and Nichols (1974), while Green and Ringwood (1968) have shown that quartz is the liquidus phase only for dacitic and rhyodacitic anhydrous melts under high pressure.

One of the constituents for which we should be able to infer behavior at depth is SiO_2. The reason for this is that the high- and low-pressure assemblages differ markedly in their contents of this oxide. This may be illustrated by the following reactions:

$$\underset{\textit{plagioclase}}{NaAlSi_3O_8} \rightleftharpoons \underset{\textit{pyroxene}}{NaAlSi_2O_6} + \underset{\textit{melt}}{SiO_2} \qquad (13.g)$$

$$\underset{\textit{plagioclase}}{CaAl_2Si_2O_8} + \underset{\textit{pyroxene}}{4(Mg,Fe)SiO_3} \rightleftharpoons \underset{\textit{pyroxene}}{Ca(Mg,Fe)Si_2O_6} + \underset{\textit{garnet}}{(Mg,Fe)_3Al_2Si_3O_{12}} + \underset{\textit{melt}}{SiO_2} \qquad (13.h)$$

$$\underset{\textit{plagioclase}}{2CaAl_2Si_2O_8} + \underset{\textit{pyroxene}}{Ca(Mg,Fc)Si_2O_8} + \underset{\textit{pyroxene}}{2(Mg,Fe)SiO_3} \rightleftharpoons \underset{\textit{garnet}}{Ca_3Al_2Si_3O_{12}} + \underset{\textit{garnet}}{(Mg,Fe)_3Al_2Si_3O_{12}} + \underset{\textit{melt}}{2SiO_2} \qquad (13.i)$$

$$\underset{\textit{plagioclase}}{CaAl_2Si_2O_8} \rightleftharpoons \underset{\textit{pyroxene}}{CaAl(AlSi)O_6} + \underset{\textit{melt}}{SiO_2} \qquad (13.j)$$

$$\underset{\textit{plagioclase}}{NaAlSi_3O_8} + \underset{\textit{olivine}}{(Mg,Fe)_2SiO_4} \rightleftharpoons NaAlSi_2O_6 + \underset{\textit{pyroxene}}{2(Mg,Fe)SiO_3} \qquad (13.k)$$

$$\underset{\textit{plagioclase}}{CaAl_2Si_2O_8} + \underset{\textit{olivine}}{(Mg,Fe)_2SiO_4} + \underset{\textit{pyroxene}}{2(Mg,Fe)SiO_3} \rightleftharpoons \underset{\textit{pyroxene}}{Ca(Mg,Fe)Si_2O_6} + \underset{\textit{garnet}}{(Mg,Fe)_3Al_2Si_3O_{12}} \qquad (13.l)$$

$$\underset{\textit{olivine}}{(Mg,Fe)_2SiO_4} + \underset{\textit{melt}}{SiO_2} \rightleftharpoons \underset{\textit{pyroxene}}{2(Mg,Fe)SiO_3} \qquad (13.m)$$

$$\underset{\textit{plagioclase}}{CaAl_2Si_2O_6} + \underset{\textit{olivine}}{2(Mg,Fe)_2SiO_4} + \underset{\textit{melt}}{SiO_2} \rightleftharpoons \underset{\textit{pyroxene}}{Ca(Mg,Fe)Si_2O_6} + \underset{\textit{garnet}}{(Mg,Fe)_3Al_2Si_3O_{12}} \qquad (13.n)$$

$$\underset{\textit{melt}}{Al_2O_3} \rightleftharpoons \underset{\textit{pyroxene}}{Al_2O_3} \qquad (13.o)$$

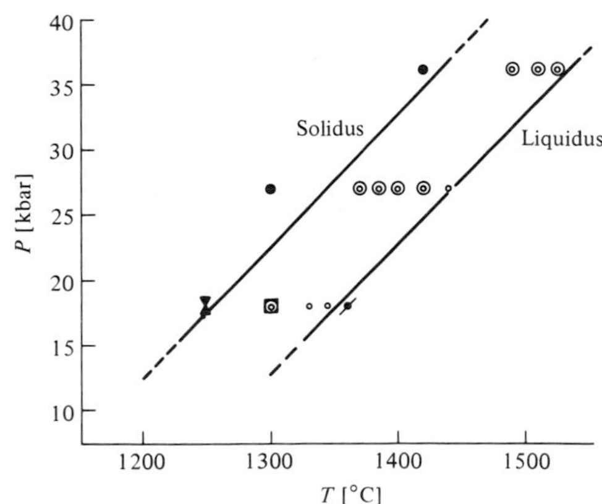

13.11 Approximate liquidus and solidus boundary curves and stable phases under anhydrous conditions for high alumina quartz tholeiite compositions. (After Green and Ringwood, 1968.)

In these reactions Mg and Fe have been combined for brevity. However in any quantitative application the end-members of the solid-solution series must be distinguished, and this will double the number of reactions in which they occur. Also all the reactions have been written in such a way that the volume is inferred to decrease toward the right.[12] Thus in the case of the first four reactions, silica is released to the melt as the low-volume (high-pressure) assemblage of garnet and pyroxene forms. However, in the case of Reactions (13.m) and (13.n) the reverse is true, and the melt is depleted in silica with the formation of the high-pressure assemblage.

Reaction (13.o) plays a crucial role in forming sub-silicic pyroxenes enriched in Al_2O_3, which characterize the high-temperature, high-pressure assemblages (Green and Ringwood, 1968). The effect of the crystallization of these pyroxenes from a melt or of forming a melt from rock containing these is to enrich the melt in silica. However, the effect is less marked where pyroxene coexists with garnet (Green and Ringwood, 1970) because Al_2O_3 enters the garnet according to the reaction

$$\underset{\textit{orthopyroxene}}{Mg_2Si_2O_6 + MgAl_2SiO_6} \rightleftharpoons \underset{\textit{garnet}}{Mg_3Al_2Si_3O_{12}} \tag{13.p}$$

The result of this equilibrium (Wood, 1974) is that although the Al content of orthopyroxene coexisting with garnet increases with the temperature, it decreases with increasing pressure at a given temperature (see also Chapter 11).

The crystallization of garnet also has the additional effect of decreasing the partitioning of iron between crystals and melt. The distribution of iron between pyroxenes and basaltic magmas is such that it is greatly concentrated in the latter relative to magnesium (Chapter 14). For garnet this effect is much less pronounced (Green and Ringwood, 1968). However, since pyroxene usually coexists with garnet on the liquidus, melts formed in the mantle should always be more iron-rich than the crystal residua.

It is clear from these reactions that high pressure will favor the appearance of pyroxene relatively enriched in Na, Ca, and Al and that garnet should be increasingly favored as the pressure is increased. Accompanying these mineral changes would be the silica enrichment or depletion implied by the stoichiometry. For example, if the system is rich enough in silica to form pyroxene instead of olivine under low-pressure conditions, then according to the first four reactions silica enrichment of the melt will occur under sufficiently high pressures. The subsolidus equivalent of this is the transformation of olivine basalt to quartz eclogite. Conversely, silica depletion of the melt will occur under high pressures if the melt is relatively undersaturated with respect to the low-pressure assemblage.

Of course the actual effect in any differentiation sequence will depend on the sum of reactions of this type. Thus it is possible for reactions such as (13.o) to bring about silica enrichment in relatively silica-depleted rocks for which Reactions (13.m) and (13.n) are operative.

One conclusion that may be drawn from reactions of this type is that they should extend over a considerable range of pressures and temperatures with both the high- and low-pressure assemblages coexisting over this range (Mueller, 1963a). This conclusion, based on the behavior of solid solutions is borne out by the results of the experiments, which show that various combinations of pyroxenes, garnet, olivine, and plagioclase can coexist in the range of 5 to 30 kbar between the liquidus and solidus.

It is interesting that the typical calc-alkaline trend of differentiation, as indicated by the MgO-FeO-alkali plots or AFM diagrams (Chapter 14), is shown under both hydrous crustal conditions (Piwinskii, 1973) and anhydrous mantle conditions (Green and Ringwood, 1968). Although the trend under hydrous conditions appears to result from increased quartzofeldspathic solubility with water content, the corresponding effect at high pressures probably results from the occurrence of low-alkali, high-iron phases such as garnet and pyroxene on the liquidus. In each case it amounts to suppression of feldspar crystallization and the precipitation of both iron- and magnesium-rich ferromagnesians rather than ferromagnesians dominantly enriched in magnesium. The result is a more rapid enrichment in alkalis than in gabbroic type differentiation.

One of the great controversies that has arisen in recent years revolves around the possibility of deriving andesitic magmas from various conceivable mantle materials by direct partial fusion rather than by differentiation of basaltic magma. Green and Ringwood (1968) showed that under anhydrous mantle conditions andesite represented the lowest melting composition within the silica-saturated calc-alkaline suite. Meanwhile it was determined (Kushiro, 1973; Nichols, 1974) that liquids of andesitic composition could also arise from

[12] The volume changes can only be roughly estimated since only STP data are available. In the case of the partial molar volume of SiO_2 in the melt the volume of pure SiO_2 may be utilized (Mueller, 1963). A better estimate might be obtained from the work of Bottinga and Weill (1970). However, no data exist for the pressures involved.

the partial melting of rocks as silica-undersaturated as peridotite. It seems that the appearance of andesite liquids versus silica-undersaturated liquids is in this case highly dependent of the partial pressure of water. Thus andesitic liquids can coexist with magnesian olivine crystals only at water pressure of less than 10 kbar, while at higher water pressure liquids in equilibrium with olivine are silica-undersaturated (Nichols, 1974). This result is regarded as reflecting unfavorably on the genesis of andesite by melting of mantle peridodite in regions overlying the Benioff fault zones, which play such an important part in current global tectonic ideas.

The effects of pressure on melting relations and differentiation trends of subcrustal materials and basaltic magma have also been widely discussed in relation to simplified synthetic systems which are extensions of classical phase equilibria studies. In particular the concern (O'Hara, 1965; O'Hara and Yoder, 1967) has been to show how increased pressure shifts the cotectic troughs and thermal divides such as those discussed in relation to Figures 13.6, 13.7, and 13.8. Such shifts would play a crucial role in determining whether or not silica-saturated magmas would result by partial melting of any given mantle material.

It is clear from the cited experiments that basaltic magma of various types can be derived by partial melting of mantle material similar to pyrolite. The precise nature of the basaltic magma—whether tholeiitic, alkaline, or more basic—can be a function of a variety of factors either related to differences in composition of the mantle source material, the P–T conditions of fusion, or to modifying factors such as volatile concentration (Ito and Kennedy, 1974).

It also seems clear that andesitic liquids can be derived by partial fusion of a variety of materials. For example (Green, 1973), when 28-percent pyrolite is fused under saturated conditions at 10 kbar, magnesian, quartz-normative basaltic andesites result; and if 33 percent of the same material is fused, quartz tholeiite melt is obtained. Similarly, Nichols (1974) showed that andesitic liquids could coexist with peridotitic material as explained previously. Such experiments do not establish, however, that andesite is an important primary magma. Rather, the experiments of Green (1973) showed that the tendency to form andesitic liquids through fractional crystallization is so strong that they are frequently formed on quenching more basic liquids. It is possible that some andesites form by direct fusion of subcrustal material. However it always must be remembered that its lower density as compared with basalt will tend to make andesite more prominent near the surface and conceal basalt (gabbro) at depth.

Subtraction and variation diagrams

So far we have been concerned largely with the physicochemical principles that govern the various mechanisms by which magmatic differentiation takes place. However, we have said little about the quantitative chemical and mineralogic changes that must be known in order to determine the extent of differentiation precisely. It is important, for example, to know what proportion of a given magma must crystallize to yield another magma of given composition on the same line of descent as the first. Then, too, it is equally important to see how the abundance of the rock-forming oxides vary along the entire differentiation sequence.

The foregoing questions can be answered very easily for synthetic systems once the phase compositions and crystallization paths have been determined (Bowen, 1928). However, such information is difficult to obtain for natural magmas. Instead, it is necessary to combine all available information on the minerals in rocks that are the products of this differentiation with certain plausible assumptions about the history of these rocks. For example, if we can assume that two such rocks represent former pure liquids on the same line of descent, and if we can infer the character of the crystallization products that separate them, then not only can we deduce the quantity of each crystal precipitated but also how much of the first magma had to crystallize to yield the second. The result of such a calculation is usually presented graphically as a *subtraction diagram.*

We shall illustrate what we have just said by making use of two volcanic rocks from the Cascade Province, one an olivine basalt and the other of basaltic andesite, which we shall assume (at least tentatively) lie on the same liquid line descent. The chemical analyses of these rocks, which were first presented by Williams (1942) and later used by Osborn (1959) in a similar illustration, are given in Table 13.1. The mole percentages of the oxides for both rocks are given in columns 2 and 4 of the table and from these Figure 13.12 has been constructed by the following method: Since the basaltic andesite does not differ greatly from the olivine basalt, it seems justified to introduce the critical assumption that the crystalline product contains no $KO_{1/2}$. We also make the second assumption, which is quite arbitrary in this case, that the crystalline product consists entirely of plagioclase, Ca-pyroxene, olivine, and magnetite. The first assumption regarding $KO_{1/2}$ enables us to establish the horizontal scale of the figure by drawing a straight line between the $KO_{1/2}$ content of the basaltic andesite and the zero-value representing the crystalline product

Table 13.1 Compositions of two rocks from the Cascade volcanic province. See Fig. 13.12

	L_1 Olivine basalt		L_2 Basaltic andesite	
	Wt. %	Mole %	Wt. %	Mole %
SiO_2	50.7	46.5	55.8	52.5
$\frac{1}{2}Al_2O_3$	18.0	19.8	18.0	19.9
CaO	9.7	9.7	7.4	7.4
MgO	7.6	10.6	5.1	7.2
FeO	7.0	5.5	4.1	3.2
$\frac{1}{2}Fe_2O_3$	1.6	1.1	2.6	1.9
$\frac{1}{2}Na_2O$	2.7	4.9	3.6	6.5
$\frac{1}{2}K_2O$	0.7	0.8	1.2	1.4

Data from Williams (1942) and Osborn (1959).

on the right. This scale, which has been arbitrarily set at 100 units, then contains the mole percent of olivine basalt crystallized. The numerical value of the latter, which may be found by fitting the basaltic $KO_{1/2}$ content to the straight-line plot, is found to be approximately 47 percent. Once the position of the olivine basalt is plotted, the oxide contents of the crystalline product are fixed by straight-line plots, except that some FeO must be oxidized to form the required magnetite. The quantity of the latter is governed by the amount of FeO + MgO that is in excess of that required for olivine and pyroxene.[13]

The most important lesson that we can learn from calculations of this type is that in the passage from one magma to another by crystallization differentiation, even when the chemical difference between them is not great, involves the crystallization of a large fraction of the parent magma. Unfortunately, it is not possible to extend this method to later members of the differentiation sequence since the assumption regarding $KO_{1/2}$ no longer holds. However, it is clear by extension of this line of reasoning that a very great proportion of basaltic parent magma must crystallize to yield dacite or rhyolite, and this effect will be all the more pronounced precisely because the late crystallization products contain large amounts of alkalis.

When the oxide contents of two or more rocks are plotted against some continuously varying parameter such as the SiO_2 content (Harker diagram), we have a type of *variation diagram* which has seen wide use in the petrologic literature. These diagrams were examined critically by Bowen (1928), who showed that the form of the variation curve for each oxide may be related to corresponding details in the crystallization path when the rocks concerned all lie on a single liquid line of descent. As in the case of the subtraction diagram, the variation diagram may be constructed in terms of a variety of parameters. For example, in the diagram introduced by Larsen (1938), the parameter $\frac{1}{3}SiO_2 + K_2O - FeO - MgO - CaO$ (in weight percent) substitutes for SiO_2 in the better-known Harker diagram. Since the SiO_2 content is usually an almost linear

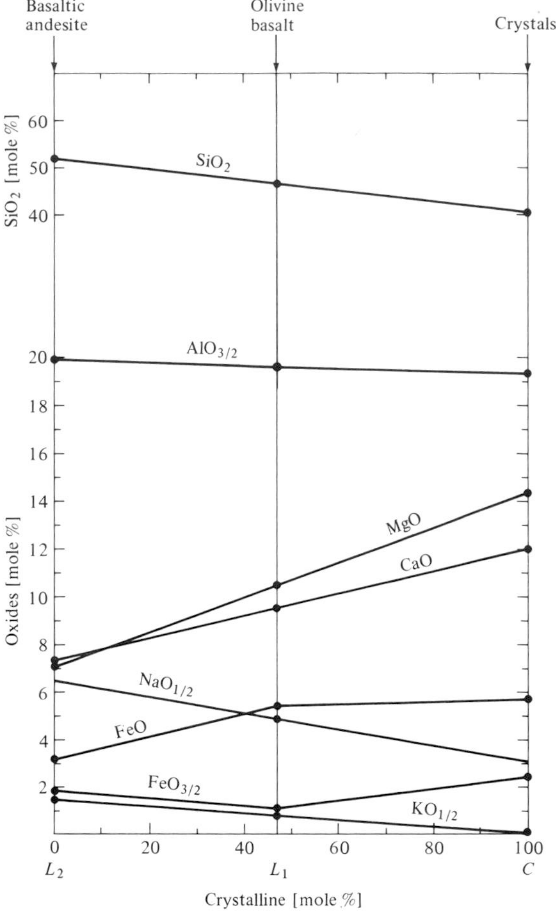

13.12 Subtraction diagram of two volcanic rocks of the Cascades province. Data are from Williams (1942) and Osborn (1959). For a detailed explanation see text.

[13] An alternative way of constructing the subtraction diagrams makes use of the quantity of a given oxide plotted on the horizontal scale. Thus Bowen (1928) and others have found the SiO_2 content a convenient index of differentiation.

function of the Larsen parameter, the oxide variation curves tend to be similar in the two types of plots.

An interesting type of variation diagram introduced by Thornton and Tuttle (1960) makes use of a parameter known as the "differentiation index," or D.I. The D.I. is defined as the sum of the weight percentages of

normative quartz + orthoclase + albite + nepheline + leucite + kalsilite.

These are the components of "petrogeny's residua system," which tend to accumulate in melts as differentiation proceeds or result from early anatexis. Since no more than three of these normative minerals will appear in any given norm, the D.I. is simply the sum of the percentages of three normative minerals. Thus in a granite this sum would be orthoclase + albite + quartz, whereas in a phonolite it could be orthoclase + albite + leucite. The D.I. has the advantage of treating each oxide in the same manner, and exhibits well the difference in silica-saturated and silica-unsaturated differentiation sequences.

Although the efficacy of the variation diagram as a petrologic tool has been challenged, it was demonstrated by Larsen in particular that significant systematic differences exist between the variation curves of different geologic provinces and different igneous episodes in the same province. Thus the rocks of the San Juan volcanic province of Colorado (Larsen and Cross, 1956), and as illustrated in Figure 13.13, are generally rich in

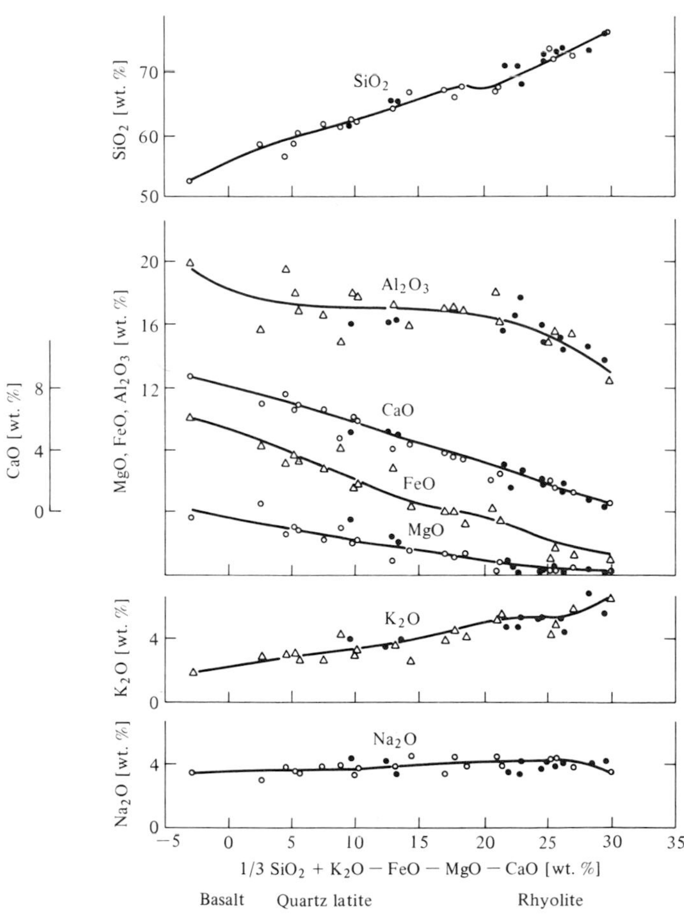

13.13 Variation diagram of Conejos quartz latite, in weight percent. Crosses and circles indicate rocks; dots are groundmasses. Curves are variation curves for the Potosi volcanic series. (After Larsen and Cross, 1956.)

K_2O as compared with other calc-alkali provinces. Of course, a great part of the uncertainty in such diagrams is immediately attributable to errors in the chemical analyses. However, it was also shown by Bowen (1928) that such diagrams are sometimes greatly influenced by admixtures of crystals so that the rocks deviate from possible pure liquids in the line of descent. If such admixtures are present in varying proportions, the variation curves tend to be straight like those of subtraction diagrams, which is, of course, in the nature of mixtures. Bowen also demonstrated that variation curves that represent a true liquid line of descent may have quite characteristic curvatures so that, for example, the Na_2O curve characteristically is concave downward while the K_2O curve is concave upward near the rhyolitic end of the diagram.

Several other interesting features are shown by Figure 13.13. One of the most important of these is the relation between a given rock and its ground mass. It is apparent that both rock and ground-mass compositions generally lie on or near the same variation curve so that we have in this case direct evidence of not only liquid lineage but also for an origin by normal crystal differentiation involving ordinary crustal minerals.

14 Igneous plutons: Their physical chemistry and mode of occurrence

The compositional fields of igneous rocks

General

Our topics here are the physical chemistry of the rocks and magmas and the structural modes of occurrence of the prominent plutonic bodies as illustrated by specific examples from the literature. However, it is desirable first to consider the observed range of chemical composition of those bodies. The primacy of composition, which we have previously emphasized for metamorphic rocks, compels us again to settle the question of the compositional fields of stabilities of magmatic minerals before it is possible to discuss meaningfully the contributions of temperature and pressure. Our concentration on plutonic rocks is dictated not only by their relative importance in the evolutionary scheme but also by their registering with greater fidelity than volcanics the physicochemical environment of the deep-seated regions in which differentiation occurs.

Although it is difficult to summarize in any succinct way the compositional variety of even the commonest igneous rocks, something approximating this may be done by the use of charts and diagrams in conjunction with stoichiometric chemical equations which display most precisely the compositional transformations between one rock and another. We already made an attempt in this direction when we introduced the simple classification scheme of Table 4.3 for igneous rocks. However, it is apparent that this chart, which is based on

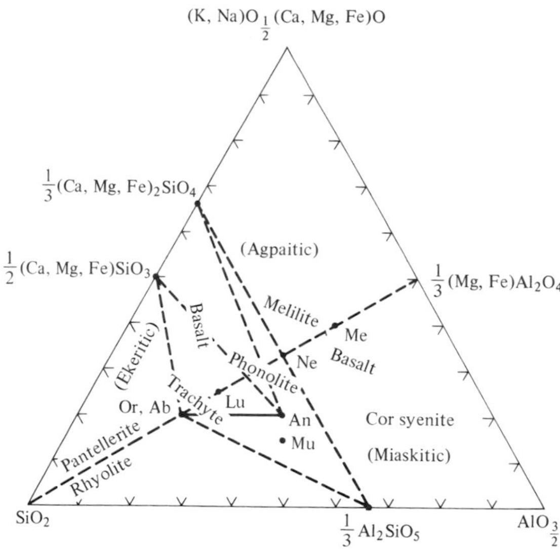

14.1 Molar plot of normative minerals to indicate degree of saturation or undersaturation with respect to SiO_2 and Al_2O_3. Other oxides have been lumped together.

the variation of quartz, feldspar, and feldspathoids, falls short of distinguishing the many important rock types which differ in Al_2O_3, K_2O, Na_2O, CaO, and other oxides.

A generalized triangular diagram, which is a useful supplement to Figure 4.3, is shown in Figure 14.1. On this diagram are plotted, on a per-cation basis, the common rock-forming minerals. Of necessity some of the mono- and di-valent cations have been lumped together to increase the scope of coverage. Most of the formulas appear regularly in normative calculations while others, such as melilite, do not.

Figure 14.1 depicts not only the degree of silica saturation (or undersaturation) but also the degree of saturation with respect to Al_2O_3. We may in fact define the line of critical Al_2O_3 undersaturation as the line which bisects the SiO_2 corner, since most rock-formers with K + Na = Al fall on this line, and only Al-deficient species fall above it. For rocks of intermediate compositions, which contain substantial CaO, MgO, FeO, etc., there is a range of undersaturation with respect to Al_2O_3 between this line and the plagioclase join albite-anorthite, since Al-rich phases ordinarily will form only after all the available CaO has been converted to anorthite. Outside this area, however, saturation with respect to Al_2O_3 is manifested as modal corundum in certain syenites and as andalusite, sillimanite, or muscovite in more siliceous rocks such as granite. The region above the line K + Na = Al is characterized by Al deficiency so that the chemical analyses show normative acmite, and Na-Fe^{3+} pyroxenes and amphiboles occur in the mode.

Because of the importance of Al_2O_3 and SiO_2 variation on the mineralogy, some authors have found it convenient to distinguish the regions of Figure 14.1 by terms which are defined as follows:

Ekeritic: silica adequate or excessive, alumina deficient

Miaskitic: silica deficient, alumina adequate or excessive

Agpaitic: silica and alumina both deficient

The origin of these terms has been discussed by Barth (1962).

Basaltic compositions

Figure 14.2 is a familiar tetrahedral plot, also on the per-cation scale, that covers most of the common basaltic compositions, although it depicts largely the degree of silica saturation. The plane albite-diopside-enstatite

14.2 Molar plot of most common minerals in the basalt system to indicate degree of saturation and undersaturation with respect to silica. The plane albite-diopside-enstatite defines silica saturation. The plane albite-diopside-forsterite is the plane of critical silica undersaturation since rocks to the left of it contain nepheline.

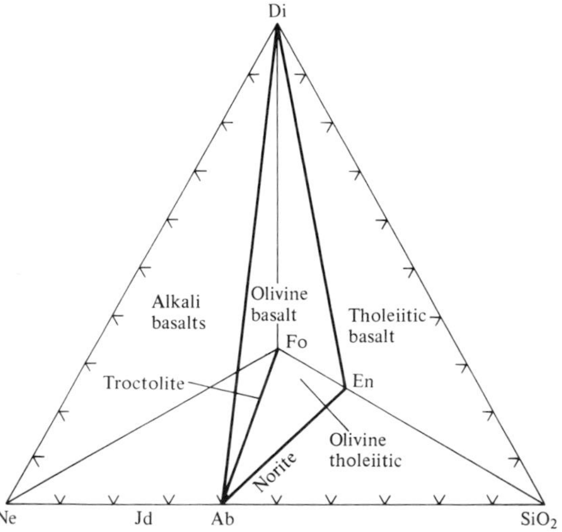

defines silica saturation as it is ordinarily understood since compositions lying to the right of this plane contain normative quartz and include the prominent basaltic type known as *quartz tholeiite*, (which is the best known example of primary magma, see Chapter 12). In spite of this normative saturation, volcanic rocks frequently contain olivine phenocrysts, since this mineral does not react completely under conditions of rapid crystallization. Also, if fractional crystallization is intense enough, free silica may occur in normatively undersaturated basalts lying to the left of this plane. However, we shall be concerned here mostly with tholeiitic magma types rather than rocks, since it is these that may give rise by differentiation to the dominant calc-alkali and gabbroic plutonic complexes.

The plane albite-diopside-forsterite of Figure 14.2 is sometimes referred to as the "plane of critical undersaturation" with respect to silica since compositions to the left of it contain normative nepheline. The natural equivalents of these compositions yield the alkaline basalts, tephrites, basanites, etc., composed of plagioclase, nepheline, and ferromagnesian silicates. On differentiation such magmas may of course give rise to even more alkaline and silica-undersaturated types.

To show the effect of varying the alumina content of a magma, we make use of a stoichiometric transformation between the normative minerals. For example, if we add Al_2O_3 to typical tholeiitic basalts near the albite-diopside-enstatite plane, the following transformations apply:

$$\underset{\text{diopside}}{CaMgSi_2O_6} + Al_2O_3 + \underset{\text{quartz}}{SiO_2} \longrightarrow \underset{\text{anorthite}}{CaAl_2SiO_8} + \underset{\text{enstatite}}{MgSiO_3} \qquad (14.a)$$

$$\underset{\text{diopside}}{CaMgSi_2O_6} + Al_2O_3 + \underset{\text{enstatite}}{MgSiO_3} \longrightarrow \underset{\text{anorthite}}{CaAl_2SiO_8} + \underset{\text{forsterite}}{Mg_2SiO_4} \qquad (14.b)$$

Reactions (14.a) and (14.b), which may be taken to apply to silica-saturated and undersaturated melts, respectively, show that the norms of nonaluminous phases are altered drastically as Al_2O_3 is added or subtracted. The effect is shown graphically by path *A* of Figure 14.3.

In the field of the alkaline or greatly undersaturated melilite basalts, variation of Al_2O_3 or CaO brings about transformations which go a step beyond Reactions (14.a) and (14.b) and involve the disappearance of normative anorthite:

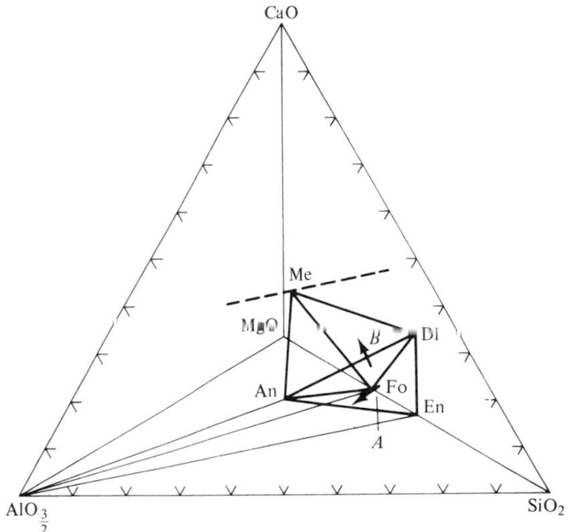

14.3 Molar plot of normative minerals illustrating certain relations in alkaline and melilite basalt systems. For explanation see text.

$$\underset{\text{anorthite}}{2CaAl_2Si_2O_8} \longrightarrow \underset{\text{melilite}}{Ca_2Al_2SiO_7} + Al_2O_3 + 3SiO_2 \qquad (14.c)$$

$$CaO + \underset{\text{anorthite}}{CaAl_2Si_2O_8} \longrightarrow \underset{\text{melilite}}{Ca_2Al_2SiO_7} + SiO_2 \qquad (14.d)$$

$$CaO + \underset{\text{diopside}}{CaMgSi_2O_6} \longrightarrow \underset{\text{melilite}}{Ca_2MgSi_2O_7} \qquad (14.e)$$

According to these equations, melilite rocks may result either through high CaO or low Al_2O_3 and SiO_2 contents. This effect is illustrated by path *B* in Figure 14.3. Analogous transformations involving nepheline are obvious and may readily be visualized in Figure 14.2.

One of the most important conclusions that may be drawn from these stoichiometric considerations is that variations in the compositions of basaltic or gabbroic magmas may arise in a variety of ways and do not necessarily involve subtraction or addition of common modal minerals such as might result from crystal settling or assimilation. Rather, the oxides, as may be seen by examination of analyses, usually vary independently. The source of this variation may of course be complex and in fact returns us again to the question of the origin of primary basaltic magmas.

Acid and alkaline compositions

The Al_2O_3 content of magmas is an important determining factor in the mineralogy of rocks that do not contain much CaO, since then the buffering effect of anorthite is not available. One of the most obvious results in such acid and alkali-rich rocks, and which has already been mentioned, is the production of normative and modal acmite in a system critically undersaturated with respect to this oxide. The stoichiometry may be written as follows:

$$\underset{\textit{albite}}{NaAlSi_3O_8} + \tfrac{1}{2}\underset{\textit{hematite}}{Fe_2O_3} \longrightarrow \underset{\textit{acmite}}{NaFeSi_2O_6} + \tfrac{1}{2}Al_2O_3 + \underset{\textit{quartz}}{SiO_2} \quad (14.f)$$

It is clear that if Al_2O_3 is withdrawn from the system or is low to start with, this reaction is displaced to the right. In silica-poor rocks this gives rise to aegerine or Na-Fe^{3+} amphibole-bearing syenites, while in the silica-rich or "ekeritic" field, aegerine granite or pantellerite volcanics will result.

If somewhat more CaO and MgO are present, the following reaction is applicable:

$$2\underset{\textit{albite}}{NaAlSi_3O_8} + \underset{\textit{hematite}}{Fe_2O_3} + \underset{\textit{diopside}}{CaMgSi_2O_6} + \underset{\textit{forsterite}}{Mg_2SiO_4} \rightleftharpoons 2\underset{\textit{acmite}}{NaFeSi_2O_6} + \underset{\textit{anorthite}}{CaAl_2Si_2O_8} + 3\underset{\textit{enstatite}}{MgSiO_3} \quad (14.g)$$

Reaction (14.g) may be regarded as a purely stoichiometric transformation or as defining an equilibrium state. In the former case it defines the normative mineral composition as given by the chemical analysis of the rock. For example, if Na = Al, the left-hand side represents the conventional norm since normative acmite and anorthite are mutually exclusive. However, if it is regarded as an equilibrium reaction, then members on both sides of the arrows could conceivably coexist in solid solutions. In fact, the acmite molecule is a prominent constituent of many pyroxenes of syenites and other rocks which show no normative acmite but contain considerable anorthite.

We omit here a detailed discussion of the more extreme chemical types of Figure 14.1 since it appears that their genesis involves processes analogous to the more common types. Such is the case, for example, of many nonfeldspathic alkaline rocks and the associated carbonatites. However, it should be mentioned that certain magmatic provinces show evidence of complete gradation between magmas of calc-alkaline and alkaline types. This was found to be the case by Larsen (1940) in the broad central Montana petrographic province. He found that the rocks of each subprovince, whether calc-alkaline or alkaline in character, exhibited smooth variation diagrams which are most easily explained as marking a dominantly liquid line of descent. He also believed that the parent magmas of these subprovinces all had a common origin. Consequently, although we shall find it necessary to concentrate our study of differentiation in the dominant gabbroic and calc-alkali series complexes, it is well to stay alert for possible alternative mechanisms of a purely magmatic character which might lead to an explanation of more unusual rock types.

General factors in the differentiation of basaltic magmas

Before we enter fully into our discussion of the differentiation of specific plutonic complexes, it is desirable to compare in the most general terms the characteristics of the two major trends as exemplified by the gabbroic and calc-alkali series rocks (Table 14.1). In so doing we should, however, caution the reader against regarding the two series as distinct from one another. Indeed, the opposite seems to be true since there is considerable evidence of complete gradation between them.

There is a rather close relationship between the different characteristics of the series as presented in Table 14.1. There is first of all the question of the compositional details of the parent magmas, which—as we have seen in Chapter 13—play such an important role in determining the course of differentiation. In this regard the calc-alkali magmas appear to be water-rich and this seems to hold for early as well as late members of the series. Whether this is a fundamental characteristic of the parent magmas at their source (presumed to be in the upper mantle) or is acquired on the way up is as yet undetermined. However, it seems highly likely now that this feature not only governs the differentiation trend but also imparts to the complex its characteristic structural-dynamic style which involves regional deformation and the upward thrust of large bouyant masses of the granodioritic plutons. By contrast, the heavy iron-enriched gabbroic bodies tend to sink out of sight since they usually are more dense than their surroundings. An exception to this behavior occurs, however, when large segregations of plagioclase form anorthosite bodies. Such bodies then appear to be of sufficiently low density to move under buoyancy forces (Martignole and Schrijver, 1970).

It is frequently possible to determine the nonvolatilic composition of the parent magmas in the gabbroic complexes because the quiescent mode of crystallization favors the preservation of quenched border-zone rocks. But in the calc-alkali complexes many factors such as synkinematic deformation, small temperature gradients, and intense local metasomatism, to mention only a few complications, work against such preservation. In spite of these difficulties, we shall see that there is much evidence that the major calc-alkali batholiths stem from a basaltic parentage not greatly different from those of the gabbros. Some of this is structural and textural evidence for the existence of liquid gabbroic magma intrusions, but most is in the form of the termination, through scattering, of the basic part of the variation diagrams. It is thought that the scatter distinguishes the crystalline cumulates from the most basic liquid of the series.

Since the gabbroic series show little or no silica enrichment during differentiation, the ordinary Harker and Larsen diagrams cannot be employed in comparing their chemistry with that of the calc-alkali rocks. In fact, silica activity remains at about 0.7 (Williams, 1971). One of the most informative diagrams for this purpose is the triangular plot shown in Figure 14.4. This plot, which compares the southern California and Idaho batholiths with the Skaergaard gabbroic complex, shows the variation of the important ferromagnesian oxides as a function of the major residue oxides in both systems. An examination of the literature will reveal that the curve *AC* is characteristic in form and location to those of most large and small calc-alkali complexes throughout the world. However, the Skaergaard curve (*ABC*) in its steep course toward the FeO corner is a somewhat extreme representative of the gabbroic complexes. It should also be kept in mind that the Skaergaard is slightly undersaturated with respect to SiO_2, whereas both the typical gabbroic and calc-alkali primary magmas tend to be saturated or oversaturated in this respect.

It is immediately apparent that although the two curves *ABC* and *AC* are significantly different, this

Table 14.1 Comparison of the features of major differentiated complexes

	Gabbroic series	Calc-alkali series
Structural form	Stratified sill, lopolith, or funnel intrusion	Complex batholiths of massive steepwalled plutons, domes, or mushroom-shaped plutons
Structural environment	Nonorogenic: faulting dominant	Orogenic; both faulting and folding prominent
Number of intrusions	One or few	Multiple
Water content of the magmas	Low	High
Mean temperatures of differentiation	High to moderate	Moderate to low
Dominant character of mineralogy	Anhydrous gabbroic	Hydrous quartzo-feldspathic
Type of differentiation	Absolute iron enrichment, no silica enrichment	Relative iron enrichment; strong silica enrichment
Contact effects	Modest thermal	Modest to profound thermal and metasomatic

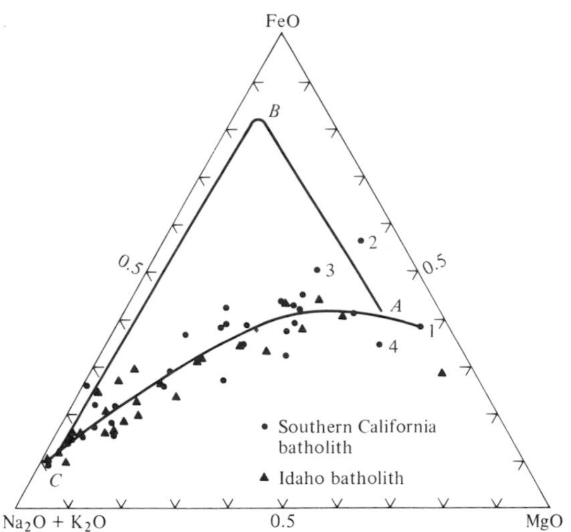

14.4 Molar plot showing fractionation trends in the southern California and Idaho batholiths (path *AC*) and the Skaergaard layered complex (path *ABC*). Numbers refer to analyses shown in Table 14.5.

difference is one of degree only. It is also obvious that curve *ABC* must represent absolute iron enrichment in the segment *AB* since there is little enrichment of alkalies until *B* is passed in going from right to left. By the same criteria, the curve *AC* cannot represent absolute iron enrichment. However, a close examination of the successive points of intersection of this curve with the lines of constant alkali content shows that there is relative iron enrichment with respect to MgO. Such behavior, which results in convexity toward the Fe corner, is characteristic of liquids from which crystals with increasing values of $Fe^{2+}/(Mg + Fe^{2+})$ are separating and, as such, points strongly to the existence of a complete family of curves between *AC* and *ABC* and which are differentiated by varying degrees of alkali enrichment.

We have seen in Chapter 13 that in an anhydrous system, such as the gabbroic complexes, the subsystems $MgO\text{-}FeO\text{-}Fe_2O_3\text{-}SiO_2$ and $Ab\text{-}An\text{-}Or\text{-}SiO_2$ are to a large degree independent. However, it is clear that in the calc-alkali series these subsystems must be strongly coupled through reactions between olivine, pyroxene, and feldspars to produce hornblende and biotite. Yet the system anorthite-albite-orthoclase, taken by itself, should exhibit many similar features in both series although details will vary. We may illustrate this point by plotting the normative feldspars of representative rocks together on the same diagram (Figure 14.5). This diagram should also be studied in relation to Figure 9.11 and the discussion in Chapter 9. In the upper left-hand part of this diagram appear the feldspar norms of the quenched border facies of the Bushveld, Stillwater, and Skaergaard gabbroic complexes which are thought to represent the primary magmas (1, *A*, and *B*). These magmas also do not differ greatly from those of the calc-alkali complexes. Further to the left, nearly on the anorthite-albite join, are plotted representative plagioclase crystals from early differentiates of the Stillwater Complex. Again these resemble the feldspar of early calc-alkali differentiates. Presumably the most anorthite-rich of these represents the primary precipitate from the parent magma *A*.

It is informative to consider differentiation under conditions when the water content of the magma increases as is thought to occur in calc-alkaline differentiation, since then the ternary feldspar solvus is intersected

14.5 Molar plot of fractionation trend in the feldspar subsystem in various complexes as illustrated by a composit from such complexes. Points 1, *A*, and *B* are thought to represent primary magmas (Table 14.2) of the Bushveld, Stillwater, and Skaergaard basic complexes, respectively. *C* represents a jotunite from Quebec and *D* the southern California batholith. ● crystals, Stillwater (Hess, 1960); ○ border facies rock, Stillwater (Hess, 1960); △ border facies rock, Bushveld (Hess, 1960); □ border facies rock, Skaergaard (Wager and Deer, 1939); ▼ jotunite rock (Philpotts, 1966); ■ jotunite crystals (Philpotts, 1966); + crystals, Woodson Mountain granodiorite (Larsen and Draisin, 1948).

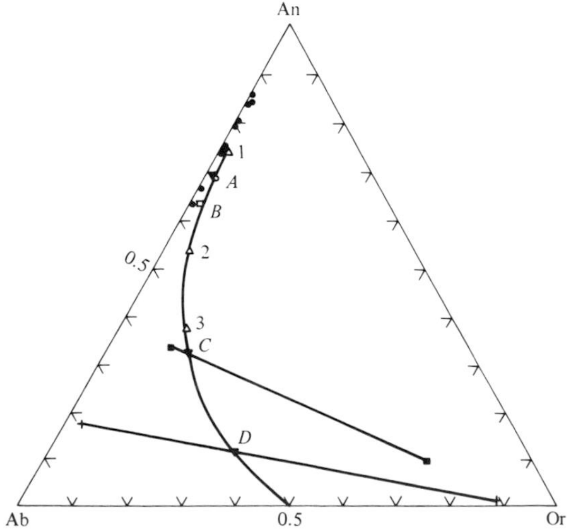

at successively lower levels and the compositions of the feldspars show maximum divergence rather than actual convergence as in high-temperature volcanic rocks (Carmichael *et al.*, 1974). The early course of differentiation may still be illustrated by the successive differentiates from the Bushveld complex (numbers 1, 2, and 3) as obtained from the quenched border-facies. The intermediate course of differentiation, corresponding to higher water content, is illustrated in the segment 3-*C*-*D*. In the general vicinity of *C*, alkali feldspar components become sufficiently enriched in the crystallizing feldspars to cause precipitation of a potassium-rich feldspar as illustrated by the pair from the jotunite rock from Quebec. The trend of alkali enrichment has been carried much further at *D*, which represents even lower temperatures than *C*. We choose to illustrate this by utilizing a feldspar pair from a granodiorite of the sourthern California batholith. The path of crystallization here is somewhat analogous to that considered in relation to Figure 14.5. Thus it is clear that a very alkali-poor parent magma can through strong fractional crystallization give rise to alkali-rich differentiates. However, the precise course of differentiation can be influenced by the coupled reactions referred to earlier. Thus if the water pressure is high enough as in the typical calc-alkali magmas, the precipitation of biotite might tend to direct the crystallization path farther from the orthoclase corner. As we have seen (Chapter 13) the temperature at each stage of differentiation is given, at least in principle, by the length and orientation of the crystal tielines assuming that equilibrium between the two crystals was attained. Unfortunately, the latter requirement is seldom satisfied. Thus although the homogeneous feldspars of the jotunite seem to indicate equilibrium (Philpotts, 1966) the highly-zoned character of the plagioclase from the granodiorite (Larsen, 1948) disproves it in that case.

The systematic increase in $Fe^{2+}/(Mg + Fe^{2+})$, which results in absolute iron enrichment in the gabbroic series but only relative iron enrichment in the calc-alkali series, is a consequence of the equally systematic change in the same parameter for all the ferromagnesian minerals. We shall see that this change in iron contents, which accompanies the enrichment of alkalies in the feldspars, implies a tendency toward reduction in both series so that oxidation-reduction phenomena do not distinguish the series as was thought by Kennedy (1955) and Osborn (1957). Rather, as was pointed out by Bowen (1913) a long time ago, the distinction must be sought in the physicochemical process that acts, on one hand, to maintain an almost constant ratio of feldspar to ferromagnesians and, on the other hand, acts to accumulate the quartzofeldspathic components in the melt.

Differentiated gabbroic complexes

General features

We have already presented a resume of characteristics by which we compared the gabbroic and calc-alkali complexes. Here we are concerned with those features that particularly characterize the gabbros and have few counterparts among the calc-alkaline rocks.

It seems clear now after the detailed studies of such complexes as the Skaergaard that the dominant chemical process involved in their formation is the co-precipitation of plagioclase and ferromagnesian silicates from a slowly cooling melt of basaltic composition. The deposition of these crystals in layers by largely mechanical sedimentation processes is accomplished at least in part through the agency of convection currents. As was pointed out by Wager and Deer (1939) the relatively low viscosity of these melts (Chapter 12), and the fact that the complexes cool principally through the roof, tend to favor convection. Then, too, the quiescent state of affairs required for orderly layer-by-layer build-up of sedimentary beds is likely to be best achieved in the non-orogenic environment, which is typical of these complexes. In general, the convection-controlled mode of crystal settling results in a *rhythmic layering* in which high concentrations of dark, heavy ferromagnesian crystals alternate with leucocratic feldspar-rich concentrations. There generally also is a mimicking of the graded bedding of aqueous sedimentation, with the ferromagnesian concentration gradually decreasing upward in each layer. Also, crystals of tabular form such as the feldspars tend to lie parallel to the bedding. This *igneous lamination* is, where the horizontal currents assert themselves, modified to an alignment of the grains in the direction of current flow.

This mechanically deposited crystal aggregate is termed the *cumulate* and has a porosity ranging from 10 to 30 percent. The pores are of course filled with liquid magma, which may or may not be in equilibrium with the crystals, depending on the rate of crystallization, as compared with the rate of diffusion and liquid composition change. Eventually the pore liquid solidifies to give rise to the *intercumulate* crystals.

Both the vertical and horizontal dimensions of the differentiated layers vary greatly, depending on the dimensions and history of the complex. In the smaller complexes, such as the Skaergaard, the thickness variation is relatively small, with the thickest beds measured in meters. But in the larger complexes some members of almost monomineralic anorthosites measure hundreds of meters in thickness.

Superimposed on the rhythmic layering is a subtler kind of variation, termed *cryptic layering* by Wager and Deer (1939), but which is actually the progressive or oscillatory vertical change in mineral composition which corresponds to iron enrichment. Where this iron enrichment has reached some critical value, at a given stratigraphic height it may, by its influence on the mineral stabilities, cause certain phases to disappear or new phases to appear. Also because the intercumulate crystals recapitulate on a minor scale the differentiation history of the complex as a whole, new crystalline phases tend to appear at a lower stratigraphic height than in the cumulate.

The absolute iron enrichment of the cryptic layering results in an increase upward in the densities of the rocks and magmas. Thus although the layering is a response to the tendency toward gravitational stability through crystal settling, the end-product is a gravitationally unstable rock mass in which dense layers overlie less dense layers.

When large bodies of basaltic magma are first intruded, they suffer chilling against the floor, walls, and roof of the intrusion. At the same time, the heat of the magma and the emitted volatiles also effect a certain amount of metamorphism. The chilling in the outer shell of magma is usually adequate to lower the temperature below the solidus range so that a rather fine-grained sample of rock approaching the bulk composition of the nonvolatile fraction is preserved. In large lopoliths where intrusion or transgression of the magma consumes appreciable time, so that differentiation occurs in the interim, samples of successive liquid differentiates may be preserved at successively higher or more advanced positions in the chill zone. Such appears to be the case with the three border-zone rocks from the Bushveld complex (Figure 14.5). On the other hand, in some small bodies such as the Skaergaard there is evidence that deposits of successive liquids comprise the inward succession of border-zone rocks of the roof and walls as well as the layered series of the floor so that the whole amounts to a box-within-a-box structure.

When the initial chilling against the wall rock has terminated and the steep temperature gradients have been much reduced, slow cooling of the main body of the magma continues. It is likely that the cooling rate and temperature are then regulated to a certain degree by the release of latent heat of crystallization. It is at the inception of this stage that the initial cumulate crystals are deposited on the floor. At present we are not yet sure where the bulk of these crystals come from, although it has most frequently been proposed that they are derived ultimately from just below the roof and are brought to their resting place either by quiescent settling or by convection currents. In any case, the nature of these initial precipitates provide at least a rough indication of the thermodynamic properties and chemical characteristics of the parent magma. To the degree then that equilibrium was attained we can obtain the coefficients for the distribution of the elements between these phases.

One of the most interesting features of the layered gabbroic complexes is the occurrence within them of thick zones of nearly monomineralic rocks such as anorthosites. As we shall see, the occurrence of such layers can be explained by variations of normal crystal fractionation processes. These occurrences should, however, be compared with the much larger anorthosite bodies of the Precambrian anorthosite complexes, such as that of the Adirondacks, where relatively much smaller quantities of gabbroic rocks occur and which usually show evidence of profound deformation. It appears that in the latter bodies buoyancy forces have brought about an even greater segregation of plagioclase than crystal fractionation acting alone (Martignole and Schrijver, 1970).

The Skaergaard gabbroic complex, East Greenland

Chiefly through the monumental work of Wager and Deer (1939),[1] the Skaergaard intrusion is the most extensively studied and best known of all differentiated complexes. Erosion has revealed an ovoid cross section of the complex approximately 5 by 10 km in dimension directly on the sea coast, so that its accessibility as well as its relatively small size have favored its study. However, this small size has brought with it certain features that are not found or at least are highly modified in the far larger lopoliths of like composition.

The complex is Tertiary in age and is intruded into older gneisses and Tertiary sediments and basalt flows. It is likewise cut by regionally developed dikes of basaltic composition. A diagrammatic cross section of the mass, which is classified as a *funnel intrusion*, is presented in Figure 14.6. In this figure account has been taken of regional tilting and bending to give the complex its approximate attitude at the time of intrusion. It is apparent that erosion has exposed three major structural units as follows:

1. The marginal border group formed by chilling and accretion of the magma on the wall rocks

[1] Most of what follows is taken directly from this work.

2. The layered series primarily of igneous-sedimentary origin

3. The upper border group, which has an origin similar to the marginal border group except that it is more complex

A fourth group of rocks, the *hidden layered series*, is inferred from the general form of the intrusion and the contrast between the composition of the complex as a whole and the composition of the chilled marginal zone, which is assumed to represent the parent magma (Tables 14.2 and 14.3).

The size and form of the intrusion are thought to have exerted considerable control on the mode and rate of cooling of its various parts as well as on the crystallization and sedimentation. This is shown, for example, by the relatively greater thickness—corresponding to more rapid heat loss of the upper border zone and the hanging wall portion of the marginal border zone as compared with the footwall (Figure 14.6). The study of the layered series has revealed a basic structure like nested saucers as well as details such as *trough banding*, which is defined by trough-shaped rhythmic layers that appear to mark former channels radiating from the center of the funnel. The latter have been interpreted as resulting from horizontal, inward-directed segments of convection cells on the floor of the intrusion. This interpretation is supported by the alignment of tabular plagioclase parallel to the axes of the troughs.

Table 14.2 Compositions of quenched-border-facies magmas of three gabbroic complexes

	Skaergaard[a]	Stillwater[b]	Bushveld[c]
SiO_2	47.92	50.68	51.45
Al_2O_3	18.87	17.64	18.67
Fe_2O_3	1.18	0.26	0.28
FeO	8.65	9.88	9.04
MgO	7.82	7.67	6.84
CaO	10.46	10.47	10.95
Na_2O	2.44	1.87	1.83[d]
K_2O	0.19	0.24	0.24[d]
H_2O^+	0.41	0.42	0.34
H_2O^-	0.10	0.06	0.03
P_2O_5	0.07	0.09	0.09
TiO_2	1.40	0.45	0.34
Cr_2O_3	0.00	0.04	—
MnO	0.11	0.15	0.47
CO_2	0.06	—	—
S	0.27	—	—
ZrO_2	0.00	—	—
SrO	0.20	—	—
BaO	0.02	—	—
CuO	0.007	—	—
NiO	0.00	0.04	—
	100.21	99.96	100.57

[a] Wager and Deer (1939).
[b] Peoples (in Hess, 1960).
[c] Daly (1928).
[d] Corrected by Hess (1960).

Table 14.3 Normative mineral compositions of border-facies magmas of three gabbroic complexes

		Skaergaard		Stillwater		Bushveld	
Quartz		0.0		0.00		2.76	
Plagioclase	orthoclase	1.11	61.55	1.39	56.17	1.11	57.82
	albite	20.69		15.72		13.62	
	anorthite	39.75		39.06		43.09	
Diopsite	wollastonite	4.99	9.74	5.04	9.95		8.47
	enstatite	2.90		2.60			
	ferrosilite	1.85		2.31			
Hypersthene	enstatite	7.50	11.92	16.70	31.82		29.52
	ferrosilite	4.42		15.12			
Olivine	forsterite	6.37	10.96	—		—	
	fayalite	4.59					
Magnetite		1.86		0.46		0.46	
Ilmenite		2.66		0.91		0.61	

Data are from Hess (1960).

Although the rhythmic layers that form the trough bands lens out within a few meters of the trough axis, the ordinary rhythmic layering frequently shows horizontal continuity for several kilometers. However the well-differentiated layers, which are usually 10 cm or so in thickness, are sometimes separated by much thicker layers of undifferentiated "average rock." This shows that the forces that produced the layers were for a time suspended so the feldspar and ferromagnesium components settled together.

From the structural features of the complex we obtain a picture of symmetrical multiple-celled convecting magma in which the upper, outward-directed horizontal component, heavily charged with crystals, is initiated by cooling from the roof. This current sweeps downward along the margin and then turns again inward along the bottom and deposits its crystals along the way. The fact that the rhythmic layering assumes a finer scale along the margin seems to support this picture.

In their monograph Wager and Deer adduced evidence that the increase of pressure with depth in the descending current would more than compensate for the rise in temperature so that the liquidus temperatures would be raised faster than the ambient temperature. Consequently, crystallization might even be enhanced in the descending current. This may be shown as follows: First let it be assumed that when convection is occurring, the rate of temperature change with height dT/dh is close to an adiabatic one. Then it may easily be shown that at any point in the magma we have

$$\frac{dT}{dh} = \frac{\alpha g T}{C} \tag{14.1}$$

where α is the coefficient of thermal expansion [see Equation (1.5)], g is the gravitational acceleration, and C is the specific heat. Now for liquid basalt $C \cong 0.3$ cal/gm/°K, or approximately 10^7 erg/gm/°K while $\alpha = 44 \times 10^{-6}$. Also for the present purposes it is adequate to assume a single temperature, which we take as 1400°K.

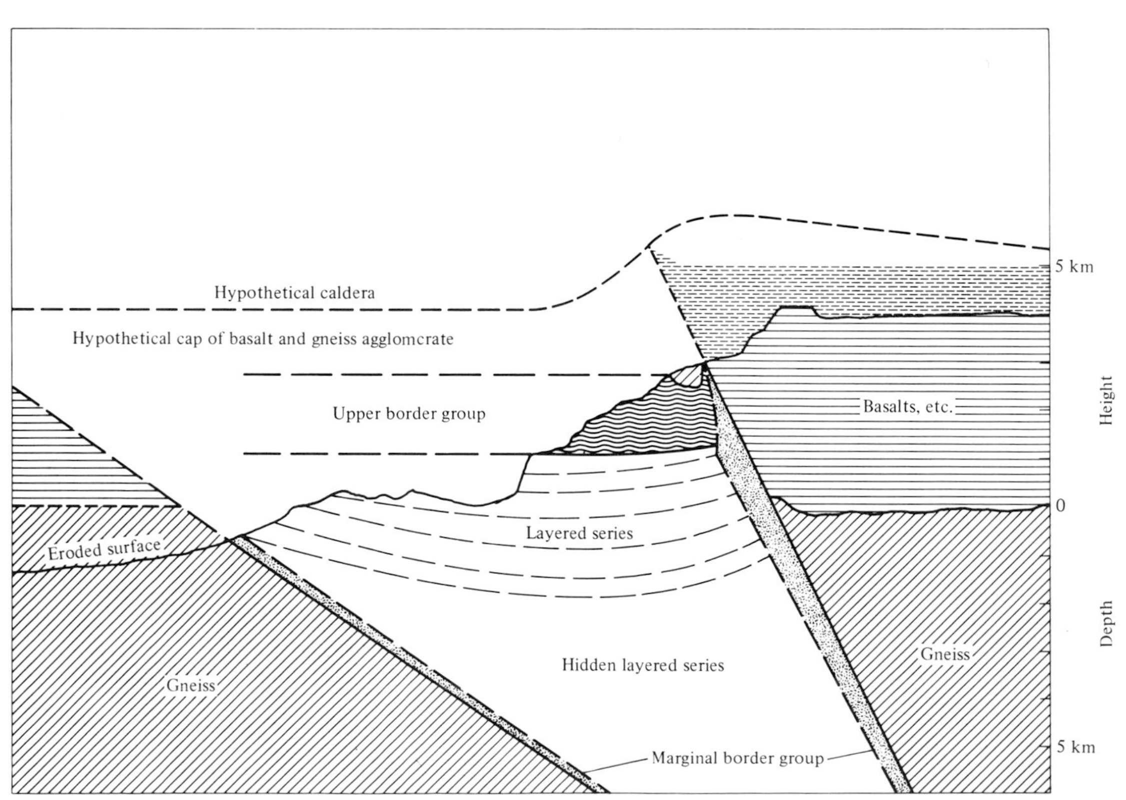

14.6 Diagrammatic cross section of the Skaergaard complex, East Greenland. (After Wager and Deer, 1939.)

We then find that

$$\frac{dT}{dh} \simeq 0.4°\text{K/km}$$

But we have already seen (Chapter 2) that liquidus temperatures increase of the order of several degrees per kilometer and this is borne out in a variety of direct experiments. Consequently the contention of Wager and Deer is well supported.

Evidence for magmatic differentiation in the Skaergaard complex is to be found particularly in the systematic change in mineral composition with stratigraphic height as shown in Figure 14.7. The steady upward increase in the albite content of plagioclase shows a reversal in the unlaminated layered series at the top of the complex, but this is easily understood as an inward accretion of the upper border group from the roof. These changes in the feldspar compositions are accompanied by equally regular increase in the fraction $Fe^{2+}/(Mg + Fe^{2+})$ of all the ferromagnesian minerals, including ilmenite and magnetite (Vincent and Phillips, 1954). In Chapter 13 in our discussion of phase relations in the system $FeO\text{-}MgO\text{-}Fe_2O_3\text{-}SiO_2$ we saw that such a trend is correlated with a steady decrease in oxygen fugacity. We can say with considerable assurance that the same conclusion applies to the Skaergaard magma, for if the oxygen fugacity did not drop with cooling we should expect hematite to be an important phase in the later differentiates. A detailed discussion of silica, oxygen, and iron activities in the Skaergaard is given by Williams (1971).

The iron enrichment of the ferromagnesian minerals has important effects on their stabilities. For example, the reaction

$$\underset{pyroxene}{2FeSiO_3} \rightleftharpoons \underset{olivine}{Fe_2SiO_4} + \underset{melt}{SiO_2} \tag{14.h}$$

is probably displaced to the right over the entire temperature range so that when the iron content of the pyroxene exceeds a certain amount, pyroxene becomes unstable relative to olivine. In the Skaergaard this occurs at an elevation of approximately 1500 m when the orthopyroxene phase has a composition of Fs_{60}. By contrast, the temporary disappearance of olivine at the level of the middle gabbros cannot be so readily explained. Such behavior does occur in the experimental system $MgO\text{-}FeO\text{-}SiO_2$ (Bowen and Schairer, 1935) when bulk

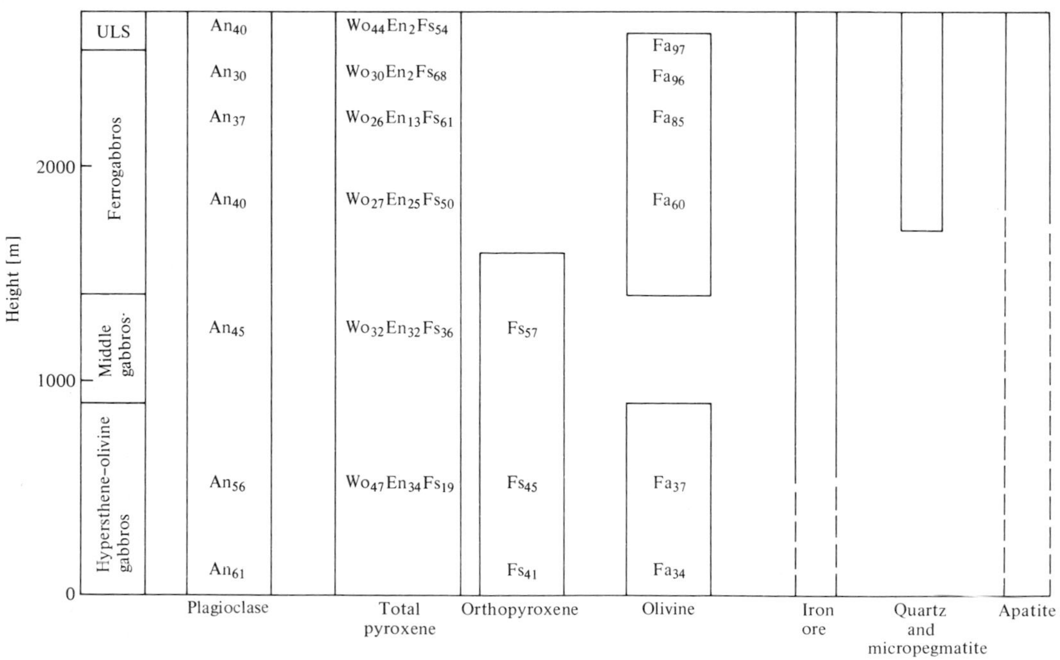

14.7 Variation of mineral compositions with height in the Skaergaard complex. ULS is the unlaminated layered series. (After Wager and Deer, 1939.)

compositions fall on the silica side of the pyroxene join, but extrapolation to the Skaergaard magma can be done with little confidence for such complex crystallization paths.

The status of the iron-titanium oxides in the scheme of Figure 14.7 is also uncertain. Wager and Deer (1939) inferred, on the basis of textural evidence, that primary precipitation of these minerals began in the upper part of the hypersthene-olivine gabbros when TiO_2 and Fe_2O_3 reached concentrations of 2.5 and 3 weight percent, respectively, in the melt. It has been shown (Vincent and Phillips, 1954; Buddington and Lindsley, 1964) that these oxides originally probably consisted of coexisting solid solutions of Fe_2TiO_4-Fe_3O_4 and $FeTiO_3$-Fe_2O_3 (ulvöspinel-magnetite and ilmenite-hematite). It may further be shown both from phase-rule considerations and direct experiment (Buddington and Lindsley, 1964) that knowledge of the composition of the two phases yields an estimate of both the temperature and oxygen fugacity. However, it must be mentioned here that there is evidence in the form of olivine reaction rims between pyroxene and iron ore for the reaction pyroxene + magnetite → olivine + O_2, which has already been discussed in Chapter 5 (Reaction 5.s). It is possible that some of this reaction involved the interstitial liquid, but much of it may have occurred under subsolidus conditions. Buddington and Lindsley (1964) found that widespread subsolidus reactions have affected the iron-titanium ores and temperatures registered by the oxide compositions are never over 1173°K.

The structural and compositional evidence indicate that the Skaergaard intrusion solidified by a combination of accretion on the walls and gravitational accumulation of sediments on the floor. The first type of solidification corresponds to the marginal and upper-border groups, while the second type is represented by the layered series. The net result of both processes is the box-within-a-box structure shown in Figure 14.8. It is clear from this figure that if an estimate can be made of the volumes and compositions of these closed shells of deposits, the compositions of the successive liquid differentiates represented by them can also be deduced. This may be done by beginning with the deposits of the last (fifth) liquid and working back toward the first, which is directly equivalent to the quenched-border facies. The results as obtained by Wager and Deer are shown in Table 14.4.

From their knowledge of the structural form, Wager and Deer also estimated that the hidden part of the complex was approximately 60 percent of the whole. If this information is combined with the known composition of the exposed complex and the parent magma, a simple calculation yields the composition of the hidden

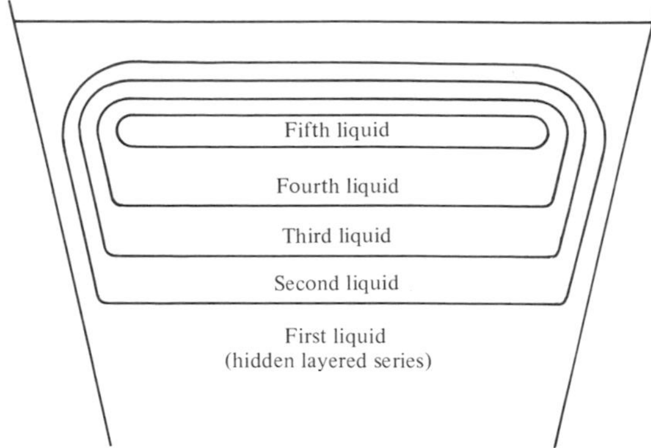

14.8 The box-within-box structure for Skaergaard magma.

part. The result, again after Wager and Deer, is given in the last column of Table 14.4.

The characteristics of the mineral assemblages, their change with height in the complex, and the major bulk chemical features of the rocks and liquid differentiates (Table 14.4) are all directly related to the characteristics of the parent magma (Table 14.1) although these have also been modified by the environment of crystallization.

As distinguished from basaltic magmas, in general this magma is quite low in K_2O, but in terms of SiO_2 content it is typical of the undersaturated olivine tholeiitic type. However, it is also quite high in alumina. It is about average in terms of Na_2O, TiO_2, and Fe_2O_3. Certainly it is not excessively high in FeO as might be thought from its late differentiates. As compared with the Stillwater and Bushveld magmas, it is relatively high in Fe_2O_3 and TiO_2, which undoubtedly favored an early appearance of iron titanium oxides. It is very likely that the low K_2O content of the primary magma contributed somewhat to the late alkali enrichment as exhibited in the variation diagram (Figure 14.4). It is thought that the latest differentiates are represented by certain transgressive granophyres. However, there is evidence, in the form of lens-shaped granophyric inclusions in the border-zone rocks, of some assimilation of alkali-rich country rocks which may have supplemented the late differentiates.[2]

More recent work on the Skaergaard (Wager and Brown, 1967) has slightly modified certain of the original findings of Wager and Deer (1939) on the structure and petrology. For example, a slightly different composition of the parent magma has been inferred, and the hidden

[2] These lens-shaped masses have been interpreted by Holgate (1954) as examples of liquid immiscibility.

layered zone is thought to be somewhat larger. Also, the behavior of the trace elements in fractionation were studied in detail. However, these modifications and additions have no major effect on the original conclusions regarding the complex, and are themselves subject to constant revision (see Chayes, 1970).

Unfortunately one of the most important characteristics of the parent magma—its water content—cannot be evaluated since this may have been lost or supplemented by later absorption. However, we have already implied, by the criterion of the type of differentiation, that this must have been quite low.

The characteristic of extreme absolute iron enrichment revealed in Figure 14.4 is equally apparent in Table 14.4. Also apparent in this table is the relatively constant ratio of quartzofeldspathic to ferromagnesian constituents in the liquids. Only in the fifth liquid, which represents the transgressive granophyre deposits, does the quartzofeldspathic constituent greatly dominate the other. It will be interesting to compare this feature with the calc-alkali differentiates.

The general characters of the minerals of the Skaergaard show that crystallization generally occurred under conditions quite far displaced from equilibrium. This is clearly shown by the common occurrence of zoned plagioclase and ferromagnesian minerals. Frequently also the crystals of pyroxene in the intercumulate fraction are far too iron-rich when compared to cumulate olivines. Thus there is a sharp compositional distinction between cumulate and intercumulate minerals.

Table 14.4 Compositions and weight norms of liquid differentiates and invisible rock of the Skaergaard complex

		First liquid		Second liquid		Third liquid		Fourth liquid		Fifth liquid		Invisible rock	
	SiO_2	47.92		46.7		45.7		46.1		49.8		48.7	
	Al_2O_3	18.86		15.3		12.7		11.5		13.7		21.2	
	Fe_2O_3	1.18		2.9		3.6		3.4		5.67		0.0	
	FeO	8.66		12.9		18.2		21.3		15.37		5.8	
	MgO	7.82		5.9		3.5		1.2		0.87		9.0	
	CaO	10.46		9.9		8.3		8.3		6.57		10.8	
	Na_2O	2.44		2.83		3.16		2.87		3.23		2.2	
	K_2O	0.18		0.29		0.44		0.51		1.08		0.10	
	TiO_2	1.35		2.18		2.36		2.10		1.32		0.80	
	MnO	0.10		0.16		0.22		0.29		0.31		0.05	
	P_2O_5	0.07		0.42		0.92		1.57		0.77		—	
	Quartz	—		—		—		—		5.28		—	
	Orthoclase	1.11	61.55	1.67	53.65	2.8	48.7	3.34	44.40	6.67	53.38	0.56	66.72
	Albite	20.69		23.58		26.7		24.10		27.25		18.34	
	Anorthite	39.75		28.40		19.2		16.96		19.46		47.82	
Diopside	wollastonite	4.99	9.74	7.66	15.28	7.1	14.6	6.26	13.20	3.48	7.34	2.44	4.70
	enstatite	2.90		3.40		1.8		0.60		0.30		1.60	
	ferrosilite	1.85		4.22		5.7		6.34		3.56		0.66	
Hypersthene	enstatite	7.50	11.92	4.7	10.40	3.1	12.6	2.40	29.46	1.90	20.25	11.00	15.22
	ferrosilite	4.42		5.7		9.5		27.06		18.35		4.22	
Olivine	forsterite	6.37	10.26	4.6	10.9	2.7	11.7	—		—		7.00	9.86
	fayalite	4.59		6.3		9.0		—		—		2.86	
	Ilmenite	2.66		4.0		4.6		3.95		2.43		1.52	
	Magnetite	1.86		4.2		5.3		4.87		8.35		—	
	Apatite	0.20		1.0		2.0		3.70		2.02		—	

Data are from Wager and Deer (1939).

In general the contact effects of the Skaergaard intrusion on the country rocks are quite modest, with notable effects extending out only a few meters. However, in some large gabbroic inclusions that were completely immersed in magma, obliteration of ophitic texture is apparent at 50 m from the contact.

The Stillwater gabbroic complex, Montana

The Stillwater complex is a variant of the stratified gabbros that differs from the Skaergaard in some significant ways. The chief source of these differences is the much greater original size of the Stillwater, but there are also important differences in the primary magmas, as was mentioned in connection with Table 14.2.

The Stillwater complex is an erosional remnant some 60 km in length which occurs in a highly faulted and dissected mountain area (Jones *et al*., 1960). In spite of its pre-Cambrian age and the mountain building it has endured, it is remarkably well preserved. In contrast to the Skaergaard, and happily for our interest here, the exposed part is approximately the lower half. In fact, in terms of rock and mineral compositions, the Stillwater stratigraphic section ends approximately where that of the Skaergaard begins.

Typically the complex shows well-developed rhythmic and cryptic layering as well as igneous lamination. However, there is little or no indication of lineation in the plane of lamination such as might be attributed to strong horizontal convection currents.

The only representatives of border-zone rocks in the exposed complex are those which occur at the contact with the floor of the intrusion, and these differ in fundamental ways from the border rocks of the Skaergaard. In the first place, they are intimately involved with and difficult to distinguish in the field from a thick zone of contact hornfelses. Secondly, they grade gradually upward into the region of more slowly cooled rocks, which contain abundant settled cumulate ferromagnesian crystals.

The mineralogic character and variation with height is shown in Figure 14.9. This figure shows that although

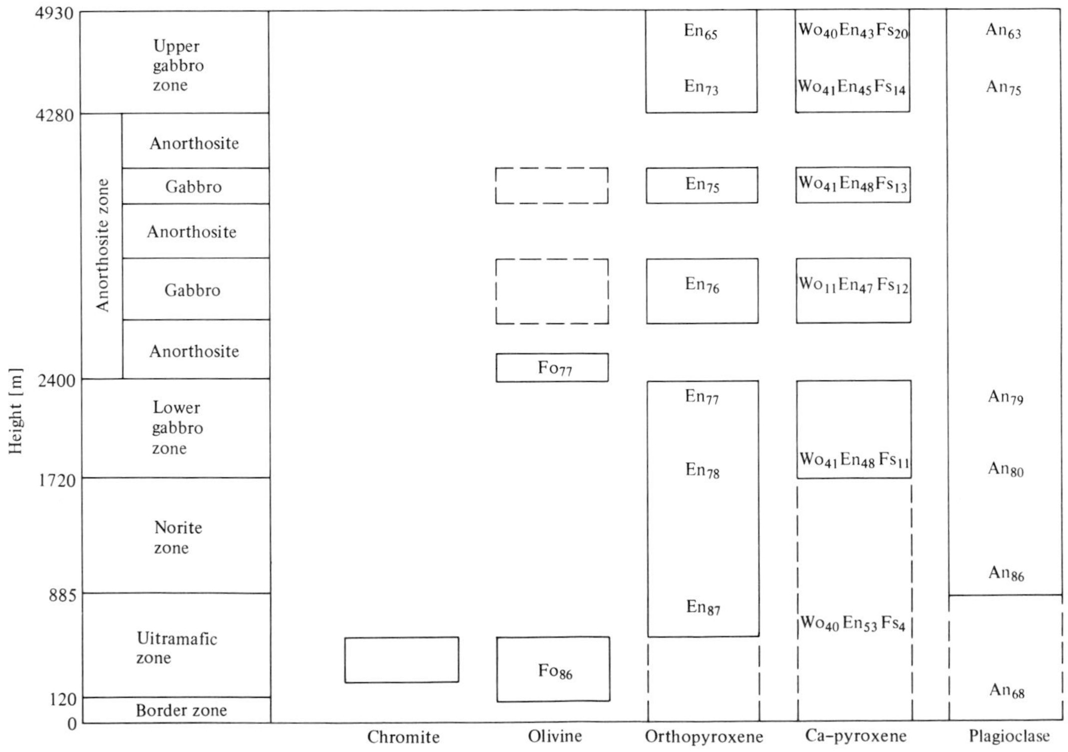

14.9 Variation of mineral compositions with height in the Stillwater complex, Montana. (After Hess, 1960.)

the first cumulate to form under the slow-cooling condition was olivine, the high normative hypersthene content (Figure 14.3) soon asserted itself so that orthopyroxene is the dominant phase throughout the ultramafic zone. By contrast, Ca-pyroxene occurs only as an intercumulate mineral below the lower gabbro. However, above this level it is found both as cumulate and intercumulate. It should be observed that plagioclase in the ultramafic zone occurs only as an intercumulate phase, and this is the reason for its relatively low anorthite content. Precipitation of this phase begins at the base of the overlying norite zone where the earliest crystals are approximately $anorthite_{86}$ as compared with normative $anorthite_{68}$ in the parent magma (Figure 14.5).

An interesting and important datum available here is the distribution of the elements between the parent magma and the primary precipitates. This distribution is shown for the system $CaSiO_3$-$MgSiO_3$-$FeSiO_3$ in Figure 14.10, where it is seen to amount to a difference of more than 0.25 in terms of $Fe^{2+}/(Mg + Fe^{2+})$. It is, of course, this factor that lies behind the strong observed iron enrichment in these magmas. Perhaps equally significant is what this means relative to the anatectic origin of the primary magma and the upward transport of iron in the Earth's crust. For it is clear that a fairly iron-rich melt can be derived from a Mg-rich crystalline aggregate and that by partial melting a continual upward enrichment of iron can occur. Not the least important consequence of this is the continually renewed gravitational instability of the crust.

The strikingly high value of the ratio Fe^{2+}/Fe^{3+} of the Stillwater parent magma (Table 14.2) indicates a somewhat more reduced state than is usual for basaltic magmas. This is also reflected in the primary crystalline precipitates which contain no free oxides other than chromite. According to McCallum (1968), even these chromites contain Fe_3O_4 only in very dilute solution. He has shown that the reaction

$$\underset{pyroxene}{2MgSiO_3} + \underset{chromite}{\tfrac{2}{3}Fe_3O_4} \rightleftarrows \underset{olivine}{Mg_2SiO_4} + \underset{olivine}{Fe_2SiO_4} + \underset{fluid}{\tfrac{1}{3}O_2}$$

may be used to estimate P_{O_2} by taking this dilution factor into account. An experimental study of the chromite system has also been made by Hill and Roedder (1975). The result is that P_{O_2} for the chromite horizon of the ultramafic zone was probably at least several orders of magnitude lower than if pure Fe_3O_4 had been precipitated. It is likely that this comparatively reduced state of the magma is a consequence of the deep-seated plutonic environment under which differentiation occurred.

In contrast to the Skaergaard, there is in the Stillwater no very clear-cut distinction between cumulate and intercumulate minerals. If, however, this distinction had been blurred by strictly local recrystallization or post-depositional reaction with the interstitial liquid, we should expect that each local composition would be intermediate between the cumulate and intercumulate and that the degree of iron enrichment would depend on the ratio of intercumulate to cumulate material. For example, in layers rich in cumulate pyroxene, the composition should resemble that of the original pyroxene precipitate at that stratigraphic level. Conversely, in an anorthositic layer in which intercumulate pyroxene greatly exceeds that in the cumulate, considerable iron enrichment might occur (Hess, 1960). Although the Stillwater rocks do exhibit this effect, it is much smaller than might be expected if the reaction had been localized as discussed above. It turns out that a number of processes were superimposed to give the observed result. For example, it appears that the rate of accumulation of the cumulate crystals was by and large slow enough for a high degree of diffusive communication to

14.10 Molar plot of the distribution of the components $MgSiO_3$, $CaSiO_3$ and $FeSiO_3$ in the Stillwater complex between the parent magma (point L) and orthopyroxene (point Opx) and olivine (point Ol). Three successive coexisting clinopyroxene-orthopyroxene pairs are also shown. The composition of the primary magma is inferred to be represented by quenched border facies (Hess, 1960) as shown in Table 14.2.

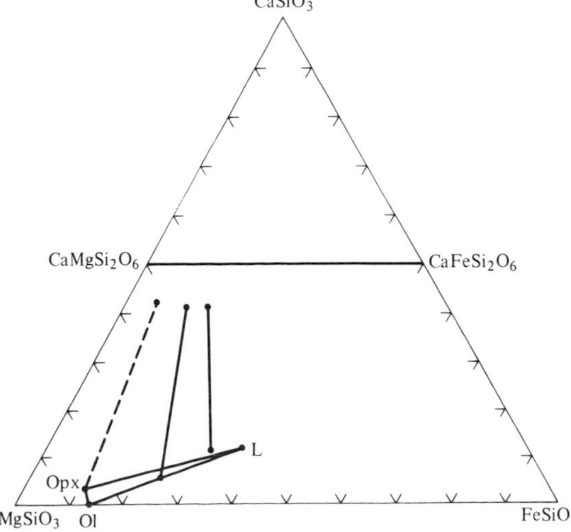

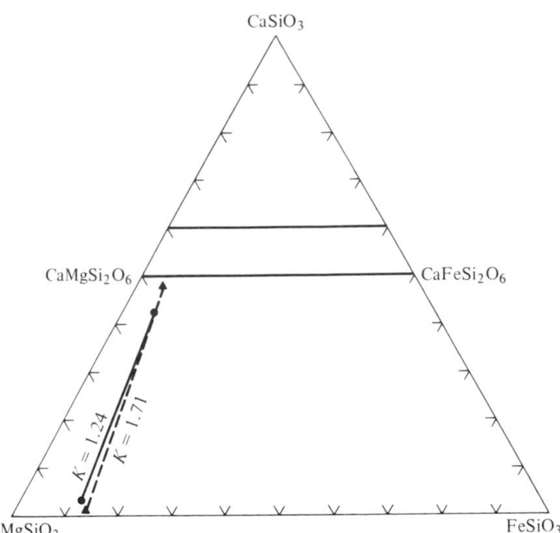

14.11 Molar plot of the distribution of the components $CaSiO_3$, $MgSiO_3$ and $FeSiO_3$ in the Stillwater complex between coexisting clino- and orthopyroxenes. The distribution constant for the cumulate and intercumulate crystals is 1.24; for the exsolution lamallae, 1.71. Data are from McCallum (1968).

be established between the interstitial liquid and the overlying magma so that something resembling equilibrium between the cumulate and intercumulate was also approached. However, it seems that there was also some reaction between this magma and the original cumulate crystals as revealed in particular by the apparent metasomatic replacement of cumulate olivine by pyroxene (Jackson, 1961). The end result of this is a quite orderly distribution of the major components as shown in Figures 14.10 and 14.11 (see McCallum, quoted in Saxena, 1973, p. 48). In the latter figure we have distinguished the mineral pairs of the cumulate and intercumulate crystals from those of exsolution lamellae. It is clear from the differences in the distribution constants for the two sets of values that exsolution occurred at temperatures considerably below the liquidus and corresponding to high-grade metamorphism.

One of the most interesting problems of the Stillwater and other large gabbroic complexes is the origin of the thick layers of practically all monomineralic rock such as the 400-m anorthosite units (Figure 14.9). This problem has been discussed by Hess (1960), who rejected the hypothesis of simple-crystal sorting of the kind that has generally been invoked to explain the rhythmic layering. He also rejected the alternative explanation that cesation of ferromagnesian deposition might be the result of a shift in the phase equilibria. The explanation finally adopted by him is a rather complicated mechanism involving

1. The preferential settling of the ferromagnesian component from crystal-charged convection currents
2. The subsequent resorption of the plagioclase component of these currents by mixing with hotter undifferentiated liquid
3. The reprecipitation solely of plagioclase from the plagioclase-enriched liquids resulting from (1) and (2)

However, consideration of the effect of water on the solubilities of the silicates and specifically the experiments of Yoder and Tilley (1962) (Figure 13.9) indicate that the second alternative—that of a shift in the phase equilibria—is not only possible but far simpler than the resorption mechanism. If, for example, the plagioclase concentration is quite high in the melt—as it is in the Stillwater—so that under low water pressure crystallization of this mineral precedes the ferromagnesians, then it seems possible that a decrease in water pressure might well shift the equilibrium from the plagioclase and ferromagnesian field to the plagioclase field. The suggested process is presented schematically in Figure 14.12.

14.12 Two hypothetical configurations of a basalt liquidus. *fm* and *pl* represent the original liquids of ferromagnesian and plagioclase components, respectively. *fm′* and *pl′* the same liquids after some differentiation has occurred. *A* and *B* are possible crystallization paths.

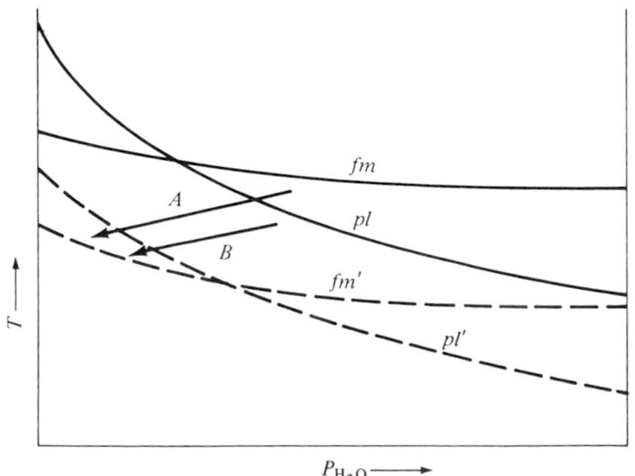

Although there is considerable evidence that plagioclase was the only mineral precipitated at the time the anorthosites were deposited, the deposition of a monomineralic cumulate is still inadequate to yield a monomineralic rock since the interstitial liquid contains ferromagnesians as well as feldspars. It was suggested by Hess that this obstacle might be overcome by the same process that tended to eliminate the compositional difference between cumulate and intercumulate crystals. This process, which we have already identified as diffusional communication with the overlying liquid, would enable the cumulate crystals to continue to grow until all the interstitial space and consequently all liquid was eliminated. The mechanism, which has been called "crystal growth differentiation" elsewhere (Mueller, 1963a) is probably of great importance in the formation of deep-seated magmatic rocks.

Structural features, such as slumping of the rhythmic layering, led Hess (1960) to conclude that the thickness of the unconsolidated crystal mush was generally in the range of a meter or so. He also attempted to estimate the time that the complex took to cool by applying Lovering's (1935) heat-conductivity equations. His result is that the complex was deposited at a rate of about 10 cm/yr. He then also made an independent estimate of the effectiveness of diffusion by applying diffusion coefficients obtained by Bowen (1921) from the system albite-anorthite-diopside, and which we referred to earlier (Chapter 13). The general agreement obtained by Hess between the effective diffusion distance from these data and the time of deposition of the 1 m thickness of crystal mush is probably fortuitous, however. This is already indicated by the many assumptions that enter into the heat-conductivity calculations, but it is virtually assured by the fact that Bowen's data refer to much higher temperatures than those which could have prevailed during crystallization of the Stillwater complex. Almost certainly the diffusion coefficients obtained by Bowen are too high for basaltic magmas, as we have already shown in our own calculations in Chapter 13. There we saw that for a typical basaltic magma, diffusion is probably effective over a distance of much less than 1 cm/yr so that a rate of deposition of 10 cm/yr would quickly overcome the effectiveness of diffusion, to maintain communication between the interstitial and overlying liquids. That there is evidence such communication was maintained seems to indicate a considerably lower rate of crystal deposition than that obtained by Hess.

It is thought that the upper part of the complex is hidden beneath overlying Paleozoic strata. It was demonstrated by Hess, however, that limits could be placed on the volume percentage of the total complex represented by these hidden rocks. It is known, for example, that the overall composition of the exposed complex is much richer in MgO than the quenched border facies that is taken to represent the parent magma. If it is assumed for the moment that the exposed portion of the complex contains all the MgO and that the hidden portion contains none, then a simple graphical solution yields a minimum of 28 percent for the hidden portion. Then by considering the compositional variation of the plagioclase, Hess also arrived at a maximum value for the hidden part—namely, 50 percent of the whole. The true value, which must fall between these extremes was considered to be approximately 40 percent. Given this value, it is a simple matter to calculate its composition from those of the border-zone rock and the exposed portion. It is found, in accord with our expectation, that this composition closely resembles the ferrogabbros of the Skaergaard complex.

The Bushveld gabbroic complex, Transvaal, South Africa

This large pre-Cambrian intrusion extends approximately 500 km in greatest dimension and bears many resemblances to the Stillwater complex. However, it is much better preserved and still reveals the typical lopolithic form. Contributions to the geology of the Bushveld complex have been numerous, as exemplified by the works of Daly (1928), Hall (1932), and Lombaard (1934). Hess made an interesting comparison with the Stillwater and other complexes.

The Bushveld complex reveals the typical stratigraphic range of rock types from a chilled-border facies closely resembling that of the Stillwater (Table 14.3) through harzburgites, norites, dunites, and anorthosites. It is also capped by a thick section of intermediate and acid rocks which some have interpreted as differentiates of the complex but which others infer to have a more complex origin. A particularly interesting feature of the Bushveld is the transgressive nature and chemical variation of its chilled border facies. Apparently the magma invaded successively more remote regions from its source as differentiation occurred (Hess, 1960). A comprehensive overview of the Bushveld and similar layered intrusion forms the subject of a special publication (Symposium on the Bushveld igneous complex, 1969).

Unfortunately we cannot dwell further on the gabbroic complexes here; however, it is felt that their salient characteristics have been adequately revealed for the interesting comparisons with other plutonic types of differentiates which follow.

Large calc-alkali plutonic complexes

General

The large calc-alkali intrusive complexes, or batholiths, are of great interest for many reasons. First, they obviously represent a course of differentiation that stands in striking contrast to that of the gabbroic complexes and that cannot be understood through classical phase-equilibria studies on anhydrous systems. Secondly, they are sufficiently large to exhibit a more or less complete series of differentiates. And third, they are unquestionably major features in the evolution of continents and the Earth's crust.

By our emphasis here on complexes of batholithic dimensions we do not mean to gloss over the fact that there is a more or less complete spectrum between these large complexes and the minor stocks and other bodies that intrude the crust at every level. Nor do we imply that little can be learned from these small intrusions. However, it seems to the writers that they can be discussed more informatively when examined in the light of processes first seen at work on this grander scale.

Because of their relatively recent age and their occurrence in accessible, highly dissected mountain areas of good exposure, the late Mezozoic batholiths of western North America have been most intensively studied. The distribution and structural trend of these batholiths, which are highly significant in the overall planetary picture, is well known and need not concern us at this point. We should, however, draw attention to the close correspondence that frequently prevails between many structural elements of the batholiths and the regional trends.

We shall concentrate our major attention on the southern California batholith since it is there that the important mineral chemical problems which bear on the environment of crystallization have been most exhaustively studied, particularly by E. S. Larsen and his co-workers. Although comparable data for the other batholiths is as yet sparse, we shall have occasion to refer to them also in terms of bulk chemical variations and the structural environment.

There are certain features that are held in common by the large batholiths such as those of southern California, Idaho, and the Sierra Nevada:

1. Their magmatic parts all consist of multiple intrusions ranging from gabbro to granite, although these differ in detail.
2. Intermediate rocks such as tonalite (southern California), quartz monzonite (Idaho), and granodiorite (Sierra Nevada) cover by far the largest areas, while gabbros and granites aggregate less than 10 percent.
3. The magmatic plutons consist of well-defined, homogeneous lithologic types, which are recognizable over large areas.
4. These plutons frequently show gneissic borders, so-called flow banding and lineation.
5. The intermediate rocks frequently contain inclusions of a basic character which appear to be cognate and traceable to early members of the series.
6. The magmatic rocks as a whole show systematic bulk chemical and mineralogic variations which appear as smooth curves on the variation diagrams. In addition, it is not unusual for individual plutons to show a distinct compositional zoning, with the latest differentiates occupying the center of the pluton.

It has not been generally stressed in the textbooks that batholiths are the products of a complex interweaving of magmatic and metamorphic processes. Instead, there has been a decided tendency in the past to stress the magmatic features, so that the student is left with the impression that batholiths always consist of well-defined, steep-walled plutons in sharp contact with their wall rocks, which have been only locally metamorphosed to rocks of the character of "hornfels." Actually the not clearly magmatic parts, which resist simple classification, have thus far been systematically excluded from study except by the unconventional few. Into this category falls the entire north-eastern side of the southern California mass as well as large parts of the northwestern margin of the Idaho batholith. However, the latter terrain has been studied by Hietanen (1963b), whose work we shall have occasion to refer to later.

At present we shall confine ourselves to what we infer to be the clearly magmatic parts of the batholiths, and save the more complicated metamorphic features for a later section.

The southern California batholith

The southern California batholith of upper Cretaceous age is the northern extension of a much larger mass which is exposed intermittently to the southern tip of

Baja California in Mexico.[3] At present we shall confine ourselves to an even more restricted part of the batholith, that which embraces the Corona, Elsinore, and San Luis Rey quadrangles as studied by Larsen (1948). According to Larsen, the average composition of this part of the batholith is tonalite with the following areal distribution of rock types:

gabbro	7 percent
tonalite	68 percent
granodiorite	28 percent
granite	2 percent

The batholith in the region mapped is complex and consists of over 20 separate intrusions. A typical part of this region is shown in Figure 14.13. It is clear from this figure that the plutons show a rough alignment parallel to the trend of the nearby coast as shown by the arrow.

Field studies clearly show that most of the contacts between individual plutons are quite sharp and that the time sequence of intrusion is gabbro-tonalite-granodiorite-granite. These plutons contain large screens of highly deformed country rock which are metamorphosed to amphibolite rank. Although no detailed study has been made of the metamorphism in these inclusions, it is known that they sometimes contain sillimanite. It is a point of some importance that the earliest

[3] The Mexican parts of the batholith are little known but appear to contain rocks similar to those of southern California (Woodford and Harris, 1938).

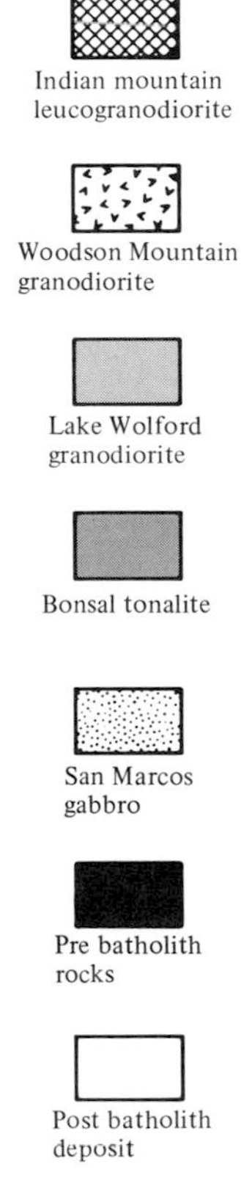

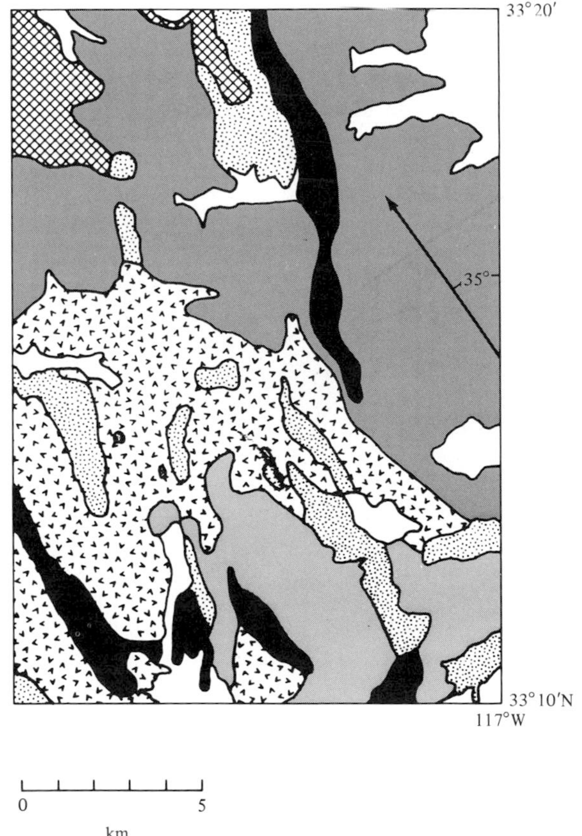

14.13 Typical area of southern California batholith, after Larsen (1948). Arrow is approximately parallel to coast and regional structure.

Table 14.5 Compositions of rocks from the southern California batholith

Sample	SLRM 354 troctolite	El 303 calcic hornblende gabbro	SLR 218 hornblende norite	SLRM 299 hornblende norite	SLR 334 norite	Average San Marcos gabbro	SLR 1016 Bonsall tonalite	Ra 135 Woodson granodiorite	El 167 Rubidoux granite
	1	2	3	4	5	6	7	8	9
SiO_2	42.86	45.78	47.22	48.16	52.12	50.78	55.14	72.58	73.60
TiO_2	0.18	5.13	1.60	0.76	0.33	0.77	0.93	0.30	0,18
Al_2O_3	24.94	15.59	18.18	19.66	20.88	20.40	17.82	14.12	13.84
Fe_2O_3	2.13	1.89	6.14	1.53	0.34	1.75	2.07	0.65	0.63
FeO	6.14	10.92	7.80	5.06	6.52	6.20	6.59	2.36	1.43
MnO	0.06	0.17	0.18	0.14	0.09	0.09	0.09	0.02	0.04
MgO	9.28	7.20	4.93	7.59	6.90	6.49	4.16	0.07	0.29
CaO	13.08	11.96	10.46	13.46	10.14	10.24	7.46	2.44	1.34
Na_2O	0.76	1.00	2.74	1.88	2.40	2.20	3.10	3.91	3.74
K_2O	0.09	0.42	0.13	0.31	0.25	0.45	1.36	2.84	4.27
H_2O^-	—	0.03	0.46	—	—	—	0.01	—	0.06
H_2O^+	0.78	0.36	—	1.48	0.22	0.65	0.75	0.32	0.17
P_2O_5	none	trace	—	0.11	none	0.05	0.11	none	0.02
S	0.03	—	0.38	0.08	none	0.04	0.15	0.07	—
CO_2	—	—	none	none	—	—	none	none	none
BaO	—	—	—	—	—	—	0.04	—	0.08
SrO	—	—	—	—	—	—	none	—	none
ZrO_2	—	—	—	—	—	—	—	0.05	—
	100.33	100.45	100.22	100.22	100.19	100.11	99.78	99.73	99.82

Data are from Larsen (1948).

member of the sequence, the San Marcos gabbro, is frequently in contact with or intrusive into the metasedimentary and metavolcanic rocks, since the hot hydrous magma which these gabbros represent should have been best able of all the batholithic magmas to assimilate these rocks.

As in the case of the screens within the batholith, there has been no detailed study of the contact relations with the outlying country rocks. Although Larsen (1948) describes this metamorphism as "mild" even within a few feet of the granitic rocks, it seems best to reserve judgment pending more detailed studies.

A prominent feature of the intermediate rocks is the common occurrence of basic inclusions. We shall see that there is evidence that these are cognate, and represent the earliest member of the series, the San Marcos gabbro. Generally these inclusions are sharply angular, and although in massive rocks they are almost equidimensional, in more gneissic rocks they are much drawn out and elongated in section. Typically also these inclusions fade out uniformly during assimilation by the magma.

The chemical and mineralogic characteristics of the southern California rocks are shown in Tables 14.5, 14.6, and 14.7, as well as in Figure 14.4 as already discussed. In these tables the first six columns represent the San Marcos gabbro. The prominence given the gabbro in the tables is due in part to the great variety it exhibits but also because it is thought to be most representative of the parent magma of the entire series. In order to better illustrate these compositional varieties, the index numbers of the rocks that appear in the tables are also shown beside their representative points in Figure 14.4.

According to Miller (1937), who made a special study of the gabbros, the troctolite (Table 14.5, No. 1) is equivalent to solidified liquid magma. This conclusion is based on a distinctive mosaic texture and lack of compositional zoning that is quite unlike the crystal cumulates. Also very similar rocks occur as fine-grained lamprophyric dikes in the region. If this conclusion is valid, comparison with ordinary basalts (Table 4.1) as well as the quenched border facies of gabbroic complexes (Table 14.2) shows this parent magma to have been one of the most Ca- and Al-rich on record. Furthermore, Tables 14.6 and 14.7 show that virtually all the Ca is bound up as anorthite. In fact, high normative feldspar content is a characteristic of these gabbros.

It should be noted at this point that the troctolite contains normative olivine rather than quartz. This feature presents some difficulty to the hypothesis that this rock represents the solidified parent magma of the southern California batholith since it is difficult to conceive of such a magma yielding a silica-rich differentiate. It is likely that this sample contains some cumulate olivine, which would have little effect on its position in Figure 14.4, since the Mg/Fe ratio does not differ greatly from other silicates. It is possible that the parent magma was more nearly like the average San Marcos gabbro, which is silica saturated (Table 14.5).

Another important characteristic of the troctolite and indeed of the San Marcos gabbros in general is the very low K_2O content, a feature that it shares with the parent magmas of the gabbroic complexes. Furthermore, Larsen

Table 14.6 Normative mineral compositions of rocks from the southern California batholith

	1	2	3	4	5	6	7	8	9
Quartz	—	2.28	1.38	—	1.26	3.18	6.96	32.10	31.20
Orthoclase	0.56	2.72	0.56	1.67	1.11	2.22	8.34	16.68	25.38
Albite	6.29	8.38	23.06	16.24	20.44	18.34	26.20	33.01	31.44
Anorthite	64.22	36.97	36.97	44.20	45.59	41.98	30.58	12.23	6.95
Corundum	—	—	—	—	—	—	—	—	—
Diopside	0.68	18.30	12.17	17.38	3.76	7.20	4.80	3.50	—
Hypersthene	4.09	19.40	13.12	9.18	26.53	21.42	17.01	—	2.55
Olivine	20.09	—	—	6.01	—	—	—	—	—
Ilmenite	0.46	9.73	3.04	1.52	0.61	1.52	1.67	0.61	0.30
Magnetite	3.02	2.78	8.82	2.09	0.46	2.55	3.02	0.93	0.93
Apatite	—	—	—	0.34	—	0.33	0.34	—	
Pyroxene	—	—	0.24	0.12	—	—		—	

Data are from Larsen (1948).

(1948) demonstrated through his variation diagrams that even the intermediate and late members of the series are slightly but systematically lower in this oxide than are many other calc-alkali complexes. Also it may be noted that although the Al_2O_3 contents of all the gabbroic varieties are high, this is never reflected as normative corundum in any rocks of the series either early or late. These chemical characteristics of the southern California batholith militate strongly against any extensive contamination of the magmas by pelitic metasediments or acidic metavolcanics of the prebatholithic screens (see also, Brooks, *et al.* 1976) and generally strengthens the view of Larsen and others that differentiation was here dominated by fractional crystallization.

The thoroughly magmatic character of the rocks of Tables 14.5 to 14.7 is revealed in manifold mineralogic characteristics as well as the large-scale structures and bulk chemical features such as those just mentioned. Thus although many of the hornblende-rich gabbros superficially resemble metamorphic amphibolites, their minerals are quite different in detail. For example, reference to Table 14.7 shows that high modal hornblende content is frequently combined with abundant and very anorthite-rich plagioclase. As we saw in Chapter 10, metamorphism of gabbroic or basaltic rocks to amphibolites almost invariably leads to a combination of calcic hornblende or epidote with rather sodic plagioclase. If, however, hornblende precipitates early from a magma, it may form cumulate rocks with co-precipitating or interstitial calcic plagioclase. Although the hornblendes do not readily reveal their primary origin by euhedral crystals or by obvious compositional zoning, these features are common among most of the gabbroic plagioclase. Most of the hornblendes in fact show a poikilitic texture with respect to plagioclase, but it is possible that this is largely of overgrowth origin. Generally also more than one type of amphibole is found in the same rock. Green-brown hornblende appears to be the primary magmatic amphibole, while pale fibrous and blue-green hornblendes are features of late replacement of olivine and pyroxene.

If we refer again to the variation diagram of Figure 14.4 we see that most of the gabbro points fall quite far off the curve, which is inferred to represent the path of compositional change of the magma. That the rocks represented by these points are very probably cumulate in origin is shown by Table 14.7. For example, Nos. 2 and 3, which lie farthest from the curve, contain unusual amounts of Fe-Ti oxides, which could have been precipitating and accumulating at the time this rock was formed. Such early precipitation of oxides distinguishes these gabbros from those of the gabbroic complexes and seems to indicate rather more oxidizing conditions in the batholithic primary magma. It is possible and even likely that this is a direct consequence of the lower liquidus temperatures, which are implied by high water content.

Next in the differentiation series and in order of intrusion to the gabbros come the tonalites, the characteristic rocks of the batholith. About a half dozen varieties were distinguished by Larsen, and one, the Bonsall tonalite, alone accounts for 38 percent of the area. A

Table 14.7 Modal minerals of the rocks from the southern California batholith

	1	2	3	4	5	7	8	9
Quartz	—	trace	1	—	trace	12	30	33
K-feldspar	—	—	—	—	trace	—	21	32
Plagioclase	67	47	61	51	65	53	42	34
Biotite	—	—	1	1	trace	14	5	2
Hornblende	10	34	12	42	1	18	1	1
Ca-pyroxene	—	5	10	5	4	—	1	—
Orthopyroxene	2	6	8	—	28	—	—	0.5
Olivine	18	—	—	—	—	—	—	—
Fe ore[a]	1	8	8	2	2	+	+	+
Spinel	1.5	—	—	—	—	—	—	—
Anorthite content of feldspar	93	86	59	70	64	47	25	21

Data are from Larsen (1948).

[a] Magnetite and Fe-Ti ores; plus indicates phase is present.

notable feature of this tonalite is the presence of abundant basic inclusions such as we have already referred to. These were studied in detail by Hurlbut (1939), who not only identified them as San Marcos gabbro but also described localities where the two rock types were in gradational contact. From this latter observation he concluded that the gabbro was still partly liquid when the Bonsall tonalite was intruded.

The tonalites are distinguished by the abundance of feldspars relative to the ferromagnesians and by the simultaneous coming into prominance of quartz and biotite as primary precipitates. As is well known, the early precipitation of biotite will impart a trend toward enrichment of Na_2O relative to K_2O in the magma. As in the case of the San Marcos gabbro, the tonalite mineralogy bears the characteristic stamp of magmatism. Many of the minerals, but particularly plagioclase, tend to assume euhedral forms, are compositionally zoned, or are corroded. Generally also there is a close resemblance between the minerals in the tonalite and the same species in the gabbroic inclusions so that it seems clear that the latter have been transformed by reaction with the tonalite magma.

The tonalite grades by imperceptible changes into granodiorite and granite by an increase of quartz, alkali feldspar, and the ratio of biotite to hornblende.

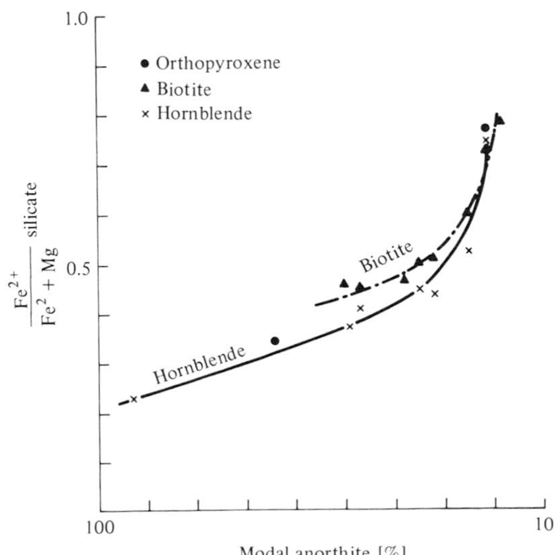

14.14 Systematic increase in the atomic fraction $Fe^{2+}/(Mg + Fe^{2+})$ with percent albite in the minerals of the southern California batholith. Data are from Larsen and Draisin (1948).

Further, the systematic decrease in anorthite content in the series gabbro to granite is accompanied by an equally systematic increase in $Fe^{2+}/(Mg + Fe^{2+})$ in all the ferromagnesian minerals (Figure 14.14). This trend reaches its culmination in the granite of Rubidoux Mountain, which contains orthopyroxenes in which this parameter reaches 0.80. We shall see that this trend in mineral composition, which parallels that of the gabbroic complexes, has important implications relative to the variation of oxygen fugacity in the series.

The Sierra Nevada batholith

This is perhaps the best known of the American batholiths, and has come to typify these complexes to many geologists. It has been, especially since the classic work of Cloos (1936), the archetype of the sharply discordant, steep-walled batholith of ultra-igneous aspect. This impression stems in part from its contacts with sediments usually being faulted. Like the southern California batholith it appears to be largely of Cretaceous age, although some individual peripheral plutons are older. Similarly it is a composite of many large relatively homogeneous plutons which have been intruded in the normal time sequence gabbro to granite. Typically also many of the plutons contain deformed or undeformed basic inclusions similar to those of the southern California rocks. However, the average composition of the Sierran rocks is nearer granodiorite than tonalite.

According to Cloos many of the plutons contain gneissic layers that are steeply dipping at the margin and nearly horizontal near the center so that they form dome-like structures. Superimposed on this essentially rheologic feature are various joint systems which also appear to be related to the broad structure of the pluton. The gneissic layers are, following Cloos, widely interpreted as representing a largely mechanical orientation of elongate or flattened elements such as hornblende and mica crystals by flowage during the last stages of solidification. In some places, as exemplified in particular by the Tioga Pass region of Yosemite Park, there is evidence that a later more acid intrusive has been discordantly emplaced in the cores of older plutons as if the more acid magma had continued to move upward after the more basic exterior had already solidified. It appears also that in some cases this mode of intrusion gives rise to marked chemical and mineralogic zoning of the plutons. Thus a study of Bateman *et al.* (1963) revealed that in one small pluton, approximately 8 km in greatest dimension, quartz and potassium feldspar increase regularly toward the center, while anorthite and ferromagnesian minerals as well as basic inclusions all increase toward the borders.

This picture is certainly consistent with continuous differentiation of the magma during intrusion.

Although to date no study comparable to that of Larsen and Draisin (1948) on the southern California batholith has been done, the bulk chemical analyses of the Sierra Nevada (Bateman *et al.*, 1963) indicate a trend very similar to but slightly displaced from those rocks (Figure 14.15). Furthermore, $Fe^{2+}/(Mg + Fe^{2+})$ shows the same increase through the series. Since iron oxides are quite generally low in these rocks, so that they have only minor influence on the curve, it is clear that the ferromagnesian minerals must also show a change in this parameter similar to that exhibited in Figure 14.14.

According to a number of field studies (Pabst, 1928; Cloos, 1936; Bateman *et al.*, 1963) the contacts of the batholith with the metasedimentary or metavolcanic country rocks are almost always sharp, and show only minor metasomatic effects. The grade of metamorphism appears to be dominantly amphibolite facies rank but again few detailed studies have been made. There have, however, been some local studies of interest (Kerrick, 1970; Morgon, 1975). Although the prebatholithic rocks appear to have formed a closely folded synclinorium (Bateman *et al.*, 1963), the above-cited authors agree that any deformation associated with the batholith is only local. It is also generally argued that the metamorphism associated with emplacement was largely of a static thermal nature so that rocks of a hornfels character resulted. However, studies to date are quite inadequate to make this argument convincing. In any case it is diametrically opposed to the hypothesis set forth by Hamilton and Myers (1967), who envision a regional flowage of metamorphosed and metasomatized country rocks beneath buoyantly rising granodioritic plutons.

14.15 Molar plot showing fractionation trends in the Sierra Nevada batholith. The trend for the southern California and Idaho batholiths is shown for comparison as the broken line. Data are from Bateman *et al.* (1963).

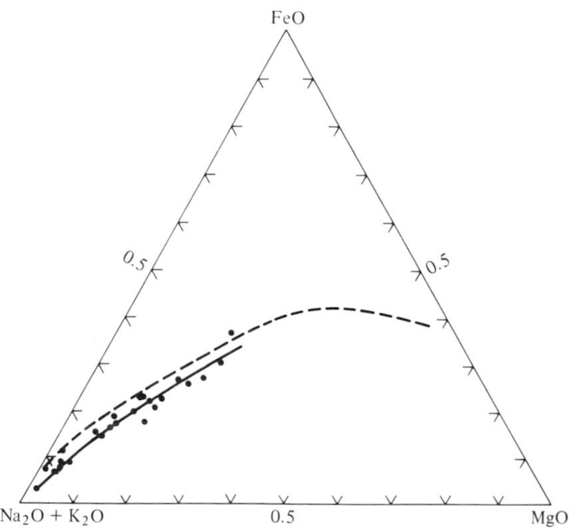

There seems to be general agreement that forceful intrusion was dominant over stoping or other permissive processes in the Sierran plutons. However, we shall see that such distinctions lose much of their meaning when considered in dynamic terms.

The Idaho batholith

The strong family resemblance of the Idaho complex to the other large western American batholiths has been made clear through the studies particularly of Larsen and Schmidt (1958) and of Hietanen (1963), whose data form the basis for Figures 14.4 and 14.15. Because of these similarities it will serve us little to discuss the magmatic aspects of this batholith in any detail. However, we shall have occasion to return to this complex again in connection with the widespread metamorphism and metasomatism encountered there.

Medium-sized calc-alkali plutons

Some of the features of emplacement and differentiation of the calc-alkali magmas are best displayed in the medium-sized plutons, which range upward from 5 to 100 km in greatest dimension but which are still much smaller than the batholiths proper.[4] This is especially true of the chemical and structural relations between the intrusion and the country rocks since these are in contact over the entire periphery. Then too many of these plutons appear to have been emplaced in one continuous process of intrusion so that they show a greater unity than the large composite masses whose different parts may differ widely in age. Examples of these medium-sized plutons are very common both in the near vicinity of the large batholiths and as quite isolated masses. Generally the range in composition they exhibit is more restricted than that of the batholiths, but sometimes the complete spectrum ranging from gabbro to granite is present.

Among the best known examples of this type of pluton are the Garabal Hill–Glen Fyne complex of Scotland, the Bald Mountain "batholith" of northeastern Oregon, and the Bald Rock "batholith" of

[4] Many of these plutons are, however, referred to as "batholiths" in the literature.

northern California. Although all three of these plutons are of calc-alkaline character, they differ greatly in their structural and chemical details.

The Garabal Hill–Glen Fyne complex (Nockolds, 1940) is the smallest of the three, with a maximum dimension of about 15 km, but it exhibits a very large range of rock types extending from ultramafic peridotites to granodiorites and pegmatites. Because of certain chemical criteria particularly as exhibited in the variation diagrams, the rock inferred to resemble the parent magma most closely is a pyroxene mica diorite. The pluton discordantly intrudes regionally metamorphosed schists of amphibolite facies rank. According to Nockolds, a metamorphic aureole, largely thermal in its effects, extends somewhat more than a kilometer from the contact. In this aureole the regionally metamorphosed schists have been transformed to hornfels facies rocks. Among the effects of the thermal metamorphism is the almost total disappearance of almandine garnet. In fact, it appears that in conformity with the reactions discussed previously (Chapters 9 and 10) the garnet-muscovite assemblages have been replaced by biotite-andalusite. The fact that peraluminous minerals such as andalusite are abundant in the aureole favors the idea that alkali metasomatism was quite limited in this case. However, evidence that some did occur locally is found in the vicinity of certain "injected schists" or migmatites very close to the contact where andalusite and sillimanite are altered to white mica.

The Garabal Hill–Glen Fyne rocks fall on the classical variation diagrams of the calc-alkali series. More particularly they exhibit the relative iron enrichment, which is so typical of the large batholiths. In this case this conclusion is also reinforced by a number of analyses of individual minerals such as pyroxene, hornblende, and cummingtonite, which also show the trend.

The Bald Mountain pluton (Taubeneck, 1957) is one of several isolated calc-alkali complexes approximately 100 km west of the Idaho batholith. It, too, is Mesozoic in age, bears many resemblances to the large batholiths of that age, and is probably a product of the same period and type of plutonic activity. The pluton abruptly transects the closely folded east–west structure of regionally metamorphosed country rocks. At the contact with the pluton, this regional trend is highly disturbed by folding locally concordant schistosity and considerable faulting and brecciation. The pluton is notable for its well-developed compositional zoning, which is similar to that already described in connection with certain Sierra Nevada plutons. This zoning takes the form of a gradation inward from a tonalitic rim to a granodioritic core. Like that of the other calc-alkali complexes, this compositional variation adheres to the classical variation diagrams.

An analogous type of zoning also occurs in the Bald Rock batholith (Compton, 1955). In this mass, which is simply an outlier of the Sierra Nevada, a tonalitic border grades inward to an intermediate zone of granodiorite and a leucotrondjhemitic core. It is interesting and perhaps significant that this trondjhemitic character is shared not only by adjacent similar plutons but also by older volcanic rocks of the area (Hietanen, 1951). In this case emplacement seems to have occurred under more deep-seated conditions than at Garabal Hill–Glen Fyne or Bald Mountain. This is shown by the generally concordant structures, evidence for plastic deformation, and intensive metasomatism and migmatization of the contact rocks. According to Compton there is a systematic increase in the anorthite content of plagioclase toward the contact in an aureole, which is several kilometers thick in some places. He has attributed these effects to the forceful intrusion and stoping action of a very energetic magma that was able to assimilate a volume of country rock equal to one-third that of the pluton itself. Although this conclusion is not at all established, it appears that metasomatism was far more active here than in some of the contact aureoles previously discussed.

The physical chemistry of calc-alkaline-type differentiation

We shall now seek a plausible explanation for the sharply contrasted differentiation trends of the gabbroic and calc-alkali series. Since we have already determined that the trend of the gabbroic series may be understood in terms of more or less classical phase-equilibrium principles, the question reduces to determining the factors at work in the calc-alkali series. To understand these we must explain the observed trend of relative iron enrichment in the ferromagnesian minerals as well as the tendency for the quartz-feldspathic constituents to be increasingly concentrated in the magma.

The question of iron enrichment can be approached through a thermodynamic analysis of the ferromagnesian compositional parameter $Fe^{2+}/(Mg + Fe^{2+})$. We already know from numerous examples that this quantity is an inverse function of the oxygen fugacity, so that we may gain some idea of the variation of the state of oxidation throughout the series by noting the mineral composition as revealed in Figure 14.14. To eliminate the complications that arise in considering H_2O and at the same time utilize phases for which thermochemical data are available, we make use of the following reaction between anhydrous phases:

$$\underset{\textit{orthopyroxene}}{FeSiO_3} + \underset{\textit{fluid}}{\tfrac{1}{6}O_2} \rightleftarrows \underset{\textit{ferrite}}{\tfrac{1}{3}Fe_3O_4} + \underset{\textit{quartz}}{SiO_2} \qquad (14.i)$$

It will be observed that two reference points for this reaction exist among the specimens represented in Figure 14.14 and Tables 14.5 to 14.7. These specimens, Nos. 5 and 9, a norite and a granite, not only exhibit large differences in $Fe^{2+}/(Mg + Fe^{2+})$ but also both contain all the necessary coexisting phases that appear in Equation (14.2). If then all these phases attained approximate equilibrium with each other during solidus conditions,[5] we may write the equation of equilibrium:

$$\log P_{O_2} = -6 \log K_{(i)} - 6 \log X_{Fe}^{Px} + 2 \log a_{Fe}^{Fer} + \frac{6P\Delta V_{(i)}}{2.303RT} \qquad (14.2)$$

In this equation we have, as usual, assumed ideality for orthopyroxene; however, no such assumption is possible regarding ferrite, and a_{Fe}^{Fer}, the activity of Fe_3O_4 in this phase, has been left in the implicit form. However, we know that the other major solution components of the latter phase are Fe_2TiO_4 and $MgFe_2O_4$, and that Fe_2TiO_4 is likely to dominate and behave quite non-ideally with respect to Fe_3O_4.

Now it may be shown that the maximum value of the fourth term of Equation (14.2) involving $\Delta V_{(i)}$, the difference in volume between the solid reactants and products, will be of the order of 1.5. Also its variation under crustal conditions is likely to be less than 0.5 in terms of $\log P_{O_2}$ so that we are safe in dropping the term to obtain

$$\log P_{O_2} = -6 \log K_{(i)} - 6 \log X_{Fe}^{Px} + 2 \log a_{Fe}^{Fer} \qquad (14.3)$$

It is desirable to consider this equation first in terms of the limiting case when all the phases of Reaction (14.2) are in their standard states corresponding to $X_{Fe}^{Px} = a_{Fe}^{Fer} = 1$. We then have

$$\log P_{O_2} = -6 \log K_{(i)} \qquad (14.4)$$

If we now make use of the available thermochemical data (Robie, 1966; Kelley, 1960; Olsen and Fuchs, 1967) we may obtain $K_{(i)}$ as a function of T. This is shown plotted as the curve (a) in Figure 14.16 along with the phase boundaries in the system Fe-O.

[5] Although it is unlikely that equilibrium was attained with respect to the distribution of Fe^{2+} and Mg between the phases, the requirements of equilibrium are not so stringent for energetic reactions of this type.

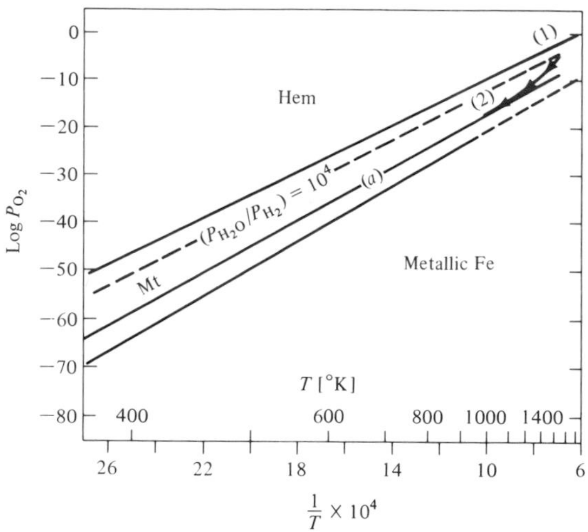

14.16 Stability fields of phases in the pure system Fe-O as a function of temperature and oxygen fugacity. The inferred path of differentiation of a typical calc-alkaline suite of magmas is shown as an arrow. The field of wüstite has been omitted.

If now MgO is added to the system, a family of curves, each corresponding to smaller values of X_{Fe}^{Px}, will lie above and parallel to curve (a). Thus if differentiation begins at some high temperature where $X_{Fe}^{Px} \sim 0.2$, as is the case for the earliest members of the series, and ends with the low-temperature granite with $X_{Fe}^{Px} \sim 0.8$, then it is clear that a curve something like curve (1)–(2) must represent the course of differentiation. However this conclusion must be modified somewhat to allow for the effect of a_{Fe}^{Fer}, which will assume a range of values less than unity depending on the quantity of Fe_2TiO_4 and other constituents dissolved in the ferrite. However, these constituents usually comprise less than 50 percent by weight of the ferrite even in such high-temperature rocks as the gabbros, while in the granites their concentrations are far lower. Consequently it seems likely that although $\log P_{O_2}$ might decrease less with temperature than the path defined by curve (1)–(2), yet it should decrease at least as much as defined by curve (a).

From the foregoing analysis we conclude that the relative iron enrichment, which generally characterizes the ferromagnesian minerals in the series gabbro-granite, corresponds to a substantial decrease in oxygen fugacity, perhaps as much as ten orders of magnitude. The few exceptions among plutonic rocks (Czamanske and Wones, 1973) only serve to reinforce this rule. Also, although we have made use of orthopyroxene to determine this trend, it is clear that it is also reflected in the compositions of hornblende and biotite since these are linked to that of orthopyroxene by the distribution relations. It is now necessary to explain how this decrease in oxygen fugacity comes about.

Let us begin with the simple gaseous system O-H and see how the oxidation state depends on the O/H ratio. We may immediately discount the case where $O/H > \frac{1}{2}$ since such mixtures have oxygen in excess to the amount present in pure water and are far more oxidizing than anything encountered under plutonic conditions. We shall therefore confine our discussion for the range $O/H \leq \frac{1}{2}$. If the quantity of H_2 derived from the decomposition of H_2O is 2α and the excess over this is β, then we obtain the following quantities of each species at equilibrium:

$$\begin{array}{ccc} 2-2\alpha & 2\alpha+\beta & \alpha \\ 2H_2O & \rightleftharpoons \; 2H_2 & + \; O_2 \end{array} \quad (14.j)$$

If we assume that the gases form a perfect mixture, we obtain for the equation of equilibrium

$$K_{(j)} = \frac{P_{H_2}^2}{P_{H_2O}^2} P_{O_2} = \left(\frac{2\alpha + \beta}{2 - 2\alpha}\right)^2 P_{O_2} \quad (14.5)$$

Then if α is negligible compared to β, as is usually the case for plutonic conditions, Equation (14.5) yields

$$P_{O_2} = \frac{4K_{(j)}}{\beta^2} \quad (14.6)$$

We see then that for this simple system the fugacity of oxygen is a function only of T and the P_{H_2O}/P_{H_2} ratio as given by β. Consequently, if $\log P_{O_2}$ is plotted as a function of T on a $1/T$ scale, a line with nearly constant slope and subparallel to curve (*a*) results for each value of P_{H_2O}/P_{H_2}. For purposes of illustration, one member of this family of curves is shown plotted in Figure 14.15.

The foregoing analysis indicates that when $O/H < \frac{1}{2}$, the degree of decomposition of water is so small that the ratio P_{H_2O}/P_{H_2} is fixed by the initial hydrogen content β quite independently of the total pressure. Under these circumstances $\log P_{O_2}$ decreases significantly with the temperature so that this simple system alone can in principle account for much of the decrease shown by this parameter during igneous differentiation. However it is desirable to see if we can also account for an even greater rate of decrease in $\log P_{O_2}$ with T such as is necessary for curve (1)-(2). For this it will be necessary to consider the interaction of the system O-H with the other components of the magma.

We might of course begin by deducing the oxidation states of more complex gaseous mixtures of the system O-H and such additional elements as C and S. However there is evidence that although such species as CO_2, CO, and SO_2 are important constituents of the gases given off by basaltic magmas, their importance tends to wane relative to water as differentiation proceeds. The occasion for this is of course readily apparent in the great tendency for H_2O to react with the silicate melt as we have already learned in Chapter 12. We shall therefore pass directly to considering how this interaction may account not only for an additional reducing tendency but also how it can bring about major changes in the crystallization path.

Regardless of the complexity of the solution components of the magma, we may characterize the oxidation state of the system by the ratio $a_{H_2O}^L/a_{H_2}^L$, the ratio of the water and hydrogen activities in the melt. Although we have considerable evidence for the strong interaction of water with the silicate melt according to reactions such as (12.b) there are no corresponding data for H_2. However, we do know that H_2 reacts with ferric iron as follows:

$$\underset{melt}{H_2} + \underset{melt}{2FeO_{3/2}} \rightleftharpoons \underset{melt}{2FeO} + \underset{melt}{H_2O} \quad (14.k)$$

so that

$$\frac{a_{H_2}^L}{a_{H_2O}^L} K_{(k)} = \left(\frac{a_{FeO}^L}{a_{FeO_{3/2}}^L}\right)^2 \quad (14.7)$$

Thus considerable hydrogen might be consumed in a melt that contains substantial ferric iron. However, it is clear that the capacity for such absorption is continuously diminished as differentiation proceeds and the absolute quantity of ferric iron is greatly reduced. In the case of water, on the other hand, the capacity for absorption into the melt should steadily increase. The reason for this is the increased tendency for reactions of the type (12.b) to occur. We shall presently see that the tendency for this type of reaction is continuously increased as differentiation proceeds and the melt becomes enriched in quartzofeldspathic relative to ferromagnesian constituents. Consequently there seems to be substantial theoretical and experimental support for the conclusion that during the normal course of calc-alkaline-type differentiation the ratio a_{H_2O}/a_{H_2} should steadily decline and that this effect further enhances the decrease in log P_{O_2}, which is already demanded by falling temperature according to Equation (14.6). Various aspects of these problems have also been discussed by Eggler and Burnham (1973).

We consider now what effect the interaction of water and the silicate melt has on the major phase relations other than those involving oxidation. The great enhancement by water of the solubilities of the quartzofeldspathic constituents is well known and is manifested in phase phenomena as a great lowering of the liquidus temperatures as the water content is increased (Chapter 13). What we are interested in here, however, is the *differential* increase of the solubilities of the quartzofeldspathic and ferromagnesian constituents. We have already seen a rather striking example of this effect in the previously discussed (Figure 13.9) experiments of Yoder and Tilley on the melting behavior of basalts under varying water pressure. In that example the plagioclase liquidus temperatures are depressed more than 200°K at $P_{H_2O} = 2000$ bar, while the ferromagnesian liquidus temperatures are depressed less than 100°K for the same water pressure. The effect is more concisely illustrated, however, in the much simpler system diopside-anorthite-H_2O. The phase diagram of this system as determined by Yoder and reported by Clark (1966) is shown in Figure 14.17. As the water pressure is increased in the system, the greater increase in solubility of anorthite relative to diopside affects a striking shift in the phase boundary and eutectic toward the anorthite side of the diagram. As a consequence all crystallization paths lead to the enrichment of anorthite in the liquid. The energetics involved may be

14.17 System diopside-anorthite-H_2O as determined by Yoder (Clark, 1966). Note reduction of field of anorthite as water pressure is increased.

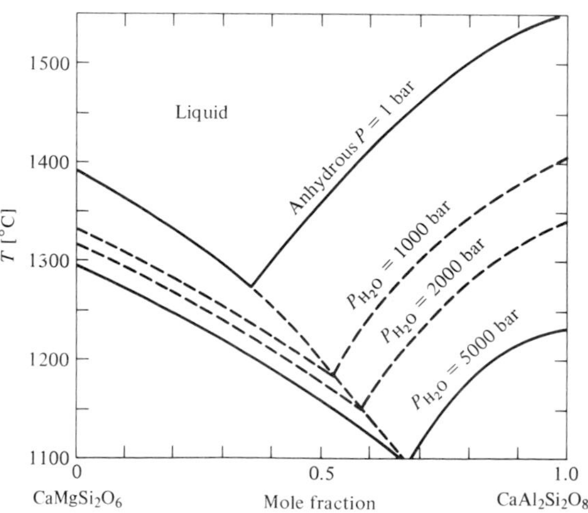

illustrated by considering the transfer reactions

$$\underset{crystal}{\text{anorthite}} \rightleftarrows \underset{liquid}{\text{anorthite}} \tag{14.l}$$

$$\underset{crystal}{\text{diopside}} \rightleftarrows \underset{liquid}{\text{diopside}} \tag{14.m}$$

at some liquidus temperature. Let us choose the temperature of the anhydrous eutectic at $X_{An}^L = 0.36$ and $T = 1548°K$. Then if the solid phases are assumed to be stoichiometric, the equations of equilibrium are

$$K_{An} = X_{An}^L \gamma_{An}^L \tag{14.8}$$

and

$$K_{Di} = X_{Di}^L \gamma_{Di}^L \tag{14.9}$$

and where the mole fractions X_{An}^L and X_{Di}^L are referred to the anhydrous system so that $X_{Dn}^L + X_{Di}^L = 1$. Then as P_{H_2O} is increased isothermally, say to 1000 bar, K_{An} and K_{Di} remain relatively constant since they are not sensitive to the pressure. However we see from Figure 14.17 that X_{An}^L changes to $(X_{An}^L)' \simeq 0.66$, while X_{Di}^L changes only to $(X_{Di}^L)'' \simeq 0.775$. We may illustrate the general process by dividing Equation (14.8) by Equation (14.9):

$$\frac{(\gamma_{Di}^L)''}{(\gamma_{An}^L)'} \frac{K_{An}}{K_{Di}} = \frac{(X_{An}^L)'}{(X_{Di}^L)''} \tag{14.10}$$

Here as above the singly and doubly primed values refer to the anorthite and diopside liquidus, respectively. Then it is clear that as P_{H_2O} is increased $(X_{An}^L)'$ always increases faster than $(X_{Di}^L)''$, so that if K_{An} and K_{Di} remain constant, the ratio $(\gamma_{Di}^L)''/(\gamma_{An}^L)'$ must steadily increase. By the conventional interpretation of activity coefficients this corresponds to a greater negative excess free energy of solution of anorthite than of diopside. In other words, anorthite has a greater affinity for water than diopside has so that water has a greater tendency to hold anorthite in solution and prevent its early precipitation. We may interpret this further to mean that water tends to form soluble species more readily with highly polymerized melts containing the small cations Si^{4+} and Al^{3+} than with the larger cations such as Ca^{2+} and Mg^{2+}. If this is true, a simple extension leads us to expect the water-melt interaction to increase steadily as differentiation proceeds and Si^{4+} increases in concentration.

The most important petrogenic consequence of all this is that increased water contents direct the crystallization paths into composition fields in which the quartzofeldspathic constituents are enriched and the ferromagnesian constituents are depleted. It is largely this effect that is thought to account for the diverse differentiation trends of the gabbroic and calc-alkali series (Mueller, 1969a). However it is likely that its effects are even more far reaching than this, so that they are felt in the alkaline and other differentiation trends as well. These results also tend to be confirmed by recent experiments on the complex system $CaAl_2\ Si_2O_8$-$NaAlSi_3O_8$-SiO_2-MgO-Fe-O_2-H_2O-CO_2 by Eggler (1974). He noted that increased water pressure in the system brought about a marked decrease in the primary phase volume of quartz. Also the fugacity of oxygen decreases markedly sub-parallel to the Ni-NiO buffer curve during fractionation as in Figure 14.16, while the ratio Fe/Mg systematically increases.

The effect of water pressure on the differentiation of a basaltic magma should also be compared with the effects of anhydrous high-pressure differentiation and partial melting discussed in Chapters 12 and 13.

Origin of calc-alkali magmas

In our discussion of the chemical variation of the batholith of southern California we noted that this appeared to be consistent with a derivation by fractional crystallization of a basaltic magma. Elsewhere we have indicated that such magmas are almost certainly derived from the upper mantle. However, recently there have been several attempts to explain the large Mesozoic batholiths of western North America as having been derived by partial melting of crustal rocks and in particular oceanic crust "subducted" under the western North American continental margin (Presnall and Bateman, 1973). Numerous attempts have also been made to explain a variety of plutonic and volcanic rocks around the world in this manner. A review has been given by Boettcher (1973). Some light is cast upon this problem by the overall compositional variation of these batholiths as discussed here and by the measured values of the initial ratio $^{87}Sr/^{86}Sr$ of constituent minerals as well. Since ^{87}Sr is derived from ^{87}Rb by radioactive decay and since ^{87}Rb is concentrated in crustal rocks, a high value of the ratio presumably reflects crustal contamination. The subject was reviewed by Kistler (1974), who found a systematic increase in initial $^{87}Sr/^{86}Sr$ from west to east in California and Nevada from low values of less than 0.704 in the west to high values of 0.709 in the east. Furthermore this increase is correlated with an increase in alkalies as well as other regional chemical variations. Since the initial values of $^{87}Sr/^{86}Sr$ of 0.704–0.706 fall in the range of this ratio in basaltic rocks derived from the mantle, it must be concluded that at least large parts of the calc-alkali batholiths of western North America have a similar origin. However, it is possible that some plutons in the eastern part of the Cordillera have been derived by partial fusion of deep-lying crustal material. According to Kistler it is difficult to fit these isotopic and chemical variations into the preferred scheme of deriving batholithic rocks by partial fusion along the hypothetical subduction zone.

Another question of great importance is the origin of the large segments of batholiths, which apparently have resulted from the metasomatic replacement of geosynclinal sediments. This problem has been largely ignored in the literature but it is quite apparent in the instances of the southern California and Idaho batholiths. Although good progress has been made by Hietanen (1963d) on the Idaho batholith detailed studies of mass balances are needed for other batholiths as well.

Alkaline rocks

Alkaline rocks comprise a minor but petrologically interesting class of igneous rocks, which have continued to attract much attention. Although we are concerned in this chapter with plutonic igneous rocks, it is convenient to discuss all alkaline rocks—whether plutonic, hyabyssal, or volcanic—together. In any case, alkaline complexes are well known for their gradation to hyabyssal and volcanic environments.

Alkaline rocks are almost all silica-undersaturated or agpaitic and miaskitic in character, although peralkaline granites are known. The most common and most ordinary members of the class are the alkaline basalts, whose compositions fall to the left of the plane of silica undersaturation in Figure 14.2. These basalts are most common in oceanic islands such as St. Helena in the Atlantic and Samoa in the Pacific, although they also occur in association with continental flood basalts.

Closely related to alkaline basalts are the lamprophyres, which consist of dark-colored dike rocks that frequently cut granite bodies. Characteristically, lamprophyres are rich in biotite, hornblende, titaniferrous pyroxene, and olivine, which form euhedral phenocrysts in a groundmass of feldspars and feldspathoids. Apparently there is also a genetic relation between alkaline rocks and kimberlites (Heinrich, 1966), which form ring structures or pipes and which, like carbonatites, frequently contain abundant carbonates.

Classic occurrences of potassic alkaline rocks are found in the Highwood Mountains of Montana, in

Italy (Mt. Vesuvius), in the African rift zone, and in Australia. Outstanding examples of sodic alkaline rocks are the nepheline syenites, ijolites, etc., which occur in the Kola peninsula of Russia, in Norway, and elsewhere. An integral part of many alkaline complexes are carbonatites consisting of a mixture of calcite, dolomite, ankerite, siderite, and sodium carbonates as well as carbonates of rarer elements. Characteristically, but not invariably, alkaline rocks occur as ring complexes in stable continental regions of tensional faulting. In such complexes the carbonatite is usually one of the latest members in the magmatic sequence.

In addition to the clearly magmatic features of alkaline-carbonatite complexes there is usually also a zone of pervasive metasomatism of the wall rocks. This zone of contact metamorphism (fenitization) has as its product the characteristic rock fenite, which consists of a development in the country rock of alkali and ferric iron-rich minerals similar to those of the alkaline intrusives.

Although it was originally thought that alkaline rocks were products of limestone assimilation, there is mounting evidence that even the carbonatites are of magmatic origin. This conclusion is strongly supported by experimental evidence that mixed carbonate melts can exist at temperatures below 600°C under only moderate water pressure (Wyllie, 1965). As in the case of basaltic magmas, carbonatites appear to originate in the upper mantle. In particular, this is indicated by their $^{87}Sr/^{86}Sr$ ratios, which fall in the range of typical mantle-derived materials (Powell *et al.*, 1966; Hyndman, 1972). The magmatic origin of at least some carbonatites is proved by the discovery of sodium-rich carbonate lava flows at the Oldoinyo Lengai volcano (Chapter 15).

The origin and evolution of alkaline rocks may be traced most rapidly and conveniently by reference to the principles we have already applied in the case of the calc-alkali sequence of differentiates. If we ignore the apparently discredited limestone syntexis hypothesis, we can sketch a crystal fractionation scheme which leads directly from alkaline basalts to nepheline syenites, phonolites, and other leucocratic differentiates which fall in the compositional field of undersaturated residual liquids shown in Figures 15.2 and 15.3. However, there is almost unanimous concurrence that there is no easy way to obtain these alkaline differentiates from silica-saturated basaltic liquids (Edgar, 1974). Furthermore, there seems to be considerable evidence that carbonatite magmas may result from the differentiation of extremely alkali-rich magmas as well as by liquid immiscibility (Wyllie, 1974; Koster van Groos, 1975).

In addition to the analogous differentiation paths of alkaline and calc-alkaline rocks, there is also a parallelism in the behavior of oxygen fugacity during differentiation. Thus, as in the calc-alkaline sequence, a drop in oxygen fugacity along the differentiation sequence of alkaline magmas is indicated by the increase in $Fe^{2+}/(Mg + Fe^{2+})$ of the ferromagnesian minerals. This has been documented in the case of the Shonkin Sag laccolith, for example (Nash and Wilkinson, 1970), where it was found that the oxygen fugacity decreased in a manner very similar to that shown in Figure 14.15 for the calc-alkaline batholiths as the laccolith differentiated from a shonkinitic magma to a nepheline syenite magma. However, in the last stages of differentiation, where olivine was no longer a precipitate, the decrease in oxygen fugacity with temperature became less marked. The decrease in oxygen fugacity with temperature along the differentiation sequence in the Shonkin Sag laccolith was also accompanied by a decrease in silica activity as determined by the coexistence of sanidine and leucite (Nash and Wilkinson, 1970). As in the case of the oxygen fugacity, this is partly a thermal effect and partly related to the compositions of the melts.

It has generally been assumed that the appearance of ferric-iron-rich acmitic pyroxenes in alkaline rocks indicates a higher value of the oxygen fugacity than does the presence of the ferrous iron silicates. However, this is not necessarily true and is not confirmed by experimental evidence (Bailey, 1969). One explanation is that the ferric iron is stabilized by the presence of the alkalis Na_2O and K_2O, which interact with it in the melt to form stable complexes which are the precursers of acmite. Various aspects of the relation of the Fe^{3+}/Fe^{2+} ratio in melts have also been discussed by Carmichael *et al.* (1974).

Assimilation and alteration of inclusions

General principles

Many controversies center around the role of assimilation reactions in petrogenesis, and particularly the extent to which crystallization-differentiation is influenced by assimilation. As is well known, magmas react directly with their wallrocks, and these reaction zones are transitional to contact metamorphism in all its variety of thermal and metasomatic manifestations. Commonly, however, we see evidence of reaction between the magma and inclusions that were completely immersed in it. These fall into two broad categories:

1. Inclusions of igneous, metamorphic, or sedimentary origin with no previous genetic relation to the magma
2. Inclusions of igneous rocks that are either solidified equivalents of cogenetic magmas earlier in the liquid line of descent or are crystal cumulates from such magmas (referred to as *cognate inclusions*)

The groundwork for a systematic physicochemical approach to assimilation effects was laid by Bowen (1928) who discussed the above-mentioned categories in considerable detail. He also discussed two categories of igneous inclusions from the same liquid line of descent as the magma:

1. Cognate, or other inclusions of a more basic nature than the magma, which are chemically and mineralogically equivalent to early differentiates
2. Inclusions equivalent to later differentiates than the magma

He demonstrated from physicochemical arguments that inclusions of the first type could not be dissolved directly by the magma since it was effectively supersaturated with their constituent minerals. These minerals could by exothermic reactions be transformed into others identical to those with which the magma itself is saturated. However, this reaction decreases the total quantity of the liquid since the exothermic heat must be supplied by precipitation of crystals from the melt. For the transformation of calcic to sodic plagioclase, the overall reaction is calcic plagioclase + liquid magma → more plagioclase of somewhat more sodic composition + a smaller quantity of liquid + ΔH, where $\Delta H < 0$. Conversely when inclusions corresponding to late differentiates are immersed in a more basic magma, these inclusions may be dissolved directly. Although this can be done only with the absorption of heat, the quantity of liquid is thereby slightly increased.

Bowen also provided a broad-ranging discussion of the more difficult problem of the reaction and assimilation of inclusions—sedimentary or otherwise—that do not stem from the same liquid line of descent as the magma. He was able to show by stoichiometric arguments that the assimilation of inclusions that differ greatly in composition from the magmatic differentiates could be regarded as being composed of a "magmatic fraction" and an "excess" fraction. By a variety of reactions with the magma, the excess fraction could give rise to wholly new phases or change the proportions of phases already crystallizing from the magma. Thus if inclusions of Al-rich pelitic sediments are immersed in a gabbroic magma, they will accentuate the precipitation of anorthite and olivine or Mg-Fe pyroxene from the magma as governed by Reactions (14.a) and (14.b). But within that part of the inclusion that may remain unassimilated and where the bulk composition might approach Al_2SiO_5, such minerals as corundum might form. Similarly when carbonate-rich inclusions are immersed in a basic magma, the bulk composition on assimilation may be silica-deficient enough to give rise to alkaline differentiates. Bowen also pointed out that limestone *syntexis*, as discussed above, could not be expected to desilicate a granitic melt directly since it could only accentuate the precipitation of minerals such as Ca-pyroxene hornblende and feldspars, which would not accomplish this end. This argument, which is supported by recent experimental evidence (Wyllie, 1974), has important implications for the hypothesis for the origin of alkaline rocks by limestone syntexis.

Most of Bowen's closely argued thesis still holds up well today. However, we will find it useful to examine some of the more important reactions in somewhat greater detail in the following sections.

Assimilation of igneous inclusions by magmas from the same liquid line of descent

The case of assimilation of anhydrous inclusions consisting of such minerals as quartz, feldspar, olivine, and pyroxene has been discussed in considerable detail by Bowen (1928). Equally interesting, however, are the assimilation reactions that involve the hydrous phases of typical calc-alkali rocks. Although we have already discussed metamorphic versions of these reactions we write them again here:

$$\underset{pyroxene}{3(Mg,Fe)SiO_3} + \underset{melt}{KAlSi_3O_8} + \underset{melt}{H_2O} \rightleftharpoons \underset{biotite}{K(Mg,Fe)_3AlSi_3O_{10}(OH)_2} + \underset{melt}{3SiO_2} \quad (14.n)$$

$$\underset{pyroxene}{3(Mg,Fe)SiO_3} + \underset{plagioclase}{2CaAl_2Si_2O_8} + \underset{melt}{H_2O} \rightleftharpoons \underset{hornblende}{Ca_2(Mg,Fe)_3Al_4Si_6O_{22}(OH)_2} + \underset{melt}{SiO_2} \quad (14.o)$$

From Reaction (14.n) we see that if anhydrous inclusions of gabbroic composition are present in, let us say, a tonalite magma, the alkali feldspar component of the magma can react with the pyroxene of the inclusion to produce biotite similar to that which will also be precipitating from the magma. Similarly Reaction (14.o) states that pyroxene and basic plagioclase of the inclusion can be converted to hornblende. It should be noted that since both of these reactions involve an absorption of water, they are bound to be exothermic, as we have seen is the case with the accompanying anhydrous reactions.

In addition to Reactions (14.n) and (14.o), both the hydrous and anhydrous minerals will undergo reactions of the type Bowen discussed. These reactions will act in such a way as to continuously adjust the anorthite content and the atomic fraction $Fe^{2+}/(Mg + Fe^{2+})$ of the inclusion minerals to the same parameters of the minerals precipitating from the magma. We shall see that while studies show that striking adjustments of this type do occur—particularly in basic cognate inclusion of calc-alkaline rocks—they are seldom adequate for the attainment of complete equilibrium between inclusions and magma precipitates.

Assimilation of inclusions—sedimentary or otherwise—that are not on the liquid line of descent of the magmas

As in the case for cognate inclusions, it is desirable to trace the assimilation of hydrous constituents in somewhat more detail than Bowen did. For example an important reaction which occurs when pelitic schist reacts with basic magma is the following:

$$\underset{\textit{muscovite}}{KAl_2(Si_3Al)O_{10}(OH)_2} + \underset{\textit{quartz}}{SiO_2} \longrightarrow \underset{\textit{melt}}{KAlSi_3O_8} + \underset{\textit{melt}}{Al_2SiO_5} + \underset{\textit{melt}}{H_2O} \quad (14.p)$$

In basic magmas this reaction always runs to the right since the temperatures of such magmas lie above the stability range of muscovite + quartz. If adequate normative diopside is present—as is always the case for normal basaltic compositions—Reaction (14.p) will be followed by the reaction

$$\underset{\textit{melt}}{CaMgSi_2O_6} + \underset{\textit{melt}}{Al_2SiO_5} \longrightarrow \underset{\textit{plagioclase}}{CaAl_2Si_2O_8} + \underset{\textit{pyroxene}}{MgSiO_3} \quad (14.q)$$

so that the precipitation of a noritic assemblage will be favored. Although this reaction has frequently been proposed to explain norites, it should be remembered that $KAlSi_3O_8$ is at the same time added to the melt. Actually this evidence for K_2O assimilation is seldom present in norites. More particularly it is totally lacking in those of the southern California batholith as we have just noted (Table 14.5).

Because of the high thermal stability of Mg-rich biotites, a certain minimum value of P_{H_2O} will bring about the following reaction in pelitic inclusions:

$$\underset{\textit{muscovite}}{KAl_2(AlSi_3)O_{10}(OH)_2} + \underset{\textit{melt}}{CaMgSi_2O_6} + \underset{\textit{melt}}{2MgSiO_4} \longrightarrow \underset{\textit{biotite}}{KMg_3(AlSi_3)O_{10}(OH)_2} + \underset{\textit{plagioclase}}{CaAl_2Si_2O_8} + \underset{\textit{melt}}{2SiO_2} \quad (14.r)$$

As a result of this reaction the thermally unstable assemblage of the schist is replaced by one in equilibrium with the magma. The biotite, pyroxene, and feldspar so derived may of course precipitate directly from the melt surrounding the inclusions so that this represents one mechanism for assimilation and dispersal. However, it may happen that these phases are nucleated within the inclusion so that little actual dispersion occurs.

When pelitic sediments of a somewhat more complex character react with magmas that contain little or no normative diopside, a reaction of the following type is possible:

$$\underset{\textit{muscovite}}{3KAl_2(Si_3Al)O_{10}(OH)_2} + \underset{\textit{biotite}}{KMg_3(AlSi_3)O_{10}(OH)_2} + \underset{\textit{quartz}}{\tfrac{15}{2}SiO_2} \longrightarrow \underset{\textit{K-feldspar}}{4KAlSi_3O_8} + \underset{\textit{cordierite}}{\tfrac{3}{2}Mg_2Al_4Si_5O_{18}} + \underset{\textit{melt}}{4H_2O} \quad (14.s)$$

If the inclusion contains a higher proportion of muscovite, the cordierite may be accompanied by sillimanite, andalusite, or other Al-rich phases either as crystallization products of the melt or within the inclusions themselves.

It should be noted that reactions such as (14.p) and (14.s) are likely to be endothermic since they involve the decomposition of hydrous phases. However, they may take place concurrently with other strongly exothermic reactions. Whether or not these reactions proceed with falling temperature will then depend on the net thermal effect and whether or not the water released can be absorbed by the melt or can in some way escape from the system.

Reactions of the type (14.p), as well as the normal crystallization processes in granodioritic and granitic magmas, frequently result in the feldspathization or "granitization" of the inclusions, which takes the form of a growth of alkali feldspar porphyroblasts. However, as was pointed out by Nockolds (1940) in his discussion of the Garabal Hill–Glen Fyne complex, the precipitation of these feldspars is seldom accompanied by a corresponding precipitation of quartz within the inclusions although this mineral may be a major precipitate in the magma.

Interpretation of the chemistry and structure of deformed cognate inclusions

When the cognate inclusions, which are so common in intermediate and acidic plutons, are examined it is found that the relation between the chemistry and structure has been quite generally misinterpreted. The observational facts of this relation were described clearly by Hurlbut (1935) for the gabbroic inclusions of the Bonsall tonalite from the southern California batholith. He provided evidence for reactions of the type (14.n) and (14.o) by which the minerals of the high-temperature gabbroic assemblages are converted to those compatible with the tonalite. That these reactions frequently did not go to equilibrium is shown by the ferromagnesian minerals of the inclusions, which usually retain higher values of $Mg/(Mg + Fe^{2+})$ than those of the surrounding tonalite. That these reactions were accompanied by a pervasive recrystallization of the entire inclusions is shown by such features as the tendency for compositional zoning in plagioclase to be obliterated as well as the parallel crystallographic alignment of minerals in the inclusions and the tonalite. For example it was found that not only the [010] faces of plagioclase are parallel in the tonalite and inclusions but also that the *a* axes of these grains are in good alignment. Furthermore the *c* axes of hornblende in the inclusion are also well aligned.

In view of the features just described it is difficult to see why the deformation of inclusions and the orientation of mineral grains within the magma are so frequently attributed to a mechanical orientation process in the flowing magma. For if this were the case, why should the crystals of the inclusion, which must have been solid, also be parallel to those in the magma? An alternative explanation is that the alignment, both within the inclusion and in the magma, is a postconsolidation feature in which the orientation was achieved by solid-state recrystallization in the differential stress field. Although this type of flow may also be "fluid" in nature (Chapter 3), it acts similarly[6] on all the crystal components and involves a minimum of physical rotation of the type encountered in pure liquid flowage alignment such as affects crystals in lavas. It is clear that this mechanism of solid-state recrystallization is highly compatible with the observed tendency toward compositional zone elimination. It is further likely that this type of recrystallization is greatly facilitated by the heterogeneous reactions that accompany it.

Structural environment and mode of emplacement of igneous plutons

Igneous plutons are emplaced into the terrain in which we observe them in essentially two different ways: they are emplaced as dominantly liquid magmas or as essentially solid bodies. However, we must also allow for transitional conditions between these two extremes, when dominantly crystalline bodies still possess a residue of melt, which although highly viscous, may impart to the bodies a degree of mobility considerably greater than if they are completely solid.

It is not our intention here to give an exhaustive review of all that has been written about the emplacement of igneous magmas. Much of the classical view of this subject is covered adequately in such books as those of Daly (1933) and of Balk (1937). Although the intrusion of solid bodies is little discussed in the classical literature, much attention is given to the question whether the mode of emplacement of a magma in any given case was "forceful" or "permissive." Daly tends to stress the permissive mode of intrusion such as is represented by *cauldron subsidence* of blocks into a lighter, upwelling magma, or the closely related *piecemeal stoping* by which the magma fractures its wall and roof rocks through thermal and buoyant stresses. Balk, on the other hand (having been a student of Hans Cloos), believed that forceful intrusion overwhelmingly dominated the igneous scene. According to his thesis the magma was generally intruded upward from below with considerable force so that it thrust aside the wallrocks and left signs of this movement in well-defined structural features of the solidified pluton. According to Cloos (1925) these structural indicators of motion consist of linear elements such as elongate crystals or stretched inclusions which define *flow lines*, or as platy mica or feldspar crystals which form *platy flow structures*. Also many joints in the margin of the pluton and the adjacent wall rocks were

[6] This "similarity" is a matter of degree (see Chapter 8).

thought to register movement that was either post-consolidation or due to viscous drag of the magma or adjacent rocks. The general picture is one of steep-sided, upward-expanding, dome-like bodies in which the marginal motion is nearly vertical but in which the central motion may be virtually horizontal. These classical models have provided us with many insights into the geometry and kinematics of intrusion; however, they are at the same time severely limited in quantitative applications since they totally ignore the energetics and dynamics of the processes involved.

An advance in the direction of the dynamical analysis of the intrusive process, particularly with respect to the gravitational effects on solid bodies, has been made by Ramberg (1967). His analysis is based largely on theoretical fluid dynamics and the application of centrifuged scale modes to duplicate the effect of gravity. Ramberg has shown that it is instructive to consider the mechanical potential change $\Delta\varphi_{\mathrm{mech}}$ of an intruding body as the sum of a *spatial potential change* $\Delta\varphi_{\mathrm{spat}}$ and a *gravitational potential change* $\Delta\varphi_{\mathrm{grav}}$, so that

$$\Delta\varphi_{\mathrm{mech}} = \Delta\varphi_{\mathrm{spat}} + \Delta\varphi_{\mathrm{grav}} \tag{14.11}$$

For example, a cross section of a pluton, which is defined by the solid line of Figure 14.18, may after some movement be defined by the broken line of the same figure. It is obvious that if the body is incompressible (which corresponds rather closely to rocks and magmas) the volume V^- lost at the end of the pluton at which contraction occurs is equal to the volume V^+ gained at the end at which expansion occurs. The spatial component of change of the mechanical potential then is

$$\Delta\varphi_{\mathrm{spat}} = \iint P\,dA\,dr + \iint P'\,dA'\,dr' \tag{14.12}$$

Here P is the pressure (taken as hydrostatic for simplicity), A is the bounding area, and r is the normal radius of displacement on the expanding end while P', A', and r' are the corresponding quantities on the contracting end. Both A and A' are positive, but whereas r is positive, r' is negative. Also P and P' are both positive for compression but negative for tension.

The change in gravitational potential is then

$$\Delta\varphi_{\mathrm{grav}} = V^+\rho g h + V^-\rho g h' \tag{14.13}$$

Here ρ is the density of the pluton and h and h' are the heights of the centers of gravity of the volumes V^+ and V^-, respectively, above a common datum plane.

It is clear that in the situation depicted in Figure 14.18, which corresponds to an upward expansion, $\Delta\varphi_{\mathrm{grav}}$, for the pluton itself must always increase. However, it is obvious that if the pluton is to move up, an equal volume of surrounding material must move down to occupy the space evacuated. If the surroundings are of the same density ρ as the pluton, then such movement would result in no net change in the total $\Delta\varphi_{\mathrm{grav}}$ for the pluton and its surroundings. In this event the driving force of the upward motion must come from a decrease in $\Delta\varphi_{\mathrm{spat}}$. In other words, the pluton expands upward into a region of low pressure, lower in fact than that which is demanded by hydrostatic equilibrium. If on the other hand the surrounding rock density is greater than ρ then the upward motion also results in a decrease in total $\Delta\varphi_{\mathrm{grav}}$ for the pluton and its surroundings.

Seen in terms of $\Delta\varphi_{\mathrm{mech}}$, much of the distinction between forceful and permissive intrusion vanishes, becoming largely a matter of the magnitude of this quantity. In the case of piecemeal stoping or cauldron subsidence, the strongest element is probably $\Delta\varphi_{\mathrm{grav}}$, but since the difference in density between the magma and country rock is seldom greater than 0.5, the forces do not reach a magnitude sufficient to greatly deform the wall rock. However when solid plutons rise, the surrounding rocks will always show distortion even though the density difference between the pluton and country rock may be considerably smaller than 0.5. The reason for this

14.18 Movement of a body of rock or magma under various forces in Earth's lithosphere. For explanation see text. (After Ramberg, 1967.)

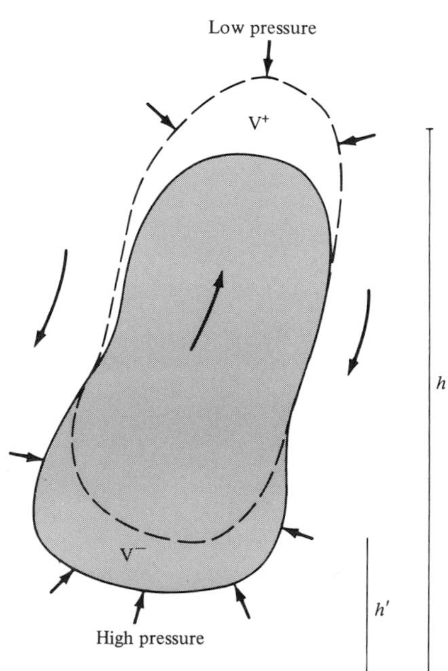

is that stoping except for salt or limestone intrusions is insignificant so that movement can occur only through external deformation, either by faulting or by plastic or fluid flow.

Consideration of gravitational effects has very broad implications in petrology in general; in particular it holds the key to the classical problem of the abundance of rock types as determined from exposures. For example, petrologists were long baffled by the relatively small quantities of basic and ultrabasic rocks in the calc-alkali plutonic complexes, as we have noted for the southern California batholith. In the past this was frequently taken to mean that the great quantities of intermediate tonalites and granodiorites could not have been derived by crystallization-differentiation acting alone but that this had to be supplemented or completely supplanted by some such mechanism as partial melting or large-scale assimilation of acid crustal rocks. However, when full account is taken of the density differences of these rock types, this apparent discrepancy is not only accounted for but it is difficult to imagine any other state of affairs. The great volumes of basic and ultrabasic rocks and magmas, which have given rise to the acidic plutons, could never have found their way to the upper crust. Rather it was only isolated segments and apophyses of them that had become entrapped in far larger volumes of more buoyant rock and were thus lifted into view.

Although we can roughly categorize rocks as to depth of origin, one of the most difficult problems in igneous petrology is the determination of the depth of emplacement of a given exposure of igneous plutonic rock. One obvious reason for this difficulty is that much further movement and even reemplacement may have occurred since the original magma was intruded. Even when the wallrocks of a pluton contain high-pressure assemblages, we cannot be certain that these were not metastable with respect to the depth at which intrusion occurred.

It has been proposed (Hamilton and Myers, 1967) that many batholithic plutons were emplaced at shallow depths and actually are covered by roof deposits of their own volcanic ejecta.[7] However, this thesis may confuse the present position of the plutons with their actual depth at the time of intrusion. Thus according to Hietanen (1963a, b) the northwest margin of the Idaho batholith exhibits synkinematic plutons of quartz diorite and tonalite as well as a broad belt of high-grade (and high-pressure) metamorphism and extensive metasomatism that extends out from the main batholith as much as 50 km. It is difficult to imagine that such features could have been acquired except under conditions of very deep-seated emplacement. Yet we know from their present levels in the crust and from the tectonic patterns that such plutons do move upward after they have solidified—according to Ramberg (1967) chiefly as a response to buoyancy forces. Martignole and Schrijver (1970) have identified the same forces at work in the Morin anorthosite of Quebec.

We can in fact see evidence in different plutons for this upward movement at every stage from a highly liquid condition to the stage of complete solidification. Some of the most convincing evidence for highly fluid magma is to be found in those plutons showing evidence for cauldron subsidence, such as has been described in the Ossipee Mountains of New Hampshire (Kingsley, 1931). Almost equally good evidence is present in some of the minor tertiary intrusives of the western United States. For example, the Notch Peak quartz monzonite of western Utah (Gehman, 1958) has displaced almost horizontal Cambrian limestones with only minor disturbance and up-arching of the strata. Also, sills are well developed. These types of structures are best explained by an upwelling of the low-density fluid magma in a block-faulted terrain. But the porphyritic character and the up-arching of the strata also indicate that the last stages of intrusive motion occurred while the magma was already quite viscous. In striking contrast to this is the pluton described by MacColl (1964) at Rattlesnake Mountain in the northern part of the San Bernardino Mountains of southern California. In this case the quartz monzonite body appears to have moved upward by shearing of largely solid blocks in a funnel-shaped structure. Comparison of its features with Ramberg's centrifuged models of buoyancy domes show many similarities.

Plutonic rocks of complex origin

General

We are concerned here chiefly with complexes of rocks known throughout the world, which appear to have been derived by processes not entirely igneous nor entirely metamorphic but containing strong elements of both. In particular the metamorphic element usually consists of evidence of large-scale metasomatism or "granitization," although the latter term is not always appropriate since the rocks concerned are frequently far from granites. Some of these controversial rocks, such as the great

[7] See Klepper *et al.*, 1974.

anorthositic massifs of the pre-Cambrian shield areas, contain clear evidence of a magmatic history which may link them to the layered gabbroic complexes. However, they are also products of considerable deformation and recrystallization in the solid state, and some authors regard them as being products of metamorphism.

Generally the replacement crystallization of the metasomatic complexes has been synkinematic or contemperaneous with intense folding and shearing. Generally, also, they are confined to rocks of amphibolite to granulite facies rank. Frequently the end product is granite, tonalite, quartz diorite, or, as in the case of the Idaho batholith, even more basic rocks. In many cases the replaced rocks are part of a miogeosynclinal pile which includes beds of carbonates. The latter serve as excellent marker horizons since they are resistant to metasomatism.

The metasomatic complexes have always been controversial as is attested by a number of classical works such as "The granite controversy" by Read (1957). We do not intend here to review this voluminous literature but present as examples a number of classic areas that have been subjected to detailed and careful investigations. Since we wish to distinguish the metasomatic rocks from the products of ordinary metamorphism and magmatism, it is well to consider the following criteria:

1. Evidence of chemical replacement of preexisting structures and minerals
2. Compatibility or lack of compatibility with possible metamorphic differentiation processes which involve only local mass transport
3. Compatibility or lack of compatibility with possible anatectic processes which involve only local mass transport
4. Compatibility or lack of compatibility of minerals with subsolidus temperatures
5. Evidence or lack of evidence of fluid magma as distinguished from solid plastic or fluid flow

These criteria should be reviewed in relation to the chemical mass-transport properties discussed in Chapters 3 and 8.

Nanga Parbat area

This area (Misch, 1949), located in the northwestern Himalayas, is one of the classic areas of granitization. A thick sequence of Precambrian argillites with calcareous and basic layers underwent progressive metamorphism that resulted in rocks ranging from slates and phyllites on the periphery to sillimanite paragneisses in the core of the complex. The dominant-introduced component was K_2O, and metasomatism by this component became important in the kyanite zone of metamorphism. Although the paragneisses of the complex have been transformed to granitic augen gneisses, the calcareous marbles have resisted transformation and serve as marker horizons. Metamorphism and metasomatism were essentially synkinematic, as is shown by such features as the porphyroblasts of the augen gneisses.

North Cascades, Washington State

This area, also studied by Misch (1968), consists of regionally developed migmatitic gneisses, biotite schists, and amphibolites. In contrast with the Nanga Parbat area, metasomatism involved the introduction of Na_2O as well as SiO_2, resulting in leucocratic trondhjemitic and quartz diorite gneisses. An epidote-bearing subfacies[8] contains sodic andesine, while an epidote-free subfacies contains plagioclase ranging from oligoclase to bytownite. The presence of staurolite and other minerals indicates Barrovian-style metamorphism at approximately 600°C according to Misch. The major criteria for large-scale metasomatism listed by Misch include the following:

1. There is no wholesale basification of schists and amphibolite remnants such as might result from anatexis or pervasive local metamorphic differentiation.
2. Generally, schist and amphibolite-derived leucocratic gneiss sequences contain only subordinate remnants of their parent rocks.
3. Plagioclase compositions range widely depending on the parent rock (Figure 14.19), while bulk compositions of the leucocratic layers do not correspond to what one would expect from anatectic melts.
4. In addition, the ratios of leucocratic gneiss to remnant material are too high for the former to have split off from the latter.
5. Finally, most of the mineral assemblages indicate temperatures of crystallization too low for melts of the compositions of the gneissic layers.

Although these characteristics also appear to rule out large-scale anatexis, they are consistent with some metamorphic differentiation as well as large-scale metasomatic introduction of material.

[8] This is merely a descriptive term as used here.

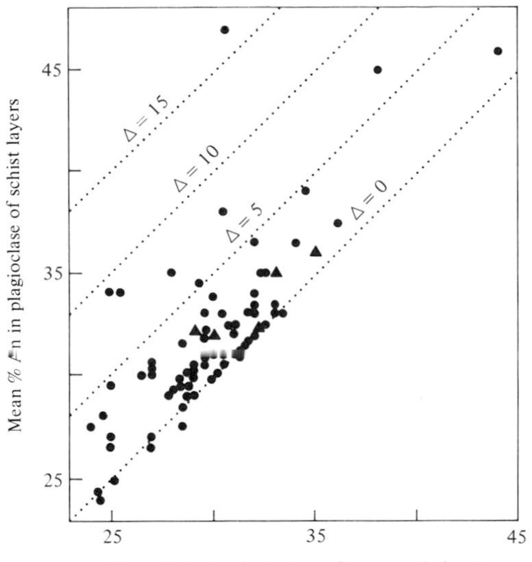

14.19 Correlation between mean plagioclase compositions (mole % An) in schist and leucocratic gneiss layers of 76 lit-par-lit–type samples from Skagit gneiss, epidote-free amphibolite facies. Dark layers (●) are (garnet)-biotite schists, a few hornblende-bearing, and a few (▲) cummingtonite-bearing. Δ = mean percentage An of plagioclase in dark layers minus mean percentage An of plagioclase in leucocratic layers. (After Misch, 1968.)

Northwest Adirondacks

This area of intensely folded and thoroughly metamorphosed metasediments (Grenville geologic province) has been studied in some detail by Engel and Engel. In their study of the major paragneiss (Engel and Engel, 1958), they found that the dominent metasomatic effect was the enrichment of biotite-plagioclase gneisses in K_2O and a depletion in TiO_2, FeO, MgO, CaO, and H_2O. As in the previously cited examples, the development of migmatites and porphyroblastic feldspar gneisses is prominent. Also abundant are calcitic marbles, which tend to resist the metasomatism. The Grenville terrain is also noted for its well-developed alaskitic granitic phacoliths described in the classic studies of Buddington (1929), who believed them to be of strictly magmatic origin. However, Engel and Engel (1963) concluded on the basis of the structurally continuous gneissic, amphibolitic, and marble remnants in and around the phacoliths that the alaskitic granites were largely of metasomatic origin. An almost universal characteristic of the Grenville rocks as observed in several seasons of field work by one of us (Mueller), is the highly fluid flow structure shown by the marble beds and the almost total lack of such structures in the feldspathic rocks (usually syenites) where these are in contact with the marbles. These features, in agreement with the conclusions of Engel and Engel, seem to indicate that the feldspathic rocks only rarely, if at all, were in a molten state in such locations.

Idaho batholith

Very widespread and complex metasomatic phenomena have been described from metasedimentary rocks in the vicinity of the Idaho batholith. In this area it was found by Hietanen (1962) that pelitic schists and quartzites underwent basification near the batholithic plutons. FeO, MgO, CaO, Al_2O_3, and Na_2O were introduced, while SiO_2 and K_2O were lost. On the basis of structural and textural criteria it was inferred by Hietanen that metasomatism was both synkinematic and postkinematic. Especially interesting is the occurrence of large bodies (10 to 20 km in length and 2.5 to 5 km in width) of anorthosite (Hietanen, 1963d) within highly aluminous pelitic schists and associated with quartzite, amphibolites, and lime silicate rocks. Schist layers, which occur as inclusions in the anorthosite, contain kyanite, andalusite, sillimanite, muscovite, staurolite, and garnet. The anorthosite itself contains two plagioclases—an early bytownite (present in small grains) and a later andesine (present in larger grains). Although some of the minerals present are consistent with metamorphism of an igneous anorthosite, it is difficult to explain the occurrence of the plagioclase in this way, as was pointed out by Hietanen. According to her the most likely origin for the anorthosite is as a product of the metamorphism of shaly limestone to yield andesine. While some metasomatic effects in the metasediments appear to be directly related to the igneous intrusives, much metasomatism is also evident at a distance from these bodies. However, there appears to be a general tendency for the end-products of metasomatism to resemble the igneous quartz diorites and tonalites of the batholith proper.

Southern California batholith

Northeast of the classical magmatic province described by Larsen (1948), the plutonic rocks of the southern California batholith assume a different face. There, between the North Pinyon Mountains and the Santa Rosa Mountains, occur regionally developed oligoclase-biotite quartz gneisses with interlayered remnants of

Paleozoic carbonaceous pelitic schists, calcite marbles, and lime-magnesian silicate skarns. Every stage of transition between the gneisses and metasediments can be observed. Here, as in the other examples discussed, calcite marbles resisted the metasomatism although the calcite has been depleted of all but traces of $MgCO_3$ (Mueller and Condie, 1964). As usual, the metasediments show evidence of intense folding and plastic deformation, and augen gneisses and migmatites are well developed. As in the case of the Idaho rocks, Na_2O rather than K_2O appears to have been introduced, and there is considerable resemblance to the tonalite magmatic rocks to the southwest. However, to date no detailed assessment of chemical mass balances has been made to establish the extent and type of metasomatism involved.

Anorthosite-mangerite complexes

These rocks are widespread throughout the Precambrian shield areas, particularly in eastern Canada, but occur throughout the world. The best-known example is the Adirondack massif in the eastern U.S., which is discussed in an important memoir by Buddington (1939).

The anorthosites consist almost entirely of intermediate plagioclase in the range of oligoclase-andesine but also contain minor amounts of ferromagnesian minerals. They are associated with subordinate gabbros and abundant quartz syenite (mangerite). Characteristically they are the loci of magnetite-ilmenite ore bodies. Profound deformation as shown by isoclinal folds and catoclastic textures is the rule. The mineralogy conforms to amphibolite-granulite facies rank while the compositions of certain coexisting minerals such as pyroxenes and feldspars indicate that these have survived from an early magmatic period. (Philpotts, 1966). According to structural studies by Martignole and Schrijver (1970), the Morin anorthosite of Quebec shows evidence of emplacement by buoyancy forces. Some authors (Hargraves, 1962) believe the mangeritic rocks to be of metasomatic origin, although there is evidence in the form of variation diagrams (Philpotts, 1966) that the entire suite is related and the product of fractional crystallization. Although some authors have been puzzled by the high ratio of acid and intermediate rocks to gabbro, the explanation of this appears to be similar to that already given for the same phenomenon in calc-alkali batholiths. It appears that in this case also the rocks found their way to the surface in some inverse proportion to their densities.

It is interesting that the members of this suite studied by Philpotts (1966) show fractionation paths that fall between those of the gabbroic complexes and the calc-alkali batholiths (Figure 14.4). It is thus possible that the parent magma of the anorthosite-mangerite suite had a water content greater than the gabbroic magma but less than the calc-alkali magma.

It seems possible that the anorthosite-mangerite suite may represent a deep-seated equivalent of the layered gabbroic complexes. If the parent magma of the former were intruded at a depth corresponding to the lower crust, the lack of strength of the wall rocks might prevent the stabilization to a form (lopolith or funnel) where quiescent differentiation into layers could occur. Instead the production of virtually monomineralic anorthosites by a crystal fractionation system outlined previously (Figure 14.12) would result in a gravitational instability. The anorthositic layers would move upward under buoyant forces creating plutons similar to batholiths. The mangeritic rocks formed later in the sequence would respond in a similar way. Philpotts (1966) suggested that the anorthosite-mangerite rocks are deep-seated equivalents of the calc-alkali suite. As explained here, it is possible that they are related to gabbros as well.

15 Volcanism and volcanic fluids

Introduction

It is only in volcanic processes that we directly experience magmatism and the fluid emanations of magmas. Consequently, it is of the greatest importance to petrology to extrapolate what we learn from these processes to the fossil products of magmatism in geologic deposits. Yet volcanism differs greatly from plutonism, and its products usually bear the stamp of that geologically unique discontinuity between the lithosphere and the relatively cool and oxidizing atmosphere or hydrosphere.

We previously concluded (Chapter 12) that basaltic magmas have their ultimate source within the mantle. However, we also have evidence from the field that magmas may undergo differentiation not only within deeper regions of the crust but also in chambers that approach closely to the surface under volcanic edifices. Differentiation under these circumstances follows a quite different course than it would for the same magma at depth.

There are, however, certain invariant characteristics of magma that are common to both extrusive and intrusive rocks. For example, it appears that the non-volatile bulk chemistry of the average volcanic rock does not differ greatly from its deep-seated equivalent. In particular, basic lavas in general and acid lavas quite frequently retain oxidation states not too dissimilar from those deduced for plutonic magmas. However, where elements of limited solubility, such as H, S, C, Cl, and F, are concerned, it is likely that extensive loss or exchange with the surroundings occurs long before the surface is

15 Volcanism and volcanic fluids

attained by the magma. Frequently, evidence for this volatile loss is apparent in the breakdown and alteration of volatile-bearing phenocrysts and in the formation of vesicles.

Although the products of volcanism are of almost infinite variety, they may be interpreted with considerably greater ease than those of plutonic magmatism. This is of course a direct consequence of the fact that they are formed under the low-pressure and short-time-scale conditions familiar to us from ordinary experience and from relatively unsophisticated laboratory experiments. Typical of these products is the abundant glass which may show to varying degrees signs of gas evolution and simple volume expansion in the form of vesicles. Other typical products of volcanism are those of mechanical brecciation and liquid flow. By contrast, these same features in plutonic rocks are obscured by postconsolidation plastic and fluid deformation, recrystallization, and metasomatism.

It is an elementary deduction of field geology that volcanic structures vary greatly in response to the chemistry that imparts to the magma characteristic viscosity, density, and consolidation features. Pyroclastics are generally associated with acid or alkaline volcanism, and highly fluid lavas appear to occur only in the basic range of compositions. However, certain products, such as ash and pumice, result from a wide range of magma compositions, although their relative volumes and modes of deposition vary greatly.

Because of the abundance of free H_2O, N_2, and O_2 in the upper crust and atmosphere, the opportunity of contamination of magmas and magmatic fluids by these constituents is always high. In addition, there is also a high probability that magmatic gases may be contaminated by their own condensation or distillation products. We shall see that both types of contamination are evident even in those gases given off directly by fresh lavas in volcanic vents. It is apparent then that in the cooler fumaroles and hot springs more remote from the vents, the contaminants may greatly dominate over original magmatic fluids. In fact, many studies have made clear that the chemistry of such waters is apt to be more a function of the rock types through which they have passed than of the magma from which their thermal energy was derived. Finally, because of their complex involvement with surficial crustal waters (including those of the oceans) and because they are frequently in contact with flows and pyroclastics for long periods, the magmatic fluids may bring about profound secondary changes in rocks that originally represented congealed magmas on the liquid line of descent. Consequently, such secondary effects must always be taken into account when attempts are made to decipher trends of magmatic evolution in volcanic deposits. Two of the most frequently encountered products of these alterations are the welded tuffs which are enriched in K_2O and the spilitic suite of volcanics which appear to result from the interaction of magma and seawater. However, such alterations are more properly considered as forms of metamorphism.

Volcanic and hypabyssal mineralogy

The primary mineral species that regularly occur in volcanic rocks are, with few exceptions, the same as those in plutonic equivalents. However, the volcanic *varieties* of these minerals may differ greatly in detail.

The characteristic of volcanism that especially influences the crystal structures and forms of volcanic minerals is the rapid rate of cooling from high temperatures. The characteristics that primarily govern the stabilities of certain minerals are the low volatile pressures and high oxidation states that result from volatile escape and interaction with surface fluids. Obviously only the hottest of magmas reach the surface, since low temperature melts can exist only under high volatile pressures. Rapid cooling from these high temperatures then tends to quench in the high-temperature structural states of such minerals as feldspar and pyroxene so that monoclinic sanidine, "high plagioclase," and supersolvous feldspars and pyroxenes are characteristic. Many volcanic pyroxenes also show a disordering of Mg^{2+} and Fe^{2+}, which reflects high temperatures (Saxena, 1973). Finally, the high-temperature forms of silica—cristobalite and tridymite—are found in the hottest rocks. However, evidence of the high-temperature β form of quartz is retained only in the morphology of phenocrysts since inversion to the low-temperature form occurs far too rapidly for quenching of the β phase to be possible.

The oscillatory compositional zoning of plagioclase already noted in plutonic rocks is even more highly developed in the phenocrysts of volcanics, where it may register a complex series of events. An example of this is the phenocryst from a volcanic ash sequence of the Taupo area, New Zealand, as shown in Figure 15.1. According to Ewart (1963), the sharp changes in anorthite content of the zones of this crystal can be correlated with the diminished water pressure attending eruption of the ash showers.

The fractionation trend in the subsystem albite-anorthite-orthoclase of volcanic rocks is strikingly different from the corresponding trend in plutonic rocks.

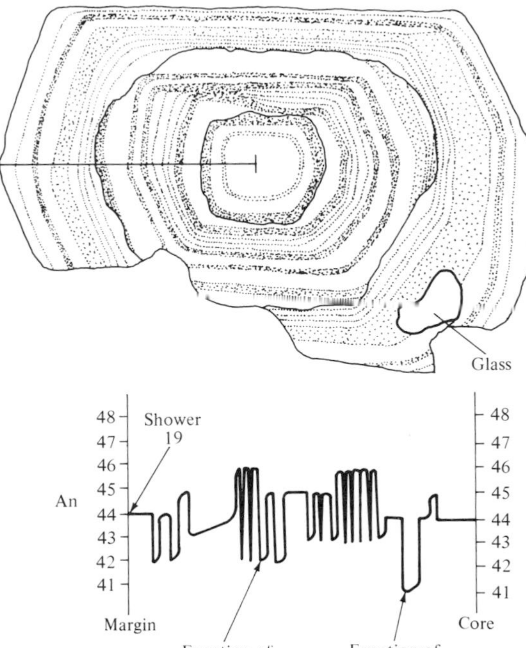

15.1 Plagioclase phenocryst from volcanic ash of Toupo area New Zealand. The changes in anorthite content registered can be correlated with other phenocrysts and probably reflect fluctuations in water pressure. (After Ewart, 1963.)

In the latter, increasing water content of the melt, as differentiation proceeds, causes the intersection of the solidus with the solvus to occur at successively lower temperatures and the feldspars diverge in composition (Chapter 14). In volcanic rocks, on the other hand, the temperature drop is much less and the compositions of coexisting feldspars may converge. The latter effect has been discussed by Carmichael *et al.* (1974).

However the most striking effects of the volcanic environment are seen in the hydrous ferromagnesian silicates such as the amphiboles and micas which are subject to simultaneous dehydration and oxidation. Some pertinent chemical reactions of these phases are as follows:

$$\underset{\textit{hornblende}}{3Ca_2Fe_5Si_8O_{22}(OH)_2} + \underset{\textit{hornblende}}{2Ca_2Mg_5Si_8O_{22}(OH)_2} + \underset{\textit{fluid}}{\tfrac{5}{2}O_2} \rightleftharpoons \underset{\textit{pyroxene}}{10CaMgSi_2O_6} + \underset{\textit{ferrite}}{5Fe_3O_4} + \underset{\textit{silica}}{20SiO_2} + \underset{\textit{fluid}}{5H_2O} \qquad (15.a)$$

$$\underset{\textit{hornblende}}{3Ca_2Fe_5Si_8O_{22}(OH)_2} + \underset{\textit{hornblende}}{2Ca_2Mg_5Si_8O_{22}(OH)_2} \rightleftharpoons \underset{\textit{pyroxene}}{10CaMgSi_2O_6} + \underset{\textit{ferrite}}{5Fe_3O_4} + \underset{\textit{silica}}{20SiO_2} + \underset{\textit{fluid}}{5H_2} \qquad (15.b)$$

$$\underset{\textit{hornblende}}{CaFe_5^{2+}Si_8O_{22}(OH)_2} \rightleftharpoons \underset{\textit{oxyhornblende}}{CaFe_3^{2+}Fe_2^{3+}Si_8O_{24}} + H_2 \qquad (15.c)$$

$$\underset{\textit{biotite}}{KFe_3AlSi_3O_{10}(OH)_2} + \underset{\textit{fluid}}{\tfrac{1}{2}O_2} \rightleftharpoons \underset{\textit{sanidine}}{KAlSi_3O_8} + \underset{\textit{ferrite}}{Fe_3O_4} + \underset{\textit{fluid}}{H_2O} \qquad (15.d)$$

$$\underset{\textit{biotite}}{KFe_3AlSi_3O_{10}(OH)_2} \rightleftharpoons \underset{\textit{sanidine}}{KAlSi_3O_8} + \underset{\textit{ferrite}}{Fe_3O_4} + \underset{\textit{fluid}}{H_2} \qquad (15.e)$$

$$\underset{\textit{biotite}}{KFe_3^{2+}AlSi_3O_{10}(OH)_2} \rightleftharpoons \underset{\textit{oxybiotite}}{KFe^{2+}Fe_2^{3+}AlSi_3O_{12}} + \underset{\textit{fluid}}{H_2} \qquad (15.f)$$

These reactions may be regarded as either equilibrium or disequilibrium processes. However, we should note that Reactions (15.a) and (15.b) are related by the reaction for the decomposition of water, and the same holds for Reactions (15.d) and (15.e). Analytical expressions for the equations of equilibrium for these reactions are difficult to write down, and this is especially true for Reactions (15.c) and (15.f), which involve the complex oxyhornblende and oxybiotite formula units. However, for purposes of illustration we may assume that Fe_3O_4 and $KAlSi_3O_8$ occur as pure phases, and that the ferromagnesian silicates are ideal solutions of their Mg^{2+} and Fe^{2+} end members. Then for Reactions (15.b) and (15.e) we may write

$$K_{(b)} = \frac{(X_{Mg}^{Cpx})^{10} f_{H_2}^5}{(X_{Fe}^{Hb})^{15}(X_{Mg}^{Hb})^{10}} \qquad (15.1)$$

$$K_{(e)} = \frac{f_{H_2}}{(X_{Fe}^{Bi})^3} \qquad (15.2)$$

Since generally $X_{Mg}^{Cpx} \simeq X_{Mg}^{Hb}$, Equation (15.1) simplifies to

$$K_{(b)}^{1/5} = \frac{f_{H_2}}{(X_{Fe}^{Hb})^3} \qquad (15.3)$$

which has the same form as Equation (15.2).

Reaction (15.c) has been subject to experimental studies by Wones and Eugster (1965), who concluded that Reaction (15.f) is also of importance, so that a quantity of oxybiotite, which depends on the temperature, fluid pressure, and oxygen fugacity, is always present as a solution component of biotite. This is in accord with the earlier observations of Larsen *et al.* (1937) on biotite and hornblende phenocrysts from the lavas of the San Juan volcanic province of Colorado.

If the foregoing reactions and equations are accepted, their consequences may be summarized as follows:

1. A decrease of the total fluid pressure on the extrusion of lavas results in all the reactions being displaced to the right so that the net effect is one of oxidation, even when no oxygen is introduced into the system.
2. If equilibrium is maintained, both hydrous and anhydrous ferromagnesian minerals, which crystallize under low fluid pressures or high oxygen pressure, will be enriched in MgO and Fe_2O_3 relative to their plutonic equivalents.
3. A drop in fluid pressure and an increase in f_{O_2} (at constant fluid pressure) leads from a hydrous to an anhydrous Mg-rich assemblage. This is the reverse of Bowen's well-known reaction series for plutonic rocks.

The disequilibrium versions of these reactions are frequently at work in volcanic rocks where they are revealed in the striking changes which phenocrysts undergo. Thus, although fresh ordinary biotites and hornblendes may be preserved in the dense, glassy parts of flows, in the broken and vesiculated parts these are likely to be replaced by their "oxy" counterparts. Where the alteration has only begun, the latter are distinguished by their aberrant optical properties, but where it has gone far enough, the entire hornblende or biotite has given way to a mixture of the anhydrous reaction products feldspar, magnetite, and pyroxene.

The affinity of the silicate melt for water aids in maintaining the iron in the ferrous state. Indeed, this effect appears to be so strong that virtually pure fayalite may crystallize from rhyolitic melts. However, just as in the cases of hornblende and biotite, these crystals frequently show evidence of oxidation according to the reaction

$$\underset{\textit{fayalite}}{\tfrac{3}{2}Fe_2SiO_4} + \underset{\textit{fluid}}{H_2O} \rightleftharpoons \underset{\textit{ferrite}}{Fe_3O_4} + \underset{\textit{quartz}}{\tfrac{3}{2}SiO_2} + \underset{\textit{fluid}}{H_2} \tag{15.g}$$

The occurrence of fayalite and its alteration products has been noted by Boyd (1961) in the welded tuffs and obsidian flows of Yellowstone Park, Wyoming.

Phase relations of the major leucocratic minerals of volcanic rocks are best understood in terms of Figures 15.2, 15.3, and certain diagrams already presented in Chapters 12 and 14. A characteristic of both Figures 15.2 and 15.3 is the thermal divide of the feldspar join which separates the depressions into which liquids descend by fractional crystallization. A consequence of this divide is as follows: It may be imagined that liquids originally of basaltic composition approach the triangle $NaAlSiO_4$-$KAlSiO_4$-SiO_2 from multidimensional space in which such components as anorthite are important. The vicinity at which these liquids approach the triangle depend on their SiO_2 contents, and it is clear that small variations of this component may throw the liquids on one or the other side of the thermal divide, giving rise either to rhyolitic or phonolitic resdua. Sometimes, however, it happens that the melt ends up directly above the divide, or, in other words, turns out to be trachytic As might be expected, such end-products usually contain substantial amounts of anorthite.

An extensive and illuminating discussion of equilibrium and fractional crystallization paths in the system $NaAlSiO_4$-$KAlSiO_4$-SiO_2-H_2O has been given by Fudali (1963). In the illustrated crystallization paths shown in Figure 15.2 the path *M-N-O* illustrates equilibrium crystallization, while path *S–T* illustrates fractional

15.2 Molar plot of the system $NaAlSiO_4$-$KAlSiO_4$-SiO_2, Paths of equilibrium and fractional crystallization are also shown. (After Schairer, 1957.) Isotherms are in °C, P = 1 bar.

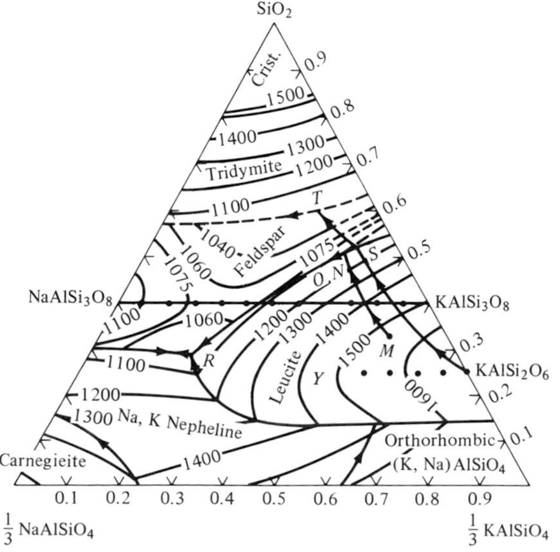

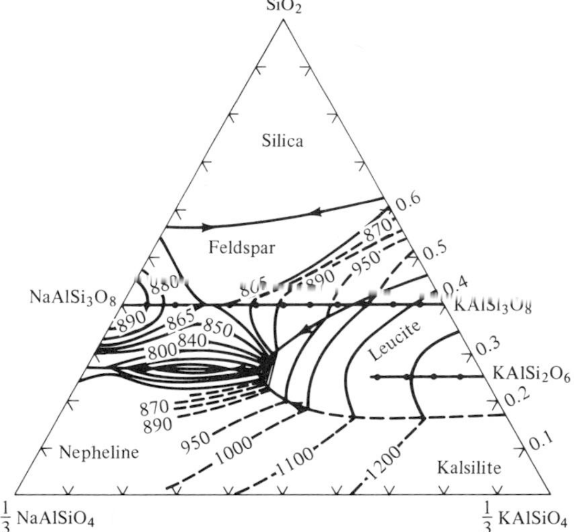

15.3 Molar plot of the system $NaAlSiO_4$-$KAlSiO_4$-SiO_2-H_2O. (After Fudali, 1963.) Paths of equilibrium and fractional crystallization are also shown. Isotherms in °C, P = 1 bar.

crystallization in which a liquid that originally precipitates leucite finally reaches the feldspar quartz cotectic and precipitates quartz.

It should be observed that while the positions of the thermal depressions are not shifted greatly by 1 kbar water pressure, the field of leucite is greatly restricted thereby. Also, we know from field experience that although leucite is a common primary crystallization product in volcanic and hypabyssal rocks, its former existence is usually manifested only by mixtures of alkali feldspar and nepheline, known as *pseudoleucite*. Furthermore, leucite in any form is unknown from plutonic rocks. This behavior of leucite cannot be understood in terms of its liquidus relations as presented in Figures 15.2 and 15.3, but must be considered in terms of subsolidus reactions (Fudali, 1963). In fact, both pseudoleucite and the absence of leucite in plutonic rocks can be explained by a reaction of the type

$$2KAlSi_2O_6 \rightleftharpoons KAlSi_3O_8 + KAlSiO_4 \quad (15.h)$$

leucite — *feldspar* — K-*nepheline*

A variant of this reaction was studied experimentally by Scarfe *et al.* (1966), who found that at 650°C, 5 kbar of pressure stabilized feldspar + hexagonal kalsilite, the latter being the equivalent of K-nepheline in natural systems. This effect of pressure might have been anticipated from the fact that the volume decrease of Reaction (15.h) is almost 7 cm^3 under STP conditions. Furthermore, Scarfe *et al.* (1966) also found that the addition of Na_2O to the system further stabilized the right-hand side of the reaction so that the field of leucite might be eliminated by only modest pressures in these systems.

In the formation of pseudoleucite the homogeneous sodium-rich leucite that crystallizes from the melt breaks up into even more sodium-rich nepheline and a potassium-rich feldspar; however, in the process the leucite becomes increasingly enriched in the $KAlSi_2O_6$ component.

Fractionation trends in a broad spectrum of silica-undersaturated volcanic rocks from the East Otago volcanic province of New Zealand have been summarized by Coombs and Wilkinson (1969). In this province they found evidence for several lineages depending on $(FeO + Fe_2O_3)/MgO$ and K_2O/Na_2O ratios, as well as the degree of silica undersaturation. Fractionation trends in both sodic and potassic series indicated that the degree of undersaturation of the salic liquids is dependent on that of the parent magma, which could be slightly undersaturated alkali basalt or more undersaturated basonite. Additional factors such as f_{O_2} are also inferred to affect the course of fractional crystallization.

In magmas greatly deficient in silica and high in alkalis, extreme rock types such as nephelenite, ijolite, and carbonatite are products of volcanism. A well-known example from Tanganyika is the volcano Oldoinyo Lengai (Dawson, 1962a, b). Especially interesting is the sodium carbonate, which in combination with calcite forms lava flows. Sodium carbonate is unknown in deep-seated carbonate rocks. According to Koster van Groos (1975) it is likely that only alkali carbonatite magmas can be stable at shallow depths within the crust, while derivation of such melts from the mantle would require that they had contained substantial quantities of Mg and Fe^{2+} at their point of origin. These constituents might however be lost in transit by reaction with various silicates.

Volcanoes and volcanic deposits

General

The lavas and pyroclastics of continental deposits are derived largely from calc-alkali series magmas similar to those that give rise to the batholiths. Where the series is best developed the vents and centers of activity are aligned along the major orogenic trends, but particularly along the island arcs and continental borders of the circumpacific region. The products of volcanism in such regions are only occasionally of an alkaline character as in Eastern Otago, New Zealand (Benson, 1941) and in central Montana (Larsen, 1940). The major fields of alkaline volcanism occur largely in regions of non-orogenic rifting, as in eastern Africa.

The character of volcanism in the ocean basins and on the oceanic ridges that cross them is still poorly known, but information is rapidly being accumulated. The major basins of the Pacific and the Atlantic appear to be dominated by vast outpourings of tholeiitic basalts which resemble rather closely those of continental deposits, except that they may be somewhat lower in alkalies (Table 15.1). By contrast, the oceanic alkaline rocks appear to be more restricted, although they received much attention in early petrographic descriptions of oceanic rocks. Descriptions and occurrences of alkaline basalts have been summarized by Wilkinson (1967). It has been suggested (Engel and Engel, 1964) that the oceanic alkaline rocks are largely confined to the tops of volcanic edifices and that they are the products of differentiated tholeiitic magma.

In the continental orogenic regions the dominant member of the calc-alkali series is andesite, and this is especially true of the geologically young circumpacific

Table 15.1 Relatively unaltered volcanic rocks of diverse origins and common types

	Porphyrite bytownite basalt[a]	Glassy to micro-crystalline basalt[b]	Labradorite olivine basalt[c]	Labradorite alkali basalt[d]	Oligoclase anorthoclase trachyte[e]	Basaltic pumice[f]	Basalt[g]	Porphyritic augite olivine basalt[h]
	1	2	3	4	5	6	7	8
SiO_2	48.53	49.80	47.13	48.00	60.30	50.21	50.87	50.17
Al_2O_3	22.30	14.88	12.54	17.42	18.08	13.36	12.92	19.65
Fe_2O_3	0.69	1.55	4.72	6.17	4.21	1.39	2.85	2.54
FeO	4.82	10.24	6.26	4.64	1.17	9.88	13.19	7.92
MgO	7.14	6.74	13.19	4.55	1.38	8.34	5.40	4.54
CaO	12.86	10.72	10.17	9.60	3.14	10.81	9.02	11.29
Na_2O	2.18	2.91	2.25	4.00	6.80	2.34	2.46	1.66
K_2O	0.06	0.24	0.65	1.30	2.91	0.55	0.76	0.29
H_2O^+	0.38	0.54	0.35	0.42	0.39	0.11	0.17	0.41
H_2O^-	0.01	0.06	0.30	0.15	0.38	0.00	0.04	0.20
TiO_2	0.76	2.20	1.84	3.20	0.94	2.63	2.95	0.87
P_2O_5	0.07	0.28	0.36	0.54	0.29	0.27	—	0.10
MnO	0.16	0.21	0.19	0.13	0.18	0.17	0.26	0.17
CO_2	—	—	—	—	—	0.01	—	—
Cl	—	—	—	—	—	0.02	—	—
F	—	—	—	—	—	0.04	—	—
BaO	—	—	—	—	—	—	—	—
ZrO_2	—	—	—	—	—	—	—	—
SO_3	—	—	—	—	—	—	—	—
	99.96	100.19	99.95	100.12	100.17	100.11	100.89	99.81

[a] Porphyritic bytownite basalt with glassy to microcrystalline groundmass, East Pacific Rise, depth 1700 m (Engel and Engel, 1964).

[b] Glassy to microcrystalline basalt, East Pacific Rise, depth 2300 m (Engel and Engel, 1964).

[c] Labradorite olivine basalt flow, Guadalupe Island, East Pacific (Engel and Engel, 1964).

[d] Labradorite, alkali basalt flow, Guadalupe Island, East Pacific (Engel and Engel, 1964).

[e] Oligoclase-anorthoclase, trachytic flow, Guadalupe Island, East Pacific Rise (Engel and Engel, 1964).

[f] Basaltic pumice, Makaopuhi Crater, Kilauea Volcano, Hawaii (Wright *et al.*, 1968).

[g] Basaltic lava, Askja Volcano, Iceland (Thorarinsson and Sigvaldason, 1962).

[h] Augite-olivine basalt (with microphenocrysts of hypersthene). Taga Volcano near Niraymа, Japan (Tsuy, 1937; as quoted by Kuno, 1950).

[i] Olivine-bearing hypersthene-augite andesite, Hakone Volcano, Japan (Kuno, 1936; as quoted by Kuno, 1950).

belt. However, even within the andesitic chains, basalt is an ever-present and abundant constituent. There it not only participates in building the large strato volcanoes but also contributes numerous valley flows and cinder cones. Frequently such basaltic flows and cones are the last lava outbreaks in a region of declining activity.

Although andesitic volcanism dominates present activity, there is much evidence that in certain parts of the orogenic belts, such as in New Zealand, Sumatra, and the Yellowstone and Nevada regions of the USA, rhyolitic magmas were more prevalent in the past. This activity is recorded for us in numerous thick and widespread deposits of welded and unwelded tuffs, and as flows of rhyolite and obsidian. By contrast, active rhyolitic volcanism is more restricted today, and although the violent outbursts, such as at Mt. Katmai, have attracted much attention, the volume of their deposits is small compared with the prehistoric pyroclastics.

Lava flows

We must distinguish here between the general term *lava*, which refers to extrusive magma, and *lava flows*, which result from the flow of lava as liquid or liquid-crystal

Hypersthene augite andesite[i]	Hypersthene hornblende quartz dacite[j]	Posphyrite adamellite[k]	Ground mass of (11)[l]	Quartz latite tuff[m]	Rhyolite tuff[n]	Rhyolite lapilli[o]	Glassy welded tuff[p]	Obsidian flow[q]
9	10	11	12	13	14	15	16	17
57.07	76.05	65.31	75.58	68.7	76.6	71.2	73.15	76.77
17.53	12.79	15.62	12.18	16.7	12.7	13.2	12.13	11.71
2.59	1.47	1.46	1.27	1.4	0.86	0.75	1.12	0.64
5.44	0.31	3.51	0.61	0.74	0.07	1.35	0.66	1.05
3.87	0.08	1.79	0.02	0.45	0.49	0.30	0.13	0.33
8.77	1.60	3.99	1.06	1.1	0.63	1.35	0.71	0.49
2.80	3.89	3.36	2.86	4.3	3.4	4.75	4.01	4.01
0.52	1.82	3.31	6.00	6.0	5.1	2.85	5.27	5.10
0.27	0.74	0.82	0.54	—	—	2.63	1.91	0.10
0.12	0.52	nil	nil	—	—	0.74	0.21	0.16
0.77	0.19	0.59	0.11	0.41	0.10	0.25	0.14	0.15
0.08	tr	0.24	0.22	0.10	0.02	0.04	0.00	0.00
0.14	0.05	0.11	0.02	0.10	0.06	0.10	0.04	0.05
—	—	—	—	—	—	<0.05	0.00	0.00
—	—	—	—	—	—	0.095	0.06	0.06
—	—	—	—	—	—	0.043	—	—
—	—	—	—	—	—	—	0.08	0.02
—	—	—	—	—	—	—	0.04	0.04
—	—	—	—	—	—	—	0.00	0.03
99.97	99.51	100.11	100.47	100.0	100.0	99.61	99.66	100.71

[j] Hypersthene-hornblende-quartz-dacite, dacite dome near Kamitaga (Kuno, 1936; as quoted by Kuno, 1950).

[k] Adamellite-porphyrite (border phase D16) New South Wales, Australia (Wilkinson *et al.*, 1964).

[l] Microcrystalline groundmass of 11.

[m] Quartz-latite tuff (Topopah Spring member of Paintbrush tuff, southern Nevada) (Lipman, 1966).

[n] Rhyolite tuff (Topopah Spring member of Paintbrush tuff, southern Nevada) (Lipman, 1966).

[o] Rhyolite Lapilli Quaternary pumice ash, Taupo area, New Zealand (Ewart, 1963).

[p] Glassy welded tuff, base of Yellowstone tuff, Yellowstone Park, Wyoming (Boyd, 1961).

[q] Obsidian flow, Obsidian Cliff, Yellowstone Park, Wyoming (Boyd, 1961).

suspensions over the surface of the ground. This distinction is necessary because we frequently encounter forms of lava that are not extruded as liquids but are fragmented and congealed within the vent as they are ejected in either a violent explosion (*nuée ardente*) or as more quiescent pumice flows. The products of such activity are classed among the *pyroclastics*.

In terms of composition, the lavas—both flow and pyroclastic, as well as hypabyssal magmas—form series that are closely parallel to the plutonic magma series. This has already been pointed out in Chapter 4, where a direct comparison of average plutonic and volcanic rocks was made. Further examples of certain select flows and pyroclastics are given in Tables 15.1 and 15.2.

It appears that magmas of virtually any composition may be extruded as freely flowing liquid-crystal suspensions, although the possibility of this happening seems to diminish greatly with a decrease in the mafic constituents and an increase of highly polymerized leucocratic constituents such as silica, feldspar, and the feldspathoids. It has been pointed out by Verhoogen (1946) that phenocrysts are almost always present in lavas, indicating that liquidus temperatures are rarely exceeded. Yet there are examples of almost phenocryst-free rocks representing lavas ranging from basalt to rhyolite, and it is not uncommon to find obsidian flows that are virtually pure glass. Because of the generally low volatile contents of these obsidians (Table 15.1), the lava temperatures must have been quite high to enable them to flow over the surface.

Unfortunately, reliable lava-temperature measurements are scarce and are almost confined to basaltic varieties. The maximum temperatures recorded are around 1200°C, although Zies (1946) found an extreme of 1250°C at Paricutin. It has been shown, however, that usually high temperatures may result from exothermic gas reactions at the surface.

Attempts have also been made to measure temperatures of various lavas and pyroclastics by indirect methods based on the mineralogy. Ewart *et al.* (1971) utilized the compositions of coexisting titanomagnetite–ilmenite solid solutions to estimate the temperatures of formation of glassy acid pumices, lavas, and ignimbrites of the Pleistocene–Recent Central volcanic region of New Zealand, and obtained temperatures ranging from 695°C to 890°C, which is in the range to be expected from such rocks. Similar results were obtained by Heming and Carmichael (1973) for pyroclastics of the Roboul Caldera in New Guinea.

As we have noted in connection with viscosity measurements, anhydrous basalts begin to lose their fluidity between 1100°C and 1200°C when crystallization sets in; but as was shown by Minakami (1951), those with a high volatile content may retain fluidity at much lower temperatures. On the other hand, Sahama and Meyer (1958) recorded a fountaining nephelinite lava at only 980°C. While there are no corresponding measurements for intermediate or acid lavas, minimum liquidus temperatures for rhyolitic compositions may be estimated from liquidus temperatures in the system albite-orthoclase-SiO_2-H_2O if it can be inferred that no significant quantities of volatiles such as F and Cl were dissolved in the melt. It then appears, from Figure 15.2, that such melts could not remain fluid below 1000°C if they contain the amount of water found in many obsidians.

Lava that flows on the surface must not only have a certain minimal fluidity but must also possess a limited potential for violent disruption, which could prevent the flow from retaining coherence. Thus, if the lava is too charged with volatiles, it may be so fragmented and undercooled by rapid loss of these volatiles so that only pumice or other pyroclastics result.

It has been noted by some volcanologists that a correlation exists between the mode or position of eruption of lava and its volatile content. As an example, Krauskopf (1948) found that the basaltic andesite lavas of the Mexican volcano Paricutin emitted different proportions of volatiles depending on whether they were discharged from the central vent or from fissures at the base of the cone. The gases from the main vent retained a higher proportion of the relatively insoluble species SO_2, while those from the basal fissures contained more HCl. An interpretation of the structure leading to this situation is depicted in Figure 15.4. It is likely that the

Table 15.2 Normative mineral compositions of volcanic rock from the Hakone Area, Japan

Numbers are the same as for Table 15.1

	8	9	10
Q	5.70	13.68	42.96
Or	1.67	2.78	10.56
Ab	14.15	23.85	33.01
An	45.31	33.92	8.06
Wo	4.18	3.60	1.43
En	11.40	9.70	0.20
Fs	11.22	6.73	0.00
Mt	3.71	3.71	0.46
Il	1.67	1.52	0.46
Hem	—	—	1.12
Ap	0.34	0.34	—

small edifice of Paricutin is fairly typical of the type of activity that builds cinder cones. In this situation the lava of the main column is buoyed up by the vesiculation of gases, which decreases its density substantially. On the other hand, the denser, volatile-poor lavas can only find their way out at the base, where they form valley flows.

It might be expected that the holocrystalline flows, in common with plutonic and hypabyssal rocks, would retain only that part of the volatile fraction of the magma that is dissolved in or reacts with the crystals themselves, or that is included as bubbles. Detailed studies (Roedder, 1965; Roedder and Coombs, 1967) have shown that although bubble inclusions within crystals may comprise only a minute fraction of the rock, they contain a variety of fluid phases such as glass (congealed silicate melt), saturated saline solution, and liquid CO_2. However, in volcanic phenocrysts such inclusions are more likely to reflect conditions at depth than at the surface.

Normally it would be expected that glassy lava flows would provide the best opportunity for retention of the small quantities of equilibrium volatiles since those should be trapped in a structural environment similar to the original melt. However, silicate glasses have a great affinity for water at all temperatures (Chapter 12) so that this constituent may be absorbed later, and greatly dilute the juvenile volatiles. This was indeed shown to be the case by Ross and Smith (1955), who discovered that the hydrous glass *perlite* is derived by postmagmatic hydration of obsidian, with the amount of water added usually ranging from 2 to 5 weight percent.

Lava flows exhibit a great variety of features that reflect their extrusive and cooling histories. Although they usually have little effect on rocks with which they come into contact, they sometimes bake unconsolidated sediments and weld glassy ash and tuff over which they flow. A somewhat idealized cross section of a rhyolite flow from Nevada is shown in Figure 15.5. In this case the flow eroded, incorporated, and altered previously

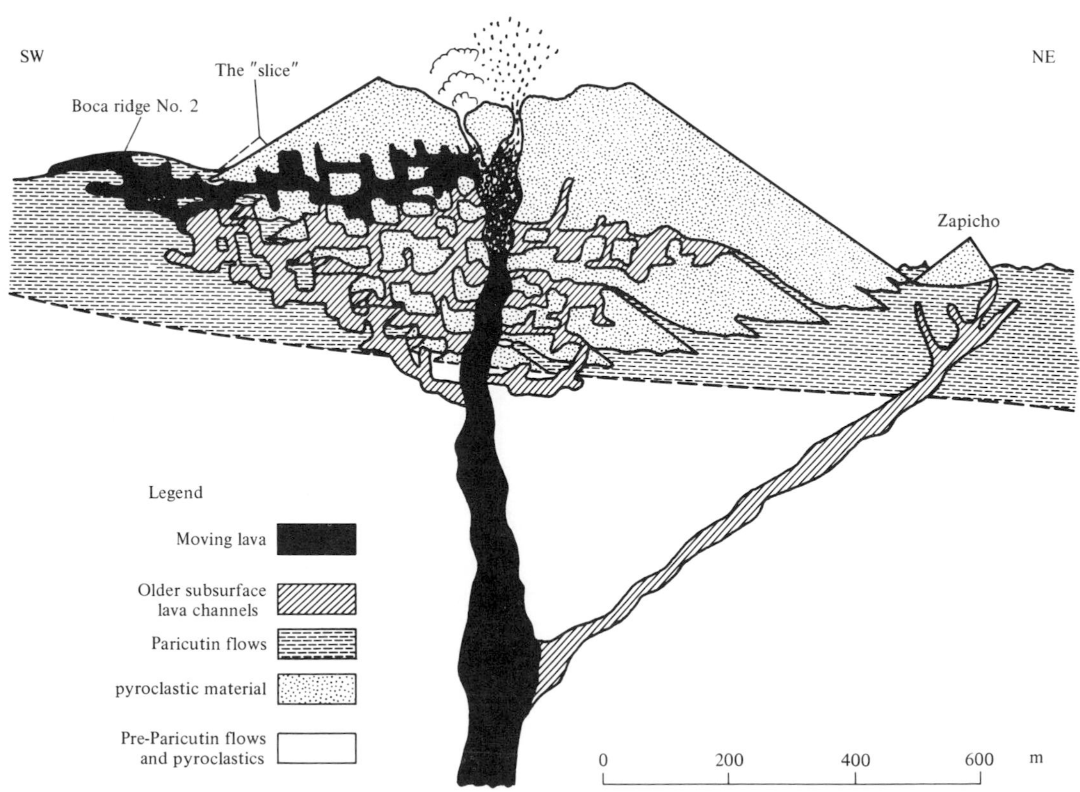

15.4 Interpretation of the structure of Paricutin volcano, Mexico, a cinder cone. (After Krauskopf, 1948.)

deposited tuffs, breccias, and agglomerates. High temperatures are indicated: the flow contains only 5 to 10 percent phenocrysts and the basal tuffs are welded. The upper part of the flow shows evidence of rapid chilling in the form of a vesiculated and almost pumaceous glass rind. Below this and completely enclosing the interior is a zone of perlitic vitrophyre which originated both by rapid cooling of magma and by welding of the basal tuff. By contrast, the central part of the flow cooled slowly enough so that thorough crystallization occurred by precipitation of microlite crystals, devitrification, and by the formation of lithophysae and open crystal-lined cavities.

The formation of glass in basaltic flows is much more infrequent, although the crystals of the aphanitic groundmass may be so small as to be invisible to the unaided eye. However, as in the case of rhyolite, the interior of flows show a more marked crystallinity than the edges, and in some cases possess a coarseness of grain approaching that of plutonites.

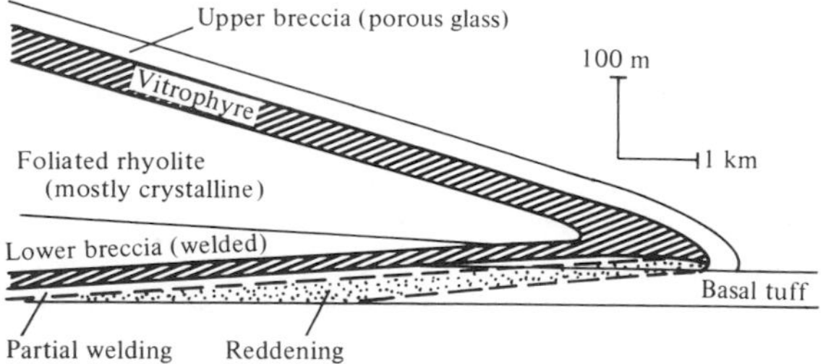

15.5 Idealized relationships in longitudinal section between emplacement, alteration, and cooling and crystallization zones in rhyolite of Comb Peak, near Fortymile Canyon, Southern Nevada. (After Christiansen and Lipman, 1966).

Pyroclastics

The term *pyroclastic* is somewhat misleading in that it seems to imply a preexisting rock has been broken up. This is of course true in the case of many breccias and agglomerates, but it is not for the most common pyroclastic material—the dust and ash that result largely from the explosive disintegration of frothing liquid magma. While it is difficult to categorize all pyroclastic material, the more common varieties may be classified as follows:

Original Deposit	Size (mm)	Consolidated equivalent
Pumice	—	
Dust	<0.05	Tuff or welded tuff (ignimbrites)
Ash	4	
Lapilli	32	
Bombs	>32	
Blocks, agglomerate, etc.	—	Agglomerates of breccias

Although lava that gives rise to pyroclastics does not flow over the surface as such, it does move as a solid-gas emulsion. When explosive disintegration of the magma is minimal, a coherent froth or pumice results. Of course every gradation exists between boulder-sized blocks of pumice and the finest ash. However, ash so derived comes largely from the attrition of larger pumice particles. By contrast, lapilli and bombs are simply congealed clots of magma, and although lapilli are frequently spheroidal, the larger bombs assume characteristic spindle-shaped forms in their flight through the air. Lapilli and bombs are more typical of fluid, basic magmas, whereas the corresponding congealed acid magmas result in blocks that are shaped by postconsolidation fracturing. Frequently avalanches of such blocks form variants of ordinary ash flows.

It is important to note that although all tuffaceous rocks are by definition consolidated, *welded tuffs* are products of consolidation under magmatic heat. Unwelded tuffs, which presumably were deposited at lower temperatures, are sometimes called *sillar* after the widespread deposits of this nature in the Andes Mountains of Peru (Fenner, 1948; Jenks and Goldich, 1956). According to current opinion, welded tuffs are deposited by *glowing avalanches* or *nuées ardentes* such as have been observed on Mt. Pelee and elsewhere in historic times. These avalanches consist of high-temperature suspensions of dust, ash, and other pyroclastics. The solids are suspended in magmatic gases that are in the process of exsolving from the solidifying magma. Such avalanches may result when frothy pumice overflows the vent, but in their most characteristic forms they break out of the side of the stratocone near the summit and flow down to the valley bottoms at high speeds under the force of gravity. Less frequently they assume the form of great horizontally directed blasts of devastating power. It is thought that in the thickest parts of such avalanche deposits, enough heat is retained to weld the fragments of ash into solid rocks and even dense obsidians. However, it should be kept in mind that no examples of welded tuffs are known that have formed during historic times. It is thought that this is a consequence of all known modern tuff flows being relatively small.

Chemical analyses of some pyroclastic lavas are shown in Table 15.1. This table includes an example of basaltic pumice from the Kilauea volcano, Hawaii, which differs very little chemically from the associated lava flows. However, extensive studies reveal that one of the salient features of the more acid tuffs is their high Fe_2O_3/FeO ratios as compared with the corresponding flow rocks. However, as we have seen, such oxidation may also affect flows that are highly vesiculated.

Numerous authors have speculated about the precise conditions that lead to the formation of pyroclastics and the attendant eruptive phenomena. Of the different deposits, pumice is perhaps the best understood since its structure and the properties of water (where this is the major volatile) set limiting conditions for its formation. Although the volume of the vesicles in pumice frequently exceeds that of the glass, it does not do so by more than a factor of four or so (Ewart, 1963). How much dissolved water this represents may be seen by reference to P–V–T data. Making use of these data, Ross (1964) gives the following illustration: If 1 percent water is released from a magma at 900°C and $P = 1$ atm, the P–V–T data for water indicate that the volume occupied by the vapor is 123 times that of the magma from which it was released. At higher pressures this volume is of course reduced, but even if $P = 100$ atm, the volume would still be of the order of the volume of the melt or greater. Thus, it appears that the quantity of dissolved water involved in pumice formation is less, and perhaps much less, than 1 percent by weight. However, the water content of unaltered obsidians and other natural glasses (Table 15.1) indicates that the amount required is probably greater than 0.1 percent by weight.

When we consider the formation of lava dust and ash, such as accompanies the ash flows, we encounter greater difficulties. It seems evident, first of all, that these products require a greater quantity of water than is required for pumice formation since the volume of vesiculation is clearly so great as to rapidly destroy the coherence of the melt. The difficulty is in explaining how this water is maintained in solution during the ascent to the surface, for if the evolution of gas occurred deep within the conduit, we should expect more vertical blasts than horizontal avalanches. Verhoogen (1951a) attempted to explain such problems by suggesting that ash formation is a question of kinetics or, more precisely, of bubble nucleation in an unstable supersaturated melt. However, Roedder and Coombs (1967) have presented evidence that suggests that many magmas, even at considerable depths, might contain bubbles of CO_2-rich fluids. Thus, there could be no supersaturation since some bubbles would always be present. Another possibility presents itself, however—that the solubility of water might be greatly increased by the presence of certain other volatiles. Later we shall see that both Cl and F are candidates for this role.

When the evolution of gases occurs slowly at depth, especially within the fluid basic lavas, the bubbles may coalesce within the conduit and rise to the surface. The rate of rise of these bubbles will be governed by their size, and this will steadily increase because of accretion and expansion. Thus, it is likely that as these bubbles reach the surface, minor but quite violent local explosions would result, and that these explosions are the major sources of the energy required to eject the clots of magma that become lapilli and bombs. We shall see that the gases emitted from liquid basaltic magma are rich in CO_2 and SO_2, so that it is possible that the formation of large bubbles is dependent on the evolution of these comparatively insoluble gases at depth rather than as the evolution of water vapor.

It is natural to turn to the active volcanic processes of the present to better understand those of the past. Thus, if we wish to interpret the ash flows and ignimbrites in the geologic column, we would do well to note studies such as those that were made by Peret (1937) on the *nuée ardentes* of Mt. Pelee. His major observations and deductions are as follows:

1. *Nuées ardentes* or glowing avalanches result when an autoexplosive (water-supersaturated) magma finds egress from a volcano and rushes downhill under the force of gravity.
2. As it descends, the *nuée* maintains its general coherence but also gives rise to high vertically rising dust clouds.
3. The *nuée* proper consists of a suspension of solidified or partially solidified lava which emits gases as it moves. These gases, consisting largely of water vapor, are thought to keep the solid particles separated, and so confer great lubrication and buoyancy to the mass.
4. The *nuées* descended Mt. Pelee at about 60 km/hr. They usually were silent, but the more powerful ones were also accompanied by a variety of sounds, including detonations.
5. *Nuées* with much of their autoexplosive force spent were sometimes reactivated by the shock of collisions with obstacles. Also, although they sometimes possessed enough momentum to override low hills, the role of gravity was well shown by their frequent deflection by such obstacles.

6. The *nuée* emission caused but slight damage to the volcanic edifice, and their deposits generally contained little or no old material.
7. The coherence of the *nuées* is thought to have been maintained by downdrafts of cool air and perhaps by electrical forces between particles.
8. The general volcanic activity was observed to begin with violent explosions, continued through a period of combined *nuées* and dome building and ended with the latter. It is thought that doming resulted after lava stood so long in the conduit that it became almost gas-free to a considerable depth.

A diagrammatic illustration of the *nuée ardente* type of activity is given in Figure 15.6. Detailed chemical analyses of the Pelean lavas have been presented by Gunn *et al.* (1974).

One aspect of processes occurring in *nuée ardente* activity was pointed out by Walker (1972). According to him there is a strong tendency for crystals to concentrate in ignimbrites by the selective loss of pumice above the volcanic vent or moving pyroclastic flow. A corollary of this study is that chemical analyses of ignimbrites probably are not representative of the magma.

Ash-flow deposits, either welded or unwelded, are common in most regions of acid volcanism, and are clearly recognizable on many of the large stratovolcanoes of the andesitic ranges. For example, they are prominent features of the more acid stratovolcanoes of the southern Cascade range of Oregon and California. In this region Crater Lake occupies the former position of the pre-

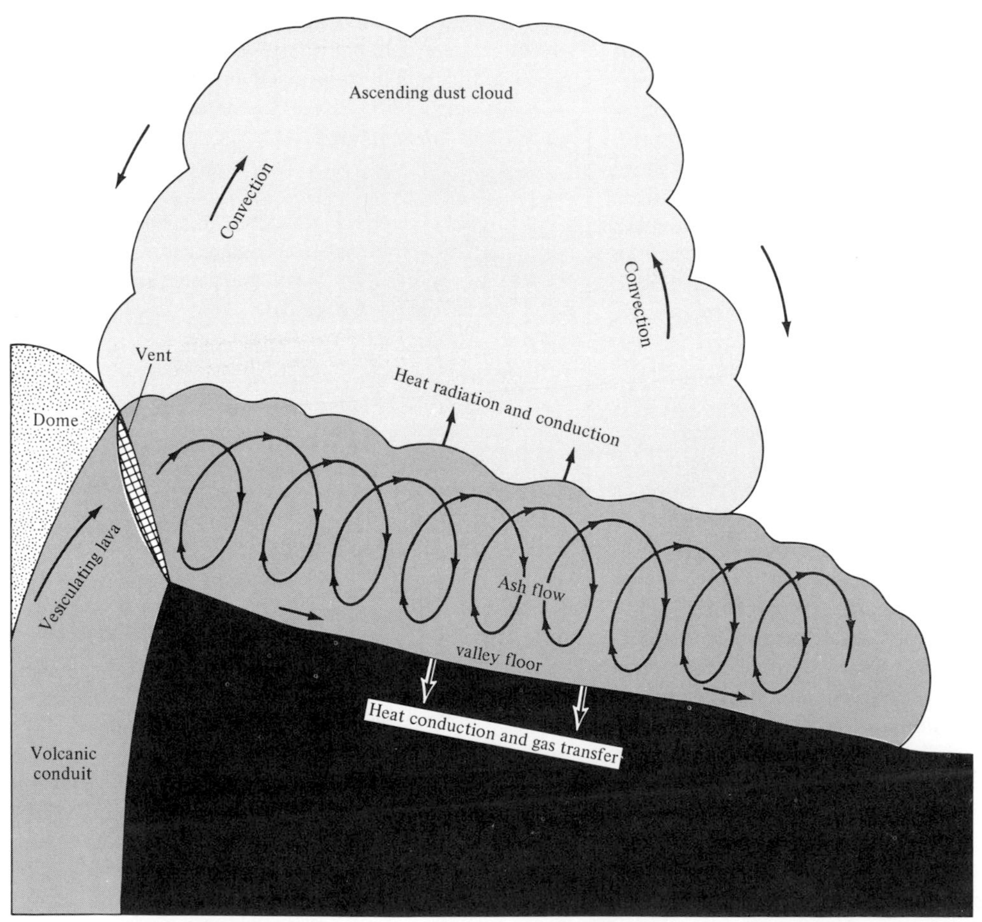

15.6 Diagrammatic illustration of *nuée ardente*–type volcanic activity. For further explanation see text.

historic but geologically recent stratocone called "Mt. Mazama." According to Moore (1934), Williams (1942), and others, the formation of the large present-day crater was preceded by activity that began with violent outbursts of pumice and ash that were carried downward to form deposits that were still more than 10 cm deep at a distance of 100 km to the northeast. These outbursts culminated in glowing avalanches that rushed down the radial valleys for some tens of kilometers and deposited about 30 km^3 of pumice and ash. The ejection of this volume of magma undermined Mt. Mazama to the extent that the large caldera of collapse was formed. As is the case with many other such deposits, the Crater Lake ash flows show no welding except in the walls of former fumarole vents. Although most of Mt. Mazama consists of andesite, the explosive phase appears to have been related to the production of more dacitic magma by crystal settling. The same appears to be true of the historic eruption of Mt. Lassen in California. In this case also, several *nuée ardente* type of eruptions accompanied the vertical ejection of ash. According to Day and Allen (1925), *nuées* like those of Mt. Pelee were emitted from beneath a plug of solidified lava near the summit.

The highly welded tuffs that characterize some of the older volcanic deposits appear to have occurred on a more massive scale than those associated with recent stratocones. They also appear to be more rhyolitic than dacitic in composition. Typical representatives are very widespread in the Great Basin of the western USA (Cook, 1963), in the Yellowstone Region of Wyoming, and in many other parts of the world.

The Yellowstone region contains a series of volcanic deposits ranging in age from early Eocene to late Pliocene. The oldest rocks, which are predominantly breccias and agglomerates, range in composition from andesite to dacite (Rouse, 1937; Boyd, 1961); they are separated by an erosion interval from the later rhyolites and minor basalts and a few rocks of intermediate composition. This late series forms the "rhyolite plateau," which consists of a complex of rhyolitic ash flows, rhyolite and obsidian lava flows, and rhyolite domes. The earliest members of the series consist of the dome extrusives. These are overlain by breccia and the thick Yellowstone tuff, which is highly welded and is interpreted as having resulted from large *nuées ardentes*. The tuff is in turn separated by an erosion interval from the plateau rhyolite lava flows ranging from holocrystalline varieties to obsidians virtually free of phenocrysts.

The Yellowstone tuff apparently extends far beyond the Yellowstone Park boundaries and may cover thousands of square kilometers. In some places it attains 300 m in thickness. According to Boyd (1961), it was deposited in "a single rapid series of eruptions" with single pyroclastic flow units more than 100 m in thickness. As no well-defined vents are known, it is inferred that eruption occurred from a number of dispersed fissures.

In this section, the Yellowstone tuff is seen to consist chiefly of aligned fragments of glass (shards) which form the matrix for collapsed pumice fragments and phenocrysts of sanidine ($\sim Or_{50}$), plagioclase ($\sim An_{10}$), quartz, ferroaugite, fayalite, and magnetite. The high albite content of the sanidine indicates high liquidus temperatures, while the occurrence of ferroaugite and fayalite with magnetite is consistent with quite reducing conditions immediately before extrusion.

As might be expected, the Yellowstone tuff is most thoroughly welded at the base, which is marked by a bed of dense obsidian averaging more than 1 m in thickness. However, the top is not welded.

Of considerable interest is the attempt made by Boyd (1961) to estimate the physical conditions and energy requirements during emplacement of the Yellowstone tuff. His approach was to calculate the maximum degree of cooling during emplacement and to compare this with the difference between the magma temperature and the temperature required for welding. He assumed that the minimum magma temperature could be approximated by the line of minimum temperatures corresponding to different water pressures in the system albite-orthoclase-SiO_2-H_2O (Tuttle and Bowen, 1958). The temperature of welding was experimentally determined by subjecting volcanic glass powders to heat and water pressure in a bomb. In order to relate these experiments to the natural situation, it was assumed that the water pressure attained a maximum value of 30 bar, corresponding to a 100-m-thick flow with a specific gravity of 2.0. Since this water pressure corresponds to a solubility of 0.4 weight percent water in the melt, glass with this water content was used as starting material. The results of these experiments suggested that on a time scale of the order of several weeks the minimum welding temperature was fairly well defined at about 600°C. If this value is adopted for the welding temperature, the allowable temperature range of cooling was found to be 50°C to 350°C.

The major processes that were inferred to have contributed to the cooling of the tuff during emplacement are

1. Exsolution and expansion of the gas and the associated magma in the conduit
2. Radiation from the top of the flow
3. Conduction and convection of heat into the ground and into the air above the flow

This list does not include certain energy expenditures and gains involved in opening the conduit, through dissipation as shock waves and in potential energy changes. The expenditure of energy in shock waves and in opening the conduit are impossible to evaluate, and the potential energy contributions can be shown to be small or as negating each other. For example, potential energy gained on extrusion will be converted into heat by turbulence in the down-rushing avalanche, and this will add a small increment to that derived from strictly magmatic sources.

Unfortunately the great uncertainties of the assumptions in Boyd's calculations for the major cooling process (item 1 above) make it unprofitable to present them here. Also, it is likely that his model would be altered significantly by the presence of Cl, F, or other volatile constituents. However, one of the major mechanisms for heat loss during flow—that from radiation—is easy to evaluate. The amount of heat that could be lost from the top of a static flow by this mechanism is negligible. However, if the flow is assumed to be in turbulent motion (as shown in Figure 15.6) heat from the interior would constantly be supplied to the surface and cooling would be much more rapid. If then it is assumed that heat was transported to the flow top as rapidly as it was radiated away, and if $\dot{Q}$ is the total energy radiated by 1 cm^2 of flow surface per minute we may write

$$\dot{Q} = GT^4 \tag{15.4}$$

where G is the Stefan–Boltzmann constant and is equal to 8.17×10^{-11} cal/cm^2/min/°K. If now we assume $T = 1123°K$ to be a safe upper limit for the temperature, we find that $\dot{Q} = 130$ cal/cm^2/min^1. According to Boyd, a minimum thickness for a single tuff flow is 30 m, so that if the density of the collapsed tuff is taken as 2.0, a vertical column 1 cm^2 in cross section would contain 6×10^3 gm of tuff. Consequently, each gram of this tuff must have lost 1.3 cal/h. If then the specific heat is taken as 0.3 cal/gm/°K, the tuff should have cooled at a rate of 4°K/h. Since this calculation corresponds to higher temperatures and a thinner flow than expected, we may conclude that even under the most favorable conditions no significant cooling can occur by radiation unless the emplacement of the tuff took many hours. Although evaluation of heat losses by conduction is more difficult, they must have been even less than those attributable to radiation.

While Boyd's analysis offers few conclusive answers, it does indicate that retention of sufficient heat for welding requirements may be possible for tuff flows of the magnitude found in the Yellowstone region.

Differentiation of hypabyssal and volcanic magmas

The mineralogy of some volcanic rocks tells us that differentiation in the corresponding magmas probably occurred at considerable depth. This is indicated in particular by the presence of phenocrysts of iron-rich hydrous silicates, which demand high water pressures and quite reducing conditions. The alterations undergone by such phenocrysts on extrusion have already been discussed.

Such bodies as the Palisade sill (Walker, 1940) provide evidence that substantial differentiation through crystal settling can also occur under hypabyssal conditions, although as compared with the plutonic complexes these sills are but poor representatives of the process. In some cases also, as in a number of diabasic sills of northern Minnesota, differentiation appears to have produced far too much granitic rock as compared with the visible parent diabase (Schwartz and Sandberg, 1940). These generally occur as granophyric concentrations near the tops of the sills and are sometimes complexly intruded into the diabase. In the instance of the Endion sill of the Duluth area, these relations have been ascribed to an up-dip migration of granophyric liquid while the sill was intruded as an inclined sheet (Ernst, 1960a). In any case, it appears that differentiation in such bodies took place largely before they reached their present locale. This conclusion is also supported by the bulk chemistry of the diabase fractions, which in most cases show higher $(FeO + Fe_2O_3)/MgO$ ratios than is typical for basaltic magma. For example, in the basal part of the Lester River sill (Schwartz and Sandberg, 1940) this (weight) ratio is approximately 10, which is greater than for the third Skaergaard liquid (Table 14.4). This seems to indicate that differentiation through absolute iron enrichment may already have been substantial before intrusion.

The differentiation of basic sills illustrates that even under shallow conditions the latest residue of crystallization is feldspathic or quartzofeldspathic in composition. In fact, the nature of this residue is, as expected, closely dependent on the chemistry of the parent magma so that melts with much normative olivine give rise to trachytic residues while quartz tholeiite magmas yield granophyres (Kennedy, 1933). MgO-FeO-alkali variation curves for the differentiated diabase sills usually lie between those for the plutonic gabbro and the calc-alkali complexes as shown in Figure 14.4. Yet this does not necessarily imply intermediate physicochemical conditions; rather, it may simply reflect the imperfect crystal settling within the smaller more rapidly cooled sill.

In their description of differentiated diabase sills many authors refer to the small quantity of rocks intermediate in composition between diabase and granophyre. A similar gap has been observed in the magmas of many volcanic provinces such as the Yellowstone region already discussed, in Iceland (Thorarinsson and Sigvaldason, 1962) and elsewhere. In fact, this characteristic was also recognized in the great Duluth gabbro by Grout (1918). Since it has subsequently been recognized that the Duluth gabbro follows a differentiation trend similar to the other gabbroic complexes (Hess, 1960), we can attribute it to the same differentiation mechanism already assigned to those rocks. It therefore seems likely that the shallow basaltic magmas that formed sills and many volcanic deposits differentiated under essentially anhydrous conditions, and that those conditions suppressed the formation of large quantities of feldspathic rocks during the middle stage of differentiation.

Diverse trends of differentiation in volcanic rocks have been suggested as explanations for mineralogic differences. Thus, Kuno (1950), in his study of the Hakone volcanic province of Japan (Table 15.1), distinguished a "hypersthene rock series" in which the ground mass is characterized by orthopyroxene and a "pigeonite rock series" in which it is characterized by pigeonite. Kuno interpreted the occurrence of hypersthene as indicative of lower temperatures of crystallization consequent to the assimilation of volatile-rich xenoliths. However, his hypersthene suite is also distinguished by the presence of hornblende, and this may simply indicate differentiation under higher water pressure as has already been suggested for the hypersthene and hornblende-bearing calc-alkali batholiths (Chapter 14).

The Hakone volcanic series also exhibit the usual trend toward relative iron enrichment, which is common to all known plutonic series. The end product of this enrichment results in the occurrence of fayalitic olivine and other ferrous iron pyroxenes as druse fillings of the late hypersthene dacite.

It has also been possible in some cases to trace differentiation trends within a series of acid volcanics. Lipman (1966) recognized a number of such series in the Tertiary ash flows of southern Nevada. According to him, the individual ash flow sheets become more mafic and phenocryst-rich with stratigraphic height so that they represent in inverse order the compositional layering within the magma chamber. Chemical analyses for two members of one of these series are shown in Table 15.1 and the entire series is shown plotted in Figure 15.7. As was pointed out by Lipman, the trend rather closely approximates the lines of fractional crystallization within

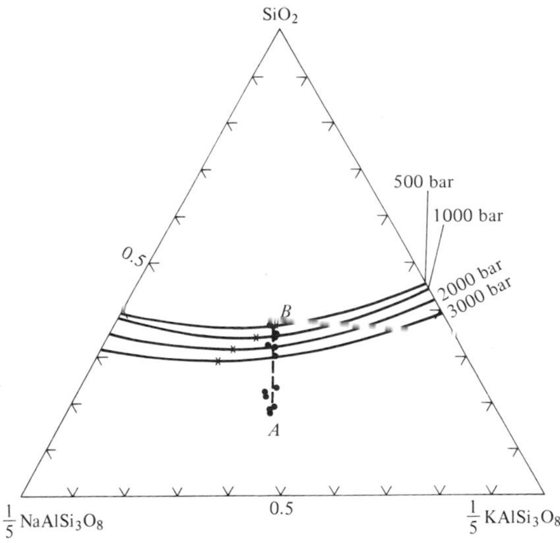

15.7 Molar plot of compositional variation of Tertiary volcanic ash flows from central Nevada. (After Lipman, 1966.)

the system albite-orthoclase-SiO_2-H_2O as given by Tuttle and Bowen (1958). It is obvious that if fractional crystallization was indeed operative in these ash-flow magmas, and if they can be represented by the simple quaternary system, then differentiation must have occurred under quite low water pressures and terminated in the range of 500 to 1000 bar. Of course, the experimental reference system implies water-saturated melts so that the water pressure might well have been even lower.

If Figure 15.7 is compared with Figure 13.5 it is seen that the early part of the differentiation path of these lavas, leading from the hypothetical basaltic parent magma (if such indeed played a role!), must have been like the path M'–N', with the line A–B of Figure 15.7 corresponding to N'–O', due account being taken of the lower water pressure. This general interpretation is supported by the observed phenocrysts which consist largely of plagioclase and alkali feldspar with little or no quartz.

Of course an interpretation such as the above must be tempered with caution. They are valid only if the rocks in question approximate the quaternary system closely. In particular, the anorthite content must be low since this component has the effect of shifting the feldspar-quartz phase boundary toward the SiO_2 corner and could yield spuriously low water pressures. Then too we must assume that no significant quantities of such volatiles as F or Cl were present since they may have large effects on the liquidus. Unfortunately, evidence of their former presence is particularly difficult to recognize.

An example of an acidic volcanic suite that follows a quite different trend than the foregoing is that of the Taupo ash showers of New Zealand (Ewart, 1963) (Table 15.1). In this case no alkali feldspar phenocrysts occur and those of quartz are virtually absent. Also, the normative anorthite contents of the glasses are all high. If we again refer to Figure 13.5 it is apparent that the crystallization path is best represented by a curve similar to M–N, but with N lying close to the base of the tetrahedron. Obviously the projection of these data onto the quaternary system cannot be used to estimate the water pressure.

In summary, it might be said that volcanic and hypabyssal magmas show evidence of the same differentiation trends as plutonic rocks, depending on the precise conditions prevailing in the magma chamber. Thus they may reflect either the calc-alkali of gabbroic trends. However, a superimposition of shallow chamber conditions may at times modify these trends greatly, an example being the restricted crystal settling in hypabyssal sills on the loss of volatile pressures in volcanic eruptions.

Volcanic fluids

General

The term *volcanic fluids* as used here includes all the emanations—gaseous or liquid—that are at least in part derived from shallow magmas and may be so identified.

Because of the high temperatures and low pressures all direct emanations of volcanic magmas are gaseous. This fact is of some importance since the solubilities of many constituents such as the refractory metals are largely functions of the temperature and fluid density. The low densities of volcanic emanations thus greatly restrict their chemical transporting capacities in spite of their high temperatures.

The orifices from which the emanations of shallow magmas issue are conveniently classified as either *fumaroles* or as *hot springs*. Fumaroles are orifices or vents close to the magma source and their gases may be as high or even higher in temperature than the magma itself if exothermic reactions take place.

Many of the chemical analyses of magmatic gases seem to point to water as the overwhelmingly dominant species. But we shall see that for at least some magmas this may not be true and that other gases such as CO_2 and SO_2 appear to be on a par with H_2O. In fact, water is the most consistantly important contaminant even at the very source of magmatic gases, and it is easy to see why this must be even more the case as the distance from the source is increased.

Because of the contaminating effect of ground water, the self-contaminating effect in magmatic gases[1], and reactions with rocks, the abundance of the dissolved constituents may be only remotely related to the chemistry of the source. Indeed, many volcanologists now favor the idea that the composition of hot spring waters simply reflects the rocks through which they have passed, although they may still contain 5 to 10 percent of a magmatic component (White, 1957).

However, the gases of the direct emanations of magmas are likely to at least approach an equilibrium with their dissolved parent species within the melt. Consequently, they not only contain potential informa tion about the "juvenile" volatiles which come from the planet's interior, but also of the physicochemical mechanisms and energetics of solution. It is therefore worth examining the magmatic gases in some detail.

Magmatic gases

The direct emanations of volcanic magmas are available from a number of studies made in the past. Although they are now more than 50 years old, the most impressive and useful set of such data are still those presented by T. A. Jaggar and E. S. Shepherd. Chemical analyses of these gases, which were collected from the vents of the basaltic volcanoes in Hawaii during the period 1917 to 1919, were carried out by Shepherd. However, the great significance of these analyses was revealed only recently through the elegant mathematical treatment of Nordlie (1967, 1971). It is the latter interpretation that is summarized here.

The gases selected for study by Nordlie comprise the J-series (for Jaggar) since these appear to be superior to the others both in terms of the collecting conditions and the chemical analytical techniques employed. The analyses had originally been corrected for what Shepherd assumed to be atmospheric contamination. However, Nordlie's calculations, which take into account several different types of contamination, required the restoration of the original values; it is these restored original gas compositions that are presented in Table 15.3.

It should be pointed out immediately that of the 14 J-series specimens, J-8 occupies a unique position in that it appears to be the least contaminated of the series.

[1] This occurs through distillation and condensation processes.

This unique quality of J-8 is attributable largely to its being the only specimen for which the glass collecting tube was sealed by melting the tip in the glowing cavity before it was withdrawn.

Some of the most obvious effects of contamination are immediately apparent from a glance at the table. These are the presence of O_2 and N_2 in approximately the atmospheric ratio, of substantial quantities of the highly oxidized constituent SO_3, and the dominant but highly variable quantity of water. We shall see that these features are qualitatively incompatible with such original emanations of magma as CO, H_2, S_2, and SO_2. It is obvious, therefore, that these gases have suffered certain modifications subsequent to their emission from the melt.

Nordlie distinguished the following major modifying effects on the magmatic gas:

I. An addition and loss of H_2O and SO_2 which apparently occurred at high temperatures even before the gases were trapped in the vacuum tube
II. A contamination of the trapped gases by atmospheric N_2, O_2, Ar, and CO_2

In his analysis the first task he set himself was the correction for the atmospheric contaminant; to this end he tested the following types of corrective procedure:

1. Based on total O_2 present
2. Based on total N_2 present
3. Based on total O_2 + total SO_3
4. Combination of cases 1, 2, and 3

In the type-1 correction, all the free oxygen of each specimen of Table 15.3 was removed with the corresponding proportion of N_2, Ar, and CO_2. In type 2, N_2 was removed with the atmospheric proportions of O_2, Ar, and CO_2. In type 3 it was assumed that all the SO_3 resulted from reaction with atmospheric oxygen, and all the reacted and unreacted oxygen was removed with the atmospheric proportions of the other gases as before.

It was found that the type-1 correction corresponded to a minimal degree of contamination, while by type 2 some specimens ended up far too reduced. Without going into further details we shall anticipate that the type-4 correction yielded the most satisfactory results in terms of the internal consistency attained. In fact, when this type of correction was applied, the data were found to correspond to a highly unique and interesting theoretical curve. This "H_2O addition curve," which is shown in Figure 15.8, has the following meaning: If water is continuously added to a specimen such as J-8, which is low in water and which plots as shown in terms of O/H and %O + %H, then the composition must change precisely along the curve as plotted and end up at 100 percent O + H and with (O/H) = 0.5. The fact that all the other specimens of the J-series fall on or very near this curve is persuasive evidence that the series is essentially one of varying degrees of water and atmospheric gas contamination.

Table 15.3 Restored volume percentages of J-series samples

No.[a]	CO	CO_2	H_2	H_2O	S_2	SO_2	SO_3	Cl_2[b]	N_2	Ar	O_2	Total
J-2 (P)	0.000	0.336	0.000	4.35	0.000	0.276	0.140	0.239	75.9	0.009	19.0	100.0
J-3 (P)	0.150	4.51	0.102	54.7	0.000	2.20	3.75	0.756	26.6	0.683	6.73	100.0
J-4 (P)	0.111	5.36	0.134	67.1	0.118	1.09	2.71	0.490	18.2	0.195	4.39	99.9
J-6 (P)	0.115	0.626	0.0504	56.7	0.000	0.00719	0.0935	0.0216	36.2	0.262	5.88	100.0
J-8 (E)	1.32	43.2	0.435	32.8	0.0362	10.2	0.0381	0.0362	9.67	0.315	2.00	100.0
J-10 (E)	0.0736	11.0	0.0669	34.1	0.0334	9.10	2.38	0.0201	35.8	0.448	6.90	99.9
J-11 (E)	0.508	18.0	0.276	53.0	0.215	9.82	0.473	0.000	14.3	0.396	2.90	99.9
J-12 (P)	0.0413	1.18	0.0662	80.4	0.0580	0.422	0.000	0.0248	14.0	0.202	1.60	98.0
J-13 (E)	0.471	13.8	0.780	54.8	0.0730	6.42	2.00	0.0812	17.4	0.711	3.95	100.0
J-14 (F)	0.218	6.89	0.0792	35.8	0.0465	1.70	0.480	0.000	43.1	0.501	11.2	100.0
J-15 (F)	0.0858	7.60	0.0660	48.7	0.0198	4.05	1.12	0.0660	30.5	0.413	7.13	99.8
J-16 (E)	0.449	14.4	0.538	53.2	0.120	6.85	2.03	0.0642	17.9	0.248	4.15	100.0
J-17 (G)	0.276	8.67	0.433	59.2	0.179	4.84	0.000	0.0374	20.8	0.267	5.33	100.0
J-18 (E)	0.576	13.7	0.646	48.2	0.171	8.41	2.51	0.101	20.7	0.299	4.64	100.0

[a] E = excellent; G = good; F = fair; P = poor.

[b] Probably mostly HCl.

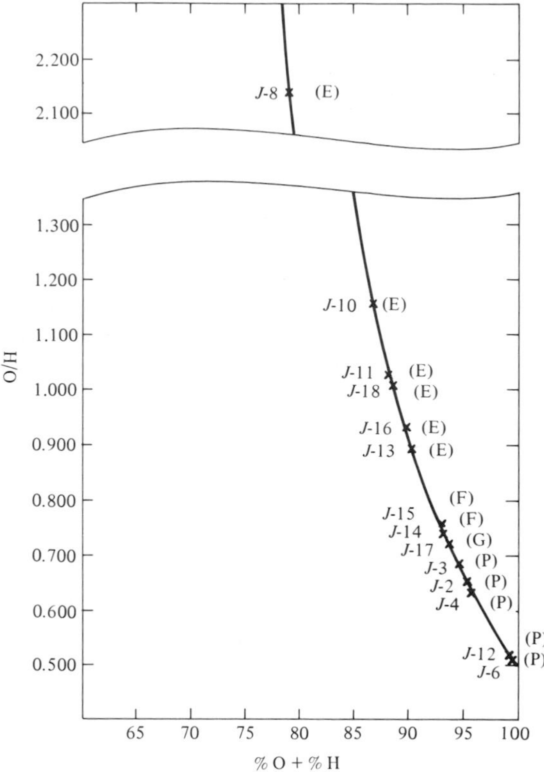

15.8 Plot of atomic ratio O/H against the sum percent of O + OH for the *J*-series of the Kilauea gas samples. (After Nordlie, 1971.) Atmospheric contamination has been removed on basis of case 4 (see text). The solid curve is the "H_2O addition curve." Letters indicate quality of sample collection: E = excellent, G = good, F = fair, P = poor. (After Jagger, 1940.)

Actually, the problem is somewhat more complicated than this, and if this is to be taken into account the analysis must be carried one step further. In this final step Nordlie took account also of the variation of carbon and sulfur. He did this by considering that there was little evidence in volcanism for any systematic loss of carbon from the gases, but that there was much evidence for the loss of sulfur in the form of deposits of the native element. Consequently, it was assumed that no carbon had been added or subtracted, and that variations in the C/S ratio was entirely due to a loss of sulfur. Furthermore, analysis showed that the sulfur loss occurred as SO_3. Thus all the analyses could be recalculated on an "equal carbon basis." Again without pursuing the subject further it may be stated that the effect of this calculation was merely to move the points of Figure 15.8 along the water-addition curve without changing the form of their distribution.

If the type-4 correction for atmospheric contamination is applied to the specimens of Table 15.3, and if *J*-8 is assumed to possess only magmatic water and oxygen, then the major gas components are fixed for each C/S ratio. If the total pressure is taken as 1 atm it is possible to calculate the variation of the molecular abundances with the temperature and this ratio. For example, if C/S is assigned values of 0.5 and 1.0, the

15.9 Possible equilibrium composition of the magmatic gas from the Kilauea volcano, Hawaii. Partial pressures of molecular species are given as a function of temperature at a total pressure of 1 atm and a C/S = 0.5. (After Nordlie, 1971.)

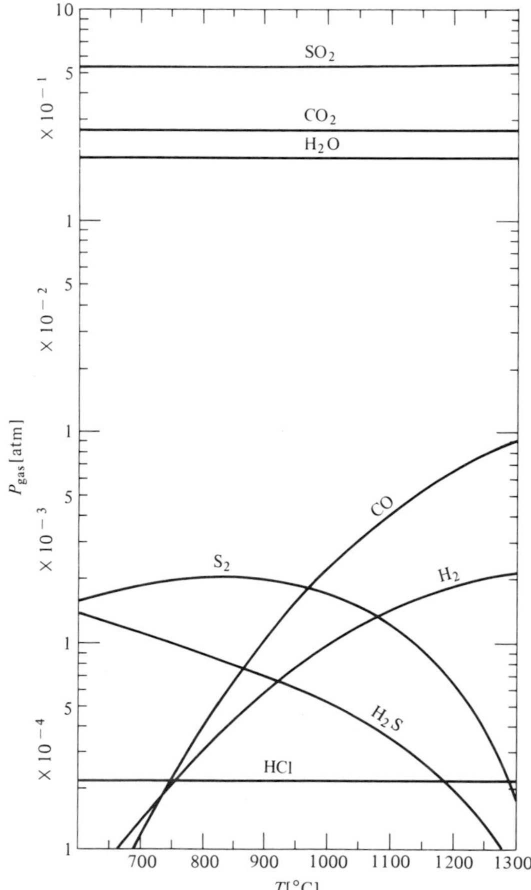

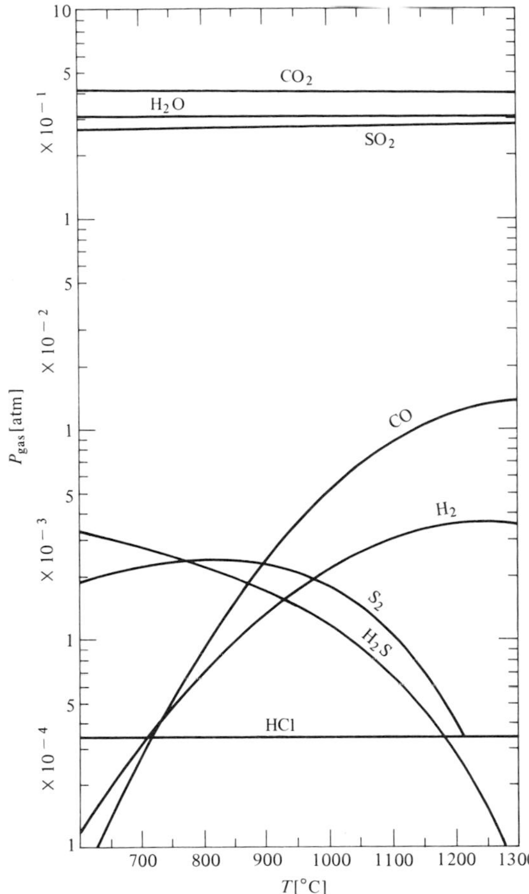

15.10 Possible equilibrium composition of the magmatic gas from the Kilauea volcano, Hawaii. Partial pressures of molecular species are given as a function of temperature at a total pressure of 1 atm and a C/S = 1.0. (After Nordlie, 1971.)

corresponding temperature variation is shown by Figures 15.9 and 15.10, respectively. It was concluded by Nordlie (1971) that a value of 0.5 for C/S probably corresponded most closely to the unmodified magmatic gas.

The state of oxidation and the variation of the oxygen fugacity with the temperature is also determined for each fixed value of C/S. In each case f_{O_2} falls in the range of 10^{-9} to 10^{-8} atm at the liquidus temperatures of 1100°C to 1200°C. These values compare rather well with those measured directly by Sato and Wright (1966).

Nordlie was able to show that the corrected gas samples reflect thermodynamic equilibrium only to the extent that quenching from magmatic temperatures was possible. He found a definite correlation between the kinds of gases and the stages of cooling at which quenching occurred. His general picture of the processes involved is shown in Figure 15.11.

It is not known to what extent the compositions of the Hawaiian gases are representative of the volatiles of basaltic magmas in general or how they differ from those of more acid magmas. It is not even known to what extent the least contaminated specimens such as *J*-8 with high CO_2 and SO_2 contents actually represent the volatiles of the Hawaiian magmas since some magmatic constituents such as SO_2 have been lost. It is possible that these compositions are really the result of the filtering of the least soluble gases into bubbles formed at depth. Presumably this early enrichment in CO_2 and SO_2 was again partially compensated for by the enrichment of water through condensation at the orifice.

An interesting comparison with the Hawaiian gases is provided by analyses for fumarolic gases collected from a dacite dome of the Usu volcano of Japan. These gases, shown in Table 15.4, are relatively rich in CO and H_2 so that for specimen No. 1 at 760°C we find that

$$\frac{H_2O}{H_2} = 158 \qquad \frac{CO_2}{CO} = 358$$

Figures 15.9 and 15.10 show that in the corrected Hawaiian gases both of these ratios would fall in the range of 500 to 1000 at the same temperature if equilibrium were attained. Thus if equilibrium is assumed in both cases the Usu gases are the more reduced of the two. This result is compatible with the low quantities of atmospheric contaminants in the latter and also argues for the magmatic origin of most of their water. However, it is again possible that this water was concentrated by condensation.

The analyses of the first three specimens of Table 15.4 are presented in more detail in Table 15.5. For convenience these analyses may be discussed in terms of two groups of species. The first group consists of those common "volatiles" such as CO_2, CO, SO_2, and H_2. The relation between these species within the gas phase are well known, as are certain of their reactions with crystalline components such as Fe_2SiO_4 and FeS. In particular these reactions may be utilized to define the oxygen fugacity of an equilibrated magma.

The second group consists of the "refractory" metals and other substances which usually are only minor components of the gas but which also bear definite reaction relations to either magmatic crystalline precipitates or the wall rocks. However, the solution mechanisms and solubility constants in the melt are only poorly known for both groups. Some of the vaporization reactions of the refractory metals and oxides were already discussed in previous chapters dealing with metasomatism and magmatic gas transfer. Here these reactions again play an important role in relating the concentrations of the refractory species to the physical and chemical conditions of volatilization.

Of the refractory constituents listed in Table 15.5, only SiO_2 and NaCl bear reasonably simple relations to mineral phases. As a result these relations have been subject to experimental investigation as shown in Figures 15.12 and 15.13. If we compare the high-temperature specimens of Table 15.5 with Figure 15.12 we see that the 200 to 300 ppm SiO_2 in the fumarolic gases corresponds to water pressures of less than 500 bar if equilibrium was established with quartz. Similarly, the approximately 50 ppm of NaCl in these gases implies water pressures of less than 100 bar if the vapor was saturated with NaCl. However, saturation with this substance is unlikely, so that the pressure might well correspond to the higher limit implied by the SiO_2 content.

It is obvious that by far the greatest amount of chloride present in the Usu gases is not associated with Na^+, K^+, etc., but with hydrogen as HCl. The abundance of this gas as well as of HF may be evaluated through certain interesting reactions that relate these acid gases to the silicates and halogen salts,[2] which may be present within the wall rock or in the magma itself.

[2] Although such phases as $CaCl_2$ and NaF are stable in certain chemically restricted environments, the presence of both Ca^{2+} and Na^+ stabilizes NaCl and CaF_2 since ΔG for the reaction $CaCl_2 + 2NaF \rightarrow 2NaCl + CaF_2$ is negative.

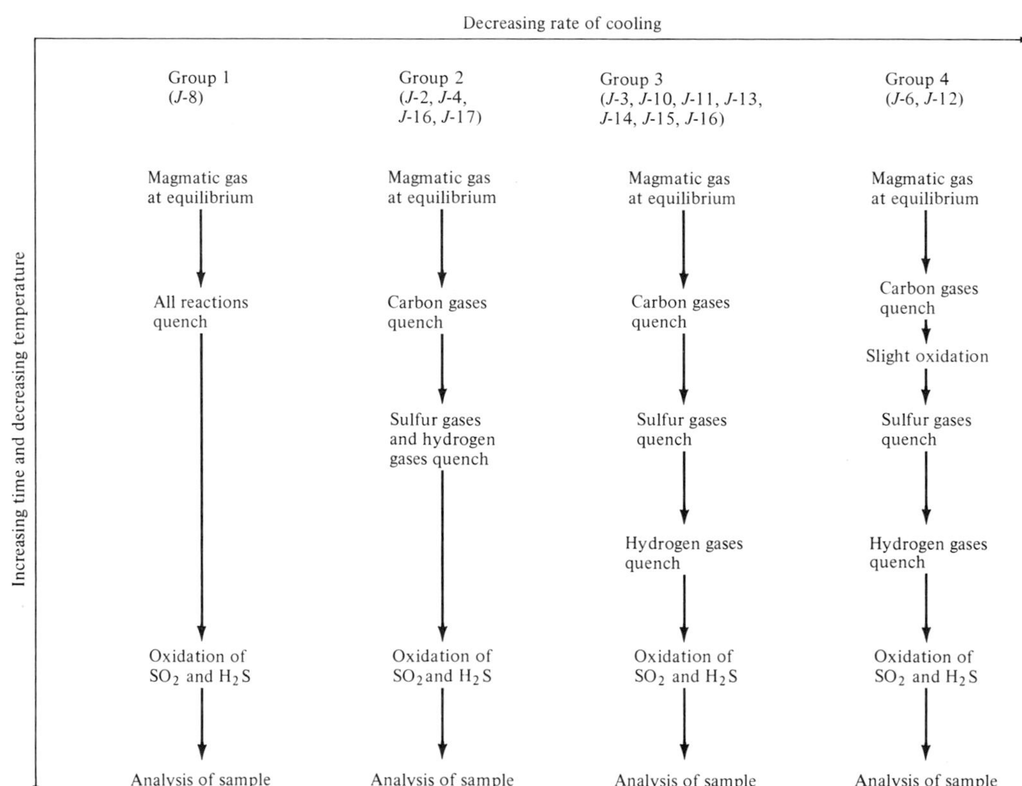

15.11 General scheme of processes effecting various groups of gas samples from the Kiluea volcano, Hawaii. (After Nordlie, 1971.)

Table 15.4 Analyses of fumarolic gases from the Usu volcano, Showa-Shinzan, Japan

The extrusive is a hypersthene dacite dome.

Sample location	Reference number	Temp. (°C)	Percent total active gases	Gross composition				"Active" gases recalculated to 100 percent										
				O_2	N_2	Ar	H_2O	CO_2	CO	CH_4	NH_3	H_2	HCl	HF	H_2S	S	SO_2	SO_3
A-1 (9081)	1	760	1.879	0.0032	0.057	2.5×10^{-5}	98.1	64.4	0.18	0.090	*a*	33.0			0.022	0.011	2.3	0.026
A-3 (9051)	2	525	1.442	0.027	0.054	—	98.5	74.2	0.15	0.15	*a*	23.0			0.16	0.007	1.41	0.019
C-3 (9063)	3	220	0.632	0.0013	0.089	6.6×10^{-5}	99.3	84.8	—	—	*a*	2.8			9.1	—	3.2	0.009
"East side"	4	655	0.516	0.0000	0.014	5×10^{-5}	99.48	56.2	0.12	0.17	—	27.2	8.7	3.5	0.39	—	3.7	—

Data are from Nemoto *et al.*, (1957), as reported by White and Waring (1963).

[a] Values given in Table 15.5.

Table 15.5 Fumorolic gases of Showa-Shinzan, Japan in weight ppm

Reference No.	1	2	3
Location No.	A-1 (9081)	A-3 (9051)	C-3 (9063)
Temp. (°C)	760	525	220
SiO_2	253	289	48
Al	15	14	1.3
Fe	1.3	1.2	5.9
Ca	4.6	4.3	21
Mg	32	14	7.9
Na	22	22	13
K	15	11	1.7
CO_2	29,200	25,800	13,000
CO	50	34	—
SO_2	1,490	716	716
SO_3	21	11	2.7
H_2S	8.0	42	1,080
S	3.7	1.8	—
Cl	728	420	433
F	238	169	35
Br	1.1	0.9	1.2
B	39	21	5.6
PO_4	2.8	3.0	0.8
NO_2	0.01	0.001	0.008
O_2	51	47	23
H_2	685	381	20
NH_3	1.3	0.8	17
N_2	567	676	1,250
Ar^{40}	0.6	—	1.5
CH_4	1.5	18	—
Ni[a]	<0.01	0.0007	0.0007
Cu[a]	0.03	0.03	0.004
Zn[a]	0.5	0.4	0.02
Ge[a]	<0.01	0.03	0.0007
As[a]	0.7	0.3	0.007
Mo[a]	trace	trace	trace
Ag[a]	0.003	0.0007	<0.0001
Sn[a]	0.3	0.006	0.001
Sb[a]	<0.1	<0.01	<0.005
Pb[a]	0.03	0.03	0.005
Bi[a]	<0.05	<0.005	<0.002
Rn	(1.3)	(2.1)	(12.6)
Total reported	33,432	28,698	16,685

Data are from Nemoto *et al.* (1957; as presented by White and Waring, 1963).

[a] Semiquantitative analyses of solids in condensate.

Some of the simplest of these reactions are as follows:

$$\underset{\textit{halite}}{2\,NaCl} + \underset{\textit{andalusite}}{Al_2SiO_5} + \underset{\textit{silica}}{5\,SiO_2} + \underset{\textit{gas}}{H_2O} \rightleftharpoons \underset{\textit{plagioclase}}{2\,NaAlSi_3O_8} + \underset{\textit{gas}}{2\,HCl} \qquad (15.i)$$

$$\underset{\textit{fluorite}}{CaF_2} + \underset{\textit{silica}}{SiO_2} + \underset{\textit{gas}}{H_2O} \rightleftharpoons \underset{\textit{wollastonite}}{CaSiO_3} + \underset{\textit{gas}}{2\,HF} \qquad (15.j)$$

$$\underset{\textit{fluorite}}{CaF_2} + \underset{\textit{andalusite}}{Al_2SiO_5} + \underset{\textit{silica}}{SiO_2} + \underset{\textit{gas}}{H_2O} \rightleftharpoons \underset{\textit{plagioclase}}{CaAl_2Si_2O_8} + \underset{\textit{gas}}{2\,HF} \qquad (15.k)$$

$$\underset{\textit{fluorite}}{CaF_2} + \underset{\textit{plagioclase}}{2\,NaAlSi_3O_8} + \underset{\textit{gas}}{2\,HCl} \rightleftharpoons \underset{\textit{halite}}{2\,NaCl} + \underset{\textit{silica}}{4\,SiO_2} + \underset{\textit{plagioclase}}{CaAl_2Si_2O_8} + \underset{\textit{gas}}{2\,HF} \qquad (15.l)$$

If all the solid phases are present as stoichiometric crystals the equations of equilibrium for these reactions are:

$$(f_{H_2O}K_{(i)})^{1/2} = f_{HCl} \qquad (15.5)$$

$$(f_{H_2O}K_{(j)})^{1/2} = f_{HF} \qquad (15.6)$$

$$(f_{H_2O}K_{(k)})^{1/2} = f_{HF} \qquad (15.7)$$

$$(K_{(l)})^{1/2} = \frac{f_{HF}}{f_{HCl}} \qquad (15.8)$$

If the crystalline phases are not stoichiometric or if certain of the reactants or products occur only as solution components of magma or vapor, then the acid gas fugacities will either be reduced or increased in comparison with those calculated from the equations.

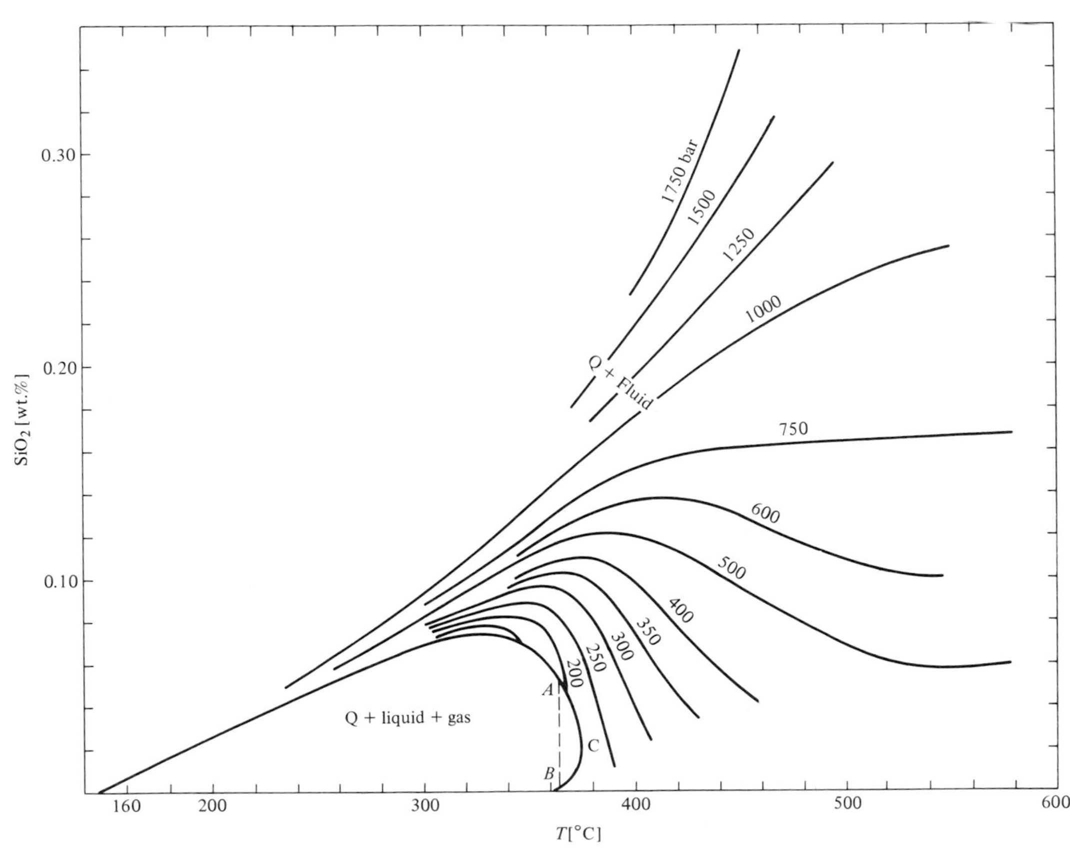

15.12 Isobaric curves of solubility of quartz in H_2O. (After Kennedy, 1950.)

15.13 Isobaric curves of solubility of crystalline NaCl in H_2O. (After Soupirajan and Kennedy, 1962.)

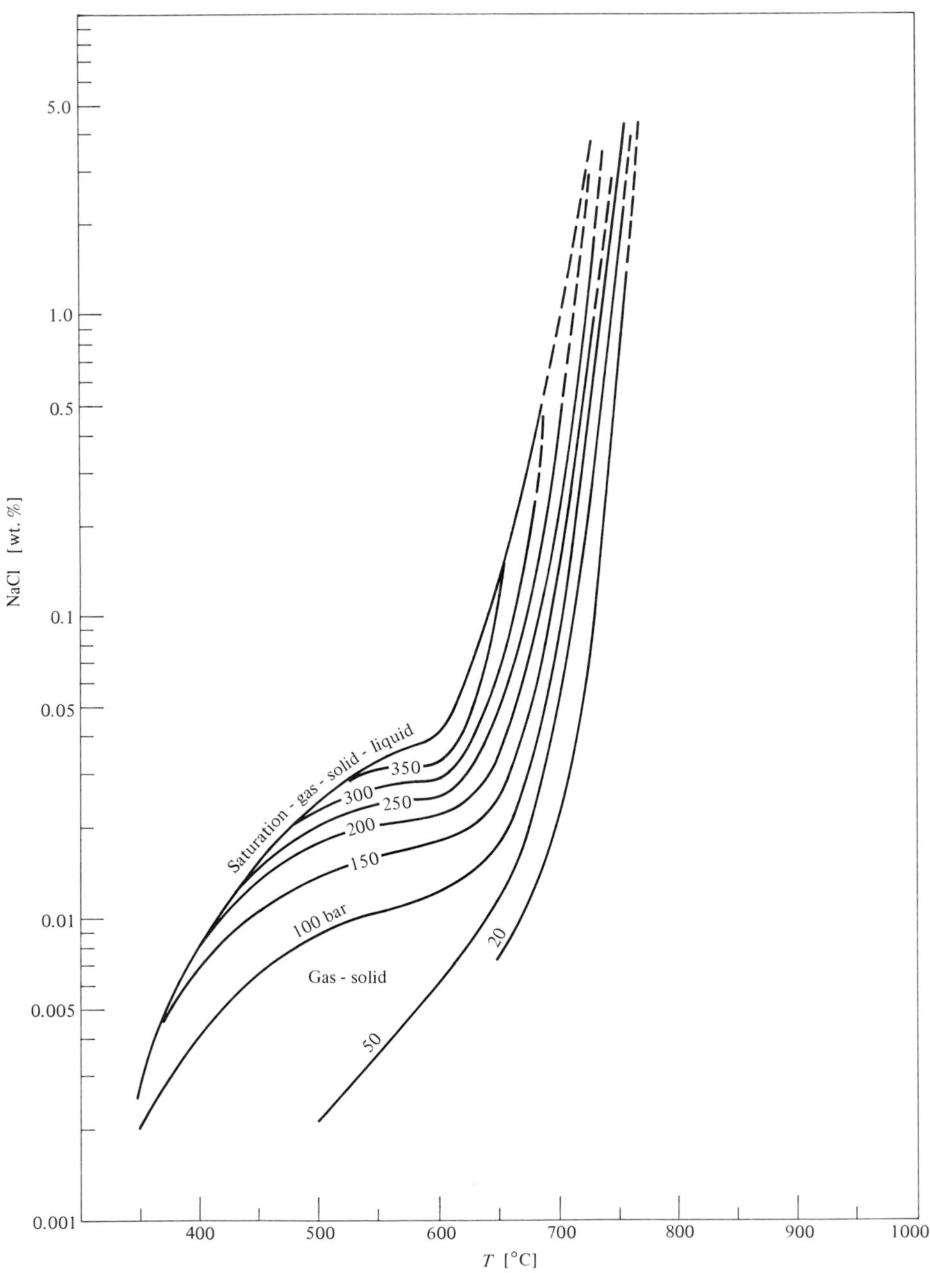

It is also apparent that reactions such as (15.l) are not independent but are obtained through combinations of the others. Equations (15.5) to (15.7) may also be written as

$$f_{H_2O}^{-1/2}K_{(i)}^{1/2} = \frac{f_{HCl}}{f_{H_2O}} \tag{15.9}$$

$$f_{H_2O}^{-1/2}K_{(j)}^{1/2} = \frac{f_{HF}}{f_{H_2O}} \tag{15.10}$$

$$f_{H_2O}^{-1/2}K_{(k)} = \frac{f_{HF}}{f_{H_2O}} \tag{15.11}$$

from which we see that the mole fraction of the acid gas decreases as the pressure increases.

The extent to which these reactions apply to fumarolic gases will depend on the degree of equilibrium attained and their approximation to the actual reactions with the rock or magma. In terms of validity, Reaction (15.1) seems the most appropriate since plagioclase, free silica, and fluorite are common phases in acid to intermediate rocks and NaCl-rich brines are common as mineral inclusions. However, there is no record of NaCl as a primary magmatic precipitate so that we should expect this reaction to be shifted somewhat to the right of the values yielded by Equation (15.8).

If we consider Reactions (15.i) and (15.k) we should expect even greater deviations from the simple equations than for Reaction (15.l). The reason for this is that neither Al_2SiO_5 nor NaCl saturate normal magmas, so that the acid gas fugacities should be smaller than those obtained from Equations (15.5) and (15.7).

The deviations from the simple Equations (15.5) to (15.11) may be checked against existing chemical analyses of fumarolic gases. The curves calculated from Equations (15.5), (15.6), and (15.7) with $f_{H_2O} = 1$ atm are shown in Figure 15.14. From Table 15.5 we note that for specimen No. 1 collected at 760°C (1033°K) the equilibrium value of f_{HCl}/f_{H_2O} as given by Equation (15.5) or (15.9) is $10^{-0.5}$. Although this ratio decreases as $f_{H_2O}^{-1/2}$ with increasing water pressure, it is still $10^{-2.0}$ when $f_{H_2O} = 1000$ atm. However, Table 15.5 shows that the observed value is approximately $728/10^6$ or approximately 10^{-3}, a result which is compatible with undersaturation of Al_2SiO_5 and NaCl.

Figure 15.15 shows the calculated curves corresponding to Equation (15.8) [Reaction (15.l)] and the analogous equation which results from combining Reactions (15.i) and (15.j) (Mueller, 1970). Also shown are a number of observed acid–gas ratios from volcanoes from Alaska, Japan, and Kamchatka as taken from the compilation of White and Waring (1963). It is perhaps important to a detailed interpretation of these data to note that the Katmai, Usu, and Kliuchevskii volcanoes are rhyolitic, dacitic, and basaltic, respectively. An interesting feature of the data is their convergence with the thermochemical curves at approximately the liquidus temperatures of anhydrous melts. Even the slight displacement of the bulk of the observational points to the right of the curves is explicable in terms of the inferred undersaturation of NaCl. It is of course also possible that the feldspar activity function $a_{Ab}/a_{An}^{1/2}$, which is assumed to be unity in Equation (15.8), was not actually so. However, it might be argued that this effect should be small if the plagioclase does not deviate greatly from an ideal solution. To see this we note that $X_{Ab}/X_{An}^{1/2}$ will be unity if $X_{An} = 0.38$, since then $0.38 = (0.62)^2$. While it is unlikely that the feldspars of acid or intermediate rocks will cause this ratio to differ greatly from unity, the same is not true of the basaltic feldspars. However, in this case the effect will be counteracted by the tendency for silica undersaturation to occur.

Although the convergence of the observational points of Figure 15.15 with the theoretical curves is consistent with an approach to equilibrium at high temperatures, the almost equally high values of f_{HF}/f_{HCl} shown by the low-temperature gases indicate that equilibrium was not

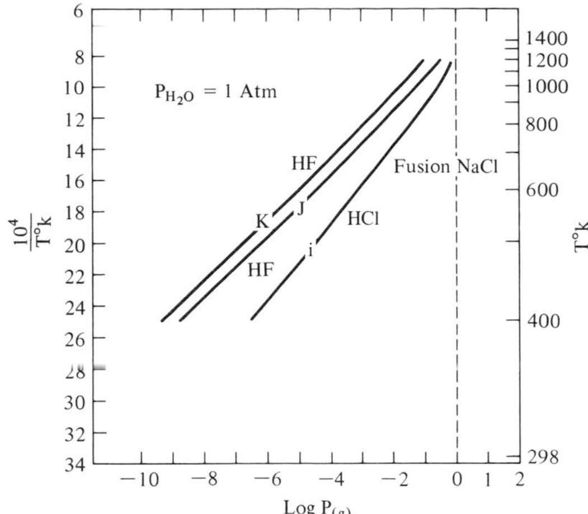

15.14 Logarithms of gas fugacities of HCl and HF as a function of the temperature and at constant fugacities of water and solid constituents. The curves, with letters referring to the corresponding reactions (see text) were calculated from thermochemical data. (After Mueller, 1969d.)

maintained on cooling. It appears that the gases were prevented from back-reacting with the vent rocks and so retained their high-temperature compositions with varying fidelity. This behavior is consistent with the deductions of Nordlie for the Hawaiian gases. It should be noted, however, that recently Stoiber and Rose (1974) reported apparent agreement between observed values of f_{HF}/f_{HCl} from certain Central American volcanoes and the corresponding value from Equation (15.8). Thus it is possible that equilibrium between fumarolic gases and minerals is sometimes attained.

The streaming and concentration of magmatic gases in volcanic vents has many implications for the origin of such extreme rock types as occur in the highly silica-deficient and alkali- and carbonate-rich magmas of carbonatite complexes. Since the low fluid pressures of

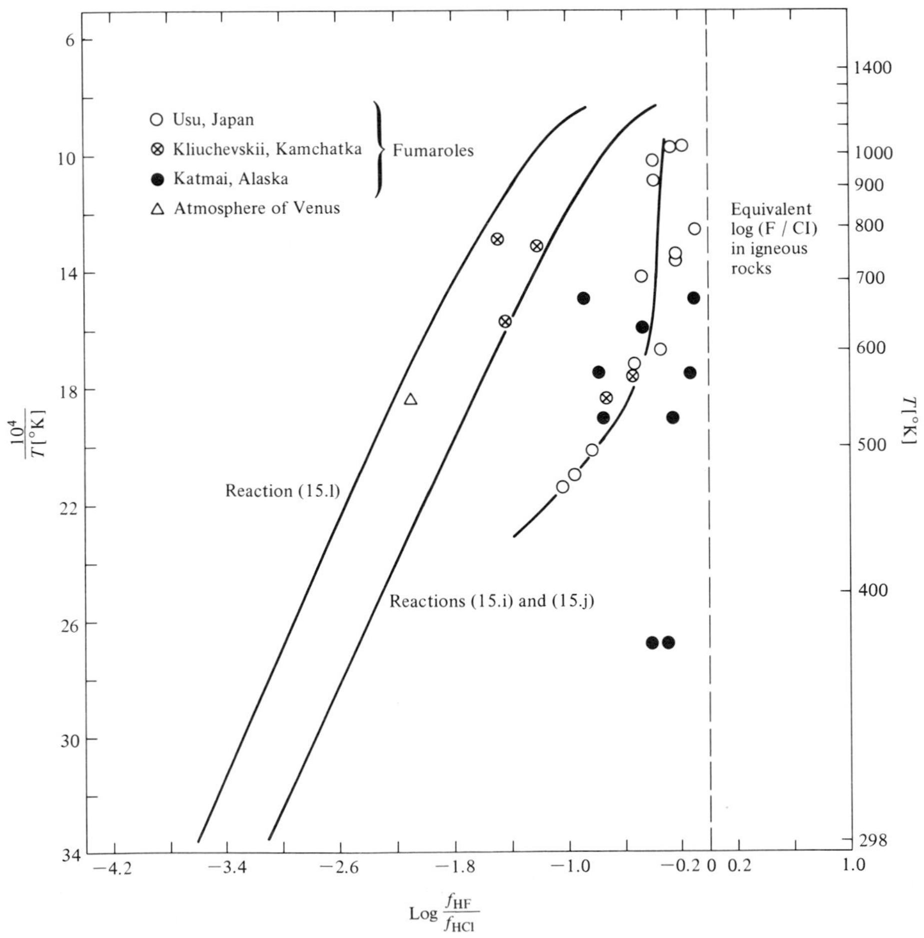

15.15 Logarithms of the fugacity ratios f_{HF}/f_{HCl} (solid lines) and the abundance ratios HF/HCl (data points) as a function of temperature. The fugacity ratios were calculated from thermochemical data and correspond to Reaction (15.l) and combined Reaction (15.i) and (15.j) (see text). The data points represent the HF/HCl abundance ratios from fumaroles and are assumed to be the same as the fugacity ratios. The corresponding value for the Venusian atmosphere is also plotted. The normal range of the equivalent ratio for igneous rocks falls to the right of the unit ratio line which is dashed. (After Mueller, 1969d.)

the volcanic environment do not favor the solubility of either alkali silicates or carbonate minerals in gases, it is likely that rapid precipitation of these constituents would occur in near surface parts of volcanic vents. This may explain the concentration of sodium carbonate in such volcanoes as Oldoinyo Lengai.

Dissolved forms of the magmatic volatiles

The chemical elements that comprise the molecular emissions of magmas must once have been dissolved in the silicate melt. However, the precise chemical combinations they may have assumed therein are still largely mysteries. As yet we have only meager experimental and theoretical knowledge to guide us in determining what these combinations might be (Chapter 12). Certainly the associations of the different elements within the gas and magma need bear no close relationship to each other. It is more likely that the specific gas molecules come into being only as their constituents are liberated. This conclusion becomes all the more apparent if we again consider the melt structure as in the following reactions:

$$\underset{melt}{2FeO} + \underset{melt}{[-Si-OH\ HO-Si-]} \rightleftharpoons \underset{melt}{[-Si-O-Si-]} + \underset{melt}{Fe_2O_3} + \underset{gas}{H_2} \quad (15.m)$$

$$\underset{melt}{[-SiOH\ F-Si-]} \rightleftharpoons \underset{melt}{[-Si-O-Si-]} + \underset{gas}{HF} \quad (15.n)$$

$$\underset{villiaumite}{2NaF} + \underset{melt}{[-Si-O-Si-]} \rightleftharpoons \underset{melt}{Na_2O} + \underset{melt}{[-Si-F\ F-Si-]} \quad (15.0)$$

$$\underset{villiaumite}{NaF} + \underset{gas}{H_2O} + \underset{gas}{[-Si-O-Si-]} \rightleftharpoons \underset{melt}{Na(OH)} + \underset{melt}{[-Si-OH\ F-Si-]} \quad (15.p)$$

$$\underset{villiaumite}{NaF} + \underset{gas}{H_2O} \rightleftharpoons \underset{melt}{Na(OH)} + \underset{melt}{HF} \quad (15.q)$$

$$\underset{fluorite}{CaF_2} + \underset{melt}{[-Si-O-Si-]} \rightleftharpoons \underset{melt}{CaO} + \underset{melt}{[-Si-F\ F-Si-]} \quad (15.r)$$

$$\underset{fluorite}{CaF_2} + \underset{melt}{2H_2O} + 2\underset{melt}{[-Si-O-Si-]} \rightleftharpoons \underset{melt}{Ca(OH)_2} + 2\underset{melt}{[-Si-OH\ F-Si-]} \quad (15.s)$$

$$\underset{melt}{NaCl} + \underset{gas}{H_2O} \rightleftharpoons \underset{melt}{Na(OH)} + \underset{gas}{HCl} \quad (15.t)$$

$$\underset{melt}{NaCl} + \underset{gas}{H_2O} + \underset{melt}{[-Si-O-Si-]} \rightleftharpoons \underset{melt}{Na(OH)} + \underset{melt}{[-Si-OH\ Cl-Si-]} \quad (15.u)$$

$$\underset{melt}{[-Si-OH\ HS-Si-]} \rightleftharpoons \underset{melt}{[-Si-O-Si-]} + \underset{gas}{H_2S} \quad (15.v)$$

$$\underset{melt}{SiO_2} + \underset{melt}{Na_2CO_3} \rightleftharpoons \underset{melt}{Na_2SiO_3} + \underset{gas}{CO_2} \quad (15.w)$$

$$\underset{melt}{SiO_2} + \underset{melt}{CaCO_3} \rightleftharpoons \underset{melt}{CaSiO_3} + \underset{gas}{CO_2} \quad (15.x)$$

$$\underset{melt}{SiO_2} + \underset{melt}{MgCO_3} \rightleftharpoons \underset{melt}{MgSiO_3} + \underset{gas}{CO_2} \quad (15.y)$$

The first of these reactions [Reaction (15.m)] is inferred by analogy with the dehydrogenation of hornblende and biotite and the viscosity data that point to a splitting of the Si-O bond by water. Thus there should be a strong relation between the escape of volatiles and the oxidation state of the melt even when no ferrous silicate crystals are present. Loss of pressure and consequent exsolution of water should bring about the oxidation of at least some dissolved FeO.

We find evidence for the leftward displacement of Reaction (15.n) in the profound lowering of liquidus temperatures and viscosities of silicate systems by HF (Wyllie and Tuttle, 1961). It is likely that its displacement to the right accounts for some of the primary fumarolic emissions of this gas. Furthermore, there is evidence that Reaction (15.o), which involves the same interaction of fluorine and the Si-O chains, occurs even in the absence of water. This is shown for example, by the considerable lowering of the melting temperature of albite by NaF and by evidence for low-viscosity melts in this system (Koster van Groos, 1966). It has also been demonstrated by Koster van Groos that when water is added to the system $NaAlSi_3O_8$-NaF there is not only a further lowering of the liquidus temperatures but also a greatly increased solubility of water over that in the system $NaAlSi_3O_8$-H_2O at corresponding water pressures. This enhancement of the solubility of water by NaF may be explained by reactions such as (15.p) and (15.q).

The affinity of F for the silicate lattice also appears to apply when this element occurs in CaF_2, as was demonstrated by Weymouth and Williamson (1957). When they heated fluorite-bearing granites under nearly anhydrous conditions, it was observed that melt readily formed where fluorite was in contact with mica, microperthite, and albite between 800°C and 900°C. The inferred melting reaction, which occurs at more than 100°C below the normal melting range of anhydrous granites, may be explained by a reaction such as (15.r). By analogy with NaF, we also expect a reaction with water such as (15.s).

Koster van Groos (1966) also found that in the system $NaAlSi_3O_8$-NaCl-H_2O, NaCl had an effect similar to NaF in that it enhanced the solubility of water in the silicate melt, so that one might anticipate a reaction such as (15.t). However, there was little or no evidence for a reaction of the type (15.u) since the effect of NaCl on the melting point of albite was observed to be quite small.

Reaction (15.v) may be regarded as entirely speculative, but it may account for some sulfur solubility of iron-poor magmas under relatively high pressures. In the presence of iron, sulfur probably occurs largely in association with this element. However, if the sulfur pressure is high and iron abundant, exsolution as a immiscible sulfide melt readily occurs.

Reactions (15.w) to (15.y) are inferred to play an important role in governing the solubility of CO_2 in magmas. They show that there should be a close relation between the thermal stability of the carbonate and the activity of silica in the melt. The carbonates Na_2CO_3 and $CaCO_3$ with high thermal stabilities should be able to exist in relatively siliceous melts as compared with $MgCO_3$. It is thus possible that Mg-rich basaltic magmas exsolve much of their CO_2 at relatively high pressures when the magma is still at considerable depth, while silica-poor alkali basalts and undersaturated alkalic rocks maintain this gas in solution at much shallower levels. These processes play a crucial role, in the evolution of alkaline rocks and carbonatites (Wyllie, 1974; Koster van Groos, 1975).

Hot spring waters

The two most easily distinguishable types of hot spring waters are (1) the *sodium chloride* or *alkaline* type and (2) the *acid sulfate* type. In addition, *acid sulfate chloride*, *sodium bicarbonate*, and *calcium bicarbonate* types have also been recognized (White, 1957). The predominant type of water in thermal areas is the sodium chloride type, and it appears that these waters are rather closely related to the magmatic source. White (1957) regarded them as mixtures of highly saline magmatic emanations modified by reactions with the wall rocks and by precipitation of the least soluble constituents. Representative chemical analyses of these different types of waters are shown in Table 15.6.

The sodium chloride waters contain only modest quantities of NaCl as compared with brines, and their concentrations of Fe, Al, Ca, Mg, and F are low. But, like other thermal waters they contain quite high concentrations of SiO_2. According to White *et al.* (1956) this dissolved silica is present in the monomeric form [probably as $Si(OH)_4$]. The observed concentrations correspond rather closely with those inferred from solubility equilibria with amorphous SiO_2 (Krauskopf, 1956). In contrast, the acid sulfate waters have rather high concentrations of Fe and Al. As was stressed particularly by Ellis and Mahon (1964), the chemistry of spring waters is closely related to the wall rock mineralogy so that we can expect their element concentrations to be governed by certain dissolution equilibria involving these minerals. A particularly interesting example of this is the composition of the feldspars that are deposited by the waters or that result from the alteration of the original volcanic feldspars. The result of this alteration is usually an adularia or other potassium-rich variety as found in the Yellowstone Park region and at Wairakei, New Zealand. According to Fenner the potassium is derived in part from magmatic waters and in part from the country rocks. As we have already seen in our discussions of metamorphism and metasomatism, the dissolution of K^+ at high temperatures and its deposition at lower temperatures is well substantiated by the laboratory experiments of Orville (1963). As a result of this process the very potassium-rich feldspars can be deposited from solutions that contain K/Na ratios no larger than those of the hot spring waters.

The ion-exchange reactions of the feldspars are also accompanied by others that result in a net release of cations. For example, feldspar may be converted to clay or mica:

$$\underset{\textit{feldspar}}{3\,NaAlSi_3O_8} + \underset{\textit{solution}}{2\,H_2O} \rightleftharpoons \underset{\textit{mica}}{NaAl_3Si_3O_{10}(OH)_2} + \underset{\textit{solution}}{6\,SiO_2} + \underset{\textit{solution}}{2\,Na(OH)} \quad (15.z)$$

Reactions of this type maintain the neutral or slightly alkaline character of the sodium chloride waters as was

Table 15.6 Representative hot spring waters

Composition is given in ppm

	Sodium chloride[a]	Sodium chloride[a]	Sodium chloride[b]	Acid sulfate[b]	Sodium bicarbonate[a]	Calcium bicarbonate[a]
	Steamboat Springs, Nevada	Morgan Springs, California	Yellowstone Park, Wyoming	Yellowstone Park, Wyoming	Well 5 Wairakei, New Zealand	Mammoth, Yellowstone Park, Wyoming
Temp. (°C)	89.2	95.4	87	73	>100	70.5
pH:	7.9	7.83	7.32[d]	2.45[d]	6.7	6.9
SiO_2	293	233	256.0	230.0	364	56
Fe	—[c]	—	0.05	0.25	2.9	—
Al	—	—	ND[e]	5.25	1.6	—
Ca	5.0	79	1.6	0.20	2.6	209
Mg	0.8	0.8	ND	ND	1.0	78
Sr	1	10	—	—	—	—
Na	653	1400	395.6	158.0	27	129
K	71	196	38.0	14.4	24	56
Li	7.6	9.2	—	—	0.1	1.4
NH_4	<1	<1	—	—	10	—
As	2.7	2.2	—	—	—	—
Sb	0.4	0.0	—	—	—	—
CO_3	0	0	—	—	—	—
HCO_3	305	52	—	—	nil	526
SO_4	100	79	145.0	404.0	482	529
Cl	865	2430	512.0	166.0	6.0	169
F	1.8	1.5	—	—	—	—
Br	0.2	0.8	—	—	—	—
I	0.1	<0.1	—	—	—	—
B	49	88	—	—	1.5	4.4
H_2S	4.7	0.7	—	—	—	2.4
PO_4	—	—	0.55	0.57	—	—

[a] Data are from White (1957).
[b] Data are from Raymahashay (1968).
[c] — indicates no determination.
[d] pH measured at the temperature recorded.
[e] ND indicates not detectable.

noted by Day and Allen (1925) in their classic study of the Mt. Lassen thermal region of California.

In the acid sulfate springs we encounter a situation quite different from the alkaline springs. Since the sulfate springs are acid they must result from reactions other than with silicates. According to White (1957), the acid character results from the condensation of steam boiled from sodium chloride waters at depth. This steam would contain greater concentrations of volatile H_2S than the parent waters, and this would react with oxygen near the surface to produce sulfuric acid. Another oxidation product is native sulfur, which results from the reaction $H_2S + \frac{1}{2}O_2 \rightleftharpoons S + H_2O$. Of course, highly acid sulfate springs can exist only where the production of H_2SO_4 is at least as rapid as its destruction or loss. The destruction of H_2SO_4 comes about by reactions that result in the distinctive mineralogy of the acid sulfate springs. One of these produces alunite as follows:

$$\underset{\textit{feldspar}}{3KAlSi_3O_8} + \underset{\textit{solution}}{3H_2SO_4} \longrightarrow \underset{\textit{alunite}}{KAl_3(SO_4)_2(OH)_6} + \underset{\textit{solution}}{9SiO_2} + \underset{\textit{solution}}{K_2SO_4} \quad (15.aa)$$

while another important product is kaolinite:

$$\underset{\textit{feldspar}}{2KAlSi_3O_8} + \underset{\textit{solution}}{(H_2SO_4 + H_2O)} \longrightarrow \underset{\textit{kaolinite}}{Al_2Si_2O_5(OH)_4} + \underset{\textit{silica}}{4SiO_2} + \underset{\textit{solution}}{K_2SO_4} \quad (15.bb)$$

The last two reactions are regarded as irreversible because the feldspar attacked was formed under other conditions. Also it is clear from all the foregoing reactions that silica is one of the most important products of the degradation of rocks by hot spring waters. This conclusion is different from that of Raymahashay (1968), who concluded that quartz in the altered rocks is largely residual!

The high aluminum content of the acid sulfate waters is attributable to the further equilibrium decomposition of kaolinite and alunite in the acid waters. Following Raymahashay we may write:

$$\underset{\textit{kaolinite}}{Al_2Si_2O_5(OH)_4} + \underset{\textit{solution}}{6H^+} \rightleftharpoons \underset{\textit{solution}}{2Al^{3+}} + \underset{\textit{solution}}{2Si(OH)_4} + \underset{\textit{solution}}{H_2O} \quad (15.cc)$$

$$\underset{\textit{alunite}}{KAl_3(SO_4)_2(OH)_6} + \underset{\textit{solution}}{6H^+} \rightleftharpoons \underset{\textit{solution}}{K^+} + \underset{\textit{solution}}{3Al^{3+}} + \underset{\textit{solution}}{2SO_4^{2-}} + \underset{\textit{solution}}{6H_2O}. \quad (15.dd)$$

In addition to the dissolved major rock constituents, hot spring waters also transport a variety of minor elements in detectable quantities (Table 15.6). Particularly common among the dissolved trace metals are As, Sb, and Hg. It was shown by Krauskopf (1951) that mercury can be transported at hot spring temperatures by alkaline sulfide solutions, as a volatile chloride, or as vapor of metallic mercury. In the Steamboat Springs thermal area of Nevada, considerable mercury is transported by the vapor emitted by the spring waters, while cinnabar, stibnite, pyrite, and other sulfides are deposited in siliceous sinter and gravel beds. Also some siliceous muds deposited by the springs contain concentrations of gold, antimony, silver, mercury, copper, and arsenic (Brannock *et al.*, 1948). With proper caution such deposits may provide many clues to the transportation and deposition of metals under deep-seated conditions.

References

Adams, L. H. (1953). A note on the stability of jadeite. *Am. J. Sci. 251*, 299–308.

Adler, I., Tromka, J., Gerard, V., Lowman, P., Schmadebeck, R., Blodget, H., Eller, E., Yin, L., Lamothe, R., Osswald, O., Gorenstein, P., Bjorkholm, P., Gursky, H., and Harris, B. (1972). Apollo 15 geochemical fluorescence experiment. Preliminary report. X-641-72-198, Goddard Space Flight Center, Greenbelt, Md.

Akella, J., and Winkler, H. G. F. (1966). Orthorhombic amphibole in some metamorphic reactions. *Contr. Mineral. Petrol. 12*, 1–12.

Albee, A. L. (1965a). A petrogenetic grid for the Fe-Mg silicates of pelitic schists. *Am. J. Sci. 263*, 512–536.

Albee, A. L. (1965b). Distribution of Fe, Mg, and Mn between garnet and biotite in natural mineral assemblages. *J. Geol. 73*, 155–164.

Albee, A. L. (1965c). Phase equilibria in three assemblages of kyanite-zone pelitic schists, Lincoln mountain quadrangle, central Vermont. *J. Petrol. 6*, 246–301.

Albee, A. L., and Zen, E-an. (1969). Dependence of the zeolitic facies on the chemical potentials of CO_2 and H_2O. In Zharikov, V. A. (ed.), Korzhinskii vol. pp. 249–260. Moscow.

Albee, A. L. (1972). Metamorphism of pelitic schists: Reaction relations of chloritoid and staurolite. *Geol. Soc. Am. Bull. 83*, 3249–3268.

Alfvén, H. (1954). *On the Origin of the Solar System.* Oxford: Clarendon.

Alfvén, H. (1962). On the mass distribution in the solar system. *Astrophys. J. 136*, 1005–1015.

Alfvén, H. (1963). On the early history of the sun and the formation of the solar system. *Astrophys. J. 137*, 981–990.

Allen, J. C., Modreski, P. J., Haygood, C., and Boettcher, A. L. (1972). The role of water in the mantle of the earth: The stability of amphiboles and micas. *Proc. 24th Intern. Geol. Cong., Sec. 2*, 231–240.

Aller, L. H. (1961). *The Abundance of Elements.* New York: Interscience.

Althaus, E. (1966). Der stabilitatsbereich des Pyrophyllits unter dem Einfluss von Sauren. I. Mitteilung, Experimentelle Untersuchungen. *Contr. Mineral. Petrol. 13*, 31–50.

Althaus, E. (1967). The triple point andalusite-sillimanite-kyanite. *Contr. Mineral. Petrol. 16*, 29–44.

Althaus, E., Nitsch, K. H., and Winkler, H. G. F. (1970). An experimental re-examination of the stability limits of muscovite plus quartz. *Neues. Jb. Mineral. 7*, 325–336.

Anders, E. (1962). Meteorite ages. *Rev. Mod. Phys. 34*, 287–325.

Anders. E. (1964). Origin, age and composition of meteorites. *Space Sci. Rev. 3*, 583–714.

Anderson, D. L. (1973). Composition and origin of the Moon. *Earth Planet. Sci. Lett. 18*, 301–316.

Anderson, G. M. (1964). The calculated fugacity of water to 1000°C and 10,000 bars. *Geochem. Cosmochim. Acta 28*, 713–715.

Annersten, H. (1968). A mineral chemical study of a metamorphosed iron formation in northern Sweden. *Lithos 1*, 374–397.

Apollo 17 preliminary examination team (1973). Apollo 17 Lunar Samples: Chemical and petrographic description. *Science 182*, 659–672.

Atherton, M. P. (1968). The variation in garnet, biotite and chlorite composition in medium grade pelitic rocks from the Dalradian, Scotland, with particular reference to the zonation in garnet. *Contr. Mineral. Petrol. 18*, 347–371.

Ave 'Lallemont, H. G., and Carter, N. L. (1970). Syntectonic recrystallization of olivine and modes of flow in the upper mantle. *Geol. Soc. Am. Bull. 81*, 2203–2220.

Baedecker, P. A., and Wasson, J. T. (1975). Elemental fractionations among enstatite chondrites. *Geochim. Cosmochim. Acta 39*, 735–765.

Bailey, A. (1971). Comparison of low-temperature with high-temperature diffusion of sodium in albite. *Geochim. Cosmochim. Acta 35*, 1073–1081.

Bailey, D. K. (1969). The stability of acmite in the presence of H_2O. *Am. J. Sci. 267A*, 1–16.

Bailey, E. H., and Blake, M. C., Jr. (1969). Late Mesozoic sedimentation and deformation in western California. *Geotektonika 3*, 17–34; *4*, 24–34.

Baldwin, R. (1949). *The Face of the Moon*. Chicago: University of Chicago Press.

Balk, R. (1937). Structural behavior of igneous rocks. *Geol. Soc. Am. Mem. 5*, 1–177.

Barrow, G. (1912). On the geology of lower Dee-side and the southern Highland border. *Geol. Assoc. Proc. 23*,268–284.

Barth, T. F. W. (1962). *Theoretical Petrology* (2nd ed.). New York: Wiley.

Barth, T. F. W. (1969). *Feldspars*. New York: Wiley-Interscience.

Barton, P. B., Jr., and Toulmin, P., III. (1966). Phase relations involving sphalerite in the Fe-Zn-S system. *Econ. Geol. 61*, 815–849.

Bateman, P. C., Clark, L. D., Huber, N. K., Moore, J. G., and Reinhart, C. D. (1963). The Sierra Nevada batholith, a synthesis of recent work across the central part. U.S. Geol. Survey prof. paper 414-D.

Becke, F. (1913). Uber Mineralbestand und Structur der Krystallinischen Schiefer. *Ksehr. Akad. Wiss. Wien, 78*, 1–53.

Beloussov, V. V. (1960). Tectonophysical investigations. *Bull. Geol. Soc. Am. 71*, 1255.

Benson, W. N. (1941). Cainozoic petrographic provinces of New Zealand and their residual magmas. *Am. J. Sci. 239*, 537–552.

Berman, R., and Simon, F. (1955). On the graphite-diamond equilibrium. *Z. Elektrochem. 59*, 333.

Berner, R. A. (1971). *Principles of Chemical Sedimentology*. New York: McGraw-Hill.

Bernstein, F. (1960). Distribution of water and electrolyte between homoionic clays and saturating NaCl solutions. *Proceedings of the Eighth National Conference on Clays and Clay Minerals*. Oxford: Pergamon, pp. 122–149.

Bhattacharji, S. (1967). Mechanics of flow differentiation in ultramafic and mafic sills. *J. Geol. 75*, 101–112.

Bhattacharji, S., and Nehru, C. E. (1972). Igneous differentiation models for the origin of Mount Johnson, a zoned monteregion intrusion, Quebec, Canada. *24th Intern. Geol. Cong. Sec. 14*, 3–17.

Bhattacharji, S., and Smith, C. S. (1964). Flowage differentiation. *Science 145*, 150–153.

Birch, F., and Bancroft, D. (1942). The elasticity of glass at high temperatures and the vitreous basaltic stratum. *Am. J. Sci. 240*, 457–490.

Bird, G. W., and Fawcett, J. J. (1973). Stability relations of Mg-chlorite-muscovite and quartz between 5 and 10 kb. water pressure. *J. Petrol. 14*, 415–428.

Blake, M. C., Jr., Irwin, W. P., and Coleman, R. G. (1969). Blueschist facies metamorphism related to regional thrust faulting. *Tectonophysics 8*, 237–246.

Blander, M., and Katz, J. L. (1967). Condensation of primordial dust. *Geochim. Cosmochim. Acta 31*, 1025–1034.

Blander, M., and Fuchs, L. H. (1975). Calcium-aluminum rich inclusions in the Allende meteorite: evidence for a liquid origin. *Geochim. Cosmochim. Acta 39*, 1605–1619.

Bloxam, T. W. (1966). Jadeite-rocks and blueschists in California. *Geol. Soc. Am. Bull. 77*, 781–786.

Boettcher, A. L. (1970). The system $CaO\text{-}Al_2O_3\text{-}SiO_2\text{-}H_2O$ at high pressures and temperatures. *J. Petrol. 11*, 337–379.

Boettcher, A. L. (1973). Volcanism and orogenic belts—The origin of andesites. *Tectonophysics 17*, 223–240.

Boettcher, A. L., and Wyllie, P. J. (1968). Melting of granite with excess water to 30 kilobars pressure. *J. Geol. 76*, 235–244.

Boettcher, A. L., and Wyllie, P. J. (1969). Phase relationships in the system $NaAlSiO_4\text{-}SiO_2\text{-}H_2O$ to 35 kilobars pressure. *Am. J. Sci. 267*, 875–909.

Bolam, T. R. (1932). *The Donnan Equilibria and Their Applications to Chemical, Physiological and Technical Processes*. London: Bell.

Bottinga, Y. A., and Weill, D. F. (1970). Densities of liquid silicate systems calculated from partial molar volumes of oxide components. *Am. J. Sci. 269*, 169–182.

Bottinga, Y. A., and Weill, D. F. (1972). The viscosity of magmatic silicate liquids: A model for calculation. *Am. J. Sci. 272*, 438–475.

Bowen, N. L. (1913). The melting phenomena of the plagioclase feldspars. *Am. J. Sci. 35*, 577–599.

Bowen, N. L. (1921). Diffusion in silicate melts. *J. Geol. 29*, 295–317.

Bowen, N. L. (1928). *The Evolution of the Igneous Rocks*. Princeton, N.J.: Princeton University Press.

Bowen, N. L., and Schairer, J. F. (1935). The system $MgO\text{-}FeO\text{-}SiO_2$. *Am. J. Sci. 29*, 151–217.

Bowen, N. L., and Tuttle, O. F. (1949). The system MgO-SiO_2-H_2O. *Geol. Soc. Am. Bull. 60*, 439–460.

Boyd, F. R. (1959). Hydrothermal investigations of amphiboles. In *Researches in Geochemistry* (P. H. Ableson, ed.). New York: Wiley, pp. 377–396.

Boyd, F. R. (1961). Welded tuffs and flows. The rhyolite plateau of Yellowstone Park, Wyoming. *Geol. Soc. Am. Bull. 72*, 387–426.

Boyd, F. R. (1970). Garnet peridotites and the system $CaSiO_3$-$MgSiO_3$-Al_2O_3. *Mineral. Soc. Am., Spec. Paper 3*, 63–67.

Boyd, F. R. (1973). A pyroxene geothermometer. *Geochim. Cosmochim. Acta 37*, 2533–2546.

Boyd, F. R., and England, J. L. (1959). Pyrope. *Carnegie Inst. Geophys. Lab. Yearbook*, 1958–1959, 83–87.

Bragg, W. L., and Williams, E. J. (1934). Effect of thermal agitation on atomic arrangement in alloys. *Proc. Roy. Soc.* (Lond.) *145A*, 699–730.

Brannock, W. W., Fix, P. F., Gianella, V. P., and White, D. E. (1948). Preliminary geochemical results at Steamboat Springs, Nevada. *Trans. Am. Geophys. Union 29*, 211–226.

Braun, G., and Stout, J. H. (1975). Some chemographic relationships in *n*-component systems. *Geochim. Cosmochim. Acta 39*, 1259–1267.

Brett, R. (1966). Cohenite in meteorites, a proposed origin. *Science 153*, 60–62.

Brewer, L. (1951). The equilibrium distribution of elements in the earth's gravitational field. *J. Geol. 59*, 490–497.

Bricker, O. P., Nesbitt, H. W., and Gunter, W. D. (1973). The stability of talc. *Am. Mineral. 58*, 64–72.

Brindley, G. W., and Hayami, R. (1965). Kinetics and mechanism of formation of forsterite (Mg_2SiO_4) by solid state reaction of MgO and SiO_2. *Philosoph. Mag. 12*, 505–514.

Brothers, R. N. (1954). Glaucophane schists from the North Berkeley Hills, California. *Am. J. Sci. 252*, 614–626.

Brown, H. (1947). An experimental method for the estimation of the age of the elements. *Phys. Rev. 72*, 348–349.

Brown, H., and Patterson, C. (1948). The composition of meteoritic matter. 3. Phase equilibria, genetic relationships and planet structure. *J. Geology 56*, 85–111.

Brown, E. H. (1967). The greenschist facies in part of eastern Otago, New Zealand, *Contr. Mineral. Petrol. 14*, 259–292.

Brown, E. H. (1975). A petrogenetic grid for reactions producing biotite and other Al-Fe-Mg silicates in the greenschist facies. *J. Petrol. 16*, 258–271.

Brown, G. M., Emeleus, C. H., Holland, J. G., Peckett, A., and Phyllips, R. (1971). Picrite basalts, ferrobasalts, feldspathic norites and rhyolites in a strongly fractionated lunar crust. *Proceedings of the Second Lunar Science Conference* Boston: M.I.T. Press, Vol. 1, pp. 583–605.

Bryhni, I., Green, D. H., Heier, K. S., and Fyfe, W. S. (1970). On the occurrence of eclogite in western Norway. *Contr. Mineral. Petrol. 26*, 12–19.

Buchwald, V. F. (1966). The iron-nickel-phosphorus system and the structure of meteorites. *Acta Polytech. Scand. Chem. Met. Ser. 51*, 1–46.

Buddington, A. F. (1929). Granite phocoliths and their contact zones in the northwest Adirondacks. *N.Y. State Museum Bull. 281*, 51–107.

Buddington, A. F. (1939). Adirondack igneous rocks and their metamorphism. *Geol. Soc. Am. Mem. 7.*

Buddington, A. F., and Lindsley, D. H., (1964). Iron-titanium oxide minerals and synthetic equivalents. *J. Petrol. 5*, 310–357.

Buerger, M. J., and Washken, E. (1947). Metamorphism of minerals. *Am. Mineral. 32*, 296–308.

Buerger, M. J. (1948). The structural nature of the mineralizer action of fluorine and hydroxyl. *Am. Mineral. 33*, 744–747.

Bunch, T. E., Keil, K., and Olsen, E. J. (1970). Mineralogy and petrology of silicate inclusions in iron meteorites. *Contr. Mineral. Petrol. 25*, 297–340.

Bunch, T. E., and Olsen, E. (1974). Restudy of pyroxene-pyroxene equilibration temperatures for ordinary chondritic meteorites. *Contr. Mineral. Petrol. 43*, 83–90.

Bundy, F. P. (1968). Direct phase transformations in carbon. In *Reactivity of Solids* (J. W. Mitchell, *et al.*, ed.). New York: Wiley-Interscience.

Burke, J. E. (1948). Metals Tech. T.P. 2472.

Burnham, C. W. (1967). Hydrothermal fluids at the magmatic stage. In *Geochemistry of hydrothermal ore deposits* (H. L. Barnes, ed.), New York: Holt, Rinehart and Winston, p. 34–76.

Burnham, C. W. (1975). Water and magmas; a mixing model. *Geochim. Cosmochim. Acta 39*, 1077–1084.

Burnham, C. W., and Davis, N. F. (1971). The role of H_2O in silicate melts. I. *P-V-T* relations in the system $NaAlSi_3O_8$-H_2O to 10 kilobars and 1000°C. *Am. J. Sci. 270*, 54–79.

Burnham, C. W., and Davis, N. F. (1974). The role of H_2O in silicate melts: II. Thermodynamic and phase relations in the system $NAlSi_3O_8$-H_2O to 10 kilobars and 700°C to 1100°C. *Am. J. Sci. 274*, 902–940.

Burnham, C. W., Holloway, J. R., and Davis, N. F. (1969). Thermodynamic properties of water to 1000°C and 10,000 bars. *Geol. Soc. Am., Spec. Pap. 132*, 1–96.

Burnham, C. W., and Jahns, R. H. (1962). A method for determining the solubility of water in silicate melts. *Am. J. Sci.* V. 260, p. 721–745.

Butler, B. C. M. (1967). Chemical study of minerals from the Moine schists of the Ardnamurchan area, Argyllshire, Scotland. *J. Petrol. 8*, 233–267.

Butler, P. Jr. (1969). Mineral compositions and equilibria in the metamorphosed iron formation of the Gagnon Region, Quebec, Canada. *J. Petrol. 10*, 56–101.

Cameron, A. G. W. (1968). A new table of abundances. In *Origin and Distribution of the Elements* (L. H. Ahrens, ed.), New York: Pergamon, pp. 125–143.

Cameron, E. N. (1963). Structure and rock sequences of the critical zone of the eastern Bushveld Complex. *Mineral. Soc. Am. Spec. Paper 1*, 93–107.

Cameron, K. L. (1975). An experimental study of actinolite-cummingtonite phase relations with notes on the synthesis of Fe-rich anthophyllite. *Am. Mineral. 60*, 375–390.

Card, K. D. (1964). Metamorphism in the Agnew Lake Area, Sudbury District, Ontario, Canada. *Bull. Geol. Soc. Am. 75*, 1011–1030.

Carmichael, I. S. E., Turner, F. J., and Verhoogen, J. (1974). *Igneous Petrology*. New York: McGraw-Hill.

Carron, J.-P. (1969). Vue d'ensemble sur la rhéologie des magmas silicates naturels. *Soc. franc. Mineral. crystallographie. Bull. 92*, 435–446.

Carswell, D. A. (1968). Picrite Magma-residual dunite relationships in garnet peridotite at Kalskaret near Tafjord, South Norway. *Contr. Mineral. Petrol. 19*, 97–124.

Chamberlin, T. C. (1928). *The Two Solar Families*. Chicago: University of Chicago Press.

Chao, E. C. T., Shoemaker, E. M., and Madsen, B. M. (1960). First natural occurrence of coesite. *Science 132*, 220–222.

Charles, R. W. (1975). The phase equilibria of richterite and ferrorichterite. *Am. Mineral. 60*, 367–374.

Chatterjee, N. D. (1970). Synthesis and upper stability of paragonite, *Contr. Mineral. Petrol. 27*, 244–257.

Chatterjee, N. D. (1972). The upper stability limit of the assemblage paragonite + quartz and its natural occurrences. *Contr. Mineral. Petrol. 34*, 288–303.

Chatterjee, N. D., and Froese, E. (1975). A thermodynamic study of the pseudobinary join muscovite-paragonite in the system $KAlSi_3O_8$-$NaAlSi_3O_8$-Al_2O_3-SiO_2-H_2O. *Am. Mineral. 60*, 985–993.

Chayes, F. (1970). On estimating the magnitude of the hidden zone and compositions of the residual liquids in the Skaergaard layered series. *J. Petrol. 11*, 1–14.

Chernosky, J. V., Jr., (1974). The upper stability of clinochlore at low pressure and the free energy of formation of Mg-cordierite. *Am. Mineral. 59*, 496–507.

Chinner, G. A. (1960). Pelitic gneisses with varying ferrous/ferric ratios from Glen Clova, Angus, Scotland. *J. Petrol. 1*, 178–217.

Chinner, G. A. (1966). The distribution of temperature and pressure during Dalradian metamorphism. *Geol. Soc. Lond. Quart. J. 486*, 159–186.

Chinner, G. A. (1967). Chloritoid, and the isochemical character of Barrow's zones. *J. Petrol. 8*, 268–82.

Christiansen, R. L., and Lipman, P. W. (1966). Emplacement and thermal history of a rhyolite lava flow near Fortymile canyon, Southern Nevada. *Geol. Soc. Am. Bull. 77*, 671–684.

Clark, B. C., Toulmin, P. III, Baird, A. K., Keil, K., and Rose, H. J. Jr. (1976). Argon content of the Martian atmosphere at Viking I landing site: Analysis by X-ray fluorescence spectroscopy. *Science 193*, 804–805.

Clark, S. P., Jr. (1957). A note on calcite-aragonite equilibrium. *Am. Mineral. 42*, 564–566.

Clark, S. P., Jr., and Kullerud, G. (1959). Iron meteorites. Yearbook 58, The Carnegie Institution of Washington, Geophysical Laboratory, pp. 167–170.

Clark, S. P., Jr. (1966). High pressure phase equilibria. In *Handbook of Physical Constants, Geol. Soc. Am., Mem. 97*, 345–370.

Clavan, W., McNabb, W. M., and Watson, E. H. (1954). Some hypersthenes from southeastern Pennsylvania and Delaware. *Am. Mineral. 39*, 566–570.

Clayton, R. N., and Epstein, S. (1958). The relationship between O^{18}/O^{16} ratios in coexisting quartz, carbonate, and iron oxides from various geologic deposits. *J. Geol. 66*, 352–373.

Clements, J. M. (1903). The Vermilion iron-bearing district of Minnesota. U.S. *Geol. Survey* monog. V. XLV.

Cloos, E. (1936). Der Sierra-Nevada-Pluton in Californien. *Neues Jahrb. Mineral. Geol., Palaont., Beil. Bd., Abt. B., 76*, 355–450.

Cloos, H. (1925). *Einfuhrung in die tektonische Behandlung magmatischer Erscheinungen* (*Granittektonik*). Berlin: Gebruder Bronbraeger.

Cobel, R. L. (1963). A model for boundary diffusion controlled creep in polycrystalline materials. *J. Appl. Phys. 34*, 1679–1682.

Coleman, R. G., and Lee, D. E. (1962). Metamorphic aragonite in the glaucophane schists of Cazadero, California. *Am. J. Sci. 260*, 577–595.

Coleman, R. G., and Lee, D. E. (1963). Glaucophane-bearing metamorphic rocktypes of the Cazadero area, California. *J. Petrol. 4*, 260–301.

Coleman, R. G., Beatty, L. B., and Brannock, W. W. (1965). Eclogites and eclogites: Their differences and similarities. *Bull. Geol. Soc. Am. 76*, 483–508.

Compton, R. R. (1955). Trondhjemite batholith near Bidwell Bar, California. *Bull. Geol. Soc. Am. 66*, 9–44.

Connes, P., Connes, J., Benedict, W. S., and Kaplan, L. D. (1967). Traces of HCl and HF in the atmosphere of Venus. *Astrophys. J. 147*, 1230–1237.

Coombs, D. S., Ellis, A. J., Fyfe, W. S., and Tayler, A. M. (1959). The zeolite facies, with comments on the interpolation of hydrothermal synthesis. *Geochim. Cosmochim. Acta 17*, 53–107.

Coombs, D. S., and Wilkinson, J. F. G. (1969). Lineages and fractionation trends in undersaturated volcanic rocks from the east Otago province (New Zealand) and related rocks. *J. Petrol. 10*, 440–501.

Coombs, D. S., Horodyski, R. J., and Naylor, R. S. (1970). Occurrence of prehnite-pumpellyite facies metamorphism in northern Maine. *Am. J. Sci. 268*, 142–156.

Cook, E. F. (1963). Ignimbrites of the Great Basin, U.S.A. *Bull. Volcanol. 25*, 89–96.

Crawford, W. A., and Fyfe, W. S. (1965). Calcite-aragonite equilibrium at 100°C. *Science 144*, 1569–1570.

Cross, W., Iddings, J. P., Pirsson, L. V., and Washington, H. S. (1902). A quantitative chemicomineralogical classification and nomenclature of igneous rocks. *J. Geol. 10*, 555–690.

Crutcher, R. M. (1973). Observation of molecules in interstellar dust clouds. *Astrophys. J. 185*, 857–868.

Currie, K. L. (1971). The reaction 3 cordierite = 2 garnet + 4 sillimanite + 5 quartz as a geological thermometer in the Opinicon Lake Region, Ontario. *Contr. Mineral. Petrol. 33*, 215–226.

Czamanske, G. K., and Wones, D. R. (1973). Oxidation during magmatic differentiation, Finnmarka complex, Oslo area, Norway. II. The mafic silicates. *J. Petrol. 14*, 349–380.

Dallmeyer, R. D., and Dodd, R. T. (1971). Distribution and significance of cordierite in paragneisses of the Hudson Highlands, southeastern New York. *Contr. Mineral. Petrol. 33*, 289–308.

Dallmeyer, R. D. (1974). Metamorphic history of the northeastern Reading Prog, New York and northern New Jersey. *J. Petrol. 15*, 325–329.

Daly, R. A. (1928). Bushveld igneous complex of the Transvaal. *Geol. Soc. Am. Bull. 39*, 703–768.

Daly, R. A. (1933). *Igneous Rocks and the Depths of the Earth*. New York: McGraw-Hill.

Danielson, A. (1950). Das Calcit-Wollastonitgleichgewicht. *Geochim. Cosmochim. Acta 1*, 55–69.

Darken, L. S., and Gurry, R. W. (1953). *Physical Chemistry of Metals*. New York: McGraw-Hill.

Dawson, J. B. (1962a). The geology of Oldoinyo Lengai. *Bull. Volcanol. 24*, 349–387.

Dawson, J. B. (1962b). Sodium carbonate lavas from Oldoinyo Lengai, Tanganyika. *Nature, 195*, 1075–1076.

Day, A. L., and Allen, E. T. (1925). The volcanic activity and hot springs of Lassen Peak. Washington D.C.: Carnegie Institution, Publ. 360.

Day, H. W. (1972). Geometrical analysis of the phase equilibria in ternary systems of six phases. *Am. J. Sci. 272*, 711–734.

Day, H. W. (1973). The high temperature stability of muscovite plus quartz. *Am. Mineral. 58*, 255–262.

De Waard, D. (1967). The occurrence of garnet in the granulite-facies terrain of the Adirondack highlands. *J. Petrol. 8*, 210–232.

Dienes, G. J. (1955). Kinetics of order-disorder transformations. *Acta Met. 3*, 549–557.

Dodd, R. T. (1969). Metamorphism of the ordinary chondrites. A review. *Geochim. Cosmochim. Acta 33*, 161–203.

Donnan, F. G., and Guggenheim, E. A. (1932). Die Genaue Thermodynamik der Membrangleichgewicht. *Z. Physik. Chem. 162(A)*, 346–360.

Drake, J. C. (1968). The composition of almandine and coexisting minerals in the mica schists of the Errol Quadrangle, New Hampshire–Maine. Ph.D. Thesis, Harvard University.

Drake, M. J., and Weill, D. F. (1975). The partition of Sr, Ba, Ca, Y, Eu^{2+}, Eu^{3+} and other REE between plagioclase feldspar and magmatic silicate liquid: An experimental study. *Geochim. Cosmochim. Acta 39*, 689–712.

Drake, M. J. (1976). Plagioclase-melt equilibria. *Geochim. Cosmochim. Acta 40*, 457–466.

Dufay, J. (1957). *Galactic Nebulae and Interstellar Matter.* New York: Philosophical Library.

DuFresne, E. R., and Anders, E. (1962). On the chemical evolution of the carbonaceaous chondrites. *Geochim. Cosmochim. Acta 26*, 1085–1114.

Dunning, W. J. (1955). Theory of crystal nucleations from vapor, liquid and solid systems. In *Chemistry of the Solid State* (W. E. Garner, ed.). London: Butterworths, pp. 159–183.

Edgar, A. D. (1974). Experimental studies. *The Alkaline Rocks* (H. Sorensen, ed.). New York: Wiley, pp. 355–389.

Edmunds, W. M., and Atherton, M. P. (1971). Polymetamorphic evolution of garnet in the Fanad Aureole, Donegal, Eire. *Lithos 4*, 147–161.

Eggler, D. H., and Burnham, C. W. (1973). Crystallization and fractionation trends in the system andesite-H_2O-CO_2-O_2 at pressures to 10 kb. *Geol. Soc. Am. Bull. 84*, 2517–2532.

Eggler, D. H. (1974). Application of a portion of the system $CaAl_2Ai_2O_8$-$NaAlSi_3O_8$-SiO_2-MgO-Fe-O_2-H_2O-CO_2 to genesis of the calc-alkaline suite. *Am. J. Sci. 274*, 297–315.

Eggleton, R. A. (1972). The crystal structure of stilpnomelane. Part II. The full cell. *Mineral. Mag. 38*, 693–711.

Einstein, A. (1906). Eine neue Bestimmung der Moleculdimensionen. *Ann. Phys. 19*, 289–306.

Einstein, A. (1911). Berichtigung zu meiner Arbeites Eine neue Bestimmung der Moleculdimensionen. *Ann. Phys. 34*, 591–592.

Elliot, D. (1973). Diffusion flow laws in metamorphic rocks. *Geol. Soc. Am. Bull. 84*, 2645–2664.

Ellis, A. J., and Mahon, W. A. J. (1964). Natural hydrothermal systems and experimental hot-water/rock interactions. *Geochim. Cosmochim. Acta 28*, 1323–1357.

Engel, A. E. J., and Engel, C. G. (1958, 1960). Progressive metamorphism and granitization of the major paragneiss, northwest Adirondack Mountains, New York. I, II. *Bull. Geol. Soc. Am. 69*, 1369–414; *71*, 1–58.

Engel, A. E. J., and Engel, C. E. (1962). Progressive metamorphism of amphibolite, northwest Adirondack mountains, New York. *Bull. Geol. Soc. Am., A–F* (Buddington vol.), 37–82.

Engel, A. E. J., and Engel, C. G. (1963). Metasomatic origin of large parts of the Adirondack phacoliths. *Bull. Geol. Soc. Am. 74*, 349–352.

Engel, A. E. J., and Engel, C. G. (1964). Igneous rocks of the East Pacific Rise. *Science 146*, 477–485.

Engel, A. E. J., Engel, C. G., and Havens, R. G. (1964). Mineralogy of amphibolite interlayers in the gneiss complex, Northwest Adirondack Mountains, New York. *J. Geol. 72*, 131–156.

Engel, A. E. J., Itson, S. P., Engel, C. G., Stickney, D. M., and Gray, E. J., Jr. (1974). Crustal evolution and global tectonics, a petrogenetic view. *Geol. Soc. Am. Bull. 85*, 843–858.

Ernst, W. G. (1959). Alkali Amphiboles. *Carnegie Inst. of Washington, Yearbook 58*, 1921–1926.

Ernst, W. G. (1960a). Diabase-granophyre relations in the Endion sill, Duluth, Minnesota. *J. Petrol. 1*, 286–303.

Ernst, W. G. (1960b). The stability relations of magnesioriebeckite. *Geochim. Cosmochim. Acta 19*, 1–40.

Ernst, W. G. (1961). Stability relations of glaucophane. *Am. J. Sci. 259*, 735–765.

Ernst, W. G. (1962). Synthesis, stability relations, and occurrence of riebeckite and riebeckite-arfvedsonite solid solutions. *J. Geol. 70*, 689–736.

Ernst, W. G. (1963a). Petrogenesis of glaucophane schists. *J. Petrol. 4*, 1–30.

Ernst, W. G. (1963b). Polymorphism in alkali amphiboles. *Am. Mineral. 48*, 241–260.

Ernst, W. G. (1963c). Significance of phengitic micas from low-grade schists. *Am. Mineral. 48*, 1357–1373.

Ernst, W. G. (1964). Petrochemical study of coexisting minerals from low-grade schists, Eastern Shikoku, Japan. *Geochim. Cosmochim. Acta. 28*, 1631–1668.

Ernst, W. G. (1966). Synthesis and stability relations of ferrotremolite. *Am. J. Sci. 264*, 37–65.

Ernst, W. G. (1968). *Amphiboles, Crystal Chemistry, Phase Relations and Occurrence.* New York: Springer-Verlag.

Ernst, W. G. (1972a). Occurrence and mineralogic evolution of blue schist belts with time. *Am. J. Sci. 272*, 657–668.

Ernst, W. G. (1972b). Ca-amphibole paragenesis in the Sharataki district, central Shikoku, Japan. *Mem. Geol. Soc. Am. 135*, 73–94.

Ernst, W. G., and Calvert, S. E. (1969). An experimental study of the recrystallization of porcelanite and its bearing on the origin of some bedded cherts. *Am. J. Sci. 267-A*, 114–133.

Eskola, P. (1915). On the relations between the chemical and mineralogical composition in the metamorphic rocks of the Orijarvi region. *Bull. Comm. geol. Finlande 44*.

Eskola, P. (1920). The mineral facies of rocks. *Norsk Geol. Tidskr. 6*, 143–194.

Eskola, P. (1921). On the eclogites of Norway. *Vidensk. Skrifter I. Mat.-naturv. Kl., Kristiania, 8*.

Eskola, P. (1939). Die metamorphen Gesteine. In *Die Entstehung der Gesteine* (Tom, F. W. Barth, C. W. Correns, and Eskola, eds.). Berlin: Springer, pp. 263–407.

Essene, E. J., Fyfe, W. S., and Turner, F. J. (1965). Petrogenesis of Franciscan glaucophane schists and associated metamorphic rocks, California. *Contr. Mineral. Petrol. 11*, 695–704.

Essene, E. J., Hensen, B. J., and Green, D. H. (1970). Experimental study of amphibolite and eclogite stability. *Phys. Earth Planet. Int. 3*, 378–384.

Eucken, A. (1944). Physikalisch-chemische Betractungen über früheste Entwicklungs geschichte der Erde. *Nachr. Akad. Wiss. Göttingen. Math.-Phys. Kl. Heft. 1*, 1–25.

Eugster, H. P. (1959). Reduction and oxidation in metamorphism. In *Researches in Geochemistry* (P. H. Abelson, ed.). New York: Wiley, pp. 397–426.

Eugster, H. P., and Wones, D. R. (1962). Stability relations of the ferruginous biotite, annite. *J. Petrol. 3*, 82–125.

Eugster, H. P., Albee, A. L., Bence, A. E., Thompson, J. B., Jr., and Waldbaum, D. R. (1972). The two-phase region and excess properties of paragonite-muscovite crystalline solutions. *J. Petrol. 13*, 147–179.

Evans, B. W. (1965). Application of a reaction-rate method to the breakdown equilibria of muscovite and muscovite plus quartz. *Am. J. Sci. 263*, 647–667.

Evans, B. W., and Guidotti, C. V. (1966). The sillimanite-potash feldspar isograd in western Maine, U.S.A. *Contr. Mineral. Petrol. 12*, 25–62.

Evans, B. W., and Trommsdorff, V. (1970). Regional metamorphism of ultramafic rocks in the Central Alps. Parageneses in the system CaO-MgO-SiO_2-H_2O. *Schweiz. Min. Pet. Mitt. 50*, 481–492.

Evans, B. W., and Trommsdorff, V. (1974). Stability of enstatite + talc, and CO_2-metasomatism of metaperidotite, Val d'efro, Lepontine Alps. *Am. J. Sci. 272*, 274–296.

Evernden, J. F., and Kistler, R. W. (1970). Chronology of the emplacement of Mesozoic batholithic complexes of California and western Nevada. *U.S. Geol. Surv. Prof. Paper 623*, 1–42.

Ewart, A. (1963). Petrology and petrogenesis of the quaternary pumic ash in the Taupo Area, New Zealand. *J. Petrol. 4*, 392–431.

Ewart, A., Green, D. C., Carmichael, I. S. E., and Brown, F. H. (1971). Voluminous low temperature rhyolitic magmas in New Zealand. *Contr. Mineral. Petrol. 32*, 128–144.

Fairbairn, H. W. (1950). Synthetic quartzite. *Am. Mineral. 35*, 735–748.

Farmer, C. B., Davis, D. W., and LaPorte, D. D. (1976). Viking: Mars atmospheric water mapping experiment-Preliminary report of results. *Science 193*, 776–780.

Fawcett, J. J., and Yoder, H. S., Jr. (1966). Phase relationships of chlorites in the system MgO-Al_2O_3-SiO_2-H_2O. *Am. Mineral. 51*, 353–380.

Feltham, P. (1957). Grain growth in metals. *Acta Metall. 5*, 97–105.

Fenner, C. N. (1926). The Katmai magmatic province. *J. Geol. 34*, 673–772.

Fenner, C. N. (1948). Incandescent tuff flows in southern Peru. *Geol. Soc. Am. Bull. 59*, 879–893.

Ferguson, I., and Currie, K. L. (1971). Evidence of liquid immiscibility in alkaline ultrabasic dykes at Collander Bay, Ontario. *J. Petrol. 12*, 566–574.

Fireman, E. L. (1958). Distribution of helium-3 in the Carbo meteorite. *Nature 181*, 1725.

Fish, R. A., Goles, G. G., and Anders, E. (1960). The record in the meteorites III. On the development of meteorites in asteroidal bodies. *Astrophys. J, 132*, 243–258.

Fisher, J. R., and Zen, E-An. (1971). Thermochemical calculations from hydrothermal phase equilibrium data and the free energy of H_2O. *Am. J. Sci. 270*, 297–314.

Fisher, G. W. (1974). Nonequilibrium thermodynamics as a model for diffusion controlled metamorphic processes. *Am. J. Sci. 273*, 897–924.

Flory, B. (1953). *Principles of Polymer Chemistry*. Ithaca, N.Y.: Cornell University Press.

Fodor, R. V., Keil, K., Jarosewich, E., and Huss, G. I. (1971). Mineralogy, petrology, and chemistry of the Burdett Kansas, Chondrite. *Chemie der Erde 30*, 103–113.

Folweiler, R. C. (1961). Creep behavior of pore-free polycrystalline aluminum oxide. *J. Appl. Phys. 32*, 773–778.

Forbes, W. C. (1971). Synthesis and stability relations of richterite $Na_2CaMg_5Si_8O_{22}(OH)_2$ or iron content of talc in the system $Mg_3Si_4O_{10}(OH)_2$-$Fe_3Si_4O_{10}(OH)_2$. *Am. Mineral. 56*, 997–1004.

Forbes, R. B., and Swainbank, R. C. (1974). Garnet-clinopyroxenite from the Red Mountain Pluton, Alaska. *Geol. Soc. Am. Bull. 85*, 285–292.

Fredriksson, K. (1963). Chondrules and the meteorite parent bodies. *Trans. N.Y. Acad. Sci. 25*, 756–769.

Fredriksson, K., and de Carli, P. (1964). Shockemplaced argon in a stony meteorite. *J. Geophys. Res. 69*, 1403–1406.

French, B. M., and Short, N. M. (eds.). (1968). *Shock Metamorphism of Natural Materials*. Baltimore: Mone Press.

French, B. M. (1971). Stability relations of siderite ($FeCO_3$) in the system Fe-C-O. *Am. J. Sci. 271*, 37–78.

Friedman, I., Long, W., and Smith, R. L. (1963). Viscosity and water content of rhyolite glass. *J. Geophys. Res. 68*, 6523–6535.

Froese, E. (1973). The oxidation of almandine and iron cordierite. *Can. Mineral. 11*, 991–1002.

Fudali, R. F. (1963). Experimental studies bearing on the origin of pseudoleucite and associated problems of alkalic rock systems. *Geol. Soc. Am. Bull. 74*, 1101–1126.

Fullman, R. L. (1957). The equilibrium form of crystalline bodies. *Acta Metall. 5*, 639–648.

Fyfe, W. S., Turner, F. J., and Verhoogen, J. (1958). Metamorphic reactions and metamorphic facies. *Geol. Soc. Am. Mem. 73*, 1–259.

Fyfe, W. S., and Turner, F. J. (1966). Reappraisal of the concept of metamorphic facies. *Contr. Mineral. Petrol. 12*, 354–364.

Gable, D. J., and Sims, P. K. (1970). Geology and regional metamorphism of some high grade cordierite gneisses, Front Range, Colorado. *Geol. Soc. Am., Spec. Paper 128*, 1–87.

Galwey, A. K., and Jones, K. A. (1963). An attempt to determine the mechanism of a natural mineral-forming reaction from examination of the product. *Chem. Soc. J. 1963* (Lond.) 5681–5686.

Ganguly, J. (1969). Chloritoid stability and related parageneses: Theory, experiments, and applications. *Am. J. Sci. 267*, 910–944.

Ganguly, J. (1972). Staurolite stability and related parageneses: Theory, experiments and applications. *J. Petrol. 13*, 335–365.

Ganguly, J. (1973). Activity-composition relation of jadeite in omphacite pyroxene. *Earth & Planet. Sci. Lett. 19*, 145–153.

Ganguly, J. (1974). Personal communication.

Ganguly, J., and Kennedy, G. C. (1974). The energetics of natural garnet solid solution: I. Mixing of the aluminosilicate end-members. *Contr. Mineral. Petrol. 48*, 137–148.

Ganguly, J., and Newton, R. C. (1968). Thermal stability of chloritoid at high pressure and relatively high oxygen fugacity. *J. Petrol. 9*, 444–466.

Garner, W. E. (ed.). (1955). *Chemistry of the Solid State*. London: Butterworths.

Garrels, R. M. (1960). *Mineral Equilibria at Low Temperature and Pressure*. New York: Harper & Row.

Garrels, R. M., and Christ, C. L. (1965). *Solutions, Minerals, and Equilibria*. New York: Harper & Row.

Garrels, R. M., and Howard, P. (1959). Reactions of feldspar and mica with water at low temperature and pressure. *Proceedings of the Sixth National Conference on Clays and Clay Minerals*. Oxford: Pergamon, pp. 68–88.

Gehman, H. M., Jr. (1958). Notch Peak Intrusive, Millard County, Utah: Geology, petrogenesis and economic deposits. *Utah Geol. & Mineral. Surv. Bull. 62*.

Gilbert, M. C. (1966). Synthesis and stability relationships of ferroparagasite. *Am. J. Sci. 264*, 698–742.

Glasstone, S., Laidler, K. J., and Eyring, H. (1941). *The Theory of Rate Processes*. McGraw-Hill, New York.

Goldberg, L., Muller, E. A., and Aller, L. H. (1960). The abundances of the elements in the solar atmosphere. *Astrophys. J., Suppl. Ser. 5*, 1–138.

Goldsmith, J. R. (1952). Diffusion in plagioclase feldspars. *J. Geol. 60*, 288–291.

Goldsmith, J. R., and Heard, H. C. (1961). Subsolidus phase relations in the system $CaCO_3$-$MgCO_3$. *J. Geol. 69*, 45–74.

Goldsmith, J. R., and Newton, R. C. (1969). *P-T-X*-relations in the system $CaCO_3$-$MgCO_3$ at high temperatures and pressures. *Am. J. Sci. 267-A*, 160–190.

Goldstein, J. I., and Doan, A. S. J. (1972). The effect of phosphorous on the formation of the Widmanstätten pattern in iron meteorites. *Geochim. Cosmochim. Acta 36*, 51–69.

Goldstein, J. I., and Ogilvie, R. E. (1965). A reevaluation of the Fe-rich portion of the Fe-Ni system. *Trans. A.I.M.E. 233*, 2083–2087.

Goldstein, J. I., and Yakowitz, H. (1971). Metallic inclusions and metal particles in the Apollo 12 lunar soil. In *Proceedings of the Second Lunar Science Conference* (A. A. Levinson, ed.). Houston: pp. 177–191. Pergamon Press.

Goodspeed, G. E. (1940). Dilation and replacement dykes. *J. Geol. 48*, 175–195.

Goranson, R. W. (1931). The solubility of water in granite magmas. *Am. J. Sci. 22*, 481–502.

Goranson, R. W. (1936). Silicate-water systems. The solubility of water in albite-melt. *Am. Geophys. Union Trans. 17*, 257–259.

Goranson, R. W. (1938). Silicate-water systems; phase equilibria in the $NaAlSi_3O_8$-H_2O and $KAlSi_3O_8$-H_2O systems at high temperatures and pressures. *Am. J. Sci. 35-A*, 71–91.

Gordon, T. M., and Greenwood, H. J. (1970). The reaction dolomite + quartz + water = talc + calcite + carbon dioxide. *Am. J. Sci. 268*, 225–242.

Gordon, T. M., and Greenwood, H. J. (1971). The stability of grossularite in H_2O-CO_2 mixtures. *Am. Mineral. 56*, 1674–1688.

Graf, D. F., and Goldsmith, J. R. (1955). Dolomite-magnesian calcite relations at elevated temperatures and CO_2 pressures. *Geochim. Cosmochim. Acta 7*, 109–128.

Graf, D. F., and Goldsmith, J. R. (1958). The solid solubility of $MgCO_3$ in $CaCO_3$: A revision. *Geochim. Cosmochim. Acta 13*, 218–219.

Grant, J. A. (1973). Phase equilibria in high grade metamorphism and partial melting of pelitic rocks. *Am. J. Sci. 273*, 289–317.

Green, D. H. (1966). The origin of the "eclogites" from Salt Lake Crater, Hawaii. *Earth & Planet. Sci. Letts. 1*, 414–420.

Green, D. H. (1973). Experimental melting studies on a model upper mantle composition at high pressure under water-saturated and undersaturated conditions. *Earth & Planet. Sci. Letts. 19*, 37–53.

Green, D. H., and Ringwood, A. E. (1967). An experimental investigation of the gabbro to eclogite transformation and its petrological applications. *Geochim. Cosmochim. Acta 31*, 767–833.

Green, D. H., and Ringwood, A. E. (1970). Mineralogy of peridotitic compositions under upper mantle conditions. *Phys. Earth Planet Interiors 3*, 359–371.

Green, D. H., and Ringwood, A. E. (1972). A comparison of recent experimental data on the gabbro-garnet granulite-eclogite transition. *J. Geol. 80*, 277–288.

Green, J. C. (1963). High-level metamorphism of pelitic rocks in northern New Hampshire. *Am. Mineral. 48*, 991–1023.

Green, T. H. (1967). An experimental investigation of subsolidus assemblages formed at high pressure in high alumina basalt, kyanite eclogite and grospydite compositions. *Contr. Mineral. Petrol. 16*, 84–114.

Green, T. H. (1969). High-pressure experimental studies on the origin of anorthosite. *Can. J. Earth Sci. 6*, 427–440.

Green, T. H. (1972). Crystallization of calc-alkaline andesite under controlled high-pressure hydrous conditions. *Contr. Mineral. Petrol. 34*, 150–166.

Green, T. H., and Ringwood, A. E. (1966). Origin of the calc-alkaline igneous rock suite. *Earth & Planet. Sci. Letts. 1*, 307–316.

Green, T. H., and Ringwood, A. E. (1967). Crystallization of basalt and andesite under high pressure hydrous conditions. *Earth & Planet. Sci. Letts. 3*, 481–489.

Green, T. H., and Ringwood, A. E. (1968). Genesis of the calc-alkaline igneous rock suite. *Contr. Mineral. Petrol. 18*, 105–162.

Green, T. H., and Vernon, R. H. (1974). Cordierite breakdown under high-pressure, hydrous conditions. *Contr. Mineral. Petrol. 46*, 215–226.

Greenwood, H. J. (1963). The synthesis and stability of anthophyllite. *J. Petrol. 4*, 317–351.

Greenwood, H. J. (1967a). The *N*-dimensional tie-line problem. *Geochim. Cosmochim. Acta 31*, 465–490.

Greenwood, H. J. (1967b). Wollastonite: Stability in H_2O-CO_2 mixtures and occurrence in a contact-metamorphic aureole near Salmo, British Columbia, Canada. *Am. Mineral. 52*, 1669–1680.

Greenwood, H. J. (1969). The compressibility of gaseous mixtures of carbon dioxide and water between 0 and 500 bars pressure and 450°C and 800°C. *Am. J. Sci. 267a*, 191–208.

Greenwood, H. J. (1971). Anthophyllite corrections and comments on its stability. *Am. J. Sci. 271*, 151–154.

Greenwood, H. J. (1972). Al^{IV}-Si^{IV} disorder in sillimanite and its effect on phase relations of the aluminum silicate minerals. *Geol. Soc. Am. Mem. 132*, 553–571.

Greenwood, H. J. (1973). Thermodynamic properties of gaseous mixtures of H_2O-CO_2 between 450°C and 800°C and 0 to 500 bars. *Am. J. Sci. 273*, 561–571.

Gregg, S. J., and Razouk, R. I. (1949). The kinetics of the thermal decomposition of magnesium hydroxide. *J. Chem. Soc., Part V*, 536–544.

Gresens, R. L. (1969). Blueschist alteration during serpentinization. *Contr. Mineral. Petrol. 24*, 93–113.

Gresens, R. L. (1972). Do mineral paragenesis reflect unusually high-pressure conditions of Franciscan metamorphism? *Am. J. Sci. 271*, 311–316.

Griffen, D. T., and Ribbe, P. H. (1973). The crystal chemistry of staurolite. *Am. J. Sci. 273A*, 479–495.

Griggs, D. T., Paterson, M. S., Heard, H. C., and Turner, F. J. (1960). Annealing recrystallization in calcite crystals and aggregates. In *Rock Deformation* (D. Griggs and J. Handin, eds.). *Geol. Soc. Am. Mem. 79*, 21–38.

Griggs, D. T., Turner, F. J., and Heard, H. C. (1958). Deformation of rocks at 500° to 800°C. In *Rock Deformation* (D. Griggs and J. Handin, eds.). *Geol. Soc. Am. Mem. 79*, 39–104.

Grim, R. E. (1953). *Clay Mineralogy*. New York: McGraw-Hill.

Grossman, L. (1972). Condensation in the primitive solar nebula. *Geochim. Cosmochim. Acta 36*, 597–619.

Grossman, L., and Olsen, E. (1974). Origin of high-temperature fraction of C_2 chondrites. *Geochim. Cosmochim. Acta 38*, 173–187.

Grout, F. F. (1918). The pegmatites of the Duluth gabbro. *Econ. Geol. 13*, 185–197.

Guidotti, C. V. (1968). Prograde muscovite pseudomorphs after staurolite in the Rangeley-Oquossoc areas, Maine. *Am. Mineral. 48*, 772–791.

Guidotti, C. V. (1969). A Comment on "Chemical study of minerals from the Moine schists of the Ardnamurchan area, Argyllshire, Scotland" by B. C. M. Butler, and its implications for the phengite problem. *J. Petrol. 10*, 164–170.

Guidotti, C. V. (1970). The mineralogy and petrology of the transition from the lower to upper sillimanite zone in the Oquossoc area, Maine. *J. Petrol. 11*, 277–336.

Guidotti, C. V. (1974). Transition from staurolite to sillimanite zone, Rangeley Quadrangle, Maine. *Geol. Soc. Am. Bull. 85*, 475–490.

Guidotti, C. V., Herd, H. H., and Tuttle, C. L. (1973). Composition and structural state of K-feldspars from K-feldspar + sillimanite grade rocks in northwestern Maine. *Am. Mineral. 58*, 705–716.

Guitard, G. (1965). Associations minérales subfaciès et types de metamorphisms dans les mica-schist es et les gneiss pélitiques du Massif du Canigou. *Bull. Geol. France 7*, 356–382.

Gunn, B. M., Roobol, M. J., and Smith, A. L. (1974). Petrochemistry of the Pelean-type volcanoes of Martinique. *Bull. Geol. Soc. Am. 85*, 1023–1030.

Gurney, R. W. (1949). *Introduction to Statistical Mechanics*. New York: McGraw-Hill.

Hahn, S. J., Ree, T., and Eyring, H. (1967). Mechanism for the plastic deformation of Yule marble. *Bull. Geol. Soc. Am. 78*, 773–782.

Halfterdahl, L. B. (1961). Chloritoid: Its composition, X-ray and optical properties, stability and occurrence. *J. Petrol. 2*, 49–135.

Hall, A. L. (1932). The Bushveld igneous complex of the central Transvaal. *Geol. Surv. S. Africa Mem. 28*.

Hall, R., and Marcus, J. (1952). On the thermal decomposition of dolomite, IV. Thermogravimetric investigation of dolomite decomposition. *J. Appl. Chem. 2*, 298–301.

Hamilton, D. L., Burnham, C. W., and Osborn, E. F. (1964). The solubility of water and effects of oxygen fugacity and water content in crystallization in mafic magmas. *J. Petrol. 5*, 21–39.

Hamilton, W., and Myers, W. B. (1967). The nature of batholiths. *U.S. Geol. Surv. Prof. Paper 554-C*.

Hamilton, W., and Myers, W. B. (1974a). Nature of the Boulder Batholith of Montana. *Geol. Soc. Am. Bull. 85*, 365–378.

Hamilton, W., and Myers, W. B. (1974b). Reply. *Geol. Soc. Am. Bull. 85*, 1958–1960.

Hanel, R., Conrath, B., Hovis, W., Kunde, V., Lowman, P., McGuire, W., Pearl, J., Pirraglia, J., Prabhakara, C., Schlachman, B., Levin, G., Straat, P., and Burke, T. (1972). Investigation of the Martian environment by infrared spectroscopy on Mariner 9. *Icarus 17*, 423–442.

Hargraves, R. B. (1962). Petrology of the Allard Lake anorthosite suite, Quebec. *Geol. Soc. Am., Buddington vol. 73*, 163–189.

Hariya, Y., and Kennedy, G. C. (1968). Equilibrium study of anorthite under high pressure and high temperature. *Am. J. Sci. 266*, 193–203.

Harker, A. (1932). *Metamorphism: A Study of the Transformations of Rock Masses*. New York: Dutton.

Harker, R. I., and Tuttle, O. F. (1955). Studies in the system CaO-MgO-CO_2. Part 2: Limits of solid solutions along the binary join, $CaCO_3$-$MgCO_3$, *Am. J. Sci. 253*, 274–282.

Harker, R. I., and Tuttle, O. F. (1956). Experimental data on the P_{CO_2}-T curve for the reaction: calcite + quartz $\rightleftharpoons$ wollastonite + carbon dioxide. *Am. J. Sci. 254*, 239–56.

Haskell, N. A. (1935). The motion of viscous fluid under a surface load. *Physics 6*, 265–269.

Hays, J. F. (1967). Lime-alumina-silicate. *Carnegie Institution Washington Year Book 65*, 234–239.

Heard, H. C. (1963). Effect of large changes in strain rate in the experimental deformation of Yule marble. *J. Geol. 71*, 162–195.

Heinrich, E. W. (1966). *The Geology of Carbonatites*. Chicago: Rand McNally.

Helgesson, H. C. (1969). Thermodynamics of hydrothermal systems at elevated temperatures and pressures. *Am. J. Sci. 267*, 729–804.

Helgesson, H. C. (1971). Kinetics of mass transfer among silicates and aqueous solutions. *Geochim. Cosmochim. Acta 35*, 421–469.

Helgesson, H. C. (1975). *Geol. Soc. Am. Meeting*.

Hellner, E., Hinrichsen, Th., and Seifert, F. (1965). The study of mixed crystals of minerals in metamorphic rocks. In *Controls of Metamorphism* (W. S. Pitcher and G. W. Flinn, eds.). New York: Wiley.

Heming, R. F., and Carmichael, I. S. E. (1973). High temperature pumice flows from the Robaul caldera, Papua, New Guinea. *Contr. Mineral. Petrol. 38*, 1–20.

Hemley, J. J. (1967). Stability relations of pyrophyllite, andalusite and quartz at elevated pressures and temperatures. [abs.]. *Am. Geophys. Union Trans. 48*, 224.

Hensen, B. J. (1971). Theoretical phase relations involving cordierite and garnet in the system MgO-FeO-Al_2O_3-SiO_2. *Contr. Mineral. Petrol. 33*, 191–214.

Hensen, B. J., and Green, D. H. (1973). Experimental study of the stability of cordierite and garnet in pelitic compositions at high pressures and temperatures. *Contr. Mineral. Petrol. 38*, 151–166.

Herd, C. W., Barth, C. A., and Steward, A. I. (1972). Mariner ultraviolet spectrometer experiment; photometry and topography of Mars. *Icarus 17*, 443–456.

Herring, C. (1951). Some theorems on the free energies of crystal surfaces. *Phys. Res. 82*, 87–93.

Herzog, G. F., Anders, E., Alexander, E. C., Jr., Davis, P. K., and Lewis, R. S. (1973). Iodine-129/Xenon-129 age of magnetite from the Orguiel meteorite. *Science 180*, 489–491.

Hess, H. H. (1960). Stillwater Igneous Complex, Montana. *Geol. Soc. Am. Mem. 80.*

Hess, P. C. (1969). The metamorphic paragenesis of cordierite in pelitic rocks. *Contr. Mineral. Petrol. 24*, 191–207.

Hess, P. C. (1971). Polymer model of silicate melts. *Geochim. Cosmochim. Acta 35*, 289–306.

Hewitt, D. A., and Wones, D. R. (1971). Experimental metamorphic petrology. *EOS, Trans. Am. Geophys. Union 52*, 73–82.

Hietanen, A. (1951). Metamorphic and igneous rocks of the Merrimac area, Plumas National Forest, California. *Bull. Geol. Soc. Am. 62*, 565–608.

Hietanen, A. (1956). Kyanite, andalusite and sillimanite in the schist in Boehls Butte quadrangle, Idaho. *Am. Mineral. 41*, 1–27.

Hietanen, A. (1959). Kyanite-garnet gedritite near Orofino, Idaho. *Am. Mineral. 44*, 539–564.

Hietanen, A. (1962). Metasomatic metamorphism in western Clearwater County, Idaho. *U.S. Geol. Surv. Prof. Paper 344-A.*

Hietanen, A. (1963a). Anorthosite and associated rocks in the Boehls Butte quadrangle and vicinity, Idaho. *U.S. Geol. Surv., Prof. Paper 344-B.*

Hietanen, A. (1963b). Metasomatic metamorphism in western Clearwater county Idaho. *U.S. Geol. Surv., Prof. Paper 344-A.*

Hietanen, A. (1963c). Metamorphism of the Belt Series in the Elk-River-Clarkia area, Idaho. *U.S. Geol. Surv., Prof. Paper 344-C.*

Hietanen, A. (1963d). Idaho batholith near Pierce and Bungalow Clearwater County, Idaho. *U.S. Geol. Surv., Prof. Paper 344-D.*

Hietanen, A. (1967). On facies series in various types of metamorphism. *J. Geol. 75*, 187–214.

Hietanen, A. (1969). Distribution of Fe and Mg between garnet, staurolite, and biotite in aluminum-rich schist in various metamorphic zones north of Idaho Batholith. *Am. J. Sci. 267*, 422–456.

Hildebrand, J. H. (1929). Solubility XII. Regular solutions. *J. Am. Chem. Soc. 51*, 66.

Hill, R., and Roedder, P. (1975). The crystallization of spinel from basaltic liquid as a fraction of oxygen fugacity. *J. Geol. 82*, 709–729.

Hinrichsen, Th.J. (1966). Hydrothermal investigation and stability relations of orthorhombic amphiboles [abs.]. *Intern. Mineral. Soc. Meeting London.*

Holdaway, M. J. (1967). Stability of epidote. *Am. Geophys. Union Trans. 48*, 225.

Holdaway, M. J. (1971). Stability of andalusite and the aluminum silicate phase diagram. *Am. J. Sci. 271*, 97–131.

Holdaway, M. J. (1972). Thermal stability of Al-Fe epidote as a function of f_{O_2} and Fe content. *Contr. Mineral. Petrol. 37*, 307–340.

Holgate, N. (1954). The role of liquid immiscibility in igneous petrogenesis. *J. Geol. 62*, 439–480.

Hollister, L. S. (1966). Garnet zoning: An interpretation based on the Rayleigh fractionation model. *Science 154*, 1647–1651.

Hollister, L. S., and Bence, A. E. (1967). Staurolite: Sectoral compositional variations. *Science 158*, 1053–1056.

Hoschek, G. (1969). The stability of staurolite and chloritoid and their significance in metamorphism of pelitic rocks. *Contr. Mineral. Petrol. 22*, 208–232.

Hounslow, A. W., and Moore, J. M., Jr. (1967). Chemical petrology of Grenville Schists near Fernleigh, Ontario. *J. Petrol. 8*, 1–28.

Howard, H. T., Tyler, G. L., Fjeldbo, G., Kliore, A. J., Levi, G. S., Bruno, D. L., Dickinson, R., Edelson, R. E., Martin, W. L., Postal, R. B., Seidel, B., Sesplaukis, T. T., Shirley, D. L., Stelzried, C. T., Sweetham, D. N., Zygielbaun, A. I., Esposito, P. B., Anderson, J. D., Shapiro, I. I., and Reaserberg, R. D. (1974). Venus: Mass, Gravity field, atmosphere and ionosphere as measured by Mariner 10 Dual-frequency radio system. *Science 183*, 1297–1301.

Howie, R. A. (1955). The geochemistry of the charnockite series of Madras, India. *Roy. Soc. Edinburgh Trans. 62*, Pt. 3, No. 18, 725–768.

Hoyle, F. (1960). *Quart. J. Roy. Astron. Soc. 1*, 28.

Hsu, L. C. (1968). Selected phase relationships in the system Al-Mn-Fe-Si-O; a model for garnet equilibria. *J. Petrol. 9*, 40–83.

Hsu, L. C., and Burnham, C. W. (1969). Phase relations in the system $Fe_3Al_2Si_3O_{12}$-$Mg_3Al_2Si_3O_{12}$-H_2O at 2.0 kilobars. *Geol. Soc. Am. Bull. 80*, 2393–2408.

Huang, W. L., Robertson, J. K., and Wyllie, P. J. (1973). Melting relations of muscovite to 30 kilobars in the system $KAlSi_3O_8$-Al_2O_3-H_2O. *Am. J. Sci. 273*, 415–427.

Huckenholz, H. G., and Yoder, H. S., Jr. (1971). Andradite stability relations in the $CaSiO_3$-Fe_2O_3 join up to 30 Kb. *Neues Jb. Mineral. Abh. 114*, 246–280.

Huebner, J. S., and Sato, M. (1970). The oxygen fugacity-temperature relationships of manganese oxide and nickel oxide buffers. *Am. Mineral. 55*, 934–952.

Hurlbut, C. S., Jr. (1935). Dark inclusions in a tonalite of southern California. *Am. Mineral. 20*, 609–630.

Hutton, C. O. (1938). The stilpnomelane group of minerals. *Min. Mag. 25*, 172–206.

Hyndman, D. W. (1972). *Petrology of Igneous and Metamorphic Rocks.* New York: McGraw-Hill.

Irving, R. D., and Van Hise, C. R. (1892). The Penokee iron-bearing series of Michigan and Wisconsin. *U.S. Geol. Surv. Monogr. V. XIX.*

Irvine, T. N. (1967). The ultramafic rocks of the Muskox intrusion, Northwest Territories, Canada. In *Ultramafic and Related Rocks* (P. J. Wyllie, ed.). New York: Wiley, pp. 38–49.

Isacks, B., Oliver, J., and Sykes, L. R. (1968). Seismology and the new global tectonics. *J. Geophys. Res. 73*, 5855–5899.

Ito, K., and Kennedy, G. C. (1971). An experimental study of the basalt-garnet granulite-eclogite transition. In *The Structure and Physical Properties of the Earth's Crust* (J. G. Heacock, ed.). *Am. Geophys. Union monogr. 14*, 303–314.

Ito, K., and Kennedy, G. C. (1974). The composition of liquids formed by partial melting of eclogites at high temperatures and pressures. *J. Geol. 82*, 383–392.

Ivanov, I. P., and Gusynin, V. F. (1970). Stability of paragonite in the system SiO_2-$NaAlSi_3O_8$-Al_2O_3-H_2O. *Geochem. Intern. 7*, 578–587.

Jackson, E. D. (1961). Primary textures and mineral associations in the ultramafic zone of the Stillwater Complex Montana. *U.S. Geol. Surv. Prof. Paper 358.*

Jacobs. P. W., and Tompkins, F. C. (1955). Classification and theory of solid reactions. In *Chemistry of the Solid State* (W. E. Garner, ed.). London: Butterworths, pp. 184–212.

James, H. L. (1955). Zones of regional metamorphism in the Precambrian of northern Michigan. *Bull. Geol. Soc. Am. 66*, 1455–1488.

Jamieson, J. C. (1953). Phase equilibrium in the system calcite-aragonite. *J. Chem. Phys. 21*, 1385–1390.

Jenks, W. F., and Goldich, S. S. (1956). Rhyolitic tuff flows in southern Peru. *J. Geol. 64*, 156–172.

Johannes, W. (1968a). Experimental investigation of the reaction forsterite + H_2O = serpentine + brucite. *Contr. Mineral. Petrol. 19*, 309–315.

Johannes, W. (1968b). Experimentelle Sederitbildung aus Calcit + $FeCl_2$. *Contr. Mineral. Petrol. 17*, 155–164.

Johannes, W. (1969). An experimental investigation of the system MgO-SiO_2-H_2O-CO_2. *Am. J. Sci. 267*, 1083–1104.

Johannsen, A. (1939). *A Descriptive Petrography of the Igneous Rocks*, Vol. 1. Chicago: University of Chicago Press.

Johnson, A. M., and Pollard, D. D. (1973). Mechanics of growth of some laccolithic intrusions in the Henry Mountains, Utah. I and II. *Tectonophysics 18*, 261–354.

Johnson, R. E. (1959). Conflicts between Gibbsian thermodynamics and recent treatments of interfacial energies in solid-liquid-vapor systems. *J. Phys. Chem. 63*, 1655–1658.

Jokipii, J. R. (1964). The distribution of gases in the post-planetary nebula. *Icarus 3*, 248–252.

Jones, K. A., and Galwey, A. K. (1964). A study of possible factors concerning garnet formation in rocks from Ardara, Co. Donegal, Ireland. *Geol. Mag. 101*, 76–92.

Jones, W. R., Peoples, J. W., and Howland, A. L. (1960). Igneous and tectonic structures of the Stillwater Complex, Montana. *U.S. Geol. Surv. Bull. 1071-H*, 281–340.

Kani, K. (1934). The measurement of the viscosity of basalt glass at high temperature II. *Imperial Acad. Tokyo Proc. 10*, 82.

Kamb, W. B. (1959). Theory of preferred crystal orientation. *J. Geol. 67*, 153–170.

Kaufman, L., and Cohen, M. (1956). *Trans. A.I.M.E. 207*, 1393.

Keesman, I., Matthes, S., Schreyer, W., and Seifert, F. (1971). Stability of almandine in the system FeO-(Fe_2O_3)-Al_2O_3-SiO_2-(H_2O) at elevated pressures. *Contr. Mineral. Petrol. 31*, 132–144.

Keil, K. (1969). Meteorite composition. In *Handbook of Geochemistry*, Vol. 1 (K. H. Wedepohl, ed.). Berlin: Springer-Verlag.

Keil, K., and Andersen, C. A. (1965). Electronmicroprobe study of the Jajh deh Kot Lalu enstatite chondrite. *Geochim. Cosmochim. Acta 24*, 621–632.

Keil, K., and Fredriksson, K. (1964). The Fe, Mg, and Ca distribution in coexisting olivines and rhombic pyroxenes in chondrites. *J. Geophys. Res. 69*, 3487–3515.

Kelly, K. K. (1960). Contributions to the data on theoretical metallurgy XIII. High temperature heat-content, heat-capacity, and entropy data for the elements and inorganic compounds. *U.S. Bureau Mines Bull. 584.*

Kennedy, W. Q. (1933). Trends of differentiation in basaltic magmas. *Am. J. Sci. 25*, 239–256.

Kennedy, G. C. (1950). A portion of the system silica-water. *Econ. Geol. 45*, 629–653.

Kennedy, G. C. (1955). Some aspects of the role of water in rock melts. *Geol. Soc. Am., Special Paper 67*, 489–503.

Kerrick, D. M. (1968). Experiments on the upper stability limit of pyrophyllite at 1.8 kilobars and 3.9 kilobars water pressure. *Am. J. Sci. 266*, 204–214.

Kerrick, D. M. (1970). Contact metamorphism in some areas of the Sierra Nevada, California. *Geol. Soc. Am. Bull. 81*, 2913–2938.

Kerrick, D. M. (1972). Experimental determination of muscovite + quartz stability with $P_{H_2O} < P_{total}$. *Am. J. Sci. 272*, 946–958.

Kerrick, D. M. (1974). Review of metamorphic mixed volatile (H_2O-CO_2) equilibria. *Am. Mineral. 59*, 729–762.

Khitarov, N. I., Lebedev, E. B., and Kadik, A. A. (1963). Solubility of water in granitic melt at pressures to 7000 atmospheres. *Geochemistry 10*, 992–994.

Kingsley, L. (1931). Cauldron-subsidence of the Ossipee Mountains. *Am. J. Sci. 22*, 139–168.

Kistler, R. W. (1974). Phaneozoic batholiths in western North America. *An. Rev. Earth & Planet. Sci. 2*, 403–418.

Klein, C. (1966). Mineralogy and petrology of the metamorphosed Wabush iron formation, southwestern Labrador. *J. Petrol. 7*. 246–305.

Klepper, M. R., Robinson, G. D., and Smedes, H. W. (1974). Nature of the Boulder Batholith of Montana: Discussion. *Geol. Soc. Am. Bull. 85*, 1953–1958.

Klotz, I. M. (1950). *Chemical Thermodynamics.* New York: Prentice-Hall.

Koenig, L. R., Murray, F. W., Michaux, C. M., and Hyatt, H. A. (1967). *Handbook of the Physical Properties of the Planet Venus.* Washington, D.C.: Office of Technology Utilization, National Aeronautics and Space Administration.

Komar, P. D. (1972a). Mechanical interactions of phenocrysts and flow differentiation of igneous dikes and sills. *Geol. Soc. Am. Bull. 83*, 973–988.

Komar, P. D. (1972b). Flow differentiation in igneous dikes and sills. Profiles of velocity and phenocryst concentration. *Geol. Soc. Am. Bull. 83*, 3443–3448.

Korzhinskii, D. S. (1957). *Physicochemical Basis of the Analysis of the Paragenesis of Minerals.* Moscow: Academy of Science (Eng. trans., New York Consultants Bureau, 1959).

Koster van Groos, A. F. (1966). The effect of NaF, NaCl and Na_2CO_3 on the phase relationships in selected joins of the system Na_2O-CaO-Al_2O_3-SiO_2-H_2O at elevated temperatures and pressures. Ph.D. Thesis, University of Leiden, Netherlands.

Koster van Groos, A. F. (1969). Melting relationships in the system $NaAlSi_3O_8$-$NaCl$-H_2O at 1 kilobar pressure. *J. Geol. 77*, 581–605.

Koster van Groos, A. F. (1975). The effect of high CO_2 pressures on alkali rocks and its bearing on the formation of alkalic ultrabasic rocks and the associated carbonatites. *Am. J. Sci. 275*, 163–185.

Koster van Groos, A. F., and Wyllie, P. J. (1966). Liquid immiscibility in the system Na_2O-Al_2O_3-SiO_2-CO_2 at pressures to 1 kilobar. *Am. J. Sci. 264*, 234–255.

Koster van Groos, A. F., and Wyllie, P. J. (1968a). Melting relationships in the system $NaAlSi_3O_8$-NaF-H_2O to 4 kilobars pressure. *J. Geol. 76*, 50–70.

Koster van Groos, A. F., and Wyllie, P. J. (1968b). Liquid immiscibility in the join $NaAlSi_3O_8$-Na_2CO_3-H_2O and its bearing on the genesis of carbonatites. *Am. J. Sci. 266*, 932–967.

Koster van Groos, A. F., and Wyllie, P. J. (1973). Liquid immiscibility in the join $NaAlSi_3O_8$-$CaAlSi_2O_8$-$NaCO_3$-H_2O. *Am. J. Sci. 273*, 465–487.

Kracek, F. C. (1930). Binary systems Li_{20}-SiO_2, *J. Phys. Chem., 34*, 2641–2650.

Kranck, S. H. (1961). A study of phase equilibria in a metamorphic iron formation. *J. Petrol. 2*, 137–184.

Krauskopf, K. B. (1948). Mechanism of eruption at Pericutin Volcano, Mexico. *Bull. Geol. Soc. Am. 59*, 711–731.

Krauskopf, K. B. (1951). Physical chemistry of quicksilver transportation in vein fluids. *Econ. Geol. 46*, 498–523.

Krauskopf, K. B. (1956). Dissolution and precipitation of silica at low temperatures. *Geochim. Cosmochim. Acta 10*, 1–26.

Krauskopf, K. B. (1959). The use of equilibrium calculations in finding the composition of a magmatic gas phase. In *Researches in Geochemistry* (P. H. Abelson, ed.), p. 260–278.

Kretz, R. (1959). Chemical study of garnet, biotite and hornblende from gneisses of southwestern Quebec, with emphasis on distribution of elements in coexisting minerals. *J. Geol. 67*, 371–402.

Kretz, R. (1961). Preliminary examination of quartz-plagioclase layers and veins in amphibolite facies gneisses, southwestern Quebec. *Geol. Assoc. Can. Proc. 13*, 25–43.

Kretz, R. (1963). Distribution of magnesium and iron between orthopyroxene and calcic pyroxene in natural mineral assemblages. *J. Geol. 71*, 773–785.

Kretz, R. (1964). Analysis of equilibrium in garnet-biotite-sillimanite gneisses from Quebec. *J. Petrol. 5*, 1–20.

Kretz, R. (1966a). Interpretation of the shape of mineral grains in metamorphic rocks. *J. Petrol. 7*, 68–94.

Kretz, R. (1966b). Grain-size distribution for certain metamorphic minerals in relation to nucleation and growth. *J. Geol. 74*, 147–173.

Kretz, R. (1968). Study of pegmatite bodies and enclosing rocks, Yellowknife-Beaulieu Region, District of MacKenzie. *Geol. Surv. Can. Bull. 159*.

Kretz, R. (1970). Variation in the composition of muscovite and albite in a pegmatite dike near Yellowknife. *Can. J. Earth Sci. 7*, 1219–1235.

Kretz, R. (1973). Kinetics of the crystallization of garnet at two localities near Yellowknife. *Can. Mineral. 12*, 1–20.

Kridelbaugh, S. J. (1973). The kinetics of the reaction: calcite + quartz = wollastonite + carbon-dioxide at elevated temperatures and pressures. *Am. J. Sci. 273*, 757–777.

Kudo, A. M., and Weill, D. F. (1970). An igneous plagioclase thermometer. *Contr. Mineral. Petrol. 25*, 52–65.

Kuiper, G. P., (ed.) (1952). *Atmospheres of the Earth and Planets*. Chicago: University of Chicago Press.

Kuno, H. (1950). Petrology of Hakone volcano and the adjacent areas, Japan. *Bull. Geol. Soc. Am. 61*, 957–1020.

Kurepin, V. A. (1970). Conditions of stability of the $MgSiO_3$-$FeSiO_3$ pyroxenes. *Geokhimiya 8*, 1000–1004.

Kushiro, I. (1969). Clinopyroxene solid solutions formed by reactions between diopside and plagioclase at high pressures. *Mineral. Soc. Am. Spec. Paper 2*, 179–191.

Kushiro, I. (1973). Origin of some magmas in oceanic and circum-oceanic regions. *Tectonophysics 17*, 211–222.

Kushiro, I., and Yoder, H. S., Jr. (1966). Anorthite-forsterite and anorthite-enstatite reactions and their bearing on the basalt-eclogite transformation. *J. Petrol. 7*, 337–362.

Kushiro, I., and Yoder, H. S., Jr. (1969). Stability field of iron-free pigeonite in the system $MgSiO_3$-$CaMgSi_2O_6$. *Carnegie Inst. Yearbook 68*, 226–229.

Lal, R. K., and Moorehouse, W. W. (1969). Cordierite-gedrite rocks and associated gneisses of Fishtail Lake, Harcourt Township, Ontario. *Can. J. Earth Sci. 6*, 145–165.

Landau, L. D., and Lifshitz, E. M. (1958). *Statistical Physics* (E. Peierls and R. F. Peierls, trans.). London: Pergamon.

Lappin, M. A. (1966). The field relationships of basic and ultrabasic masses in the basal gneiss complex of Stadlandet and Almklovdalen, Nordfjord, Southwestern Norway. *Norsk Geol. Tidskr. 46*, 439–496.

Larimer, J. W. (1967). Chemical fractionation in meteorites. I. Condensation of the elements. *Geochim. Cosmochim. Acta 31*, 1215–1238.

Larimer, J. W. (1968). Experimental studies on the system Fe-MgO-SiO_2-O_2 and their bearing on the petrology of chondritic meteorites. *Geochim. Cosmochim. Acta 32*, 1187–1207.

Larimer, J. W. (1973). Chemistry of the solar nebula. *Space Sci. Rev. 15*, 103–119.

Larimer, J. W., and Buseck, P. R. (1974). Equilibration temperatures in enstatite chondrites. *Geochim. Cosmochim. Acta 38*, 471–477.

Larsen, E. S. (1938). Some new variation diagrams for groups of igneous rocks. *J. Geol. 46*, 506–520.

Larsen, E. S. (1940). Petrographic province of central Montana. *Bull. Geol. Soc. Am. 51*, 887–948.

Larsen, E. S., Jr. (1948). Batholith and associated rocks of Corona, Elsinore and San Luis Rey quadrangles Southern California. *Geol. Soc. Am. Mem. 29*.

Larsen, E. S., Jr., and Draisin, W. M. (1948). Composition of the minerals in the rocks of the southern California batholith. *Intern. Geol. Cong. Rept., 18th Sess.*, Great Britain, Pt. 2, pp. 66–79.

Larsen, E. S., Jr., and Cross, W. (1956). Geology and petrology of the San Juan Region, southwestern Colorado. *U.S. Geol. Surv., Prof. paper 258*.

Larsen, E. S., Jr., Irving, J., Gonyer, F. A., and Larsen, E. S. (1937). Petrologic results of a study of the minerals from the Tertiary volcanic rocks of the San Juan region. Colorado. *Am. Mineral. 22*, 889–905.

Larsen, E. S., Jr., and Schmidt, R. G. (1958). A reconnaissance of the Idaho batholith and comparison with the southern California batholith. *U.S. Geol. Surv. Bull. 1070A*.

Latimer, W. M. (1950). *Science*, 112, 101.

Laves, F., and Goldsmith, J. R. (1961). Polymorphism, order, disorder, diffusion and confusion in the feldspars. Cursiles Y. Conferences, Fusc. VIII, 71–80.

Lebedev, E. B., and Khitarov, N. I. (1964). The dependence of the beginning of melting and the electrical conductivity of granite melt on high pressure of water. *Geochem. Intern. 1964*, 193–197.

Leelanandam, C. (1967). Chemical study of pyroxenes from the charnockitic rocks of Kondapalli (Andhra Pradesh), India, with emphasis on the distribution of elements in coexisting pyroxenes. *Mineral. Mag. 36*, 153–179.

Lepezin, G. C. (1969). Importance of H_2O in cordierite in natural mineralogenesis. *Dokl. Akad. Nauk. SSR 186*, 122–125.

Lever, R. F., and Mandel, G. (1962). Diffusion and vapotransport of solids. *J. Phys. Chem. Solids 23*, 599–600.

Levin, B. Y., and Slonimskii, G. L. (1957). On the problem of the origin of chondrules in meteorites (in Russian). *Dokl. Akad. Nauk. SSR 113*, 62–64.

Lewis, J. S. (1968). Geochemistry of volatile elements on Venus. *Icarus 11*, 367.

Lindsley, D. H. (1965). Ferrosilite. *Carnegie Inst. Yearbook 64*, 148.

Liou, J. G. (1970). Synthesis and stability relations of wairakite, $CaAl_2Si_4O_{12}$-$2H_2O$. *Contr. Mineral. Petrol. 27*, 259–282.

Liou, J. G. (1971a). Synthesis and stability relations of prehnite, $Ca_2Al_2Si_3O_{10}(OH)_2$. *Am. Mineral. 56*, 507–531.

Liou, J. G. (1971b). Stilbite-laumontite equilibrium. *Contr. Mineral. Petrol. 31*, 171–177.

Liou, J. G. (1973). Synthesis and stability relations of epidote, $Ca_2Al_2FeSi_3O_{12}$. *J. Petrol. 14*, 381–413.

Liou, J. G., Ito, K., and Kuniyoshi, S. (1971). Experimental studies of the phase relations between greenschist and amphibolite in a basaltic system. *Abs. Geol. Soc. Am. Meeting*, 634–635.

Lipman, P. W. (1966). Water pressures during differentiation and crystallization of some ash-flow magmas from southern Nevada. *Am. J. Sci. 264*, 810–826.

Lipschutz, M. E., and Anders, A. (1961). The record in the meteorites. IV. Origin of diamonds in iron meteorites. *Geochim. Cosmochim. Acta 24*, 83–105.

Lombaard, B. V. (1934). On the differentiation and relationships of the rocks of the Bushveld complex. *Geol. Soc. S. Africa Trans. 35*, 126–189.

Lord, H. C., III (1965). Molecular equilibria and condensation in a solar nebula and cool stellar atmospheres. *Icarus 4*, 279–288.

Lovering, J. F. (1957). Pressures and temperatures within a typical parent meteorite body. *Geochim. Cosmochim. Acta 12*, 253–261.

Lovering, J. F., and White, A. J. R. (1969). Granulitic and eclogitic inclusions from basic pipes at Delegate, Australia. *Contr. Mineral. Perol. 21*, 9–52.

Lovering, T. S. (1935). Theory of heat conduction applied to geological problems. *Geol. Soc. Am. Bull. 46*, 69–94.

Lowman, P. D., Jr. (1972). The geologic evolution of the moon. *J. Geol. 80*, 125–166.

Lowman, P. D. (1973). Evolution of the Earth's crust: evidence from comparative planetology. *NASA X-document*, X-644-73-322.

Lowman, P. D., Jr. (1976). Crustal evolution in silicate planets: Implications for the origin of continents. *J. Geol. 84*, 1–30.

Luth, W. C., Jahns, R. H., and Tuttle, O. F. (1964). The granite system at pressures of 4 to 10 kilobars. *J. Geophys. Res. 69*, 759–773.

Luth, W. C., and Tuttle, O. F. (1966). The alkali feldspar solvus in the system Na_2O-K_2O-Al_2O_3-SiO_2-H_2O. *Am. Mineral. 51*, 1359–1373.

MacColl, R. S. (1964). Geochemical and structural studies in batholithic rocks of southern California. Part I: Structural geology of the Rattlesnake Mountain pluton. *Geol. Soc. Am. Bull. 75*, 805–822.

MacDonald, G. J. F. (1957). Thermodynamics of solids under nonhydrostatic stress with geologic applications. *Am. J. Sci. 255*, 266–281.

MacGregor, I. D. (1974). The system MgO-Al_2O_3-SiO_2: Solubility of Al_2O_3 in enstatite for spinel and garnet peridotite compositions. *Am. Mineral. 59*, 110–119.

MacGregor, I. D., and Basu, A. R. (1974). Thermal structure of the lithosphere: A petrologic model. *Science 185*, 1007–1011.

Mandel, G. (1962). Vapor transport of solids by vaporphase reactions. *J. Phys. Chem. Solids 23*, 587–598.

Marfunin, A. (1966). *The Feldspars* (translated from Russian, 1962). Jerusalem: Israel Scientific Translations.

Marshall, C. E. (1955). Thermodynamic, quasi-thermodynamic and non-thermodynamic methods as applied to the electro-chemistry of clays in clays and clay minerals. *Fourth National Conference of Clays and Clay Minerals* (A. Swineford, ed.). Washington, D.C.: National Academy of Sciences.

Martignole, J., and Schrijver, K. (1970). Tectonic setting and evolution of the Morin anorthosite, Grenville province, Quebec. *Bull. Geol. Soc. Finland 42*, 165–209.

Mason, B. (1960). *Meteorites*. New York: Wiley.

Mason, B., and Wiik, H. B. (1961). The composition of the Ottawa, Chateau-Renard, Mocs and New Concord meteorites. *Am. Museum Novitates 2069*, 1–25.

Mason, B. (1962a). The carbonaceous chondrites. *Space Sci. Rev. 1*, 621–646.

Mason, B. (1962b). Extraterrestrial mineralogy. *Am. Mineral. 52*, 307–325.

Mason, B. (1965). The enstatite chondrites. *Geochim. Cosmochim. Acta 30*, 23–39.

Mason, B. (1972). The mineralogy of meteorites. *Meteoretics 7*, 309–326.

Mather, J. D. (1970). The biotite isograd and the lower greenschist facies in the Dalradian rocks of Scotland. *J. Petrol. 11*, 253–275.

Matthes, S. (1961). Ergebnisse zur Granatsynthese und ihre Beziehungen zur naturlichen Granatbildung innerhalb der Pyralspit-Gruppe. *Geochim. Cosmochim. Acta 23*, 233–294.

McCallum, I. S. (1968). Equilibrium relationships among the coexisting minerals in the Stillwater Complex, Montana. Ph.D. Thesis, University of Chicago.

McCauley, J. F., Carr, M. H., Cutts, J. A., Hartmann, W. K., Masursky, H., Milton, D. V., Sharp, R. P., and Wilhelms, D. E. (1972). Preliminary Mariner 9 report on the geology of Mars. *Icarus 17*, 289–327.

McConnell, R. K. (1968). Viscosity of the mantle from relaxation time spectra of isostatic adjustment. *J. Geophys. Res. 73*, 7089–7105.

McDowell, S. D. (1974). Emplacement of the Little Chief Stock Paramint Range, California. *Geol. Soc. Am. Bull. 85*, 1539–1546.

McNamara, M. J. (1965). The lower greenschist facies in the Scottish Highlands. *Geol. Foren. Stockholm Forh. 87*, 347–389.

Medaris, L. G., Jr. (1969). Partitioning of Fe^{2+} and Mg^{2+} between coexisting synthetic olivine and orthopyroxene. *Am. J. Sci. 267*, 945–968.

Merill, R. B., Robertson, J. K., and Wyllie, P. J. (1970). Melting reaction in the system $NaAlSi_3O_8$-$KAlSi_3O_8$-SiO_2-H_2O at 20 kilobars compared with results for other feldspar-quartz-H_2O and rock-H_2O systems. *J. Geol. 78*, 558–569.

Merrihue, C. M. (1963). Excess Xenon-129 in chondrules from the Bruderheim meteorite, *J. Geophys. Res. 68*, 325–330.

Metz, P. (1970). Experimental investigation of the metamorphism of siliceous dolomites, 2. The conditions of diopside formation. *Contr. Mineral. Petrol. 28*, 221–250.

Metz, P., and Puhan, D. (1970). Experimentelle Untersuchung der Metamorphose von kieselig dolomitischen Sedimenten. *Contr. Mineral. Petrol. 26*, 302–314.

Metz, P., and Puhan, D. (1971). Korrektur zur Arbeit. Experimentalle Untersuchung der Metamorphose von Kieslig Dolomitischen Sedimenten. *Contr. Mineral. Petrol. 31*, 169–170.

Metz, P., and Trommsdorff, V. (1968). On phase equilibria in metamorphosed siliceous dolomites. *Contr. Mineral. Petrol. 18*, 305–309.

Meyer, C., Jr. (1971). An experimental approach to circumstellar condensation. *Geochim. Cosmochim. Acta 35*, 531–565.

Miller, F. S. (1937). Petrology of the San Marcus gabbro, southern California. *Geol. Soc. Am. Bull. 48*, 1397–1426.

Millhollen, G. L., and Wyllie, P. J. (1974). Melting relations of brown hornblende mylonite from St. Pauls rocks under water saturated and water-undersaturated conditions to 30 kilobars. *J. Geol. 82*, 589–606.

Minakami, T. (1951). On the temperature and viscosity of the fresh lava extruded in the 1951 Oo-sima eruption: *Earthquake Res. Inst. Bull.* (Tokyo) *29*, pt. 3, 487–498.

Misch, P. H. (1949). Metasomatic granitization of batholithic dimensions. Part I. *Am. J. Sci. 247*, 209–245.

Misch, P. H. (1964a). Stable association wollastonite-anorthite, and other calc-silicate assemblages in amphibolite-facies crystalline schists of Nanga Parbat, northwest Himalayas. *Beitr. Mineral. Petrogr. 10*, 315–356.

Misch, P. H. (1964b). Radial epidote glomeroblasts, formed under conditions of synkinematic metamorphism—a new mechanism of collective crystallobastesis. *Geolog. Rundschau 54*, 944–956.

Misch, P. H. (1968). Plagioclase compositions and non-anatectic origin for migmatitic gneisses in northern Cascade Mountains of Washington State. *Contr. Mineral. Petrol. 17*, 1–70.

Misch, P. H. (1969). Paracrystalline microboudinage of zoned grains and other criteria for synkinematic growth of metamorphic minerals. *Am. J. Sci. 267*, 43–63.

Miyashiro, A. (1953a). Progressive metamorphism of the calcium-rich rocks of the Gosaisyo-Takanuki district, Abu-Kuma Plateau, Japan. *Japan J. Geol. Geogr. 23*, 81–107.

Miyashiro, A. (1953b). Calcium-poor garnet in relation to metamorphism. *Geochim. Cosmochim. Acta 4*, 179–208.

Miyashiro, A. (1960). Thermodynamics of reactions of rock-forming minerals with silica. Part IV. Decomposition reactions of muscovite. *Japan J. Geol. Geogr. 31*, 113–120.

Miyashiro, A. (1961). Evolution of metamorphic belts. *J. Petrol. 2*, 277–318.

Miyashiro, A. (1967a). Aspects of metamorphism in the circum-Pacific region. *Tectonophysics 4*, 519–521.

Miyashiro, A. (1967b). Orogeny, regional metamorphism, and magmatism in the Japanese Islands. *Medd. Dansk. Geol. Forening 17*, 390–446.

Miyashiro, A. (1967c). Metamorphism of mafic rocks. In *Basalts*: *Treatise on Rocks of Basaltic Composition*, Vol. 2 (H. H. Hess and A. Poldervaart, eds.). New York: Wiley-Interscience, pp. 799–834.

Miyashiro, A. (1972a). Pressure and temperature conditions and tectonic significance of regional and ocean floor metamorphism. *Tectonophysics 13*, 141–159.

Miyashiro, A. (1972b). Metamorphism and related magmatism in plate tectonics. *Am. J. Sci. 272*, 629–656.

Miyashiro, A. (1973). *Metamorphism and Metamorphic Belts.* New York: Halsted.

Miyashiro, A., and Banno, S. (1958). Nature of glaucophane metamorphism. *Am. J. Sci. 256*, 97–110.

Miyashiro, A., and Seki, Y. (1958). Enlargement of the composition field of epidote and piedmontite with rising temperature. *Am. J. Sci. 256*, 423–430.

Moore, B. N. (1934). Deposits of possible *Nuée Ardente* origin in the Crater Lake region, Oregon, *J. Geol. 42*, 358–385.

Morelock, C. R. (1962). Sub-micro whiskers by vapor deposition. *Acta. Metal. 10*, 161–167.

Morey, G. W. (1922). The application of thermodynamics to heterogeneous equilibria. *J. Franklin Inst. 194*, 425–484.

Morey, G. W., and Fleischer, M. (1940). Equilibrium between vapor and liquid phases in the system CO_2-H_2O-K_2O-SiO_2. *Geol. Soc. Am. Bull. 51*, 1035–1058.

Morey, G. W., and Hesselgesser, J. M. (1951). The solubility of some minerals in superheated steam at high pressures. *Econ. Geol. 46*, 821–835.

Morgan, B. A. (1970). Petrology and mineralogy of eclogite and garnet amphibolite from Puerto Cabella, Venezuela. *J. Petrol. 11*, 101–145.

Morgan, B. A. (1975). Mineralogy and origin of skarns in the Mount Morrison pendant, Sierra Nevada, California. *Am. J. Sci. 275*, 119–142.

Morse, S. A. (1968). Feldspars. *Carnegie Inst. Yearbook 67*, 120–126.

Morse, S. A. (1970). Alkali feldspars with water at 5 kb pressure. *J. Petrol. 11*, pt. 2, 221–251.

Mosesman, M. A., and Pitzer, K. S. (1941). Thermodynamic properties of crystalline forms of silica. *J. Am. Chem. Soc. 63*, 356–438.

Muan, A., and Osborn, E. F. (1956). Phase equilibria at liquidus temperatures in the system MgO-FeO-Fe_2O_3-SiO_2. *Am. Ceramic Soc. J. 39*, 121–140.

Mueller, R. F. (1960). Compositional characteristics and equilibrium relations in mineral assemblages of a metamorphosed iron formation. *Am. J. Sci. 258*, 449–497.

Mueller, R. F. (1961a). Oxidation in high temperature petrogenesis. *Am. J. Sci. 259*, 460–470.

Mueller, R. F. (1961b). Analysis of relations among Mg, Fe, Mn in certain metamorphic minerals. *Geochim. Cosmochim. Acta 25*, 267–296.

Mueller, R. F. (1962). Energetics of certain silicate solutions. *Geochim. Cosmochim. Acta 26*, 581–598.

Mueller, R. F. (1963a). Interaction of chemistry and mechanics in magmatism. *J. Geol. 71*, 759–772.

Mueller, R. F. (1963b). A comparison of oxidative equilibria in meteorites and terrestrial rocks. *Geochim. Cosmochim. Acta 27*, 273–278.

Mueller, R. F. (1964). Theory of the equilibria between complex silicate melts and crystalline solutions. *Am. J. Sci. 262*, 643–652.

Mueller, R. F. (1965). Chemistry in planetology. *J. Chem. Educ. 42*, 294–301.

Mueller, R. F. (1966). Stability relations of the pyroxenes and olivine in certain high grade metamorphic rocks. *J. Petrol. 7*, 363–374.

Mueller, R. F. (1967a). Mobility of the elements in metamorphism. *J. Geol. 75*, 565–582.

Mueller, R. F. (1967b). Model for order-disorder kinetics in certain quasi-binary crystals of continuously variable composition. *J. Phys. Chem. Solids 28*, 2239–2243.

Mueller, R. F. (1967c). High temperature and pressure aqueous solubility data bearing on the stability of the silicates. *J. Geol. 75*, 121–126.

Mueller, R. F. (1969a). Hydration, oxidation and the origin of the calc-alkali series. Natl. Aeronautics & Space Admin. Tech. Note D-5400.

Mueller, R. F. (1969b). Effect of temperature on the strength and composition of the upper lithosphere of Venus. *Nature 224*, 354–356.

Mueller, R. F. (1969c). Kinetics and thermodynamics of intracrystalline distributions. *Mineral. Soc. Am. Spec. Paper 2*, 83–93.

Mueller, R. F. (1969d). Energetics of HCl and HF in volcanic emanations. Natl. Aeronautics & Space Admin. Document X-644-69-254.

Mueller, R. F. (1970). Energetics of HCl and HF in volcanic emanations. *Geochim. Cosmochim. Acta 34*, 737–744.

Mueller, R. F. (1972). Stability of biotite: A discussion. *Am. Mineral. 57*, 300–316.

Mueller, R. F. (1973). System CaO-MgO-FeO-SiO_2-C-H_2-O_2: Some correlations from nature and experiment. *Am. J. Sci. 273*, 152–170.

Mueller, R. F. (1975). Chemical aspects of viscous flow in planetary mantles. *Geochem. J. 9*, 185–199.

Mueller, R. F., and Condie, K. C. (1964). Stability relations of carbon mineral assemblages in the southern California batholith. *J. Geol. 72*, 400–411.

Mueller, R. F., and Kridelbaugh, S. J. (1973). Kinetics of CO_2 production on Venus. *Icarus 19*, 531–541.

Murase, T., and McBirney, A. (1970). Viscosity of lunar lavas. *Science 167*, 1491–1492.

Murase, T., and McBirney, A. R. (1973). Properties of some common igneous rocks and their melts at high temperatures. *Geol. Soc. Am. Bull. 84*, 3563–3592.

Murdock, T. L., and Ney, E. P. (1970). Mercury: The dark side temperature. *Science 170*, 535–537.

Murry, B. C., Soderblom, L. A., Cutts, J. A., Sharp, R. P., Milton, D. J., and Leighton, R. B. (1972). Geological framework of the South Polar region of Mars. *Icarus 17*, 328–345.

Murry, B. C., Belton, M. J. S., Danielson, G. E., Duis, M. E., Gault, D., Hoke, B., O'Leory, B., Strom, R. G., Suomi, V., and Trask, N. (1974). Mariner 10 pictures of Mercury: First results. *Science 184*, 459–461.

Mutch, T. A., Binder, A. B., Huck, F. O., Levinthal, E. C., Liebes, S. Jr., Morris, E. C., Patterson, W. R., Pollack, J. B., Sagan, Carl, and Taylor, G. R. (1976). The surface of Mars; The view from the Viking I Lander. *Science 193*, 791–801.

Mysen, B. O. (1976). The role of volatiles in silicate melts: Solubility of carbon dioxide and water in feldspar, pyroxene and feldspathoid melts to 30 kb and 1625°C. *Am. J. Sci. 276*, 969–996.

Nafziger, R. H., and Muan, A. (1967). Equilibrium phase compositions and thermodynamic properties of olivines and pyroxenes in the system MgO-"FeO"-SiO_2. *Am. Mineral. 52*, 1364–1385.

Nash, W. P., and Wilkinson, J. F. G. (1970). Shonkin Sag laccolith, Montana I. Mafic minerals and estimates of temperature, pressure, oxygen fugacity and silica activity. *Contr. Mineral. Petrol. 25*, 241–269.

Nemoto, T., Hayakawa, M., Takahashi, K., and Oana, S. (1957). Report on the geological, geophysical and geochemical studies of Showa-Shinzan, Usu Volcano. *Geol. Surv., Japan Rept. 170* (in Japanese).

Newton, M. S., and Kennedy, G. C. (1968). Jadeite, analcite, nepheline and albite at high temperatures and pressures. *Am. J. Sci. 266*, 728–735.

Newton, R. C. (1966a). Some calc-silicate equilibrium relations. *Am. J. Sci. 264*, 204–222.

Newton, R. C. (1966b). Kyanite-andalusite equilibrium from 700–800°C. *Science* (N.Y.) *153*, 170–172.

Newton, R. C. (1972). An experimental determination of the high-pressure stability limits of magnesian cordierite under wet and dry conditions. *J. Geol. 80*, 398–420.

Newton, R. C., Goldsmith, J. R., and Smith, J. V. (1969). Aragonite crystallization from strained calcite at reduced pressures and its bearing on aragonite in low-grade metamorphism. *Contr. Mineral. Petrol. 22*, 335–348.

Newton, R. C., and Kennedy, G. C. (1963). Some equilibrium reactions in the join $CaAl_2Si_2O_8$-H_2O. *J. Geophys. Res. 68*, 2967–2983.

Newton, R. C., and Smith, J. V. (1967). Investigations concerning the breakdown of albite at depth in the earth. *J. Geol. 75*, 268–286.

Nichols, I. A. (1974). Liquids in equilibrium with peridotite mineral assemblages at high water pressures. *Contr. Mineral. Petrol. 45*, 289–316.

Niggli, P. (1936). Die Magmentypen. *Schweiz. Min. Petr. Mitt. 16*.

Nitsch, K. H. (1971). Stabilititätsbezichungen von Prehnit und Pumpellyit-haltigen Paragenesen. *Contr. Min. Pet. 30*, 240–260.

Nitsch, K. H., and Winkler, H. G. F. (1965). Bildungsbedingungen von Epidot und Orthozoisit. *Beitr. Miner. Petrogr. 11*, 470–486.

Nockolds, S. R. (1940). The Garabal Hill-Glen Fyne igneous complex. *Geol. Soc. Lond. Quart. J. 96*, 451–511.

Nockolds, S. R. (1954). Average chemical composition of some igneous rocks. *Geol. Soc. Am. Bull. 65*, 1007–1032.

Nordenskjold, K. E. (1878). On the composition and common origin of the meteorites. *Nature 10*, 510–511.

Nordlie, B. E. (1967). The composition of the basaltic gas phase. Ph.D. thesis. University of Chicago.

Nordlie, B. E. (1971). The composition of the magmatic gas of Kilouea and its behavior in the near surface environment. *Am. J. Sci. 271*, 417–463.

Norton, D. A., and Clavan, W. S. (1959). The optical mineralogy, chemistry and X-ray crystallography of ten clinopyroxenes from the Pennsylvania and Delaware Piedmont Province. *Am. Mineral. 44*, 844–874.

Nussbaumer, H., and Swings, V. P. (1970). (Fe II) Magnetic dipole transition probabilities and the problem of the solar iron abundance. *Astron. Astrophys. 7*, 455–458.

O'Hara, M. J. (1965). Primary magmas and the origin of basalts. *Scot. J. Geol. 1*, 19–40.

O'Hara, M. J., and Mercy, E. L. P. (1963). Petrology and petrogenesis of some garnetiferrous peridotites. *Trans. Roy. Soc. Edinb. 65*, 251–314.

O'Hara, M. J., and Yoder, H. S. (1967). Formation and fractionation of basic magmas at high pressures. *Scot. J. Geol. 3* 67–113.

Olsen, E. (1963). Equilibrium calculations in the system Mg, Fe, Si, O, H, and Ni. *Am. J. Sci. 261*, 943–956.

Olsen, E. (1964). Some calculations concerning the effect of nickel on the stability of cohenite in meteorites. *Geochim. Cosmochim. Acta 28*, 609–617.

Olsen, E. J., and Fuchs, L. (1966). The state of oxidation of some iron meteorites. *Icarus 6*, 242–253.

Olsen, E., and Frederiksson, K. (1966). Phosphate in iron and pallasite meteorites. *Geochim. Cosmochim. Acta 30*, 459–470.

Olsen, E. J., and Mueller, R. F. (1964). Silicates in some iron meteorites. *Nature 201*, 596–597.

Olsen, E., and Mueller, R. F. (1966). Stability of orthopyroxenes with respect to pressure and composition. *J. Geol. 74*, 620–625.

Onuma, N., Clayton, R. N., and Mayeda, T. K. (1972). Oxygen isotope temperatures of "equilibrated" ordinary chondrites. *Geochim. Cosmochim. Acta 36*, 157–168.

Öpik, E. J. (1962). *Progress in the Astronautical Sciences*, Vol. I. Amsterdam: North-Holland.

Öpik, E. J., and Singer, S.F. (1960). Escape of gases from the moon. *J. Geophys. Rev. 65*, 3065–3070.

Orville, P. M. (1963). Alkali ion exchange between vapor and feldspar phases. *Am. J. Sci. 261*, 201–237.

Osann, A. (1899). Versuch einer Chemischen Klassification der Eruptivgesteine. *Tscherm. Mineral. Petrograph. Mitt. XIX*, 351–469; *XX* (1901), 399–559; *XXI* (1902), 365–448; *XXII* (1903), 322–356, 403–436.

Osborn, E. F. (1959). Role of oxygen pressure in the crystallization and differentiation of basaltic magma. *Am. J. Sci. 257*, 609–647.

Owen, E. A., and Sully, A. H. (1939). The equilibria diagram of iron-nickel alloys. *Phil. Mag. 27*, 614–636.

Owen, E. A., and Liu, Y. H. (1949). Further x-ray study of the equilibrium diagram of the iron-nickel system. *J. Iron, Steel Inst. 163*, 132–138.

Owen, Tobias and Biemann, K. (1976). Composition of the atmosphere at the surface of Mars, Detection of Argon-36 and preliminary analysis. *Science 193*, 801–803.

Pabst, A. (1928). Observations on inclusions in the granitic rocks of the Sierra Nevada. *Calif. Univ. Dept. Geol. Sci. Bull. 17*, 325–386.

Pauling, L. (1948). *The Nature of the Chemical Bond*. Ithaca, N.Y.: Cornell University Press.

Perchuk, L. L. (1965). The paragenesis of nepheline with alkali feldspar as the indicator of mineral equilibrium thermodynamic conditions. *Dokl. Akad. Nauk. SSSR 161* (4).

Peret, F. A. (1937). The eruption of Mt. Pelee 1929–1932. *Carnegie Inst. Pub. No. 458*.

Perry, K., Jr. (1967). An application of linear algebra to petrologic problems. *Geochim. Cosmochim. Acta 31*, 1043–1078.

Philpotts, A. R. (1966). Origin of the anorthosite-mangerite rocks in Southern Quebec. *J. Petrol. 7*, 1–64.

Philpotts, A. R. (1972). Density surface tension and viscosity of the immiscible phase of a basic alkaline magma. *Lithos 5*, 1–18.

Pickering, J. S. (1961). *Captives of the Sun*. New York: Dodd, Mead.

Pistorius, C. W. F. T., and Kennedy, G. C. (1960). Stability relations of grossularite and hydrogrossularite at high temperatures and pressures. *Am. J. Sci. 258*, 247–257.

Piwinskii, A. J. (1968). Experimental studies of igneous rock series, central Sierra Nevada batholith, California. *J. Geol. 76*, 548–570.

Piwinskii, A. J. (1973). Experimental studies of igneous rock series, central Sierra Nevada batholith, California. Part II. *N. Jb. Miner. Mh. H5*, 193–215.

Piwinskii, A. J., and Wyllie, P. J. (1968). Experimental studies of igneous rock series; A zoned pluton in the Wallowa batholith, Oregon. *J. Geol. 76*, 205–234.

Popp, R. K., and Gilbert, M. C. (1972). Stability of acmite-jadeite pyroxenes at low pressure. *Am. Mineral. 57*, 1210–1231.

Post, R. H., Jr., and Griggs, D. T. (1973). The Earth's mantle; evidence of non-Newtonian flow. *Science 181*, 1242–1244.

Powell, J. L., Hurley, P. M., and Fairbairn, H. W. (1966). The strontium isotopic composition and origin of carbonatites. In *Carbonatites* (O. F. Tuttle and J. Gittins, eds.). New York: Wiley, pp. 365–378.

Presnall, D. C. (1969). The geometrical analysis of partial fusion. *Am. J. Sci. 267*, 1178–1194.

Presnall, D. C., and Bateman, P. C. (1973). Fusion relations in the system $NaAlSi_3O_8$-$CaAl_2Si_2O_8$-$KAlSi_3O_8$-SiO_2-H_2O and generation of granitic magmas in the Sierra Nevada Batholith. *Geol. Soc. Am. Bull. 84*, 3181–3202.

Prigogine, I., and Defay, R. (1954). *Chemical Thermodynamics*. London: Longmans Green.

Prior, G. T. (1916). On the genetic relationship and classification of meteorites. *Mineral. Mag. 18*, 28–44.

Prior, G. T. (1920). The classification of meteorites. *Mineral. Mag. 19*, 51–63.

Prior, G. T. (1953). *Catalogue of Meteorites*, (2nd ed., rev. by M. H. Hoy). London: British Museum.

Puhan, D., and Hoffer, E. (1973). Phase relations of talc and tremolite in metamorphic calcite-dolomite sediments in the southern portion of the Damara Belt (Southwest Africa). *Contr. Mineral. Petrol. 40*, 207–214.

Rabbitt, J. C. (1948). A new study of the anthophyllite series. *Am. Mineral. 33*, 263–323.

Ramberg, H. (1944a). Petrologic significance of subsolidus phase transitions in mixed crystals. *Norsk Geol. Tidskr. 24*, 42–74.

Ramberg, H. (1944b). The thermodynamics of the earth's crust. I, II. *Norsk Geol. Tidskr. 24*, 98–111; *25*, 307–326.

Ramberg, H. (1946). Kjemisk likevekt i gravitasjonsfeldet og dens betydning for jordskorpens differentiasjon. *Medd. Dansk. Geol. Foren. 11*, 13–18.

Ramberg, H. (1947). The force of crystallization as a well definable property of crystals. *Geol. Foren. Stockholm Förh. 69*, 189–194.

Ramberg, H. (1948a). On sapphirine-bearing rocks in the vicinity of Sukkertoppen (West Greenland). *Medd. Gronland, 142* (5), 1–33.

Ramberg, H. (1948b). Radial diffusion and chemical stability in the gravitational field. *J. Geol. 56*, 448–458.

Ramberg, H. (1949). The facies classification of rocks: A clue to the origin of quartzofeldspathic massifs and veins. *J. Geol. 57*, 18–54.

Ramberg, H. (1952). *The Origin of Metamorphic and Metasomatic Rocks*. Chicago: University of Chicago Press.

Ramberg, H. (1956). Pegmatites in west Greenland. *Bull. Geol. Soc. Am. 67*, 185–214.

Ramberg, H. (1959). The Gibbs' free energy of crystals under anisotropic stress, a possible cause for preferred mineral orientation. Arais du Escola de Minas de Ouro Preto (Brazil). *Ancto ao. 72*, 1–13.

Ramberg, H. (1961). A study of veins in Caledonian rocks around Trondheim Fjord, Norway. *Norsk Geol. Tidskr. 41*, 1–43.

Ramberg, H. (1963). Experimental study of gravity tectonics by means of centrifuged models. *Bull. Geol. Uppsala 42*.

Ramberg, H. (1964). A model for the evolution of continents, oceans and orogens. *Tectonophysics 2*.

Ramberg, H. (1967). *Gravity, Deformation and the Earth's Crust*. London: Academic.

Ramberg, H., and DeVore, G. W. (1951). The distribution of Fe^{2+} and Mg^{2+} in coexisting olivines and pyroxenes. *J. Geol. 59*, 193–210.

Ramsay, J. G. (1967). *Folding and Fracturing of Rocks*. New York: McGraw-Hill.

Raymahashay, B. C. (1968). A geochemical study of rock alteration by hot springs in the Paint Pot Hill area, Yellowstone Park. *Geochim. Cosmochim. Acta. 32*, 499–522.

Read, H. H. (1957). *The Granite Controversy*. New York: Interscience.

Read, W. T., and Shockley, W. (1950). Dislocation models of crystal grain boundaries. *Phys. Rev. 78*, 275–289.

Reinhardt, E. W. (1968). Phase relations in cordierite bearing gneisses from the Gananoque area, Ontario. *Can. J. Earth Sci. 5*, 455–483.

Reverdatto, V. V., and Sharapov, V. N. (1970). The controls and selected peculiarities of the origin of contact metamorphic zonation. *Contr. Mineral. Petrol. 29*, 310–337.

Reynolds, J. H. (1960). Determination of the age of the elements. *Phys. Rev. Letts. 4*, 8–10.

Richardson, S. W. (1968). Staurolite stability in a part of the system Fe-Al-Si-O-H. *J. Petrol. 9*, 468–488.

Richardson, S. W., Bell, P. M., and Gilbert, M. C. (1968). Kyanite-sillimanite equilibrium between 700 and 1500°C. *Am. J. Sci. 266*, 513–541.

Richardson, S. W., Gilbert, M. C., and Bell, P. M. (1969). Experimental determination of kyanite-andalusite and andalusite-sillimanite equilibria; the aluminum silicate triple point. *Am. J. Sci. 267*, 259–272.

Riebling, E. F. (1966). Structure of sodium alumino-silicate melts containing at least 50 mole % SiO_2 at 1500°C. *J. Chem. Phys. 44*, 2857–2865.

Ringwood, A. E. (1960). Cohenite as a pressure indicator in iron meteorites. *Geochim. Cosmochim. Acta 20*, 155–158.

Ringwood, A. E. (1961a). Silicon in the metal phase of enstatite chondrites and some geochemical implications. *Geochim. Cosmochim. Acta 25*, 1–13.

Ringwood, A. E. (1961b). Chemical and genetic relationship among meteorites. *Geochim. Cosmochim. Acta 24*, 159–197.

Ringwood, A. E. (1966). Genesis of chondritic meteorites. *Rev. Geophys. 4*, 113–175.

Ringwood, A. E., and Essene, E. (1970). Petrogenesis of lunar basalts and the internal constitution and origin of the Moon. *Science 167*, 607–610.

Ringwood, A. E., and Green, D. H. (1966a). An experimental investigation of the gabbro-eclogite transformation and some geophysical implications. *Tectonophysics 3*, 383–427.

Ringwood, A. E., and Green, D. H. (1966b). Petrological nature of the stable continental crust. In *The Earth Beneath the Continents* (J. S. Steinhart and T. J. Smith, eds.). Washington, D.C.: American Geophysical Union, Monogr. *10*, pp. 611–619.

Robie, R. A., and Bethke, P. M. (1966). Molar volumes and densities of minerals. *Geol. Soc. Am. Mem. 97*, 60–73.

Robie, R. A., Bethke, P. M., and Beardsley, K. M. (1967). Selected X-ray crystallographic data; molar volumes, and densities of minerals and related substances. *U.S. Geol. Surv. Bull. 1248*.

Robie, R. A., and Waldbaum, D. R. (1968). Thermodynamic properties of minerals and related substances at 298.15 K (25.0°C) and one atmosphere (1.013 bars) pressure and at higher temperature. *U.S. Geol. Surv. Bull. 1259*.

Robinson, P. (1963). Gneiss domes of the Orange Area, Mass. and N.H.Ph.D. thesis, Harvard University.

Robinson, P., and Jaffe, H. W. (1969a). Aluminous enclaves in gedrite-cordierite gneiss from southwestern New Hampshire. *Am. J. Sci. 267*, 389–421.

Robinson, P., and Jaffe, H. W. (1969b). Chemographic exploration of amphibole assemblages from Central Massachusetts and southwestern New Hampshire. *Mineral. Soc., Spec. Paper 2*, 251–274.

Robinson, P., Ross, M., and Jaffe, H. W. (1971). Composition of the anthophyllite-gedrite series, comparison of gedrite and hornblende, and the anthophyllite-gedrite solvus. *Am. Mineral. 56*, 1005–1041.

Roedder, E. (1951). Low temperature liquid immiscibility in the system K_2O-FeO-Al_2O_3-SiO_2. *Am. Mineral. 36*, 282–286.

Roedder, E. (1965). Liquid inclusions in olivine-bearing nodules and phenocrysts from basalts. *Am. Mineral. 50*, 1746–1782.

Roedder, E., and Coombs, D. S. (1967). Immiscibility in granite melts, indicated by fluid inclusions in ejected granitic blocks from Ascension Island. *J. Petrol. 8*, 417–451.

Roedder, E., and Weiblen, P. W. (1970). Silicate immiscibility in lunar magmas evidenced by melt inclusions in lunar rocks. *Science 167*, 641–644.

Roeder, P. L. (1974). Paths of crystallization and fusion in systems showing ternary solid solution. *Am. J. Sci. 274*, 48–60.

Roscoe, R. (1952). The viscosity of suspensions of rigid spheres. *Br. J. Appl. Phys. 3*, 267–269.

Rosenfeld, J. L. (1968). Garnet rotations due to major Palaeozoic deformations in southeast Vermont. In *Studies of Appalachian Geology, Northern and Maritime* (E-An Zen *et al.*, eds.). New York: Wiley, pp. 185–202.

Rosenfeld, J. L. (1970). Rotated garnets in metamorphic rocks. *Geol. Soc. Am., Spec. Paper 129*.

Ross, C. S., and Smith, J. (1955). Water and other volatiles in volcanic glasses. *Am. Mineral. 40*, 1071–1089.

Ross, C. S. (1964). Volatiles in volcanic glasses and their stability relations. *Am. Mineral. 49*, 258–271.

Ross, M., Papike, J. J., and Shaw, K. W. (1969). Exsolution textures in amphiboles as indicators of subsolidus thermal histories. *Mineral. Soc. Paper 2*, 275–299.

Rouse, J. R. (1937). Genesis and structural relationships of the Absaroka volcanic rocks. *Geol. Soc. Am. Bull. 48*, 1257–1296.

Roy, D. M., and Roy, R. (1957). Synthesis and stability of minerals in the system MgO-Al_2O_3-SiO_2-H_2O. *Am. Mineral. 40*, 147–178.

Runcorn, S. K. (1973). On the implications of the shape of Mars. *Icarus 18*, 109–112.

Rutherford, M. J. (1973). The phase relations of aluminous iron biotites in the system $KAlSi_3O_8$-$KAlSiO_4$-Al_2O_3-Fe-O-H. *J. Petrol. 14*, 159–180.

Sahama, T. G., and Meyer, A. (1958). A study of the Volcano Nyirangongo: Exploration du Parc National Albert: Mission d'études vulcanologiques, fasc. *2*.

Sander, B. (1930). *Gefügekunde der Gesteine*. Vienna: Springer.

Sato, M., and Wright, T. L. (1966). Oxygen fugacities directly measured in magmatic gases. *Nature 153*, 1103–1105.

Sauramo, M. (1958). Die Geschichte der Ostsee. *Ann. Acad. Sci. Fennicae, A(3)*, 44.

Saxena, S. K. (1968). Chemical study of phase equilibria in charnockites, Varberg, Sweden. *Am. Mineral. 53*, 1674–1695.

Saxena, S. K. (1971). Mg^{2+}-Fe^{2+} order-disorder in orthopyroxene and the Mg^{2+}-Fe distribution between coexisting minerals. *Lithos 4*, 345–354.

Saxena, S. K. (1973). *Thermodynamics of Rock-Forming Crystalline Solutions*. New York: Springer-Verlag.

Saxena, S. K. (1976). The two-pyroxene geothermometer. *Am. Mineral. 61*, 643–652.

Saxena, S. K., and Ghose, S. (1970). Order-disorder and the activity-composition relation in a binary crystalline solution. Part 1. Metamorphic orthopyroxene. *Am. Mineral. 55*, 1219–1225.

Saxena, S. K., and Ghose, S. (1971). Mg^{2+}-Fe^{2+} order-disorder and the thermodynamics of the orthopyroxene crystalline solution. *Am. Mineral. 56*, 532–559.

Scarfe, C. M., Luth, W. C., and Tuttle, O. F. (1966). An experimental study bearing on the absence of leucite in plutonic rocks. *Am. Mineral. 51*, 726–735.

Scarfe, C. M., and Wyllie, P. J. (1967). Serpentine dehydration curves and their bearing on serpentine deformation orogenesis. *Nature 215*, 945–946.

Scarfe, C. M. (1973). Viscosity of basaltic magmas at varying pressures. *Nature 241*, 101–102.

Schairer, J. F. (1957). Melting relations of the common rock-forming silicates. *Am. Ceram. Soc. J. 40*, 215–235.

Schwarcz, H. P., Scott, S. D., and Kissin, S. A. (1975). Pressure of formation of iron meteorites from sphalerite compositions. *Geochim. Cosmochim. Acta 39*. 1457–1466.

Schwartz, G. M., and Sandberg, A. E. (1940). Rock series in diabase sills at Duluth, Minnesota. *Bull. Geol. Soc. Am. 51*, 1135–1172.

Schreyer, W. (1968). A reconnaissance study of the system MgO-Al_2O_3-SiO_2-H_2O at pressures between 10 and 20 kbar. *Carnegie Inst. Geophys. Lab. Yearbook, 1967–1968*, pp. 380–392.

Schreyer, W., and Chinner, G. A. (1966). Staurolite-quartzite bands in Kyanite quartzite at Big Rock, Rio Arriba County, New Mexico. *Contr. Mineral. Petrol. 12*, 233–244.

Schreyer, W., and Seifert, F. (1967). Metastability of an osumilite end member in the system K_2O-MgO-Al_2O_3-SiO_2-H_2O and its possible bearing on the rarity of natural osumilites. *Contr. Mineral. Petrol. 14*, 343–358.

Schreyer, W., and Seifert, F. (1968). Synthetic amphiboles in the system Na_2O-MgO-SiO_2-H_2O and their significance for the chemistry of natural amphiboles (abst.). *Proc. 5th Meet. Int. Mineral. Assoc., Cambridge*, 337–338.

Schreyer, W., and Seifert, F. (1969). High-pressure phases in the system MgO-Al_2O_3-SiO_2-H_2O. *Am. J. Sci. 267A*, 407.

Scott, S. D., and Barnes, H. L. (1971). Sphalerite geothermometry and geobarometry. *Econ. Geol. 66*, 653–669.

Seck, H. A. (1971a). Koexistierende Alkalifeldspate und Plagioclase in System $NaAlSi_3O_8$-$KAlSi_3O_8$-$CaAl_2Si_2O_8$-H_2O bei Temperaturen von 650°C bis 900°C. *Neues Jahrb. Mineral. Abhandl. 115*, 315–345.

Seck, H. A. (1971b). Der Einfluss des Drucks auf die Zusammensetzung koexistierender Alkalifeldspate und Plagioclase. *Contr. Mineral. Petrol. 31*, 67–86.

Seifert, F., and Schreyer, W. (1970). Low temperature stability limit of Mg cordierite in the range 1–7 kbar water pressure: A redetermination. *Contr. Mineral. Petrol. 27*, 225–238.

Seifert, F. (1970). Low-temperature compatibility relations of cordierite in haplopelites of system K_2O-MgO-Al_2O_3-SiO_2-H_2O. *J. Petrol. 11*, 73–99.

Seifert, F., and Virgo, D. (1975). Kinetics of the Fe^{2+}-Mg order-disorder reaction in anthophyllite: Quantitative cooling rates. *Science 188*, 1107–1109.

Seki, Y. (1972). Lower grade stability limit of epidote in the light of natural occurrences. *J. Geol. Soc. Japan 78*, 405–413.

Shafer, H. (1956). Über der transport der Bodenkörpers im Temperatur gefälle mit Hilfe hetergeser Gleichgewichte. *Z. Anorg. Chem. 286*, 27–41.

Shaw, H. R. (1963). Obsidian-H_2O viscosities at 1000 and 2000 bars in the temperature range 700 to 900°C. *J. Geophys. Res. 68*, 6337–6343.

Shaw, H. R. (1964). Theoretical solubility of H_2O in silicate melts: quasi-crystalline models. *J. Geol. 72*, 601–617.

Shaw, H. R. (1965). Comments on viscosity, crystal settling and convection in granitic magmas. *Am. J. Sci. 263*, 120–152.

Shaw, H. R. (1969). Rheology of basalt in the melting range. *J. Petrol. 10*, 510–535.

Shaw, H. R. (1972). Viscosities of magmatic silicate liquids: An empirical method of prediction. *Am. J. Sci. 272*, 870–893.

Shaw, H. R., and Wones, D. R. (1964). Fugacity coefficients for hydrogen gas between 0° and 1000°C, for pressures to 3000 atm. *Am. J. Sci. 262*, 918–929.

Shimizu, M. (1963). Vertical distribution of neutral gases on Venus. *Planet Space Sci. Letts. 11*, 269–273.

Sippel, R. F. (1963). Sodium selfdiffusion in natural minerals. *Geochim. Cosmochim. Acta 27*, 107–120.

Skippen, G. B. (1971). Experimental data for reactions in siliceous marbles. *J. Geol. 79*, 451–481.

Skippen, G. (1974). An experimental model for low pressure metamorphism of siliceous dolomitic marble. *Am. J. Sci. 274*, 487–509.

Smith, C. S. (1948). Grains, phases and interfaces; an interpretation of microstructure. *Trans. Am. Inst. Min. Metall. Engrs. 175*, 15–51.

Smith, D. (1971). Stability of the assemblage iron-rich orthopyroxene-olivine-quartz. *Am. J. Sci. 271*, 370–382.

Smith, J. V. (1968). The crystal structure of staurolite. *Am. Mineral. 53*, 1139–1155.

Smith, J. V. (1973). Critical review of synthesis and occurrence of plagioclase feldspars and a possible phase diagram. *J. Geol. 80*, 505–523.

Smith, J. V. (1974). *Feldspar Minerals*, Parts 1 and 2. New York: Springer-Verlag.

Sobolev, N. V., Jr., Kuznetsova, I. K., and Zyuzin, N. I. (1968). The petrology of grospydite zenoliths from the Zagadochnaya kimberlite pipe in Yakutia. *J. Petrol. 9*, 253–280.

Sorby, H. C. (1877). On the structure and origin of meteorites. *Nature 15*, 495–498.

Sourirajan, S., and Kennedy, G. C. (1962). The system H_2O-NaCl at elevated temperatures and pressures. *Am. J. Sci. 260*, 115–141.

Speidel, D. H., and Osborn, E. F. (1967). Element distribution among coexisting phases in the system MgO-FeO-Fe_2O_3-SiO_2 as a function of temperature and oxygen fugacity. *Am. Mineral. 52*, 1139–1152.

Spengler, C. J., and Burnham, C. W. (1962). Compositions in the upper three-phase region of the system $KAlSi_3O_8$-H_2O at pressures up to 6 kilobars (Abs.). *Geol. Soc. Am., Spec. Paper 68*, 277.

Spitzer, L. Jr. (1952). The terrestrial atmosphere above 300 km. In *Atmospheres of the Earth and Planets*. (G. Kuiper, ed.). Chapt. 7, Chicago, University of Chicago Press, Illinois.

Spry, A. (1963). The origin and significance of snowball structure in garnet. *J. Petrol. 4*, 211–222.

Staatz, M. H., Murata, K. J., and Glass, J. (1955). Variation in composition and physical properties of tourmaline with its position in the pegmatite. *Am. Mineral. 40*, 789–804.

Stecher, T. P., and Williams, D. A. (1966). Interstellar molecule formation. *Astrophys. J. 146*, 88–103.

Stern, C. R., and Wyllie P. J. (1973). Water-saturated and undersaturated melting relations of a granite to 35 kilobars. *Earth Planet Sci. Letts. 18*, 163–167.

Stewart, D. B. (1958). System $CaAl_2Si_2O_8$-SiO_2-H_2O (Abs.). *Geol. Soc. Am. Bull. 69*, 1648.

Stoiber, R. E., and Rose, W. I., Jr. (1974). Fumarole incrustations at active Central American volcanoes. *Geochim. Cosmochim. Acta 38*, 495–516.

Storre, B. (1972). Dry melting of muscovite + quartz in the range $P_s = 7$ kb to $P_s = 20$ kb. *Contr. Mineral. Petrol. 37*, 87–89.

Storre, B. (1970). Stabilitatsbedingungen Grossular-furender Paragenesen im System CaO-Al_2O_3-SiO_2-CO_2-H_2O. *Contr. Mineral. Petrol. 29*, 145–162.

Storre, B., and Karotke, E. (1971). An experimental determination of the upper stability limit of muscovite + quartz in the range 7–20 kb water pressure. *Neues Jahrb. Mineral. Monatsh.* 237–240.

Struve, O. (1950). *Stellar Evolution*. Princeton, N.J.: Princeton University Press.

Studier, M. H., Hayatsu, R., and Anders, E. (1972). Origin of organic matter in early solar system. 2. Further studies of meteorite hydrocarbons and a discussion of their origin. *Geochim. Cosmochim. Acta 36*, 189–215.

Suess, H. E. (1949). Zur chemie der Planeten und Meteoritenbildung. *Z. Electrochemie 53*, 237–241.

Suess, H. E., and Urey, H. C. (1956). Abundances of the elements. *Rev. Mod. Phys. 28*, 53–74.

Sundius, N. (1933). Uber die Mischungslücken Zwischen Anthophyllit-Gedrit, Cummingtonit-Grünerit and Tremolit-Aktinolith. *Tschermak Mineralog. Petrog. 43*, 422–440.

Surdam, R. C. (1973). Low grade metamorphism of tuffaceous rocks in the Karmutsen Group, Vancouver Island, British Columbia. *Geol. Soc. Am. Bull. 84*, 1911–1925.

Swalin, R. A. (1962). *Thermodynamics of Solids*. New York: Wiley.

Sylvester, A. G., and Christie, J. M. (1968). The origin of cross-girdle orientation of optic oxes in deformed quartzites. *J. Geol. 76*, 571–580.

Symposium on the Bushveld igneous complex and other layered intrusions. (1969). D. J. L. Visser and G. von Gruenewaldt, eds. Publication No. 1, *Geol. Soc. of South Africa, Johannesburg, S.A.*

Tatsumoto, M., Knight, R. J., and Allegre, C. J. (1973). Time difference in the formation of meteorites as determined from the ratio of lead-207 to lead-206. *Science 180*, 1279–1283.

Taubeneck, W. H. (1957). Geology of the Elkhorn Mountains, Northeastern Oregon: Bald Mountain batholith. *Bull. Geol. Soc. Am. 68*, 181–238.

Taylor, S. R. (1975). *Lunar Science: A Post-Apollo View*. New York: Pergamon.

Thomas, G. E. (1974). Mercury: Does its atmosphere contain water? *Science 183*, 1197–1198.

Thompson, A. B. (1970a). A note on the kaolinite-pyrophyllite equilibrium. *Am. J. Sci. 268*, 454–458.

Thompson, A. B. (1970b). Laumontite equilibria and the zeolite facies. *Am. J. Sci. 269*, 267–275.

Thompson, A. B. (1971a). Analcite-albite equilibria at low temperatures. *Am. J. Sci. 271*, 79–92.

Thompson, A. B. (1971b). P_{CO_2} in low-grade metamorphism; zeolite, carbonate; clay mineral, prehnite relations in the system CaO-Al_2O_3-SiO_2-CO_2-H_2O. *Contr. Mineral. Petrol. 33*, 145–161.

Thompson, J. B., Jr. (1957). The graphical analysis of mineral assemblages in pelitic schists. *Am. Mineral. 42*, 842–858.

Thompson, J. B., Jr. (1967). Thermodynamic properties of simple solutions. In *Researches in Geochemistry*, Vol. II. (P. H. Abelson, ed.). New York: Wiley, pp. 340–361.

Thompson, J. B., Jr. (1970). Geochemical reaction and open systems. *Geochim. Cosmochim. Acta 34*, 529–551.

Thompson, J. B., and Norton, S. A. (1968). Paleozoic regional metamorphism in New England and adjacent areas. In *Studies of Appalachian Geology: Northern and Maritime*. (E-an Zen *et al.*, eds.). New York: Interscience, pp. 319–327.

Thompson, J. B., Jr., and Waldbaum, D. R. (1969). Mixing properties of sanidine crystalline solutions. III. Calculations based on two-phase data. *Am. Mineral. 54*, 811–838.

Thorarinsson, S., and Sigvaldason, G. E. (1962). The eruption in Askja, 1961, a preliminary report. *Am. J. Sci. 260*, 641–651.

Thornton, C., and Tuttle, O. F. (1960). Chemistry of igneous rocks, I. Differentiation index. *Am. J. Sci. 258*, 664–684.

Tilley, C. E. (1922). Density, refractivity and composition relation of some natural glasses. *Mineral. Mag. 19*, 275–294.

Tilley, C. E. (1924). Contact metamorphism in the Comrie area of the Perthshire Highlands. *Quart. J. Geol. Soc. Lond. 80*, 22–70.

Tilley, C. E. (1937). Anthophyllite-cordierite granulites of the Lizard. *Geol. Mag. 74*, 300–309.

Tomasson, J., and Kristmansdottir, H. (1972). High temperature alteration minerals and thermal brines, Reykjanes, Iceland. *Contr. Mineral. Petrol. 36*, 123–134.

Tomlinson, J. W. (1953). Some aspects of the constitution of liquid oxides. In *Physical Chemistry of Melts*. London: Institution of Mining and Metallurgy.

Toulmin, Priestley, III, and Barton, P. B., Jr. (1964). A thermodynamic study of pyrite and pyrrhotite. *Geochim. Cosmochim. Acta 28*, 641–671.

Trommsdorff, V., and Evans, B. W. (1969). The stable association enstatite-forsterite-chlorite in amphibolite facies ultramafics of the Lepontine Alps. *Schweiz. Mineral. Petrograph. Mitteil. 49/2*, 326–332.

Trommsdorff, V., and Evans, B. W. (1972). Progressive metamorphism of antigorite schist in the Bergell Tonalite Aureole (Italy). *Am. J. Sci. 272*, 423–437.

Turnbull, D., and Fisher, J. C. (1949). Rate of nucleation in condensed systems. *J. Chem. Phys. 17*, 71–73.

Turner, F. J., and Verhoogen, J. (1960). *Igneous and Metamorphic Petrology*. New York: McGraw-Hill.

Turner, F. J. (1968). *Metamorphic Petrology*. New York: McGraw-Hill.

Turnock, A. C. (1960). The stability of iron chlorites. *Carnegie Inst. Yearbook 59*, 98–103.

Tuttle, O. F., and Bowen, N. L. (1958). Origin of granite in the light of experimental studies in the system $NaAlSi_3O_8$-$KAlSi_3O_8$-SiO_2-H_2O. *Mem. Geol. Soc. Am. 74.*

Uhlig, H. H. (1954). Contribution of metallurgy to the study of meteorites. Part I. Structure of metallic meteorites, their composition and the effect of pressure. *Geochim. Cosmochim. Acta 6*, 282–301.

Urey, H. C. (1952). *The Planets: Their Origin and Development.* New Haven: Yale University Press.

Urey, H. C., and Craig, H. (1953). The composition of the stone meteorites and the origin of the meteorites. *Geochim. Cosmochim. Acta 4*, 36–82.

Urey, H. C. (1956). Diamonds, meteorites and the origin of the solar system. *Astrophys. J. 124*, 623–637.

Urey, H. C. (1959). Primary and secondary objects. *J. Geophys. Res. 64*, 1721–1722.

Urey, H. C. (1963). The origin and evolution of the solar system. In *Space Science* (P. P. LeGalley, ed.). New York: Wiley, pp. 123–168.

Urey, H. C. (1964). A review of atomic abundances in chondrites and the origin of meteorites. *Rev. Geophys. 2*, 1–34.

Urey, H. C., and Mayeda, T. (1959). The metal particles of some chondrites. *Geochim. Cosmochim. Acta 17*, 113–124.

Uruno, K., and Kanisawa, S. (1965). Staurolite-bearing rocks in the Abukuma metamorphic belt, Japan. *J. Assoc. Geol. Collab. Japan 81*, 1–12.

Van Bemmelen, R. W. (1933). On the geophysical foundation of the undation theory. *Koninbl. Ned. Akad. Vetenskap. Proc. Ser. 36*, 686.

Van Schmus, W. R., and Koffman, D. M. (1967). Equilibrium temperatures of iron and magnesium in chondritic meteorites. *Science 155*, 1009–1011.

Van Schmus, W. R., and Wood, J. A. (1967). A chemical petrologic classification for the chondritic meteorites. *Icarus 31*, 747–765.

Velde, B. (1966). Upper stability of muscovite. *Am. Mineral. 51*, 924–929.

Velde, B., and Kornprobst, J. (1969). Stabilite des silicates d'alumine hydrates. *Contr. Mineral. Petrol. 21*, 63–74.

Verhoogen, J. (1946). Volcanic heat. *Am. J. Sci. 244*, 745–771.

Verhoogen, J. (1951a). Mechanics of ash formation. *Am. J. Sci. 249*, 729–739.

Verhoogen, J. (1951b). The chemical potential of a stressed solid. *Am. Geophys. Union 32*, 251–258.

Vernon, R. H. (1968). Micro-structures of high grade metamorphic rocks at Broken Hill, Australia. *J. Petrol. 9*, 1–22.

Vernon, R. H. (1970). Comparative grain-boundary studies of some basic and ultrabasic granulites, nodules and cumulates. *Scot. J. Geol. 6*, 337–351.

Vincent, E. A., and Phillips, R. (1954). Iron-titanium oxide minerals in layered gabbros of the Skaergaard intrusion, East Greenland. Part I. Chemistry and ore-microscopy. *Geochim. Cosmochim. Acta 6*, 1–26.

Virgo, D., and Hafner, S. S. (1969). Fe^{2+}, Mg order-disorder in heated orthopyroxenes. *Mineral. Soc. Am., Spec. Paper 2*, 67–81.

Volarovic, M. P., Tolstoj, T. M., and Korcemkin, L. I. (1936). A study of the viscosity of molten lavas from Mount Alaghez. *Compt. Rend.* (*Doklady*) *Acad. Sci. URSS 1*, No. 8 (85), 333–336.

von Weizäcker, C. F. (1944). Über die Entstehung des Planetensystems. *Zeitschr. Astrophys. Bd. 22*, 319.

Vinogradov, A. P., Surkov, Yu. Aij., and Kirnozov, F. F. (1973). U, Th and K content in the Venusian rock measured by Venera 8: Preprint. *Inst. Geochemistry and Analytical Geochemistry, Acad. Sci. U.S.S.R.*

Wager, L. R., and Deer, W. A. (1939). Geological investigations in east Greenland. Part III. The petrology of the Skaergaard intrusion, Kangerdlugssuaq, east Greenland. *Medd, Grönland 105* (4), 1–352.

Wager, L. R., and Brown, G. M. (1967). *Layerd Igneous Rocks.* San Francisco: Freeman.

Wahl, W. (1910). Beitrage zur Chemi der Meteoriten. *Z. Anorg. Chem. 69*, 52–96.

Waldbaum, D. R., and Thompson, J. B., Jr. (1969). Mixing properties of sanidine crystalline solutions; IV. Phase diagrams from equations of state. *Am. Mineral. 54*, 1274–1298.

Waldbaum, D. R., and Robie, R. A. (1971). Calorimetric investigation of Na-K mixing and polymorphism in the alkali feldspars. *Zeit. Kristallogr. 134*, 381–420.

Walker, F. (1940). Differentiation of the Palisade diabase, New Jersey. *Bull. Geol. Soc. Am. 51*, 1059–1106.

Walker, G. P. L. (1972). Crystal concentration in ignombrites. *Contr. Mineral. Petrol. 36*, 135–146.

Wasserburg, G. J. (1957). The effects of H_2O in silicate systems. *J. Geol. 65*, 15–23.

Wasserburg, G. J. (1958). The solubility of quartz in supercritical water as a function of pressure. *J. Geol. 66*, 559–578.

Wasson, J. T. (1974). *Meteorites, Classification and Properties.* New York: Springer-Verlag.

Watts, B. J. (1973). Relationship between fluid-bearing and fluid-absent invariant points and a petrogenetic grid for a greenschist facies assemblage in the system CaO-MgO-Al_2O_3-SiO_2-CO_2-H_2O. *Contr. Mineral. Petrol. 40*, 225–238.

Weeks, W. F. (1956). A thermochemical study of equilibrium relations during metamorphism of siliceous carbonate rocks. *J. Geol. 64*, 245–1270.

Weidner, J. R. (1968). Phase equilibria in a portion of the system Fe-C-O from 250 to 10,000 bars and 400°C to 1200°C and its petrologic significance. Ph.D. dissertation, Pennsylvania State University.

Weidner, J. R. (1972). Equilibria in the system Fe-C-O. Part I; Siderite-magnetite-carbon-vapor equilibrium from 500 to 10,000 bars. *Am. J. Sci. 272*, 735–775.

Weisbrod, A. (1973). Refinements of the equilibrium conditions of the reaction Fe cordierite $\rightleftharpoons$ almandine + quartz + sillimanite (+H_2O). *Carnegie Inst. Geophys. Lab., Yearbook 72*, 515–522.

Welch, A. J. E. (1955). Solid-solid reactions. In *Chemistry of the Solid State* (W. E. Garner, ed.). London: Butterworths, pp. 297–310.

Weyl, P. K. (1959). Pressure solution and the force of crystallization—A phenomenological theory. *J. Geophys. Res. 64*, 2001–2025.

Weymouth, J. H., and Williamson, W. O. (1957). Some effects of artificial heating on fluorite-bearing albite-granite from St. Austell, Cornwall. *Geol. Mag. 44*, 69–80.

White, D. E., Brannock, W. W., and Murata, K. J. (1956). Silica in hot spring waters, *Geochim. Cosmochim. Acta 10*, 27–59.

White, D. E. (1957). Thermal waters of volcanic origin. *Bull. Geol. Soc. Am. 68*, 1637–1658.

White, D. E., and Waring, G. A. (1963). Volcanic emanations. In *Data of Geochemistry* (M. Fleischer, ed.). *U.S. Geol. Surv. Prof. Paper 440-K.*

Wiik, H. B. (1956). The chemical composition of some stony meteorites. *Geochim. Cosmochim. Acta 9*, 279–289.

Wilkinson, J. F. G., Vernon, R. H., and Shaw, S. E. (1964). The petrology of an adamellite porphyrite from the New England Bathylith (New South Wales). *J. Petrol. 5*, 461–488.

Wilkinson, J. F. G. (1967). Petrography of basaltic rocks. In *Basalts* (H. H. Hess and Aric Poldervaart, eds.). New York: Wiley.

Williams, H. (1942). Geology of Crater Lake National Park, Oregon. *Carnegie Inst. Publ. 540*.

Williams, H., Turner, F. J., and Gilbert, C. M. (1954). *Petrography: An Introduction to the Study of Rocks in Thin Sections*. San Francisco: Freeman.

Williams, R. J. (1971). Equilibrium temperatures, pressure, and oxygen fugacities of the equilibrated chondrites. *Geochim. Cosmochim. Acta 35*, 407–411.

Winkler, H. G. F. (1965, 67, 74). *Petrogenesis of Metamorphic Rocks*, 1st and 2nd eds. New York: Springer-Verlag.

Wood, J. A. (1963). On the origin of chondrules and chondrites. *Icarus 2*, 152–180.

Wood, B. J. (1974a). The solubility of alumina in orthopyroxenes coexisting with garnet. *Contr. Mineral. Petrol. 46*, 1–15.

Wood, B. J. (1974b). Fe^{2+}-Mg^{2+} partition between coexisting cordierite and garnet—a discussion of the experimental data. *Contr. Mineral. Petrol. 40*, 253–258.

Woodford, A. O., and Harris, T. F. (1938). Geological reconnaissance across Sierra San Pedro Martir, Baja California. *Geol. Soc. Am. Bull. 49*, 1297–1336.

Wones, D. R., and Eugster, H. P. (1965). Stability of biotite: Experiment, theory and application. *Am. Mineral. 50*, 1228–1272.

Wones, D. R., and Gilbert, M. E. (1969). The fayalite-magnetite-quartz assemblage between 600°C and 800°C. *Am. J. Sci. 267-A*. (Schairer Volume), 480–488.

Wright, T. L., Kinoshita, W. T., and Peek, D. L. (1968). March 1965 eruption of Kilauea Volcano and the formation of Ma Kaopuhi Lava Lake. *J. Geophys. Res. 73*, 3181–3205.

Wulff, G. (1901). Zur Frage der Geschwindigkeit des Wachsthums und der Auflosung der Krystallflachen. *Z. Kristallogr. Mineral. 34*, 449–530.

Wyllie, P. J. (1965). Heating relationship in the system CaO-MgO-CO_2-H_2O with petrologic applications. *J. Petrol. 6*, 101–123.

Wyllie, P. J. (1971a). Experimental limits of melting in the earth's crust and upper mantle. In *The Structure and Physical Properties of the Earth's Crust* (J. G. Heacock, ed.). Washington, D.C.: American Geophysical Union.

Wyllie, P. J. (1971b). Role of water in magma generation and initiation of diapiric uprise in the mantle. *J. Geophys. Res. 76*, 1328–1338.

Wyllie, P. J. (1974). Limestone assimilation. In *The Alkaline Rocks* (H. Sorensen, ed.). New York: Wiley, pp. 459–474.

Wyllie, P. J., and Tuttle, O. F. (1959). Effect of carbon dioxide on the melting of granite and feldspars. *Am. J. Sci. 257*, 648–655.

Wyllie, P. J., and Tuttle, O. F. (1961). Experimental investigations of silicate systems containing two volatile components. Part II: The effects of NH_3 and HF, in addition to H_2O on the melting temperatures of albite and granite. *Am. J. Sci. 259*, 128–143.

Wyllie, P. J., and Tuttle, O. F. (1964). Experimental investigations of silicate systems containing two volatile components. Part III: The effects of SO_3, P_2O_5, HCl and Li_2O in addition to H_2O, on the melting temperatures of albite and granite. *Am. J. Sci. 262*, 930–939.

Wyllie, P. J., and Hass, J. L., Jr. (1966). The system CaO-SiO_2-CO_2-H_2O. II—The petrogenetic model. *Geochim. Cosmochim. Acta 30*, 525–543.

Wynne-Edwards, H. R. (1967). Westport map-area, Ontario, with special emphasis on the Precambrian rocks. *Geol. Surv. Canada Mem. 346*.

Yoder, H. S., Jr. (1950). Stability of grossularite. *J. Geol. 58*, 221–253.

Yoder, H. S., Jr. (1955). Almandine stability range (abs.). *Am. Mineral. 40*, 342.

Yoder, H. S., Jr. (1973). Contemporaneous basaltic and rhyolitic magmas. *Am. Mineral. 58*, 153–171.

Yoder, H. S., Jr., and Eugster, H. P. (1954). Phlogopite synthesis and stability range. *Geochim. Cosmochim. Acta 6*, 157–185.

Yoder, H. S., Jr., and Eugster, H. P. (1955). Synthetic and natural muscovites. *Geochim. Cosmochim. Acta 8*, 225–280.

Yoder, H. S., Jr., Stewart, D. B., and Smith, J. R. (1957). Ternary Feldspars. Annual Report 1956–1957. *Carnegie Inst. Geophys. Lab. Yearbook*, pp. 206–214.

Yoder, H. S., Jr., and Tilley, C. E. (1962). Origin of basalt magmas; an experimental study of natural and synthetic rock systems. *J. Petrol. 3*, 342–532.

Yoder, H. S., Jr., and Weir, C. E. (1951). Change of free energy with pressure of the reaction nepheline + albite = 2 jadeite. *Am. J. Sci. 249*, 683–694.

Young, L. D. G. (1972). High resolution spectra of Venus. A review. *Icarus 17*, 632–658.

Zavaritsky, A. N. (1950). On some structural peculiarities of meteorites. *Meteoritika 8*, 100–115.

Zen, E-an (1960). Metamorphism of Lower Paleozoic rocks in the vicinity of the Taconic Range in west-central Vermont. *Am. Mineral. 45*, 129–175.

Zen, E-an (1966). Construction of pressure-temperature diagrams for multicomponent systems after the method of Schreinemakers—a geometric approach. *U.S. Geol. Surv. Bull. 1225*.

Zen, E-an (1967). Some topological relationships in multisystems of $n + 3$ phases. II. Unary and binary metastable sequences. *Am. J. Sci. 265*, 871–897.

Zen, E-an (1969). The stability relations of the polymorphs of aluminum silicate; a survey and some comments. *Am. J. Sci. 267*, 297–309.

Zen, E-an, and Roseboom, E. H., Jr. (1972). Some topological relationships in multisystems of $n + 3$ phases III. Ternary systems. *Am. J. Sci. 272*, 677–710.

Zies, E. G. (1946). Temperature measurements at Paricutin Volcano. *Am. Geophys. Union Trans. 27*, 178–180.

Subject index

Subject index

Author index

Springer-Verlag New York Heidelberg Berlin

Petrogenesis of Metamorphic Rocks
Fourth Edition

By **H. G. F. Winkler**

1976. xi, 334p. 100 illus. paper
Springer Study Edition
ISBN 0-387-07473-2

The fourth edition of *Petrogenesis of Metamorphic Rocks* is an expanded and updated version of the extensively revised third edition, which provided a completely new approach to the study of metamorphic transformations.

From a review of the third edition:

"The third edition of Winkler's already classic work on metamorphism is 35 percent longer, extensively revised and considerably improved. Among major changes, the use of metamorphic facies for purposes of classification has been abandoned as obsolete and the descriptive aspects of metamorphic rocks and reactions are instead developed using groups of rocks related by common bulk composition. The book is a good standard reference and, supplemented by some thermodynamics, would make an excellent graduate text. . . . the book is an important one and deserves to be read carefully by anyone seriously studying metamorphic petrology."

American Mineralogist

Stress and Strain
Basic Concepts of Continuum Mechanics for Geologists

By **W. D. Means**

1976. xi, 339p. 223 illus. paper
ISBN 0-387-07556-9

Appropriate as a text for a semester lecture course or as a self-instruction manual, *Stress and Strain* is divided into four parts and twenty-seven short chapters, each followed by solved problems. Part One introduces such fundamental ideas as the distinction between instantaneous quantities (i.e. stress) and two-state quantities (i.e. finite strain). Part Two discusses stress in detail. Part Three treats deformation and strain, introduces infinitesimal and finite strain tensors, and discusses strain history. The relations between stress and strain for ideally elastic and ideally viscous materials are examined in Part Four.

Students of structural geology and tectonics and working geologists who wish to improve their understanding of mechanical topics are given an extensive, yet concise introduction to the theories of stress and strain.